Lecture Notes in Computer Science 1470

Edited by G. Goos, J. Hartmanis and J. van Leeuwen

Springer-Verlag Berlin Heidelberg GmbH

David Pritchard   Jeff Reeve   (Eds.)

Euro-Par'98
Parallel Processing

4th International Euro-Par Conference
Southampton, UK, September 1-4, 1998
Proceedings

Springer-Verlag
Berlin Heidelberg GmbH

Series Editors

Gerhard Goos, Karlsruhe University, Germany
Juris Hartmanis, Cornell University, NY, USA
Jan van Leeuwen, Utrecht University, The Netherlands

Volume Editors

David Pritchard
Jeff Reeve
University of Southampton
Department of Electronics and Computer Science
Southampton SO17 1BJ, UK
E-mail: {djp,jsr}@ecs.soton.ac.uk

Cataloging-in-Publication data applied for

Die Deutsche Bibliothek - CIP-Einheitsaufnahme

Parallel processing : proceedings / Euro-Par '98, 4th International
Euro-Par Conference, Southampton, UK, September 1 - 4, 1998.
David Pritchard ; Jeff Reeve (ed.). - Berlin ; Heidelberg ; New York ;
Barcelona ; Budapest ; Hong Kong ; London ; Milan ; Paris ;
Singapore ; Tokyo : Springer, 1998
  (Lecture notes in computer science ; Vol. 1470)
  ISBN 978-3-540-64952-6

CR Subject Classification (1991): C.1-4, D.1-4, F.1-2, G.1-2, E.1, H.2

ISSN 0302-9743
ISBN 978-3-540-64952-6   ISBN 978-3-540-49920-6 (eBook)
DOI 10.1007/978-3-540-49920-6

Typesetting: Camera-ready by editors from LaTeX by authors
SPIN 10638643   06/3142 – 5 4 3 2 1 0   Printed on acid-free paper

Preface

Euro-Par

Euro-Par is an international conference dedicated to the promotion and advancement of all aspects of parallel computing. The major themes can be divided into the broad categories of hardware, software, algorithms and applications for parallel computing. The objective of Euro-Par is to provide a forum within which to promote the development of parallel computing both as an industrial technique and an academic discipline, extending the frontier of both the state of the art and the state of the practice. This is particularly important at a time when parallel computing is undergoing strong and sustained development and experiencing real industrial take up. The main audience for and participants in Euro-Par are seen as researchers in academic departments, government laboratories and industrial organisations. Euro-Par's objective is to become the primary choice of such professionals for the presentation of new results in their specific areas. Euro-Par is also interested in applications which demonstrate the effectiveness of the main Euro-Par themes.

There is now a permanent Web site for the series which can be reached at http://brahms.fmi.uni-passau.de/cl/europar where the history of the conference is described. Euro-Par is now sponsored by the Association of Computer Machinery and the International Federation of Information Processing.

Euro-Par'98

The format of Euro-Par'98 follows that of the past two conferences and consists of a number of workshops each individually monitored by a committee of four. There were 23 original workshops for this year's conference. The call for papers attracted 238 submissions of which 129 were accepted. Of the papers accepted 5 were judged as distinguished, 70 as regular and 54 as short papers. Distinguished papers are allowed 12 pages in the proceedings and 30 minutes for presentation, regular papers are allowed 8 pages and 15 minutes for presentation, short papers are allowed 4 pages and 15 minutes for presentation. Two extra pages could be purchased. There were on average 3.6 reviews per paper. Submissions were received from 33 countries 26 of which are represented at the conference. The principal contributors by country are the UK with 23 papers, Germany 22, France 19 and the USA 16.

The Web site for the conference is at http://www.europar98.ecs.soton.ac.uk.

Acknowledgements

Knowing the quality of past Euro-Par conferences makes the task of organising one daunting indeed and we have many people to thank. Ron Perrott, Christian

Lengauer and Luc Bougé have given us the benefit of their experience and helped us generously throughout the past 18 months. The workshop structure of the conference means that we must depend on the goodwill and enthusiasm of all the 77 programme committee members listed below. Their professionalism makes this the most academically rigorous conference in the field worldwide. The programme committee meeting at Southampton in April was well attended and thanks to sound preparation by everyone and Ron Perrott's guidance resulted in a coherent, well structured conference. The smooth running of the organisation of the conference can be attributed to a few individuals. Firstly the software for the submission and refereeing of the papers that we inherited from Lyons via Passau was significantly enhanced by Flavio Bergamaschi. This attracted many compliments from those who benefited. Panagiotis Melas ably assisted by Duncan Simpson spent copious hours checking, printing and correcting papers. Finally Lesley Courtney, secretary to the conference and the research group, has been invaluable in monitoring the conference organisation and seeing to the myriad of tasks that invariably arise, including organising the social programme.

Southampton, June 1998   Jeff Reeve and David Pritchard.

Euro-Par Steering Committee

Chair
Ron Perrott   Queen's University Belfast, UK
Vice Chair
Emilio Zapata   University of Malaga, Spain
Committee
Luc Bougé   ENS Lyon, France
Agnes Bradier   EC, Belgium
Helmar Burkhart   University of Basel, Switzerland
Paul Feautrier   University of Versailles, France
Ian Foster   Argonne National Lab, USA
Seif Haridi   SICS, Sweden
Peter Kacsuk   KFKI, Hungary
Christian Lengauer   University of Passau, Germany
Jeff Reeve   University of Southampton, UK
Paul Spirakis   CTI, Greece
Marian Vajtersic   Slovak Academy, Slovakia
Jens Volkert   Johannes Kepler University, Austria
Makoto Amamiya   Kyushu University, Japan

Euro-Par'98 Local Organisation

The conference has been organised by the Concurrent Computation Group of the Department of Electronics & Computer Science, University of Southampton

Chairs
Tony Hey
Jeff Reeve
Committee
Alistair Dunlop
Hugh Glaser
Luc Moreau
Mark Papiani
David Pritchard
Secretary Lesley Courtney
Technical Support
Flavio Bergamaschi
Panagiotis Melas
Duncan Simpson

Euro-Par'98 Programme Committee

Workshop 1: Support Tools and Environments

Global Chair
Helmar Burkhart   University of Basel, Switzerland
Local Chair
Chris Wadsworth   Rutherford Appleton Lab, UK
Vice Chairs
Peter Kacsuk   KFKI, Hungary
Vaidy Sunderam   Emory University, USA

Workshop 2+8: Performance Evaluation and Prediction

Global Chairs
Allen Malony   University of Oregon
Rajeev Alur   University of Pennsylvania and Bell Labs
Local Chairs
Wolfgang Gentzsch   GENIAS Software, Germany
Eric Rogers   University of Southampton
Vice Chairs
Daniel Reed   University of Illinois, Urbana-Champaign
Aad van der Steen   University of Utrecht, The Netherlands
Hans-J.Siegert   TU Munich, Germany
Guenter Hommel   University of Berlin, Germany

Workshop 3: Scheduling and Load Balancing

Global Chair
Susan Flynn Hummel   IBM T.J.Watson Research Center
Local Chair
Graham Riley   University of Manchester
Vice Chairs
Rizos Sakellariou   University of Manchester
Wolfgang Gentzsch   GENIAS Software, Germany

Workshop 4: Automatic Parallelization and High-Performance Compilers

Global Chair
Jean-François Collard   CNRS and U. of Versailles, France
Local Chair
Thomas Brandes   IASC, Germany
Vice Chairs
Martin Rinard   MIT, USA
Martin Griebl   University of Passau, Germany

Workshop 5+9+15: Distributed Systems and Database Systems

Global Chairs

Andreas Reuter	International University in Germany
Ernst Mayr	TU Munich, Germany

Local Chairs

Lionel Brunie	ENS Lyon, France
Pavlos Spirakis	CTI, Greece
Kam-Fai Wong	Hong Kong University

Vice Chairs

Harald Kosch	University of Klagenfurt, Austria
Usama Fayyad	Microsoft
Arbee Chen	National Tsing Hua University, Taiwan
Friedemann Mattern	TU Darmstadt, Germany
Marios Mavronicolas	University of Cyprus, Cyprus

Workshop 6+16+18: Languages

Global Chairs

Denis Caromel	University de Nice - INRIA Sophia Antipolis
Christian Lengauer	University of Passau, Germany
Mike Quinn	Oregon State University

Local Chairs

Antonio Corradi	University of Bologna, Italy
Henk Sips	Technical University, Delft, Netherlands
Murray Cole	University of Edinburgh, UK

Vice Chairs

Gul Agha	University of Illinois
Geoffrey Fox	University of Syracuse
Phil. Hatcher	University of New Hampshire, USA
Luc Bougé	ENS Lyon, France
Beverly Sanders	University of Florida, Gainesville
Gaétan Hains	University of Orléans, France

Workshop 7+20: Numerical and Symbolic Algorithms

Global Chairs

Wolfgang Küchlin	University of Tubingen, Germany
Maurice Clint	Queen's University Belfast, UK

Local Chairs

Ali Abdallah	University of Reading, UK
Marian Vajtersic	Slovak Academy, Slovakia

Vice Chairs

Krzysztof Apt	CWI, Netherlands
Kevin Hammond	University of St. Andrews, Scotland
Michael Thune	University of Upsala, Sweden
Peter Arbenz	Institute for Sci. Computing, Switzerland

Workshop 10+17+21+22:
Theory and Applications of Parallel Computation

Global Chairs

Bill McColl	Oxford University, UK
Mike Brady	University of Oxford, UK
Paul Messina	California Institute of Technology
Freidel Hossfeld	Forschungszentrum Julich GmbH, Germany

Local Chairs

Michel Cosnard	INRIA, Nancy, France
Paul Lewis	University of Southampton, UK
Ed Zaluska	University of Southampton, UK

Vice Chairs

Andrea Pietracaprina	University of Padova, Italy
Frank Dehne	Carlton University, Canada
Patrice Quinton	IRISA-CNRS, Rennes, France
Hartmut Schmeck	University of Karlsruhe, Germany
Rajeev Thakur	Argonne National Laboratory, USA
David Walker	University of Wales, UK

Workshop 13+14: Architectures and Networks

Global Chairs

Abhiram Ranade	Indian Institute of Technology, Bombay
Mateo Valero	UPC Barcelona, Spain

Local Chairs

Kieran Herley	University College Cork, Eire
David Snelling	FECIT, UK

Vice Chairs

Sanguthevar Rajasekaran	University of Florida
Geppino Pucci	University of Padova, Italy
Olivier Temam	University of Versailles, France
Nigel Topham	University of Edinburgh, UK
Rupert Ford	University of Manchester, UK

Esprit Workshop

Global Chair

Ron Perrott	Queen's University Belfast

Local Chair

Colin Upstill	PAC,University of Southampton

Vice Chairs

Jesus Labarta	Universitat Politecnica de Catalunya, Spain
Karl Solchenbach	PALLAS, Germany

Other Referees

Arbab, Farhad
Barthou, Denis
Bischof, Stefan
Buffo, Mathieu
Detert, Ulrich
Ellmenreich, Nils
Erlebach, Thomas
Esser, Ruediger
Flammini, Michele
Foisy, Christian
Formenti, Enrico
Friedetzky, Tom
Gerndt, Michael
Gorlatch, Sergei
Harmer, Terence
Hascoet, Laurent
Hege, Hans-Christian
Herrmann, Christoph
Hogstedt, Karin
Irigoin, Francois
Jerraya, Ahmed
Lazure, Dominique
Lim, Amy
Ludwig, Thomas

Massingill, Berna
Mattern, Friedemann
McKendrick, Rob
Mehaut, Jean-Francois
Mery, Dominique
Mountjoy, Jon
Namyst, Raymond
Pazat, Jean-Louis
Perez, Christian
Rauber, Thomas
Redon, Xavier
Robert, Yves
Scales, Dan
Schwabe, Eric
Stadtherr, Hans
Stewart, Alan
Surridge, Mike
Thompson, Simon
Trinder, Phil
Utard, Gil
Vivien, Frederic
Wedler, Christoph
Wonnacott, David
Zait, Mohamed

Contents

Workshop 2+8

Workshop 3

Workshop 4

Workshop 6+16+18
Languages **625**

Workshop 7+20
Numerical and Symbolic Algorithms 747

Workshop 10+17+21+22
Theory and Algorithms for Parallel Computation 863

Workshop 13+14
Architectures and Networks 967

Workshop 23

Random Number Generation and Simulation on Vector and Parallel Computers

Richard P. Brent

Oxford University Computing Laboratory,
Wolfson Building, Parks Road,
Oxford OX1 3QD, UK
rpb@comlab.ox.ac.uk
http://www.comlab.ox.ac.uk/oucl/people/richard.brent.html

Abstract. Pseudo-random numbers are often required for simulations performed on parallel computers. The requirements for parallel random number generators are more stringent than those for sequential random number generators. As well as passing the usual sequential tests on each processor, a parallel random number generator must give different, independent sequences on each processor. We consider the requirements for a good parallel random number generator, and discuss generators for the uniform and normal distributions. We also describe a new class of generators for the normal distribution (based on a proposal by Wallace). These generators can give very fast vector or parallel implementations. Implementations of uniform and normal generators on vector and vector/parallel computers are discussed.

1 Introduction

Pseudo-random numbers have been used in Monte Carlo calculations since the earliest days of digital computers [32]. In this paper we are concerned here with random number generators (RNGs) on fast, modern computers – typically either vector processors or parallel computers using vector or pipelined RISC processors. What we say about vector processors often applies to pipelined RISC processors with a memory hierarchy (the vector registers of a vector processor corresponding to the first-level cache of a RISC processor).

With the increasing speed of vector processors and parallel computers, considerable attention must be paid to the quality of random number generators. A program running on a supercomputer might use 10^8 random numbers per second over a period of many hours or even months in the case of QCD calculations, so 10^{14} random numbers might contribute to the result. Small correlations or other deficiencies in the random number generator could easily lead to spurious effects and invalidate the results of the computation.

Applications require random numbers with various distributions (uniform, normal, exponential, binomial, Poisson, etc.) but the algorithms used to generate these random numbers usually require a good uniform random number generator – see for example [2, 5, 14, 24, 34, 39]. In this paper we consider the generation of uniformly and normally distributed numbers.

Pseudo-random numbers generated in a deterministic fashion on a digital computer can not be truly random. What is required is that finite segments of the sequence behave in a manner indistinguishable from a truly random sequence. In practice, this means that they pass all statistical tests which are relevant to the problem at hand. Since the problems to which a library routine will be applied are not known in advance, random number generators in subroutine libraries should pass a number of stringent statistical tests (and not fail any) before being released for general use.

A sequence $u_0, u_1, \cdots$ depending on a finite state must eventually be periodic, i.e. there is a positive integer p such that $u_{n+p} = u_n$ for all sufficiently large n. The minimal such p is called the *period*.

Following are some of the more important requirements for a good uniform pseudo-random number generator and its implementation in a subroutine library (the modifications for a normal generator are obvious) –

- *Uniformity.* The sequence of random numbers should pass statistical tests for uniformity of distribution. In one dimension this is easy to achieve. Most generators in common use are provably uniform (apart from discretisation due to the finite wordlength) when considered over their full period.
- *Independence.* Subsequences of the full sequence $u_0, u_1, \cdots$ should be independent. For example, members of the even subsequence $u_0, u_2, u_4, \cdots$ should be independent of their odd neighbours $u_1, u_3, \cdots$. Thus, the sequence of pairs (u_{2n}, u_{2n+1}) should be uniformly distributed in the unit square. More generally, random numbers are often used to sample a d-dimensional space, so the sequence of d-tuples $(u_{dn}, u_{dn+1}, \ldots, u_{dn+d-1})$ should be uniformly distributed in the d-dimensional cube $[0, 1]^d$ for all "reasonable" values of d (certainly for all $d \le 6$).
- *Long Period.* As mentioned above, a simulation might use 10^{14} random numbers. In such a case the period p must exceed 10^{14}. For many generators there are strong correlations between $u_0, u_1, \cdots$ and $u_m, u_{m+1}, \cdots$, where $m = p/2$ (and similarly for other simple fractions of the period). Thus, in practice the period should be *much* larger than the number of random numbers which will ever be used.
- *Repeatability.* For testing and development it is useful to be able to repeat a run with *exactly* the same sequence of random numbers as was used in an earlier run [22]. This is usually easy if the sequence is restarted from the beginning (u_0). It may not be so easy if the sequence is to be restarted from some other value, say u_m for a large integer m, because this requires saving the state information associated with the random number generator.
- *Portability.* Again, for testing and development purposes, it is useful to be able to generate *exactly* the same sequence of random numbers on two different machines, possibly with different wordlengths. In practice it will be expensive to simulate a long wordlength on a machine with a short wordlength, but the converse should be easy – a machine with a long wordlength (say

$w = 64$) should be able to simulate a machine with a smaller wordlength without loss of efficiency.

- *Disjoint Subsequences.* If a simulation is to be run on a machine with several processors, or if a large simulation is to be performed on several independent machines, it is essential to ensure that the sequences of random numbers used by each processor are disjoint. Two methods of subdivision are commonly used. Suppose, for example, that we require 4 disjoint subsequences for a machine with 4 processors. One processor could use the subsequence $(u_0, u_4, u_8, \cdots)$, another the subsequence $(u_1, u_5, u_9, \cdots)$, etc. This partitioning method is sometimes called "decimation" or "leapfrog" [11]. For efficiency each processor should be able to "skip over" the terms which it does not require. Alternatively, processor j could use the subsequence $(u_{m_j}, u_{m_j+1}, \cdots)$, where the indices m_0, m_1, m_2, m_3 are sufficiently widely separated that the (finite) subsequences do not overlap. This requires some efficient method of generating u_m for large m without generating all the intermediate values $u_1, \ldots, u_{m-1}$.

- *Efficiency.* It should be possible to implement the method efficiently so that only a few arithmetic operations are required to generate each random number and all vector/parallel capabilities of the machine are used. To minimise subroutine call overheads, the random number routine should return an array of (optionally) several numbers at a time.

Several recent reviews [4, 6, 11, 16, 22, 24, 28, 33] of uniform random number generators are available. The most important conclusion regarding uniform generators is that good ones may exist, but are hard to find [33]. Linear congruential generators with a "short" period (less than say 2^{48}) are certainly to be avoided. Generalised (or "lagged") Fibonacci generators using the "exclusive or" operation are also to be avoided; other generalised Fibonacci generators may be satisfactory if the lags are sufficiently large (if they use the operation of addition then the lags should probably be at least 1000). See, for example, [12, Table 2]. Our recommendation, implemented as RANU4 on Fujitsu VP2200 and VPP300 vector/parallel processors, is a generalised Fibonacci generator with very large lags, e.g. (79500, 132049) (see [21]), and careful initialisation which avoids any initial atypical behaviour and ensures disjoint sequences on parallel processors. For further details see [6].

In the interests of conserving space, we refer the reader to the reviews cited above for uniform generators, and concentrate our attention on the less often considered, but still important, case of normal random number generation on vector/parallel processors. "Classical" generators are considered in §2, and an interesting new class of "Wallace" generators [40] is considered in §3.

We do not attempt to cover the important topic of *testing* random number generators intended for use on vector/parallel computers. A good, recent survey of this topic is [12]. The user should always remember that a deterministic sequence of pseudo-random numbers can not truly be random; all that testing can do is inspire confidence that a generator is indistinguishable from random

4

in a particular application [37, 38]. In practice, testing is essential to cull bad generators, but can not provide any guarantees.

2 Normal RNGs Based on Uniform RNGs

In this section we consider some "classical" methods for generating normally distributed pseudo-random numbers. The methods all assume a good source of uniform random numbers which is transformed in some manner to a sequence of normally distributed random numbers. The transformation is not necessarily one to one.

The most well-known and widely used methods for generating normally distributed random variables on sequential machines [2, 5, 14, 20, 24, 26] involve the use of different approximations on different intervals, and/or the use of "rejection" methods [14, 24], so they often do not vectorise well. Simple, "old-fashioned" methods may be preferable. In §2.1 we describe two such methods, and in §§2.2–2.3 we consider their efficient implementation on vector processors, and give the results of implementations on a Fujitsu VP2200/10. In §§2.4–2.5 we consider some other methods which are popular on serial machines, and show that they are unlikely to be competitive on vector processors.

2.1 Some Normal Generators

Assume that a good uniform random number generator which returns uniformly distributed numbers in the interval $[0, 1)$ is available, and that we wish to sample the normal distribution with mean μ and variance σ^2. We can generate two independent, normally distributed numbers x, y by the following old algorithm due to Box and Muller [31] *(Algorithm B1):*

1. Generate independent uniform numbers $u, v \in [0, 1)$.
2. Set $r \leftarrow \sigma\sqrt{-2\ln(1 - u)}$.
3. Set $x \leftarrow r\sin(2\pi v) + \mu$ and $y \leftarrow r\cos(2\pi v) + \mu$.

The proof that the algorithm is correct is similar to the proof of correctness of the Polar method given in Knuth [24].

Algorithm B1 is a reasonable choice on a vector processor if vectorised square root, logarithm and trigonometric function routines are available. Each normally distributed number requires 1 uniformly distributed number, 0.5 square roots, 0.5 logarithms, and 1 sin or cos evaluation. Vectorised implementations of the Box-Muller method are discussed in §2.2.

A variation of Algorithm B1 is the *Polar* method of Box, Muller and Marsaglia described in Knuth [24, Algorithm P]:

1. Generate independent uniform numbers $x, y \in [-1, 1)$.
2. Set $s \leftarrow x^2 + y^2$.
3. If $s \in (0, 1)$ then go to step 4 else go to step 1 (i.e. *reject* x and y).
4. Set $r \leftarrow \sigma\sqrt{-2\ln(s)/s}$, and return $rx + \mu$ and $ry + \mu$.

It is easy to see that, at step 4, (x, y) is uniformly distributed in the unit circle, so s is uniformly distributed in $[0, 1)$.

A proof that the values returned by Algorithm P are independent, normally distributed random numbers (with mean μ and variance σ^2) is given in Knuth [24]. On average, step 1 is executed $4/\pi$ times, so each normally distributed number requires $4/\pi \simeq 1.27$ uniform random numbers, 0.5 divisions, 0.5 square roots, and 0.5 logarithms. Compared to Algorithm B1, we have avoided the sin and cos computation at the expense of more uniform random numbers, 0.5 divisions, and the cost of implementing the acceptance/rejection process. This can be done using a vector gather. Vectorised implementations of the Polar method are discussed in §2.3.

2.2 Vectorised Implementation of the Box-Muller Method

We have implemented the Box-Muller method (Algorithm B1 above) and several refinements (B2, B3) on a Fujitsu VP2200/10 vector processor at the Australian National University. The implementations all return double-precision real results, and in cases where approximations to sin, cos, sqrt and/or ln have been made, the absolute error is considerably less than 10^{-10}. Thus, statistical tests using less than about 10^{20} random numbers should not be able to detect any bias due to the approximations. The calling sequences allow for an array of random numbers to be returned. This permits vectorisation and amortises the cost of a subroutine call over the cost of generating many random numbers.

Our method B2 is the same as B1, except that we replace calls to the library sin and cos by an inline computation, using a fast, but sufficiently accurate, approximation (for details see [7]).

Times, in machine cycles per normally distributed number, for methods B1, B2 (and other methods described below) are given in Table 1. In all cases the generalised Fibonacci random number generator RANU4 (described in [6]) was used to generate the required uniform random numbers, and a large number of random numbers were generated, so that vector lengths were long. RANU4 generates a uniformly distributed random number in 2.2 cycles on the VP2200/10. (The cycle time of the VP2200/10 at ANU is 3.2 nsec, and two multiplies and two adds can be performed per clock cycle, so the peak speed is 1.25 Gflop.)

The Table gives the total times and also the estimated times for the four main components:

1. ln computation (actually 0.5 times the cost of one ln computation since the times are per normal random number generated).
2. sqrt computation (actually 0.5 times).
3. sin or cos computation.
4. other, including uniform random number generation.

The results for method B1 show that the sin/cos and ln computations are the most expensive (65% of the total time). Method B2 is successful in reducing the sin/cos time from 33% of the total to 19%.

Table 1. Cycles per normal random number

component	B1	B2	B3	P1	P2	R1
ln	13.1	13.1	7.1	13.1	7.1	0.3
sqrt	8.8	8.8	1.0	8.8	1.0	0.0
sin/cos	13.8	6.6	6.6	0.0	0.0	0.0
other	5.9	5.6	11.6	11.9	13.8	35.1
total	41.6	34.1	26.3	33.8	21.9	35.4

In Method B2, the computation of $\sqrt{-\ln(1-u)}$ consumes 64% of the time. An obvious way to reduce this time is to use a fast approximation to the function

$$f(u) = \sqrt{-\ln(1-u)},$$

just as we used a fast approximation to sin and cos to speed up method B1. However, this is difficult to accomplish with sufficient accuracy, because the function $f(u)$ is badly behaved at both endpoints of the unit interval. Method B3 overcomes this difficulty in the following way.

1. We approximate the function

$$g(u) = u^{-1/2}f(u) = \sqrt{\frac{-\ln(1-u)}{u}},$$

rather than $f(u)$. Using the Taylor series for $\ln(1-u)$, we see that $g(u) = 1 + u/4 + \cdots$ is well-behaved near $u = 0$.
2. The approximation to $g(u)$ is only used in the interval $0 \le u \le \tau$, where $\tau < 1$ is suitably chosen. For $\tau < u < 1$ we use the slow but accurate library ln and sqrt routines.
3. We make a change of variable of the form $v = (\alpha u + \beta)/(\gamma u + \delta)$, where $\alpha, \ldots, \delta$ are chosen to map $[0, \tau]$ to $[-1, 1]$, and the remaining degrees of freedom are used to move the singularities of the function $h(v) = g(u)$ as far away as possible from the region of interest (which is $-1 \le v \le 1$). To be more precise, let ρ be a positive parameter. Then we can choose

$$\tau = 1 - \left(\frac{\rho}{\rho+2}\right)^2,$$

$$v = (\rho+1)\left(\frac{(\rho+2)u - 2}{2(\rho+1) - (\rho+2)u}\right),$$

and the singularities of $h(v)$ are at $\pm(\rho+1)$.

For simplicity, we choose $\rho = 1$, which experiment shows is close to optimal on the VP2200/10. Then $\tau = 8/9$, $v = (6u-4)/(4-3u)$, and $h(v)$ has singularities at $v = \pm 2$, corresponding to the singularities of $g(u)$ at $u = 1$ and $u = \infty$. A polynomial of the form $h_0 + h_1 v + \cdots + h_{15}v^{15}$ can be used to approximate

$h(v)$ with absolute error less than 2×10^{-11} on $[-1, 1]$. About 30 terms would be needed if we attempted to approximate $g(u)$ to the same accuracy by a polynomial on $[0, \tau]$. We use polynomial approximations which are close to minimax approximations. These may easily be obtained by truncating Chebyshev series, as described in [10].

It appears that this approach requires the computation of a square root, since we really want $f(u) = u^{1/2}g(u)$, not $g(u)$. However, a trick allows this square root computation to be avoided, at the expense of an additional uniform random number generation (which is cheap) and a few arithmetic operations. Recall that u is a uniformly distributed random variable on $[0, 1)$. We generate *two* independent uniform variables, say u_1 and u_2, and let $u \leftarrow \max(u_1, u_2)^2$. It is easy to see that u is in fact uniformly distributed on $[0, 1)$. However, $u^{1/2} = \max(u_1, u_2)$ can be computed without calling the library sqrt routine. To summarise, a non-vectorised version of method B3 is:

1. Generate independent uniform numbers $u_1, u_2, u_3 \in [0, 1)$.
2. Set $m \leftarrow \max(u_1, u_2)$ and $u \leftarrow m^2$.
3. If $u > 8/9$ then
   3.1. set $r \leftarrow \sigma\sqrt{-\ln(1 - u)}$ using library routines, else
   3.2. set $v \leftarrow (6u - 4)/(4 - 3u)$, evaluate $h(v)$ as described above, and set $r \leftarrow \sigma m h(v)$.
4. Evaluate $s \leftarrow \sin(2\pi u_3 - \pi)$ and $c \leftarrow \cos(2\pi u_3 - \pi)$ as in [7].
5. Return $\mu + cr\sqrt{2}$ and $\mu + sr\sqrt{2}$, which are independent, normal random numbers with mean μ and standard deviation σ.

Vectorisation of method B3 is straightforward, and can take advantage of the "list vector" technique on the VP2200. The idea is to gather those $u > 8/9$ into a contiguous array, call the vectorised library routines to compute an array of $\sqrt{-\ln(1 - u)}$ values, and scatter these back. The gather and scatter operations introduce some overhead, as can be seen from the row labelled "other" in the Table. Nevertheless, on the VP2200, method B3 is about 23% faster than method B2, and about 37% faster than the straightforward method B1. These ratios could be different on machines with more (or less) efficient implementations of scatter and gather.

Petersen [35] gives times for normal and uniform random number generators on a NEC SX-3. His implementation *normalen* of the Box-Muller method takes 55.5 nsec per normally distributed number, i.e. it is 2.4 times faster than our method B1, and 1.51 times faster than our method B3. The model of SX-3 used by Petersen has an effective peak speed of 2.75 Gflop, which is 2.2 times the peak speed of the VP2200/10. Considering the relative speeds of the two machines and the fact that the SX-3 has a hardware square root function, our results are encouraging.

2.3 Vectorised Implementation of the Polar Method

The times given in Table 1 for methods B1–B3 can be used to predict the best possible performance of the Polar method (§2.1). The Polar method avoids the computation of sin and cos, so could gain up to 6.6 cycles per normal random number over method B3. However, we would expect the gain to be less than this because of the overhead of a vector gather caused by use of a rejection method. A straightforward vectorised implementation of the Polar method, called method P1, was written to test this prediction. The results are shown in Table 1. 13.8 cycles are saved by avoiding the sin and cos function evaluations, but the overhead increases by 6.0 cycles, giving an overall saving of 7.8 cycles or 19%. Thus, method P1 is about the same speed as method B2, but not as fast as method B3.

Encouraged by our success in avoiding most ln and sqrt computations in the Box-Muller method (see method B3), we considered a similar idea to speed up the Polar method. Step 4 of the Polar method (§2.1) involves the computation of $\sqrt{-2\ln(s)/s}$, where $0 < s < 1$. The function has a singularity at $s = 0$, but we can approximate it quite well on an interval such as $[1/9, 1]$, using a method similar to that used to approximate the function $g(u)$ of §2.2.

Inspection of the proof in Knuth [24] shows that step 4 of the Polar method can be replaced by

4a. Set $r \leftarrow \sigma\sqrt{-2\ln(u)/s}$,
 and return $rx + \mu$ and $ry + \mu$

here u is any uniformly distributed variable in $(0, 1]$, provided u is independent arctan(y/x). In particular, we can take $u = 1 - s$. Thus, omitting the constant ctor $\sigma\sqrt{2}$, we need to evaluate $\sqrt{-\ln(1 - s)/s}$, but this is just $g(s)$, and we can use exactly the same approximation as in §2.2. This gives us method P2. To summarise, a non-vectorised version of method P2 is:

1. Generate independent uniform numbers $x, y \in [-1, 1)$.
2. Compute $s \leftarrow x^2 + y^2$.
3. If $s \geq 1$ then go to step 1 (i.e. *reject* x and y) else go to step 4.
4. If $s > 8/9$ then
    4.1. set $r \leftarrow \sigma\sqrt{-\ln(1 - s)/s}$ using library routines, else
    4.2. set $v \leftarrow (6s - 4)/(4 - 3s)$, evaluate $h(v)$ as described in §2.2, and set $r \leftarrow \sigma h(v)$.
5. Return $xr\sqrt{2} + \mu$ and $yr\sqrt{2} + \mu$, which are independent, normal random numbers with mean μ and standard deviation σ.

To vectorise steps 1-3, we simply generate vectors of x_j and y_j values, compute $s_j = x_j^2 + y_j^2$, and compress by omitting any triple (x_j, y_j, s_j) for which $s_j \geq 1$. This means that we can not predict in advance how many normal random numbers will be generated, but this problem is easily handled by introducing a level of buffering.

The second-last column of Table 1 gives results for method P2. There is a saving of 11.9 cycles or 35% compared to method P1, and the method is 17% faster than the fastest version of the Box-Muller method (method B3). The cost of logarithm and square root computations is only 37% of the total, the remainder being the cost of generating uniform random numbers (about 13%)

and the cost of the rejection step and other overheads (about 50%). On the VP2200/10 we can generate more than 14 million normally distributed random numbers per second.

2.4 The Ratio Method

The Polar method is one of the simplest of a class of rejection methods for generating random samples from the normal (and other) distributions. Other examples are given in [2, 5, 14, 24]. It is possible to implement some of these methods in a manner similar to our implementation of method P2. For example, a popular method is the Ratio Method of Kinderman and Monahan [23] (also described in [24], and improved in [26]). In its simplest form, the Ratio Method is given by *Algorithm R:*

1. Generate independent uniform numbers $u, v \in [0, 1)$.
2. Set $x \leftarrow \sqrt{8/e}(v - \frac{1}{2})/(1 - u)$.
3. If $-x^2 \ln(1 - u) > 4$ then go to step 1 (i.e. reject x) else go to step 4.
4. Return $\sigma x + \mu$.

Algorithm R returns a normally distributed random number using on average $8/\sqrt{\pi e} \simeq 2.74$ uniform random numbers and 1.37 logarithm evaluations. For the proof of correctness, and various refinements which reduce the number of logarithm evaluations, see [23, 24, 26]. The idea of the proof is that x is normally distributed if the point (u, v) lies inside a certain closed curve C which in turn is inside the rectangle $[0, 1] \times [-\sqrt{2/e}, +\sqrt{2/e}]$. Step 3 rejects (u, v) if it is outside C.

The function $\ln(1 - u)$ occurring at step 3 has a singularity at $u = 1$, but it can be evaluated using a polynomial or rational approximation on some interval $[0, \tau]$, where $\tau < 1$, in much the same way as the function $g(u)$ of §2.2.

The refinements added by Kinderman and Monahan [23] and Leva [26] avoid most of the logarithm evaluations. The following step is added:

2.5. If $P_1(u, v)$ then go to step 4
   else if $P_2(u, v)$ then go to step 1
   else go to step 3.

Here $P_1(u, v)$ and $P_2(u, v)$ are easily-computed conditions. Geometrically, P_1 corresponds to a region R_1 which lies inside C, and P_2 corresponds to a region R_2 which encloses C, but R_1 and R_2 have almost the same area. Step 3 is only executed if (u, v) lies in the borderline region $R_2 \backslash R_1$.

Step 2.5 can be vectorised, but at the expense of several vector scatter/gather operations. Thus, the saving in logarithm evaluations is partly cancelled out by an increase in overheads. The last column (R1) of Table 1 gives the times for our implementation on the VP2200. As expected, the time for the logarithm computation is now negligible, and the overheads dominate. In percentage terms the times are:

1% logarithm computation (using the library routine),
17% uniform random number computation,
23% scatter and gather to handle borderline region,
59% step 2.5 and other overheads.

Although disappointing, the result for the Ratio method is not surprising, because the computations and overheads are similar to those for method P2 (though with less logarithm computations), but only half as many normal random numbers are produced. Thus, we would expect the Ratio method to be slightly better than half as fast as method P2, and this is what Table 1 shows.

2.5 Other Methods

On serial machines our old algorithm GRAND [5] is competitive with the Ratio method. In fact, GRAND is the fastest of the methods compared by Leva [26]. GRAND is based on an idea of Von Neumann and Forsythe for generating samples from a distribution with density function $c \exp(-h(x))$, where $0 \leq h(x) \leq 1$:

1. Generate a uniform random number $x \in [0, 1)$, and set $u_0 \leftarrow h(x)$.
2. Generate independent uniform random numbers $u_1, u_2, \ldots \in [0, 1)$ until the first $k > 0$ such that $u_{k-1} < u_k$.
3. If k is odd then return x,
   else reject x and go to step 1.

A proof of correctness is given in Knuth [24].

It is hard to see how to implement GRAND efficiently on a vector processor. There are two problems –

1. k is not bounded, even though its expected value is small. Thus, a sequence of gather operations seems to be required. The result would be similar to Petersen's implementation [35] of a generator for the Poisson distribution (much slower than his implementation for the normal distribution).
2. Because of the restriction $0 \leq h(x) \leq 1$, the area under the normal curve $\exp(-x^2/2)/\sqrt{2\pi}$ has to be split into different regions from which samples are drawn with probabilities proportional to their areas. This complicates the implementation of the rejection step.

For these reasons we would expect a vectorised implementation of GRAND to be even slower than our implementation of the Ratio method. Similar comments apply to other rejection methods which use an iterative rejection process and/or several different regions.

3   Vectorisation of Wallace's Normal RNG

Recently Wallace [40] proposed a new class of pseudo-random generators for normal variates. These generators do not require a stream of uniform pseudo-random numbers (except for initialisation) or the evaluation of elementary functions such as log, sqrt, sin or cos (needed by the Box-Muller and Polar methods).

The crucial observation is that, if x is an n-vector of normally distributed random numbers, and A is an $n \times n$ orthogonal matrix, then $y = Ax$ is another n-vector of normally distributed numbers. Thus, given a pool of nN normally distributed numbers, we can generate another pool of nN normally distributed numbers by performing N matrix-vector multiplications. The inner loops are very suitable for implementation on vector processors. The vector lengths are proportional to N, and the number of arithmetic operations per normally distributed number is proportional to n. Typically we choose n to be small, say $2 \le n \le 4$, and N to be large.

Wallace implemented variants of his new method on a scalar RISC workstation, and found that its speed was comparable to that of a fast uniform generator, and much faster than the "classical" methods considered in §2. The same performance relative to a fast uniform generator is achievable on a vector processor, although some care has to be taken with the implementation (see §3.6).

In §3.1 we describe Wallace's new methods in more detail. Some statistical questions are considered in §§3.2–3.5. Aspects of implementation on a vector processor are discussed in §3.6, and details of an implementation on the VP2200 and VPP300 are given in §3.7.

3.1 Wallace's Normal Generators

The idea of Wallace's new generators is to keep a pool of nN normally distributed pseudo-random variates. As numbers in the pool are used, new normally distributed variates are generated by forming appropriate combinations of the numbers which have been used. On a vector processor N can be large and the whole pool can be regenerated with only a small number of vector operations[1].

The idea just outlined is the same as that of the generalised Fibonacci generators for uniformly distributed numbers – a pool of random numbers is transformed in an appropriate way to generate a new pool. As Wallace [40] observes, we can regard the uniform, normal and exponential distributions as maximum-entropy distributions subject to the constraints:

$0 \le x \le 1$ (uniform)
$E(x^2) = 1$ (normal)
$E(x) = 1,\ x \ge 0$ (exponential).

We want to combine $n \ge 2$ numbers in the pool so as to satisfy the relevant constraint, but to conserve no other statistically relevant information. To simplify notation, suppose that $n = 2$ (there is no problem in generalising to $n > 2$). Given two numbers x, y in the pool, we could satisfy the "uniform" constraint by forming

$$x' \leftarrow (x + y) \bmod 1,$$

and this gives the family of generalised Fibonacci generators [6].

---

[1] The process of regenerating the pool will be called a "pass".

We could satisfy the "normal" constraint by forming

$$\begin{pmatrix} x' \\ y' \end{pmatrix} \leftarrow A \begin{pmatrix} x \\ y \end{pmatrix},$$

where A is an orthogonal matrix, for example

$$A = \frac{1}{\sqrt{2}} \begin{pmatrix} 1 & 1 \\ -1 & 1 \end{pmatrix}$$

or

$$A = \frac{1}{5} \begin{pmatrix} 4 & 3 \\ -3 & 4 \end{pmatrix}.$$

Note that this generates two new pseudo-random normal variates x' and y' from x and y, and the constraint

$$x'^2 + y'^2 = x^2 + y^2$$

is satisfied because A is orthogonal.

Suppose the pool of previously generated pseudo-random numbers contains $x_0, \ldots, x_{N-1}$ and $y_0, \ldots, y_{N-1}$. Let $\alpha, \ldots, \delta$ be integer constants. These constants might be fixed throughout, or they might be varied (using a subsidiary uniform random number generator) each time the pool is regenerated.

One variant of Wallace's method generates $2N$ new pseudo-random numbers $x'_0, \ldots, x'_{N-1}$ and $y'_0, \ldots, y'_{N-1}$ using the recurrence

$$\begin{pmatrix} x'_j \\ y'_j \end{pmatrix} = A \begin{pmatrix} x_{\alpha j + \gamma \bmod N} \\ y_{\beta j + \delta \bmod N} \end{pmatrix} \tag{1}$$

for $j = 0, 1, \ldots, N-1$. The vectors x' and y' can then overwrite x and y, and be used as the next pool of $2N$ pseudo-random numbers. To avoid the copying overhead, a double-buffering scheme can be used.

3.2 Desirable Constraints

In order that all numbers in the old pool (x, y) are used to generate the new pool (x', y'), it is essential that the indices

$$\alpha j + \gamma \bmod N$$

and

$$\beta j + \delta \bmod N$$

give permutations of $\{0, 1, \ldots, N-1\}$ as j runs through $\{0, 1, \ldots, N-1\}$. A necessary and sufficient condition for this is that

$$\mathrm{GCD}(\alpha, N) = \mathrm{GCD}(\beta, N) = 1. \tag{2}$$

For example, if N is a power of 2, then any odd α and β may be chosen.

The orthogonal matrix A must be chosen so each of its rows has at least two nonzero elements, to avoid repetition of the same pseudo-random numbers. Also, these nonzeros should not be too small.

For implementation on a vector processor it would be efficient to take $\alpha = \beta = 1$ so vector operations have unit strides. However, statistical considerations indicate that unit strides should be avoided. To see why, suppose $\alpha = 1$. Thus, from (1),

$$x_j' = a_{0,0} x_{j+\gamma \bmod N} + a_{0,1} y_{\beta j + \delta \bmod N} ,$$

where $|a_{0,0}|$ is not very small. The sequence (z_j) of random numbers returned to the user is

$$x_0, \ldots, x_{N-1}, y_0, \ldots, y_{N-1},$$
$$x_0', \ldots, x_{N-1}', y_0', \ldots, y_{N-1}', \cdots$$

so we see that z_n is strongly correlated with $z_{n+\lambda}$ for $\lambda = 2N - \gamma$.

Wallace [40] suggests a "vector" scheme where $\alpha = \beta = 1$ but γ and δ vary at each pass. This is certainly an improvement over keeping γ and δ fixed. However, there will still be correlations over segments of length $O(N)$ in the output, and these correlations can be detected by suitable statistical tests. Thus, we do not recommend the scheme for a library routine, although it would be satisfactory in many applications.

We recommend that α and β should be different, greater than 1, and that γ and δ should be selected randomly at each pass to reduce any residual correlations.

For similar reasons, it is desirable to use a different orthogonal matrix A at each pass. Wallace suggests randomly selecting from two predefined 4×4 matrices, but there is no reason to limit the choice to two[2]. We prefer to choose "random" 2×2 orthogonal matrices with rotation angles not too close to a multiple of $\pi/2$.

3.3 The Sum of Squares

As Wallace points out, an obvious defect of the schemes described in §§3.1–3.2 is that the sum of squares of the numbers in the pool is fixed (apart from the effect of rounding errors). For independent random normal variates the sum of squares should have the chi-squared distribution χ_ν^2, where $\nu = nN$ is the pool size.

To overcome this defect, Wallace suggests that one pseudo-random number from each pool should not be returned to the user, but should be used to approximate a random sample S from the χ_ν^2 distribution. A scaling factor can be introduced to ensure that the sum of squares of the ν values in the pool (of which $\nu - 1$ are returned to the user) is S. This only involves scaling the matrix A, so the inner loops are essentially unchanged.

---

[2] Caution: if a finite set of predefined matrices is used, the matrices should be multiplicatively independent over $GL(n, R)$. (If $n = 2$, this means that the rotation angles (mod 2π) should be independent over the integers.) In particular, no matrix should be the inverse of any other matrix in the set.

There are several good approximations to the χ_ν^2 distribution for large ν. For example,

$$2\chi_\nu^2 \simeq \left(x + \sqrt{2\nu - 1}\right)^2 , \tag{3}$$

where x is $N(0, 1)$. More accurate approximations are known [1], but (3) should be adequate if ν is large.

3.4 Restarting

Unlike the case of generalised Fibonacci uniform random number generators [8], there is no well-developed theory to tell us what the period of the output sequence of pseudo-random normal numbers is. Since the size of the state-space is at least 2^{2wN}, where w is the number of bits in a floating-point fraction and $2N$ is the pool size (assuming the worst case $n = 2$), we would expect the period to be at least of order 2^{wN} (see Knuth [24]), but it is difficult to guarantee this. One solution is to restart after say 1000N numbers have been generated, using a good uniform random number generator with guaranteed long period combined with the Box-Muller method to refill the pool.

3.5 Discarding Some Numbers

Because each pool of pseudo-random numbers is, strictly speaking, determined by the previous pool, it is desirable not to return all the generated numbers to the user[3]. If $f \geq 1$ is a constant parameter[4], we can return a fraction $1/f$ of the generated numbers to the user and "discard" the remaining fraction $(1 - 1/f)$. The discarded numbers are retained internally and used to generate the next pool. There is a tradeoff between independence of the numbers generated and the time required to generate each number which is returned to the user. Our tests (described in §3.7) indicate that $f \geq 3$ is satisfactory.

3.6 Vectorised Implementation

If the recurrence (1) is implemented in the obvious way, the inner loop will involve index computations modulo N. It is possible to avoid these computations. Thus $2N$ pseudo-random numbers can be generated by $\alpha + \beta - 1$ iterations of a loop of the form

```
do j = low, high
xp(j) = A00*x(alpha*j + jx) + A01*y(beta*j + jy)
yp(j) = A10*x(alpha*j + jx) + A11*y(beta*j + jy)
enddo
```

---

[3] Similar remarks apply to some uniform pseudo-random number generators [24, 27].
[4] We shall call f the "throw-away" factor.

where **low, high, jx,** and **jy** are integers which are constant within the loop but vary between iterations of the loop. Thus, the loop vectorises. To generate each pseudo-random number requires one load (non-unit stride), one floating-point add, two floating-point multiplies, one store, and of order

$$\frac{\alpha + \beta}{N}$$

startup costs. The average cost should is only a few machine cycles per random number if N is large and $\alpha + \beta$ is small.

On a vector processor with interleaved memory banks, it is desirable for the strides α and β to be odd so that the maximum possible memory bandwidth can be achieved. For statistical reasons we want α and β to be distinct and greater than 1 (see §3.2). For example, we could choose

$$\alpha = 3, \quad \beta = 5,$$

provided $\mathrm{GCD}(\alpha\beta, N) = 1$ (true if N is a power of 2). Since $\alpha + \beta - 1 = 7$, the average vector length in vector operations is about $N/7$.

Counting operations in the inner loop above, we see that generation of each pseudo-random $N(0,1)$ number requires about two floating-point multiplications and one floating-point addition, plus one (non-unit stride) load and one (unit-stride) store. To transform the $N(0,1)$ numbers to $N(\mu, \sigma^2)$ numbers with given mean and variance requires an additional multiply and add (plus a unit-stride load and store) [5]. Thus, if f is the throw-away factor (see §3.5), each pseudo-random $N(\mu, \sigma^2)$ number returned to the user requires about $2f + 1$ multiplies and $f + 1$ additions, plus $f + 1$ loads and $f + 1$ stores.

If performance is limited by the multiply pipelines, it might be desirable to reduce the number of multiplications in the inner loop by using fast Givens transformations (i.e. diagonal scaling). The scaling could be undone when the results were copied to the caller's buffer. To avoid problems of over/underflow, explicit scaling could be performed occasionally (e.g. once every 50-th pass through the pool should be sufficient).

The implementation described in §3.7 does not include fast Givens transformations or any particular optimisations for the case $\mu = 0$, $\sigma = 1$.

3.7 RANN4

We have implemented the method described in §§3.5–3.6 in Fortran on the VP2200 and VPP300. The current implementation is called **RANN4**. The implementation uses **RANU4** [6] to generate uniform pseudo-random numbers for initialisation and generation of the parameters $\alpha, \ldots, \delta$ (see (1)) and pseudo-random orthogonal matrices (see below). Some desirable properties of the uniform random number generator are inherited by **RANN4**. For example, the processor id is appended to the seed, so it is certain that different pseudo-random sequences

---

[5] Obviously some optimisations are possible if it is known that $\mu = 0$ and $\sigma = 1$.

will be generated on different processors, even if the user calls the generator with the same seed on several processors of the VPP300.

The user provides RANN4 with a work area which must be preserved between calls. RANN4 chooses a pool size of $2N$, where $N \geq 256$ is the largest power of 2 possible so that the pool fits within part (about half) of the work area. The remainder of the work area is used for the uniform generator and to preserve essential information between calls. RANN4 returns an array of normally distributed pseudo-random numbers on each call. The size of this array, and the mean and variance of the normal distribution, can vary from call to call.

The parameters $\alpha, \ldots, \delta$ (see (1)) are chosen in a pseudo-random manner, once for each pool, with $\alpha \in \{3,5\}$ and $\beta \in \{7,11\}$. The parameters γ and δ are chosen uniformly from $\{0, 1, \ldots, N-1\}$. The orthogonal matrix A is chosen in a pseudo-random manner as

$$A = \begin{pmatrix} \cos\theta & \sin\theta \\ -\sin\theta & \cos\theta \end{pmatrix},$$

where $\pi/6 \leq |\theta| \leq \pi/3$ or $2\pi/3 \leq \theta \leq 5\pi/6$. The constraints on θ ensure that $\min(|\sin\theta|, |\cos\theta|) \geq 1/2$. We do not need to compute trigonometric functions: a uniform generator is used to select $t = \tan(\theta/2)$ in the appropriate range, and then $\sin\theta$ and $\cos\theta$ are obtained using a few arithmetic operations. The matrix A is fixed in each inner loop (though not in each complete pass) so multiplications by $\cos\theta$ and $\sin\theta$ are fast.

For safety we adopt the conservative choice of throw-away factor $f = 3$ (see §3.5), although in most applications the choice $f = 2$ (or even $f = 1$) is satisfactory and significantly faster.

Because of our use of RANU4 to generate the parameters $\alpha, \ldots, \delta$ etc, it is most unlikely that the period of the sequence returned by RANN4 will be shorter than the period of the uniformly distributed sequence generated by RANU4. Thus, it was not considered necessary to restart the generator as described in §3.4. However, our implementation monitors the sum of squares and corrects for any "drift" caused by accumulation of rounding errors.

On the VP2200/10, the time per normally distributed number is approximately $(6.8f + 3.2)$ nsec, i.e. $(1.8f + 1.0)$ cycles. With our choice of $f = 3$ this is 23.6 nsec or 6.4 cycles. The fastest version, with $f = 1$, takes 10 nsec or 2.8 cycles. For comparison, the fastest method of those considered in [7] (the Polar method) takes 21.9 cycles. Thus, we have obtained a speedup by a factor of about 3.2 in the case $f = 3$.

Times on a single processor of the VPP300 are typically faster by a factor of about two, which is to be expected since the peak speed of a processor on the VPP300 is 2.285 GFlop (versus 1.25 Gflop on the VP2200/10). On the VPP300 with P processors, the time per normally distributed number is $11.4/P$ nsec if $f = 3$ and $5.4/P$ nsec if $f = 1$.

Various statistical tests were performed on RANN4 with several values of the throw-away factor f. For example:

- If (x, y) is a pair of pseudo-random numbers with (supposed) normal $N(0, 1)$ distributions, then $u = \exp(-(x^2 + y^2)/2)$ should be uniform in $[0, 1]$, and $v = \arctan(x/y)$ should be uniform in $[-\pi/2, +\pi/2]$. Thus, standard tests for uniform pseudo-random numbers can be applied. For example, we generated batches of (up to) 10^7 pairs of numbers, transformed them to (u, v) pairs, and tested uniformity of u (and similarly for v) by counting the number of values occurring in $1,000$ equal size bins and computing the χ^2_{999} statistic. This test was repeated several times with different initial seeds etc. The χ^2 values were not significantly large or small for any $f \geq 1$.
- We generated a batch of up to 10^7 pseudo-random numbers, computed the sample mean, second and fourth moments, repeated a number of times, and compare the observed and expected distributions of sample moments. The observed moments were not significantly large or small for any $f \geq 3$. The fourth moment was sometimes significantly small (at the 5% confidence level) for $f = 1$.

A possible explanation for the behaviour of the fourth moment when $f = 1$ is as follows. Let the maximum absolute value of numbers in the pool at one pass be M, and at the following pass be M'. By considering the effect of the orthogonal transformations applied to pairs of numbers in the pool, we see that (assuming $n = 2$),

$$M/\sqrt{2} \leq M' \leq \sqrt{2}M .$$

Thus, there is a correlation in the size of outliers at successive passes. The correlation for the subset of values returned to the user is reduced (although not completely eliminated) by choosing $f > 1$.

4   Summary and Conclusions for Normal RNG

We showed that both the Box-Muller and Polar methods for normally distributed random numbers vectorise well, and that it is possible to avoid and/or speed up the evaluation of the functions (sin, cos, ln, sqrt) which appear necessary. On the VP2200/10 our best implementation of the Polar method takes 21.9 machine cycles per normal random number, slightly faster than our best implementation of the Box-Muller method (26.3 cycles).

We considered the vectorisation of some other popular methods for generating normally distributed random numbers, and showed why such methods are unlikely to be faster than the Polar method on a vector processor.

We showed that normal pseudo-random number generators based on Wallace's ideas vectorise well, and that their speed on a vector processor is close to that of the generalised Fibonacci uniform generators, i.e. only a small number of machine cycles per random number.

Because Wallace's methods are new, there is little knowledge of their statistical properties. However, a careful implementation should have satisfactory statistical properties provided distinct non-unit strides α, β satisfying (2) are used, the sums of squares are varied as described in §3.3, and the throw-away

18

factor f is chosen appropriately. The pool size should be fairly large (subject to storage constraints), both for statistical reasons and to improve performance of the inner loops. Wallace uses 4×4 orthogonal transformations, but a satisfactory generator is possible with 2×2 orthogonal transformations.

It may appear that we have concentrated on vector rather than parallel implementations. If this is true, it is because vectorisation is the more interesting and challenging topic. Parallelisation of random number generators is in a technical sense "easy" since no communication is required after the initialisation on different processors. However, care has to be taken with this initialisation to ensure independence (see §1), and testing of parallel RNGs should not ignore this important requirement.

Acknowledgements

Thanks are due to:

- Don Knuth for discussions regarding the properties of generalised Fibonacci methods and for bringing some references to my attention.
- Wes Petersen for his comments and helpful information on implementations of random number generators on Cray and NEC computers [34, 35].
- Chris Wallace for sending me a preprint of his paper [40] and commenting on my attempts to vectorise his method.
- Andy Cleary, Bob Gingold, Markus Hegland and Peter Price for their assistance on the Vector/Parallel Scientific Subroutine Library ("area 4") project.

This work was supported in part by a Fujitsu-ANU research agreement. The ANU Supercomputer Facility provided computer time for development and testing on Fujitsu VP2200 and VPP300 computers at the Australian National University.

References

1. M. Abramowitz and I. A. Stegun: Handbook of Mathematical Functions. Dover, New York, 1965, Ch. 26.
2. J. H. Ahrens and U. Dieter: Computer Methods for Sampling from the Exponential and Normal Distributions. Comm. ACM **15** (1972), 873–882.
3. S. Aluru, G. M. Prabhu and J. Gustafson: A Random Number Generator for Parallel Computers. Parallel Computing **18** (1992), 839.
4. S. L. Anderson: Random Number Generators on Vector Supercomputers and Other Advanced Architectures, SIAM Review **32** (1990), 221–251.
5. R. P. Brent: Algorithm 488: A Gaussian Pseudo-Random Number Generator (G5). Comm. ACM **17** (1974), 704–706.
6. R. P. Brent: Uniform Random Number Generators for Supercomputers. Proc. Fifth Australian Supercomputer Conference, Melbourne, December 1992, 95–104. ftp://nimbus.anu.edu.au/pub/Brent/rpb132.dvi.gz
7. R. P. Brent: Fast Normal Random Number Generators for Vector Processors. Report TR-CS-93-04, Computer Sciences Laboratory, Australian National University, March 1993. ftp://nimbus.anu.edu.au/pub/Brent/rpb141tr.dvi.gz

8. R. P. Brent: On the Periods of Generalized Fibonacci Recurrences, Math. Comp. **63** (1994), 389–401.
9. R. P. Brent: A Fast Vectorised Implementation of Wallace's Normal Random Number Generator. Report TR-CS-97-07, Computer Sciences Laboratory, Australian National University, Canberra, April 1997. `ftp://nimbus.anu.edu.au/pub/Brent/rpb170tr.dvi.gz`
10. C. W. Clenshaw, L. Fox, E. T. Goodwin, D. W. Martin, J. G. L. Michel, G. F. Miller, F. W. J. Olver and J. H. Wilkinson: Modern Computing Methods. 2nd edition, HMSO, London, 1961, Ch. 8.
11. P. D. Coddington: Random Number Generators for Parallel Computers. The NHSE Review **2** (1996). `http://nhse.cs.rice.edu/NHSEreview/RNG/PRNGreview.ps`
12. P. D. Coddington and S-H. Ko: Techniques for Empirical Testing of Parallel Random Number Generators. Proc. International Conference on Supercomputing (ICS'98), Melbourne, Australia, July 1998, to appear.
13. S. A. Cuccaro, M. Mascagni and D. V. Pryor: Techniques for Testing the Quality of Parallel Pseudo-Random Number Generators. Proc. 7th SIAM Conf. on Parallel Processing for Scientific Computing, SIAM, Philadelphia, 1995, 279–284.
14. L. Devroye: Non-Uniform Random Variate Generation. Springer-Verlag, New York, 1986.
15. P. L'Ecuyer: Efficient and Portable Combined Random Number Generators. Comm. ACM **31** (1988), 742–749, 774.
16. P. L'Ecuyer: Random Numbers for Simulation. Comm. ACM **33**, 10 (1990), 85–97.
17. P. L'Ecuyer and S. Côté: Implementing a Random Number Package with Splitting Facilities. ACM Trans. Math. Software **17** (1991), 98–111.
18. W. Evans and B. Sugla: Parallel Random Number Generation. Proc. 4th Conference on Hypercube Concurrent Computers and Applications (ed. J. Gustafson), Golden Gate Enterprises, Los Altos, CA, 1989, 415.
19. A. M. Ferrenberg, D. P. Landau and Y. J. Wong: Monte Carlo Simulations: Hidden Errors From "Good" Random Number Generators. Phys. Rev. Lett. **69** (1992), 3382–3384.
20. P. Griffiths and I. D. Hill (editors): Applied Statistics Algorithms. Ellis Horwood, Chichester, 1985.
21. J. R. Heringa, H. W. J. Blöte and A. Compagner: New Primitive Trinomials of Mersenne-Exponent Degrees for Random-Number Generation. Internat. J. of Modern Physics C **3** (1992), 561–564.
22. F. James: A Review of Pseudo-Random Number Generators. Computer Physics Communications **60** (1990), 329–344.
23. A. J. Kinderman and J. F. Monahan: Computer Generation of Random Variables Using the Ratio of Uniform Deviates. ACM Trans. Math. Software **3** (1977), 257–260.
24. D. E. Knuth: The Art of Computer Programming. Volume 2: Seminumerical Algorithms. 3rd edn. Addison-Wesley, Menlo Park, 1997.
25. D. H. Lehmer: Mathematical Methods in Large-Scale Computing Units. Ann. Comput. Lab. Harvard Univ. **26** (1951), 141–146.
26. J. L. Leva: A Fast Normal Random Number Generator. ACM Trans. Math. Software **18** (1992), 449–453.
27. M. Lüscher:, A Portable High-Quality Random Number Generator for Lattice Field Theory Simulations. Computer Physics Communications **79** (1994), 100–110.
28. G. Marsaglia: A Current View of Random Number Generators. Computer Science and Statistics: Proc. 16th Symposium on the Interface, Elsevier Science Publishers B. V. (North-Holland), 1985, 3–10.

29. M. Mascagni, S. A. Cuccaro, D. V. Pryor and M. L. Robinson: A Fast, High-Quality, and Reproducible Lagged-Fibonacci Pseudorandom Number Generator. J. of Computational Physics **15** (1995), 211–219.

30. M. Mascagni, M. L. Robinson, D. V. Pryor and S. A. Cuccaro: Parallel Pseudorandom Number Generation Using Additive Lagged-Fibonacci Recursions. Springer-Verlag Lecture Notes in Statistics **106** (1995), 263–277.

31. M. E. Muller: A Comparison of Methods for Generating Normal Variates on Digital Computers. J. ACM **6** (1959), 376–383.

32. J. von Neumann: Various Techniques Used in Connection With Random Digits. The Monte Carlo Method, National Bureau of Standards (USA) Applied Mathematics Series **12** (1951), 36.

33. S. K. Park and K. W. Miller: Random Number Generators: Good Ones are Hard to Find. Comm. ACM **31** (1988) 1192–1201.

34. W. P. Petersen: Some Vectorized Random Number Generators for Uniform, Normal, and Poisson Distributions for CRAY X-MP. J. Supercomputing **1** (1988), 327–335.

35. W. P. Petersen: Lagged Fibonacci Series Random Number Generators for the NEC SX-3. Internat. J. High Speed Computing **6** (1994), 387–398.

36. D. V. Pryor, S. A. Cuccaro, M. Mascagni and M. L. Robinson: Implementation and Usage of a Portable and Reproducible Parallel Pseudorandom Number Generator. Proc. Supercomputing '94, IEEE, New York, 1994, 311–319.

37. I. Vattulainen, T. Ala-Nissila and K. Kankaala: Physical Tests for Random Numbers in Simulations. Phys. Rev. Lett. **73** (1994), 2513.

38. I. Vattulainen, T. Ala-Nissila and K. Kankaala: Physical Models as Tests of Randomness. Phys. Rev. E **52** (1995), 3205.

39. C. S. Wallace: Transformed Rejection Generators for Gamma and Normal Pseudorandom Variables. Australian Computer Journal **8** (1976), 103–105.

40. C. S. Wallace: Fast Pseudo-Random Generators for Normal and Exponential Variates. ACM Trans. Math. Software **22** (1996), 119–127.

Heterogeneous HPC Environments

Marco Vanneschi

Department of Computer Science, University of Pisa, Italy

Abstract. Directions of software technologies for innovative HPC environments are discussed according to the industrial user requirements for heterogeneous multidisciplinary applications, performance portability, rapid prototyping and software reuse, integration and interoperability of standard tools. The various issues are demonstrated with reference to the PQE2000 project and its programming environment SkIE (Skeleton-based Integrated Environment). Modules developed by a variety of standard languages and tools are encapsulated into SkIECL (SkIE *Coordination Language*) structures to form the global application. A performance model associated to SkIECL allows the static and dynamic tools to introduce a large amount of global optimizations without the direct intervention of the programmer. The paper discusses also some of the most critical issues in matching architecture and software technology, showing how the SkIE environment, and its evolutions, can act as a solid framework in which innovative hardware-software systems for HPC can be studied and experimented.

1 Introduction

Broadening the use of High Performance Computing beyond its current research applications requires correction of some shortcomings in the technology at all levels: hardware architecture, software environments, and their relationships. A very critical role is played by the *software technology* of integrated programming environments for the modular and efficient development of industrial and commercial applications. Traditionally, the target of research and development in HPC has been mainly directed towards the scientific users and the Grand Challenge applications. However, it is important to recognize [1] that the widespread diffusion of HPC technology depends on its ability to satisfy the needs of industrial users: their main goal is to exploit HPC technology to significantly reduce the time-to-market of products. In the same way, HPC must be able to satisfy the requirements for critical public services and government decision making strategies.

These requirements represent the basic goals of the *PQE2000 project* [3]. PQE2000 is a joint initiative of the main Italian research agencies - CNR, ENEA, INFN - and of Finmeccanica's QSW (Quadrics Supercomputers World Ltd) for the realization of innovative HPC general-purpose systems and their applications for industry, commerce and public services. As a consequence of PQE2000 emerging technology, a Research Programme on HPC has been proposed aiming at the definition of a large European initiative in HPC technology and applications. The results of PQE2000 activities are transferred into industrial products by QSW, as in the case of the programming environment discussed in this paper. A research road map towards PetaFlops architectures and programming environments has been established and, at the same time, a first innovative version of PQE2000 technology has been realized by integrating stable existing products. In this paper, the basic ideas on the technologies for innovative HPC software environments are discussed and described with reference to the PQE2000 project and its programming environment SkIE (Skeleton-based Integrated Environment) [2, 4]. In SkIE an application is developed with the assistance of a coordination language (SkIECL), based on the *skeletons model* [9–11, 5, 12], in order to integrate, and to globally optimize the usage, of independent software modules: these last are, possibly very complex, programs independently developed by existing standard languages (host languages) and tools. Currently, the host languages include C, C++, F77, F90, and HPF. Work is in progress to integrate Java and other standard object-oriented tools. The coordination language SkIECL is an industrial version of P^3L, a prototype research vehicle designed at the Department of Computer Science, University of Pisa [5–8]. SkIECL allows the software designers to express, in a primitive and structured way, both *data parallelism* and *task parallelism*. Such structures can be combined hierarchically at any depth according to the principles for modular software development. The variety of parallelism strategies and their structured composition are the key issues for achieving global optimizations in parallel programs and to overcome the typical difficulties and inefficiencies of SPMD/data parallel languages. In section 2 we discuss the basic issues in the development of heterogeneous HPC environments. Section 3 contains a description of the SkIE programming model. Section 4 is dedicated to the discussion of critical issues in matching architecture and software technology.

2   Heterogeneous Multidisciplinary Environments

The introduction of software engineering methodologies in HPC global applications development requires that the current "in the small" approach evolves rapidly towards "in the large" development tools and environments. A *parallel application* is not merely a parallel algorithm: the parallelization problems must be solved along with the efficient integration of heterogeneous objects such as existing software modules, files and data base management tools, visualization tools, application libraries and packages. More in general, a comprehensive solution to the industrial needs in HPC application development implies to deal with several *heterogeneity* issues as discussed in the following.

Multidisciplinary Environments. Multidisciplinary environments for parallel machines are needed, by which an entire development team can easily take part in a complex industrial project. As an example, in many industrial areas - e.g. motor-car, energy and environment, aerospace, biotechnology and medicine, finance, new media services - application development implies, besides the utilization of different specific simulation techniques, the exploitation of techniques and tools for data mining, virtual reality, multipurpose analysis and optimization, interactive visualization, and so on. Achieving the goal of multidisciplinary applications requires the realization of *integrated* programming environments for parallel machines. As in the general context of Information Technology, this means that all the current and future standard languages, tools and packages for application design, each of which is possibly specialized towards an application area, are made available in the context of a uniform and well-structured development environment. However, the integration must also be characterized by *high performance* and *efficiency*, and this is much more important for multidisciplinary applications aiming to improve the industrial time-to-market and productivity: often, the combined utilization of several different tools, to form a multidisciplinary application, needs high performance systems, since it involves the execution of large amounts of tasks, each one structurable as a (massively) parallel program. The goal of high performance in multidisciplinary applications cannot be achieved by low-level machine-dependent mechanisms and tools and by mere "porting/rewriting" techniques: *high-level, automatic* tools must be available in order to restructure and to optimize a, possibly very large, set of heterogeneous modules individually developed by means of different standard languages and tools.

Performance Portability. Because of the requirements for reliability and standardization, as well as the continuous evolution of systems and of simulation techniques, industry privileges the utilization of commercial, general purpose systems and products that guarantee the portability of applications across different machine platforms. This is especially needed in a situation in which a great deal of uncertainty exists among industrial users about the characteristics of the next generation machines. Thus, the development environment we advocate is a high-level, multidisciplinary, *multi-platform* one. All the known HPC platforms are of interest: distributed memory and shared memory MPP, SMP, clusters of SMP, PIM, combined architectures (e.g. MPP+SMP, MPP+PIM), as well as metacomputers. Moreover, the programming environment must be able to address the effective utilization of future PetaFlops machines.

Especially in HPC, portability implies *performance portability*, i.e. the ability of a programming environment to support restructuring and re-optimization of applications passing from one machine platform to another one, or simply from one configuration/model of the same machine to the next one. This has to be done mainly in a *(semi-)automatic* manner, by means of powerful restructuring compiler tools and/or by suitable run-time supports. In SkIE the run-time support consists of a collection of very efficient implementation *templates* (fully invisible to the programmer), where more than one template exists for each skeleton with respect to the variety of underlying platforms. The design of implementation templates is done on top of a standard, portable abstract machine: an MPI machine has been adopted in the current version.

Rapid Prototyping and Software Reuse. The programming environment must allow the design team to develop new applications by a rapid prototyping approach. In successive design phases the application prototype could be refined and restructured for the sake of performance and/or modularity and reliability, by means of explicit and/or semiautomatic techniques. Very often, this feature implies the ability *to reuse* existing software modules or objects, without changing them, i.e. the ability to encapsulate existing modules into structured parallel programs, or with minimum modifications. This concept should be applied to modules written in any sequential language or even to already parallelized modules. Analyzing the software reuse feature in terms of achievable performance, two approaches to parallel program optimization can be distinguished :

- *global optimizations*, in which a computation is viewed as a parallel collection of modules, each of which is considered as a "black box" : optimizations are done at the level of the whole structure, without affecting the optimizations done by the native compiler of the imported modules;
- *local optimizations*, in which new opportunities for parallelism detection are exploited inside the modules according to the characteristics of the whole application and/or of the underlying machine.

Global optimizations are fundamental for rapid prototyping and software reuse, with the performance portability goal in mind. However, in terms of performance refinement, software reuse may imply also the adoption of local optimizations of the imported modules. Currently, in SkIE only global optimizations are implemented, relying on the existing standard compilers for the local optimizations. Research is in progress along the line of *global + local* optimizations: a notable example is the optimization of HPF modules in the context of a skeleton global structure that includes task parallelism [13].

3 Programming Models for Heterogeneous Environments

3.1 Modular HPC Environments

The issues discussed in sect. 2 emphasize the viewpoint that, despite several current approaches, HPC methodologies and tools should fully respect the Computer Science principles of modular software development. We believe that, even in this context, it is feasible to achieve high performance and performance portability all the same, i.e. the find satisfactory solutions to the so called PPP (Programmability, Portability, Performance) problem.

Principles of modular software development have been often, e.g. [17], enunciated by J. Dennis: *information hiding*: the internal structure of a module must not be known to the user; *invariant behavior*: the functional behavior of a module must be independent of the context in which it is invoked; *data generality*: it must be possible to pass any data object onto the interface to a module; *secure arguments*: side-effects on arguments must not be allowed at the interface of a module; *recursive construction*: any, possibly complex, computation must be usable as a module in building larger computations; *system resource management*:

resource, e.g. storage and communication, management must be performed by the system, invisibly to the program modules.

3.2 The Coordination Language Approach

The requirements stated till now have led to the idea of a coordination language to design complex HPC applications. The *constructs* of the coordination language must allow the designer to express the global structure of the parallel application in a high-level, fully machine independent fashion (in-the-large development of HPC programs). The constructs represent *the only forms of parallelism* (mechanisms + strategies) that can be used to express a parallel computation. Independently developed modules are easily inserted in to the global application structure according to the rules for combining and interfacing the parallel constructs.

The programming model of SkIE , i.e. the coordination language SkIECL of the integrated development environment, is based on the *skeletons* model [9–11, 5, 12]. Potentially, the skeletons approach has all the interesting properties advocated in the previous sections. In fact, some of the proposed skeletons systems are characterized by the compositionality property and by the existence of a formal performance model to drive the compiler optimizations: the term *structured parallel programming* has been introduced to denote those skeletons models that possess such powerful features. Among them : BMF [20], SCL [10], BACS [21] and P^3L [5]. In SkIECL the following skeletons (second order functions) have been chosen for the parallel constructs:

- *stream parallel skeletons*:
  seq: sequential module with well defined in-out interfaces
  pipe: pipelined composition of skeletons
  farm: self-load balancing process farm, where the generic (functionally replicated) worker can be any skeleton
  loop: data-flow loop, where the body can be any skeleton
- *data parallel skeletons:*
  map: independent data-parallel computations without communication
  reduce: parallel reduction by binary associative and commutative operators
  comp: sequential function composition to express data parallel computation with communication (stencils)

Figure 1 and 2 illustrate the graphical representation and the textual syntax of stream parallel skeleton and data parallel skeleton respectively. Specific issues to understand the approach are discussed in the following.

Compositionality. Compositionality of the coordination language constructs is, of course, fundamental to build modular program structures and, at the same time, to exploit the desired degree of parallelism. Figure 3 shows a SkIECL program computing a ray tracing algorithm on stream of images. In particular the

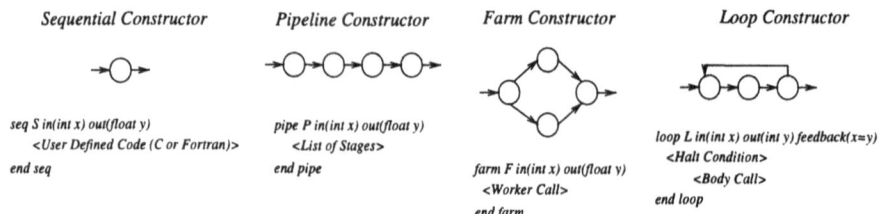

Fig. 1. Stream parallel skeletons representation

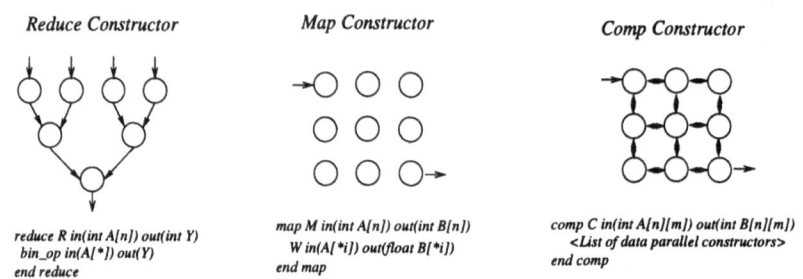

Fig. 2. Data parallel skeletons representation

computation of the ray tracer on each image I is parallelized using a **map** constructor decomposing I by row and applying the function **worker** to each row. Furthermore the map **trace** is nested in the farm **disp** that is in charge of dispatching the images to the first **map** ready to start a new computation on the next stream image.

The composite parallel computation can be viewed as a *construct tree*, where the leafs correspond to modules assumed as "black box" (information hiding, invariant behavior) and the remaining nodes to instances of the constructs of the coordination language (recursive construction). The root-construct corresponds to the outermost view of the parallel program. Figure 4 shows the tree associated to the program computing the parallel ray tracer on a stream of images.

Host Languages and Interfacing. In principle, the leaf-modules can be written in any language, called *host languages*; such modules are, possibly complex, programs independently developed by means of their native tools (in-the-large development of multidisciplinary heterogeneous applications, integration of tools).

The coordination language defines the *interfacing rules* and *mechanisms* for the composition, and the *data objects* which can be passed at the module interfaces [4]. In our approach, these mechanisms and objects must respect the data generality and secure arguments principles for modular construction. The basic semantics of the composition is *functional* in nature, though any language can be used as host language. The computation is *determinate*, except where spe-

```
#include "SceneType.h"
////////////////////////  SkIECL MAIN ///////////////////////////
pipe main in(char filename[STR_LENGTH]) out()
  read_data  in(filename) out(stream of Scene image)
  disp       in(image)    out(Pixl color[(XRESOLUTION + 1)][(YRESOLUTION +1)])
  write_data in(stream color) out()
end pipe
////////////////////  First Sequential Module  ///////////////////////////
read_data in(char filename[STR_LENGTH]) out(stream of Scene image)
$c{ /* Read the scene from the file filename */}c$
end
////////////////////  FARM Module  ///////////////////////////
farm disp in(Scene image) out(Pixl color[(XRESOLUTION + 1)][(YRESOLUTION + 1)])
    trace in(image) out(color)
end farm
////////////////////  MAP module Nested in a Farm  /////////////////
map trace in(Scene image) out(Pixl color[(XRESOLUTION + 1)][(YRESOLUTION + 1)])
    worker in(*i, *j, image) out(color[*j][*i])
end map
////////////////////  MAP Worker  ///////////////////////////
worker in(int x, int y, Scene image) out(Pixl col)
$c{ /* Compute the ray tracing algorithm on
    each row of the image */}c$
end
////////////////////  Last Module  ///////////////////////////
write_data in(stream of Pixl img[(XRESOLUTION + 1)][(YRESOLUTION + 1)]) out()
$c{ /* write the computed RGB image on file */ }c$
end
```

Fig. 3. A SkIECL program computing a ray tracer on a stream of images

cific constructs are used to control constrained forms of nondeterminism (e.g. feedbacks in data-flow loops). As in functional parallel programming, the *stream* data type is a very powerful mechanisms to define clear and powerful interfaces. Streams are SkIECL data types that allow programmers to manage the input as an unbounded sequence of items. A declaration **stream of int a** defines a sequence of items each one having type integer and name **a**. Streams can be *nested*. Stream can be *specialized*, that is programmers can specify the constant part that has to be replicated in each element of the stream. Only sequential constructors can create or delete a stream by means of the stream library functions provided by SkIECL language.

Forms of Parallelism: efficiency in irregular computations. The choice of the forms of parallelism corresponding to the skeletons is critical for achieving the best trade-off between modularity, predictability and performance.
The combined usage of data and task parallelism constructs in SkIE is considered fundamental to overcome the limitations of existing SPMD or data parallel standards, notably HPF, in designing irregular computations, i.e. computations in which the synchronization and data patterns, as well as data dependencies, vary dynamically, and dynamic load balancing is often explicitly required to achieve a satisfactory performance level [18]. In SkIECL a very frequent program structure for irregular/dynamic problems consists of parallel *loops* including stream

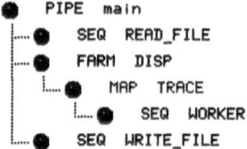

Fig. 4. The Constructor tree to the SkIECL parallel ray tracing

parallel substructures, e.g. *farms* that guarantee automatic load balancing ("on demand" strategy), and/or *pipes of farms*; shortly, this kind of computation will be denoted by *loop-farm*. Loop-farms emulate divide-and-conquer strategies efficiently [23]. In some cases, this structure can be improved by introducing *comp* and *map/reduce* in place of *farm* or at a deeper nesting level inside *pipe* or *farm* skeletons. Besides stream parallelism and data partitioning, another feature frequently used in the implementation of irregular/dynamic computations is *replication* of objects in farming ("specializer" constructor) and in map structures.

A programming model with a performance model. A cost calculus, or performance model, is associated to the coordination language SkIECL. This feature is fundamental in order to implement tools, and in particular the compiler, able to introduce as many global optimizations as possible. In the current implementation, the cost model consists of a set of functions predicting the performance of the template network, according to a set of parameters characterizing the target architecture and the user-defined sequential module code.

An important issue is the *predictability* of the underlying machine mechanisms, e.g. bandwidth/latency of interconnection structures, performance of memory subsystems, performance of I/O and file system. For example the current architecture of PQE2000 adopts an interconnection structure (fat tree) which is highly predictable. When predictability of machine mechanisms is low, the utilization of more powerful performance evaluation tools [22] is of great benefit, provided that they can be really integrated with the tools of the programming environment.

Global Optimizations. In SkIE all the low level issues and the very hard (NP-hard) problems of parallel processing [12] are left to the static and dynamic tools of the coordination language, that applies *approximate* heuristics to the composition of skeletons. The reader is referred to the literature on P^3L for the detailed description of the kind of optimizations adopted by the SkIE compiler. The main features can be summarized as follows:

– the same skeleton can have *different implementation templates*, among which the compiler can choose the best one for the specific program and target machine;

- while values of the architecture-dependent parameters are known for a "predictable" abstract/physical machine, the application-dependent parameters are estimated by *profiling* sequential modules at compile time;
- the construct tree can be optimized according to *formal transformations* (rewriting rules) applied to the skeleton composition. This is a very promising field in which substantial research has still to be done [19];
- the *template/resource optimization* phase applies the set of approximate heuristics by descending the construct tree and consulting the performance models and the local optimization rules for each template. This phase assumes an unbounded amount of computing resources;
- to meet the actual amount of computing resources, the implementation is reduced by applying the so called *Limited Resources Algorithm* [7] that "shrinks" the template network iteratively; at each step the amount of eliminated resources is the one that produces the minimum degradation of performance with respect to the unbounded resources version.

Figure 5 shows the results obtained running the parallel ray tracer, presented in section 3.2, on a CS2 with 24 nodes. Using the performance models, the SkIECL compiler is able to fix the optimal number of **map** and **farm** workers in order to reach the best trade off between efficiency and speed up, with a very good approximation (error less than 10%)with respect to the measured performances.

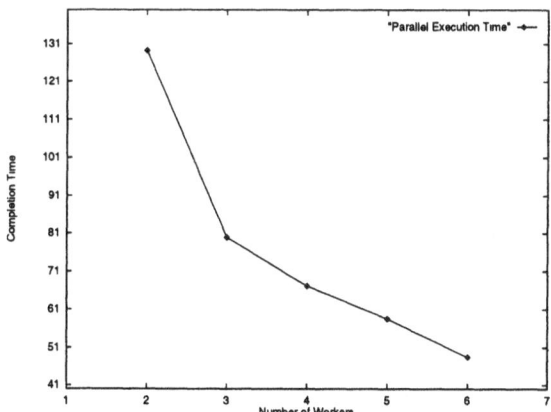

Fig. 5. Execution time of parallel ray tracer (speed up 21, 38 processes generated)

4 Matching Architecture and Software Technology

The approach in which the programming model is rigidly derived from the architectural model (or even from a specific architecture) has been, unfortunately, adopted for many years, leading to non-portable, non-modular, very expensive and error-prone parallel programs for SIMD and MIMD specific machines. In our opinion, the correct approach is different: the programming model must be *able to efficiently exploit the new directions in computer architecture*; at the same time, *the new directions in computer architecture should be influenced by the general characteristics of the programming models*.

Distributed Virtual Shared Memory. Especially in irregular/dynamic computations, skeletons implementations based on a form of *shared memory* is potentially able to increase the application performance significantly. In the current version of PQE2000, shared memory is emulated, on top of the native hardware, by a component of the Intermediate Virtual Machine called *Distributed Virtual Shared Memory (DVSM)* [24, 3]. Its goal is to allow the compiler to choose and to implement the most efficient strategies for shared objects in complex cooperation tasks. As an example, the DVSM implementation of MPI-1 collective operations achieves an improvement by a factor from 3 to 12 with respect to MPI-CH: this is very significant especially if we consider that DVSM is entirely implemented in software. In [25] the performance of irregular/dynamic computations has been evaluated, by using the *loop-farm* structure with a shared memory implementation using a uniform distribution strategy and excess parallelism in each node to hide memory access latency. It has been shown that significant improvements, of at least one order of magnitude, can be achieved with respect to message passing implementations. The opportunities for optimizations depends on the "by need" access patterns to objects (also a certain amount of fine grain accesses is efficiently tolerated) and by the proper management of the excess parallelism feature. The optimal number of processes per node can be approximated.

Based in these first evaluations, and exploiting the shared memory facilities of PQE2000, work is in progress to deeply evaluate the impact of shared memory implementation of SkIECL ; static and dynamic optimizations will be evaluated with respect to various architectural features (caching, prefetching, interconnection network, multi-threading, dynamic job scheduling, and so on) and language features (e.g., visibility of shared objects at the programmer level).

Memory-Oriented Models. An issue of growing importance is the set of problems caused by the increasing gap between processors and main memory [14] and, consequently, by the fact that access to memory is becoming the main parameter affecting performance. The problem can be dealt with, either on the side of the programming model, or on the side of the architectural model, or both. The SkIE programming environment is, by its own nature, defined in terms of a "performance-driven" programming model. According to the results achieved till now, and to the experiments on new run-time solutions, we are confident that, given a specific architecture, intensive optimizations can be realized to give a reasonable solution to the memory gap problem. This is due to the features of the coordination language explained in section 3.2.

If we study the architecture-software technology matching problem from the viewpoint of the parallel architecture, a general trend seems to be common to the most interesting architectural models for the next generation: *the exploitation of larger amounts of fine grain parallelism.* The skeleton model, and SkIECL in particular, has this potential, and this feature doesn't increase significantly the complexity in the application development. Future architectures will be able to exploit the high parallelism of the skeleton model even better. Two main solutions, not mutually exclusive, can be individuated: *multi-threading* [17] and *Processors-In-Memory (PIM)* [26, 3, 2, 28]. Their effectiveness to support the efficient implementation of structured parallel programs has to be evaluated accurately.

Another interesting solution to match architecture and software technology is *dynamic scheduling* in a multiuser context [15]. Dynamic scheduling techniques are fully compatible with the skeleton programming model, and SkIECL in particular. In fact, as sketched in sect. 3.2, the compilation of a SkIE program provides to the allocation and optimization of the available resources through the application of the Limited Resources Algorithm: the same algorithm can be applied at run-time in order to modify the amount of resources allocated to a program, provided that the scheduling strategy maintains the "processor working set" for each job in execution.

Seamless integration of parallel machines into distributed environments. The need for development environments that are much more oriented towards industrial requirements imposes that the characterization of MPP includes distributed systems at local/geographical level, clusters and metacomputers. Many complex industrial processes can be solved by exploiting the full power of all the computational resources available within a corporation, thus adding significant tools for the cooperation in a multidisciplinary environment. Some interesting powerful environments for managing distributed computing resources are emerging, notably at PAC [1] and Imperial College [16].The programming environment we advocate for parallel machines, being based on the concept of in-the-large application development, is compatible with the requirement of seamless integration of distributed heterogeneous resources. The current version of SkIE contains only basic, state-of-the art features for such integration. Work is in progress to integrate, in the same environment, object-based software infrastructures for integrating distributed MPP servers in complex multidisciplinary applications.

5   Conclusion

We have discussed the basic ideas for the development of heterogeneous, multidisciplinary, multi-platform environments for industrial HPC applications. These ideas have been illustrated with reference to the SkIE programming environment of the PQE2000 project, where the research results on parallel programming tools and MPP architectures have been transferred into industrial products at the European level. This experience has demonstrated the feasibility of an approach aiming to achieve a satisfactory trade-off between high-level modular software development, high performance, performance portability, rapid prototyping and software reuse, and productivity in a multidisciplinary environment. Moreover, SkIE can act as a solid framework in which the complex set of relationships between new directions in MPP architecture and programming environments can be studied and experimented in the next few years, along with the fundamental issues of integration between object-oriented distributed systems and MPP resources.

Acknowledgments

The PQE2000 project is the result of the joint efforts of many people I wish to thank collectively. In particular, the following people have done an enormous and qualified work to realize the SkIE programming environment: Marco Danelutto and Susanna Pelagatti, Department of Computer Science, University of Pisa, have coordinated the scientific project of SkIE ; Bruno Bacci, QSW Pisa, has coordinated the design and implementation team composed by: Barbara Cantalupo, Paolo Pesciullesi, Roberto Ravazzolo and Alessandro Riaudo, QSW Pisa, and Marco Aldinucci, Massimo Coppola and Massimo Torquati, Department of Computer Science. Silvia Ciarpaglini and Laura Folchi, INFN fellows at the Department of Computer Science, have contributed to the prototype evaluation of SkIECL compiler. Fabrizio Baiardi, Department of Computer Science, has coordinated the activity on the Intermediate Virtual Machine; in this context, Davide Guerri has realized the first version of DVSM. Domenico Laforenza, CNUCE-CNR Pisa, has coordinated the activity on HPF, with the collaboration of Salvatore Orlando (University of Venice) and Raffaele Perego. I have no words to express my gratitude to all of them. The activity on the software tools of PQE2000 have been supported by the PQE-1 program of QSW and ENEA-HPCN leaded by Agostino Mathis. Thanks are due to Lidia Arcipiani, Paolo Novelli, Vittorio Rosato, Paolo Palazzari, Roberto Iacono of the ENEA Labs, and to Roberto Marega, QSW Rome. Finally, a special thank to Antonio Sgamellotti, University of Perugia, president of the Scientific Committee of PQE2000, to Armando Orlandi, manager director of QSW, and to Roberto Vaccaro, IRSIP-CNR in Naples, for his continuous work to pursue the idea of PQE2000.

References

1. A.Hey, C.J. Scott, M. Surridge, C. Upstill, "Integrating computation and information resources - an MPP perspective". *3rd International Working Conference on Massively Parallel Programming Models*, London, Nov. 12-14, 1997. To be published by IEEE Press.
2. M. Vanneschi, "Variable grain architectures for MPP computation and structured parallel programming. *3rd International Working Conference on Massively Parallel Programming Models*, London, Nov. 12-14, 1997. To be published by IEEE Press.

3. M. Vanneschi, *The PQE2000 Project on General Purpose Massively Parallel Systems*, PQE2000 Report 96/1, 1996. Also published in Alta Frequenza, IEEE, and in Ercim News.

4. B. Bacci, B. Cantalupo, P. Pesciullesi, R. Ravazzolo, A. Riaudo, "SkIECL User Manual". PQE200 Report, QSW, Pisa, Dec. 997.

5. B. Bacci, M. Danelutto, S. Orlando, S. Pelagatti, M. Vanneschi, "P3L : a structured parallel programming language and its structured support", *Concurrency : Practice and Experience*, 7 (3), 225-255, May 1995.

6. B. Bacci, B. Cantalupo, M. Danelutto, S. Orlando, D. Pasetto, S. Pelagatti, M. Vanneschi, "An environment for structured parallel programming", *Advances in High Performance Computing*, Kluwier, The Netherlands, 219-234, 1997.

7. B. Bacci, M. Danelutto, S. Pelagatti, "Resources optimization via structured parallel programming", *Programming Environments for Massively Parallel Distributed Systems*, Birkhauser-Verlag, 13-25, 1994.

8. S. Pelagatti, "Compiling and supporting skeletons on MPP, *3rd International Working Conference on Massively Parallel Programming Models*, London, Nov. 12-14, 1997. To be published by IEEE Press.

9. M. Cole, *Algorithmic skeletons : structured management of parallel computation*, MIT Press, Cambridge, Mass., 1989.

10. J. Darlington, Y. Guo, H.W. To, Y. Jing, "Skeletons for Structured Parallel Composition", in *Proc. of the 15th ACM SIGPLAN Symposium on Principles and Practice of Parallel Programming*, 1995.

11. D.B. Skillicorn, *Foundations of Parallel Programming*, Cambridge University Press, 1994.

12. S. Pelagatti, *Structured development of parallel programs* "Taylor&Francis","London",1998

13. S. Orlando, R. Perego, *COLT-HPF, a coordination layer for HPF tasks*, University of Venice, Department of Mathematics and Computer Science, TR CS-98-4, March 1998.

14. W.D. Gropp, "Performance driven programming models", *3rd International Working Conference on Massively Parallel Programming Models*, London, Nov. 12-14, 1997. To be published by IEEE Press

15. E. Rosti, E. Smirini, L.W. Dowdy, G. Serazzi, K.C. Sevcik, "Processor saving scheduling policies for multiprocrssor systems", *IEEE Transaction on Computers* 47 (2) 178-189 Febrary 1998 .

16. J. Chattratichat, J. Darlington, Y. Guo, S. Hedvall, M. Koehler, A. Saleem, J. Sutiwaraphun, D. Yang, "A software architecture for deploying high performance solution tio Internet", *HPCN Conference*, Amsterdam, 1998.

17. J.B. Dennis, "A parallel program execution model supporting modular software construction", *3rd International Working Conference on Massively Parallel Programming Models*, London, Nov. 12-14, 1997. To be published by IEEE Press.

18. P. Mehrotra, J. Van Rosendale, H. Zima, "Solving irregular problems with High Performance Fortran", *3rd International Working Conference on Massively Parallel Programming Models*, London, Nov. 12-14, 1997. To be published by IEEE Press.

19. S. Gorlatch, C. Lengauer, "Homomorphism (de)composition for parallel programs", *3rd International Working Conference on Massively Parallel Programming Models*, London, Nov. 12-14, 1997. To be published by IEEE Press.

20. D. Skillicorn, "Categorical data types", *2nd Workshop on Abstract Machine Models for Highly Parallel Computers*, Leeds, UK, April 1993.

21. W. Kuhn, H. Burkhart, "The Alpstone Project : an overview of a performance modeling environment", *2nd Int. Conf. On HiPC'96*, 1996.
22. A. Dunlop, A. Hey, "PERFORM - A fast simulator for estimating program execution time", *3rd International Working Conference on Massively Parallel Programming Models*, London, Nov. 12-14, 1997. To be published by IEEE Press.
23. R. Di Meglio, *Dynamic computations in massively parallel programming models*, PhD dissertation, Department of Computer Science, University of Pisa, March 1994.
24. F. Baiardi, D. Guerri, *Op-Memory Library*, PQE2000 report, October 1997.
25. M. Torquati, *Shared memory support to structured parallel programming paradigms*, Computer Science dissertation, Department of Computer Science, University of Pisa, February 1997.
26. T. Sterling, P. Messina, P.H. Smith, *Enabling Technologies for Peta(FL)OPS Computing*, Caltech Concurrent Supercomputing Facilities, California Institute of Technology, CCSF-45, July 1994.
27. P.M. Kogge, T. Sunaga, H. Miyataka, K. Kitamura, E. Retter, "Combined DRAM and Logic for Massively Parallel Systems", *Conference on Advanced Research in VLSI*, Chapel Hill, NC, March 1995, pp. 4-16.
28. P. Tenti, *Variable grain MIMD/PIM architectures for MPP systems*, Computer Science dissertation, Department of Computer Science, University of Pisa, April 1998.

Quantum Cryptography on Optical Fiber Networks

Paul D. Townsend

BT Laboratories,
B55-131D, Martlesham Heath, Ipswich, IP5 3RE, U.K.
paul.townsend@bt-sys.bt.co.uk

Abstract. The security of conventional or classical cryptography systems relies upon the supposed (but often unproven) difficulty of solving certain classes of mathematical problem. Quantum cryptography represents a new paradigm for secure communications systems since its security is based not on computational complexity, but instead on the laws of quantum physics, the same fundamental laws that govern the behaviour of the universe. For brevity, this paper concentrates solely on providing a simple overview of the practical security problems that quantum cryptography addresses and the basic concepts that underlie the technique. The accompanying talk will also cover this introductory material, but the main emphasis will be on practical applications of quantum cryptography in optical fiber systems. In particular, I will describe a number of experimental systems that have been developed and tested recently at BT Laboratories. The experimental results will be used to provides some insights about the likely performance parameters and application opportunities for this new technology.

1 Introduction

The information revolution of the latter half of the twentieth century has been driven by rapid technological developments in the fields of digital communications and digital data processing. A defining feature of today's digital information systems is that they operate within the domain of classical physics. Each bit of information can be fully described by a classical variable such as, for example, the voltage level at the input to a transistor or microprocessor. However, information can also be processed and transported quantum-mechanically, leading to a range of radically new functionalities [1]. These new features occur because the properties of quantum and classical information are fundamentally different. For example, while classical bits must take on one of the two mutually exclusive values of zero or one, quantum superposition enables a bit of quantum information to have the strange property of being both zero and one simultaneously. Furthermore, it is always possible, in principle, to read classical information and make a perfect copy of it without changing it. In contrast, quantum systems can be used to carry information coded in such a way that it is impossible to read it without changing it and impossible to make a perfect copy of it [2]. It is these properties that are directly exploited in quantum cryptography [3–5] to provide secure communications on optical fiber systems [6–10]. The technique enables the secrecy of information transmitted over public networks to be tested in a fundamental way, since an eavesdropper will inevitably introduce readily detectable errors in the transmission. Quantum cryptography offers the intriguing prospect of certifiable levels of security that are guaranteed by fundamental physical laws. This possibility arises from the quantum nature of the communication system and cannot be realised with any conventional classical system.

2 Classical Cryptograhpy

In order to understand the problems that quantum cryptography addresses it is necessary to first review some background ideas from conventional or classical cryptography. Encryption is the process of taking a message or plaintext and scrambling it so that it is unreadable by anyone except the authorised recipient. This process is perhaps best described by a simple analogy. Encryption is the mathematical analogue of taking a message and placing it inside a lockable box. The box is then locked with a key held only by the sender and the authorised recipients of the message. Any unauthorised person intercepting the locked box will be unable to retrieve the message from the box without possessing a copy of the key.

Mathematically encryption can be described as an invertible mapping of a message $\mathbf{m}$ into a ciphertext $\mathbf{c}$. The mapping is achieved by an encryption algorithm, E, which takes as input the secret key, $\mathbf{k}$, and the message so that $\mathbf{c} = E(\mathbf{m}, \mathbf{k})$. The message, key and ciphertext are expressible as binary data strings. The decryption is achieved with the use of the inverse algorithm, D, so that $\mathbf{m} = D(\mathbf{c}, \mathbf{k})$.

In terms of the lockable box analogy the algorithm is the locking mechanism which is activated by the key. As with the case of a mechanical lock it is the secrecy of the key that keeps the system secure. A well-designed lock should be difficult to open without the key even when the details of the lock are known. It is an accepted design principle of cryptographic algorithms that the security should not be dependent on the secrecy of the algorithm but should reside *entirely* in the secrecy of the key. A graphic example of this idea is provided by the 'one-time pad' cipher system [11] that was proposed by Vernam in 1926. In the binary version of the one-time pad encryption is performed by modulo 2 addition (equivalent to a bit by bit Boolean XOR) of the message and a random key of equal length. Despite the simplicity (and public availability!) of the XOR algorithm the one-time pad offers theoretically perfect secrecy as long as the key is kept secret and used only once [12]. In practice, the one-time pad is not widely used because of the requirement for large volumes of key data. However, most commercial algorithms, such as the Data Encryption Standard (DES), also use publicly available algorithms [13].

The requirement that the secrecy be entirely dependent upon the key also imposes another good design principle. In a practical system such as DES, that does not offer the perfect secrecy of the one-time pad, an eavesdropper can always attack the system simply by trying each possible key in turn. A good algorithm will be designed such that this least efficient strategy is, nevertheless, the fastest method of attack. DES, for example, uses a key length of 56 bits so that an attacker performing an exhaustive key search will have to test, on average 2^{55} keys. Although this is a huge number, future increases in computational power will necessitate the use of longer keys.

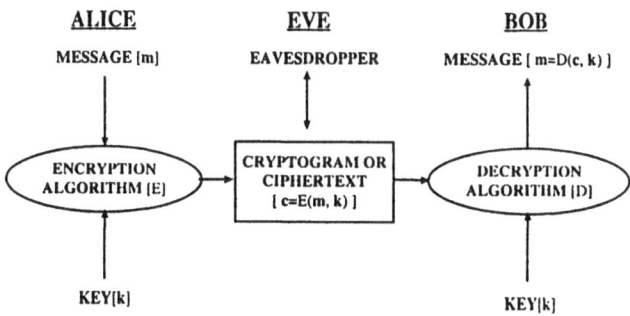

Fig. 1. The basic elements of a cipher system.

Figure 1 shows a schematic representation of a cryptography system, where the standard terminology of Alice, Bob and Eve is used to describe the transmitter, receiver and an unauthorised eavesdropper respectively. In order to operate the scheme Alice and Bob must exchange keys and, most importantly, must ensure that these keys do not fall into Eve's hands. Furthermore, if the system is to remain secure Alice and Bob must regularly repeat this process ensuring that the updated keys are also secret. How to achieve this goal is one of the central problems in secure communications; keys must be generated and distributed and the whole process managed in such a way as to ensure the secrecy of the system.

In essence quantum cryptography is a communication protocol that enables cryptographic keys to be securely distributed over communication channels that are physically insecure and hence potentially subject to eavesdropping. The technique achieves the desirable goal of an automated key establishment procedure between two or more parties in such a way that any unauthorised interception can be discovered and dealt with. Some modern cryptography systems attempt to solve the key management problems using a technique known as public-key cryptography [13]. In public-key cryptography each user has two keys, a private key, and a public key. The public key is widely distributed; the private key is kept secret. To send a confidential message using a public-key scheme Alice obtains Bob's public key from a public directory and uses this to encrypt her message to Bob. Once this has been done only Bob can decrypt the message using his private key. In terms of the lockable box analogy this is like having a box whose lock is operated by two keys. Everyone has access to the key to lock the box but cannot unlock it. Only one key will open the box and that is the private key held by the message recipient.

The security of public-key cryptography systems is based on 'one-way functions' that is readily computable mathematical problems for which the solution of the inverse problem is thought to be computationally infeasible. The classic example of such a function is multiplication and its inverse factorisation. For example, it easy to multiply two large prime numbers, but extremely difficult to factorise the resulting product into its two prime factors. Cryptographers using public-key cryptography have to make the assumption that advances in mathematics and computing will not dramatically change the ability of an attacker to compute these difficult problems. This issue has received a great deal of attention recently because of Shor's discovery of a high-speed factorisation algorithm for quantum computers [14]. If such computers can ever be built, the security of public-key cryptosystems may be dramatically reduced.

3 Quantum Cryptography

3.1 Background

The following section describes the original 'BB84' quantum key distribution protocol developed by Bennett and Brassard [3]. This protocol is used exclusively in the BT experiments that will be described in the talk. In general, quantum information systems require the use of some suitable two-state quantum objects to provide quantum-level representations of binary information (qubits). In the BB84 scheme Alice and Bob employ the linear and circular polarization states of single photons of light for this purpose. Figure 2 shows schematic representations

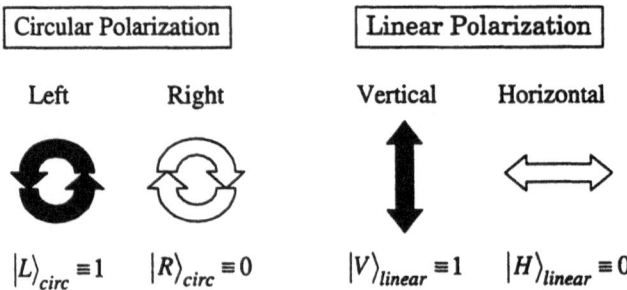

Fig. 2. Schematic representation and notation used to describe the linear and circular polarization states of single photons of light. In the BB84 quantum cryptography scheme the linear and circular states are used to provide two different quantum level representations of zero and one.

of these states together with the notation used to represent them and their associated binary values. The linear and circular 'bases' are used to provide two different quantum level representations of zero and one. Before describing how the properties of these single photon polarization states are exploited in the key distribution protocol we will consider the outcomes and interpretation of various possible measurements that can be performed on them.

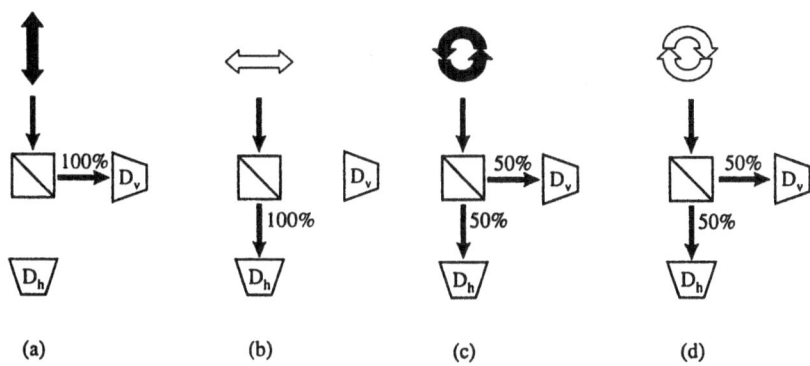

Fig. 3. A receiver uses a polarizer and two single photon detectors to perform linear polarization measurements on the linear (a and b) and circular (c and d) single photon polarization states. The probabilities of the various outcomes are indicated as percentage values. The linear measurement is incompatible with the circular representation and the photon is equally likely to be repolarized as either a horizontal or vertical state and detected at the appropriate detector.

Figures 3(a)-3(d) illustrate the possible outcomes of measurements on each of the four polarization states shown in figure 2. In each case the receiver has arranged a polarizer and two single photon detectors to perform a linear polarization measurement on the incoming photon (the apparatus is assumed to be perfect). For the two linear states the outcome of the measurement is deterministic, i.e. the $|V\rangle_{linear}$ photon is registered at detector D_V and the $|H\rangle_{linear}$ photon is registered at detector D_h, both with 100% accuracy. Of course similar results would also be obtained for classical input fields in the vertical and horizontal polarization states. In contrast, a classical input state with circular polarization would generate equal-intensity signals at the two detectors. Single photons, however, are elementary excitations of the electro-magnetic field and they cannot split in two. Instead the $|R\rangle_{circ}$ and $|L\rangle_{circ}$ states behave in a random fashion and have a 50% probability of being repolarized in the state $|V\rangle_{linear}$ and registered at D_V and, similarly, a 50% probability of being repolarized in the state $|H\rangle_{linear}$ and registered at D_h. In quantum mechanics terminology the photon is said to be projected into an eigen-state of the measurement operator, namely either $|V\rangle_{linear}$ or $|H\rangle_{linear}$. Taking the example of the $|R\rangle_{circ}$ state, the probability of each possible outcome is given by the squared modulus of the amplitude coefficients in (1),

$$|R\rangle_{circ} = \frac{1}{\sqrt{2}(|V\rangle_{linear} + i|H\rangle_{linear})} \tag{1}$$

the expansion of $|R\rangle_{circ}$ in the linear representation. In the case of the current example it can be seen that the measurement provides no information at all on a photon's state of circular polarization. Of course, the receiver could have included a 1/4-wave retardation plate in front of the polarizer in order to perform a perfect circular measurement. However, this would only be obtained at the expense of no information on the photon's linear polarization state since the $|V\rangle_{linear}$ and $|H\rangle_{linear}$ states behave in a random fashion when subjected to a circular polarization measurement. This situation arises because linear and circular polarization are complementary quantum observables that are incompatible in the sense that a measurement of one property will always disturb the other. Consider now the situation where one of the four possible polarization states is chosen at random and the receiver is asked to distinguish which one has been sent. Evidently the receiver must choose a measurement to perform on the photon. If he happens to choose an incompatible measurement the outcome will be random, but this fact cannot be deduced from the measurement itself. ence, the receiver cannot determine unambiguously the polarization state of e photon unless he is supplied with additional information on which type of state has been sent. All that can be said after a particular measurement is that the photon is now in the polarization state measured. Although we have used the example of a specific type of measurement here, there is in fact no single measurement or sequence of measurements that will allow the receiver to accurately determine the state of the photon when he is told only that it has been coded in one of two incompatible representations, but not which one. As we shall see, this peculiarly quantum phenomenon can be directly exploited to provide security in a quantum cryptography scheme. The reader might be tempted to suggest at this point that a solution to the measurement problem could be to copy the photon so that the receiver could measure different properties on the identical copies. Fortunately, at least from the quantum cryptographer's point of view, this type of 'cloning' is forbidden by quantum mechanics [2].

3.2 Quantum Key Distribution Protocol

The aim of quantum key distribution is not to take a specific predetermined key and send it to the recipient but rather to generate a random bit sequence in two separate locations that can then be used as a key. The key only comes into existence when the entire protocol has been executed. In order to carry out this process the transmitter Alice and the receiver Bob employ a pair of communication channels: a quantum channel that is used for sending the photons and a classical channel that is used for a subsequent public discussion about various aspects of the quantum transmission. It is assumed that Eve may have access to both of these channels at any time. She may make any measurement that she chooses on the quantum channel and is free to learn the contents of all the public channel messages although, as discussed later, she cannot change their content. The key distribution process is illustrated schematically in figure 4. Alice begins by choosing a random bit sequence and for each bit randomly selects either the linear or the circular representation. She then prepares a regular-clocked sequence of photons in the appropriate polarization states and transmits them to Bob over the quantum channel. For the example shown in the figure Alice has chosen to send a one in the first timeslot using the linear representation i.e. a vertically polarized photon, a one in the second timeslot using the circular representation i.e. a right polarized photon, and so on. At the other end of the channel Bob makes an independent random choice of whether to perform a linear or a circular measurement in each time period and records the result. As discussed in the example above, the type of measurement is selected by applying a 1/4-wave retardation before the polarizer for circular and no retardation for linear, and the result depends on which of the detectors registers the photon. The optical transmission system will also inevitably suffer from a variety of loss processes that randomly delete photons from the system and in these cases neither of Bob's detectors fires and he denotes this null event by an 'X'. In the timeslots where Bob did receive a photon and happened to choose the same representation as Alice (as in timeslot 1, for example) the bit that he registers agrees with what Alice sent. In timeslots where Bob used a different representation to Alice (as in time-slot 2, for example) the outcome of the measurement is unpredictable and he is equally likely to obtain a one or a zero.

Alice and Bob now carry out a discussion using the classical public channel that enables them to generate a shared random bit sequence from the data and test its secrecy. Bob announces the timeslots when he received a photon and the *type* of measurement he performed in each but, crucially, not the *result* of the measurement as this would give away the bit value if Eve were monitoring the discussion. In the case of the transmission shown in figure 4 Bob reveals that he received photons in timeslots 1, 2, 6 and 8 using linear, and timeslots 3 and 5 using circular. Alice then tells Bob in which of these timeslots they both used the same representation (1, 3 and 8) and they discard all other data. In this way they identify the photons for which Bob's measurements had deterministic outcomes and hence distil the shared subset 101 from Alice's initial random bit sequence. This is the 'raw' key data. In practice, the process would of course be continued until a much longer sequence had been established.

Why do Alice and Bob go through this tedious and inefficient process? Recall that the goal is to establish a secret random bit sequence for use as a cryptographic key. Consider the situation illustrated in Fig 5 where Eve attempts to intercept the quantum transmission, make a copy, and then resend it on to Bob. In each timeslot Eve (like Bob) has to make a choice of what measurement to perform. Without any aprioi knowledge of the representation chosen by Alice, Eve (like Bob) has a 50% chance of performing an incompatible measurement as shown in the figure. This leads to a random outcome and a change of polariza-

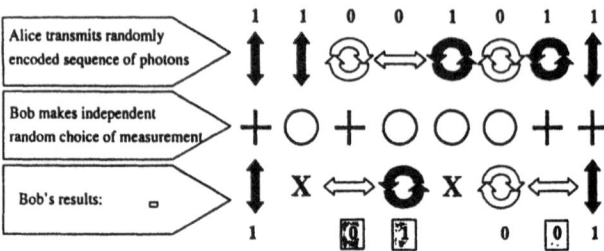

Fig. 4. Schematic representation of the quantum transmission phase of the BB84 quantum cryptography protocol. The top row shows Alice's outputs in 8 time slots that run sequentially from right to left. Bob's choice of measurements (cross=linear, circle =circular) and his results for the 8 timeslots are shown in the lower rows (X=no photon detected). During the public discussion Alice and Bob agree to discard the results where they used different representations (greyed boxes) and retain the shared bit sequence 101.

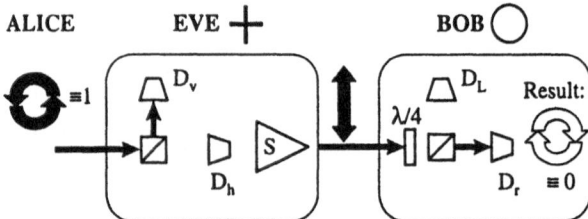

Fig. 5. Eve tries to intercept, make a copy, and then resend the bit sequence on to Bob. For each photon Eve has a 50% chance of choosing an incompatible measurement that leads to a random outcome. This is the case shown here where Eve has chosen to perform a linear measurement on Alice's circular photon. As a result Eve uses her source S to send on a vertical photon to Bob who now has a 50% chance of obtaining an error even though he has performed a compatible (as far as Alice and Bob are concerned) measurement.

tion state for the photon (note that we assume for simplicity that Eve makes the same type of polarization measurements as Bob, but the system is designed to be secure for any measurements that Eve may make). Eve has no way of knowing that she has performed an incompatible measurement because, unlike Bob, she does not have the benefit of the post transmission public discussion. The latter only takes place after the photons have arrived at Bob's end of the channel. Consequently, Eve sends a repolarized photon on to Bob who now has a 50% chance of observing an error in a timeslot where he has used the same representation as Alice and therefore expects a deterministic outcome. Consequently, if Eve makes continuous measurements on the channel she will generate a substantial bit-error rate of the order of 25% in the raw key. Of course, Eve may make measurements on only some of the photons or she may use other measurement techniques that yield less information but produce less disturbance on the channel [4, 15, 16]. Nevertheless, the crucial point is that Eve can only reduce the disturbance that she causes by also reducing the amount of information that she obtains; quantum mechanics ensures that these two quantities are directly related. For the specific example given above, Eve has a 50% chance per interception of making the correct choice of measurement and hence obtaining the correct bit and a 50% chance of making an incorrect choice of measurement that yields no information and causes disturbance on the channel. If Eve listens to the subsequent public discussion she can identify the instances in which her choice of measurement was correct and hence identify the bits (approximately 50%) in her data string that are shared with Alice and Bob. Hence if Eve intercepts 20% of the photons, for example, she will learn 10% of Alice and Bob's bits and generate an error-rate of around 5%. This simple example is unrealistic, because it does not describe all the measurement strategies that are available to Eve in a practical system, or the relative 'values' of the different types of information that she can obtain. Nevertheless, it does emphasise the basic principle of the technique which is that Alice and Bob can set an upper limit on the amount of information that Eve has obtained on their shared bit string simply by evaluating the error rate for their transmission [4]. As discussed below, if this is sufficiently low, they can generate a final highly secret key, if not they identify that the channel is currently insecure and cannot be used for secure key distribution. Quantum cryptography thus prevents the situation, which can always arise in principle with any classical key distribution system, where Alice and Bob's encrypted communications are compromised because they are unaware that Eve has surreptitiously obtained a copy of the key.

A full discussion of the final stages of the protocol that achieve this goal is beyond the scope of this paper, however, the main procedures that are involved will be briefly outlined. After discarding the inconclusive results from their bit strings, Alice and Bob publicly compare the parities of blocks of their data, and where these do not match, performing a bisective search within the block to identify and discard the error [4]. In this way they can derive a shared, error free, random bit sequence from their raw data together with a measurement of the error rate. This is achieved at the cost of leaking some additional information to

Eve in the form of the parities of the data subsets. In real systems transmission errors occur even when no eavesdropper is present on the channel. These errors can arise, for example, from the finite polarization extinction ratios of the various optical components in the system or from noise sources such as detector dark counts. In principle there is no way in to distinguish these errors from those caused by eavesdropping, hence all errors are conservatively assumed to have been generated by Eve. If the error rate is sufficiently low (typically of the order of a few percent in practice), then a technique known as 'privacy amplification' [17, 18] can be employed to distil from the error-corrected (but possibly only partially secret) key a smaller amount of highly secret key about which Eve is very unlikely to know even one bit. If, for example, Alice and Bob calculate that Eve may know e bits of their error-corrected n-bit string, they can generate from this data a highly secret m-bit key (where m(n-e) simply by computing (but not announcing as in error-correction) the parities of m publicly agreed-on random subsets of their data [4, 17, 18]. The final stage of the process is for Alice and Bob to authenticate their public channel discussions using an information-theoretically secure authentication scheme [19]. This is to prevent the powerful attack in which Eve breaks in to the public channel and impersonates Bob to Alice and visa-versa [4]. If this were possible Eve could share one key with Alice and another with Bob who would be unaware that this were the case. Like the one-time pad cipher, the authentication scheme requires Alice and Bob to possess in advance a modest amount of shared secret key data, part of which is used up each time a message is authenticated. However, this initial key is only required at system turn-on time because each implementation of the quantum key distribution protocol generates a fresh volume of key data some of which can be used for authentication. After privacy amplification and authentication Alice and Bob are now in possession of a shared key that is certifiably secret. They can now safely use this key together with an appropriate algorithm for data encryption purposes.

4 Conclusions

In the accompanying talk I will demonstrate how the theoretical ideas discussed above can be implemented and tested in a number of scenarios of real practical interest. In particular I will describe a scheme based on photon interference that we have developed over the last few years at BT [7, 20–24]. This system operates in the 1.3μm-wavelength fiber transparency window over point-to-point links up to $\sim$50km in length [23] and on multi-user optical networks [6, 25]. I will also discuss how this technology performs on fiber links installed in BT's public network and discuss issues such as cross-talk with conventional data channels propagating at different wavelengths in the same fiber [24]. The experiments demonstrate that quantum cryptography can provide, at least from a technical point of view, a practical solution for secure key distribution on optical fiber links and networks. In summary, quantum cryptography has now moved a long way from the original fascinating theoretical concept towards practical applications in optical communication systems.

Acknowledgements I would especially like to thank Keith Blow for his continued advice, support and encouragement and Simon Phoenix for his theoretical contributions. I would also like to acknowledge the important contributions of two graduate students, Christophe Marand and Guilhem Ensuque, to the experimental work mentioned above.

References

1. C. H. Bennett, 'Quantum information and computation', *Physics Today*, October (1995), for a review of this topic.
2. W. K. Wootters and W. H. Zurek, 'A single quantum cannot be cloned', *Nature*, **299** 802-803 (1982)
3. C. H. Bennett and G. Brassard, 'Quantum cryptography: public-key distribution and coin tossing', in *Proceedings of IEEE International Conference on Computers, Systems and Signal Processing*, Bangalore, India, 175-179 (1984).
4. C. H. Bennett, F. Bessette, G. Brassard, L. Salvail and J. Smolin, 'Experimental quantum cryptography', *Journal of Cryptology*, **5** 3-28 (1992).
5. A. K. Ekert, 'Quantum cryptography based on Bell's Theorem', *Physical Review Letters*, **67** 661-663 (1991)
6. P. D. Townsend, 'Quantum cryptography on multi-user optical fiber networks', *Nature*, **385**, 47-49 (1997)
7. C. Marand and P. D. Townsend, 'Quantum key distribution over distances as long as 30km', *Optics Letters*, **20** 1695-1697 (1995)
8. J. D. Franson and B. C. Jacobs, 'Operational system for quantum cryptography', *Electronics Letters*, **31** 232-234 (1995)
9. R. J. Hughes, G. G. Luther, G. L. Morgan and C. Simmons, 'Quantum cryptography over 14km of installed optical fiber', *Proc. 7th Rochester Conf. on Coherence and Quantum Optics (eds J. H. Eberly, L. Mandel and E. Wolf)*, 103-112 (Plenum, New York, 1996)
10. H. Zbinden, J. D. Gautier, N. Gisin, B. Huttner, A. Muller, and W. Tittel, 'Interferometry with Faraday mirrors for quantum cryptography', *Electronics Letters*, **33**, 586-587 (1997)
11. G. S. Vernam, *J. Amer. Inst. Electr. Engrs.*, **45**, 109-115 (1926)
12. C. E. Shannon, 'Communication theory of secrecy systems', *Bell Syst. Tech. J.*, **28**, 656-715 (1949)
13. H. Beker and F. Piper, Cipher Systems : *the Protection of Communications*, (Northwood Publications, London, 1982). See also G. Brassard, Modern *Cryptology, Lecture Notes in Computer Science*, eds G. Goos and J. Hartmanis (Springer-Verlag, Berlin, 1988)
14. P Shor, 'Algorithms for quantum computation: Discrete logarithm and factoring', Proc. 35th *Annual IEEE Symposium on Foundations of Computer Science* (IEEE Computer Society Press, 1994), 124-134
15. A. K. Ekert, B. Huttner, G. M. Palma and A. Peres, 'Eavesdropping on quantum cryptosystems', *Physical Review A* **50**, 1047-1056 (1994)
16. B. Huttner and A. K. Ekert, 'Information gain in quantum eavesdropping', *J. Mod. Opt.*, **41**, 2455-2466 (1994)

17. C. H. Bennett, G. Brassard and J.-M. Robert, 'Privacy amplification by public discussion', *SIAM Journal on Computing*, **17** 210-229 (1988).

18. C. H. Bennett, G. Brassard, C. Crepeau and U. Maurer, 'Generalized privacy amplification', *SIAM Journal on Computing*, **17** 210-229 (1988).

19. M. N. Wegman and J. L. Carter, 'New hash functions and their use in authentication and set equality', *J. Computer and System Sciences*, **22**, 265-279 (1981)

20. P. D. Townsend, J. G. Rarity and P. R. Tapster, 'Single-photon interference in a 10km long optical fiber interferometer', *Electronics Letters*, **29** 634-635 (1993)

21. P. D. Townsend, 'Secure key distribution system based on quantum cryptography', *Electronics Letters*, **30** 809-810 (1994)

22. S. J. D. Phoenix and P. D. Townsend, 'Quantum cryptography: how to beat the code breakers using quantum mechanics', *Contemporary Physics*, **36** 165-195 (1995)

23. P. D. Townsend, 'Quantum cryptography on optical fiber networks', *Optical Fiber Technology*, (In Press)

24. P. D. Townsend, 'Simultaneous quantum cryptographic key distribution and conventional data transmission over installed fiber using wavelength division multiplexing', *Electronics Letters*, **33** 188-189 (1997)

25. P. D. Townsend, S. J. D. Phoenix, K. J. Blow and S. M. Barnett, 'Design of quantum cryptography systems for passive optical networks', Electronics Letters, **30** 1875-1877 (1994). See also S. J. D. Phoenix, S. M. Barnett, P. D. Townsend and K. J. Blow, 'Multi-user quantum cryptography on optical networks', *Journal of Modern Optics*, **42**, 1155-1163 (1995) 18 19

Very Distributed Media Stories: Presence, Time, Imagination

Glorianna Davenport

Interactive Cinema Group,
MIT Media Laboratory,
Massachusetts Institute of Technology,
Cambridge, MA, USA 02139,
gid@media.mit.edu

"...the uncommon vista raises a pleasure in the imagination because it fills the soul with an agreeable surprise, gratifies its curiosity and gives it an idea of which it was not before possessed."

–Addison
The Spectator 1712

1 Introduction

The action of stories is always grounded and contextualized in a specific place and time. For centuries, artists seeking places worthy of representation have found inspiration in both the natural landscape and in man-made surrounds. This inspiration traditionally dwells on the scenic aspects of place and situation, in styles ranging from photorealistic to impressionistic. Sometimes, as in Australian aboriginal "dreamtime maps," the real, the historical, and the spiritual components of a place are simultaneously depicted with equal weightings. Sometimes, as in road maps and contour maps, super-simplified representations are enhanced with integrated or overlaid technical measurements; constructed artifacts, such as roads and airports, share equal billing with natural landmarks, such as lakes and rivers. The scale, focus, point-of-view, and narrative content of landscapes are chosen and manipulated to suit the artist's (and the audience's) specific purposes: they embody affordances which exert great influence over a work's final use.

When we view the image of a broad, sweeping vista, we seldom notice every leaf on every tree: indeed, artists seldom provide us with this level of detail. Instead, our attention is drawn through and across a collection of landmarks, consciously and unconsciously brought together to serve as sensually satisfying iconic and symbolic representations of a more complex whole. This notion of bringing together and arranging selected elements into an integrated environment – customized and personalized for specific uses – has profound relevance to the digital systems of tomorrow.

2   Touchstones from the Past

In the 20th century, artists have increasingly moved away from strict representations of what they see in the world to formal and informal explorations of form
and space. Many artists have challenged the limits of art as object by extending
expression into the natural or man-made surround; often, fanciful landscape is
used to proffer metaphors for engagement and imagination.

Towards the end of his life, Matisse – no longer able to hold a paintbrush –
developed a technique by which he could produce flat paper cut-outs. By past-
ing these flat elements onto his hotel walls, Matisse was able to explore anew
luminous dimensions of light in space. In these explorations, the artist inverts
the objectness of sculpture and negates the boundaries of painting. In the flat
frieze of bathers swimming in a band around his dining room, Matisse created
an immersive space which is no less substantial and emotionally charged than
his earlier works on canvas. The experience of actually dining in this customized
and privileged environment – individually and collectively – is in itself a transformation of the familiar into a joyous, ethereal, unfamiliar collision of art and
life.

Marcel Duchamp's "Etant Donnes" – the last major work of his life – at
first appears to be a large, ramshackle wooden door. On closer examination,
one discovers a peep-hole through which an allegorical 3D tableau (involving a
nude, a waterfall, and a gas burner) is visible. Duchamp's widow has steadfastly
upheld his request that no-one be allowed to photograph or otherwise reproduce
this voyeuristic vision; only those who physically travel to the museum and
peek through the door cracks are rewarded. Even then, the visitor meets "Etant
Donne" on a reticent footing: the piece is situated at the end of a corridor,
and look as if workmen may be doing construction behind it; to peer through
cracks at a spectacular nude, while other people watch you, seems a bit tawdry.
The limited views of a 3D scene present a shifting dynamic were information is
selectively concealed or revealed.

In the early 70s, beleaguered by minimalism and America's struggle with their
values relative to the environment, artists such as Michael Heizer and Robert
Smithson created mystical, monumental Earth Art. These enormous sculptural
reworkings of the landscape ranged from huge gouges in the earth, to enormous
decorative patterns of ecological flows, to fences across grand grazing terrains.
Over time, these earthworks would erode and decay until they became indistinguishable from the natural surround.

These and other examples from the past provide a relevant touchstone for
modern makers, particularly those who wish to create their own CD-ROMs,
immersive interactive environments, or "virtual reality" installations.

3   Art in Transition

Whatever its wellspring, the artistic imagination is shaped by skills, exposure
to pre-existing conventions, the available materials, and worldly beliefs. In this

sense, the work is never precisely as we imagined; rather, it is a running compromise which mediates that inner thought with the outer reality of contemporary ideas, available technology, practical collaborations, and expressive powers of creative artists. Today's digital technology is particularly empowering; unlike the passive, pastoral beauty of painted landscapes, digital environments are computational entities seething with activity, change, and movement. The hardware / software duality of these creations offers new forms of dynamic control over both the physical and metaphysical aspects of space, time, form, and experience. The marriage of content with production and delivery systems extends far beyond the technological; as we gain more experience with these systems, new sets of aesthetics, expectations, and challenges are emerging.

A typical "interactive multimedia" installation ties together devices which sense and track human activity; powerful computational engines; graphical displays; and, increasingly, high-capacity communications networks which interconnect people and resources across vast distances. At the core of these systems is some combination of software and hardware which attempts to glean the desires and motivations of the participant audience, remap them to the intentions of the primary author, and respond in meaningful ways. Many of these exploratory systems empower the audience to enter and traverse vast, sophisticated information landscapes via a "driving" or "flying" metaphor; many offer constructionist environments which invite the audience to add their own stories, objects, and other manifestations of desire and personality to the surround. The usefulness and expressive power of these systems depends not only on the technology – which is often "invisible" to the audience – but also on the choice of overarching metaphor which conjoins and energizes the parts, relating them to specific story content.

Today, we are no longer required to sit in front of a small computer screen, typing, pointing, and clicking our way through every kind of locally-stored content. Instead, we can use our voice, gaze, gesture, and body motion – whatever expressive form is most appropriate – to communicate our desires and intentions to the computational engine. These engines can reach across vast networks to find whatever material is needed, bring it back, and massage it on-the-fly to form personalized, customized presentations. Modern sensing devices and displays range from the familiar personal computer to large- scale, immersive environments of superb quality. A complex series of transactions underlies this process of interface, information retrieval, and presentation; as interactive systems rise to dominance, the problems and opportunities presented by these transactions – drawing from a broad base of networked resources owned by others – must ultimately be addressed by e-commerce.

4 Kinesics, Kinesthesia, and the Cityscape

In the physical world, natural vistas engage the human eye, stimulate the brain, and generate emotional responses such as joy, anxiety, or fear. These experiences also produce a degree of kinesthesia: the body's instinctive awareness and deep understanding of how it is moving through space.

Film directors as well as makers of virtual reality scenarios seek to create transformational imagery which will spark a sensational journey through a constructed or simulated reality. Cinema captures or synthetically provides many sensory aspects of genuine body motion and audiovisual perception, but fails to represent others (such as muscle fatigue). As a result, the empathetic kinesthesia of film has proven to be extremely elusive and transitory; there is a limit to what the audience can experience emotionally and intellectually as they sit at the edge of the frame. By experimenting with lens-based recording technology, multilayered effects, and a variety of story elements, cinema presents an illusion of a world which offers sufficient cues and clues to be interpreted as "real." More recent experiments with remote sensing devices, theme-park "thrill rides," networked communities, constructionist environments, and haptic input/output devices have greatly expanded the modern experience-builder's toolkit.

In cinema, the perceiver is transported into a manufactured reality which combines representations of space or landscape with the actions of seemingly human characters. This reality must be conceived in the imagination of the film's creators before it can be realized for the mind and heart of the spectator. The attachment of any director to a particular landscape – be it the desert in Antonioni's "The Passenger," the futuristic city and under-city of Fritz Lang's "Metropolis," or Ruttmann's poetic "Berlin: Symphony of a Great City" – is a function of personal aesthetic and symbolic intention. The "holy grail" of cinematography may well be the realization of satisfying kinesthetic effects as we glide over "Blade Runner's" hyperindustrialized Los Angeles or descend into the deep caverns of Batman's ultra-gothic Gotham City. Note that all of these examples draw upon intellectual properties subject to licensing fees for use: however, the rules and models of commerce governing the on-line delivery of this art remain substantially undefined.

5 The Landscape Transformed

Lured by the potential of a near infinite progression of information, artists are constructing digital environments which reposition the perceiver within the synthetic landscape of the story frame itself. Awarded agency and a role, the Arthur-like explorer sets out on a quest for adventure and knowledge. Positioned at the center of a real-time dynamic display, the perceiver navigates a landscape of choice in which information – text, HTML pages, images, sound, movies, or millions of polygons – is continuously rendered according to explicit rules to move in synch with the participant's commands.

In 1993-1994, Muriel Cooper, Suguru Ishizaki, and David Small of MIT's Visual Language Workshop designed an "Information Landscape" in which local information (such as news) was situated on a topological map of the world. As the explorer pursued particular information, they found themselves hurtling across a vast, familiar landscape, or zooming down into an almost infinite well of detail. In this work, the illusion of driving and flying are produced through the smooth scaling of the information. The perceiver's body is at rest but, as she comes to understand the metaphor of motion, her brain makes the analogy to a phenomenon which is well known in our 20th century world.

Fig. 1. Flavia Sparacino's "City of News"

What new possibilities for navigational choice are offered when a system is able to parse full-body motion and gestures? Recently, Flavia Sparacino has created the "City of News," a prototype 3D information browser which uses human gesture as the navigational input. "City of News" explores behavioral information space in which HTML pages are mapped onto a familiar urban landscape – a form of "memory palace" where the location of particular types of information remains constant, but the specific content is forever changing as fresh data feeds in from the Internet. Still in development, this work raises interesting questions about organizational memory structures and the economics of information space. Over the past year, Sparacino has demonstrated "City of News" in a desktop environment, in a small room-sized environment, and in a larger space which could accommodate floor and wall projection surfaces. As this environment develops, Sparacino will consider how two or more people might build and share such a personal information landscape.

6 Landscape as Harbinger of Story Form

Stories help us to make sense of the chaotic world around us. They are a means – perhaps the principal human cognitive means – by which we select, interpret, reshape, and share our experiences with others. In order to evolve audiovisual stories in which the participant-viewer has significant agency, the artist seeks to conjoin story materials with relevant aspects of real-world experience. Sets, props, settings, maps, characters, and "plot points" are just a few of the story elements which can be equipped with useful "handles" which allow the audience to exert control. As these computer-assisted stories mature – just as cinema has – they will in turn serve as metaphoric structures against which we can measure the real world.

Today's networking technology allows us to bring an audience "together" without requiring them to be physically present in the same place at the same time: we call this art form "very distributed story." Regardless of its underlying spatio-temporal structure, the participant audience always perceives the playout of story as a linear experience. Thus, "real time" is a crucial controller of the audience's perceptions of "story time."

In networked communications, the passing-on of information from one person to the next is characterized by time delays and differences in interpretation: what effect does this have on the communal experience of story? Long-term story structures such as suspense, expectation, and surprise depend on the linear, sequential revelation of knowledge over time. Context and the "ticking clock" reveal absolute knowledge of a story situation – but only after a significant period of involvement with the narrative.

Beyond its consequential responses to the remotely-sensed desires and commands of its audience, very distributed story must embark upon a metaphysics of landscape. This is the focus of the "Happenstance" system, currently under development by Brian Bradley, another of my graduate students at MIT.

"Happenstance" is flexible storytelling testbed which expands the traditional literary and theatrical notions of Place and Situation to accommodate interactive, on-the-fly story construction. Important aspects of story content and context are made visible, tangible, and manipulable by systematically couching them within the metaphors of ecology, geology, and weather. Information-rich environments become conceptual landscapes which grow, change, and evolve over time and through use. Current information follows a natural cycle modeled after the Earth's water cycle. Older information, history, and complex conceptual constructs – built up by the flow of data over time – are manifested in the rock and soil cycles. Directed inquiries, explorations of theory, and activities associated with the audience's personal interests are captured and reflected by plant growth. As a result, information itself is imbued with sets of systemic, semi-autonomous behaviors which allow it to move and act intelligently within the story world or other navigable information spaces in ways which are neither tightly scripted nor random.

In Bradley's work, landscape and the broad forces of weather which sweep across it are the carriers of information as well as scenic elements: they are bound by the rules, cycles, and temporal behavior of ecological systems. Landscape, context, and new information flowing through the system work together to provide a "stage" where story action can take place.

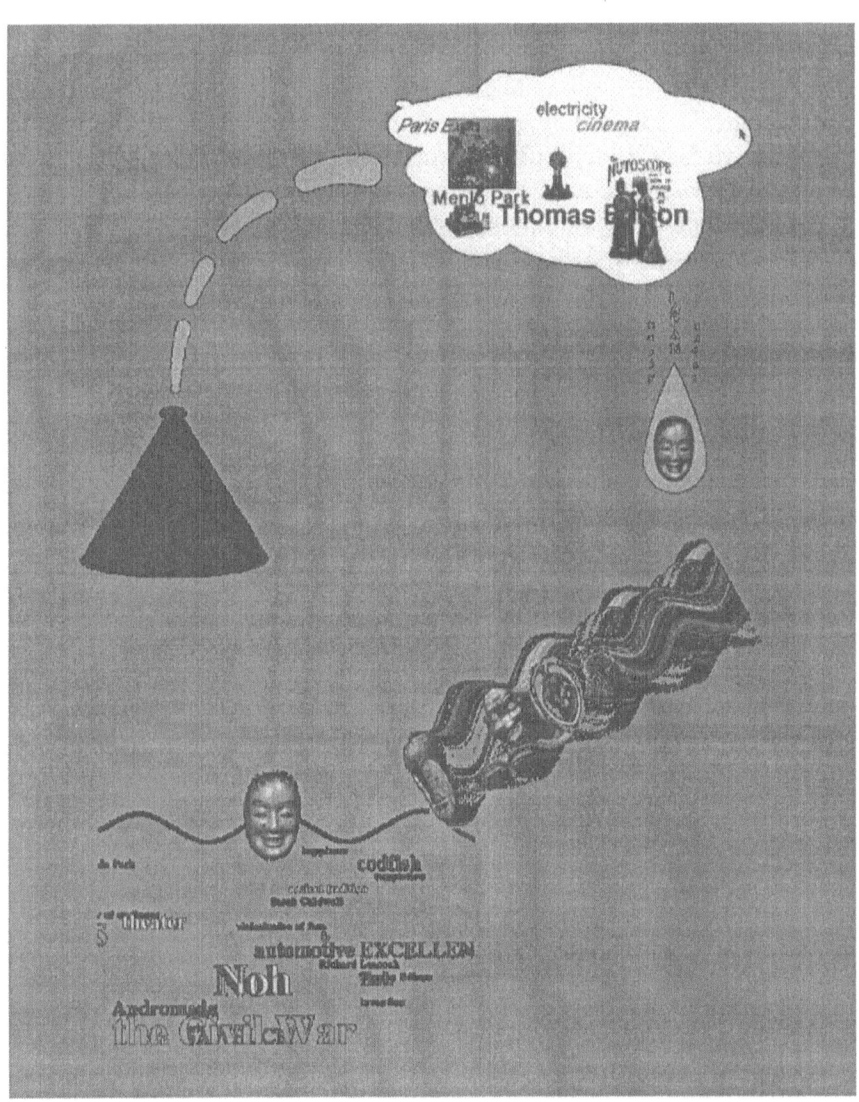

Fig. 2. The information "water cycle" of Brian Bardley's "Happenstance"

As precious computational resources – known in the jargon as "real estate" – meet specialized, continuous Information Landscapes, new and intuitive forms of navigation and content selection are beginning to emerge. The symbology of landscapes built by the gathering-together of scattered resources must hook seamlessly into a personalized, customizable transactional model capable of resolving higher-level ambiguities and vaguenesses. The movement of information through these story spaces must have consequence over time, just as the interactions of the participant audience must always elicit promt and meaningful responses from the storytelling system.

As narrative finds its appropriate form in immersive electronic environments, the traditional classification of story – myth, history, satire, tragedy, comedy – must be extended to include hybridized forms which reflect the moment-to-moment desires and concerns of the audience. Rather than strictly mapping events to prototypical story action, we will evolve new story forms based on navigation of choice, constrained and guided by authorial will. Hopefully, in time and with practice, these metaphoric structures and maps can be – as Addison foretold – "as pleasing to the Fancy as the Speculations of Eternity or Infinitude are to the Understanding."

HPcc as High Performance Commodity Computing on Top of Integrated Java, CORBA, COM and Web Standards

G.C. Fox, W. Furmanski, T. Haupt, E. Akarsu and H. Ozdemir

Northeast Parallel Architectures Center, Syracuse University, Syracuse NY, USA
gcf@npac.syr.edu, furm@npac.syr.edu
http://www.npac.syr.edu

Abstract. We review the growing power and capability of commodity computing and communication technologies largely driven by commercial distributed information systems. These systems are built from CORBA, Microsoft's COM, JavaBeans, and rapidly advancing Web approaches. One can abstract these to a three-tier model with largely independent clients connected to a distributed network of servers. The latter host various services including object and relational databases and of course parallel and sequential computing. High performance can be obtained by combining concurrency at the middle server tier with optimized parallel back end services. The resultant system combines the needed performance for large-scale HPCC applications with the rich functionality of commodity systems. Further the architecture with distinct interface, server and specialized service implementation layers, naturally allows advances in each area to be easily incorporated. We illustrate how performance can be obtained within a commodity architecture and we propose a middleware integration approach based on JWORB (Java Web Object Broker) multi-protocol server technology. Examples are given from collaborative systems, support of multidisciplinary interactions, proposed visual HPCC ComponentWare, quantum Monte Carlo and distributed interactive simulations.

1 Introduction

We believe that industry and the loosely organized worldwide collection of (freeware) programmers is developing a remarkable new software environment of unprecedented quality and functionality. We call this DcciS - Distributed commodity computing and information System. We believe that this can benefit HPCC in several ways and allow the development of both more powerful parallel programming environments and new distributed metacomputing systems. In the second section, we define what we mean by commodity technologies and explain the different ways that they can be used in HPCC. In the third and critical section, we define an emerging architecture of DcciS in terms of a conventional 3 tier commercial computing model, augmented by distributed object and component technologies of Java, CORBA, COM and the Web. This is followed in sections four and five by more detailed discussion of the HPcc core technologies and high-level services.

In this and related papers [5], we discuss several examples to address the following critical research issue: can high performance systems - called HPcc or High Performance Commodity Computing - be built on top of DcciS. Examples include integration of collaboration into HPcc; the natural synergy of distribution simulation and the HLA standard with our architecture; and the step from object to visual component based programming in high performance distributed computing. Our claim, based on early experiments and prototypes is that HPcc is feasible but we need to exploit fully the synergies between several currently competing commodity technologies. We refer to our approach towards HPcc, based on integrating several popular distributed object frameworks as Pragmatic Object Web and we describe a specific integration metodology based on multi-protocol middleware server, JWORB (Java Web Object Request Broker).

2 Commodity Technologies and Their Use in HPCC

The last three years have seen an unprecedented level of innovation and progress in commodity technologies driven largely by the new capabilities and business opportunities of the evolving worldwide network. The web is not just a document access system supported by the somewhat limited HTTP protocol. Rather it is the distributed object technology which can build general multi-tiered enterprise intranet and internet applications. CORBA is turning from a sleepy heavyweight standards initiative to a major competitive development activity that battles with COM, JavaBeans and new W3C object initiatives to be the core distributed object technology.

There are many driving forces and many aspects to DcciS but we suggest that the three critical technology areas are the web, distributed objects and databases. These are being linked and we see them subsumed in the next generation of "object-web" [1] technologies, which is illustrated by the recent Netscape and Microsoft version 4 browsers. Databases are older technologies but their linkage to the web and distributed objects, is transforming their use and making them more widely applicable.

In each commodity technology area, we have impressive and rapidly improving software artifacts. As examples, we have at the lower level the collection of standards and tools such as HTML, HTTP, MIME, IIOP, CGI, Java, JavaScript, Javabeans, CORBA, COM, ActiveX, VRML, new powerful object brokers (ORB's), dynamic Java clients and servers including applets and servlets, and new W3C technologies towards the Web Object Model (WOM) such as XML, DOM and RDF.

At a higher level collaboration, security, commerce, multimedia and other applications/services are rapidly developing using standard interfaces or frameworks and facilities. This emphasizes that equally and perhaps more importantly than raw technologies, we have a set of open interfaces enabling distributed modular software development. These interfaces are at both low and high levels and the latter generate a very powerful software environment in which large preexisting components can be quickly integrated into new applications. We believe that there are significant incentives to build HPCC environments in a way that naturally inherits all the commodity capabilities so that HPCC applications can also benefit from the impressive productivity of commodity systems. NPAC's HPcc activity is designed to demonstrate that this is possible and useful so that one can achieve simultaneously both high performance and the functionality of commodity systems.

Note that commodity technologies can be used in several ways. This article concentrates on exploiting the natural architecture of commodity systems but more simply, one could just use a few of them as "point solutions". This we can term a "tactical implication" of the set of the emerging commodity technologies and illustrate below with some examples:

- Perhaps VRML,Java3D or DirectX are important for scientific visualization;
- Web (including Java applets and ActiveX controls) front-ends provide convenient customizable interoperable user interfaces to HPCC facilities;
- Perhaps the public key security and digital signature infrastructure being developed for electronic commerce, could enable more powerful approaches to secure HPCC systems;
- Perhaps Java will become a common scientific programming language and so effort now devoted to Fortran and C++ tools needs to be extended or shifted to Java;
- The universal adoption of JDBC (Java Database Connectivity), rapid advances in the Microsoft's OLEDB/ADO transparent persistence standards and the growing convenience of web-linked databases could imply a growing importance of systems that link large scale commercial databases with HPCC computing resources;
- JavaBeans, COM, CORBA and WOM form the basis of the emerging "object web" which analogously to the previous bullet could encourage a growing use of modern object technology;
- Emerging collaboration and other distributed information systems could allow new distributed work paradigms which could change the traditional teaming models in favor of those for instance implied by the new NSF Partnerships in Advanced Computation.

However probably more important is the strategic implication of DcciS which implies certain critical characteristics of the overall architecture for a high performance parallel or distributed computing system. First we note that we have seen over the last 30 years many other major broad-based hardware and software developments – such as IBM business systems, UNIX, Macintosh/PC desktops, video games – but these have not had profound impact on HPCC software. However we suggest the DcciS is different for it gives us a world-wide/enterprise-wide distributing computing environment. Previous software revolutions could help individual components of a HPCC software system but DcciS can in principle be the backbone of a complete HPCC software system – whether it be for some global distributed application, an enterprise cluster or a tightly coupled large scale parallel computer.

In a nutshell, we suggest that "all we need to do" is to add "high performance" (as measured by bandwidth and latency) to the emerging commercial concurrent DcciS systems. This "all we need to do" may be very hard but by using DcciS as a basis we inherit a multi-billion dollar investment and what in many respects is the most powerful productive software environment ever built. Thus we should look carefully into the design of any HPCC system to see how it can leverage this commercial environment.

3 Three Tier High Performance Commodity Computing

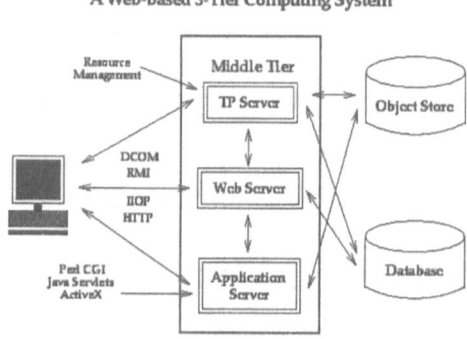

Fig. 1. Industry 3-tier view of enterprise Computing

We start with a common modern industry view of commodity computing with the three tiers shown in fig 1. Here we have customizable client and middle tier systems accessing "traditional" back end services such as relational and object databases. A set of standard interfaces allows a rich set of custom applications to be built with appropriate client and middleware software. As indicated on figure, both these two layers can use web technology such as Java and Javabeans, distributed objects with CORBA and standard interfaces such as JDBC (Java Database Connectivity). There are of course no rigid solutions and one can get "traditional" client server solutions by collapsing two of the layers together. For instance with database access, one gets a two tier solution by either incorporating custom code into the "thick" client or in analogy to Oracle's PL/SQL, compile the customized database access code for better performance and incorporate the compiled code with the back end server. The latter like the general 3-tier solution, supports "thin" clients such as the currently popular network computer.

The commercial architecture is evolving rapidly and is exploring several approaches which co-exist in today's (and any realistic future) distributed information system. The most powerful solutions involve distributed objects. Currently, we are observing three important commercial object systems - CORBA, COM and JavaBeans, as well as the ongoing efforts by the W3C, referred by some as WOM (Web Object Model), to define pure Web object/component standards. These have similar approaches and it is not clear if the future holds a single such approach or a set of interoperable standards.

CORBA is a distributed object standard managed by the OMG (Object Management Group) comprised of 700 companies. COM is Microsoft's distributed object technology initially aimed at Window machines. JavaBeans (augmented with RMI and other Java 1.1 features) is the "pure Java" solution - cross platform but unlike CORBA, not cross-language! Finally, WOM is an emergent Web model that uses new standards such as XML, RDF and DOM to specify respectively the dynamic Web object instances, classes and methods.

Legion is an example of a major HPCC focused distributed object approach; currently it is not built on top of one of the three major commercial standards. The HLA/RTI [2] standard for distributed simulations in the forces modeling community is another important domain specific distributed object system. It appears to be moving to integration with CORBA standards.

Although a distributed object approach is attractive, most network services today are provided in a more ad-hoc fashion. In particular today's web uses a "distributed service" architecture with HTTP middle tier servers invoking via the CGI mechanism, C and Perl programs linking to databases, simulations or other custom services. There is a trend toward the use of Java servers with the servlet mechanism for the services. This is certainly object based but does not necessarily implement the standards implied by CORBA, COM or Javabeans. However, this illustrates an important evolution as the web absorbs object technology with the evolution from low- to high-level network standards:

- from HTTP to Java Sockets to IIOP or RMI
- from Perl CGI Script to Java Program to JavaBean distributed object

As an example consider the evolution of networked databases. Originally these were client-server with a proprietary network access protocol. In the next step, Web linked databases produced a three tier distributed service model with an HTTP server using a CGI program (running Perl for instance) to access the database at the backend. Today we can build databases as distributed objects with a middle tier JavaBean using JDBC to access the backend database. Thus a conventional database is naturally evolving to the concept of managed persistent objects.

Today as shown in fig 2, we see a mixture of distributed service and distributed object architectures. CORBA, COM, Javabean, HTTP Server + CGI, Java Server and servlets, databases with specialized network accesses, and other services co-exist in the heterogeneous environment with common themes but disparate implementations. We believe that there will be significant convergence as a more uniform architecture is in everyone's best interest.

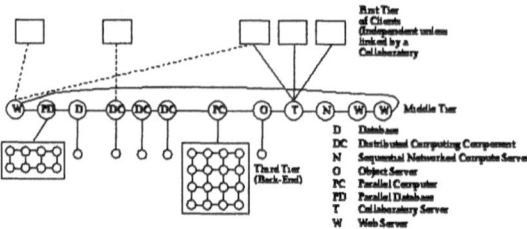

Fig. 2. Today's Heterogeneous Interoperating Hybrid Server Architecture. HPcc involves adding to this system, high performance in the third tier.

We also believe that the resultant architecture will be integrated with the web so that the latter will exhibit distributed object architecture shown in fig 3.

More generally the emergence of IIOP (Internet Inter-ORB Protocol), CORBA2, rapid advances with the Microsoft's COM, DCOM, and COM+, and the realization that both CORBA and COM are naturally synergistic with Java is starting a new wave of "Object Web" developments that could have profound importance. Java is not only a good language to build brokers but also Java objects are the natural inhabitants of object databases. The resultant architecture in fig 3 shows a small object broker (a so-called ORBlet) in each browser as in Netscape's current plans. Most of our remarks are valid for all these approaches to a distributed set of services. Our ideas are however easiest to understand if one assumes an underlying architecture which is a CORBA or Javabean distributed object model integrated with the web.

We wish to use this service/object evolving 3-tier commodity architecture as the basis of our HPcc environment. We need to naturally incorporate (essentially) all services of the commodity web and to use its protocols and standards wherever possible. We insist on adopting the architecture of commodity distribution systems as complex HPCC problems require the rich range of services offered by the broader community systems. Perhaps we could "port" commodity services to a custom HPCC system but this would require continued upkeep with each new upgrade of the commodity service.

By adopting the architecture of the commodity systems, we make it easier to track their rapid evolution and expect it will give high functionality HPCC systems, which will naturally track the evolving Web/distributed object worlds. This requires us to enhance certain services to get higher performance and to incorporate new capabilities such as high-end visualization (e.g. CAVE's) or massively parallel systems where needed. This is the essential research challenge for HPcc for we must not only enhance performance where needed but do it in a way that is preserved as we evolve the basic commodity systems.

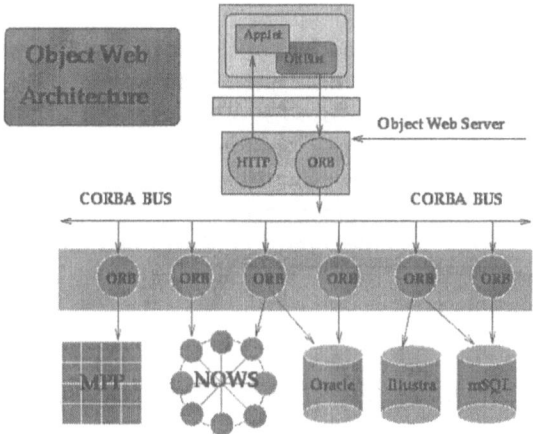

Fig. 3. Integration of Object Technologies (CORBA) and the Web

We certainly have not demonstrated clearly that this is possible but we have a simple strategy that we will elaborate in ref. [5] and sec. 5. Thus we exploit the three-tier structure and keep HPCC enhancements in the third tier, which is inevitably the home of specialized services in the object-web architecture. This strategy isolates HPCC issues from the control or interface issues in the middle layer. If successful we will build an HPcc environment that offers the evolving functionality of commodity systems without significant re-engineering as advances in hardware and software lead to new and better commodity products.

Returning to fig 2, we see that it elaborates fig 1 in two natural ways. Firstly the middle tier is promoted to a distributed network of servers; in the "purest" model these are CORBA/ COM/ Javabean object-web servers, but obviously any protocol compatible server is possible. This middle tier layer includes not only networked servers with many different capabilities (increasing functionality) but also multiple servers to increase performance on an given service.

The use of high functionality but modest performance communication protocols and interfaces at the middle tier limits the performance levels that can be reached in this fashion. However this first step gives a modest performance scaling, parallel (implemented if necessary, in terms of multiple servers) HPcc system which includes all commodity services such as databases, object services, transaction processing and collaboratories. The next step is only applied to those services with insufficient performance. Naively we "just" replace an existing back end (third tier) implementation of a commodity service by its natural HPCC high performance version. Sequential or socket based messaging distributed simulations are replaced by MPI (or equivalent) implementations on low latency high bandwidth dedicated parallel machines. These could be specialized architectures or "just" clusters of workstations.

Note that with the right high performance software and network connectivity, workstations can be used at tier three just as the popular "LAN consolidation" use of parallel machines like the IBM SP-2, corresponds to using parallel computers in the middle tier. Further a "middle tier" compute or database server could of course deliver its services using the same or different machine from the server. These caveats illustrate that as with many concepts, there will be times when the relatively clean architecture of fig 2 will become confused. In particular the physical realization does not necessarily reflect the logical architecture shown in fig 2.

4 Core Technologies for High Performance Commodity Systems

4.1 Multidisciplinary Application

We can illustrate the commodity technology strategy with a simple multidisciplinary application involving the linkage of two modules A and B – say CFD and structures applications respectively. Let us assume both are individually parallel but we need to link them. One could view the linkage sequentially as in fig 4, but often one needs higher performance and one would "escape" totally into a layer which linked decomposed components of A and B with high performance MPI (or PVMPI). Here we view MPI as the "machine language" of the higher-level commodity communication model given by approaches such as WebFlow from NPAC.

There is the "pure" HPCC approach of fig 5, which replaces all commodity web communication with HPCC technology. However there is a middle ground between the implementations of figures 4 and 5 where one keeps control (initialization etc.) at the server level and "only" invokes the high performance back end for the actual data transmission. This is shown in fig 6 and appears to obtain the advantages of both commodity and HPCC approaches for we have the functionality of the Web and where necessary the performance of HPCC software. As we wish to preserve the commodity architecture as the baseline, this strategy implies that one can confine HPCC software development to providing high performance data transmission with all of the complex control and service provision capability inherited naturally from the Web.

4.2 JavaBean Communication Model

We note that JavaBeans (which are one natural basis of implementing program modules in the HPcc approach) provide a rich communication mechanism, which supports the separation of control (handshake) and implementation. As shown below in fig 7, Javabeans use the JDK 1.1 AWT event model with listener objects and a registration/call-back mechanism.

Simple Server Approach

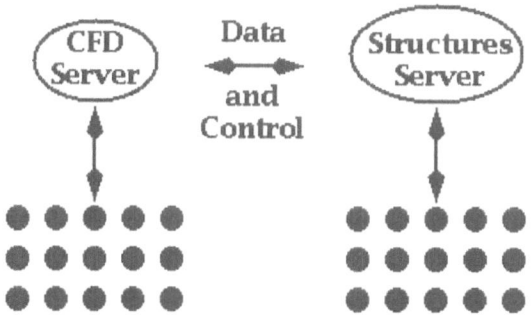

Fig. 4. Simple sequential server approach to Linking Two Modules

Classic HPCC Approach

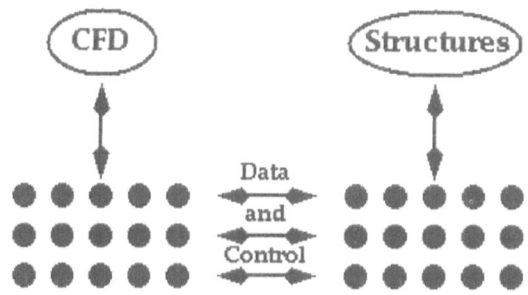

Fig. 5. Full HPCC approach to Linking Two Modules

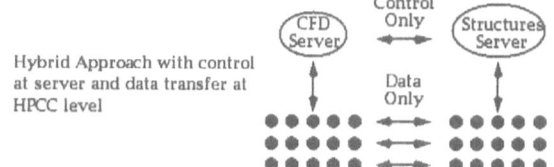

Fig. 6. Hybrid approach to Linking Two Modules

64

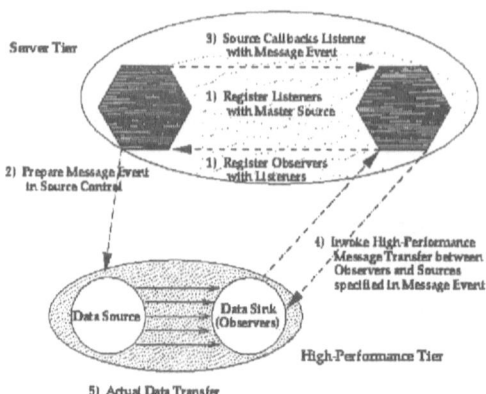

Server Tier

3) Source Callbacks Listener
with Message Event

1) Register Listeners
with Master Source

1) Register Observers
with Listeners

2) Prepare Message Event
in Source Control

4) Invoke High-Performance
Message Transfer between
Observers and Sources
specified in Message Event

Data Source

Data Sink
(Observers)

High-Performance Tier

5) Actual Data Transfer

Fig. 7. JDK 1.1 Event Model used by (inter alia) Javabeans

JavaBeans communicate indirectly with one or more "listener objects" acting
as a bridge between the source and sink of data. In the model described above,
this allows a neat implementation of separated control and explicit communication with listeners (a.k.a. sink control) and source control objects residing in
middle tier. These control objects decide if high performance is necessary or
possible and invoke the specialized HPCC layer. This approach can be used to
advantage in "run-time compilation" and resource management with execution
schedules and control logic in the middle tier and libraries such as MPI, PCRC
and CHAOS implementing the determined data movement in the high performance (third) tier. Parallel I/O and "high-performance"" CORBA can also use
this architecture. In general, this listener model of communication provides a
virtualization of communication that allows a separation of control and data
transfer that is largely hidden from the user and the rest of the system. Note
that current Internet security systems (such as SSL and SET) use high functionality public keys in the control level but the higher performance secret key
cryptography in bulk data transfer. This is another illustration of the proposed
hybrid multi-tier communication mechanism.

4.3 JWORB based Middleware

Enterprise JavaBeans that control, mediate and optimize HPcc communication
as described above need to be maintained and managed in a suitable middleware
container. Within our integrative approach of Pragmatic Object Web, a CORBA
based environonment for the middleware management with IIOP based control
protocol provides us with the best encapsulation model for EJB components.
Such middleware ORBs need to be further integrated with the Web server based
middleware to assure smooth Web browser interfaces and backward compatibility
with CGI and servlet models. This leads us to the concept of JWORB (Java Web

Object Request Broker)[6] - a multi-protocol Java network server that integrates several core services (so far dispersed over various middleware nodes as in fig 2) within a single uniform middleware management framework.

An early JWORB prototype has been recently developed at NPAC. The base server has HTTP and IIOP protocol support as illustrated in fig 8. It can serve documents as an HTTP Server and it handles the IIOP connections as an Object Request Broker. As an HTTP server, JWORB supports base Web page services, Servlet (Java Servlet API) and CGI 1.1 mechanisms. In its CORBA capacity, JWORB is currently offering the base remote method invocation services via CDR based IIOP and we are now implementing higher level support such as the Interface Repository, Portable Object Adapter and selected Common Object Services.

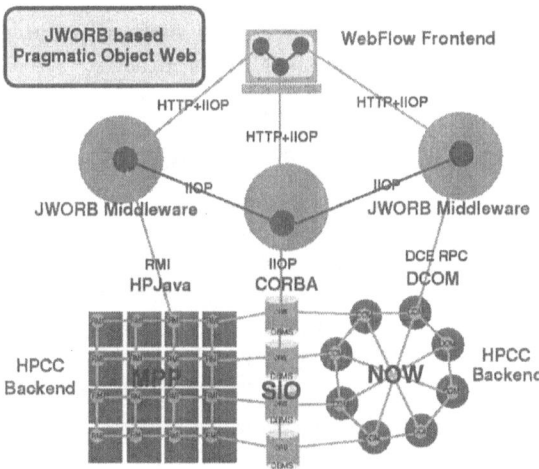

Fig. 8. Overall architecture of the JWORB based Pragmatic Object Web middleware

During the startup/bootstrap phase, the core JWORB server checks its configuration files to detect which protocols are supported and it loads the necessary protocol classes (Definition, Tester, Mediator, Configuration) for each protocol. Definition Interface provides the necessary Tester, Configuration and Mediator objects. Tester object inpects the current network package and it decides how to interpret this particular message format. Configuration object is responsible for the configuration parameters of a particular protocol. Mediator object serves the connection. New protocols can be added simply by implementing the four classes described above and by registering a new protocol with the JWORB server.

After JWORB accepts a connection, it asks each protocol handler object whether it can recognize this protocol or not. If JWORB finds a handler which can serve the connection, is delegates further processing of the connection stream to this protocol handler. Current algorithm looks at each protocol according to their order in the configuration file. This process can be optimized with randomized or prediction based algorithm. At present, only HTTP and IIOP messaging is supported and the current protocol is simply detected based on the magic anchor string value (GIOP for IIOP and POST, GET, HEAD etc. for HTTP). We are currently working on further extending JWORB by DCE RPC protocol and XML co-processor so that it can also act as DCOM and WOM/WebBroker server.

5 Commodity Services in HPcc

We have already stressed that a key feature of HPcc is its support of the natural inclusion into the environment of commodity services such as databases, web servers and object brokers. Here we give some further examples of commodity services that illustrate the power of the HPcc approach.

5.1 Distributed Collaboration Mechanisms

The current Java Server model for the middle tier naturally allows one to integrate collaboration into the computing model and our approach allow one to "re-use" collaboration systems built for the general Web market. Thus one can without any special HPCC development, address areas such as computational steering and collaborative design, which require people to be integrated with the computational infrastructure. In fig 9, we define collaborative systems as integrating client side capabilities together. In steering, these are people with analysis and visualization software. In engineering design, one would also link design (such as CATIA or AutoCAD) and planning tools. In both cases, one would need the base collaboration tools such as white-boards, chat rooms and audio-video conferencing.

If we are correct in viewing collaboration (see Tango [10, 11] and Habanero [12]) as sharing of services between clients, the 3 tier model naturally separates HPCC and collaboration and allows us to integrate into the HPCC environment, the very best commodity technology which is likely to come from larger fields such as business or (distance) education. Currently commodity collaboration systems are built on top of the Web and although emerging facilities such as Work Flow imply approaches to collaboration, are not yet defined from a general CORBA point of view. We assume that collaboration is sufficiently important that it will emerge as a CORBA capability to manage the sharing and replication of objects. Note CORBA is a server-server model and "clients" are viewed as servers (i.e. run Orb's) by outside systems. This makes the object-sharing view of collaboration natural whether application runs on "client" (e.g. shared Microsoft Word document) or on back-end tier as in case of a shared parallel computer simulation.

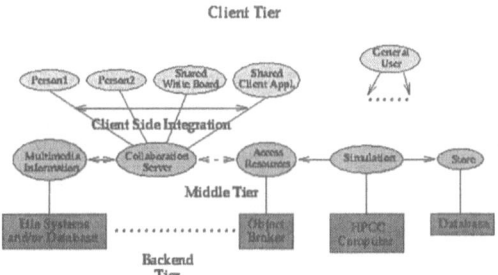

Fig. 9. Collaboration in today's Java Web Server implementation of the 3 tier computing model. Typical clients (on top right) are independent but Java collaboration systems link multiple clients through object (service) sharing

5.2 Object Web and Distributed Simulation

The integration of HPCC with distributed objects provides an opportunity to link the classic HPCC ideas with those of DoD's distributed simulation DIS or Forces Modeling FMS community. The latter do not make extensive use of the Web these days but they have a commitment to CORBA with their HLA (High Level Architecture) and RTI (Runtime Infrastructure) initiatives. Distributed simulation is traditionally built with distributed event driven simulators managing C++ or equivalent objects. We suggest that the Object Web (and parallel and distributed ComponentWare described in sec. 5.3) is a natural convergence point for HPCC and DIS/FMS. This would provide a common framework for time stepped, real time and event driven simulations. Further it will allow one to more easily build systems that integrate these concepts as is needed in many major DoD projects – as exemplified by the FMS and IMT DoD computational activities which are part of the HPCC Modernization program.

HLA is a distributed object technology with the object model defined by the Object Model Template (OMT) specification and including the Federation Object Model (FOM) and the Simulation Object Model (SOM) components. HLA FOM objects interact by exchanging HLA interaction objects via the common Run-Time Infrastructure (RTI) acting as a software bus similar to CORBA. Current HLA/RTI follows a custom object specification but DMSO's longer term plans include transferring HLA to industry via OMG CORBA Facility for Interactive Modeling and Simulation.

At NPAC, we are anticipating these developments are we are building a prototype RTI implementation in terms of Java/CORBA objects using the JWORB middleware [7]. RTI is given by some 150 communication and/or utility calls, packaged as 6 main managment services: Federation Management, Object Management, Declaration Managmeent, Ownership Management, Time Management, Data Distribution Management, and one general purpose utility service. Our de-

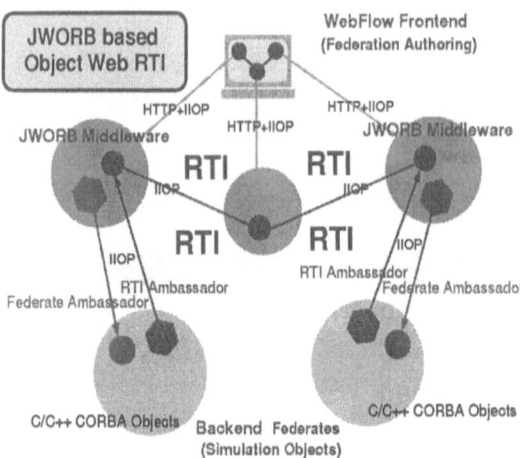

Fig. 10. Overall architecture of the Object Web RTI - a JWORB based RTI prototype recently developed at NPAC

sign shown in fig 10 is based on 9 CORBA interfaces, including 6 Managers, 2 Ambassadors and RTIKernel. Since each Manager is mapped to an independent CORBA object, we can easily provide support for distributed management by simply placing individual managers on different hosts.

The communication between simulation objects and the RTI bus is done through the RTIambassador interface. The communication between RTI bus and the simulation objects is done by their FederateAmbassador interfaces. Simulation developer writes/extends FederateAmbassador objects and uses RTIambassador object obtained from the RTI bus.

RTIKernel object knows handles of all manager objects and it creates RTI-ambassador object upon the federate request. Simulation obtains the RTIambassador object from the RTIKernel and from now on all interactions with the RTI bus are handled through the RTIambassador object. RTI bus calls back (asynchronously) the FederateAmbassador object provided by the simulation and the federate receives this way the interactions/attribute updates coming from the RTI bus.

Although coming from the DoD computing domain, RTI follows generic design patterns and is applicable to a much broader range of distributed applications, including modeling and simulation but also collaboration, on-line gaming or visual authoring. From the HPCC perspective, RTI can be viewed as a high level object based extension of the low level messaging libraries such as PVM or MPI. Since it supports shared objects management and publish/subscribe based multicast channels, RTI can also be viewed as an advanced collaboratory framework, capable of handling both the multi-user and the multi-agent/multi-module distributed systems.

5.3 Visual HPCC ComponentWare

HPCC does not have a good reputation for the quality and productivity of its programming environments. Indeed one of the difficulties with adoption of parallel systems, is the rapid improvement in performance of workstations and recently PC's with much better development environments. Parallel machines do have a clear performance advantage but this for many users, this is more than counterbalanced by the greater programming difficulties. We can give two reasons for the lower quality of HPCC software. Firstly parallelism is intrinsically hard to find and express. Secondly the PC and workstation markets are substantially larger than HPCC and so can support a greater investment in attractive software tools such as the well-known PC visual programming environments. The DcciS revolution offers an opportunity for HPCC to produce programming environments that are both more attractive than current systems and further could be much more competitive than previous HPCC programming environments with those being developed by the PC and workstation world. Here we can also give two reasons. Firstly the commodity community must face some difficult issues as they move to a distributed environment, which has challenges where in some cases the HPCC community has substantial expertise. Secondly as already described, we claim that HPCC can leverage the huge software investment of these larger markets.

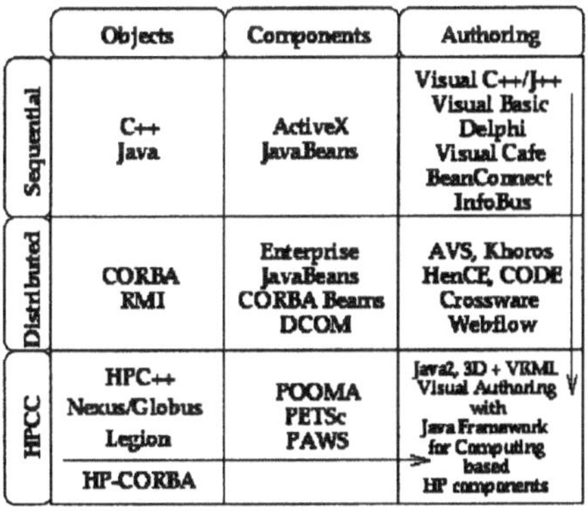

	Objects	Components	Authoring
Sequential	C++ Java	ActiveX JavaBeans	Visual C++/J++ Visual Basic Delphi Visual Cafe BeanConnect InfoBus
Distributed	CORBA RMI	Enterprise JavaBeans CORBA Beans DCOM	AVS, Khoros HenCE, CODE Crossware Webflow
HPCC	HPC++ Nexus/Globus Legion HP-CORBA	POOMA PETSc PAWS	Java2, 3D + VRML Visual Authoring with Java Framework for Computing based HP components

Fig. 11. System Complexity (vertical axis) versus User Interface (horizontal axis) tracking of some technologies

In fig 11, we sketch the state of object technologies for three levels of system complexity – sequential, distributed and parallel and three levels of user (programming) interface – language, components and visual. Industry starts at the top left and moves down and across the first two rows. Much of the current commercial activity is in visual programming for sequential machines (top right box) and distributed components (middle box). Crossware (from Netscape) represents an initial talking point for distributed visual programming. Note that HPCC already has experience in parallel and distributed visual interfaces (CODE and HenCE as well as AVS and Khoros). We suggest that one can merge this experience with Industry's Object Web deployment and develop attractive visual HPCC programming environments as shown in fig 12.

Currently NPAC's WebFlow system [9][12] uses a Java graph editor to compose systems built out of modules. This could become a prototype HPCC ComponentWare system if it is extended with the modules becoming JavaBeans and the integration with CORBA. Note the linkage of modules would incorporate the generalized communication model of fig 7, using a mesh of JWORB servers to manage a recourse pool of distributedHPcc components. An early version of such JWORB based WebFlow environment, illustrated in fig 13 is in fact operational at NPAC and we are currently building the Object Web management layer including the Entperprise JavaBeans based encapsulation and communication support discussed in the previous section.

Returning to fig 1, we note that as industry moves to distributed systems, they are implicitly taking the sequential client-side PC environments and using them in the much richer server (middle-tier) environment which traditionally had more closed proprietary systems.

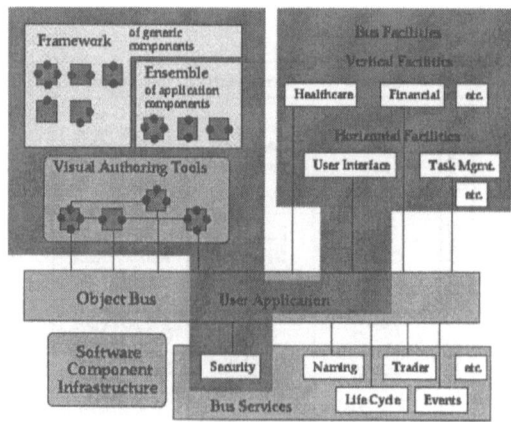

Fig. 12. Visual Authoring with Software Bus Components

We will then generate an environment such as fig 12 including object broker services, and a set of horizontal (generic) and vertical (specialized application) frameworks. We do not have yet much experience with an environment such as fig 12, but suggest that HPCC could benefit from its early deployment without the usual multi-year lag behind the larger industry efforts for PC's. Further the diagram implies a set of standardization activities (establish frameworks) and new models for services and libraries that could be explored in prototype activities.

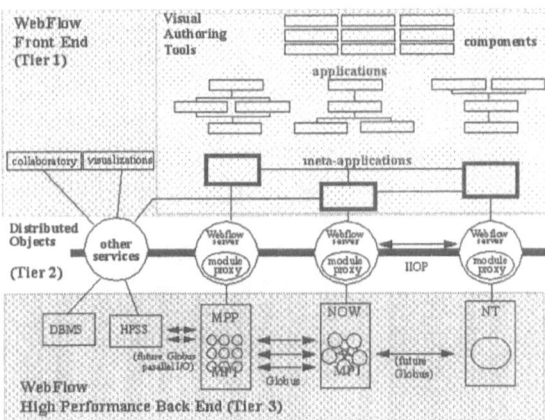

Fig. 13. Top level view of the WebFlow environment with JWORB middleware over Globus metacomputing or NT cluster backend

5.4 Early User Communities

In parallel with refining the individual layers towards production quality HPcc environment, we start testing our existing prototypes such as WebFlow, JWORB and WebHLA for the selected application domains.

Within the NPAC participation in the NCSA Alliance, we are working with Lubos Mitas in the Condensed Matter Physics Laboratory at NCSA on adapting WebFlow for Quantum Monte Carlo simulations [13]. This application is illustrated in figures 14 and 15 and it can be characterized as follows. A chain of high performance applications (both commercial packages such as GAUSSIAN or GAMESS or custom developed) is run repeatedly for different data sets. Each application can be run on several different (multiprocessor) platforms, and consequently, input and output files must be moved between machines. Output files are visually inspected by the researcher; if necessary applications are rerun with modified input parameters. The output file of one application in the chain is the input of the next one, after a suitable format conversion.

The high performance part of the backend tier in implemented using the GLOBUS toolkit [14]. In particular, we use MDS (metacomputing directory services) to identify resources, GRAM (globus resource allocation manager) to allocate resources including mutual, SSL based authentication, and GASS (global access to secondary storage) for a high performance data transfer. The high performance part of the backend is augmented with a commodity DBMS (servicing

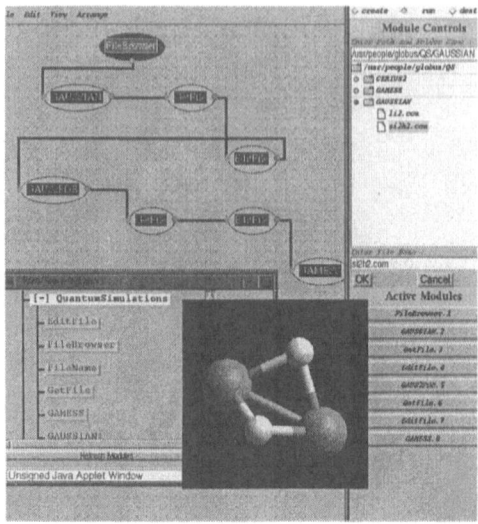

Fig. 14. Screendump of an example WebFlow session: running Quantum Simulations on a virtual metacomputer. Module GAUSSIAN is executed on Convex Exemplar at NCSA, module GAMESS is executed on SGI Origin2000, data format conversion module is executed on Sun SuperSparc workstation at NPAC, Syracuse, and file manipu-
tion modules (FileBrowser, EditFile, GetFile) are run on the researcher's desktop

Permanent Object Manager) and LDAP-based custom directory service to maintain geographically distributed data files generated by the Quantum Simulation project. The diagram illustrating the WebFlow implementation of the Quantum Simulation is shown in fig 15.

Another large application domain we are currently adressing is DoD Modeling Simulation, approached from the perspective of FMS and IMT thrusts within the DoD Modernization Program. We already described the core effort on building Object Web RTI on top of JWORB. This is associated with a set of more application- or component-specific efforts such as: a) building distance training space for some mature FMS technologies such as SPEEDES; b) parallelizing and CORBA-wrapping some selected computationally intense simulation modules such as CMS (Comprehensive Mine Simulator at Ft. Belvoir, VA); c) adapting WebFlow to support visual HLA simulation authoring. We refer to such Pragmatic Object Web based interactive simulation environment as WebHLA [8] and we believe that it will soon offer a powerful modeling and simulation framework, capable to address the new challenges of DoD computing in the areas of Simulation Based Design, Testing, Evaluation and Acquisition.

References

[1] Client/Server Programming with Java and CORBA by Robert Orfali and Dan Harkey, Wiley, Feb '97, ISBN: 0-471-16351-1

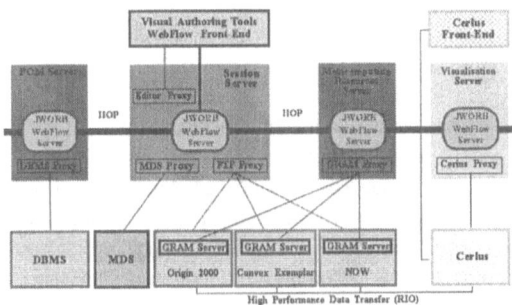

Fig. 15. WebFlow implementation of the Quantum Simulations problem

[2] High Level Architecture and Run-Time Infrastructure by DoD Modeling and Simulation Office (DMSO), http://www.dmso.mil/hla

[3] Geoffrey Fox and Wojtek Furmanski, "Petaops and Exaops: Supercomputing on the Web", IEEE Internet Computing, 1(2), 38-46 (1997); http://www.npac.syr.edu/users/gcfpetastuff/petaweb

[4] Geoffrey Fox and Wojtek Furmanski, "Java for Parallel Computing and as a General Language for Scientific and Engineering Simulation and Modeling", Concurrency: Practice and Experience 9(6), 415-426(1997).

[5] Geoffrey Fox and Wojtek Furmanski, "Use of Commodity Technologies in a Computational Grid", chapter in book to be published by Morgan-Kaufmann and edited by Carl Kesselman and Ian Foster.

[6] Geofrey C. Fox, Wojtek Furmanski and Hasan T. Ozdemir, "JWORB - Java Web Object Request Broker for Commodity Software based Visual Dataflow Metacomputing Programming Environment", NPAC Technical Report, Feb 98, http://tapetus.npac.syr.edu/iwt98/pm/documents/

[7] G.C.Fox, W. Furmanski and H. T. Ozdemir, "Java/CORBA based Real-Time Infrastructure to Integrate Event-Driven Simulations, Collaboration and Distributed Object/Componentware Computing", In Proceedings of Parallel and Distributed Processing Technologies and Applications PDPTA '98, Las Vegas, Nevada, July 13-16, 1998, http://tapetus.npac.syr.edu/iwt98/pm/documents/

[8] David Bernholdt, Geoffrey Fox and Wojtek Furmanski, B. Natarajan, H. T. Ozdemir, Z. Odcikin Ozdemir and T. Pulikal, "WebHLA - An Interactive Programming and Training Environment for High Performance Modeling and Simulation", In Proceedings of the DoD HPC 98 Users Group Conference, Rice University, Houston, TX, June 1-5 1998, http://tapetus.npac.syr.edu/iwt98/pm/documents/

[9] D. Bhatia, V. Burzevski, M. Camuseva, G. Fox, W. Furmanski and G. Premchandran, "WebFlow – A Visual Programming Paradigm for Web/Java based coarse grain distributed computing", Concurrency Practice and Experience 9,555-578 (1997), http://tapetus.npac.syr.edu/iwt98/pm/documents/

[10] L. Beca, G. Cheng, G. Fox, T. Jurga, K. Olszewski, M. Podgorny, P. Sokolowski and K. Walczak, "Java enabling collaborative education, health care and comput-

ing", Concurrency Practice and Experience 9,521-534(97).
http://trurl.npac.syr.edu/tango

[11] Tango Collaboration System, http://trurl.npac.syr.edu/tango

[12] Habanero Collaboration System,
http://www.ncsa.uiuc.edu/SDG/Software/Habanero

[13] Erol Akarsu, Geoffrey Fox, Wojtek Furmanski, Tomasz Haupt, "WebFlow - High-Level Programming Environment and Visual Authoring Toolkit for High Performance Distributed Computing", paper submitted for Supercomputing 98,
http://www.npac.syr.edu/users/haupt/ALLIANCE/sc98.html

[14] Ian Foster and Carl Kessleman, Globus Metacomputing Toolkit,
http://www.globus.org

Workshop 1
Support Tools and Environments

Chris Wadsworth and Helmar Burkhart

Co-chairmen

Support Tools and Environments

Programming tools are a controversial topic: on the one hand, everyone agrees that good tools are needed for productive software development, and a plethora of tools was built in past; on the other hand, many of the tools developed are not used (may be not even by their developers) and it is not easy for companies offering tools for high-performance computing to survive. In order to improve the situation, both scientific activities such as workshops and conferences discussing technical challenges, as well as organizational activities such as consortia and software tool repositories are needed.

The one-day workshop *Support Tools and Environments* consists of a special session on PSPTA projects (see the separate introduction by Chris Wadsworth) and five additional papers concentrating on *Performance-oriented Tools* and *Debugging and Monitoring*.

Peter Kacsuk (KFKI, Hungary), Vaidy Sunderam (Emory University, USA), and Chris Wadsworth (Rutherford Appleton Lab, UK) have been co-organizers of this workshop. I would like to thank them and the reviewers for all help and work, which only made this event possible.

Introduction to PSTPA Session

Background and Aims

Despite the forecast demand for the high performance computing power of parallel machines, the fact remains that programming such machines is still an intellectually demanding task. While there have been some significant recent advances (e.g. the evolution and widespread adoption of the PVM and MPI message passing standards, and the development of High Performance Fortran), there remains an urgent need for a greater understanding of the fundamentals, methods, and methodology of parallel software development to support a broad application base and to engender the next generation of parallel software tools.

The PSTPA programme seeks in particular:

- to capitalise on UK strengths in research into the foundations of generic software tools, languages, application generators, and program transformation techniques for parallel computing,

- to widen the application programmer base and to improve productivity by an order of magnitude,
- to maximise the exploitation potential of tools developed in the programme,
- to maximise architectural independence of the tools and to maximise the leverage of open systems standards,
- to focus on easy-to-use tools which can provide particular benefit to the engineering, industrial, and information processing user sectors, and
- to prove the effectiveness of the tools and the value of the standards to potential users with appropriate demonstrations.

A long term goal is to hide the details of parallelism from the user, and this programme also aims to address this. In the short term, however, it remains necessary to provide tools which facilitate the exploitation of parallelism explicitly at an appropriate level for the community of application programmers. These tools will help in the expression, modelling, measurement, and management of parallel software, as much as possible in a portable fashion.

The scope of the programme thus covers the spectrum from incremental advances in the present generation of tools to research that will engender the next generation of tools. In the medium term (five years) it is envisaged that tools will become increasingly 'generative', e.g. application generators, and 'performance-oriented' at least for particular application domains, e.g. for databases, embedded systems, or signal and image processing. The effectiveness of the tools in practice is a key objective.

15 projects (listed below) have been supported by the programme, including both projects to aid the porting of existing software and ones that are researching methods and tools to reduce the effort in developing new applications software. Each project is academically led with one or more (up to 7 in one case) industrial or commercial partners (not listed). At the time of writing (May 1998) 9 of the projects have finished, with the remaining 6 finishing over the next 12 months.

Papers

The six papers in this PSTPA Special Session are from different projects and have been refereed and accepted as regular submissions for EuroPar'98. It is pleasing to note that two of these — by Hill et al (Oxford) and by Delaitre et al (Westminster) — have been chosen as Distinguished Papers for EuroPar'98.

The first paper "Process Migration and Fault Tolerance of BSPlib Programs Running on a Network of Workstations" by Hill, Donaldson and Lanfear builds on the achievements of the BSP (Bulk Sychronous Parallelism) model over recent years. The PSTPA project has been developing a programming and run-time environment for BSP to extend the parctical methodology. Hill et al describe a checkpointing technique that enables a parallel BSP job to adapt itself continually to run on the least loaded machines in the network and present experimental results for a network of workstations. Distributed techniques for determining global load are also addressed. It is seen that the 'superstep' structure of BSP provides natural times when communication is quiescent for taking checkpoints.

The authors conclude that fault tolerance and process migration are achievable in a transparent way for BSP programs.

The next paper "A Parallel-System Design Toolset for Vision and Image Processing" by Fleury, Sarvan, Downton and Clarke is targetted at real-time, data-dominated systems such as those typically found in embedded systems, in vision, and in signal and image processing. Many systems in these areas are naturally built as software pipelines each stage of which can contain internal parallelism. The authors identify a generic form of pipelines of processor farms (PPF) as of wide interest. The paper summarises the PPF methodology and goes on to describe the development of a toolset for the construction, analysis, and tuning of pipeline farms. Performance prediction for the tuning phase is by simulation, using a simulation visualizer, augmented by analytic prediction for some known standard distributions. It is seen that a key benefit is that pipeline farms may be crafted quantitatively for desired performance characteristics (minimum latency, maximum throughput, minimum cost etc).

Image processing is also the target domain for "Achieving Portability and Efficiency through Automatic Optimisation: an Investigation in Parallel Image Processing" by Crookes et al. They present the EPIC model, a set of portable programming abstractions for image processing, and outline an automatically optimising implementation. This uses a coprocessor structure and an optimiser which extends the coprocessor's instruction set for compositions of built-in instructions as they are used on a program by program basis. The technique removes most forms of inefficiency that otherwise arise from a straightforward abstract-machine based implementation. The paper concludes that the EPIC model is a portable software platform for its domain and the implementation demonstrates portability without loss of effieciency for this domain.

The design and construction of a performance-oriented environment is addressed in "EDPEPPS: A Toolset for the Design and Performance Evaluation of Parallel Applications" by Delaitre et al. Their environment is designed for a rapid prototyping approach based on a design-simulate-analyse cycle, including graphical tools for design, simulation, and performance visualisation and prediction. The simulator has been developed and refined over several projects, with results in the paper showing good comparisons against measured execution times. The present toolset is built around PVM, however the modular strucure of the simulation model makes it reconfigurable.

Databases are an area in which parallel processing is already being exploited but remains in need of performance-oriented tools for accurate modelling and prediction. In "Verifying a Performance Estimator for Parallel DBMSs" Dempster et al address issues in predicting the performance characteristics of parallel database systems, in verifying the results by both simulation and process algebra. There is particular application to topics such as application sizing, capacity planning, and data placement. The paper describes a performance estimation tool STEADY designed to predict maximum transaction throughput, resource utilisation, and response times. Close accuracy is verified for predictions for throughput and resource utilisation. Predictions for response times are less accurate as yet, particular as the transaction workload increases.

Finally, "Generating Parallel Applications of Spatial Interaction Models" by Davy and Essah characterises application generators as a high-level approach to generating programs, typically here by specifying input model and parameters, usually for a particular application domain. Davy and Essah point to a wide range of applications of Spatial Interaction Models (SIM) in the social sciences and describe opportunities for parallelism in the model evaluation and in the application. Practical experience is presented building an application generator for SIM applications, with results presented showing that the run-time overheads (compared to hand-coded versions) are less than 2% for a variety of model types and number of processors. Parallelism is encapsulated within the application generator, hiding the need for knowledge of it by the user who needs only to specify a SIM application by defining its parameters.

List of Projects

The 15 projects with title, leader, and email contact are as follows:

A Portable Coprocessor Model for Parallel Image Processing
Prof Danny Crookes, Queen's University Belfast; `email: d.crookes@qub.ac.uk`

Portable Software Tools for the Parallelisation of Computational Mechanics Software
Prof Mark Cross, University of Greenwich; `email: m.cross@greenwich.ac.uk`

An Application Generator for Spatial Interaction Modelling on a Scalable Computing Platform
Prof Peter Dew, University of Leeds; `email: dew@scs.leeds.ac.uk`

Portable Software Components
Dr Peter Dzwig, QMW College London; `email: Heather.Liddell@dcs.qmw.ac.uk`

Parallel Software Tools for Embedded Signal Processing Applications
Prof Andy Downton, University of Essex; `email:A.C.Downton@essex.ac.uk`

A Distributed Application Generator in the Search/Optimisation Domain
Dr Hugh Glaser, University of Southampton; `email: hg@ecs.soton.ac.uk`

Automatic Generation of Parallel Visualisation Modules
Dr Terry Hewitt, University of Manchester; `email: w.t.hewitt@mcc.ac.uk`

An Integrated Environment for Modelling High-Performance Parallel Database Systems
Prof Tony Hey, University of Southampton; `email: C.J.Scott@pac.soton.ac.uk`

A BSP Programming Environment
Prof Bill McColl, University of Oxford; `email: bob.mclatchie@comlab.ox.ac.uk`

Automatic Checking of Message-Passing Programs
Dr Denis Nicole, University of Southampton; `email: D.A.Nicole@ecs.soton.ac.uk`

The Inference of Data Mapping and Scheduling Strategies from Fortran 90 Programs on the Cray T3D
Dr Alex Shafarenko, University of Surrey; `email: A.Shafarenko@ee.surrey.ac.uk`

A Framework for Distributed Application
Prof Philip Treleaven, University College London; `email: P.Treleaven@ucl.ac.uk`

occam for All

Prof Peter Welch, University of Kent; email: `P.H.Welch@ukc.ac.uk`

Application Sizing, Capacity Planning and Data Placement for Parallel Database Systems

Prof Howard Williams, Heriot-Watt University; email: `howard@cee.hw.ac.uk`

An Environment for the Design and Performance Evaluation of Portable Parallel Software

Prof Steve Winter, University of Westminster; email: `wintersc@westminster.ac.uk`

Process Migration and Fault Tolerance of *BSPlib* Programs Running on Networks of Workstations

Jonathan M.D. Hill[1], Stephen R. Donaldson[1] and Tim Lanfear[2]

[1] Oxford University Computing Laboratory, UK.
[2] British Aerospace Sowerby Research Centre, UK.

Abstract. This paper describes a system that enables parallel programs written using the *BSPlib* communications library to migrate processes among a network of workstations. Not only does the system provide fault tolerance of *BSPlib* jobs, but by utilising a load manager that maintains an approximation of the global load of the system, it is possible to continually schedule the migration of BSP processes onto the least loaded machines in a network. Results are provided for an industrial electromagnetics application that show that we can achieve similar throughput on a publically available collection of workstations as a dedicated NOW.

1 Introduction

The Bulk Synchronous Parallel (BSP) model [14, 10] views a parallel machine as a set of processor-memory pairs, with a global communication network and a mechanism for synchronising all processors. A BSP program consists of a sequence of *supersteps*. Each superstep involves all of the processors and consists of three phases: (1) processor-memory pairs perform a number of computations on data held locally at the start of a superstep; (2) processors communicate data into other processor's memories; and (3) all processors barrier synchronise. The globally consistent state that is available after the barrier not only helps when reasoning about parallel programs, but also suggests a programming discipline in which computation (and communication) is balanced across all the processes. As balance is so important to BSP, profiling tools have concentrated upon exposing imbalances to the programmer so that they can be eliminated [5, 7]. However, not all imbalances that arise during program execution are caused by the program. In an environment where processors are not dedicated resources, the BSP computation proceeds at the speed of the slowest processor. This would suggest that the synchronous nature of BSP is a disadvantage compared to the more lax synchronisation regime of message passing systems such as MPI. However, most programs written using collective communications, or scientific applications such as the NAS parallel benchmarks [1] are highly synchronous in nature, and are therefore limited to the performance of the slowest running process in either BSP or MPI. Therefore, if a network user logs onto a machine that is part of a BSP job, this may have an undue effect on the entire job. This paper describes a technique that ensures a p process BSP job continually adapts itself to run on the p least loaded processors in a network consisting of P machines ($p \leq P$).

Dedicated parallel machines can impose a global scheduling policy upon their user community such that, for example, parallel jobs do not interfere with each other in a detrimental manner. The environment that we describe is one where it is not possible to impose some schedule on the user community. The components of the parallel computation are invariably guests on other peoples machines and should not impose any restrictions on them for hosting the computation. Further, precisely because of this arrangement, the availability of the nodes and the available resources at these nodes is quite erratic and unpredictable. We adopt the philosophy that in such a situation it is reasonable to expect the parallel job to look after itself.

We briefly describe the steps involved in migrating a BSP job, that has been written using the *BSPlib* [6] communications library, among a set of machines and the strategy used in making check-pointing and migration decisions across all machines. The first technical challenge (Section 2) describes how we capture the state of a UNIX process and restart it in the same state on another machine of the same type and operating system. The simplicity of the superstep structure of BSP programs provides a convenient point at which local checkpoints capture the global state of the entire BSP computation. This therefore enables process migration and check-pointing to be achieved without any changes to the users program. Next we describe a strategy whereby all processes simultaneously decide that a different set of machines would provide a better service (Section 3). When the BSP job decides that processes should be migrated, all processes perform a global checkpoint, they are then terminated and restarted on the least loaded machines from that checkpoint. Section 4 describes a technique for determining the global load of a system, and Section 5 describes an experiment using an industrial electro-magnetic application on a network of workstations. This demonstrates how the scientist or engineer is allowed to concentrate on the application and not on maintaining or worrying about the choice of processors in the network. Section 6 describes some related work and Section 7 concludes the paper.

2   Check-Pointing and Restarting Single Processes

BSPlib provides a simple API for inter-processor communication in the context of the BSP model. This simple interface has been implemented on four classes of machine: (1) distributed memory machines where the implementation uses either proprietary message passing libraries or MPI; (2) Distributed memory machines where the implementation uses primitive one sided communication, for example the Cray SHMEM library of the T3E; (3) shared memory multi-processors where the implementation uses either proprietary concurrency primitives or System V semaphores; and (4) Networks of workstations where the implementation uses TCP/IP or UDP/IP. In this paper we concentrate upon check-pointing programs running on the network of workstations version of the library. Unlike other check-pointing schemes for message passing systems (See Section 6), by choosing to checkpoint at the barrier synchronisation that delimits supersteps, because there is a globally consistent state upon exiting the barrier (where all communication is quiesced), the task of performing a global checkpoint reduces to the task of check-pointing all the processes at the local process level.

All that is required to perform a local checkpoint is to save all program data that is active. Unfortunately, because data (i.e., the state) can be arbitrarily dispersed amongst the stack, heap and program text, capturing the state of a running program is not as simple as it would first appear. A relatively straight-forward solution in a UNIX environment is to capture an image of the running process and create an executable which contains the state of the modified data section (including any allocated heap storage) and a copy of the stack. When the check-pointed executable is restarted the original context is restored and all *BSPlib* supporting IPC connections (pipes and sockets) are re-established before the computation is allowed to proceed. All this activity is transparent to the programmer as it is performed as part of the *BSPlib* primitives. Furthermore, by restricting the program to the semantics of *BSPlib*, no program changes are required. The process of taking a checkpoint involves making a copy of the stack on the heap, saving the current stack pointer and frame pointer registers, and saving any additional state information (for example, on the SPARC the register windows need to be flushed onto the stack before it is saved, and the subroutine return address needs to be saved as it is stored in a register; in contrast, on the Intel X86 architecture, all necessary information is already stored on the stack). The executable that captures this state information is built using the `unexec()` function which is distributed as part of Emacs [11]. The use of `unexec()` is similar to its use (or the use of `undump`) in Emacs, LaTeX (which build executables containing initialised data structures) and Condor which also performs check-pointing [4]. However, the state saving in Condor captures the point of execution and the stack height using the standard C functions `setjmp()` and `longjmp()` which only guarantee far jumping into activation records already on the stack and within the same process instance. Instead, we capture the additional required information based on the concrete semantics of the processor. To restart a process and restore its context, the restart routine adjusts the stack pointer to create enough space on the stack so that the saved stack can be copied to its original address and restores any saved registers.

3 Determining when to Migrate Processes

As mentioned above, our philosophy is that the executing job be sensitive to the environment in which it is executing and it is the job, and not an external scheduler, that makes appropriate scheduling decisions. For the job to make an informed decision, some global resource information needs to be maintained. The solution we have adopted is that there are daemons running on each machine in the network which maintain local approximations to the global load. The accuracy of these approximations is discussed in the next section. Here we assume that each machine contains information on the entire network which is no more than a few minutes old with a high probability.

When a BSP job requests a number of processes, the local daemon is queried for a suitable list of machines on which to execute the job (the daemon responds so that the job may be run on the least loaded machines). In order that the decisions are not too fickle, the five minute load averages are used. Also, it is a requirement that not too much network traffic be generated to maintain a reasonable global state. Since the five minute load averages are being used, it is not too important that entries in the load table become slightly out of date as the wildest swings in the load averages take a couple of minutes to register in the five minute load average figures.

Let G_i be the approximation to the global load on machine i, then given P machines, the true global load is $G = G_1 \sqcup \cdots \sqcup G_P$; where $\sqcup$ is used to merge the approximations from two machines. Given a BSP job running on p processors with machine names[1] j in the set $\{i_1, \ldots, i_p\}$, we use the approximation $G' = G_{i_1} \sqcup \cdots \sqcup G_{i_p}$ which is better than any of the individual approximations with a high probability. G' is a sequence of machine names sorted in decreasing priority order (based on load averages, number of CPUs and their speeds). If the top set of p entries of G' is not $\{i_1, \ldots, i_p\}$ then an alternate and better assignment of processes to machines exists (call this predicate fn). In order not to cause processes to thrash between machines, a measure x of the load of the job (where $0 \leq x \leq 1$) is added to the load averages of all machines not involved in the BSP computation before the predicate fn is applied. This anticipates the maximum increase in the load of a machine when a process is migrated to it. Any observed increase in load greater than x is therefore caused by additional external load.

Our aim is to ensure that the only overhead in process migration is the time taken to write p instances of the check-pointed program to disk. Therefore, we require that the calculation that determines when to perform process migration does not unduly impact the computation or communication performance of *BSPlib*. We need to solve $fn(G') = fn(G_{i_1} \sqcup \cdots \sqcup G_{i_p})$ either on a superstep by superstep basis or every N supersteps. However, the result can be obtained without first merging the global load approximations. This can be done by each processor independently computing its load approximations G_i and checking that it is amongst the top p after adding x_j to the loads of machines not involved in the current BSP job; where $0 \leq x_j \leq 1$ is the contribution of the BSP process on machine j, to the load average on that machine. This calculation can be performed on entry to the superstep T seconds after the last checkpoint (i.e., this checking does not have to be performed in synchrony). The Boolean result from each of the processors is reduced with the or-operator to ensure that all processors agree to checkpoint during the same superstep. In the TCP/IP and UDP/IP implementations of *BSPlib*, this is piggy-backed onto a reduction that is used to globally optimise communication [3]. Therefore if a checkpoint is not necessary, there is no substantial impact on the performance of *BSPlib*.

---

[1] we distinguish between machines and processors as each machine may contain a number of processors, each of which runs multiple processes.

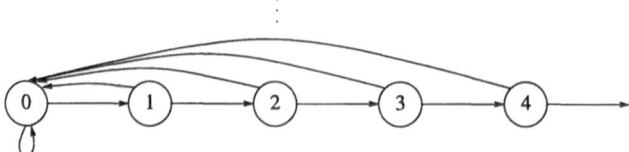

Fig. 1. Markov chain considering only single direct updates

4 Determining the Global Load of a System

The load daemons use a protocol in which the locally held global load states are periodically sent to k randomly chosen daemons running on other machines. This update period is uniformly and randomly distributed with each machine being independent. When a load daemon receives an update, it responds by merging the locally held load table and sending the resultant table back to the sender. The sender then merges the entries of the returned table with the entries of the locally held table. For purposes of simplifying the analysis, the updates are assumed to happen at fixed intervals and in a lockstep fashion. We also assume that the processors do not respond with the merged table, but merely update their tables by merging in the update requests. The analysis that follows always provides an upper bound for the actual protocol used.

If each processor sent out a message at the end of each interval to all the other processors, this would require P^2 messages to maintain the global state. A job requiring $p \leq P$ processes could contact P machines and arrive at an optimal choice of processors with considerably fewer messages provided that jobs did not start very often. However, with a large network of machines, the rate of jobs starting and the number of machines P would quickly lengthen the delay in scheduling a BSP job. By maintaining a global processor utilisation state at each of the machines, starting a job only involves contacting the local daemon when choosing a set of p processors and thus need not contribute to network traffic. Even once the ordering of processors has been selected, the problem of over assigning work to the least loaded machines can be avoided by having those machines reject the workload request based on knowledge built up locally. The algorithm then simply tries the machine with the next highest priority.

The quality of the decision for the optimal set of machines depends on the ages of the entries in the distributed load averages tables. If each machine uniformly and randomly chooses a partner machine at the end of each interval and sends its load average value to that machine, then the mean age of each entry in a distributed load average table can be calculated by considering the discrete time Markov chain shown in Figure 1. In this case there would only be p messages at the end of each interval, but the age distribution $\{\pi_i : i \in \mathbb{N}\}$, and the mean age μ are given by:

$$\pi_i = \frac{1}{p-1} \left(\frac{p-2}{p-1} \right)^i \tag{1}$$

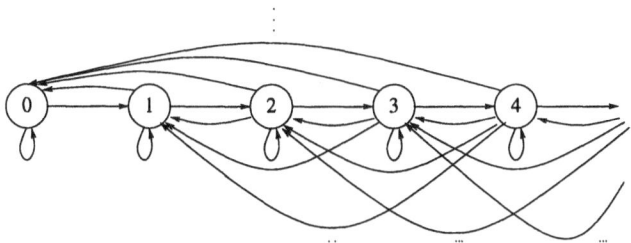

Fig. 2. Markov chain when indirect updates are allowed

$$\mu = \sum_{i=1}^{\infty} i\pi_i = p - 2 \tag{2}$$

By sending out messages more frequently, or sending out k messages at the end of each interval, the mean age can be reduced to $\mathcal{O}(p/k)$, but this increases the traffic and required number of virtual circuits to pk.

By allowing each machine to exchange *all* their load average information with k other machines at the end of each interval, a significant reduction in the mean age of the entries can be achieved with pk circuits. This scheme allows machines to choose between the most current load average figures, even if they were indirectly obtained from other machines. Figure 2 shows the corresponding Markov chain. In this stochastic process, transitions into state 0 can only arise out of direct updates, that is, a machine directly updating its entry in a remote table. The distribution, $\{\pi_i' : i \in \mathbb{N}\}$, of the ages of the entries in the load average table is given by the recurrence:

$$\pi_i' = \begin{cases} k/(p-1), & \text{if } i = 0 \\ \pi_{i-1}' P\{no\ useful\ updates\} + \sum_{j=i}^{\infty} \pi_j' P\{min\ age\ is\ i\}, & \text{otherwise} \end{cases} \tag{3}$$

Figure 3 shows the mean table entry ages of the three strategies when $k = 1, 3, 6$ against P. As P is given as a log scale, it is clear from the figure that while the first two strategies give a mean age μ as $\mathcal{O}(P)$, the third strategy (allowing indirect updates) gives a mean age of μ' as $\mathcal{O}(\log P)$.

If we replace the discrete times of the Markov chains with update intervals, t, the distributions above give the mean age at the beginning of each of the intervals. The figure shows that in order to bound the mean age to, say, five minutes we must ensure that:

$$t(\mu + \frac{1}{2}) \leq 5 \text{ minutes, or}$$

$$t \leq \frac{10}{2\mu + 1} \tag{4}$$

Therefore when $p = 32$, t should be less than or equal $3\frac{1}{3}$ minutes. The line marked "experimental results" shows the actual bounds on the algorithm for $t = 4$ minutes for all values of P. The experimental results are better than the upper bounds of the analysis as the updates don't occur in lock-step fashion, and a shorter sequence of updates are therefore possible. Also the actual protocol re-uses the established circuit to send the merged tables back to the sender daemon;

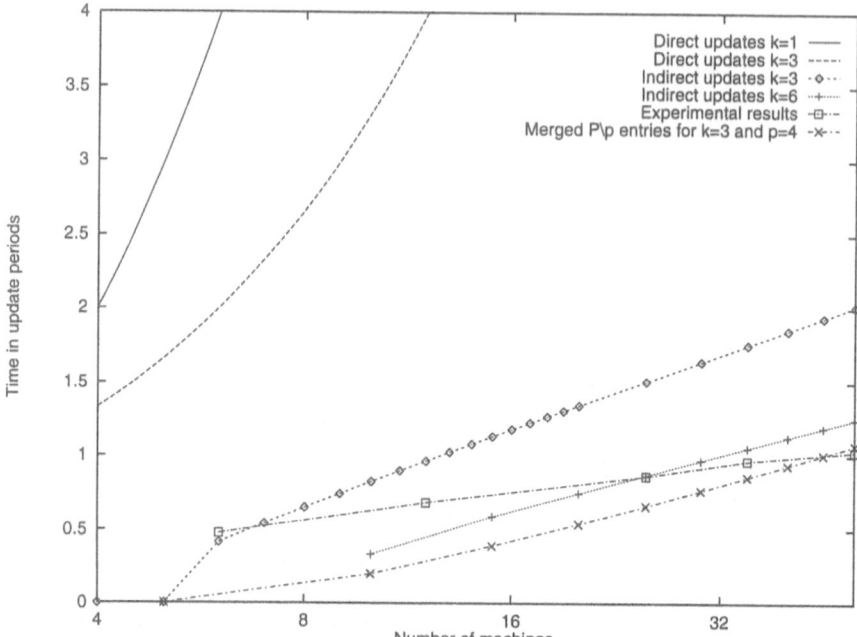

Fig. 3. Mean ages achieved by each of the three strategies and the merged $P \backslash p$ entries

this in effect allows the system to perform twice as many updates with fewer circuits.

As described in Section 3, by having the p processes involved in a BSP computation merge their load average tables before choosing where to migrate the processes, the *current* five minute load averages for the p processors executing the job is obtained and the ages of the load average table entries for the other $P - p$ machines have a distribution $\{\pi_i'' : i \in \mathbb{N}\}$ where

$$\pi_i'' = \sum_{x=1}^{p} \binom{p}{x} (\pi_i')^x (1 - \sum_{t=0}^{i} \pi_t')^{p-x} \tag{5}$$

Figure 3 includes the resulting mean age from this distribution for $p = 4$ against the total number of machines P, and compares it with the mean derived from the original distribution $\{\pi_i' : i \in \mathbb{N}\}$. It is clear from the figure that as p increases, the mean age of the data from the merged tables decreases until the mean age is zero when $p = P$.

5   Experimental Results for an Electro-Magnetics Application

The code EMMA T:FE3D (part of the British Aerospace EMMA electro-magnetic software suite), uses the finite element time domain method for solving Maxwell's

Table 1. Execution time in seconds for the electro-magnetics simulation

p	ypcat order	daemon order	daemon order + forced migration	daemon order + selective migration
2	3255	623	841	658
4	3855	1155	1916	1092
8	2968	1161		

equations in three dimensions. The finite element approach offers several advantages over other full wave solution methods (e.g., finite difference time domain, method of moments). A volume of space around the target is filled with an unstructured mesh of tetrahedra which can conform accurately to the geometry of the object being analysed and, because of the unstructured nature of the mesh, many small elements can be introduced in regions where the solution has rapid spatial variation. A time marching algorithm using a Taylor-Galerkin method is used to advance the fields through time to simulate the propagation of a wave through the mesh. The CFD community have developed considerable knowledge and expertise in unstructured mesh generation and finite element solvers which have been exploited for solving electro-magnetic problems. Applications areas for electro-magnetic solvers in the aerospace industry include electro-magnetic scattering, analysis of electro-magnetic compatibility and hazards, antenna design, and modelling of effects of lightning strike.

Table 4 shows the results from the electro-magnetics application running on various numbers of processors. The four columns of execution time show the following: (1) a job running on a random choice of machines. These may be highly loaded or may not have fast processors; (2) a job that is initiated on the p best machines (i.e., the fastest least loaded machines); (3) a job that is started on the best p machines, but is check-pointed every two minutes; and (4) a job that is started on the best p machines and checks every two minutes whether it is beneficial to check-point and migrate.

The first experiment is an approximation to a job that encounters poor service during execution. As can be seen from the dramatic decrease in execution times in the second experiment it is always beneficial to start a job on the most powerful unloaded machines. In the situation where the chosen processors have the same power, and the job is long running, then the second experiment will degrade to the performance of the first. The third experiment quantifies the overhead in check-pointing. As ten check-points were performed at $p = 4$ the increase in execution time shows that a local checkpoint takes 19 seconds to write a seven megabyte image to disc. The fourth experiment shows that there is little overhead in checking whether a check-point is necessary.

As it costs $19p$ seconds to perform a migration, it is only beneficial to migrate in situations where the processor usage is not too erratic, thus allowing the job to recoup the cost of the migration on the set of processors that were migrated to. If jobs are long running and compute bound then there is a lot of potential for regaining lost ground due to having to migrate from a loaded machine.

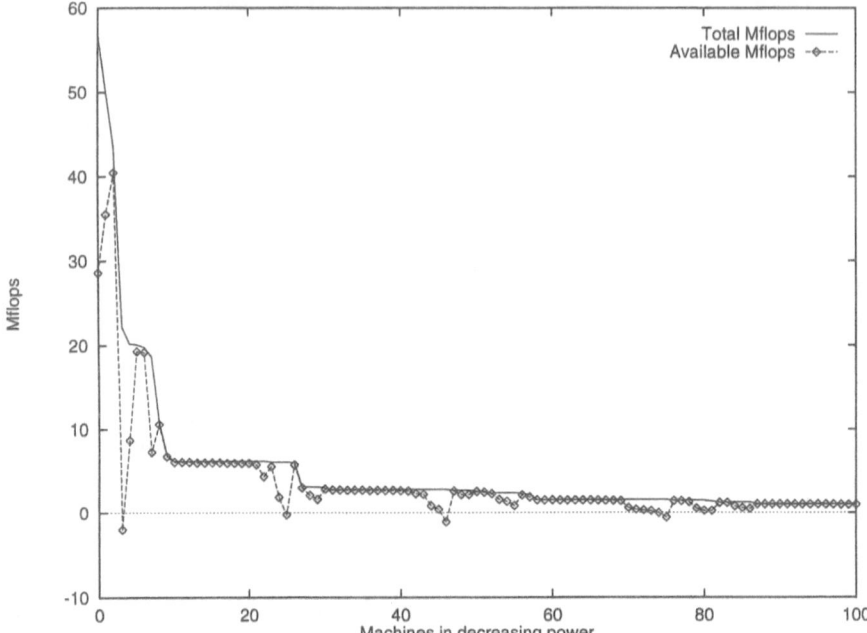

Fig. 4. Moore's law: a profile of the computational power of a collection of workstations

The results shown in Table 4 are from an electro-magnetics experiment that simulates a field around a sphere. As can be seen from the figure, no parallel speedup was achieved when increasing the size of the NOW; this was due to the dominance of communication over computation in this small test case. Larger realistic test cases at BAe have shown linear speedup up-to sixteen processes as computation begins to dominate.

This work is based on the assumption that the available cycles on the machines changes continuously over time. However, with a large resource acquired over time, the machines are unlikely to be homogeneous in available power. Figure 4 shows a graph of available computing power on each workstation in Oxford University Computing Laboratory against available power at a particular instance in time. We express available power by the formula $(n - L)s$; where n is the number of processors in a machine, L is the load average on that machine and s is the Mflops/s rating of a single processor. The graph demonstrates Moore's Law in the purchase of workstations over time, i.e., to the right of the graph, there are a large number of low powered aging workstations, whereas toward the left of the graph, there are few high performance machines procured over the last six months. When choosing to schedule a p process parallel job, either a homogeneous set of unloaded (slow) machines can be used, or the best p. The policy we adopt is that although the unloaded machines ensure little interference, they only provide a fraction of the power of the best machines. However, running on the popular powerful machines, can also make a job susceptible to low throughput as the available cycles at a node can vary dramatically over time.

For example, the figure shows that the fourth machine from the left has a peak performance of 50 Mflops/s, yet it is so loaded that there are no free cycles for parallel jobs. In summary, the erratic, but powerful machines, to the left of the graph are most suited to computational intensive parallel jobs, yet they are the very machines that require process migration.

6   Related Work

The process migration work described in this paper, also provides fault tolerance if the mean time between failure is greater than the rate at which processes migrate/checkpoint. Existing work in this area has tended to concentrate on fault tolerance using redundant computation. For example Nibhanupudi and Szymanski [9] minimise the slowdown of a BSP job when external loads are applied to machines in a network by replicating computation on a number of machines. At the end of a superstep, the results of the fastest of the replicated jobs is used to form the global state of the system. Although their system allows the mean time between failure to be less than ours due to the replicated jobs, their prototype system requires user annotation of the data structures to be included in a checkpoint, and they assume that there won't be a machine failure during communication. In contrast, the fault tolerance in our system is transparent to the user, and has no restrictions on when faults can occur. Also, our approach doesn't suffer from the considerable overhead that would be incurred to implement a process replication scheme. As already noted, the only overhead we incur is the time taken to write the p checkpoints to disk. A similar approach to ours is that of Kaashoek *et al.* [8] where fault tolerance of Orca programs is provided on top of the Amoeba distributed operating system. Their approach is complicated by the fact that they have to determine locally when communication is quiescent so that a stable checkpoint can be taken. Other parallel check-pointing systems for MPI [12] and PVM [13] also suffer from this problem, as a checkpoint can only be performed if there is no communication in transit when each process performs a local checkpoint to capture the global state [2]. Fortunately, the check-pointing regime described here is far simpler than any of the approaches used in message passing systems as opportunities for a global checkpoint naturally arise out of the superstep structure of BSP programs. The process migration facilities provided for MPI [12] and PVM have usually been developed on top of the check-pointing and batch scheduling facilities provided by Condor [4] and LSF[15].

7   Conclusions

We have shown that it is possible to perform fault tolerance and process migration of BSP programs in a transparent way on a network of workstations. By paying careful attention to the design of a distributed load manager, it is possible to determine the global load of a system with minimal impact on network traffic. This, in conjunction with the pro-active manner in which BSP jobs migrate between machines, enables a system that is unobtrusive to non-BSP users, whilst providing the best of the resource as a whole.

References

1. David Bailey, Tim Harris, William Saphir, Rob van der Wijngaart, Alex Woo, and Maurice Yarrow. Nas parallel benchmarks 2.0. Technical Report 95-020, NAS Applied Research Branch (RNR), December 1995.
2. R. Baldoni, J. M. Hélary, A. Mostefaoui, and M. Raynal. A communication-induced checkpointing protocol that ensures the rollback-dependency trackability property. In *Proc. of the 27th IEEE Symposium on Fault-Tolerant Computing Systems (FTCS)*, pages 68–77, Seattle, WA, June 1997. IEEE.
3. Stephen R. Donaldson, Jonathan M. D. Hill, and David B. Skillicorn. Predictable communication on unpredictable networks: Implementing BSP over TCP/IP. In *EuroPar'98*, LNCS, Southampton, UK, September 1998. Springer-Verlag.
4. D. H. J. Epema, M. Livny, R. van Dantzig, X. Evers, and J. Pruyne. A worldwide flock of condors : Load sharing among workstation clusters. *Future Generations of Computer Systems*, 12, 1996.
5. Jonathan M. D. Hill, Stephen Jarvis, Constantinos Siniolakis, and Vasil P. Vasilev. Portable and architecture independent parallel performance tuning using a call-graph profiling tool. In *6th EuroMicro Workshop on Parallel and Distributed Processing (PDP'98)*, pages 286–292. IEEE Computer Society Press, January 1998.
6. Jonathan M. D. Hill, Bill McColl, Dan C. Stefanescu, Mark W. Goudreau, Kevin Lang, Satish B. Rao, Torsten Suel, Thanasis Tsantilas, and Rob Bisseling. BSPlib: The BSP Programming Library. *Parallel Computing*, to appear 1998. see www.bsp-worldwide.org for more details.
7. Jonathan M.D. Hill, Stephen Jarvis, Constantinos Siniolakis, and Vasil P. Vasilev. Analysing an sql application with a bsplib call-graph profiling tool. In *EuroPar'98*, LNCS, Southampton, UK, September 1998. Springer-Verlag.
8. M. F. Kaashoek, R. Michiels, H. E. Bal, and A. S Tanenbaum. Transparent fault-tolerance in parallel orca programs. In *Proc. Symp. on Experiences with Distributed and Multiprocessor Systems III*, pages 297–312, 1992.
9. Mohan V. Nibhanupudi and Boleslaw K. Szymanski. Adaptive parallelism in the bulk synchronous parallel model. In *EuroPar'96*, number 1124 in Lecture Notes in Computer Science, pages 311–318, Lyon, France, aug 1996. Springer-Verlag.
10. David Skillicorn, Jonathan M. D. Hill, and W. F. McColl. Questions and answers about BSP. *Scientific Programming*, 6(3):249–274, Fall 1997.
11. Richard M. Stallman. Emacs: The extensible, customizable, self-documenting display editor. AI memo 519A, Artificial Intelligence Laboratory, Massachusetts Institute of Technology (MIT), 1979.
12. G. Stellner. CoCheck: checkpointing and process migration for MPI. In IEEE, editor, *Proceedings of IPPS '96. The 10th International Parallel Processing Symposium: Honolulu, HI, USA, 15–19 April 1996*, pages 526–531, 1109 Spring Street, Suite 300, Silver Spring, MD 20910, USA, 1996. IEEE Computer Society Press.
13. Kasidit Chanchio Xian-He Sun. Efficient process migration for parallel processing on non-dedicated networks of workstations. Technical Report TR-96-74, Institute for Computer Applications in Science and Engineering, December 1996.

14. Leslie G. Valiant. A bridging model for parallel computation. *Communications of the ACM*, 33(8):103–111, August 1990.
15. Jingwen Wang, Songnian Zhou, Khalid Ahmed, and Weihong Long. LSBATCH: A distributed load sharing batch system. Technical Report CSRI-286, Computer Systems Research Institute, University of Toronto, April 1993.

A Parallel-System Design Toolset for Vision and Image Processing

M. Fleury, N. Sarvan, A. C. Downton and A. F. Clark

Dept. of Electronic Systems Engineering, University of Essex, Wivenhoe Park,
Colchester, CO4 4SQ, U.K
tel: +44 - 1206 - 872795
fax: +44 - 1206 - 872900
e-mail fleum@essex.ac.uk

Abstract. This paper analyses the requirements of a toolset intended
for integrated parallel system design. Real-time, data-dominated systems
are targeted, particularly those in vision and image-processing. A constrained abstract model is represented based around software pipelines
each stage of which can contain a parallel component. The toolset is
novel in seeking to enable generalized parallelism in which the needs
of the system as a whole are given priority. The paper describes the
tool infra-structure, and reports on current progress on the performance
prediction toolset element, including simulation visualizer and analytic
support.

1 Introduction

This paper describes a parallel-system design toolset which will guide the engineer 'in-the-field' towards the construction of Pipeline Processor Farm (PPF)
applications. PPF is a stylized design and development methodology aimed at
continuous-flow, embedded systems. More precisely, the target systems are expected to be driven by soft real-time constraints such as pipeline throughput,
and traversal latency. Recently, a number of medium-scale, data-dominated, vision and image-processing applications [1–6], have been parallelized along PPF
lines. We tentatively define medium-sized systems to contain several algorithmic components, with approximately 3,000 to 50,000 lines of code. Examples of
applicable irregular, continuous-flow systems can be found in vision, radar [7],
speech processing [8], and data compression [9].

PPF promotes the notion of a software pipeline, which can subsequently be
transferred onto available hardware according to client need, analogously to the
way relational database systems are mapped to differing parallel hardware [10].
It has been observed [11] that many parallel algorithms merely form a sub-system
of a vision-processing system. A way of combining algorithmic components in
a coherent whole is required, though any subsequent changes should be self-
contained. A pipeline may be the natural architecture for vision processing [12];
indeed, it is the architecture used by the human mind, the result of engineering
through natural selection. In PPF, a pipeline is a convenient organising entity.

Each stage of the pipeline can independently cater for the differing requirements of a system's algorithmic components, either through *centralized processing*, data farming, or some other form of algorithmic parallelism.

By adjusting the system parameters differing performance goals can be achieved. For example, identification of handwritten postcodes falls into three stages: pre-processing of a digitised postcode image; classification of the characters within the postcode; and matching the postcode against a dictionary of postal addresses. By timing the algorithmic components of a system in a sequential setting, the ratio of processing power required at each stage of a pipeline can be initially assessed, though the extent of testing will determine the veracity of the results. For full generality (rather than heuristics), either second-order statistics or even identification of the statistical distribution formed by the processing times is necessary. The distribution can be estimated by fitting a distribution to the timings' histogram, for example by a chi-squared test. In the postcode example, the first stage might be achieved by farming complete postcode images to a set of worker processors, the second stage by the same or by decomposing processing to individual postcode characters, and the third stage by means of a trie search on each processor. A pipeline enables (say) easy transition to an alternative search engine, such as a syntactic neural net. The first two stages may be approximated by parameterized truncated Gaussian (normal) distributions while processing of the final stage is bi-modal, UK postcodes being either six or seven characters in length. It emerges that a reduction in latency is achievable if the second-stage processing is on a character basis and the postcode characters can be re-assembled into postcodes without significantly impeding the dictionary search [4].

As timings on sequential code give static requirements, the system designer, aiming for a reduction in hardware costs, will also want to know dynamic effects such as: the mean, variance and maximum work flow, i.e. the throughput and traversal latency metrics of a particular configuration of a parallel pipeline in the long run; the likely steady-state per-stage scheduling regime behaviour; what memory requirements will be needed if there are temporary holdups in the workflow requiring buffering; the effect of varying compute and communicate parameters; and will a particular processor employed at any one stage spend a significant time idling while a less-powerful processor would suit the granularity of the tasks more closely. The PPF performance tools progress the designer to a resolution of these issues.

2 Toolset Overview

Previous work on the PPF toolset has entailed identifying semi-manual methods of partitioning existing sequential code, i.e. identifying likely partition points, and transferring the partitioned code to pre-written high-level software templates which will enact individual processing stages. The ability to routinely parallelize algorithms in a generic manner enables algorithms to be prototyped on intermediate, general-purpose parallel hardware. Introducing an intermediate stage has several advantages: a central parallel version of the system is available for transfer to a variety of client machines according to individual processing needs; the correct working of the parallelisation can be checked; and if instrumented high-level, software templates have been used in the construction of the pipeline an iterative design cycle is set-up in which predicted performance is compared to actual through the medium of an event trace. The PPF design cycle is shown in Fig. 1. The performance predictor works by simulation and analytic means. Because one wishes to compare predicted with actual performance the simulation should set-up a visual impression of activity within a pipeline segment which the designer can compare with recorded activity on the prototyping machine, using a similar display format. The display format of the simulator is deceptively simple as: the PPF methodology restricts the degrees of freedom in the parallel system development path; and extraneous detail is avoided so that the mapping between prototype design and target system(s) is not prematurely fixed. For generality, a machine-neutral description [13] is required, which will reflect a designer's linguistically-based thought processes in developing a design. For example, it is important to present the concept of 'pipeline' and 'data-farm' which do not already exist in other common visualizers, such as ParaGraph. Analytic results are used for synchronous pipelines (Section 4) and to check the simulated results for some mathematically tractable distributions. Synchronous and asynchronous pipeline segment results are then combined. Single-stage scheduling regimes using either fixed or variable task sizes are also open to simulation and analytic prediction.

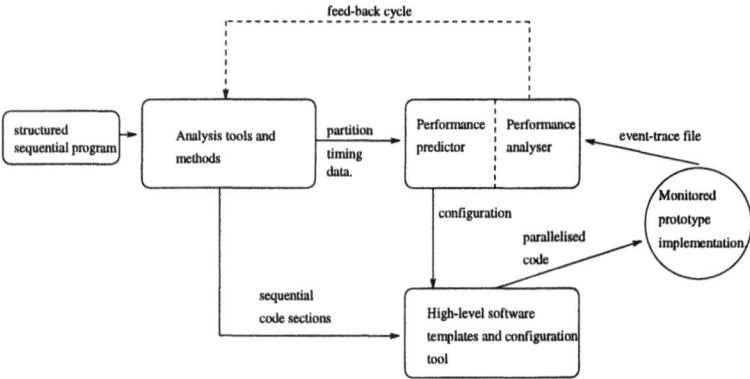

Fig. 1. The PPF Design Cycle

3 PPF Outlook

In the vision and image-processing fields, the choice of parallel hardware has narrowed, since SIMD-style machines are not generally available — not because of any particular difficulty with the design but because a critical mass in the market place has not been achieved.[1] Similarly, topology was a source of considerable academic debate when store-and-forward communication was predominant amongst first generation message-passing machines. With the advent of generalized communication methods such as the routing switch, the 'fat' tree, wormhole routing, and virtual channels the situation appears less critical.

Communication latency is also apparently becoming less of an issue. The Bulk Synchronous Parallelism (BSP) programming model, which has been widely ported to recent machines, adopts a linear model of piece-wise execution time, characterised by a single network permeability constant. Message-latency variance can be reduced by a variety of techniques, such as the message aggregation used on the IBM SP2 SMP. Routing congestion can be palliated either by a randomized routing step, or by a changing pattern of communication generated through a latin square. It appears that a single metric, such as the mean latency, may suitably characterize PPF communication performance. The advantages for the system designer were this to be the case is that the behaviour of the algorithm becomes central, with communication characteristics remaining stable and decoupled from the algorithm.

4 PPF Taxonomy

The postcode example given in Section 1 may give the impression that PPF pipelines are invariably asynchronous, though the need to collect a complete postcode before commencing the dictionary search is a form of synchronous constraint. Throughput and traversal latency depend on whether there is a synchronous constraint on a pipeline stage, which, in PPF, can be: an algorithmic constraint, i.e. completion of a given set of jobs; one or more feedback loops; a folded-back pipeline. An ordering constraint is a feedback loop that can be confined to one stage, but generally feedback constraints extend over at least one stage. A folded-back pipeline is a way of avoiding synchronisation overhead by using the same pipeline stage for more than one step in the processing, but ineluctabley this strategy will only partially succeed because partitioning is imperfect and because even a perfect partition may have variance due to data-dependency. PPF specifies the elementary types of software pipeline (Fig. 2) as: a linear pipeline; a dual pipeline; a pipeline with feedback; a folded pipeline; and combinations of any of the four previous pipeline types. Input to the complete pipeline is separately specified as either being instantaneously available or modelled by an input stream distribution, typically exponential.

A pipeline is split into asynchronous and synchronous segments. Examples of splittings are given in Fig. 3. A form of pipeline process algebra is possible. A segment is a set of adjacent stages. A segment need not be a proper subset of the whole pipeline. A synchronous segment may contain any finite number of complete asynchronous segments and optionally one additional part of an asynchronous segment. An asynchronous segment is the largest set of adjacent

---

[1] MasPar Inc. and their word-oriented machines and Cambridge Memory Systems, who support the DAP bit-serial machine, remain in active trading.

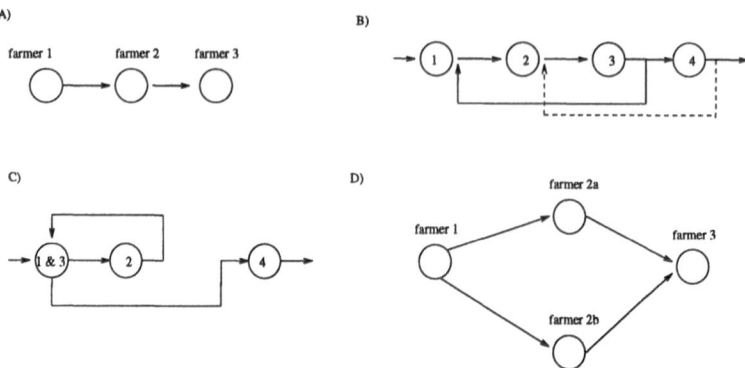

Fig. 2. Elementary Pipeline Types: A) Linear B) Simple feedback C) Folded D) Dual

stages, other than the trivial case, that can be formed at any one site in which no synchronous constraints apply. An asynchronous segment is capable of being simulated and any single stage may also be simulated. Synchronous segments must contain all synchronous constraints to ensure that the segment is self-contained but cannot include any constraints which are not needed to make the segment self-contained. Notice that a synchronous segment may contain another synchronous segment as a subset.

For the purposes of the predictor tool, a feedback-loop can be replaced by a nominal pipeline stage which models the delay distribution experienced by the succeeding stage (Fig. 4). A nominal stage, which can be synchronous or asynchronous, will not correspond to a stage in the implemented software pipeline. A folded-back pipeline can also be unwrapped again to form a linear pipeline by means of first replacing the folding and then providing a nominal stage. By alternating simulation and analytic predictions the performance of a complete pipeline can be built up in a piecewise fashion. In the pipeline of Fig. 3(C), after first replacing the feedback loop by a nominal stage, the performance metrics for the first segment are found. The throughput characteristics act as inputs to the first asynchronous segment proceeding in the direction of data flow, which is simulated. The output metrics of the first asynchronous segment are available for subsequent segments. The maximum latency is additive between pipeline segment. Notice that the ubiquitous central-limit theory in various forms suggests that the result of concatenating random variables, the service times, across a number of stages is a Gaussian distribution, which is determined solely by mean and variance.

5 Implementation of the Simulator

Fig. 5 shows a sample screen from the simulator, which is implemented for portability through the Java 1.1 AWT. The number of pipeline stages and jobs to be completed as well as the job input type and arrival rate are entered initially. Within any one stage, jobs can be grouped into tasks. The user enters: the per-stage statistical distributions and parameters as estimated from test runs; the number of processors; and their performance characteristics. The interconnect parameters are also adjustable. If algorithmic parallelism is used then each worker may output work according to a different distribution. The scrollable pipeline display animates a discrete-event simulation running as a background thread. The display includes running indication of minimum, maximum and mean pipeline traversal latency. Throughput details, including a measure of distribution spread, are also included. Simulation time, which is adjustable, is shown in counter or clock format. The display can be zoomed into to show activity on individual pipeline stages. The state of the pipeline can be displayed, e.g. the total idle time to date of any worker process is to be available by clicking on that worker.

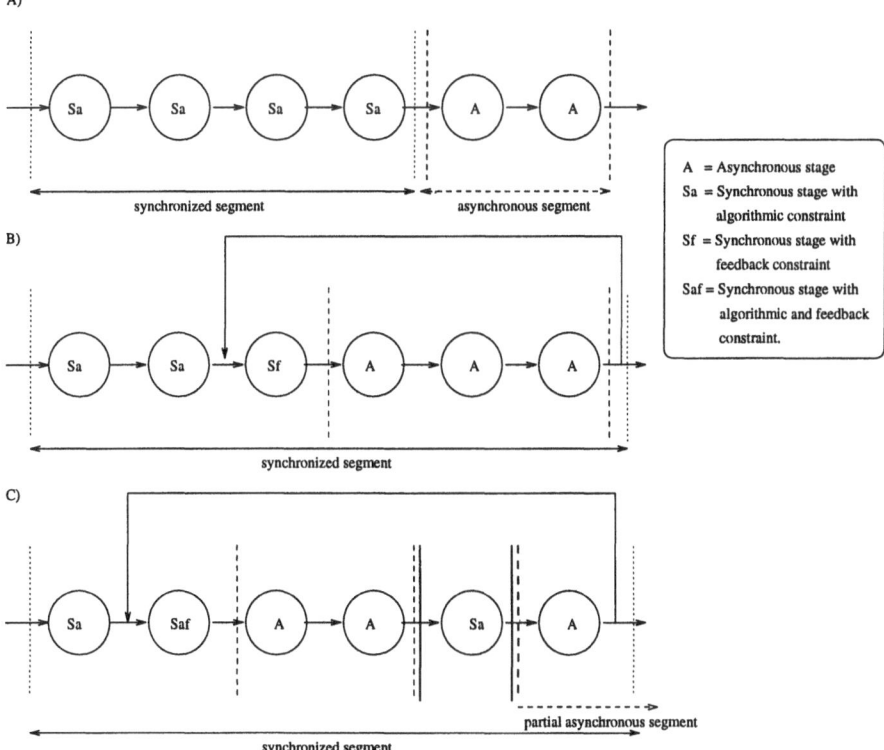

Fig. 3. Pipeline splittings: A) Disjoint segments B) Singly Nested segments C) Multiply Nested Segments

98

The simulation moves forward in time according to the global minimum remaining service time for any task. A preliminary, two-stage, version of the simulator used local minima on a stage-by-stage basis, which correctly calculated the output statistics but did not preserve correct event ordering on the screen. To keep a record of job latency a job is tagged with its time to date in the pipeline along with any remaining time at an individual stage. The simulation program also accounts for differing numbers of jobs per task at each stage of the pipeline. Interstage buffer queues are represented internally by calls to the vector class library, which enables dynamic data structures. At each time update point, the latency times of all jobs in a buffer are incremented. To give an impression of parallelism on a sequential machine using a slower semi-compiled language, communication is animated by changes in the colour and size of the links resulting in a persisting display.

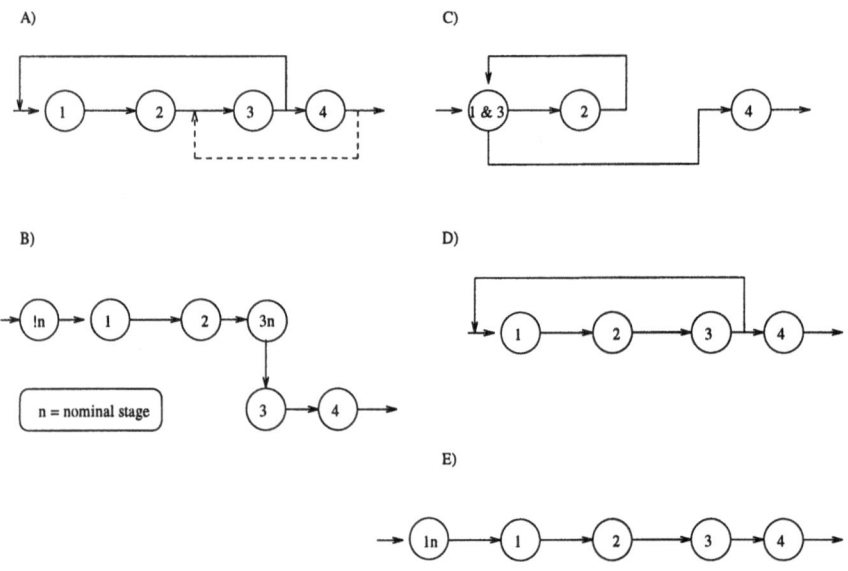

Fig. 4. A) Two simple feedback loops B) Replacement by nominal stages C) Folded-back pipeline D) After unwrapping E) After substituting a nominal stage.

6   Analytic Prediction

PPF is intended to develop systems that can guarantee or adapt to specifications. A principle problem is to determine maximum latency given data-dependent behaviour. The maximum latency distribution is the output distribution for algorithmically synchronous pipeline stages, i.e. latency and throughput are bound together. Distribution spread occurs even on regular problems within synchronous pipeline segments as there will still be system noise [14]. The obvious analytic technique, queueing theory, is generally confined to exponential arrival distributions. A further serious detraction is that the distributions of waiting times are not easily found, though we have made some progress with delay-cycle analysis which will estimate variances as well as means. The waiting time distribution is needed to find maximum latency due to buffering in asynchronous pipeline segments. However, order statistics have been shown in simulation and in timing experiments [15] to give suitable statistics for maximal behaviour.

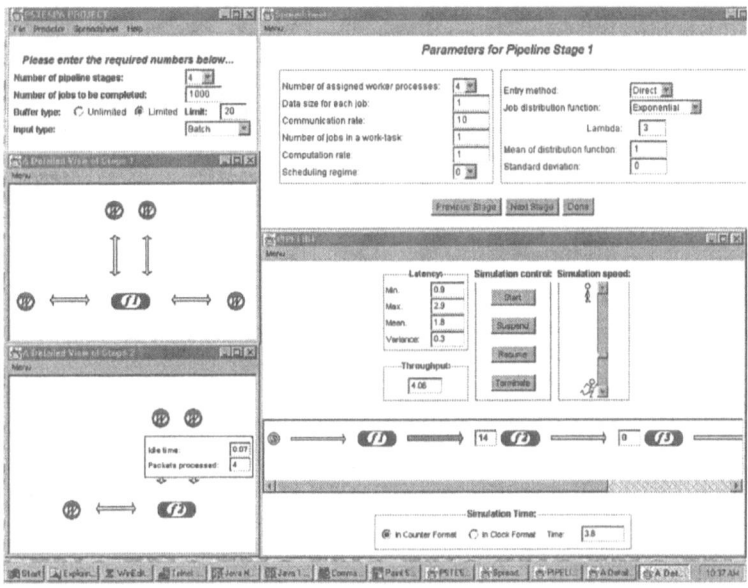

Fig. 5. Sample Screen Display for the PPF Simulator

Because of the changed nature of parallel hardware, as envisaged in Section 3, it becomes possible to use order statistics whereas previously linear programming or queueing theory or both appeared necessary [16]. This paper considers just one way of using order statistics. Taylor expansions of all order statistics [17] are available from $X_{i:p} = F^{-1}(U_{i:p}) = G(U_{i:p})$, when the $E[U_{i:p}]$ are $i/(p+1) = p_i$, resulting in:

$$X_{i:p} = G(p_i) + G'(p_i)(U_{i:p} - p_i) + \frac{1}{2}G''(U_{i:p} - p_i)^2 \cdots, \qquad (1)$$

with $G' = d(G(u))/du|_{u=p_i}$, $u = F(x)$. For example, with F a logistic distribution, $p = 20$, after taking the expectation of (1), the results are plotted in Fig. 6 for odd values of p. The advantage of this procedure is twofold: the amount of work that needs to be reserved at each scheduling round in order to minimize the final idling time, assuming a monotonically-reducing task-size duration scheduling regime, can be found by observing the shape of the cdf of $\{E[X_{i:p}]\}$; and likewise the performance degradation from any ordering constraint on task output is estimated by the expectation of the order statistics expectation cdf.

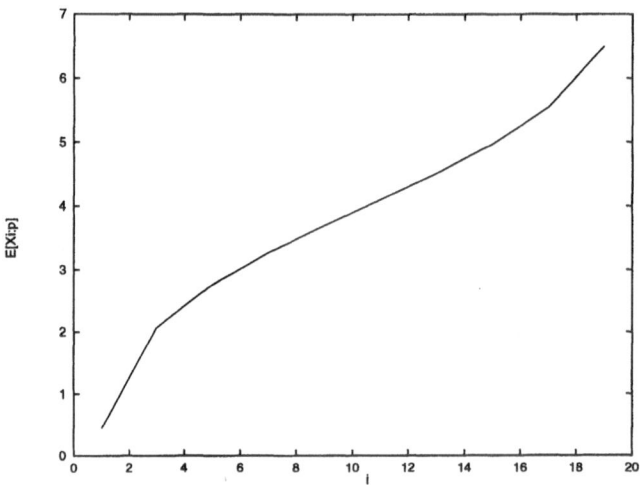

Fig. 6. Approximate Mean Order Statistics for the Logistic Distribution

7 Conclusion

A coherent scheme, PPF, for developing parallel systems has been introduced. The system design element is supported by a performance prediction tool which gives an abstract graphical representation of pipelines formed from independent parallel components. The complete-pipeline performance is built-up by combining simulated and analytic predictions for respectively asynchronous and synchronous pipeline segments. Input to the predictor is found by test runs of software developed in a sequential environment, which implies that the first stages of development can occur without purchase of parallel hardware. In fact, the path to full implementation is further eased because an intermediate step is introduced whereby the system is prototyped on general-purpose parallel kit, the results being fed-back for comparison with predictions. Order statistics are used to form estimates of maximal pipeline segment output characteristics. A discrete-event simulator, written in Java, has been animated to visualize pipeline activity. The event traces from typical runs can be animated for graphical comparison.

Acknowledgement

This work is being carried out under EPSRC research contract GR/K40277 'Parallel software tools for embedded signal-processing applications' as part of the EPSRC Portable Software Tools for Parallel Architectures directed programme.

References

1. A. C. Downton. Speed-up trend analysis for H.261 and model-based image coding algorithms using a parallel-pipeline model. *Signal Processing: Image Communications*, 7:489–502, 1995.
2. H. P. Sava, M. Fleury, A. C. Downton, and A. F. Clark. A case study in pipeline processor farming: Parallelising the H.263 encoder. In *UK Parallel '96*, pages 196–205. Springer, London, 1996.
3. A. Çuhadar, D. Sampson, and A. Downton. A scalable parallel approach to vector quantization. *Real-Time Imaging*, 2:241–247, 1996.
4. A. Çuhadar, A. C. Downton, and M. Fleury. A structured parallel design for embedded vision systems: A case study. *Microproc. and Microsys.*, 21:131–141, 1997.
5. M. Fleury, A. C. Downton, and A. F. Clark. Pipelined parallelization of face recognition. *Machine Vision and Applications*, 1998. submitted for publication.
6. M. Fleury, A. C. Downton, and A. F. Clark. Co-design by parallel prototyping: Optical-flow detection case study. In *High Performance Architectures for Real-Time Image Processing*, pages 8/1–8/13, 1998. IEE Colloquium Ref. No. 1998/197.
7. M. N. Edward. Radar signal processing on a fault tolerant transputer array. In T.S Durrani, W.A. Sandham, J.J. Soraghan, and S.M. Forbes, editors, *Applications of Transputers 3*. IOS, Amsterdam, 1991.
8. S. Glinski and D. Roe. Spoken language recognition on a DSP array processor. *IEEE Transactions on Parallel and Distributed Systems*, 5(7):697–703, 1994.
9. A-M Cheng. High speed video compression testbed. *IEEE Transactions on Consumer Electronics*, 40(3):538–548, 1994.
10. J. Spiers. Database management systems for parallel computers. Technical report, Oracle Corporation, 1990.
11. S-Y. Lee and J. K. Aggarwal. A system design/scheduling strategy for parallel image processing. *IEEE Trans. on PAMI*, 12(2):194–204, February 1990.
12. D. P. Agrawal and R. Jain. A pipeline pseudoparallel system architecture for real-time dynamic scene analysis. *IEEE Trans. on Comp.*, 31(10):952–962, 1982.
13. K. M. Nichols and J. T. Edmark. Modeling multicomputer systems with PARET. *IEEE Computer*, pages 39–47, May 1988.
14. M. Dubois and F. A. Briggs. Performance of synchronized iterative processes in multiprocessor systems. *IEEE Trans. on SW. Eng.*, 8(4):419–431, July 1988.
15. M. Fleury, A. C. Downton, and A. F. Clark. Modelling pipelines for embedded parallel processor system design. *Electronic Letters*, 33(22):1852–1853, 1997.
16. M. Fleury and A. F. Clark. Performance prediction for parallel reconfigurable low-level image processing. In *International Conference on Pattern Recognition*, volume 3, pages 349–351. IEEE, 1994.
17. F. N. David and N. L. Johnson. Statistical treatment of censored data part I fundamental formulæ. *Biometrika*, 41:228–240, 1954.

Achieving Portability and Efficiency Through Automatic Optimisation: An Investigation in Parallel Image Processing

D Crookes[1], P J Morrow[2], T J Brown[1], G McAleese[2], D Roantree[2] and I T A Spence[1]

[1] Department of Computer Science, The Queen's University of Belfast, Belfast BT7 1NN, UK
[2] Department of Computing Science, University of Ulster at Coleraine, Coleraine, BT52 7EQ, UK

Abstract. This paper discusses the main achievements of the EPIC project, whose aim was to design a high level programming environment with an associated implementation for portable parallel image processing. The project was funded as part of the EPSRC Portable Software Tools for Parallel Architectures (PSTPA) programme. The paper summarises new portable programming abstractions for image processing, and outlines the automatically optimising implementation which achieves portability of application code and efficiency of implementation on a closely coupled distributed memory parallel system. The paper includes timings for optimised and unoptimised versions of typical image processing algorithms; it draws the main conclusion that it is possible to achieve portability with efficiency, for a specific application, by adopting a high level algebraic programming model, together with a transformation-based optimiser which reclaims the loss of efficiency which an algebraic approach traditionally entails.

1 Introduction

The need for portability of program code for parallel systems is an important but elusive goal. For instance, in an anecdotal keynote address to a recent European conference on parallel computing a distinguished American visitor from a major research facility recounted how the parallel supercomputers which his laboratory uses are renewed typically every three years. After installation, some two years is then spent modifying the suite of programs which researchers at the laboratory use so that they will run on the new hardware. Useful computing can then proceed for a period of about a year, before the machines are replaced again by the next generation; and so the cycle of events repeats. His purpose in relating these events was to highlight the relative difficulty in producing parallel software and porting it between machines. Software for parallel computers is often tailored to the architecture of particular machines in order to deliver optimal performance, particularly in the case of distributed memory machines.

On the other hand, this complicates the problem of constructing programs and inhibits their portability once constructed.

The EPIC project has been undertaken to consider to what extent automatic optimising software tools can take a high level, portable algorithm description, and generate parallel code whose efficiency on a distributed machine rivals than of hand-tuned parallel code. Because of the difficulty of such a problem in a general purpose programming environment, the project has deliberately chosen an application specific approach - in our case, the domain of image processing. The result is the EPIC environment: a high level, portable image processing programming environment, with an implementation which automatically generates very efficient code running on distributed memory parallel systems[1, 2].

More specifically, the key objectives of the EPIC project were:

1. To identify a set of programming abstractions for image processing. These abstractions constitute the core of the *Extensible Parallel Image Coprocessor* (EPIC) model. In the long term, we also wished to see how far these abstractions could be pushed towards general purpose programming abstractions.
2. To construct an (object-oriented) application development environment based on the EPIC model which enables users to construct portable image processing applications.
3. To develop a rule based transformation system which generates parallel code, for distributed memory machines, whose efficiency rivals that of hand tuned parallel code.

In the rest of this paper we first summarise the new image processing programming abstractions which were developed at the outset of the project, based on extensions to Image Algebra[3]. We then describe the EPIC application development environment in general, before looking at the off-line optimiser in particular. We present performance figures illustrating the benefit which optimisation brings for a number of simple algorithms and for the three architectures on which the system has now been implemented. Finally we draw conclusions and review the scientific insights which we have gained.

2 EPIC Programming Abstractions for Parallel Image Processing

It is known that efficiency on distributed memory parallel architectures benefits from the predictability and locality of data references. For low level image processing applications this is readily provided by the neighbourhood based operations which are typically used. For these forms of operation one can identify easily any data communication needs associated with an updating operation from the neighbourhood itself. Thus an application specific neighbourhood based programming model offers an advantage over more general purpose language notations where operations would typically be defined in terms of loop constructs and subscripting, from which the extraction of communication requirements is in general non-trivial.

Another advantage of a neighbourhood based programming model is that it can readily be supported by the development of a high level algebraic notation. This has been done for image processing by the development of Image Algebra[3] whose basic operations, plus image and template data types, form the starting point for the EPIC programming notation. Our programming abstractions for image processing have been developed in two stages:

1. Because *neighbourhood* processing is particularly suited to implementation on *parallel* machines (since data access patterns are predictable), the basic operations and concepts of *Image Algebra* were used as the basis of our programming abstractions, with minor extensions.
2. To continue to exploit the locality property of neighbourhood processing, we extended considerably the concept of a neighbourhood. For instance, we introduced *sets* and *sequences* of neighbourhoods. Together with variant neighbourhoods of Image Algebra, these novel abstractions give a very powerful and flexible notation which is still nevertheless capable of parallel implementation. As a result, we can now express various complete image transforms such as the Hadamard transform at a high level[4], and obtain an implementation whose performance is comparable with hand coding. We can also express various kinds of geometric transformation, including downsampling and upsampling operations of the kind used in some forms of wavelet transform, and data permutation operations such as perfect shuffle or the bit reversal permutation which is a component of many standard image transforms such as the FFT.

Evaluation of the abstractions was carried out by re-engineering a number of existing (sequential) systems. One large one (on calculating optical flow) has highlighted the need for extending the abstractions to include 3-D and video image processing applications. (The latter is currently the subject of more recent work at QUB[5].)

In trying to extend the applicability and expressive power of the above abstractions beyond the domain of image processing, we have insisted on retaining the basic concept of neighbourhood processing (as in (2) above). We have investigated several standard numerical algebra problems, and found that, with facilities for building new compound operators from a group of primitives, sets and sequences, we can code algorithms for problems such as matrix multiplication, LU factorisation, and finding the transitive closure of directed graphs. This work has not yet been reported in detail, but results to date indicate that a number of numerical problems *can* be expressed in terms of neighbourhood processing, but whether or not it is *natural* to do so needs further investigation.

3 Implementation and the Need for Optimisation

Although the kind of high level, algebraic notation outlined above is very convenient as a programming notation, there are some forms of inefficiency which necessarily arise from a straightforward library-based implementation approach. There are two main reasons for this:

- The fact that *special cases*, in which an operands value is known and could be exploited manually, cannot be fully exploited, because the implementations of the high level operations must of necessity be general purpose.
- The fact that implementing expressions involving *compound* operations will frequently involve replication of overheads.

As a simple example of the first problem, consider a convolution operation between an image and a template (a weighted window used in neighbourhood operators). This operation is normally provided as a general purpose routine applicable to any pair of operands. However it is frequently the case that the template may contain weights of zero or one. Using a general purpose routine will therefore entail unnecessary arithmetic operations.

The second problem arises with any expression which involves compound operations using image operands. Each individual operation will normally be implemented by a routine which traverses the image domain (for example using a double loop construct). When a compound expression is implemented this overhead is replicated for each individual operation, leading to a significant loss in efficiency.

Of course, a programmer interested primarily in efficiency would manually write additional routines for all the above cases; but this would require operating at a lower level at which the parallelism and communication would be visible, thus reducing portability. The essence of the EPIC environment is that it automatically generates these additional routines, and links them in to the runtime environment; in this way it gives the same efficiency as a good programmer, but allows the application developer to continue to operate at the highest level. The environment is based on a rule based optimiser which aims to apply the same reasoning steps which a programmer would apply manually in generating efficient, tailored versions of the standard routines.

4 EPIC : A Portable Application Development Environment

The second objective defined above was the implementation of an object oriented application development environment for parallel image processing based on the programming abstractions defined within the framework of the EPIC model. We have developed a sophisticated environment, providing C++ as the users language, with the EPIC abstract machine provided as a range of C++ methods[6, 7]. In developing the EPIC environment, we have proposed and integrated several ideas which we believe could have longer term benefits for the design of future systems of this kind. The more significant ideas (or developments of previous ideas and approaches) which appear to us to have particular relevance and importance in this area of the project are now considered.

An Extensible Abstract Machine

The traditional approach to achieving portability across architectures is to define an abstract machine with an instruction set matching the application. In our case, we provide an image coprocessor with an instruction set which supports our Image Algebra-based programming abstractions. This results in a static instruction set (like a library). However this traditional approach results in inefficiencies which can sometimes make the whole approach less attractive to real application developers. As discussed above, these inefficiencies arise typically for two reasons:

1. Special cases of instructions are often implemented more efficiently by manual programmers, and
2. The array processing overheads for compound instructions are usually cumulative.

We have developed an architecture for an abstract machine in which these inefficiencies can be avoided while retaining the elegance and portability of the abstract machine model. We have proposed the concept of an Extensible abstract machine, in which the instruction set has two parts:

1. The basic, static instruction set, and
2. A set of additional, optimised instructions which avoid the inefficiencies mentioned above.

These additional instructions are program-specific, and implement special cases and compound instructions as efficiently as manual coding. As indicated below, these extended instructions are generated and called automatically.

A Self-Optimising Abstract Machine

To enable the programmer to continue to use the static EPIC algebraic programming abstractions, and thus gain clarity, portability and conciseness, while at the same time gain the performance benefits of the types of optimisation mentioned above, we have developed an off-line optimiser which carries out the optimisation and generation task which a performance-minded programmer would carry out. The EPIC environment automatically builds new, optimised instructions which are then linked in to the dynamic, extended part of the EPIC instruction set. This involves the following program execution behaviour:

1. The first time a program is run, it does so using only the static instruction set. But at the same time, syntax trees for compound operations and special cases are dumped to file.
2. Off-line, these syntax trees are transformed into new, optimised routines, and linked into the EPIC coprocessors extended instruction set.
3. On subsequent execution, the extended instructions are automatically used, with no user input. The only visible difference will be the increase in execution speed.

To demonstrate how these ideas are provided by the EPIC environment, the next section presents the architecture of the EPIC system itself.

The EPIC System Architecture

The EPIC application development environment has three principal components. These are:

1. The C++ class library which forms the users application programming environment.
2. The parallel coprocessor which is installed on the parallel machine.
3. The off-line optimiser which is the rule-based transformation system used to generate extended instructions from combinations of the basic programming abstractions. The optimiser is described in more detail in the next section of this report, but it forms an integral part of the overall system architecture.

The overall system architecture is illustrated in Figure 1 below.

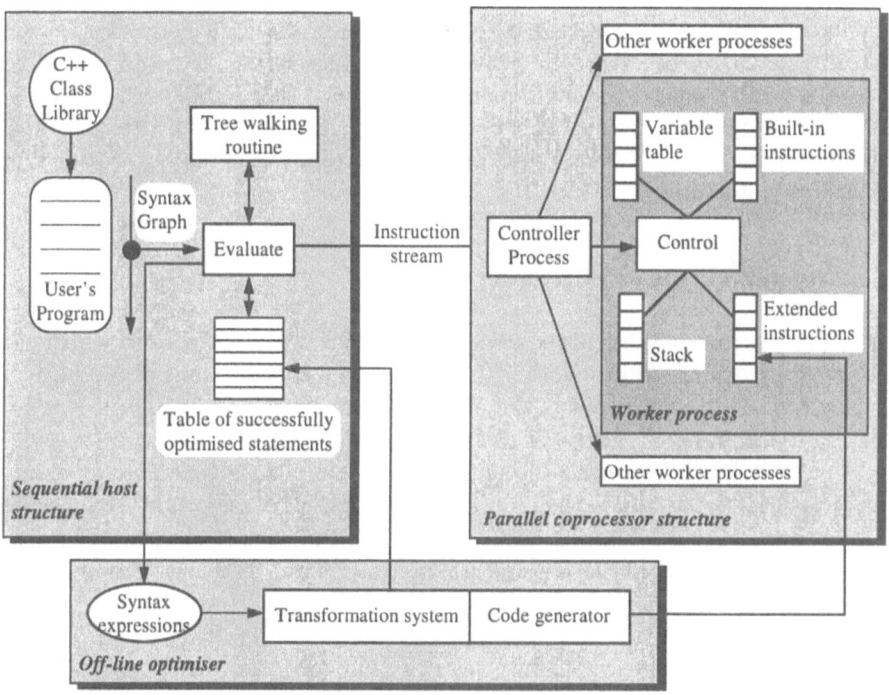

Fig. 1. Overall System Architecture

The Parallel Coprocessor

The parallel coprocessor consists of a controller process and a set of worker processes. Images are geometrically decomposed into horizontal strips or vertical columns. Dynamic redistribution at runtime is supported.

The worker processes perform the actual computations on image data. Each worker process has the internal structure illustrated in the inner box (Figure 1), with a control module, a variable table storing information about program variables including pointers to their data, an evaluation stack, and two tables of instructions. The first is the table of built-in instructions, and the second is the table of extended instructions generated by the optimiser. The control logic can handle the implementation of either form of instruction.

The coprocessor is implemented in C, using MPI for communications. In fact, only six MPI routines are used. On the C40 system, we wrote our own MPI routines, making sure they implement only the minimum necessary functionality (which proved a significant performance advantage).

Portability of the System

Our initial goal was to base the EPIC implementation on a network of Texas Instruments C40 processors. We have not only successfully met this objective, but we have also ported the implementation to two other parallel architectures. These are: a quad-processor Unix workstation, and a network of Pentium PCs running Linux. The application development environment is fully operational on all three architectures.

5   The Off-Line Optimiser

The EPIC environment includes a rule-based transformation system which can generate new optimised instructions from arbitrary compositions of the built-in instructions. This component is the off-line optimiser shown in Figure 1 and it is in reality an integral feature of the overall system architecture.

The off-line optimiser takes the syntactic representations generated by execution of the users program on the sequential host and generates optimised parallel code to implement the complete operation defined in each statement. The optimised routine, after compilation and linking, becomes a new instruction on the parallel coprocessor.

The optimiser is a rule driven system written in Prolog. It uses optimisation techniques which are not in themselves new. For the most part these are loop combining, loop unrolling and loop interchanging, along with the elimination of redundant arithmetic. Several strategies are employed to identify the most appropriate path to follow based on the contents of the syntax graph. This includes trial and error approaches when an obvious course is not evident from the nature of the syntax graph. A pattern-matching approach is used to find the best transformation applicable.

The output is initially in the form of an object code syntax graph. A code generation module is then used to generate C code with calls to library routines for such purposes as border swapping. The generated optimised instructions are independent of the number of available processors. Direct calls to MPI, which provides the communication framework, are also embedded in the code.

The capabilities built into the optimiser enable it to handle any composition of the operations provided within the class library, including operations nested to any depth. Also provided is the ability to handle the syntactic representations of variant templates operations, including sequences of variant templates.

6 Performance of the EPIC Optimiser

As an illustrative case, consider the example program statement of the form:

```
EdgeImage = (absconv(Im,Sv) + absconv(Im,Sh)) > Threshold)*255 ;
```

This performs a simple edge detection operation when `Sv` and `Sh` are vertical and horizontal gradient operators such as the Sobel operators. When we execute a statement like this without optimisation a total of 5 of the basic instructions will be needed, each entailing a pass over the image domain. Optimisation can produce a new instruction requiring only one traverse of the image domain to perform the whole operation. Our initial studies showed that a sequential unoptimised version would run up to nearly 5 times slower than a hand- optimised version for statements of this level of complexity (and a factor of around 4.5 on a parallel system).

Trials using the EPIC optimiser have now been performed on all three of the architectures on which the EPIC system is implemented. The results are tabulated (Table 1) for two such operations, namely a top hat filter operation and the edge detector example shown above. Times are in seconds and the image size in each case is 256*256.

Table 1. Execution times

	Top Hat Filter	Edge Detector
C40 - 4 Workers		
Before optimisation	0.62	0.39
After Optimisation	0.13	0.10
Improvement	4.8 fold	3.9 fold
C40 - 2 Workers		
Before optimisation	1.12	0.47
After Optimisation	0.17	0.16
Improvement	6.6 fold	2.93 fold
PC-Linux System		
Before optimisation	1.83	0.94
After Optimisation	0.75	0.53
Improvement	2.44 fold	1.77 fold
Quad-Processor Workstation		
Before optimisation	2.87	1.95
After Optimisation	1.74	1.36
Improvement	1.65 fold	1.43 fold

These are of course quite short programs but they do illustrate that on the C40 implementation we are seeing impressive performance improvements from the optimisation.

Measurement of the overheads incurred by the host processor (tree building, tree matching and host-coprocessor communication) is harder to assess. However, on a selection of tests with a (slow!) SUN host and a C40-based coprocessor, the host overheads reduced overall system performance by between 10% and 35%. This demonstrates another significant finding namely, the importance of a fast, closely-coupled host to a parallel coprocessor.

It is clear from all our experiments that workstation clusters are going to struggle in obtaining parallel efficiency for image processing, because of the communication overheads and process switching times. On such systems, the EPIC optimisation capability will probably not give sufficient speedup to make its use worthwhile (although recent work on lightweight messaging[8] could improve the situation somewhat). However, the C40 system gives very promising results, and the EPIC approach definitely pays off. This is largely because the C40 system is much more closely coupled, and because implementation of the MPI communication routines are architecture-aware. This is an important lesson for other implementors.

7 Conclusions

The EPIC project has demonstrated the main thesis of our approach to achieving portability with efficiency, namely that in this problem domain:

1. Portability over parallel architectures can be achieved by adopting an algebraic, application-specific programming model, and by retaining the traditional implementation concept of an abstract coprocessor machine - but in a more sophisticated form.
2. The inherent loss of efficiency arising from a standard abstract machine-based approach to implementing an algebraic model can be recovered, by developing a rule-based optimiser which generates the equivalent of new machine instructions dynamically, thus giving an extensible instruction set for the abstract coprocessor.

As part of the project, we have therefore developed a powerful self-optimising tool which demonstrates the feasibility of automatically generating new operations from special cases or compositions of library operations, and have provided a transformation system for program optimisation. An efficient implementation of EPIC has been developed for C40 networks and portability has been achieved through the use of an MPI communications layer. In addition the system has been ported to a quad-processor Unix workstation and a PC-based Linux network.

The work described in this paper has contributed to the ongoing debate on how to achieve portability with efficiency, though very much from an application specific viewpoint. The main contributions to this debate are:

- The concept of an *extensible* abstract machine, for obtaining both portability and efficiency.
- New sophisticated *neighbourhood-based abstractions*, which are proving capable of expressing some algorithms from other application areas (e.g. numerical problems), and which guarantee parallel efficiency.
- Automatic *identification* of extended instructions has been achieved, though we are undecided at this stage as to whether or not it is *advisable* (the programmer could identify these manually).
- The novel concept of a *self-optimising machine*, made possible by automatic extension of the instruction set by the system generating, linking and calling extended instructions on subsequent program executions.
- Portability across different distributed memory platforms, using a small subset of the MPI interface. For full efficiency, we recommend rewriting these MPI routines for a specific configuration, rather than rely on a general-purpose MPI implementation.
- For image processing applications at least, the benefits of automatic optimisation are most apparent on a *closely coupled network* of processors. They are not so apparent on a loosely coupled cluster with a general purpose MPI implementation.
- At the highest level, possibly our most significant achievement is to enable programmers to continue to operate using a set of high level primitives — and hence retain application *portability* — by reclaiming the usual inherent loss of efficiency through sophisticated optimising software tools. It is this concept which should be applicable to a wide range of other application domains.

Acknowledgements

This work has been supported by the UK EPSRC within the framework of the PSTPA (Portable Software Tools for Parallel Architectures) initiative. The support received from Transtech Parallel Systems is also gratefully acknowledged.

References

1. Crookes, D., Brown, J., Dong, Y., McAleese, G.,Morrow, P.J., Roantree D. and Spence I.: A Self-Optimising Coprocessor Model for Portable Parallel Image Processing. Proc. EUROPAR'96. Springer-Verlag (1996) 213–216
2. Crookes, D., Brown, J., Spence, I., Morrow, P., Roantree, D. and McAleese, G.: An Efficient, Portable Software Platform for Parallel Image Processing. Proc. PDP'98, Madrid (1998)
3. Ritter, G.X., Wilson, J.N. and Davidson, J.L.: Image Algebra : An overview. Computer Vision, Graphics and Image Processing **49** (1990) 297–331
4. Crookes, D., Spence, I.T.A. and Brown, T.J.: Efficient parallel image transforms: a very high level approach. Proc. 1995 World Transputer Congress, IOS Press (1995) 135-143

5. Hill, S. J., Crookes, D. and Bouridane, A.: Abstractions for 3-D and video processing. Proc. IMVIP-97, Londonderry, (1997)
6. Morrow, P.J, Crookes, D., Brown, J., Dong, Y., McAleese, G., Roantree, D. and Spence, I.: Achieving Scalability, Portability and Efficiency in a High-Level Programming Model for Parallel Architectures. Proc. UKPAR'96. Springer-Verlag (1996) 29–39
7. Morrow, P., Roantree, D., McAleese, G., Crookes, D., Spence, I., Brown, J. and Dong, Y.: A Portable Coprocessor Model for Parallel Image Processing. European Parallel Tools Meeting. ONERA, Chatillon, France (1996)
8. Nicole, D.A. et al. High performance message passing under chorus/Mix using Java. Department of Electronics and Computer Science, University of Southampton (1997)

EDPEPPS*: A Toolset for the Design and Performance Evaluation of Parallel Applications

T. Delaitre, M.J. Zemerly, P. Vekariya, G.R. Justo, J. Bourgeois,
F. Schinkmann, F. Spies, S. Randoux, and S.C. Winter

Centre for Parallel Computing,
Cavendish School of Computer Science,
University of Westminster
115 New Cavendish Street, London W1M 8JS
Email: edpepps-all@cpc.wmin.ac.uk
Web: http://www.cpc.wmin.ac.uk/~edpepps

Abstract. This paper describes a performance-oriented environment for
the design of portable parallel software. The environment consists of a
graphical design tool based on the PVM communication library for building parallel algorithms, a state-of-the-art simulation engine, a CPU characteriser and a visualisation tool for animation of program execution and
visualisation of platform and network performance measures and statistics. The toolset is used to model a virtual machine composed of a cluster
of workstations interconnected by a local area network. The simulation
model used is modular and its components are interchangeable which
allows easy re-configuration of the platform. Both communication and
CPU models are validated.

1 Introduction

A major obstacle to the widespread adoption of parallel computing in industry
is the difficulty in program development due mainly to lack of parallel programming design tools. In particular, there is a need for performance-oriented tools,
and especially for clusters of heterogeneous workstations, to allow the software
designer to choose between design alternatives such as different parallelisation
strategies or paradigms. A portable message-passing environment such as Parallel Virtual Machine (PVM) [12] permits a heterogeneous collection of networked
computers to be viewed by an application as a single distributed-memory parallel
machine. The issue of portability can be of great importance to programmers but
optimality of performance is not guaranteed following a port to another platform
with different characteristics. In essence, the application might be re-engineered
for every platform [26]. Traditionally, parallel program development methods
start with parallelising and porting a sequential code on the target machine and

---

* The EDPEPPS (Environment for the Design and Performance Evaluation of
Portable Parallel Software) project is funded by an EPSRC PSTPA Grant No.:
GR/K40468 and also by EC Contract Nos.: CIPA-C193-0251 and CP-93-5383.

running it to measure and analyse its performance. Re-designing the parallelisation strategy is required when the reached performance is not satisfactory. This is a time-consuming process and usually entails long hours of debugging before reaching an acceptable performance from the parallel program. Rapid prototyping is a useful approach to the design of (*high-performance*) parallel software in that complete algorithms, outline designs, or even rough schemes can be evaluated at a relatively early stage in the program development life-cycle, with respect to possible platform configurations, and mapping strategies. Modifying the platform configurations and mappings will permit the prototype design to be refined, and this process may continue in an evolutionary fashion throughout the life-cycle before any parallel coding takes place.

The EDPEPPS toolset described here is based on a rapid prototyping philosophy and comprises four main tools:

- A graphical design tool (PVMGraph) for designing of parallel applications.
- A simulation utility (SES/Workbench [24]) based on discrete-event simulation.
- A CPU performance prediction tool (Chronos) which characterises computational blocks within the C/PVM code based on basic operations in C.
- A visualisation tool (PVMVis) for animation of program execution using traces generated by the simulator and visualisation of platform and network performance measures and statistics.

Other tools used in the environment for integration purposes are:

- A trace instrumentation utility (Tape/PVM) [19].
- A translator (SimPVM) [7] from C/PVM code to queueing network graphical representation.
- A modified version of the SAGE++ toolkit [3] for restructuring C source code. In the original version the C files were passed through the C-preprocessor (*cpp*) and then processed by the SAGE++ parser (*pC++2dep*), which creates a parse tree containing nodes for the individual statements and expressions in the code (stored in .dep files). The modification is needed because pC++2dep does not understand preprocessor directives such as #define and #include. Therefore, rules for these directives were added to the grammar of pC++2dep and new node types for them were introduced in the parse tree.
- A translator from existing C/PVM parallel applications into PVMGraph graphical representation (C2Graph) based on the modified SAGE++ toolkit. This is provided in order to allow already written parallel applications to experiment with the toolset.

The advantage of the EDPEPPS toolset is that the cyclic process of design-simulate-visualise is executed within the same environment. Also the EDPEPPS toolset allows generation of code for both simulation and real execution to run on the target platform if required. The toolset is also modular and extensible to allow modifications and change of platforms and design as and when required.

This paper describes the various tools within the EDPEPPS environment and presents a case study for illustration and validation of the models used. In the next section we describe several modelling tools with similar aims to EDPEPPS and we highlight the differences between them. In section 3 we describe the different tools in the EDPEPPS toolset. In section 4 we present results obtained from the case study. Finally, in section 5 we present conclusions and future work.

2 Parallel System Performance Modelling Tools

The current trend in parallel software modelling tools is to support all the software performance engineering activities in an integrated environment [22]. A typical toolset should be based on at least three main tools: a graphical design tool, a simulation facility and a visualisation tool [22]. The graphical design tool and the visualisation tool should coexist within the same environment to allow information about the program behaviour to be related to its design. Many existing toolsets consist of only a subset of these tools but visualisation is usually a separate tool. In addition, the modelling of the operating system is usually not addressed.

The HAMLET toolset [23] supports the development of real-time applications based on transputers and PowerPCs. HAMLET consists of a design entry system (DES), a specification simulator (HASTE), a debugger and monitor (INQUEST), and a trace analysis tool (TATOO). However, the tools are not tightly integrated as in the case of EDPEPPS but are applied separately on the output of each other. Also no animation tool is provided.

HeNCE (Heterogeneous Network Computing Environment) [1] is an X-window based software environment designed to assist scientists in developing parallel programs that run on a network of computers. HeNCE provides the programmer with a high level of abstraction for specifying parallelism as opposed to real parallel code in EDPEPPS. HeNCE is composed of integrated graphical tools for creating, compiling, executing, and analysing HeNCE programs. HeNCE relies on the PVM system for process initialisation and communication. HeNCE displays an event-ordered animation of application execution.

The ALPSTONE project [17] comprises performance-oriented tools to guide a parallel programmer. The process starts with an abstract, BACS (Basel Algorithm Classification Scheme), description [5]. This is in the form of a macroscopic abstraction of program properties, such as process topology and execution structure, data partitioning and distribution descriptions, and interaction specifications. From this description, it is possible to generate a time model of the algorithm which allows performance estimation and prediction of the algorithm runtime on a particular system with different data and system sizes. If the prediction promises good performance implementation and verification can start. This can be helped with a skeleton definition language (ALWAN or PEMPI-Programming Environment for MPI), which can derive a time model in terms of the BACS abstraction, and a portability platform (TIANA), which translates the program to C with code for a virtual machine such as PVM.

The VPE project [21] aims to design and monitor parallel programs in the same tool. The design is described as a graph where the nodes represent sequential computation or a reference to another VPE graph. Performance analysis and graph animation are not used here, but the design aspect of this work is elaborate.

The PARADE project [27] is mainly oriented on the animation aspects. PARADE is divided into a general animation approach which is called POLKA, and specific animation developments such as PVM with PVaniM, Threads with GThreads and HPF. This work does not include any graphical design of parallel programs, thus, the predefined animations and views can decrease the user understanding. One of the most important aspects of POLKA is the ability of classification between general and specific concepts.

In the SEPP project [6] (Software Engineering for Parallel Processing) a toolset based on six types of tools has been developed. There are static design tools, dynamic support tools, debugging tools, behaviour analysis tools, simulation tools and visualisation tools [14, 10, 15]. These tools are integrated within the GRADE environment [16, 11]. The GRAPNEL application programming interface currently supports the PVM message passing library.

The TOPSYS (TOols for Parallel SYStems) project [2] aims to develop a portable environment which integrates tools that help programmers cope with every step of the software development cycle of parallel applications. The TOPSYS environment contains tools which support specification and design, coding and debugging, and optimisation of multiprocessor programs. The TOPSYS environment comprises: a CASE environment for the design and specification of applications (SAMTOP) including code generation and mapping support, a multi-processor operating system, a high level debugger (DETOP), a visualiser (VISTOP), and a performance analyser (PATOP). The tools are based on the MMK operating system which has been implemented for Intel's iPSC/2 hypercube. The tools were later ported to PVM in [18]. A more detailed review of parallel programming design tools and environments can be found in [8].

3   Description of the Integrated Toolset

The advantages of the EDPEPPS toolset over traditional parallel design methods are that it offers a rapid prototyping approach to parallel software development, allows performance analysis to be done without accessing the target platform, helps the user to take decisions about scalability and sizing of the target platform, offers modularity and extensibility through layered partitioning of the model and allows the software designer to perform the cycle of design-simulate-analysis in the same environment without having to leave the toolset.

Figure 1 shows the components of the EDPEPPS toolset. The process starts with the graphical design tool (PVMGraph), step (1) in the figure, by building a graph representing a parallel program design based on the PVM programming model. The graph is composed of computational tasks and communications. The tool provides graphical representation for PVM calls which the user can select to build the required design.

The software designer can then generate (by the click of a button) C/PVM code (.c files) for both simulation and real execution. The toolset also provides a tool (C2Graph), step (0), to translate already developed parallel applications onto graphical representation suitable for PVMGraph. The software designer can then experiment with the toolset by changing the parallelisation model or other parameters, such as the number of processors or processor types to optimise the code.

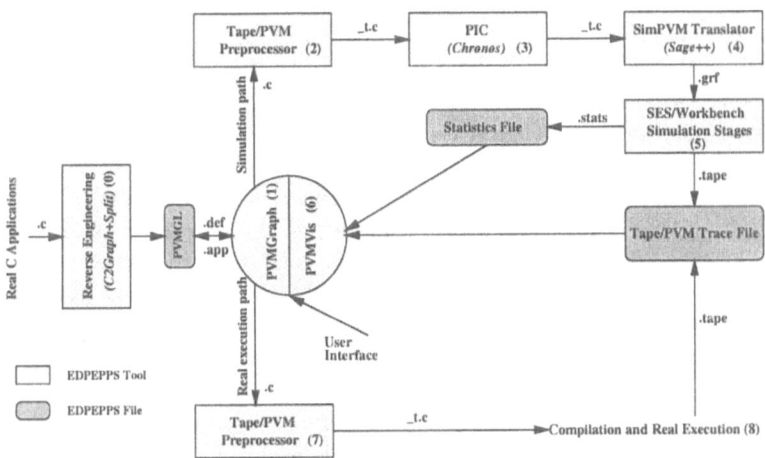

Fig. 1. The EDPEPPS Integrated Environment.

In the simulation path each C/PVM source code obtained from the PVM-Graph is instrumented using a slightly modified version of the Tape/PVM trace pre-processor, step (2), [19]. The output is then parsed using the Program Instruction Characteriser (PIC), step (3), which forms a part of the CPU performance prediction tool called "Chronos" which inserts *cputime* calls at the end of each computational block. The instrumented C source files are translated using the SimPVM Translator [7], step (4), into a queueing network representation suitable for Workbench graph (.grf file). SES/Workbench, step (5), translates the graph file into the Workbench object oriented simulation language called SES/*sim* [25] using an SES utility (*sestran*). The *sim* file is then used to generate an executable model using some SES/Workbench utilities, libraries, declarations and the PVM platform model. The simulation is based on discrete-event modelling. SES/Workbench has been used both to develop and simulate the platform models. Thus the Workbench simulation engine is an intrinsic part of the toolset. All these simulation actions are hidden from the user and are executed from the PVMGraph window by a click on the simulation button and hence shown in the EDPEPPS environment in Figure 1 in one box under "Simulation Stages". The simulation executable is carried out by using three input files containing parameters concerning the target virtual environment (e.g. number of hosts, host names, architecture, the UDP communication characteristics and the timing costs for the set of instructions used by Chronos [4]). The UDP model and the instruction costs are obtained by benchmarking (benchmarks are provided off-line) the host machines in the network.

The simulation outputs are the execution time, a Tape/PVM trace file and a statistics file about the virtual machine. These files are then used by the visualisation tool (PVMVis), step (6), in conjunction with the current loaded application to animate the design and visualise the performance of the system. The design can be modified and the same cycle is repeated until a satisfactory performance is achieved.

In the real execution path the Tape/PVM pre-processor, step (7), is used to instrument the C source files and these are then compiled and executed, step (8), to produce the Tape/PVM trace file required for the visualisation/animation process. This step can be used for validation of simulation results but only when the target machine is accessible. The visualisation tool (PVMVis) offers the designer graphical views (animation) representing the execution of the designed parallel application as well as the visualisation of its performance. The PVM-Graph and PVMVis are incorporated within the same Graphical User Interface where the designer can switch between these two possible modes. The performance visualisation presents graphical plots, bar charts, space-time charts, and histograms for performance measures concerning the platform at three levels (the message passing layer, the operating system layer and the hardware layer). The following sections describe the main tools within EDPEPPS.

3.1 PVMGraph

PVMGraph is a graphical programming environment to support the design and implementation of parallel applications. PVMGraph offers a simple but yet expressive graphical representation and manipulation for the components of a parallel application. The main function of PVMGraph is to allow the parallel software designer or programmer to develop PVM applications using a combination of graphical objects and text. Graphical objects are composed of boxes which represent tasks (which may include computation) and arrows which represent communications. The communication actions are divided into two groups: input and output. The PVM actions (calls) are numbered to represent the link between the graph and text in the parallel program. Also different types and shapes of arrows are used to represent different types of PVM communication calls. Parallel programs (PVM/C) can be automatically generated after the completion of the design. Additionally, the designer may enter PVM/C code directly into the objects. The graphical objects and textual files are stored separately to enable the designer to re-use parts of existing applications [13].

A PVMGraph on-line[1] demonstration is available on the Web.

---

[1] http://www.cpc.wmin.ac.uk/~edpepps/demo_html/ppf.html

3.2 PVMVis

The main objective of this tool is to offer the designer graphical views and animation representing the execution and performance of the designed parallel application from the point of view of the hardware, the design and the network.

The animation is an event-based process and is used to locate an undesirable behaviour such as deadlocks or performance bottlenecks. The animation view in PVMVis is similar to the design view in PVMGraph except that the pallet is not shown and two extra components for performance analysis are added: barchart view and platform view. The barchart view shows historical states for the simulation and the platform view shows some statistics for selected performance measures at three levels: the message passing layer, the operating system layer and the hardware layer.

3.3 The CPU Performance Prediction Tool: Chronos

The Program Instruction Characteriser (PIC) is called only in the simulation path to estimate the time taken by computational blocks within a parallel algorithm. PIC characterises a workload by a number of high-level language instructions (e.g. float addition) [4] taking into account the effect of instruction and data caches. Assumptions have been made to reduce the number of possible machine instructions to 43 (see [4] for more details on these assumptions). The costs associated with the various instructions are kept in a file in the hardware layer accessible by the SES utilities. These costs are obtained by benchmarking the instructions on different machines.

PIC first parses an instrumented C/PVM program using the modified pC++2dep (mpC++2dep) tool from the SAGE++ toolkit In the second stage, PIC traverses the parse tree using the SAGE++ library and inserts *cputime* calls with the number of machine instructions within each sequential C code fragment.

The *cputime* call is a simple function with a fixed number of parameters (a total of 31). This is different from the number of machine instructions because the instruction cache duplicates some of the instructions (hit or miss).

Each parameter of the *cputime* function represents the number of times each instruction is executed within the sequential C code fragment. The only exception is the last parameter, which determines whether the instruction cache is hit or miss for the code fragment in question.

3.4 C2Graph

As mentioned before, this tool allows existing PVM applications to be converted into the EDPEPPS format. The C/PVM application files are first parsed with mpC++2dep to get the .dep files. These files are then traversed by the C2Graph translator using the SAGE++ library.

The translator also takes into account the PVM calls in the original code and generates their corresponding graphical representation in the PVMGraph files.

The translator then determines the master process, positions it with the other tasks by calculating appropriate coordinates for them in the PVMGraph screen, and writes the PVMGraph definition files (.def) for each task. The translator finally writes the application file (.app) required for PVMGraph.

3.5 SimPVM Translator

From PVMGraph graphical and textual objects, executable and "simulatable" PVM programs can be generated. The "simulatable" code generated by PVM-Graph is written in a special intermediary language called SimPVM, which defines an interface between PVMGraph and SES/Workbench [7].

To simulate the application, a model of the intended platform must also be available. Thus, the simulation model is partitioned into two sub-models: a dynamic model described in SimPVM, which consists of the application software description and some aspects of the platform (e.g. number of hardware nodes) and a static model which represents the underlying parallel platform.

The SimPVM language contains C instructions, PVM and PVM group (PVMG) functions, and simulation constructs such as computation delay and probabilistic functions.

3.6 The EDPEPPS Simulation Model

The EDPEPPS simulation model consists of the PVM platform model library and the PVM programs for simulation. The PVM platform model is partitioned into four layers: the *message passing layer*, the group functions layer which sits on top of the message passing layer, the *operating system layer* and the *hardware layer*. Modularity and extensibility are two key criteria in simulation modelling, therefore layers are decomposed into modules which permit a re-configuration of the entire PVM platform model. The modelled configuration consists of a PVM environment which uses the TCP/IP protocol, and a cluster of heterogeneous workstations connected to a 10 Mbit/s Ethernet network.

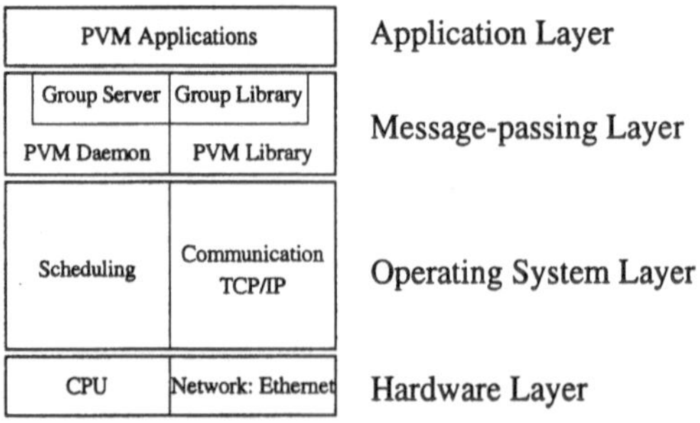

Fig. 2. Simulation model architecture.

The message-passing layer models a single (parallel) virtual machine dedicated to a user. It is composed of a daemon, which resides on each host making up the virtual machine, a group server and the libraries (PVM and PVMG), which provide an interface to PVM services. The daemon and the group server act primarily as message routers. They are modelled as automatons or state machines which are a common construct for handling events. The LIBPVM library allows a task to interact with the daemon and other tasks. The PVM library is structured into two layers. The top layer includes most PVM programming interface functions and the bottom layer models the communication interface with the local daemon and other tasks. Only this layer needs to be modified if the Message Passing Interface (MPI [20]) model is to be supported.

The major components in the operating system layer are the System Call Interface, the Process Scheduler, and the Communication Module. The Communication Module is structured into 3 sub-layers: the Socket Layer, the Transport Layer and the Network Layer. The Socket Layer provides a communications endpoint within a domain. The Transport Layer defines the communication protocol (either TCP or UDP). The Network Layer models the Internet Protocol (IP).

The Hardware Layer is comprised of hosts, each with a CPU layer, and the communications subnet (Ethernet). Each host is modelled as a single server queue with a time-sliced round-robin scheduling policy. The communications subnet is Ethernet, whose performance depends on the number of active hosts and the packet characteristics. Resource contention is modelled using the CSMA/CD (Carrier Sense Multiple Access with Collision Detection) protocol. The basic notion behind this protocol is that a broadcast has two phases: propagation and transmission. During propagation, packet collisions can occur. During transmission, the carrier sense mechanism causes the other hosts to hold their packets.

4 Case Studies

4.1 Bessel Equation

The computational intensive application chosen here to validate the CPU model is the Bessel equation which is a differential equation defined by:

$$x^2 \frac{d^2y}{dx} + x\frac{dy}{dx} + (x^2 - m^2)y = 0$$

When m is an integer, solutions to the Bessel differential equation are given by the Bessel function of the second kind [4], also called Neumann function or Weber function.

The Bessel function of the second kind is translated in C and executed 100000 times. The results for four machines are given in Figure 3. The average error is about 6.13%. The errors for the 486 machine are larger than those of other machines and this is probably due to the fact that there is only one cache used for both data and instructions in the 486 machine. However, for the superscalar machines, Pentium and SuperSparc, the results are encouraging with errors of 1.7% and 3.4% respectively.

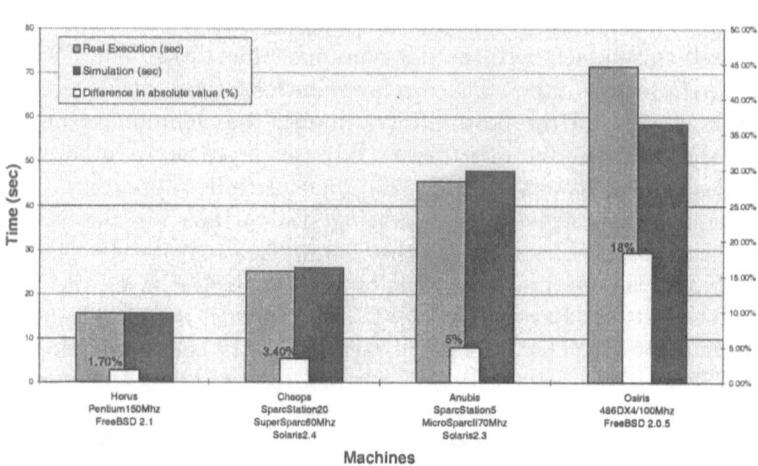

Fig. 3. Comparison between real execution and simulation.

4.2 CCITT H.261 Decoder

The application chosen here to demonstrate the capabilities of the environment in the search for the optimal design is the Pipeline Processor Farm (PPF) model [9] of a standard image processing algorithm, the CCITT H.261 decoder [9]. Figure 4 shows how the H.261 algorithm decomposes into a three-stage pipeline: frame initialisation (T1); frame decoder loop (T2) with a farm of 5 tasks; and frame output (T3).

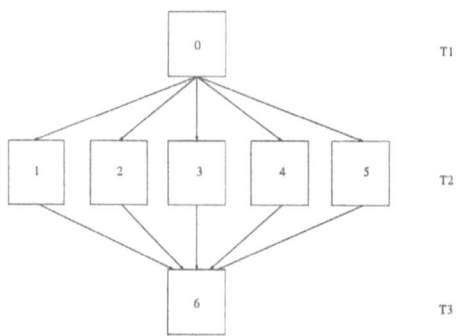

Fig. 4. PPF topology for a three-stage pipeline.

The first and last stages are inherently sequential, whereas the middle stage contains considerable data parallelism.

The same topological variation in the PPF model leads directly to performance variation in the algorithm, which, typically, is only poorly understood at the outset of design. One of the main purposes of the simulation tool in this case is to enable a designer to identify the optimal topology, quickly and easily, without resorting to run-time experimentation. Two experiments for 1 and 5 frames were carried out. The number of processors in Stage T2 is varied from 1 to 5 (T1 and T3 were mapped on the same processor). In every case, the load is evenly balanced between processors.

The target platform is a heterogeneous network of up to 6 workstations (SUN4's. SuperSparcs and PC's). Timings for the three computational stages of the algorithm were extracted from [9] and inserted as time delays. Figure 5 shows the simulated and real experimental results for speed-up.

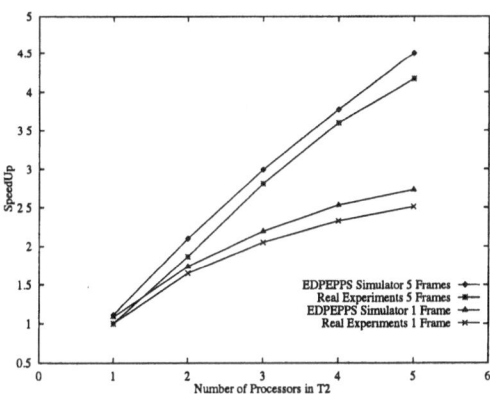

Fig. 5. Comparison between speed-ups of simulation and real experiments for the H.261 PPF algorithm.

As expected, the figure shows that the 5-frame scenario performs better than the 1-frame scenario, since the pipeline is fuller in the former case. The difference between simulated and real speed-ups is below 10% even though the PPF simulation results do not include packing costs.

5 Conclusion

This paper has described the EDPEPPS environment which is based on a performance-oriented parallel program design method. The environment supports graphical design, performance prediction through modelling and simulation and visualisation of predicted program behaviour. The designer is not required to leave the graphical design environment to view the program's behaviour, since the visualisation is an animation of the graphical program description. It is intended that this environment will encourage a philosophy of program design, based on a rapid synthesis-evaluation design cycle, in the emerging breed of parallel programmers.

Success of the environment depends critically on the accuracy of the underlying simulation system. Preliminary validation experiments showed average errors between the simulation and the real execution of less than 10%.

An important future direction of our work is to extend the simulation model to support other platforms, such as MPI. The modularity and flexibility of our model will ensure that PVM layer may be re-used where appropriate in the development of the MPI model component. Another planned extension to our environment is the integration of a distributed debugger such as the DDBG [16].

References

1. A. Beguelin, et. al. HeNCE: A Heterogeneous Network Computing Environment *Scientific Programming*, Vol. 3, No. 1, pp 49–60.
2. T. Bemmerl. The TOPSYS Architecture, In H. Burkhart, editor, *CONPAR90-VAPPIV Conf.*, Zurich, Switzerland, Springer, September 1995, Lecture Notes in Computer Science, 457, pp 732–743.
3. F. Bodin, et. al. Sage++: An Object-Oriented Toolkit and Class Library for Building Fortran and C++ Restructuring Tools, *Proc. 2nd Annual Object-Oriented Numerics Conf.*, 1994. http://www.extreme.indiana.edu/sage/docs.html.
4. J. Bourgeois. CPU Modelling in EDPEPPS EDPEPPS EPSRC Project (GR/K40468), D3.1.6, EDPEPPS/35, Centre for Parallel Computing, University of Westminster, London, June 1997.
5. H. Burkhart, et. al. BACS: Basel Algorithm Calssification Scheme, version 1.1, Tech. Report 93-3, Universität Basel, URZ+IFI, 1993.
6. T. Delaitre, et. al. Simulation of Parallel Systems in SEPP, in: A. Pataticza, ed., *The 8th Symposium on Microcomputer and Microprocessor Applications* 1 (1994), pp 294–303.
7. T. Delaitre, et. al. Final Syntax Specification of SimPVM, EDPEPPS EPSRC Project (GR/K40468) D2.1.4, EDPEPPS/22, Centre for Parallel Computing, University of Westminster, London, March 1997.
8. T. Delaitre, M.J. Zemerly, and G.R. Justo, Literature Review 2, EDPEPPS EPSRC Project (GR/K40468) D6.2.2, EDPEPPS/32, Centre for Parallel Computing, University of Westminster, London, May 1997.
9. A.C. Downton, R.W.S. Tregidgo and A. Cuhadar, Top-down structured parallelisation of embedded image processing applications, in: *IEE Proc.-Vis. Image Signal Process.* 141(6) (1994) 431-437.
10. G. Dozsa, T. Fadgyas and P. Kacsuk, A Graphical Programming Language for Parallel Programs, in: A. Pataricza, E. Selenyi and A. Somogyi, ed., *Proc. of the symposium on Microcomputer and Microprocessor Applications* (1994), pp 304–314.
11. G. Dozsa, P. Kacsuk and T. Fadgyas, Development of Graphical Parallel Programs in PVM Environments, *Proc. of DAPSYS'96*, pp 33–40
12. A. Geist, et. al. *PVM: Parallel Virtual Machine*, MIT Press, 1994.
13. G.R. Justo, PVMGraph: A Graphical Editor for the Design of PVM Programs, EDPEPPS EPSRC Project (GR/K40468) D2.3.3, EDPEPPS/5, Centre for Parallel Computing, University of Westminster, February 1996.

14. P. Kacsuk, P.Dozsa and T. Fadgyas, Designing Parallel Programs by the Graphical Language GRAPNEL, *Microprocessing and Microprogramming* 41 (1996), pp 625–643.

15. P. Kacsuk, et. al. Visual Parallel Programming in Monads-DPV, in: López Zapata, ed., *Proc. of the 4th Euromicro Workshop on Parallel and Distributed Processing* (1996), pp 344–351.

16. P. Kacsuk, et. al. A Graphical Development and Debugging Environment for Parallel Programs, *Parallel Computing*, 22:1747–1770, 1997.

17. W. Kuhn and H. Burkhart. The ALPSTONE Project: An Overview of a Performance Modelling Environment, In *2nd Int. Conf. on HiPC'96*, McGraw Hill 1996, pp 491–496.

18. T. Ludwig, et. al. The TOOL-SET - An Integrated Tool Environment for PVM, In *EuroPVM'95*, Lyon, France, September 1995. Tech. Rep. 95-02, Ecole Normale Superieure de Lyon.

19. E. Maillet, TAPE/PVM: An Efficient Performance Monitor for PVM Applications - User Guide. LMC-IMAG, ftp://ftp.imag.fr/ in pub/APACHE/TAPE, March 1995.

20. M. P. I. Forum. MPI: A Message Passing Interface Standard. *The Int. Journal of Supercomputer Applications and High-Performance Computing*, 8(3/4), 1994.

21. P. Newton, J. Dongarra, Overview of VPE: A Visual Environment for Message-Passing, *Heterogeneous Computing Workshop*, 1995.

22. C. Pancake, M. Simmons and J. Yan, Performance Evaluation Tools for Parallel and Distributed Systems, *Computer* 28 (1995), pp 16–19.

23. P. Pouzet, J. Paris and V. Jorrand, Parallel Application Design: The Simulation Approach with HASTE, in: W. Gentzsch and U. Harms, ed., *HPCN* 2 (1994), pp 379–393.

24. Scientific and Engineering Software Inc. SES/workbench Reference Manual, Release 3.1, Scientific Engineering Software Inc., 1996.

25. K. Sheehan and M. Esslinger, The SES/*sim* Modeling Language, *Proc. The Society for Computer Simulation*, San Diego CA, July 1989, pp 25–32.

26. A. Reinefeld and V. Schnecke, Portability vs Efficiency? Parallel Applications on PVM and Parix, in *Parallel Programming and Applications*, P. Fritzson and L. Finmo eds., (IOS Press, 1995), pp 35–49.

27. J.T. Stasko. The PARADE Environment for Visualizing Parallel Program Executions, Technical Report GITGVU-95-03, Graphics, Visualization and Usability Center, Georgia Inst. of Tech., 1994.

Verifying a Performance Estimator for Parallel DBMSs

E W Dempster[1], N T Tomov[1], J Lü[1], C S Pua[1], M H Williams[1], A Burger[1], H Taylor[1] and P Broughton[2]

[1] Department of Computing and Electrical Engineering, Heriot-Watt University, Riccarton, Edinburgh, Scotland, EH14 4AS, UK
[2] International Computers Limited, High Performance Technology, Wenlock Way, West Gorton, Manchester, England, M12 5DR, UK

Abstract. Although database systems are a natural application for parallel machines, their uptake has been slower than anticipated. This problem can be alleviated to some extent by the development of tools to predict the performance of parallel database systems and provide the user with simple graphic visualisations of particular scenarios. However, in view of the complexities of these systems, verification of such tools can be very difficult. This paper describes how both process algebra and simulation are being used to verify the STEADY parallel DBMS performance estimator.

1 Introduction

Database systems are an ideal application area for parallel computers. The inherent parallelism in database applications can be exploited by running them on suitable parallel platforms to enhance their performance - a fact which has attracted significant commercial interest. A number of general purpose parallel machines are currently available that support different parallel database systems, including adaptations of standard commercial DBMSs produced by vendors such as Oracle, Informix and Ingres. For a platform based on non-shared memory, such systems distribute their data (and associated query processing) amongst the available processing elements in such a way as to balance the load [1].

However, despite the potential benefits of parallel database systems, their uptake has been slower than expected. While information processing businesses have a strong interest in obtaining improved performance from current DBMSs, they have a substantial investment in existing database systems, running generally on mainframe computers. Such users need to be convinced that the benefits outweigh the costs of migrating to a parallel environment. They need assistance to assess what parallel database platform configuration is required and what improvements in performance can be achieved at what cost. They need tools to tune the performance of their applications on a parallel database platform to ensure that they take best advantage of such a platform. They need help in determining how the system should be upgraded as the load on it increases, and how it should be reorganised to meet anticipated changes in usage of database capacities.

Application sizing is the process of determining the database and machine configuration (and hence the cost) required to meet the performance requirements of a particular application. *Capacity planning*, on the other hand, is concerned with determining the impact on performance of changes to the parallel database system or its workload. *Data placement* is the process of determining how to lay out data on a distributed memory database architecture to yield good (optimal) performance for the application.

All three activities require suitable tools to predict the performance of parallel database systems for the particular application. They determine whether a given hardware/software configuration will meet a user's performance requirements, how performance will change as the load changes, and how the user's data should be distributed to achieve good performance [2]. Such tools would rely on a model of transaction throughput, response time and resource utilisation for given system configurations, workloads and data distributions. To complicate matters further, since a parallel machine may host several different DBMSs, such tools should be capable of supporting models of different parallel DBMSs and of providing facilities for describing different features of applications running on these DBMSs.

This paper presents the results from an initial verification exercise of a parallel DBMS performance estimation tool called STEADY. The predictions produced by components of STEADY are compared with results from a simulation and with those derived from a process algebra model. The rest of the paper is organised as follows. Section 2 briefly describes STEADY. Sections 3 and 4 provide a comparison between predictions of components of STEADY and those obtained from a simulation and a process algebra model. Section 5 provides a summary and conclusion.

2 STEADY

STEADY is designed to predict maximum transaction throughput, resource utilisation and response time given a transaction arrival rate. The maximum throughput value is derived by analysing the workload and identifying the system bottlenecks. Given a transaction arrival rate, lower than the maximum throughput, the response time is derived using an analytical queuing model. During the process of calculating the response time, the resource utilisation can also be obtained.

The basic STEADY system originally worked with an Ingres Cluster model, and has been extended to model the Informix OnLine Extended Parallel Server (XPS) and the Oracle7 Parallel Server with the Parallel Query Option. These two systems are representative of two different parallel DBMS architectures: shared disk and shared nothing. The underlying hardware platform for both systems consists of a number of processing elements (PEs) on which instances of the DBMS are running. PEs communicate using a fast interconnecting network. The platform which STEADY currently supports is the ICL GoldRush machine [3].

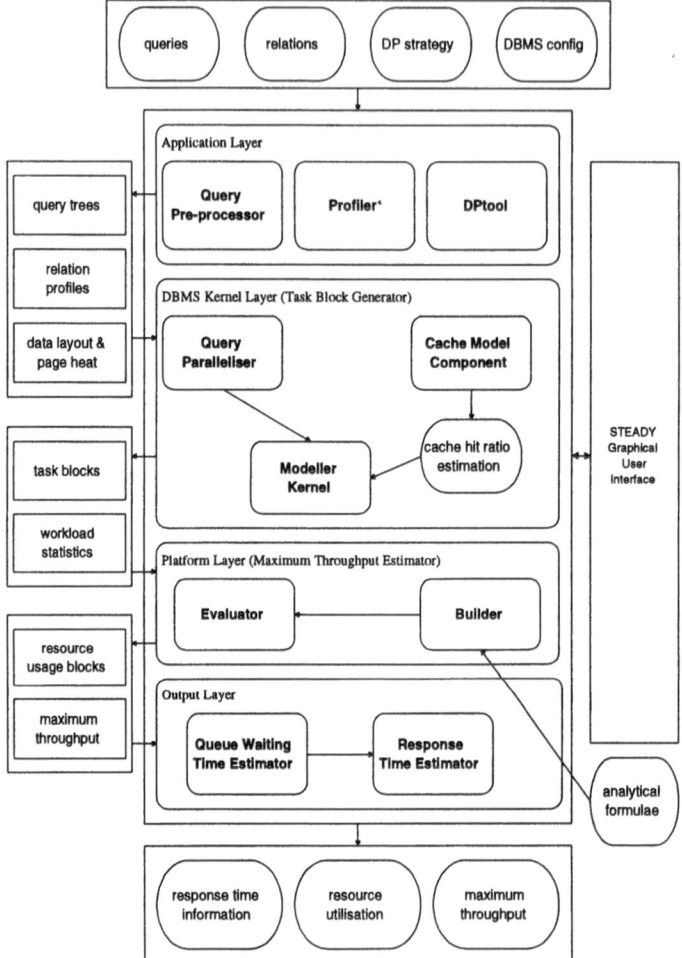

Fig. 1. STEADY Architecture

In a shared disk DBMS (Oracle 7), each of the PEs of the parallel machine may access and modify data residing on the disks of any of the PEs of the configuration. In a shared nothing DBMS (Informix XPS), the disks of a PE are not directly accessible to other PEs.

STEADY takes as input the execution plans of SQL queries, represented as annotated query trees. The query trees capture the order in which relational operators are executed and the method for computing each operator. Within STEADY the annotated query trees undergo several transformations until they reach a form - the *resource usage representation* - which allows the prediction of response time and resource utilisation. This process of transformation can be followed in Fig. 1 which illustrates the architecture of STEADY.

In addition to the graphical user interface the system comprises four parts. The application layer consists of the Profiler, DPTool and the Query Pre-Processor. The Profiler is a statistical tool, primarily responsible for generating base relation profiles and estimating the number of tuples resulting from data operations. DPTool is used to generate data placement schemes using different strategies, and estimates the access frequency (heat) of different pages in each relation. DPTool provides the necessary heat information along with the generated data layout to the Cache Model Component. Both the Profiler and DPTool make use of the annotated query tree format of the original SQL queries.

The DBMS kernel layer consists of the Cache Model Component, the Query Paralleliser and the Modeller Kernel. The Cache Model Component estimates the cache hit ratio for pages from different relations [4]. The Query Paralleliser transforms the query tree into a *task block* structure. Each task block represents one or more phases in the execution of the relational operators within the query trees. Examples of phases are the following: building a hash table in the first phase of a hash join operator; merging two sorted streams of tuples in the last phase of a merge-sort join operator; performing a full table scan of a relation. The task blocks are organised as the nodes of a tree with dependencies among blocks represented by links. The dependencies specify ordering constraints among execution phases. One form of dependency is a pipeline dependency where a sending and a receiving block communicate tuples through a pipeline. Another form is full dependency which requires the previous block to complete before the next one can start. In addition, the task block tree structure captures the inter- and intra-operator parallelism within the query. The Query Paralleliser is able to construct the tree based on knowledge of the parallelisation techniques and the load balancing mechanism employed by the parallel DBMS.

The Modeller Kernel takes as input the relation profiles, data layout, estimated cache hit ratios and the task block profile of the query, produced by the Query Paralleliser. It produces workload profiles in terms of the numbers of basic operations which are to be executed on each PE in the course of a transaction. This results in a set of workload statistics. Together with this, the Modeller Kernel fills in the details of the task blocks by expanding each execution phase within the block into a corresponding sequence of basic operations. For example, a full table scan of a relation is an execution phase process which is represented by the following sequence of operations: wait for a page lock; obtain the page lock; read the page; select tuples from the page; send selected tuples to the next phase. This sequence will be executed once for each page of the relation. The obtained locks are released within the body of a different task block which represents the commit phase of the transaction the query belongs to.

The platform layer consists of the Evaluator and the Builder. The task block profiles of the queries are mapped by the Evaluator into sequences of resource usages. The Evaluator contains a hardware platform model which is generated by the Builder from a set of analytical formulae. This enables the Evaluator to map each basic operation into the appropriate sequence of resource usages. For example, a basic page read operation may be mapped to the following sequence, which makes explicit the order of usage and time consumption associated with the read:

cpu 32µs;
disk3 150ms;
cpu 50µs

Apart from producing these resource usage profiles, the evaluator also gives an initial estimation of the maximum throughput value.

A resource usage representation is then produced which captures the pattern of resource consumption of the original query by mapping the task blocks to resource blocks. The Queue Waiting Time Estimator and the Response Time Estimator of the output layer of STEADY work with this representation to estimate individual resource utilisation and response time and, from these, overall query response time. The details of these processes is the subject of a future paper.

3 Verification by Simulation

When estimating the performance of complex systems using analytical methods, two basic approaches are available: 1) to build a high-level abstract model and find an exact solution for the model, or 2) to build a fairly detailed model and find an approximate solution for it — finding exact solutions for detailed models is in general not feasible. In the case of STEADY, the second approach has been adopted.

Part of the verification of STEADY is to determine the level of accuracy of results that can be achieved using the approximation algorithm described in the previous section. This can be achieved by finding solutions, within error bounds, for the model using discrete event simulation and then comparing these with the figures that are obtained using STEADY.

The simulation is based on the resource usage profiles of transactions, which are generated by STEADY's platform layer. Transactions are implemented as simulation processes which access shared resources according to the specifications in these profiles. New transactions are generated according to the overall arrival rate distribution and the respective transaction frequencies. Simulation output includes statistics on resources (including utilisation) and transaction response times.

A number of examples have been selected and experiments conducted to compare the results obtained from simulation with those predicted by STEADY. One example is given in Figure 2 which shows the average utilisation of resources as predicted by simulation and by STEADY. By contrast Figure 3 provides a comparison of average transaction response times. In all the experiments results for average utilisation show very good agreement whereas those for response time are more variable. This is being investigated further.

4 Verification by Process Algebra

The second way in which the STEADY approach is being "verified" is through process algebras. A process algebra is a mathematical theory which models communication and concurrent systems. In the early versions of process algebras,

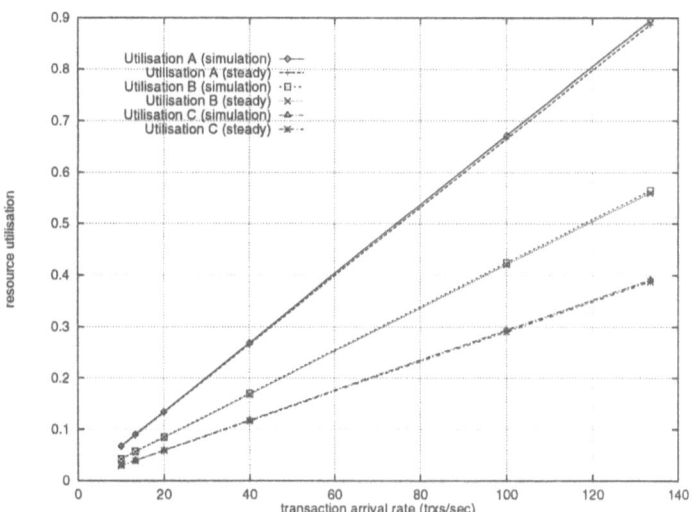

Fig. 2. Utilisation of resources A, B and C

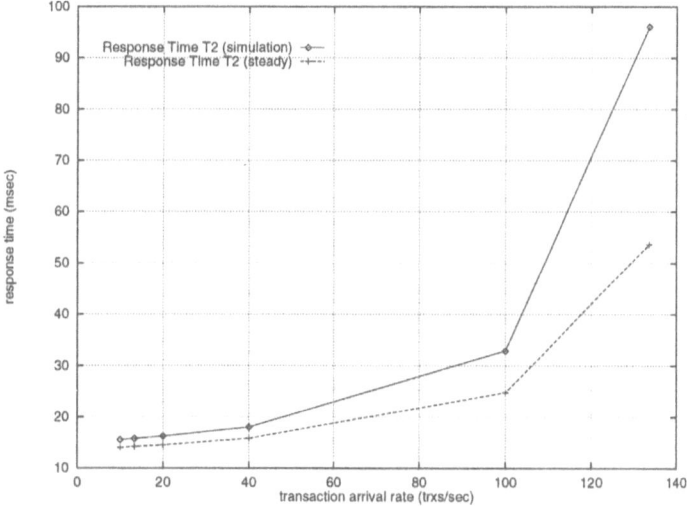

Fig. 3. Mean response time of t_2

132

time was abstracted away. A classical example of such an algebra is CCS (Calculus of Communicating Systems) [7].

The form of process algebra which has been selected for use here is known as a Performance Evaluation Process Algebra (PEPA) [6]. This is a stochastic process algebra which has been developed to investigate the impact of the computational features of process algebras upon performance modelling.

In PEPA, a system is expressed as an interaction of components, each of which engages in some specific activity. These components correspond to the parts of the system or the events in the behaviour of the system. The behaviour of each component is defined by the activities in which it can engage. Every activity has an action type and an associated duration (which is represented by a parameter known as the activity rate), and is written as (α, r) where α is an activity and r the activity rate.

To illustrate the PEPA approach, consider the simplest of models, comprising a single processing element with one processing unit and one disk (see fig 4). Suppose that transactions arrive at the transaction manager (TM) module, which on receipt of a transaction passes it to the concurrency control unit (CCU). The latter sends a message to the distributed lock manager (DLM) requesting a lock; the DLM grants the lock and replies accordingly. The CCU sends a message to the buffer manager (BM) which reads from disk and returns the data. Finally the CCU releases the lock and commits the transaction.

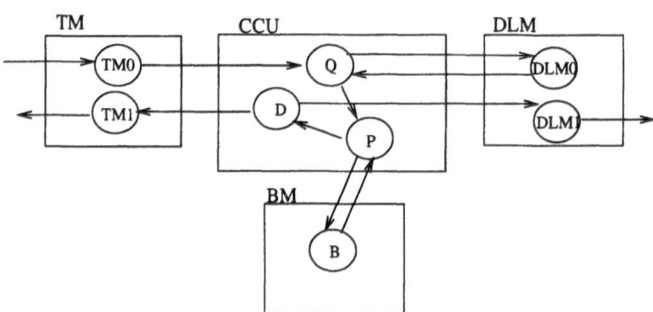

Fig. 4. Single Processing Element Configuration

This is expressed in PEPA notation as follows:

#Q0 = (tm2ccu,infty).(ccu2dlm,r_sgn0).Q0 ;
#Q1 = (dlm2ccu,infty).(lock_granted,r_sgn).Q1;
#Q = Q0 ⟨⟩ Q1;
#D = (release_lock,infty).(commit,infty).(ccu2dlm0,r_sgn0). (ccu2tm,r_sgn0).D;
#P = (lock_granted,infty).P1;
#P1 = (ccu2bm,r_sgn0).(bm2ccu,infty).(release_lock,r_sgn). (commit,r_commit).P;
#CCU = (Q ⟨⟩ D) ⟨ commit, release_lock, lock_granted ⟩ P ;
#DLM0 = (ccu2dlm,infty).(dlm2ccu,r_gnt).DLM0;

```
#DLM1 =(ccu2dlm0,infty).(release,r_sgn).DLM1;
#DLM = DLM0 ⟨⟩ DLM1;
#BM = (ccu2bm,infty).(deliver,r_deliver).(bm2ccu,r_sgn0).BM;
#TM0 = (request,r_req).(tm2ccu,r_sgn0).TM0;
#TM1 = (ccu2tm,infty).(reply, r_reply).TM1;
#TM = TM0 ⟨⟩ TM1;
#PEPA = (TM ⟨⟩ BM ⟨⟩ DLM)
⟨ tm2ccu,ccu2tm,bm2ccu,ccu2bm,ccu2dlm,ccu2dlm0,dlm2ccu ⟩ CCU;
PEPA
```

Here the notation $(\alpha, r).REST$ is the basic mechanism by which the behaviour of a component is expressed. It denotes the fact that the component will carry out activity (α, r) which has action type α and a duration of mean $1/r$, followed by the rest of the behaviour (REST). The notation $P\langle L\rangle Q$ where P and Q are components and L is a set of action types, denotes the fact that P and Q proceed independently and concurrently with any activity whose action type is not contained in L. However, for any activity whose action type is included in L, components P and Q must synchronise to achieve the activity. These shared activities will only be enabled in $P\langle L\rangle Q$ when they are enabled in both P and Q , i.e. one component may be blocked waiting for the other component to be ready to participate.

For this very simple configuration, a very simple benchmark has been adopted in which a single relation with tuple size of 100 bytes is scanned by a query performing a simple search. The page size is set to 1024 bytes and each page is assumed to be 70% full. Each fragment of the relation is assumed to consist of a single page. The average time to perform a disk read is taken as 41.11 milliseconds.

Results for this very simple example are given in Table 1.

Query arrival rate	Throughput (tps)			
	data in 1 frag	data in 2 frags	data in 3 frags	data in 4 frags
5.0	4.996	4.962	4.816	4.512
10.0	9.919	9.015	7.339	5.858
25.0	20.418	11.924	8.058	6.059
100.0	24.019	12.089	8.077	6.063
STEADY Max throughput(tps)	24.324	12.162	8.108	6.081

Table 1. System performance of PEPA model and STEADY: throughput (in tps)

In order to solve such a model, the method reduces the model to a set of states. In the case of this very simple example, the number of states generated is 1028. However, when one considers slightly more complex models, such as configurations involving more than one processing element, the number of states increases rapidly (approx 810000 states for two processing elements) and the computation required to solve it becomes too large.

In order to handle such cases we are currently experimenting with the flow equivalent aggregation method [8] to decompose the models into manageable sub-models and find solutions to these. The results so far are encouraging although no complete solution is yet available.

5 Conclusions

This paper has discussed the verification of an analytical model (STEADY) for estimating the performance of parallel relational DBMSs. Two different methods of verification were investigated. The first approach was to apply simulation to a set of problems. The same problems were also solved using STEADY. The results of the comparison between STEADY and the simulation were mixed, as some compared very favourably whereas others require further investigation. Although the simulation alone is not sufficient to verify all aspects of STEADY, it has proved to be valuable in analysing, choosing and developing approximation algorithms for mathematical solutions of detailed analytical models.

The second, more theoretical, method of verification that was investigated was a process algebra system known as Performance Evaluation Process Algebra (PEPA). This system was used to verify the components of STEADY which produce both the resource usage profiles and the first estimation of maximum system throughput. The system used to verify STEADY was a simple system as PEPA becomes unmanageable when more complex configurations are considered. Further work on PEPA is being carried out to handle more complex configurations, whilst keeping the system manageable.

The processes described are intended as an initial stage in the verification of STEADY, being performed in parallel to validation against actual measured performance.

6 Acknowledgments

The authors acknowledge the support received from the Engineering and Physical Sciences Research Council under the PSTPA programme (GR/K40345) and from the Commission of the European Union under the Framework IV programme, for the Mercury project (ESPRIT IV 20089). They also wish to thank Arthur Fitzjohn of ICL and Shaoyu Zhou of Microsoft Corp. for their contribution to this work.

References

1. K. Hua, C. Lee, and H. Young. "An efficient load balancing strategy for shared-nothing database systems" *Proceedings of DEXA'92 conference, Valencia, Spain, September 1992, pages 469-474*

2. M. B. Ibiza-Espiga and M. H. Williams. "Data placement strategy for a parallel database system" *Proceedings of Database and Expert Systems Applications 92, Spain, 1992. Springer-Verlag, pages 48-54*

3. P. Watson and G. Catlow. "The architecture of the ICL GoldRush MegaSERVER" *Proceedings of the 13th British National Conference on Databases (BNCOD 13), Manchester, U.K., July 1995, pages 250-262*

4. S. Zhou, N. Tomov, M.H. Williams, A. Burger, and H. Taylor. "Cache Modelling in a Performance Evaluator of Parallel Database Systems" *Proceedings of the Fifth International Symposium on Modelling, Analysis and Simulation of Computer and Telecommunication Systems, IEEE Computer Society Press, January 1997, pages 46-50*

5. L. Kleinrock. "Queueing Systems, Volume 1: Theory" *John Wiley & Sons, Inc., Canada, 1975*

6. J. Hilston. "A Compositional Approach for Performance Modelling", *PhD Thesis, University of Edinburgh, 1994.*

7. R. Milner. "Communication and Concurrency" *Prentice Hall International, UK, 1989*

8. K. Kant. "Introduction to Computer System Performance Evaluation" *McGraw-Hill Inc., Singapore, 1992*

Generating Parallel Applications of Spatial Interaction Models

John Davy and Wissal Essah

School of Computer Studies, University of Leeds,
Leeds. West Yorkshire, UK, LS2 9JT

Abstract. We describe a tool enabling portable parallel applications of spatial interaction modelling to be produced automatically from high-level specifications of their parameters. The application generator can define a whole family of models, and produces programs which execute only slightly more slowly than corresponding hand-coded versions.

1 Introduction

This paper describes a prototype *Application Generator* (AG) for rapid development of parallel programs based on *Spatial Interaction Models* (SIMs). The work was a collaboration between the School of Computer Studies and the Centre for Computational Geography at the University of Leeds, with GMAP Ltd, a University company, as industrial partner. It was supported by the Engineering and Physical Sciences Research Council within its Portable Software Tools for Parallel Architectures programme.

An application generator can be characterised as a tool "with which applications can be built by specifying their parameters, usually in a domain-specific language" [1]. Our AG follows precisely this definition:

- both the SIM itself and the various computational tasks which use it are specified by defining parameters in a textual Interface File;
- parallel programs for the required tasks are generated from the Interface File, using templates which implement generic forms of these tasks;
- automatically generated programs deliver identical results to hand-coded versions of the same tasks.

Portability is achieved on two counts. Firstly, the level of abstraction of the Interface File is high enough to be machine-independent. Secondly, the AG generates source-portable programs in Fortran-77 with MPI. Currently the AG has been tested on a 4-processor SGI Power Challenge and a 512-processor Cray T3D. A serial version runs on a range of Sun and SGI workstations.

A common criticism of such high-level approaches to program generation is that excessive and unacceptable overheads are introduced in comparison with equivalent hand-coded programs. For this reason, every effort was made to ensure that needless inefficencies arising from genericity were avoided; also the AG determines an efficient form of the parallel program based on the dimensions of the model and the number of processors available. With execution times for generated programs no more than 2% greater than the corresponding hand-coded programs, we have demonstrated that acceptable performance from high-level abstractions is indeed possible, at least within a limited application domain.

The work is closely related to *algorithmic skeletons* [3], in which common patterns of parallel algorithms are specified as higher-order functions, each of which has a template for parallel implementation. Our current prototype is only first order, but extensions to cover a wider range of industrial requirements are likely to lead to higher-order versions.

2 Spatial Interaction Modelling

Spatial interaction modelling was developed in the context of the social sciences, notably quantitative geography. A SIM is a set of non-linear equations which defines *flows* (people, commodities etc) between spatial zones. Such models are of importance to the business sector, academic researchers, and policy makers; they have been used in relation to a wide range of spatial interaction phenomena, such as movements of goods, money, services and persons, and spread of innovation.

Many realistic spatial interaction problems involve large data flows, together with computation which grows rapidly with the number of zones the model uses to represent geographical areas. Computer constraints have therefore limited the level of geographical detail that can practically be used for these models. Greater computing power, available by the exploitation of parallel processing, allows models to be used on a finer level with more realistic levels of detail and may enable better quality results. Despite these benefits, the exploitation of parallelism has received limited attention.

Recent research [2, 5, 6] has confirmed the reality of these benefits and the considerable potential of parallelism in this area. In particular, [5] shows the effectiveness of parallel genetic algorithms for solving the network optimisation problem, in comparison with previous heuristic methods. To date, however, such work has proceeded on an ad hoc basis, and there is still a need to provide a more uniform and consistent approach, enabling these technologies to be more readily exploited outside the specialised community of parallel processing.

The research reported here aims to rectify this deficiency by developing an application generator to:

- cover a wide range of SIM applications in the social sciences;
- enable efficient parallel implementations over a range of machines;
- be scalable to enable the use of large data sets on appropriate machines.

2.1 A Family of Models

There is a family of closely-related SIMs derived from so-called entropy-maximising methods [7]. A simple *origin-constrained* model is specified by

$$T_{ij} = O_i D_j A_i f(c_{ij}) \tag{1}$$

$$A_i = 1/\sum_j D_j f(c_{ij}) \qquad (2)$$

where T_{ij} predicts the flow from origin zone i to destination zone j, O_i, D_j represent the 'sizes' of i and j, and c_{ij} is a measure of the travel cost from i to j (often simply the distance). The *deterrence function* $f(c_{ij})$ decreases as c_{ij} increases and is commonly modelled as either the exponential function $e^{-\beta c_{ij}}$ or the exponentiation function $c_{ij}^{-\beta}$ where β is an unknown *impedance factor* to be estimated. Equation (2) ensures that the predicted flows satisfy the constraint:

$$\sum_j T_{ij} = O_i \qquad (3)$$

thus equating the 'size' of a zone to the number of flows starting there. A single *model evaluation* involves computing the set of values T_{ij} defined by (1) and (2).

Destination-constrained models are an obvious variant of origin-constrained models, in which (1) and (2) are replaced by

$$T_{ij} = O_i D_j A_j f(c_{ij}) \qquad (4)$$

$$A_j = 1/\sum_i O_i f(c_{ij}) \qquad (5)$$

thus satisfying the constraint

$$\sum_i T_{ij} = D_j \qquad (6)$$

Doubly-constrained models are rather more complex:

$$T_{ij} = O_i D_j A_i B_j f(c_{ij}) \qquad (7)$$

$$A_i = 1/\sum_j D_j B_j f(c_{ij}) \qquad (8)$$

$$B_j = 1/\sum_i O_i A_i f(c_{ij}) \qquad (9)$$

In this case the predicted flows satisfies constraints similar to both (3) and (6).

2.2 Applications Involving SIMs

SIMs are commonly used to represent human trip behaviour (such as to retail outlets) in network optimisation problems, where some 'profit' or other performance indicator of a global network is maximised. A typical problem is to locate some number N_F of facilities (e.g. shops, dealerships, hospitals) in a set of N_Z distinct zones (where $N_F < N_Z$) in such a way as to maximise the 'profit' from the facilities, which is computed from an underlying SIM by evaluating (a variant of) equation (1) for the zones in which facilities are located.

A necessary preliminary to solving this non-linear optimisation problem is the *calibration* of the model, in which the value of β is estimated from observed values T_{ij}^{obs} of trips (flows) between zones. This is a further non-linear optimisation problem which minimises an appropriate error norm $f(\beta)$ between T_{ij}^{obs} and the flows T_{ij} predicted by the model. Choice of appropriate error norms is a complex issue [7], which is beyond the scope of this paper; for simplicity we here use only maximum likelihood (10).

$$f(\beta) = \frac{\sum_i \sum_j T_{ij}^{\text{obs}} c_{ij} - \sum_i \sum_j T_{ij} c_{ij}}{\sum_i \sum_j T_{ij}} \tag{10}$$

The reliability of solutions to network optimisation problems depends on the reliability of the calibration process, which can be addressed computationally. The robustness of the computed value of β in relation to minor changes in T_{ij}^{obs} is assessed by *bootstrapping*, involving multiple recalibrations of the model, with slightly different sets of observed trip values obtained by systematic sampling and replacement from the original T_{ij}^{obs}. Thus the mean and variance of β can be derived. The number of calibrations of carried out, B is the *bootstrap size*. Clearly, greater values of B lead to more accurate estimates of the mean and variance of β; in practice, the heavy computation places practical limits on B, emphasising the potential benefits of parallel processing.

3   Parallel Implementation

We have implemented all three application components (calibration, bootstrapping, optimisation) on all three kinds of model (origin-, destination- and doubly-constrained), using Fortran77 with MPI. The aim of this part of the work was to experiment with alternative implementation techniques and derive templates for use in the AG. Details have already been reported in [4]; here there is space only to outline the principles in the origin-constrained case.

For all three components parallelism can be obtained within both the model evaluation and the application which evaluates the model: model evaluation is highly data parallel (as implied by (1) and (2)) and the applications can be parallelised by executing independent SIM evaluations in parallel.

For calibration we used a 'Golden Section' non-linear optimiser, which generates few independent model evaluations. Hence we expected parallelism to be exploited most effectively within the SIM evaluation. On the other hand, bootstrapping involves multiple independent calibrations and thus a high level of more coarsely-grained application-level parallelism was expected.

Following [5], a genetic algorithm was used to solve the non-linear network optimisation problem. Here the SIM is used to evaluate the fitness of each member of a population of possible solutions, hence multiple independent SIM evaluations are again involved, leading to plentiful application-level parallelism.

To explore the optimal combination of application- and model-based parallelism the P processors are viewed as a logical grid of dimensions $p_{\text{app}} \times p_{\text{mod}}$.

The application is parallelised between the p_{app} *application processors* of one row, each of which acts as a master processor for model evaluation, distributing its copy of the model between the p_{mod} *model processors* in its column.

3.1 Parallelism in the model

One of the main computational challenges for model evaluation is the large volume of stored data; space is required for the cost matrix c_{ij} and (for calibration and bootstrapping) the trip matrix T_{ij}^{obs}. A data parallel implementation distributes them across the processors using cyclic partitioning by rows: processor 1 receives rows 1, $P+1$, $2P+1$,..., processor 2 rows 2, $P+2$, $2P+2$,..., and so on. The matrices are treated slightly differently, as follows:

- each processor reads an (x, y) pair for each zone centroid, to compute its own rows of c_{ij}, stored in *dense* format.
- each processor stores the corresponding rows of T_{ij}^{obs} (also read from file), in *sparse* format since most trips are between physically close zones.

This distribution ensures that model evaluation is scalable from the perspective of memory usage, as long as the number of processors increases proportionately to the model size. Partitioning cyclically leads to a better load balance than partitioning contiguously because of the removal of systematic matrix patterns.

Evaluation of (1) and (2) may then proceed entirely in parallel with no further communication. It is only necessary to store one row of the T_{ij} matrix, since the error norms for calibration or bootstrapping are cumulatively evaluated in parallel from equation (10). The 'profit' in network optimisation is similarly accumulated.

3.2 Performance results

Performance results were obtained on a 512-processor Cray T3D machine, using 'journey-to-work' data derived from the 1991 UK census (see [6]). This records the numbers of journeys (T_{ij}^{obs}) out of and into all 10764 electoral wards. Costs c_{ij} are computed from distances between the centroids (x_i, y_i) of wards; in the case of intra-ward journeys the costs c_{ii} are fixed at some notional distance d_i depending on ward size. A second, smaller data set contains equivalent information aggregated into 459 electoral districts.

These data are representative of a range of other origin-constrained models, such as journeys to shopping centres or car dealerships, and can therefore be used to simulate corresponding location optimisation problems. As most journeys are between relatively close zones the trips matrix has the typical highly sparse pattern. Also, since the data sets satisfy both origin and destination constraints they may be used to assess doubly-constrained calibration and bootstrapping.

In all cases an exponential deterrence function was used. For space reasons, only a selection of the results in [4] are given.

As expected, calibration obtained best results with all parallelism in the model ($p_{app} = 1$), whereas bootstrapping and network optimisation enabled

Table 1. Optimisation times (sec) for locating 15 facilities in 10764 zones

P	p_{app}				
	1	2	4	8	16
32	3779	-	-	-	-
64	3317	1843	-	-	-
128	3300	1657	928	-	-
256	3459	1650	833	471	-
512	3736	1724	824	418	241

Table 2. Calibration times (sec) for 10764 zones

P	Time at $p_{app} = 1$	Speedup
32	274.3	-
64	137.6	1.99
128	69.2	3.96
256	35.1	7.8
512	17.9	15.3

maximum parallelism in the application. Table 1 shows optimisation times for allocating 15 car dealerships within 10764 zones, for varying values of P and p_{app}. Increasing p_{app} always decreases execution time, so the best time is obtained when the maximum value of p_{app} is used (in this case $P/32$, since the model requires 32 processors to evaluate). Similar results were obtained for the 459-zone model, which can be evaluated on a single processor (ie $P_{app} = P$). Encouragingly, the diagonal entries in Table 1 show a speedup close to linear.

Bootstrapping results show a similar pattern to network optimisation. Calibration was predictably less scalable, since only the finer-grain model parallelism was available. However, even here the results were encouraging. Table 2 shows a near-linear improvement in execution time as the number of processors increases 16-fold – the baseline of 32 processors was again the minimum necessary to evaluate the model. Even with the smaller 459-zone model useful performance improvements were obtained by parallelism up to 64 processors (see [4] for details) but performance degraded thereafter. By contrast, bootstrapping and optimisation continued to achieve near-linear speedup up to 512 processors even with the smaller model, because parallelism was exploited at the application level.

4 Specifying a Spatial Interaction Model

A SIM application can be specified by defining its parameters in a textual Interface File. We illustrate this for a simple case in Fig. 1, which defines a locational problem to optimise placement of 15 facilities within 10764 zones. (The need for MAXTRP in this file arises from using static arrays in the Fortran-77 implementation and is not inherent in the model specification.)

```
SIM                    % Model parameters
  MODTYP=0                     % origin constrained model
  DETFUN=exp                   % exponential deterrence function
CALIBRATE              % calibration parameters
  ERROR=maxlike                % maximum likelihood error norm
BOOTSTRAP              % bootstrapping parameters
  BSIZE=256                    % size of bootstrap sample
OPTIMISE               % optimisation parameters
  LOC=15                       % number of locations to be optimised
DATASIZES              % data set details for
  NZ=10764                     % number of origin zones
  MZ=10764                     % number of destination zones
  MAXTRP=586645                % total number of non-zero trips
                               % in calibration data
```

Fig. 1. Interface File for simple origin-constrained model

This Interface File format could be used as the input to an AG to generate parallel programs using templates based on the codes described in section 3. In effect we would have a generator based on SIMs defined by equations (1) to (9). While interesting as a demonstration of principle, the simplified form of the models is very restrictive. To develop a more powerful AG, we studied the requirements of a wide range of spatial location problems, including examples from retailing, agricultural production, industrial location, and urban spatial structure. From these a more generic SIM formulation was derived. Here we outline the origin-constrained version, without seeking to justify the modelling process.

4.1 A Generic Origin-Constrained Model

We assume that each origin zone produces 'activities' or 'goods' of several types and that each destination zone has several facilities' (consumers of activities/goods). This rather general terminology describes a range of different phenomena; for instance (and rather counter-intuitively) 'goods' may be m different categories of potential buyers travelling to one of r different car dealerships.

The earlier flow value, T_{ij} generalises to T_{ij}^{mr}, the flow of good type m from zone i to facility r in zone j. Origin and destination sizes, O_i and D_j become O_i^m and D_j^r and we allow for the latter to be exponentiated. Thus (1), (2) and (3) generalise to

$$T_{ij}^{mr} = O_i^m (D_j^r)^\alpha A_i f(c_{ij}) \tag{11}$$

$$A_i = 1 / \sum_j \sum_r (D_j^r)^\alpha f(c_{ij}) \tag{12}$$

$$\sum_j \sum_r T_{ij}^{mr} = O_i^m \tag{13}$$

The network optimisation problem is defined as maximising the difference between revenues (D_j^r) and costs (C_j^r) for each facility, ie for each r maximise

$$\sum_{j \in J} (D_j^r - C_j^r) \tag{14}$$

over a subset J of destination zones, subject to (13).

Revenues are computed as $\sum_i \sum_m p_i^m T_{ij}^{mr}$ where p_i^m is the revenue generated from a good of type m from zone i. In the most general case, costs at facility of type r at j have three components: maintaining the facility, transport, and the costs associated with transactions. This is modelled as

$$C_j^r = v_j^r (D_j^r)^\alpha + \left(\sum_i \sum_m c_{ij} T_{ij}^{mr} \right) + \left(q_j^r \sum_i \sum_m T_{ij}^{mr} \right) \tag{15}$$

where v_j^r, q_j^r are the unit costs of running facility r at j and making transactions there. The second and third terms are not always needed.

Finally, unit transport costs, c_{ij}, can be modelled by a term proportional to distance together with 'terminal costs', such as parking. Thus, in the most general case,

$$c_{ij} = t d_{ij} + \eta_i + \rho_j \tag{16}$$

where t is travel cost per unit distance, d_{ij} is a distance measure (we use Euclidean distance), η_i and ρ_j are the terminal costs at origin and destination. In some cases the t value may be irrelevant (ie it is only necessary to have costs proportional to distance) and the terminal costs are not always required.

Specifying this more general model requires the values of m, r and α, as well as stating whether the optional parts of the model are included. The values of ρ_j, η_i, v_j^r, q_j^r and p_i^m are data to be read from files at runtime if required.

Thus, the previous Interface File (Fig. 1) would be replaced by Fig. 2.

5 The Application Generator

The prototype AG implements the generic model above. It accepts an Interface File as input and generates the required parallel programs using templates based on the codes described in section 3. The user supplies an Interface File as above, or alternatively is prompted for parameters at a simple command-line interface. Each of the CALIBRATE, BOOTSTRAP, OPTIMISE sections is optional.

Using the results of section 3, calibration only exploits parallelism in the SIM evaluation, whereas bootstrapping and optimisation use maximum parallelism in the application, subject to limits imposed by the model size, derived from the NZ and MZ parameters. Thus, within the constraints of the program structure chosen, the values of p_{app} and p_{mod} give optimal execution time.

The use of a generic model can lead to superfluous and time-consuming computations, such as adding 0 when optional parts of the computation are not present, and needless exponentiation when ALPHA has its (common) value 1. The AG removes all such redundant computation, bringing execution times of

```
SIM
  MODTYP=0
  DETFUN=exp
  O_ACTIVITY=1   % number of activities at origins
  D_FACILITY=1   % number of facilities at destinations
  CFLAG=0        % travel costs NOT included in optimisation
  TFLAG=0        % terminal and unit travel costs NOT included
  PFLAG=0        % revenue costs NOT included in optimisation
  QFLAG=0        % transaction costs NOT included in optimisation
  ALPHA=1.0      % value of alpha
CALIBRATE
  ERROR=maxlike
BOOTSTRAP
  BSIZE=256
OPTIMISE
  LOC=7
DATASIZES
  NZ=10764
  MZ=10764
  MAXTRP=586645
```

Fig. 2. Interface file with generic SIM

generated code very close to hand-coded versions. Table 3 shows overheads of less than 2% on the Cray T3D for a variety of model types, application components and values of P (with 10764-zone model).

Source code for the AG is a mix of Fortran, Perl and Unix scripts. The templates consist of 2700 lines of code totalling some 97 kbytes, and the rest of the AG consists of 2500 lines (77 kbytes). It has been implemented and tested on an SGI Power Challenge, a Cray T3D, and various Sun and SGI workstations.

6   Evaluation and Conclusions

The prototype AG shows that it is possible to generate non-trivial parallel applications from very high-level specifications, with small overheads compared with direct coding. This is made possible by a restricted application domain, which nevertheless is useful in practice and benefits from parallelism.

The current tool is not yet of industrial-strength, on several counts. It has not been adequately evaluated by real users, does not cover the full range of spatial interaction modelling variants, and is limited in the outputs produced. Further generalisation will require more detailed inputs from domain experts, including a standardisation of data formats in a flexible way to permit greater generality. More control over the optimisation process (such as specifying the number of generations for the genetic algorithm) is desirable. Lastly, the user interface betrays the system's Fortran origins and would benefit from a graphical presentation.

Table 3. Overheads of application generator

application type	model constraint	P	overheads (%)
calibration	origin	1	1
		2	1.26
		8	1.4
calibration	double	1	1.8
		2	1.7
		8	1.6
optimisation	origin	1	1.6
		4	1.85
		16	1.9

Despite its limitations, we believe this work is valuable in showing that rapid parallel application development through a skeleton-like approach can be made to work effectively in an appropriate, well-defined application domain.

References

1. C. Barnes and C. Wadsworth. Portable software tools for parallel architectures. In *Proceedings of PPECC'95*, 1995.
2. M. Birkin, M. Clarke, and F. George. The use of parallel computers to solve non-linear spatial optimisation problems. *Environment and Planning A*, 17:1049–1068, 1994.
3. M. Cole. *Algorithmic Skeletons: Structured Management of Parallel Computation*. Pitman/MIT Press, 1989.
4. W. Essah, J. R. Davy, and S. Openshaw. Systematic exploitation of parallelism in spatial interaction models. In *Proceedings of PDPTA '97*, July 1997.
5. F. George. Hybrid genetic algorithms with immunisation to optimise networks of car dealerships. Edinburgh Parallel Computing Centre, EPCC–PAR–GMAP, 1994.
6. I. Turton and S. Openshaw. Parallel spatial interaction models. *Geographical and Environmental Modelling*, 1:179–197, 1997.
7. A. G. Wilson. A family of spatial interaction models, and associated developments. *Environment and Planning*, 3:1–32, 1971.

Performance Measurement of Interpreted Programs

Tia Newhall and Barton P. Miller

Computer Sciences Department, University of Wisconsin, Madison, WI 53706-1685

Abstract. In an interpreted execution there is an interdependence between the interpreter's execution and the interpreted application's execution; the implementation of the interpreter determines how the application is executed, and the application triggers certain activities in the interpreter. We present a representational model for describing performance data from an interpreted execution that explicitly represents the interaction between the interpreter and the application in terms of both the interpreter and application developer's view of the execution. We present results of a prototype implementation of a performance tool for interpreted Java programs that is based on our model. Our prototype uses two techniques, dynamic instrumentation and transformational instrumentation, to measure Java programs starting with unmodified Java .class files and an unmodified Java virtual machine. We use performance data from our tool to tune a Java program, and as a result, improve its performance by more than a factor of three.

1 Introduction

An *interpreted execution* is the execution of one program (the interpreted application) by another (the interpreter); the application code is input to the interpreter, and the interpreter executes the application. Examples include just-in-time compiled, interpreted, dynamically compiled, and some simulator executions. Performance measurement of an interpreted execution is complicated because there is an interdependence between the execution of the interpreter and the execution of the application; the implementation of the interpreter determines how the application code is executed and the application code triggers what interpreter code is executed. We present a representational model for describing performance data from an interpreted execution. Our model characterizes this interaction in a way that allows the application developer to look inside the interpreter to understand the fundamental costs associated with the application's execution, and allows the interpreter developer to characterize the interpreter's execution in terms of application workloads. This model allows for a concrete description of behaviors in the interpreted execution, and it is a reference point for what is needed to implement a performance tool for measuring interpreted executions. An implementation of our model can answer performance questions about specific interactions between the interpreter and the application. For example, we can represent performance data of Java interpreter activities like thread context switches, method table lookups, garbage collection, and byte-code instruction execution associated with different method functions in a Java application. We present results from a prototype implementation of our model for measuring the performance of interpreted Java applications and applets. Our prototype tool uses Paradyn's dynamic instrumentation [4] to dynamically insert and remove instrumentation from the Java virtual machine and Java method byte-codes as the byte-code is interpreted by the Java virtual machine. Our tool requires no modifications to the Java virtual machine nor to the Java source nor class files prior to execution.

The difficulties in measuring the performance of interpreted codes is demonstrated by comparing an interpreted code's execution to a compiled code's execution. A compiled code is in a form that can be executed directly on a particular operating system/architecture platform. Performance tools for compiled code provide performance measures in terms of platform-specific costs associated with executing the code; process time, number of page faults, and I/O blocking time are all examples of platform-specific measures. In contrast, an interpreted code is in a form that can be executed by the interpreter virtual machine. The interpreter virtual machine is itself an application program that executes on some operating system/architecture platform. One obvious difference between compiled and interpreted application execution is the extra layer of the interpreter program that, in part, determines the application's performance (see Figure 1). A performance tool for measuring the performance of an interpreted execution needs to measure the interaction between the Application layer (AP) and the InterpreterVM layer (VM).

There are potentially two different program developers that would be interested in performance measurement of the interpreted execution: the virtual machine developer and the application program developer. Both want performance data described in terms of platform-specific costs associated with executing parts of their program. However, each views the platform and the program as different layers of the interpreted execution. The VM developer sees the AP as input to the VM program that is run by the Platform layer (left side of the interpreted execution in Figure 1). The AP developer sees the Application layer as the program that is run on the VM (right side of the interpreted execution in Figure 1). For a VM developer, this means characterizing the virtual machine's performance in terms of Platform layer costs of the VM program's execution of

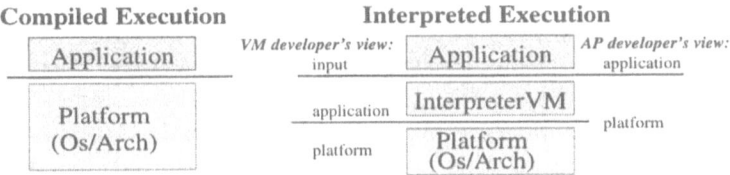

Fig. 1. Compiled application's execution vs. Interpreted application's execution. *VM and AP developers view the interpreted execution differently.*

its input (AP)—for example, the amount of process time executing VM function `invokeMethod` while interpreting AP method `foo`. The application program developer wants performance data that measures the same interaction to be defined in terms of VM-specific costs associated with AP's execution; for example, the amount of method call context switch time in AP method `foo`.

Another characteristic of an interpreted execution is that the application program has multiple execution forms. By multiple execution forms we mean that AP code is transformed into another form or other forms while it is executed. For example, a Java class is read in by the Java VM in `.class` file format. It is then transformed into an internal form that the Java VM executes. This differs from compiled code where the application binary is not modified as it executes. Our model characterizes AP's transformations as a measurable event in the code's execution, and we represent the relationship between different forms of an AP code object so that performance data measured for one form of the code object can be mapped back, to be viewed in previous forms of the object.

2 Performance Measurement Model

We present a representational model for describing performance data from an interpreted execution that explicitly represents the interaction between the application program and the virtual machine. Our model addresses the problems of representing an interpreted execution and describing performance data from the interpreted execution in a language that both the virtual machine developer and the application program developer can understand.

2.1 Representing an Interpreted Execution

Our representation of an interpreted execution is based on Paradyn's representation of a program execution as a set of resource hierarchies. A *resource* is a physical or logical component of a program (a semaphore, a function, and a process are all examples of program resources). A *resource hierarchy* is a collection of hierarchically related program resources. For example, the Process resource hierarchy views the running program as a set of processes. It consists of a root node that represents all processes in the application, and some number of child resources—one for each process in the running program. Other examples of resource hierarchies are a Code hierarchy for the code view of the program, a Machine hierarchy for the set of machines on which the application is running, and a Synchronization hierarchy for the set of synchronization objects in the application. An application's execution might be represented as the following set of resource hierarchies: {**Process, Machine, Code, SyncObj**}. An individual resource is represented by a path from its root node. For example, the function resource **main** is represented by the path **/Code/main.C/main**. Its path represents its relationship to other resources objects in the Code hierarchy.

Since both the application program and the virtual machine can be viewed as executing programs, we can represent each of their executions as a set of resource hierarchies. For example, AP's execution might be represented by {**Code,**

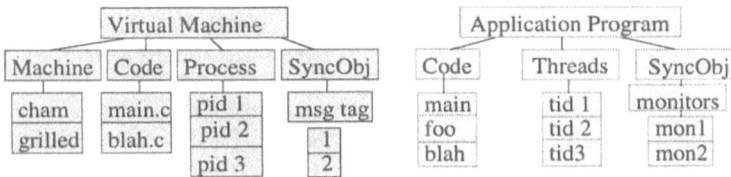

Fig. 2. Example resource hierarchies for the virtual machine and application program.

Thread, SyncObj} (right half of Figure 2), and the virtual machine's execution might be represented as **{Machine, Process, Code, SyncObj}** (left half of Figure 2). The resource hierarchies in this figure represent AP and VM's executions separately. However, there is an interaction between the execution of the two that must be represented. The relationship between the virtual machine and its input (AP) is that VM runs AP. We represent the VM *runs* AP relationship as an interaction between program resources of the two; code, process, and synchronization objects in the virtual machine interact with code, process, and synchronization objects in the application program during the interpreted execution. The interpreted execution is the union of AP and VM resource hierarchies. Figure 3 is an example of an interpreted execution represented by **{Machine, Code, Process, SyncObj, APCode, APThreads, APSyncObj}**.

The application program's multiple execution forms are represented by a set of sets of resource hierarchies–one set for each of the forms that AP takes during its execution, and a set of mapping functions that map resources in one form to resources in another form of AP. For example, in a dynamically compiled Java application, method functions may be translated from byte-code form to native code form during execution. Initially we create one AP Code hierarchy for the byte-code form of AP. As run-time transformations occur, we create a new AP code hierarchy for the native form of AP objects, and create mapping functions that map resources in one form to resources in the other form.

2.2 Representing Points in an Interpreted Execution

We can think of a program's execution as a series of events. Executing an instruction, waiting on a synchronization primitive, and accessing a memory page are examples of events. Any event in the program's execution can be represented

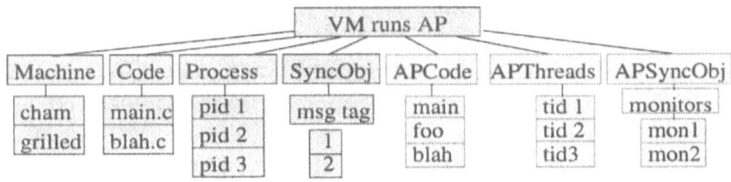

Fig. 3. Resource hierarchies representing the execution of VM *runs* AP.

as a subset of its *active* resources—those resources that are currently involved in the event. For example, if the event is "process 1 is executing code in function foo", then the resource **/Code/main.C/foo** and the resource **/Process/pid_1** are both active when this activity is occurring. These single resource selections from a resource hierarchy are called *constraints*. The activity specified by the constraint is active when its constraint function is true.

Definition 1. *A constraint is a single resource selection from a resource hierarchy. It represents a restriction of the hierarchy to some subset of its resources.*

Definition 2. *A constraint function, constrain(r), is a boolean function of resource r that is true when r is active. For example, constrain(/Process/pid_1) is true when process 1 is active (i.e. running).*

Constraint functions are *resource-specific*; all constraint functions test whether their resource argument is active, but the type of test that is done depends on the type of the resource argument. The constraint function for a process resource will test whether the specified process is running. The constraint function for a function resource will test whether the program counter is in the function. Each resource hierarchy exports the constraint functions for the resources in its hierarchy. For example, `Code.constrain(/Code/main.C)` is the constraint function applied to the Code hierarchy resource `main.C`.

Constraint functions can be combined with AND and OR to create boolean expressions containing constraints on more than one resource. By combining one constraint from each resource hierarchy with the AND operator, we represent different activities in the running program. This representation is called a *focus*.

Definition 3. *A focus is a selection of resources (one from each resource hierarchy). It represents an activity in the running program. A focus is active when all of its resources are active (i.e., the AND of the constraint functions).*

If the focus contains resources that are refined on both VM and AP resource hierarchies, then it represents a specific part of virtual machine's execution of the application program. For example,

</Machine, /Code/main.C/invokeMethod, /Process/pid_1, /SyncObj, /APCode/foo.class/foo, /APThreads, /APSyncObj>

is a focus from Figure 3 that represents when VM's process 1 is executing function `invokeMethod`, and interpreting AP method `foo`. This activity is occurring when the corresponding constraint functions are all true.

2.3 Performance Data for Interpreted Execution

To describe performance data that measure the interaction between the application program and the virtual machine, we provide a mechanism to selectively constrain performance information to any part of the execution. There are two complementary (and occasionally overlapping) ways to do this: constraints are either implicitly specified by metric functions or explicitly specified by foci. A metric function is a time-varying function that measures some aspect of a program execution's performance. Metric functions consist of time or count functions combined with boolean expressions built from constraint functions and constraint operators. For example:

- **CPUtime** = [ ]processTime/sec. The amount of process time per second.
- **methodCallTime** = [Code.constrain(/Code/main.C/invokeMethod)] processTime/sec. The time spent in VM function invokeMethod.
- **io_wait** = [Code.constrain(/Code/libc.so/read) **OR** Code.constrain(/Code/libc.so/write)] wallTime/sec. The time spent reading or writing.

The second way to constrain performance data is by specifying foci that represent restricted locations in the execution. Foci with both VM and AP resources represent an interaction between VM and AP. The focus **</Code/main.C/fetchCode, /APCode/foo.class/foo>** represents the part of the execution when the virtual machine is fetching AP code object **foo**. This part of the execution is active when `[Code.constrain( /Code/main.C/fetchCode) AND APCode.constrain( /APCode/foo.class/foo)]` is true. If we combine metric functions with this focus, we can represent performance data for the specified interaction.

We represent performance data as metric-focus pairs. The AND operator is used to combine a metric with a focus. The following are some example metric–focus pairs (we only show those components of the focus that are refined beyond a hierarchy root node):

1. **CPUtime, </APCode/foo.class/foo>**:
   [ ] `processTime/sec` **AND** [`APCode.constrain(/APCode/foo.class/foo)`]
2. **CPUtime, </Code/main.C/invokeMethod,/APCode/foo.class/foo>**:
   [ ] `processTime/sec` **AND** [`Code.constrain(/Code/main.C/invokeMethod)` `AND APCode.constrain(/APCode/foo.class/foo)`]
3. **methodCallTime, </APCode/foo.class/foo>**:
   [`Code.constrain(/Code/main.C/invokeMethod)`] `processTime/sec` **AND** [`APCode.constrain(/APCode/foo.class/foo)`]

Example 1 measures the amount of process time spent in AP function **foo**. The performance measurements in examples 2 and 3 are identical—both measure the amount of process time spent in VM function **invokeMethod** while interpreting AP function **foo**. However, example 2 is represented in a form that is more useful to a VM developer and example 3 is represented in a form that is more useful to an AP developer. Example 3 uses an *VM-specific* metric. VM-specific metric functions measure activities that are specific to a particular virtual machine. They are designed to present performance data to an AP developer who may have little or no knowledge of the virtual machine; they encode knowledge of the virtual machine in a representation that is closer to the semantics of the application language. Thus, an AP developer can measure VM costs associated with the execution of their program without having to know the details of the implementation of the VM; the **methodCallTime** metric encodes information about the VM function **invokeMethod** that is used to compute its value.

A final issue is representing performance data for foci with application program resources. An AP object may currently be in one form, while its performance data should be viewable in any of its forms. To do this, AP mapping functions are used to map performance data that is measured in one form of an AP object to a logical view of the same object in any of its other forms.

3   Measuring Interpreted Java Applications

We present a tool for measuring the performance of interpreted Java applications and applets running on Sun's version 1.0.2 of the Java VM [6]. The tool is an implementation of our model for representing performance data from an interpreted execution.

The Java VM is an abstract stack-based processor architecture. A Java program consists of a set of classes, each compiled into its own .class file. Each method function is compiled into byte-code instructions that the VM executes.

To measure the performance of an interpreted Java application or applet, our performance tool (1) discovers Java program resources as they are loaded by the VM, (2) generates and inserts SPARC instrumentation code into Java VM routines, and (3) generates and inserts Java byte-code instrumentation into Java methods and triggers the Java VM to execute the instrumentation code.

Since the Java VM performs delayed loading of class files, new classes can be loaded at any point during the execution. We insert instrumentation code in the VM that notifies our tool when a new .class file is loaded. We parse the VM's internal form of the class to create application program code resources for the class. At this point, instrumentation requests can be made for the class by specifying metric–focus pairs containing the class's resources.

We use dynamic instrumentation [4] to insert and delete instrumentation into Java method code and Java VM code at any point in the execution. Dynamic instrumentation is a technique where instrumentation code is generated in the heap, a branch instruction is inserted from the function's instrumentation point to the instrumentation code, and the function's instructions that were replaced by the branch are relocated to the heap and executed before or after the instrumentation code. Because the SPARC instruction set has instructions to save and restore stack frames, the instrumentation code and the relocated instructions can execute in their own stack frames. Thus instrumentation code will not destroy the values in the function's stack frame.

Using this technique to instrument Java methods is complicated by the fact that a method's byte-code instructions push and pop operands from their own operand stack. Java instrumentation code should use its own operand stack and have its own execution stack frame. The Java instruction set does not contain instructions to explicitly save and restore execution stack frames or to create new operand stacks. Our solution uses a technique called *transformational instrumentation*. This technique forces the Java VM to create a new operand stack and execution stack frame for our instrumentation code. The following are the transformational instrumentation steps:

1. The first time an instrumentation request is made for a method, relocate the method byte-code to the heap and expand its size by adding nop byte-code instructions around each instrumentation point. The nop instructions will be replaced with branches to instrumentation code.

2. Get the VM to execute the relocated method byte-code by replacing the first bytes in the original method with a goto_w byte-code instruction that branches to the relocated method. Since the goto_w instruction is inserted after the VM has

verified that this method byte-code is legal, the VM will execute this instruction even though it branches outside the original method function.

3. Generate instrumentation code in the heap. We generate SPARC instrumentation code in the heap, and use Java's native methods facility to call our instrumentation code from the Java method byte-code.

4. Insert method call byte-code instructions in the relocated method to call the native method function that will execute the instrumentation code. This will implicitly cause the VM to create a new execution stack frame and value stack for the instrumentation code.

4   Results

We present results from running a Java application with our performance tool. The application is a CPU scheduling simulator that consists of eleven Java classes and approximately 1200 lines of Java code. Currently, we can represent performance data in terms of VM program resources, AP code resources, and the combination of VM and AP program resources using both foci and VM-specific metrics to describe the interaction. Figure 4 shows the resource hierarchies from the interpreted execution, including the separate VM and AP code hierarchies.

We began by looking at the overall CPU utilization of the program (about 98%). We next tried to account for the part of the CPU time due to method call context switching and object creation in the Java VM. To measure these, we created two VM-specific metrics—**MethodCall_CS** measures the time for the Java VM to perform a method call context-switch, and **obj_create** measures the time for the Java VM to create a new object. Both are a result of the VM interpreting certain byte-code instructions in the AP. We measured these values for the Whole Program focus (no constraints). As a result, we found that a large portion ($\sim 35\%$) of the total CPU time is spent handling method context switching and object creation (Figure 5)[1].

Because of these results, we first tried to reduce the method call context switching time by in-lining method functions. Figure 6 shows the method functions that are accounting for the most CPU time and the largest number of

---

[1] With CPU enabled for every Java class, about 45% is due to executing Java code, 5% to method call context switching, and 40–45% to instrumentation overhead.

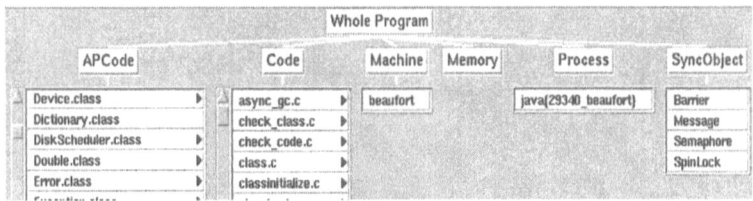

Fig. 4. Resource Hierarchies from the interpreted execution.

154

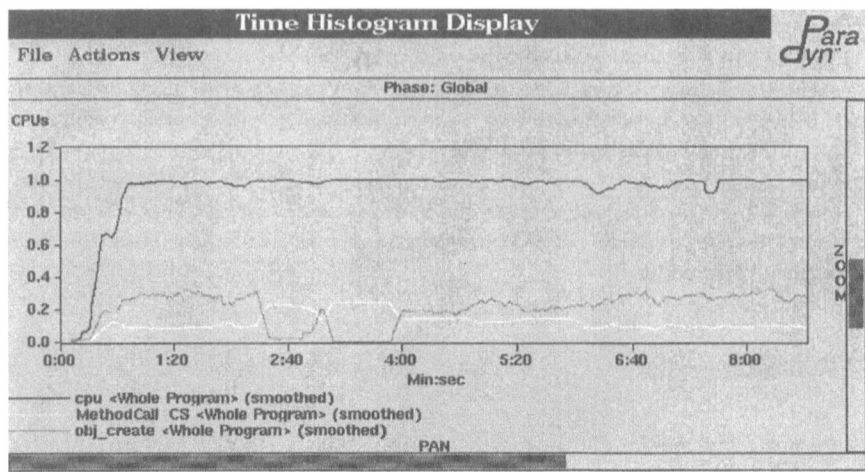

Fig. 5. Time Histogram showing CPU utilization, object creation time, and method call context switching time. *This graph shows that method call context switching time plus object creation time account for ~ 35% of the total CPU time.*

method function calls. We found that the `nextInterrupt()` and `isBusy()` methods of the `Device` class were being called often, and were accounting for a relatively large amount of total CPU time. By examining the code we found that the `Sim.now()` method was also called frequently. These three methods return the value of a private data member, and thus are good candidates for in-lining. After changing the code to in-line calls to these three method functions, the total number of method calls decreased by 31% and the total execution time decreased by 12% (second row in Table 1).

Table Visualization — Phase: phase_1	cpu CPUs
/APCode/Job.class/nextBurst()I	0.028
/APCode/Device.class/isBusy()Z	0.057
/APCode/Device.class/start(LJob;I)V	0.077
/APCode/Device.class/stop()LJob;	0.099
/APCode/Sim.class/firstInterrupt()I	0.1
/APCode/Device.class/nextInterrupt()I	0.11

Table Visualization — Phase: phase_1	proc_calls ops/sec
/APCode/Device.class/start(LJob;I)V	715
/APCode/Device.class/isBusy()Z	2,033
/APCode/StringBuffer.class	3,032
/APCode/Device.class/nextInterrupt()I	4,024
/APCode/Sim.class	6,987
/APCode/Device.class	7,487

Fig. 6. AP classes and methods that account for the largest % CPU (left) and that are called the most frequently (right). *The first column lists the focus, and the second column lists the metric value for the focus.*

We next tried to reduce the object creation time. We examined the CPU time for the new version of the AP code, and found that the Sim, Device, Job, and StringBuffer classes accounted for most of the CPU time. The time spent in StringBuffer methods is due to a large number of calls to the append and constructor methods made from the Sim and Device classes. We were able to reduce the number of StringBuffer and String objects created by removing strings that were created but never used, and by creating static data members for parts of strings that were recreated multiple times (in Device.stop() and Device.start()). With these changes we are able to reduce the total execution time by 70% (fourth row in Table 1).

Table 1. Performance results from different versions of the application.

Optimization	Number of Method Calls	Number of Object Creates	Total Execution Time (in seconds)
Original Version	13,373,200	465,140	389.29
Method in-lining	9,213,400 (31% less)	465,140	343.99 (-12% change)
Fewer Obj. Creates	9,727,800 (27% less)	17,350 (96% less)	234.13 (-40% change)
Both Changes	5,568,100 (58% less)	17,350	113.60 (-70% change)

In this example, our tool provided performance data that is difficult to obtain with other performance tools. Our tool provided performance data that described expensive interactions between the Java VM and the Java application, and accounted for these costs in terms of AP resources. With this data, we were easily able to determine what changes to make to the Java application to improve its performance.

5   Related Work

There are many general purpose program performance measurement tools [3, 5, 7, 4] that can be used to measure the performance of the virtual machine. However, these are unable to present performance data in terms of the application program or in terms of the interaction between VM and AP. There are some performance tools that provide performance data in terms of AP's execution [1, 2]. These tools provide performance data in terms of Java application code. However, they do not represent performance data in terms of the Java VM's execution or in terms of the interaction between the VM and the Java application. If the time values provided by these tools include Java VM method call, object creation, garbage collection, thread context switching, or class file loading activities in the VM, then performance data that explicitly represents this interaction between the Java VM and the Java application's execution will help an application developer determine how to tune his or her application.

6 Conclusion and Future Work

This paper describes a new approach to performance measurement of interpreted executions that explicitly models the interaction between the interpreter program and the interpreted application program so that performance data can be associated with any part of the execution. Performance data is represented in terms that either an application program developer or an interpreter developer can understand.

Currently, we are working to expand our prototype to include a larger set of VM-specific metrics, AP thread and synchronization resource hierarchies, and support for mapping performance data between different views of AP code objects. With support for AP thread and synchronization resources our tool can provide performance data for multi-threaded Java applications.

7 Acknowledgments

Thanks to Marvin Solomon for providing the CPU simulator Java application. This work is supported in part by DARPA contract N66001-97-C-8532, NSF grant CDA-9623632, and Department of Energy Grant DE-FG02-93ER25176. The U.S. Government is authorized to reproduce and distribute reprints for Governmental purposes notwithstanding any copyright notation thereon.

References

1. Intuitive Systems. Optimize It. http://www.optimizeit.com/.
2. KL Group. JProbe. http://www.klg.com/jprobe/.
3. A. Malony, B. Mohr, P. Beckman, D. Gannon, S. Yang, and F. Bodin. Performance Analysis of pC++: A Portable Data-Parallel Programming System for Scalable Parallel Computers. *Proceedings of the 8th International Parallel Processing Symposium (IPPS), Cancun, Mexico,* pages 75–85, April 1994.
4. Barton P. Miller, Mark D. Callaghan, Jonathan M. Cargille, Jeffrey K. Hollingsworth, R. Bruce Irvin, Karen L. Karavanic, Krishna Kunchithapadam, and Tia Newhall. The Paradyn Parallel Performance Measurement Tools. *IEEE Computer 28, 11,* November 1995.
5. Daniel A. Reed, Ruth A. Aydt, Roger J. Noe, Phillip C. Roth, Keith A. Shields, Bradley W. Schwartz, and Luis F. Tavera. Scalable Performance Analysis: The Pablo Performance Anlysis Environment. *Proceedings of the Scalable Parallel Libraries Conference,* pages 104–113. IEEE Computer Society, 1993.
6. Sun Microsystems Computer Corporation. *The Java Virtual Machine Specification,* August 21 1995.
7. J. C. Yan. Performance Tuning with AIMS – An Automated Instrumentation and Monitoring System for Multicomputers. *27th Hawaii International Conference on System Sciences, Wailea, Hawaii,* pages 625–633, January 1994.

Analysing an SQL Application
with a *BSPlib* Call-Graph Profiling Tool

Jonathan M.D. Hill, Stephen A. Jarvis,
Constantinos Siniolakis, and Vasil P. Vasilev

Oxford University Computing Laboratory, UK.

Abstract. This paper illustrates the use of a post-mortem *call-graph* profiling tool in the analysis of an SQL query processing application written using *BSPlib* [4]. Unlike other parallel profiling tools, the architecture independent metric of imbalance in size of communicated data is used to guide program optimisation. We show that by using this metric, *BSPlib* programs can be optimised in a portable and architecture independent manner. Results are presented to support this claim for unoptimised and optimised versions of a program running on networks of workstations, shared memory multiprocessors and tightly coupled distributed memory parallel machines.

1 Introduction

The Bulk Synchronous Parallel model [8,6] views a parallel machine as a set of processor-memory pairs, with a global communication network and a mechanism for synchronising all processors. A BSP program consists of a sequence of *supersteps*. Each superstep involves all of the processors and consists of three phases: (1) processor-memory pairs perform a number of computations on data held locally at the start of a superstep; (2) processors communicate data into other processor's memories; and (3) all processors barrier synchronise.

The BSP cost model [6] asserts that globally balancing computation and communication is the key to optimal parallel design. The rationale for balancing computation is clear as the barrier that marks the end of the superstep ensures that all processors have to wait for the slowest processor before proceeding into the next superstep. It is therefore desirable that all processes enter the barrier at about the same time to minimise idle time. In contrast, the need to balance communication is not so clear-cut. The BSP cost model implicitly asserts that the dominating cost in communicating a message is the cost of crossing the boundary of the network, rather than the cost of internal transit. This is largely true for today's highly-connected networks, as described in detail in [2]. Therefore, a good design strategy is to ensure that all processors inject the same amount of data into the network and are consequently equally involved in the communication.

This paper explores the use of a call-graph profiling tool [3] that exposes imbalance in either computation or communication, and highlights those portions of the program that are amenable to improvement. Our hypothesis is that by minimising imbalance, significant improvements in the algorithmic complexity of parallel algorithms usually follow. In order to test the hypothesis, the call-graph profiler is used to optimise a database query evaluation program [7]. We show how the profiling tool helps identify problems in the implementation of a randomised sample sort algorithm. It is also shown how the profiling tool guides the improvement of the algorithm. Results are presented which show that significant performance improvements can be achieved without tailoring algorithms to a particular machine or architecture.

2 Profiling an SQL Database Application

A walk-through of the steps involved in optimising an SQL database query evaluation program is used to demonstrate the use of the call-graph profiling tool.

A series of relational queries were implemented in BSP. This was done by transcribing the queries into C function calls and then linking them with a *BSPlib* library of SQL-like primitives. The test program was designed to take as input a sequence of relations (tables), it would then processes the tables and yield as output a sequence of intermediate relations. The input relations were distributed among the processors using a simple block-cyclic distribution. Three input relations ITEM, QNT and TRAN were defined. Six queries were evaluated which created the following intermediary relations: (1) TEMP1, an aggregate sum and a "group-by" rearrangement of the relation TRAN; (2) TEMP2, an equality-join of TEMP1 and ITEM; (3) TEMP3, an aggregate sum and group-by of TEMP2; (4) TEMP4, an equality-join of relations TEMP3 and QNT; (5) TEMP5, a less-than-join of relations TEMP4 and ITEM; and (6) a filter (IN "low 1%") of the relation TEMP5.

2.1 Using the Profiler to Optimise SQL Queries

Figure 2 shows a screen shot of a call-graph profile for the SQL query processing program running on a sixteen processor Cray T3E. The call-graph contains a series of interior and leaf nodes. The interior nodes represent procedures entered during program execution, whereas the leaf nodes represent the textual position of the end of a superstep, that is, the line of code containing a call to the barrier synchronisation function bsp_sync. The path from a leaf to the root of the graph identifies the nesting of procedure calls that were active when bsp_sync was executed. This path is termed a *call-stack* and a collection of call-stacks comprise a *call-graph*. The costs of shared procedures can be accurately apportioned to their parents via a scheme known as *inheritance* [5]. This is particularly important when determining how costs are allocated to library functions. Instead of simply reporting that all the time was spent in a parallel sorting algorithm for example, a call-graph profile also attributes costs to the procedures which contained calls to the sort routine.

In line with superstep semantics, the cost of all communication issued during a superstep is charged to the barrier that marks the end of the superstep. Similarly, all computation is charged to the end of the superstep. Leaf nodes

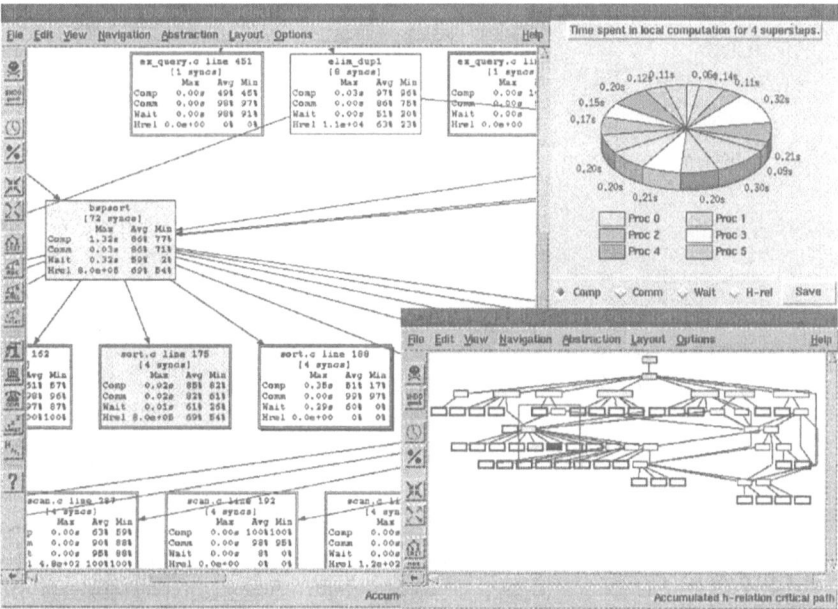

Fig. 1. Screen shot of the call-graph profiling tool

record the costs accrued during a superstep as: (i) the textual position of the bsp_sync call within the program; (ii) the number of times a particular superstep is executed; and (iii) summaries of computation cost, communication cost, idle time, and the largest amount of data either entering or leaving any process (a h-relation). These costs are given as a maximum, average, and minimum cost over p processors (the average and minimum costs are given as a percentage of the maximum). Interior nodes record similar information to leaf nodes, except that the label of the node is a procedure name and the accumulated cost is inherited from each of the supersteps executed during the lifetime of that procedure.

Critical paths are made visible by shading each of the nodes in the graph with a colour ranging from white to red. A red node corresponds to a bottleneck (or 'hot-spot'!) in the program. In the graphs that follow, the shading for each node is determined by the largest difference between the maximum and average cost, and also the largest percentage-wise deviation between maximum and average cost [3]. Terms of the form, (12% | 7%), are used to quantify this balance. This states that the average cost of executing a piece of code on p processors is 12% of the cost of the processor that spent the longest time performing the task; whereas the process with the least amount of work took only 7% of the time of the most loaded process. In the following sections we refine the algorithms used in the SQL program until there is virtually no imbalance, which is identified by terms of the form (100% | 100%).

Query Evaluation Stage 1 A portion of the call-graph for the original program (version 1) is shown in Figure 2. The three input relations were initially unevenly distributed among the processors. It is to be expected that an irregular

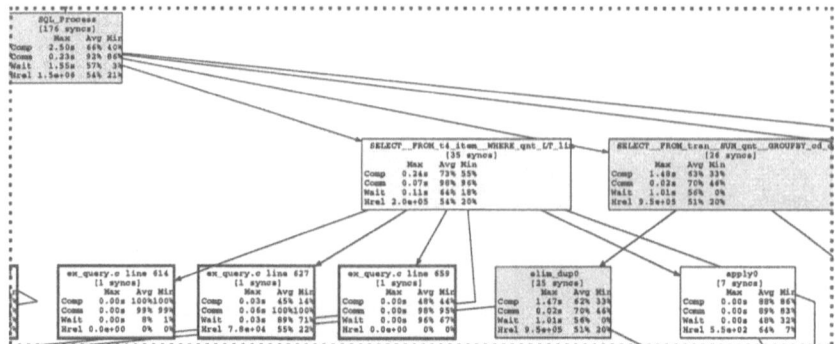

Fig. 2. Query version 1: SQL query evaluation.

distribution of input relations would produce a considerable amount of imbalance in computation and communication when performing operations using these data structures. For example, Figure 2 shows a (54%| 21%) imbalance in h-relation size. To remedy this imbalance, load balancing functions were introduced into the code which ensured that each processor contained approximately an equal sized partition of the input relation.

Query Evaluation Stage 2 Load balancing the input relations reduced the amount of communication and computation imbalance by 26%. Further profiles revealed that the imbalance had not been completely eradicated. It appeared that the SQL primitives had inherent imbalances in communication even for perfectly balanced input data. This can be seen in Figure 3.

The critical paths identifying the imbalance in communication were followed from the select function to **elim_dup0**. At this point it was easy to identify that the major cause of the communication imbalance, (69%| 54%), was attributed to the function **bspsort** at line **175**. This is shown in the screen-shot in Figure 1.

In the same way, following the imbalance in the computation critical paths, (51%| 17%), it was also clear that the primary cause of computation imbalance was attributed to the same function **bspsort**, this time at line **188** in Figure 1. Figure 1 also shows a pie-chart that gives a breakdown of the accumulated computation time at the superstep at line **188** for all the processes. To give some idea of the type of computation that may cause problems, the underlying algorithm of the **bspsort** function is (briefly) described. The function **bspsort** implements a refined variant of the optimal randomised BSP sorting algorithm of [1].

The algorithm consists of seven stages: (1) each processor locally sorts the elements in its possession; (2) each processor selects a random sample of $s \times p$ elements (where s is the oversampling factor) which are gathered onto process zero; (3) the samples are sorted and p regular pivots are picked from the $s \times p^2$ samples; (4) the pivots are broadcast to all processors; (5) each processor partitions the elements in its possession into the p blocks as induced by the

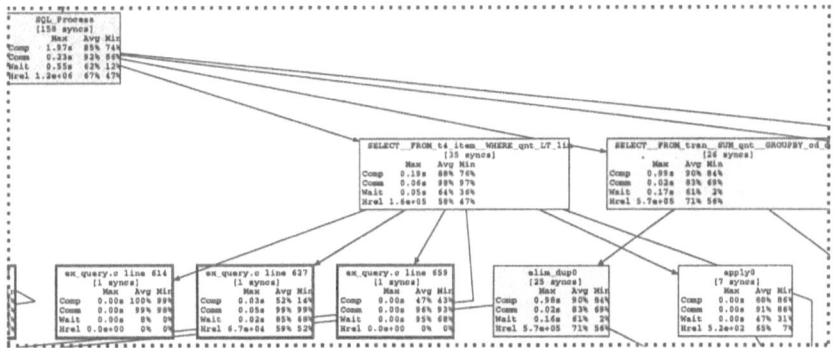

Fig. 3. Query version 2: SQL query evaluation after load balance.

pivots; (6) each processor sends partition i to processor i; and (7) a local multi-way merge produces the desired effect of a global sort.

If stages (6) and (7) are not balanced, then this can only be attributed to a poor selection of splitters in stage (2). Since the random number generator that selects the sample had been extensively tested prior to usage the only possible cause for the disappointing performance of the algorithm was the choice of the oversampling factor s. The algorithm had however been previously tested and the oversampling procedure had been fine tuned by means of extensive experimental results using simple timing functions. The experimental results had suggested that the oversampling factor established by the theoretical analysis of the algorithm had been a gross overestimate and, therefore, at the implementation level it had been decided to employ a considerably reduced factor.

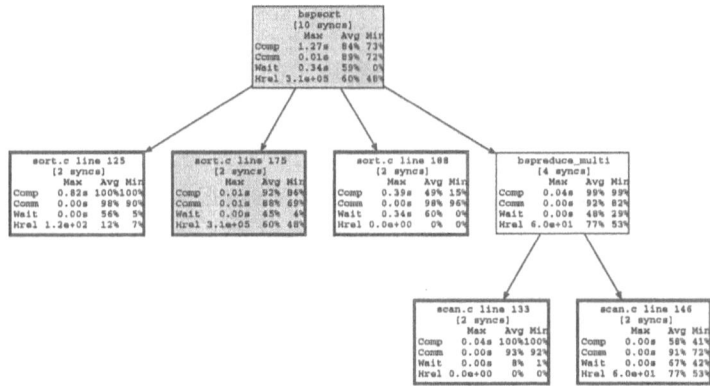

Fig. 4. Sorting version 1: experimental oversampling factor.

Query Evaluation: Improving Parallel Sorting Experiments continued by varying the oversampling factor for the sorting algorithm. A portion of the call-graphs for the optimal experimental and theoretical parameters are exhibited in figures 4 and 5 respectively. The original experimental results were confirmed by the profile, that is, the algorithm utilising the theoretical oversampling factor (sort version 2) delivered performance that was approximately 50% inferior to that of the algorithm utilising the experimental oversampling factor (sort version

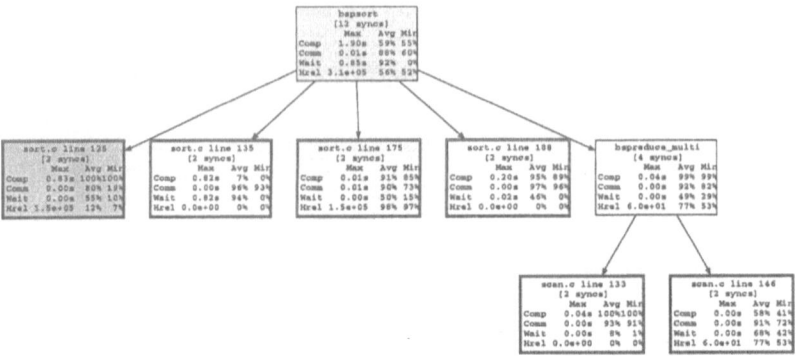

Fig. 5. Sorting version 2: theoretical oversampling factor.

1). The computation imbalance of (49%| 15%) in stage (7) of sort version 1, shown at line 188 in Figure 4, shifted to a computation imbalance of (7%| 0%) in stage (2) of sort version 2, shown at line 135 of Figure 5. Similarly, the communication imbalance of (60%| 48%) in stage (6) of sort version 1 shifted to an imbalance of (12%| 7%) in stage (3) at line 125 of sort version 2.

It was noted however that the communication and computation requirements of stages (6) and (7) in sort version 2 of Figure 5 are highly balanced, (98%| 97%) and (95%| 89%) respectively. Therefore, the theoretical analysis had accurately predicted the oversampling factor value required to achieve load balance. Unfortunately, the sustained improvement gained by balancing the communication pattern of stage (6) of the underlying sorting algorithm – and consequently, the communication requirements of the algorithm – had been largely overwhelmed by the cost of communicating and sorting a larger sample in stages (2) and (3).

To remedy this problem the ideas of [1] were adopted. In particular, the unbalanced communication and computation algorithm of stages (2) and (3), which collected and sorted a sample on a single process, were replaced with a simple and efficient parallel sorting algorithm, which sorted the sample set among all the processes. As noted in [1], an appropriate choice for such a parallel sorting algorithm is presented by an efficient variant of the bitonic-sort network.

The introduction of the bitonic sorter caused the data presented in the **bspsort** node in Figures 4 and 5 to: Computation (99%| 98%), Communication (83%| 70%), and h-relation (99%| 98%). This improved the wall-clock running time of the sorting algorithm by 8.5%.

Query Evaluation Stage 3 As the sorting algorithm is central to the implementation of most of the SQL queries, a minor improvement in the sorting algorithm brings about a marked improvement in the performance of the query evaluation as a whole. This can be clearly seen from the lack of any shading of critical paths in Figure 6 (contrast with Figure 2)–all the h-relations in the SQL queries are almost perfectly balanced.

To reinforce the conclusion that these improvements are architecture independent, Table 2.1 shows the wall-clock times for the original and optimised pro-

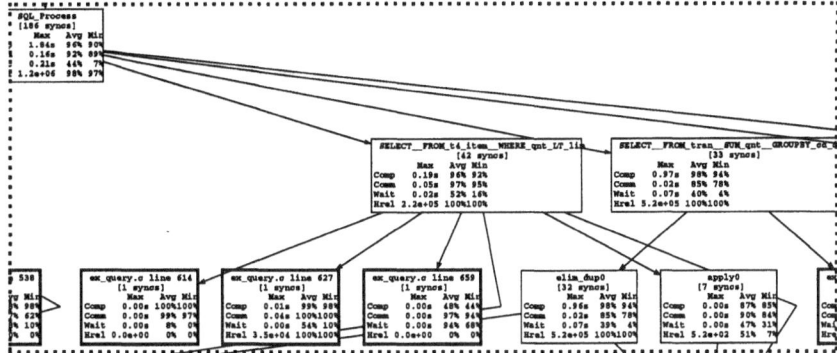

Fig. 6. Query version 3: SQL query evaluation, final version.

grams running on a variety of parallel machines. The size of the input relations was chosen to be small, so that the computation time in the algorithms would not dominate. This is particularly important if we are to highlight the improvements in communication cost. A small relation size also enabled the Network of Workstations (NOW) implementation of the algorithm to complete in a reasonable amount of time. For all but one of the results, the optimised version of the program provides an approximate 20% improvement over the original program. This evidence supports our hypothesis that the call-graph profiling tool provides a mechanism that guides architecture independent program optimisation. As an aside, the parallel efficiencies in the table are marginal for the T3E and Origin 2000, due to the communication intensive nature of this query processing application, in combination with the small amount of data held on each process (750 records per process at $p = 16$). In contrast, a combination of the slow processors, fast communication, and poor cache organisation on the T3D gives super-linear speedup even for this small data set. However, the communication intensive nature of this problem makes it unsuitable for running over loosely coupled distributed memory machines. For example, the results for a NOW, exhibit super-linear speedup at $p = 2$ in the optimised program, yet *super-linear slowdown* in all other configurations.

3 Conclusions

The performance improvements resulting from the analysis of the call-graph profiles demonstrate that the tool can be used to optimise programs in a portable and architecture independent manner. Unlike other profiling tools, the architecture independent metric – h-relation size – guides the optimisation process. The major benefit of this profiling tool is that the amount of information displayed when visualising a profile for a parallel program is no more complex than that of a sequential program.

---

[1] A 10Mbps Ethernet network of 266MHz Pentium Pro processors.

Table 1. Wall-clock time in seconds for input relations containing 12,000 records

Machine	p	Unoptimised		Optimised		gain
		time	speedup	time	speedup	
Cray T3E	1	6.44	1.00	5.37	1.00	17%
	2	4.30	1.50	3.38	1.59	21%
	4	2.48	2.60	1.85	2.90	25%
	8	1.23	5.23	1.04	5.17	15%
	16	0.68	9.43	0.67	8.02	1%
Cray T3D	1	27.18	1.00	22.41	1.00	18%
	2	13.18	2.06	10.88	2.06	17%
	4	6.89	3.94	5.70	3.93	17%
	8	3.29	8.25	3.07	7.30	7%
	16	1.66	16.34	1.89	11.88	-14%
SGI Origin 2000	1	2.99	1.00	2.42	1.00	19%
	2	1.65	1.81	1.27	1.91	23%
	4	1.26	2.37	1.11	2.16	12%
	8	0.88	3.39	0.77	3.15	13%
Intel NOW[1]	1	4.96	1.00	4.04	1.00	18%
	2	78.21	0.06	1.65	2.45	98%
	4	174.77	0.03	88.76	0.05	49%
	6	216.42	0.02	101.33	0.04	53%

References

1. A. V. Gerbessiotis and C. J. Siniolakis. Deterministic sorting and randomized median finding on the BSP model. In *Proceedings of the 8th ACM Symposium on Parallel Algorithms and Architctures*, Padova, Italy, June 1996. ACM Press.
2. J. M. D. Hill, S. Donaldson, and D. Skillicorn. Stability of communication performance in practice: from the Cray T3E to networks of workstations. Technical Report PRG-TR-33-97, Programming Research Group, Oxford University Computing Laboratory, October 1997.
3. J. M. D. Hill, S. Jarvis, C. Siniolakis, and V. P. Vasilev. Portable and architecture independent parallel performance tuning using a call-graph profiling tool. In *6th EuroMicro Workshop on Parallel and Distributed Processing (PDP'98)*. IEEE Computer Society Press, January 1998.
4. J. M. D. Hill, B. McColl, D. C. Stefanescu, M. W. Goudreau, K. Lang, S. B. Rao, T. Suel, T. Tsantilas, and R. Bisseling. BSPlib: The BSP Programming Library. *Parallel Computing*, to appear 1998. see www.bsp-worldwide.org for more details.
5. S. A. Jarvis. *Profiling large-scale lazy functional programs*. PhD thesis, Computer Science Department, University of Durham, 1996.
6. D. Skillicorn, J. M. D. Hill, and W. F. McColl. Questions and answers about BSP. *Scientific Programming*, 6(3):249–274, Fall 1997.
7. K. R. Sujithan and J. M. D. Hill. Collection types for database programming in the BSP model. In *5th EuroMicro Workshop on Parallel and Distributed Processing (PDP'97)*. IEEE Computer Society Press, Jan. 1997.
8. L. G. Valiant. A bridging model for parallel computation. *Communications of the ACM*, 33(8):103–111, August 1990.

A Graphical Tool for the Visualization and Animation of Communicating Sequential Processes

Ali E. Abdallah

Department of Computer Science
The University of Reading
Reading, RG6 6AY, UK
email: A.Abdallah@reading.ac.uk
WWW:http://www.fmse.cs.reading.ac.uk/people/aea

Abstract. This paper describes some aspects of an interactive graphical tool designed to exhibit, through animation, the dynamic behaviour of parallel systems of communicating processes. The tool, called *VisualNets*, provides functionalities for visually creating graphical representations of processes, connecting them via channels, defining their behaviours in Hoare's CSP notation and animating the evolution of their visualization with time. The tool is very useful for understanding concurrency, analysing various aspects of distributed message-passing algorithms, detecting deadlocks, identifying computational bottlenecks, and estimating the performance of a class of parallel algorithms on a variety of MIMD parallel machines.

1 Introduction

The process of developing parallel programs is known to be much harder than that of developing sequential programs. It is also not as intuitive. This is especially the case when an explicit *message passing* model is used. The user usually takes full responsibility for identifying parallelism, decomposing the system into a collection of parallel tasks, arranging appropriate communications between the tasks, and mapping them onto physical processors. Essentially, a parallel system can be viewed as a collection of independent sequential subsystems (processes) which are interconnected through a number of common links (channels) and can communicate and interact by sending and receiving messages on those links. While understanding the behaviour of a sequential process in isolation is a relatively straightforward task, studying the effect of placing a number of such processes in parallel can be very complex indeed. The behaviour of all the processes becomes inter-dependent in ways which are not at first obvious making it very difficult to comprehend, reason about, and analyse the system as a whole. Athough part of this difficulty can be attributed to the introduction of new aspects such as synchronizations and communications, the main problem lies in the inherent complexity of *parallel control* as opposed to sequential control.

Visualizations are used to assist in absorbing large quantities of information very rapidly. Animations add to their worth by allowing visualizations to evolve with time and, hence, making apparent the dynamic behaviour of complex systems. *VisualNets* is an interactive graphical tool for facilitating the design and analysis of synchronous networks of communicating systems through visualization and animation.

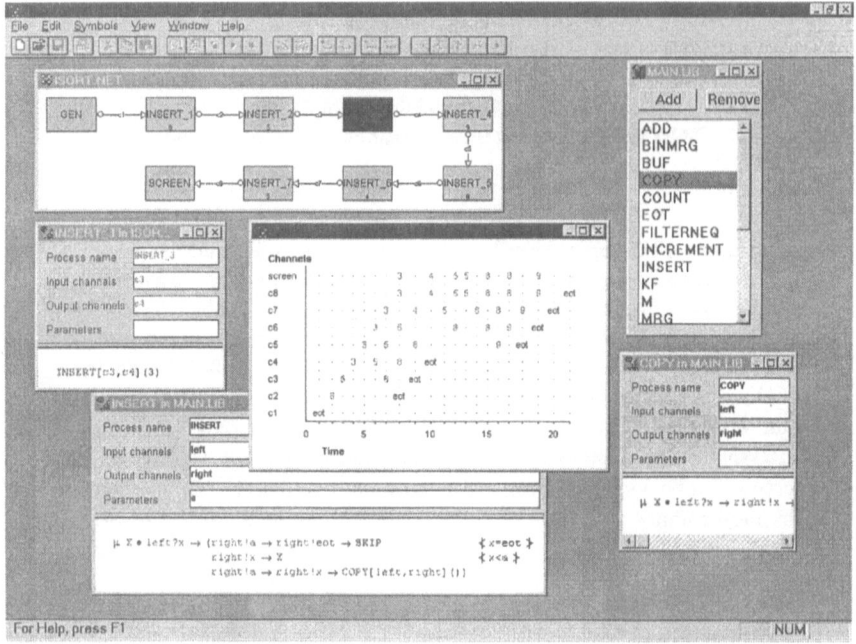

Fig. 1. A screenshot showing the animation of an insertion sort algorithm

The tool, depicted in Fig. 1, allows the user to interactively create graphical representations of processes, link them via channels to form specific network configurations, describe the behaviour of each process in the network in Hoare's CSP notation [11], and study an animated simulation of the network in execution. The CSP syntax is checked before the animation starts. The visualization of the network animates communications as they take place by flashing the values being communicated over channel links. The tool builds a timing diagram to accompany the animation and allows a more in-depth analysis to be carried out. Networks can be built either as part of an orchestrated design process, by systematic refinement of functional specifications using techniques described in [1, 2], or "on the fly" for rapid prototyping.

The benefits of using visualization as an aid for analysing certain aspects of parallel and distributed systems has long been recognised. The main focus in this area has been on tools for monitoring and visualizing the performance of parallel programs. These tools usually rely on trace data stored during the actual execution of a parallel program in order to create various performance displays. A good overview of relevant work in this area can be found in [10, 13]. A different approach aimed at understanding not only the performance of parallel and distributed systems but also their logical behaviours is based on simulation and animation for specific architectures [6, 5, 12].

The remainder of the paper is organised as follows. Section 2 introduces various features of the tool through a case study and Section 3 concludes the paper and indicates future directions.

2 A Case Study

We will give a brief description of the functionality of the tool through the development of a distributed sorting algorithm based on insertion sort. Given a non-empty list of values, the insertion sort algorithm, *isort*, starts from the empty list and constructs the final result by successively inserting each element of the input list at the correct position in an accumulated sorted list. Therefore, sorting a list, say $[a_1, a_2, ..a_n]$, can be visualized as going through n successive stages. The ith intermediate stage, say *insert_i*, holds the value a_i, takes the sorted list *isort* $[a_1, a_2, ..a_{i-1}]$ as input from the preceeding stage and returns the the longer list *isort* $[a_1, a_2, ..a_i]$ to the following stage.

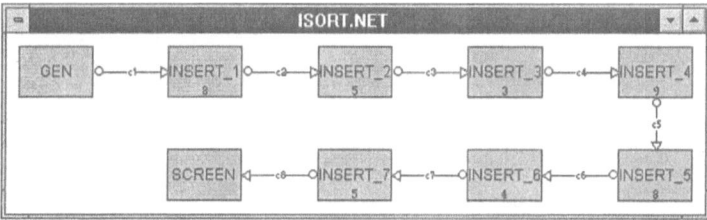

Fig. 2. Network configuration for parallel insertion sort

2.1 Creating a Graphical Network

The network configuration for sorting the list $[8, 5, 3, 9, 8, 4, 5]$ is depicted in Fig. 2. Each process in the network is represented graphically as a box with some useful information, such as values of local parameters and process name, inside it. Processes in the network can be linked via channels to form any topographical configuration. The construction of a network involves operations for adding, removing, and editing both processes and channels. The interface allows these operations to be carried out visually and interactively.

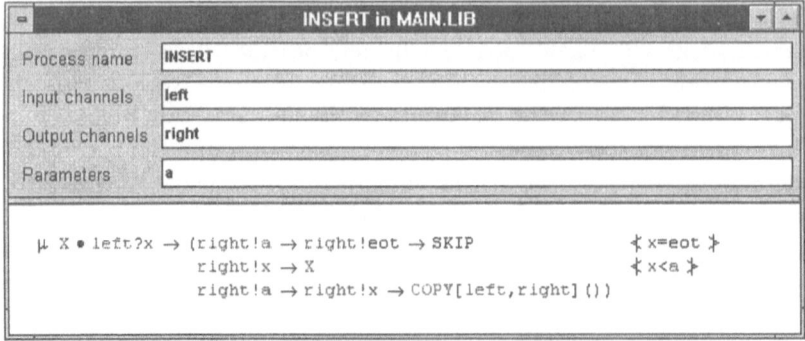

Fig. 3. A CSP definition of $INSERT(a)$

2.2 Defining Process Behaviour

Having defined the network configuration, the next stage is to define the behaviour of each process in the network. This is done using a subset of CSP. The screen shot captured in Fig. 3 shows the CSP definition for the process $INSERT(a)$ which is stored in a library of useful processes. The processes $INSERT\_i$, $1 \leq i \leq 7$, depicted in the above sorting network are all defined as specific instances of the library process $INSERT(a)$ by appropriately instantiating the value parameter a and renaming the channels *left* and *right*.

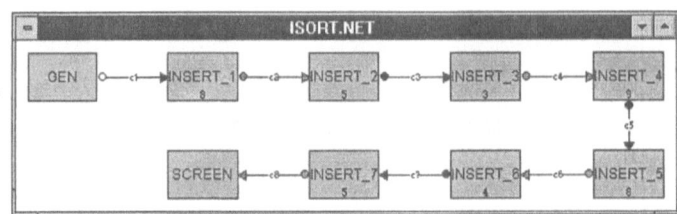

Fig. 4. Animation of the network

In CSP, outputing a specific value v on a channel c is denoted by the event $c!v$, inputing any value v on channel c and storing it in a local variable x is denoted by the event $c?x$. The arrow $\rightarrow$ denotes prefixing an event to a process. The notation $P \; \{ \; b \; \} \; Q$, where b is a boolean expression and P and Q are processes, is just an infix form for the traditional selection construct **if** b **then** P **else** Q. Note that the prefix and conditional operators associate to the right. The special message *eot* is used to indicate the end of a stream of messages transmitted on a channel. The process $COPY$ denotes a one place buffer. For any lists s, the process $Prd(s)$ outputs the values of s in the same order on channel *right* and followed by the message *eot*. The process GEN is $Prd([])$.

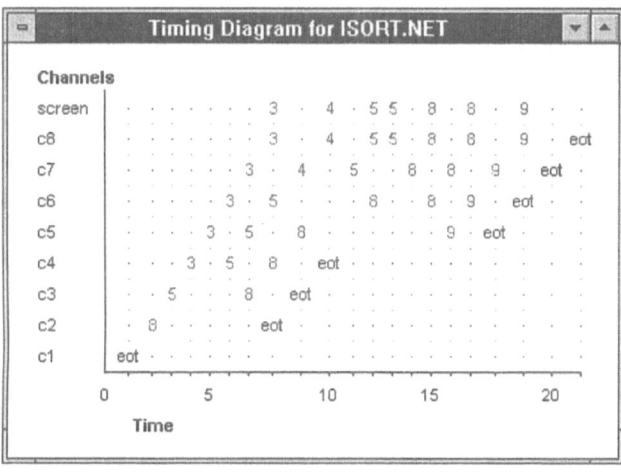

Fig. 5. Timing diagram for the network

2.3 Animation of the Network

Having created the network and defined each process, we can now graphically animate the execution of the network by selecting the run button for continuous animation, or the **step** button to show the new state of each process in the network after one time step. Channels in each process are colour coded to reflect the state of the process as follows: red for unable to communicate, green for ready to communicate, and white when the process has successfully terminated. When the indicators of a channel between two processes are both green, the communication takes place at the next time step.

Fig. 4 clearly illustrates that on the next time step, communications can only happend on channels with green (light grey) colour on both ends of the links. These are, channels c_2, c_4, c_6, and c_8. Fig. 5 illustrates the timing diagram for animating the network. It contains a record of all communications on each channel in the network coupled with the appropriate time stamp.

The evolution of each individual process in the network can be dynamically monitored during the animation of the network. An indicator highlights the exact place in the code of the process which will be executed at the next time step. For example, the cursor in Fig. 6 indicates that the process $INSERT\_2$ is willing to output the value 8 at the next time step.

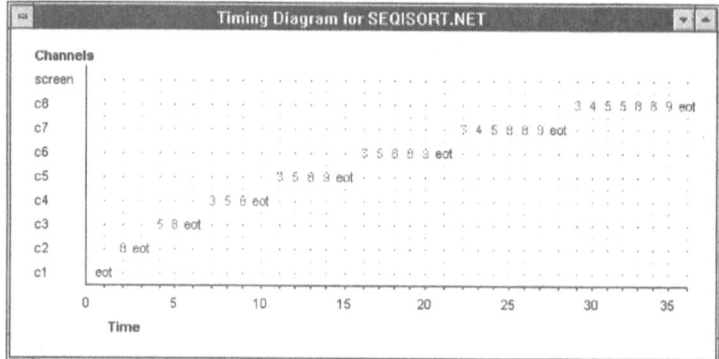

```
                          INSERT_2
  Process name     INSERT_2::INSERT
  Input channels   c2::left
  Output channels  c3::right
  Parameters       a

   μ X • left?x → (right!a → right!eot → SKIP          { x=eot }
                   right!x → X                         { x<a }
                   right!a → right!x → COPY[left,right]())
```

Fig. 6. Monitoring control within an individual process during network animation

```
                        INS in MAIN.LIB
  Process name     INS
  Input channels   left
  Output channels  right
  Parameters       a, s, t

   left?x  → ( Prd[right](s++[a]++t)            { x=eot }
               INS[left,right](a, s++[x], t)    { x<a }
               INS[left,right](a, s, t++[x])          )
```

Fig. 7. The process $INS(a, s, t)$.

```
                    Timing Diagram for SEQISORT.NET
  Channels
  screen  · · · · · · · · · · · · · · · · · · · · · · · · · · · · · · · · · ·
  c8      · · · · · · · · · · · · · · · · · · · · · · · · · · 3 4 5 5 8 8 9 eot
  c7      · · · · · · · · · · · · · · · · · · · 3 4 5 8 8 9 eot · · · · · · · ·
  c6      · · · · · · · · · · · · · · 3 5 8 8 9 eot · · · · · · · · · · · · · ·
  c5      · · · · · · · · · 3 5 8 9 eot · · · · · · · · · · · · · · · · · · · ·
  c4      · · · · · · 3 5 8 eot · · · · · · · · · · · · · · · · · · · · · · · ·
  c3      · · · 5 8 eot · · · · · · · · · · · · · · · · · · · · · · · · · · · ·
  c2      · 8 eot · · · · · · · · · · · · · · · · · · · · · · · · · · · · · · ·
  c1      eot · · · · · · · · · · · · · · · · · · · · · · · · · · · · · · · · ·

          0        5       10       15       20       25       30       35
          Time
```

Fig. 8. Timing diagram for the execution of the new network.

2.4 Alternative Designs

The process $INSERT(a)$ is just a valid implementation of the function $insert(a)$ which takes a sorted list of values and inserts the value a at the appropiate position so that the extended list is also sorted. Another process $INSERT'(a)$ which also correctly implements $insert(a)$ can be defined as depicted in Fig. 7. In this definition, the additional variable s (and t resp.) is used to accumulate the list of input values which are stricly less than a (and greater or equal to a respectively).

$$INSERT'(a) = INS(a, [], [])$$

The timed diagram of the new network is shown in Fig. 8. The behaviour of the network is completely sequential. Parallelism is only syntactic; each stage in the pipeline needs to completely terminates before the following stage starts. One of the strengths of *VisualNets* is that it allows the user to alter the network being investigated very quickly, so that variations can be tested and compared interactively. After each change the user can immediately run the network again, view the animation and the timing diagram as in Fig. 8. until a good design is reached.

3 Conclusions and Future Works

In this paper we have presented a graphical tool for the visualization, simulation, and animation of systems of communicating sequential processes. A brief overview of the functionality of the tool is described through the process of developing a distributed solution to a specific problem. Such a tool is of a great educational value in assisting the understanding of concurrency and in illustrating many distributed computing problems and the techniques underlying their solutions. Perhaps the most important aspect of the tool is the ability to visually alter a design, experiment with ideas for overcoming specific problems, and investigate, through animations, the potential consequences of certain design decisions. The tool proved very useful in detecting, through animation, undesirable behaviours such as bottlenecks deadlock [3]. However, currently *VisualNets* cannot deal with non-deterministic processes our underlying visualisation techniques are not easily scalable.

Work is presently in progress on a new version of the tool, rewritten in Sun Microsystems' Java language. The tool is platform-independant and will operate on UNIX or Windows-based systems. The new tool implements a considerably larger set of CSP operators, and can deal with networks that synchronise on events as well as on channel input or output. Internal parallelism within processes is supported, permitting a smaller network to be visualised as a single process and later zoomed to display the detail of the internal communications. The emphasis of the new project is on developing an advanced visualisation and animation tool and integrating it within an environment which allows higher level of abstractions and capabilities for systematic specification refinement and program transformation.

Acknowledgements

The work reported in this paper has evolved over several years and is the result of working very closely with several final year and postgraduate Computer Science students at the University of Reading, UK. In particular, I would like to thank Aiden Devine, Robin Singleton (for the implementation of the current tool) and Mark Green for developing the new Java implementation.

References

1. A. E. Abdallah, Derivation of Parallel Algorithms from Functional Specifications to *CSP* Processes, in: Bernhard Möller, ed., *Mathematics of Program Construction*, LNCS **947**, (Springer Verlag, 1995) 67-96.
2. A. E. Abdallah, Synthesis of Massively Pipelined Algorithms for List Manipulation, in L. Boug, P. Fraigniaud, A. Mignotte, and Y. Robert, eds, *Proceedings of the European Conference on Parallel Processing, EuroPar'96*, LNCS **1024**, (Springer Verlag, 1996), pp 911-920.
3. A. E. Abdallah, Visualization and Animation of Communicating Processes, in N. Namazi and K. Matthews, eds, *Proc. of IASTED Int. Conference on Signal and Image Processing, SIP-96*, Orlando, Florida, USA. (IASTED/ACTA Press, 1996), 357-362.
4. R. S. Bird, and P. Wadler, *Introduction to Functional Programming*, (Prentice-Hall, 1988).
5. Rajive L. Bagrodia and Chien-Chung Shen, MIDAS: Integrated Design and Simulation of Distributed Systems, *IEEE Transactions On Software Engineering*, **17** (10), 1993, pp. 1042-1058.
6. Daniel Y. Chao and David T. Wang, An Interactive Tool for Design, Simulation, Verification and Synthesis of Protocols, *Software and Experience*, **24** (8), 1994, pp. 747-783.
7. Jim Davies, *Specification and Proof in Real-Time* CSP, (Cambridge University Press, 1993).
8. H. Diab and H. Tabbara, Performance Factors in Parallel Programs, Submitted for publication, 1996.
9. Michael T. Heath and Jennifer A. Etheridge, Visualizing the Performance of Parallel Programs, *IEEE Software*, **8** (5), 1991, pp. 29-39.
10. Michael T. Heath, Visualization of Parallel and Distributed Systems, in A. Zomaya (ed), *Parallel and Distributed Computing Handbook*, (McGraw-Hill, 1996)
11. C. A. R. Hoare, *Communicating Sequential Processes*. (Prentice-Hall, 1985).
12. T. Ludwig, M. Oberhuber, and R. Wismller, An Open Monitoring System for Parallel and Distributed Programs, in L. Boug, P. Fraigniaud, A. Mignotte, and Y. Robert, eds, *Proceedings of the European Conference on Parallel Processing, EuroPar'96*, LNCS **1123** (Springer, 1996) 78-83.
13. Guido Wirtz, A Visual Approach for Developing, Understanding and Analyzing Parallel Programs, in E.P. Glinert, editor, Proc. *Int. Symp. on Visual Programming*, (IEEE CS Press, 1993) 261-266.

A Universal Infrastructure
for the Run-Time Monitoring
of Parallel and Distributed Applications*

Roland Wismüller, Jörg Trinitis, Thomas Ludwig

Technische Universität München (TUM), Informatik
Lehrstuhl für Rechnertechnik und Rechnerorganisation (LRR-TUM)
Arcisstr. 21, D-80333 München
email: {wismuell|trinitis|ludwig}@in.tum.de

Abstract. On-line tools for parallel and distributed programs require a facility to observe and possibly manipulate the programs' run-time behavior, a so called monitoring system. Currently, most tools use proprietary monitoring techniques that are incompatible to each other and usually apply only to specific target platforms. The On-line Monitoring Interface Specification (OMIS) is the first specification of a universal interface between different tools and a monitoring system, thus enabling interoperable, portable and uniform tool environments. The paper gives an introduction into the basic concepts of OMIS and presents the design and implementation of an OMIS compliant monitoring system (OCM).

1   Introduction

The development and maintenance of parallel programs is inherently more expensive than that of sequential programs. This is partly due to the lack of widely available, powerful tools supporting test and production phases of those programs, e.g. performance analyzers, debuggers, or load balancers. What these tools have in common is their need for a module that enables them to observe and possibly manipulate the parallel application, a so called *monitoring* module. Usually every tool includes a monitoring module that is specifically adapted to its needs. Since these modules typically use exclusive low level interfaces to the operating system and possibly the underlying hardware, these layers tend to be incompatible to each other. As a result, it is impossible to connect two or more tools to a running application at the same time. Even worse, if you want to use two different tools, one after the other, you often have to re-compile, re-link, or otherwise re-process the application.

In this paper we will describe the design and implementation of a universally usable on-line monitoring infrastructure that is general enough to support a wide variety of tools and their interoperability. We will first describe existing approaches, analyze the requirements for a general monitoring interface, then introduce our approach and finally describe our implementation.

* This work is partly funded by *Deutsche Forschungsgemeinschaft*, Special Research Grant SFB 342, Subproject A1.

2 State of the Art

Existing tools can be classified into interactive vs. automatic, observing vs. manipulating, and on-line vs. off-line tools. Taking a look at existing tools for parallel and distributed programming, one can state that:

- there are more off-line tools than on-line tools,
- most tools are only available on one or very few platforms,
- different on-line tools can typically not be applied concurrently,
- there are no generally available integrated tool environments.

There are several reasons for this. First, developing on-line monitoring systems is complex and expensive. Second, these systems are typically difficult to port from one platform to another. And third, they are usually specifically adapted and restricted to the tools that sit on top of them and require incompatible modifications to the application. Nevertheless, very powerful on-line tools have been developed. Two that are especially interesting concerning their monitoring parts will be mentioned here.

In the area of debugging tools, p2d2 by Cheng and Hood [6] is an important representative. Different from most earlier approaches, p2d2 clearly separates the platform specific on-line monitoring system from portable parts of the tool itself. To accomplish this, a client-server approach is used [7]. The p2d2 server contains all platform specific parts, whereas the client consists of the portable user interface. Client and server communicate via a specified protocol. Cheng and Hood encourage platform vendors to implement debugger servers that comply with the p2d2 protocol. Users could then use an existing p2d2 debugger client to debug applications on different machines.

Taking a look at performance analysis tools, the ambitious Paradyn project by Miller and Hollingsworth [10] clearly deserves a closer look. Paradyn utilizes the exceptional technique of dynamic instrumentation [5], where code snippets performing performance analysis tasks are dynamically inserted into and removed from the running process's code. The interface between the tool and the dynamic instrumentation library *dyninst* is clearly defined and published [3]. On the higher level, Paradyn uses the so called W^3-model to automatically determine performance bottlenecks [4, 10]. The W^3-model in particular profits from dyninst because intrusion can be kept low by removing unnecessary instrumentation.

The above tools are two of the very few that define an explicit interface between the monitoring system and the tool itself. By clearly defining this interface it becomes possible to separate the development of the tool from that of the on-line monitoring system. In addition, porting the tool becomes much easier, thus improving the availability of tools. Unfortunately, p2d2 and Paradyn both target only a single class of tools: p2d2 is limited to debugging, whereas Paradyn/dyninst is concentrated on performance analysis. It is impossible to build new tools or interoperable tools with either of these environments. OMIS, on the other hand, is designed with that primary goal in mind.

3 The OMIS Approach

One key idea in OMIS (*On-line Monitoring Interface Specification*) [9] is to define a standardized interface between the tools and the on-line monitoring system. This is comparable to p2d2 and Paradyn/dyninst. Our approach however is much more general as we do not concentrate on a single class of tools. OMIS supports a wide variety of on-line tools, including centralized, decentralized, automatic, interactive, observing, and manipulating tools (off-line tools can be implemented by a tiny on-line module writing trace files).

The tools we initially had in mind were those in THE TOOL-SET [12], developed at LRR-TUM: VISTOP (program flow visualization), PATOP (performance analysis) [2], DETOP (debugging) [11], CoCheck (checkpointing/process migration), LoMan (load balancing), Codex (controlled deterministic execution), and Viper (computational steering). But we do not only want to support classical stand-alone tools. OMIS is intended to also support interoperable tools (two or more tools applied at the same time), e.g. a performance analyzer covering and measuring the dynamic load balancing and process migration as well. Another example is a debugger interoperating with a checkpointing system and a deterministic execution unit, thus enabling the user to start debugging sessions from checkpoints of long lasting program runs. OMIS will also allow to build uniform tool environments. It will thus become possible to develop portable development environments for parallel and distributed programming.

In order to achieve all of the above, the monitoring system has to be programmable. OMIS supports this by specifying so called event/action-relations. An event/action-relation consists of a (possibly empty) event definition and a list of actions, where the event specification defines a condition for the actions to be invoked. Every time the event occurs, the actions are executed (in case of an empty event, they are executed immediately). The OMIS core already provides a rich set of events and actions. In addition to this, there are user defined events.

The main interface between the tool and the monitoring system consists of only a single function. To this function, requests from the tool are specified as event/action-relations in a machine independent string representation. In order to minimize intrusion, requests can be quickly disabled and re-enabled when e.g. a measurement is unnecessary for a while. When it is no longer needed, the whole request can be deleted from the monitoring system. Depending on how often the corresponding event occurs and what flags were specified, a single request can result in $0...n$ replies while it is active.

Every tool that connects to an OMIS compliant monitor is able to define its own view of the observed system. This is done by explicitly attaching to the nodes and processes of interest. Different tools can thus have different views, and are only informed about events taking place within their observed system. This is especially useful in non-exclusively used target environments.

While not being object oriented in the usual sense, objects form an important part of the interface. OMIS represents processes, threads, nodes, requests etc. as objects, which are identified by so called tokens. To improve scalability, all events and actions can operate on object *sets*. The objects are automatically localized within the observed (distributed) system and the requests are distributed according to the objects' locations (location transparency). OMIS also automatically converts object tokens where appropriate. For example, a process token can be converted into the token of the node where this process is currently located.

Although the above concepts are very powerful if wisely applied, there always remain situations where they do not suffice to support some specific environment. To overcome this, OMIS employs another concept: extendibility. Two types of extensions are defined: *Tool extensions* (e.g. for distributed tools) introduce new actions into the monitoring system, whereas *monitor extensions* can introduce new events, actions and objects. One monitor extension is, for example, the PVM extension. It introduces message buffers, barrier objects and services (events and actions) related to the PVM programming model.

A detailed description of the current version of OMIS can be found in [9] and on the OMIS home page in the WWW[1].

4 Design of an OMIS Compliant Monitoring System

Starting from the OMIS specification, we have done a first implementation of an OMIS compliant monitoring system, called OCM. In contrast to OMIS itself, which is not oriented towards specific programming libraries or hardware platforms, the OCM currently targets PVM applications on networks of UNIX workstations. However, we also plan to extend this implementation to MPI and NT workstations.

Goals of the OCM project. The implementation of the OCM serves three major goals. First, it will be used as an implementation platform for the ongoing tool developments at LRR-TUM, and also at other institutes. E.g. the MTA SZTAKI Research Institute of the Hungarian Academy of Science will use the OCM to monitor programs designed with the GRADE graphical programming environment[8]. [2] The second goal is to validate the concepts and interfaces specified by OMIS and to gain experience guiding further refinements of the specification. Finally, the OCM also serves as a reference for other implementors that want to develop an OMIS compliant monitoring system. Therefore, the OCM will be made available under the terms of the GNU license.

Global structure of OCM. Since the current target platform for the OCM is a distributed system where virtually all of the relevant information can only be retrieved locally on each node, the monitoring system must itself be distributed. Thus, we need one or more local monitoring components on each node. We decided to have one additional process per node, which is automatically started when a tool issues the first service request concerning that node. An alternative would be to completely integrate the monitoring system into the programming

---

[1] http://wwwbode.informatik.tu-muenchen.de/~omis

[2] Funded by the German/Hungarian intergovernmental project GRADE-OMIS, contract UNG-053-96.

libraries used by the monitored application, as it is done with many trace collectors. However, we need an independent process, since an OMIS compliant monitoring system must be able to autonomously react on events, independently of the application's state. In addition, the monitoring support offered by the UNIX operating system requires the use of a separate process.

A principal decision has been to design these monitor processes as independent local servers that do not need any knowledge on global states or the other monitor processes. Thus, each monitor process offers a server interface that is very similar to the OMIS interface, with the exception that it only accepts requests that can be handled completely locally.

However, since OMIS allows to issue requests that relate to multiple nodes, we need an additional component in the OCM. This component, called Node Distribution Unit (NDU), has to analyze each request issued by a tool and must split it into pieces that can be processed locally on the affected nodes. E.g. a tool may issue the request :thread_stop[p_1,p_2][3] in order to stop all threads in the processes identified by the tokens p_1 and p_2. In this case, the NDU must determine the node executing process p_1 and the node executing p_2. If these nodes are different, the request must be split into two separate requests, i.e. :thread_stop[p_1] and :thread_stop[p_2], which are sent to the proper nodes. In addition, the NDU must also assemble the partial answers it receives from the local monitor processes into a global reply sent to the tool.

In order to keep our first implementation of the OCM manageable, the NDU is currently implemented as a single central process. Thus, a typical scenario of the OCM looks as in Figure 1. The figure also shows that the NDU contains a request parser that transforms the requests specified by the tools into a preprocessed binary data structure that can be handled efficiently by all other parts of the monitoring system.

The communication between all of these components is handled by a small communication layer offering non-blocking send and interrupt-driven receive operations. We need interrupt-driven receives (or related techniques like Active Messages) to avoid blocking of the monitor processes. It would result in the inability to timely react on occurrences of other important events. A polling scheme using non-blocking receive operations is not feasible, because it consumes too much CPU time and thus leads to an unacceptable probe effect. In the current version of the OCM, the communication layer is based on PVM. However, we are working on an independent layer based on sockets, which is necessary for the MPI port. As is indicated in Figure 1, tools are linked against a library that implements the procedural interface defined by OMIS and transparently handles the connection to the NDU and the necessary message passing.

Structure of the local monitor. The local monitor process operates in an event driven fashion. This means that the process is normally blocked in a system call (either **wait** or **poll**, depending on whether we use **ptrace** or **/proc** for process

---

[3] The leading colon in the request is the separator between event definition and action list. Since the event definition is empty, the action must be executed immediately.

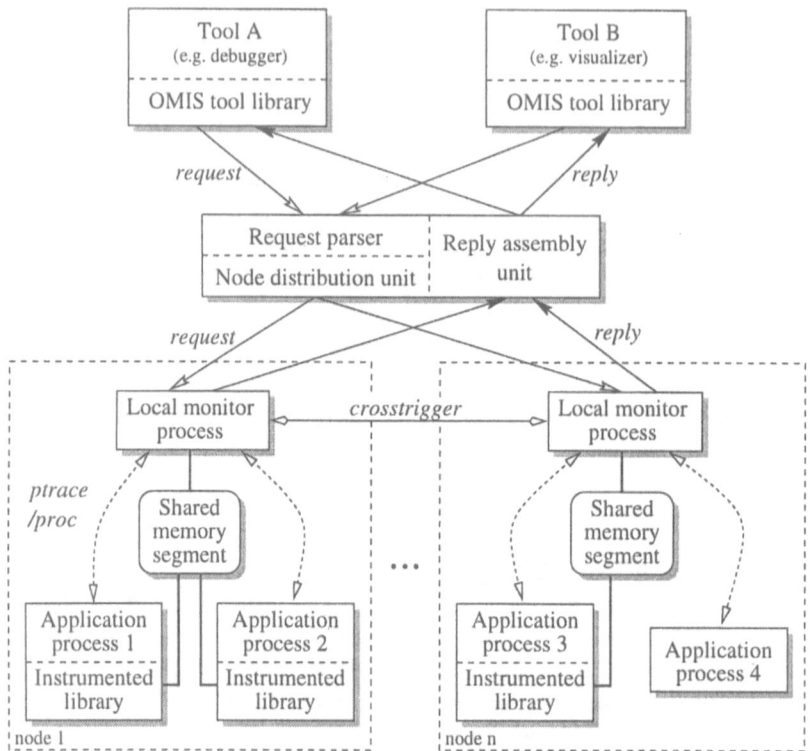

Fig. 1. Coarse Structure of the OCM

control), which returns when an event in one of the monitored processes has been detected by the operating system. In addition, the call is interrupted when a message arrives. After the monitor process has been unblocked, it analyzes the exact kind of event that has occurred and executes the list of actions that are associated with that event. If the event happens to be the arrival of a service request, the associated action causes the request to be received and analyzed. In the case of an unconditional request (i.e. a request with an empty event definition), it is executed and the results are sent back. When a conditional request is received, the monitor process dynamically inserts or activates the necessary instrumentation of the target system and stores the request's action list, so it can be executed later when the event occurs.

Unfortunately, this relatively easy execution scheme is insufficient due to efficiency constraints. For example, assume that a performance analysis tool wants to measure the time process p_1 spends in **pvm_send** calls. The tool could then issue the following requests:

```
thread_has_started_lib_call([p_1],"pvm_send"): pa_timer_start(ti_1)
thread_has_ended_lib_call([p_1],"pvm_send"): pa_timer_stop(ti_1)
```

Thus, a timer is started when **pvm_send** is called by p_1[4] and is stopped again when the call returns. If event detection and action execution are only performed in the monitor process, we introduce an overhead of at least four context switches for each library call, which renders performance analysis useless.

Thus, the OCM must be able both to detect certain events within the context of the triggering application process and to execute simple actions directly in that process, thereby removing the need for context switching. To achieve this, the monitored processes must be linked against instrumented libraries, i.e. extended versions of standard programming libraries that include additional parts of the monitoring system. In order to tell the monitoring parts in these libraries what they shall do, we use a shared memory segment that stores information on which events to monitor and which actions to execute. Note that actions in the application processes should not return results (except in the case of errors), since this would again result in context switches. Rather they update data stored in the shared memory segment (e.g. the timer value in the above example), which are read by actions executed in the monitor process upon request from a tool.

With this extension, the final structure as shown in Figure 1 results. Note that the OCM does not require that each monitored application process is linked with an instrumented library. However, if it is not, some events and actions may not be available for this process.

5   Project Status and Future Work

A first implementation of the OCM, as it is described in Section 4, has already been finished. It is currently supporting PVM 3.3.x and PVM 3.4 on networks of workstations running Solaris, Linux, and Irix. However, some of the services defined by OMIS still need to be implemented. In order to test the OCM, the TOOL-SET debugger has been adapted and is already operational.

Our current work includes the specification of an interface allowing extensions of OMIS compliant monitoring systems to be built independently of the individual implementation of these monitoring systems. This interface is supported by a stub generator that eases the programming of new events and actions.

Future work will focus on adapting the OCM to the MPI implementation **mpich**, and on porting it to the operating system Windows NT. In addition, we will use the OCM as the basis for providing interoperable tools within THE TOOL-SET project[12]. Finally, as a more long term goal, we will implement the OCM also for distributed shared memory systems, both based on hardware support (SCI coupled cluster of PCs using a hardware monitoring module [1]) and pure software implementations.

---

[4] More exactly: when **pvm_send** is called by *any thread in* process p_1. However, since PVM is not multi-threaded, there is only one thread in the process.

6 Acknowledgments

We would like to thank Manfred Geischeder, Michael Uemminghaus, and Hans-Günter Zeller for their participation in the OCM design. Klaus Aehlig, Alexander Fichtner, Carsten Isert, and Andreas Schmidt are still busy with implementing the OCM. We appreciate their support. Special thanks to Michael Oberhuber and Vaidy Sunderam who have been heavily involved in discussing OMIS, and to all other people that provided feedback.

References

1. G. Acher, H. Hellwagner, W. Karl, and M. Leberecht. A PCI-SCI Bridge for Building a PC-Cluster with Distributed Shared Memory. In *Proc. 6th Intl. Workshop on SCI-based High-Performance Low-Cost Computing*, pages 1–8, Santa Clara, CA, Sept. 1996.
2. O. Hansen and J. Krammer. A Scalable Performance Analysis Tool for PowerPC Based MPP Systems. In Mirenkov et al., editors, *The First Aizu Intl. Symp. on Parallel Algorithms / Architecture Synthesis*, pages 78–84, Aizu, Japan, Mar. 1995. IEEE Comp. Soc. Press.
3. J. K. Hollingsworth and B. Buck. *DynInstAPI Programmer's Guide Release 1.1*. Computer Science Dept., Univ. of Maryland, College Park, MD20742, May 1998.
4. J. K. Hollingsworth and B. P. Miller. Dynamic Control of Performance Monitoring on Large Scale Parallel Systems. In *Intl. Conf. on Supercomputing*, Tokio, July 1993.
5. J. K. Hollingsworth, B. P. Miller, and J. Cargille. Dynamic Program Instrumentation for Scalable Performance Tools. In *Scalable High-Performance Computing Conference*, pages 841–850, Knoxville, TN, May 1994.
6. R. Hood. The *p2d2* Project: Building a Portable Distributed Debugger. In *Proc. SPDT'96: SIGMETRICS Symposium on Parallel and Distributed Tools*, pages 127–136, Philadelphia, Pennsylvania, USA, May 1996. ACM Press.
7. R. Hood and D. Cheng. Accomodating Heterogeneity in a Debugger - A Client-Server Approach. In *Proc. 28th Annual Hawaii Intl. Conf. on System Sciences, Volume II*, pages 252–253. IEEE, Jan. 1995.
8. P. Kacsuk, J. Cunha, G. Dzsa, J. Lourenco, T. Antao, and T. Fadgyas. GRADE: A Graphical Development and Debugging Environment for Parallel Programs. *Parallel Computing*, 22(13):1747–1770, Feb. 1997.
9. T. Ludwig, R. Wismüller, V. Sunderam, and A. Bode. *OMIS — On-line Monitoring Interface Specification (Version 2.0)*, volume 9 of *LRR-TUM Research Report Series*. Shaker Verlag, Aachen, Germany, 1997. ISBN 3-8265-3035-7.
10. B. P. Miller, J. M. Cargille, R. B. Irvin, K. Kunchithap, M. D. Callaghan, J. K. Hollingsworth, K. L. Karavanic, and T. Newhall. The Paradyn parallel performance measurement tools. *IEEE Computer*, 11(28), Nov. 1995.
11. M. Oberhuber and R. Wismüller. DETOP - An Interactive Debugger for PowerPC Based Multicomputers. In P. Fritzson and L. Finmo, editors, *Parallel Programming and Applications*, pages 170–183. IOS Press, Amsterdam, May 1995.
12. R. Wismüller, T. Ludwig, A. Bode, R. Borgeest, S. Lamberts, M. Oberhuber, C. Röder, and G. Stellner. THE TOOL-SET Project: Towards an Integrated Tool Environment for Parallel Programming. In *Proc. 2nd Sino-German Workshop on Advanced Parallel Processing Technologies, APPT'97*, Koblenz, Germany, Sept. 1997.

Net-dbx: A Java Powered Tool for Interactive Debugging of MPI Programs Across the Internet

Neophytos Neophytou and Paraskevas Evripidou

Department of Computer Science
University of Cyprus
P.O. Box 537
CY-1678 Nicosia, Cyprus
skevos@turing.cs.ucy.ac.cy
Tel: +357-2-338705 (FAX 339062)

Abstract. This paper describes Net-dbx, a tool that utilizes Java and other WWW tools for the debugging of MPI programs from anywhere in the Internet. Net-dbx is a source level interactive debugger with the full power of gdb augmented with the debug functionality of LAM-MPI. The main effort was on a low overhead but yet powerful graphical interface that would be supported by low bandwidth connections. The portability of the tool is of great importance as well because it enables us to use it on heterogeneous nodes that participate in an MPI multicomputer. Both needs are satisfied a great deal by the use of Internet Browsing tools and the Java programming language. The user of our system simply points his browser to the URL of the Net-dbx page, logs in to the destination system, and starts debugging by interacting with the tool just like any GUI environment. The user has the ability to dynamically select which MPI-processes to view/debug. A working prototype has already been developed and tested successfully.

1 Introduction

This paper presents Net-dbx, a tool that uses WWW capabilities in general and Java applets in particular for portable parallel and distributed debugging across the Internet. Net-dbx is now a working prototype of a full fledged development and debugging tool. It has been tested for the debugging of both Fortran and C MPI programs.

Over the last 3-4 years we have seen an immense growth of the Internet and a very rapid development of the tools used for browsing it. The most common of form of data traveling in gigabytes all over the net is WWW pages. In addition to information formatted with text, graphics, sound and various gadgets, WWW enables and enhances a new way of accomplishing tasks: Teleworking.

Although, teleworking was introduced earlier, it has been well accepted and enhanced through the use of the web. A lot of tasks, mostly transactions, are done in the browser's screen, enabling us to order and buy things from thousands of miles away, edit a paper, move files along to machines in different parts of the world. Until recently, all these tasks were performed using CGI scripts [1] .

The CGI scripts are constrained in exchanging information only through text-boxes and radio buttons in a standard Web Form . The Java language [1][2] is now giving teleworking a new enhanced form. Java applets allow the user to manipulate graphics with his mouse and keyboard and send information, in real time, anywhere in the Internet. Graphical web applications are now able to use low bandwidth connections, since the interaction with the server does not demand huge amounts of X-Protocol [3] data. This even makes it possible to perform the task of remote program development and debugging from anywhere in the net.

Developing and debugging parallel programs has proven to be one of the most hazardous tasks of program development. When running a parallel program, many things can go wrong in many places (processing nodes). It can lead to deadlock situations when there are bugs in the message distribution (in distributed machines), or when access to the common memory is not well controlled. Some processing nodes may crash under some circumstances in which the programmer may or may not be responsible. To effectively debug parallel programs we should know what went wrong in each of these cases [4].

In order to effectively monitor the programs execution, there are two approaches: Post mortem and runtime debugging. In post mortem debugging, during execution, the compiled program, or the environment, keeps track of everything that happens and writes every event in special files called trace files. These trace files are then parsed using special tools to help the user guess when and what went wrong during execution of a buggy program. Runtime tools, on the other side, have access to the programs memory, on each machine, and using operating system calls they can stop or resume program execution on any execution node. These tools can change variables during the execution and break the program flow under certain conditions.

The prototype described in this paper is a runtime source-level debugging tool. It enhances the capabilities of the gdb debugger [5],[6] , and the monitoring facilities of LAM [7][8] into a graphical environment. Using Java, we developed a tool that integrates these two programs in a collection of applets in a single Web Page. Our tool acts as an interface to the two existing tools, which provides the user with a graphical interaction environment, and in addition, it optimizes interaction between LAM and gdb.

2 Architecture of Net-dbx

Net-dbx is a client-server package. The server side is the actual MPI-Network, that is the group of workstations of processors participating in an MPI-Multicomputer. It consists of tools that are to be installed locally on each Node and

---

[1] CGI stands for Common Gateway Interface. It consists of a server program, residing on the HTTP server, which responds to a certain form of input (usually entered in web forms) and produces an output Web page to present results to the user.

for individually debugging the processes running on that Node. These tools include the MPI runtime environment (we currently use the LAM implementation) and a source level runtime debugger (we currently use gdb). On the client side there should be a tool that integrates debugging of each Node and provides the user with a wider view of the MPI program which runs on the whole Multicomputer. The program residing in the client side is an applet written in Java. It is used to integrate the capabilities of the tools which rely on the server side. We can see Net-dbx as an environment consisting of two layers: the lower layer which is the MPI Network and the higher layer as the applet on the Web browser running on the user's workstation.

To achieve effective debugging of an MPI Parallel Program, one must follow the execution flow on every node of the multicomputer network. The status of messages and status of the MPI processes have to be monitored. Net-dbx is designed to relieve the user from the tedious procedure of manual parallel debugging. This procedure would involve him capturing PIDs as a parallel program is started, connecting to each node he wants to debug via telnet, and attaching the running processes to a debugger. There are also several synchronization issues that make this task even more difficult to achieve. In addition, Net-dbx offers a graphical environment as an integrated parallel debugging tool, and it offers means of controlling/monitoring processes individually or in groups.

2.1 Initialization Scheme

As mentioned above, to attach a process for debugging you need its PID. But that is not known until the process starts to run. In order to stall the processes, until they are all captured by the debugger, a small piece of code has to be added in the source code of the program. An endless loop, depending on a variable and two MPI_Barriers that can hold the processes waiting until they are captured. In a Fortran program the synchronization code looks like the following:

```
if ( myid .eq. 0 ) then
        dummydebug=1
        dowhile ( dummydebug .eq. 1 )
        enddo
    endif
call MPI_BARRIER( MPI_COMM_WORLD, ierr )
call MPI_BARRIER( MPI_COMM_WORLD, ierr )
```

The variable myid represents the process' Rank. As soon as all the processes are attached to the debugger, we can proceed to setting a breakpoint on the second MPI_BARRIER line and then setting the dummydebug variable on the root process to 0 so that it will get out of the loop and allow the MPI_Barriers to unlock the rest of the processes. After that the processes are ready to be debugged. The dummy code is to be added by the tool using standard searching techniques to find the main file and then compile the code transparently to the user.

2.2   Telnet Sessions to Processing Nodes

Telnet Sessions are the basic means of communication between the client and the server of the system. For every process to be debugged, the Java applet initiates a telnet connection to the corresponding node. It also has to initiate some extra telnet connections for message and process task monitoring. The I/O communication with this connection is done by exchanging strings, just like a terminal where the user sends input from the keyboard and expects output on the screen. In our case, the Telnet Session must be smart enough to recognize and extract information from the connections response. The fact that standard underlying tools are used guarantees same behavior for all the target platforms.

Standard functionality that is expected from the implementation of the Telnet Sessions role is to provide abstractions to:

- setting/unsetting breakpoints,
- starting/stopping/continuing execution of the program,
- setting up and reporting on variable watches,
- evaluating/changing value of expressions and
- providing the source code (to the graphical interface).
- Capturing the Unix PIDs of every active process on the network (for the telnet session used to start the program to be debugged).

In addition, the telnet Session needs to implement an abstraction for the synchronization procedure in the beginning. That is, if it is process 0 the session should wait for every process to attach and then proceed to release (get out of the loop) the stalled program. If it is a process ranked 1..n, then it should just attach to the debugger, set the breakpoints and wait. This can be achieved using a semaphore that will be kept by the object which will "own" all the telnet Sessions. In programming telnet session's role, standard Java procedures were imported and used, as well as third-party downloaded code, incorporated in the telnet part of our program. To the existing functionality we added interaction procedures with Unix shell, GNU Debugger (gdb), and with LAM-MPI so that the above abstractions were implemented.

One of the major security constraints posed in Java is the rule that applets can have an Internet connection (of any kind-telnet, FTP, HTTP, etc) only with their HTTP server host [9], [10]. This was overcome by having all telnet connections to the server and then rsh to the corresponding nodes. This approach assumes that the server can handle a large number of telnet connections and that the user will be able to login to the server and then rsh to the MPI network nodes.

Telnet Sessions need to run in their own thread of execution and thus need to be attached to another object which will create and provide them with that thread. A Telnet Session can be attached either to a graphical DebugWindow, or to a non-graphical global control wrapper. Both of these wrappers acquire control to the process under debugging using the abstractions offered by the telnet session.

2.3   Integration/Coordination

As mentioned above, several telnet sessions are needed in order to debug an MPI program in the framework that we use. All these sessions need a means of coordination. The coordinator should collect the starting data (which process runs on which processing node) from the initial connection and redistribute this information to the other sessions so that they can telnet to the right node and attach the right process to the debugger.

A large array holds all the telnet sessions. The size is determined by the number of processes that the program will run. Several pointers of the array will be null, as the user might not want to debug all the running processes. The other indices are pointing to the wrappers of the according telnet sessions that will be debugged. The wrapper of a telnet session can be either a graphical *DebugWindow*, described in the next subsection, or a non-graphical *TelnetSessionWrapper*, which only provides a thread of execution to the telnet session. Additionally it provides the methods required for the em coordinator to control the process together with a group of other processes. The capability of dynamically initiating a telnet session to a process not chosen from the beginning is a feature under development.

As the user chooses, using the user interface, which processes to visualize, and which are in the group controlled by the *coordinator*, the wrapper of the affected process will be changed from a *TelnetSessionWrapper* to a *DebugWindow* and vice-versa.

When LAM MPI starts the execution of a parallel program it reports which machines are the processing nodes, where each MPI process is executed and what is its Unix PID. The interpreted data is placed in an *AllConnectionData* object, which holds all the necessary startup information that every telnet session needs in order to be initialized. After acquiring all the data needed, the user can decide with the help of the graphical environment which of the processes need to be initiated for debugging.

After that, the coordinator creates *ConnectionData* objects for each of the processes to be started and triggers the start of the initialization process. It acts as a *semaphore holder* for the purposes of the initial synchronization.

Another duty of the coordinator is to keep a copy of each source file downloaded so that every source file is downloaded only once. All the source files have to be acquired using the list command on the debugger residing on the server side. These are kept as string arrays on the client side, since access to the user's local files is not allowed to a Java applet. A future optimization will be the use of FTP connections to acquire the needed files.

3   Graphical Environment

As mentioned in the introduction the user Interface is provided as an applet housed in a Web Page. Third-party packages [2] are specially formed to build a

---

[2] We modified and used the *TextView* component which was uploaded to the *gamelan Java directory* [11] by Ted Phelps at DTSC Australia. The release version will use

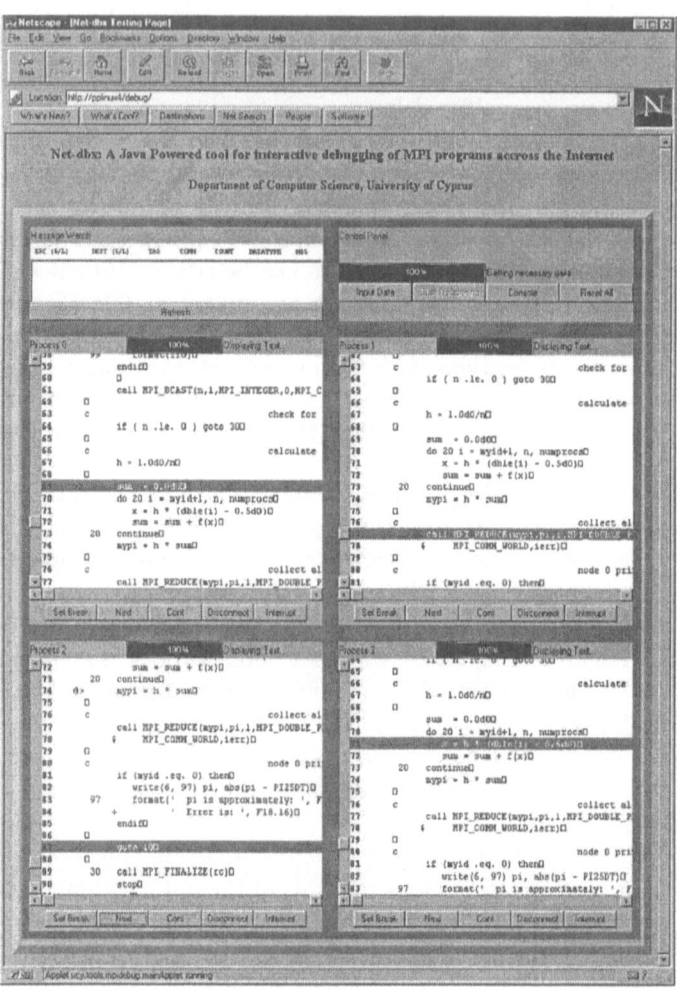

Fig. 1. A snapshot of the Net-dbx debugger in action.

fast and robust user interface. The graphical environment that is shown in the main browser screen (see figure 1) is consisted of:

- The *Process coordinator* window (control panel) which is the global control panel of the system.
- The *debugging windows*, where all the debugging work is done
- The *interaction console*, where the user can interact with the main console of the program to be debugged
- The *message window*, where all the pending messages are displayed.

The *Process coordinator window* provides a user Interface to the *coordinator* which is responsible for getting the login data from the user, starting the debugging procedure and coordinating the several telnet sessions. It provides means for entering all the necessary data to start a debugging session such as UserID, Password, Program Path which are given by the user. The user is also enabled to select which processes are going to be active, visualized, or controlled within a global process group.

The most important visual object used in this environment is the *Debug Window* in which the code of the program is shown. In addition to the code, as mentioned before, this object has to implement behaviors such as showing the current line of execution and by indicating which lines are set as breakpoints. These behaviors are shown using painted text (red color) for the breakpoints, painted text for the line numbers (blue color), and painted and highlighted text for the current line. The appropriate buttons to *set/unset breakpoints, next* and *continue* are also provided.

The *interaction console* is actually a typical telnet application. It is the user interface to a special telnet session used by the coordinator to initialize the debugging procedure. After initialization and capturing of all the needed data, the interaction console is left as a normal telnet window within which the user I/O takes place.

The *message view* is the applet responsible for showing the message status. It displays all the information that is given by the LAM environment when an mpimsg command is issued. All the pending messages are shown in a scrolling List object.

4   Related Work

A survey on all the existing graphical debugging tools would be beyond the scope of this paper. However we present the tools that appear to be most similar to Net-dbx. For an overview of most of the existing parallel tools, the interested reader can visit the Parallel Tools Consortium home page [12].

---

the *IFC Java widget set* as it appears to be one of the fastest and most robust *100% Compatibility certified* Java toolkits.

4.1 XMPI

XMPI [7] is a graphical environment for visualization and debugging, created by the makers of the LAM- MPI environment at the Ohio Supercomputer Center. It basically features graphical representation of the program's execution and message traffic with respect to what each process is doing. XMPI uses a special trace produced by the LAM-MPI environment during the execution. Although XMPI can be used at runtime it is actually a post-mortem tool and uses the LamTraces for every function that it provides. XMPI seems to be more of a visualization tool for displaying the program execution, but not controlling it. It is supported on DEC, HP, SGI, IBM and Sun platforms.

4.2 TotalView

This package is a commercial product of Dolphin Interconnect Solutions [13]. It is a runtime tool and supports source level debugging on MPI programs, targeted to the MPICH implementation by Argonne National Labs. Its capabilities are similar to those of dbx or gdb, which include changing variables during execution and setting conditional breakpoints. It is a multiprocess graphical environment, offering an individual display for each process, showing source code, execution stack etc. TotalView is so far implemented on Sun4 with SunOS 4.1.x platforms, Digital Alpha with Digital Unix system and IBM RS/6000.

4.3 p2d2

Another ongoing project is the p2d2 Debugger [14] being developed at the NASA Ames Research Center. It is a parallel debugger based on the client/server paradigm. It is based on an Object Oriented framework that uses debugging abstractions for use in graphical environments. It relies on the existence of a debugging server on each of the processing nodes. The debugging server provides abstractions, according to a predefined framework and can be attached to the graphical environment at runtime. Following this architecture the tool achieves portability and heterogeneity on the client side, whereas it depends on the implementation of the server on the server side.

5 Concluding Remarks and Future Work

In this paper we have presented Net-dbx, a tool that utilizes standard WWW tools and Java for fully interactive parallel and distributed debugging. The working prototype we developed provides proof of concept that we will soon be able to apply teleworking for the development debugging and testing of parallel programs for very large machines. A task that, up to now, is mostly confined to the very few supercomputer centers worldwide.

Net-dbx provides runtime source level debugging on multiple processes and message monitoring capabilities. The prototype provides a graphical user interface at minimal overhead. It can be run from anywhere in the Internet even using a low bandwidth dialup connection (33KBps). Most importantly, Net-dbx is superior over similar products in compatibility and heterogeneity issues. Being a Java applet, it can run with no modifications on virtually any console, requiring only the presence of a Java enabled WWW browser on the client side.

The tool at it's present implementation is being utilized in the Parallel Processing class at the University of Cyprus. After this alpha testing, it will be ready for beta testing on other sites as well. We are currently working on extending the prototype presented, to an integrated MPI-aware environment for program development, debugging and execution. For more information of the Net-dbx debugger's current implementation state, one can visit the Net-dbx home page [15].

References

1. Laura Lemay and Charlse L. Perkins, Teach yourself Java in 21 days, Sams.net Publishing, 1996
2. The JavaSoft home page, http://java.sun.com, The main java page at Sun Corporation. This address is widely known as the JAVA HOME
3. X Window System, The Open Group, http://www.opengroup.org/tech/desktop/x/, The Open Group's information page on the X-Window System.
4. Charles E. McDowell, David P. Helmbold, Debugging concurrent programs, ACM Computing Surveys Vol. 21, No. 4 (Dec. 1989), Pages 593-622
5. Richard M. Stallman and Roland H. Pesch, Debugging with GDB, The GNU Source-Level Debugger, Edition 4.09, for GDB version 4.9
6. GNU Debugger information, http://www.physik.fu-berlin.de/edv_docu/documentation/gdb/index.html,
7. LAM / MPI Parallel Computing, Ohio Supercomputer Center, http://www.osc.edu/lam.html, The info and documentation page for LAM. It also includes all the documentation on XMPI
8. Ohio Supercomputer Center, The Ohio State University, MPI Primer Development With LAM
9. J. Steven Fritzinger, Marianne Mueller, Java Security white Paper, Sun Microsystems Inc., 1996
10. Joseph A. Bank, Java Security, http://www-swiss.ai.mit.edu/~jbank/javapapaer/javapaper.html
11. The GameLan home page, http://www.gamelan.com. Here resides the most complete and well known java code collection in the Internet
12. The Parallel Tools Consortium home page, http://www.ptools.org/
13. Dolphin Toolworks, Introduction to the TotalView Debugger, http://www.dolphinics.com/TINDEX.html
14. Doreen Cheng and Robert Hood, A Portable Debugger for Parallel and Distributed Programs, Supercomputing '94
15. The Net-dbx home page, http://www.cs.ucy.ac.cy/~net-dbx/

Workshop 2+8
Performance Evaluation and Prediction

Allen D. Malony and Rajeev Alur

Co-chairmen

Measurement and Benchmarking

The core of any performance evaluation strategy lies in the methodologies used for measuring performance characteristics and comparing performance across systems. The first paper in this session describes an efficient performance measurement system created for workstation clusers that can report runtime information to application level tools. A hardware monitor approach is argued for in the second paper for measuring the memory access performance of a parallel NUMA system. The third paper demonstrates analysis methods for testing self-similarity of workload characterizations of workstations, servers, and the world-wide web. It is hard enough deciding performance metrics for parallel application benchmarking, but making the methodology portable is a real challenge. This is the subject of the fourth paper.

Modeling and Simulation

Parallel systems pose unique challenges for constructing analytical models and simulation studies because of the inter-relationship of performance factors. The first paper deals with the problem of building probabilistic models for CPU's with multilevel memory hierarchies when irregular access patterns are applied. The second paper studies the validity of the h-relation hypothesis of the BSP model and proposes an extension to the model based on an analysis of message passing programs. Simulating large-scale parallel programs for predictive purposes requires close attention to the choice of program factors and the effect of these factors on architectural analysis. This is shown effectively in the third paper for predicting data paralel performance.

Communications Modeling and Evaluation

Scalable parallel systems rely on high-performance communication subsystems and it is important to relate their characteristics to communication operations in software. The first paper assess LogP parameters for the IBM SP machine as applied to the MPI library and distinguished between sender and receiver operations. In contrast, pre-evaluating communications performance at the level of a data parallel language, independent of machine architecture, relies instead on the aspects of data distribution and communication volume and pattern. The second paper pre-evaluates communication for HPF. Targeting both a communcations model and its application to program development, the third paper applies genetic programming to the problem of finding efficient and accurate runtime functions for the performance of parallel algorithms on parallel machines.

Real-Time Systems

The real-time systems session is dedicated to the design, implementation, and verification techniques for computerized systems that obey real-time constraints. The subject comprises programming languages for real-time systems, associated compiling or synthesis technology, real-time operating systems and schedulers, and verification of real-time constraints. The first paper addresses the issue of language support for design of reliable real-time systems, proposing a methodology based on constraint programming. The second paper presents theoretical results concerning scheduling of periodic processes. In particular, the authors present sufficient conditions, weaker than previously known, for guaranteed schedulability so as to meet given age constraints. The third paper describes how the authors formulated and solved a control problem in the design of an industrial grinding plant to optimize economic criteria using parallel processin.

Configurable Load Measurement in Heterogeneous Workstation Clusters

Christian Röder, Thomas Ludwig, Arndt Bode

LRR-TUM — Lehrstuhl für Rechnertechnik und Rechnerorganisation
Technische Universität München, Institut für Informatik, D–80290 München
{roeder|ludwig|bode}@informatik.tu-muenchen.de

Abstract. This paper presents the design and implementation of NSR–
the Node Status Reporter. The NSR provides a standard mechanism
for measurement and access to status information in clusters of heterogeneous workstations. It can be used by any application that relies on
static and dynamic information about this execution environment. A
key feature of NSR is its flexibility with respect to the measurement requirements of various applications. Configurability aims at reducing the
measurement overhead and the influence on the observed cluster.

1 Motivation

During the last years, clusters of time-shared heterogeneous workstations became a popular execution platform for parallel applications. Parallel programming environments like PVM [5] and MPI [13] have been developed to implement
message passing parallel applications. Late development stages and efficient production runs of parallel applications typically require sophisticated on-line tools,
e.g. performance analysis tools, program visualization tools, or load management systems. A major drawback of existing tools like e.g. [1,7,9,12] is that all
of them implement their own monitoring technique to observe and/or manipulate the execution of a parallel application. Proprietary monitoring solutions
have to be redesigned and newly implemented when shifting to a new hardware
platform and/or a new programming environment.

In 1995, researchers at LRR-TUM and Emory University started the OMIS[1]
project. It aims at defining a standard interface between tools and monitoring
systems for parallel and distributed systems. Using an OMIS compliant monitoring system, tools can efficiently observe and manipulate the execution of a message passing based parallel application. Meanwhile, OMIS has been redesigned
and is available in version 2 [8]. At the same time, the LoMan project [11]
has been started at LRR-TUM. It aims at developing a cost-sensitive system-
integrated load management mechanism for parallel applications that execute
on clusters of heterogeneous workstations. Combining the goals of these two
projects, we designed the NSR [6] that is a separate module to observe the
status of this execution environment.

---

[1] OMIS: On-line Monitoring Interface Specification.

The goal of NSR is to provide a uniform interface for *data measurement* and *data access* in clusters of heterogeneous workstations. Heterogeneity covers two aspects: different performance indices of the execution nodes and different data representations due to different architectures. Although primarily influenced by the measurement requirements of on-line tools, the NSR serves as a powerful and flexible measurement component for any type of application that requires static and/or dynamic information about workstations in a cluster. However, the NSR does not post-process and interpret status information because the semantics of measured values is known by measurement applications only. In the following, we will use *measurement application* to denote any application that requires status information.

The document is organized as follows. The next section presents an overview of related work. In §3 we propose the reference model of NSR and discuss the steps towards its design. Implementation aspects are presented in §4. The concept of NSR is evaluated in §5. Finally, we summarize in §6 and discuss future development steps.

2 Related Work

Available results taken into consideration when designing and implementing the NSR can be grouped into three different categories:

- Standard commands and public domain tools (e.g. `top`[2] and `sysinfo`[3]) to retrieve status information in UNIX environments.
- Load modeling and load measurement by load management systems for parallel applications that execute in clusters of heterogeneous workstations (e.g. load monitoring processes [4, 7]; classification of load models [10]).
- The definition of standard interfaces to separately implement components of tools for parallel systems (e.g. OMIS [8], UMA[4] [14], SNMP[5] [2], and PSCHED [3]).

Typically, developers implement measurement routines for their own purposes on dedicated architectures or filter status information from standard commands or public domain tools. An overview on commands and existing tools is presented in [6]. The implementation of public domain tools is typically divided into three parts. A textual display presents the front-end of the tools. A controlling component connects the front-end with measurement routines that are connected by a tool-defined interface. Hence, porting a tool to a new architecture is limited to implement a new measurement routine. However, most tools only provide status information from the local workstation. Routines for transferring measured data have to be implemented if information is also needed from remote workstations.

---

[2] `top`: available at ftp://ftp.groupsys.com/pub/top.

[3] `sysinfo`: available at ftp://ftp.usc.edu/pub/sysinfo.

[4] UMA: Universal Measurement Architecture.

[5] SNMP: Simple Network Management Protocol.

The second category taken into considerations covers load modeling requirements of load management systems. A load management system can be modeled as a control loop that tries to manage the load distribution in a parallel system. Developers of load management techniques often adopt measurement routines to retrieve load values from the above mentioned public domain tools. However, it is still not yet clear which status information is best suited for load modeling in heterogeneous environments. A flexible load measurement component can serve as a basis to implement an adaptive load model [10]. Furthermore, different combinations of status information could be tested to find the values that can adequately represent the load of heterogeneous workstations.

Standard interfaces are defined to allow separate development of tool components. Tool components can subsequently cooperate on the basis of well-defined communication protocols. The exchange of one component by another component (even from different developers) is simplified if both are designed taking into account the definitions of such interfaces. We already mentioned the OMIS project. In § 4.2 we will see that OMIS serves as the basis to define the status information provided by NSR.

The UMA project addresses data collection and data representation issues for monitoring in heterogeneous environments. Status information is collected independently of vendor defined operating system implementations. The design of UMA includes data collection from various sources like e.g. operating systems, user processes, or even databases with historical profiling data. Although most measurement problems are addressed, the resulting complexity of UMA might currently hamper the development and widespread usage of UMA compliant measurement systems. Furthermore, the concept does not strictly distinguish between data collection and subsequent data processing.

The PSCHED project aims at defining standard interfaces between various components of a resource management system. The major goal is to separate resource management functionality, message passing functionality of parallel programming environments, and scheduling techniques. Monitoring techniques are not considered by PSCHED at all. The PSCHED API is subdivided into two parts: 1) a task management interface defines a standard mechanism for any message passing library to communicate with any resource management system; 2) a parallel scheduler interface defines a standard mechanism for a scheduler to manage execution nodes of a computing system regardless of the resource manager actually used. By using a configurable measurement tool, the implementation of a general-purpose scheduler can be simplified.

3 The Design of NSR

The design of a flexible measurement component has to consider the measurement requirements of various applications. A measurement component serves as an *auxiliary module* in the context of an application's global task. Hence, the NSR is designed as a *reactive system*, i.e., it performs activities on behalf of the supported applications. According to the main goal of NSR to provide a uniform interface for *data measurement* and *data access*, we consider three questions:

196

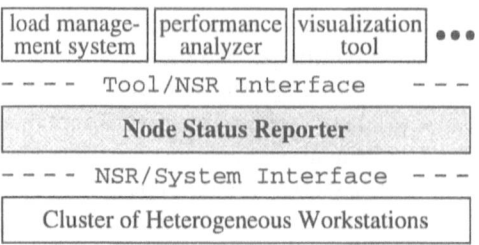

Fig. 1. The Reference Model of NSR.

1. *Which data is needed?* Different applications need different types of status information. Status information can be classified into *static* and *dynamic* depending on the time-dependent variability of its values. Static and dynamic information is provided for all major resource classes, i.e., operating system, CPU, main memory, secondary storage, and communication interfaces. Interactive usage of workstations in a time-shared environment gains increasing importance when parallel applications should be executed. Hence, the number of users and their workstation utilization are measured as well.

2. *When is the data needed?* On-line tools like e.g. load management systems, often have a cyclic execution behavior and measurements are *periodically* initiated. In this case, the period time between two successive measurements depends on the intent of measurement applications. However, measurement applications can have other execution behaviors as well. Now, status information can be measured *immediately and exactly once* as soon as it is needed.

3. *From which set of nodes is the data needed?* On-line tools can be classified regarding their implementation as distributed or centralized. In case of a central tool component (e.g. the graphical user interface of a program visualization tool), status information of multiple workstations is typically needed at once. Status information of any workstation is accessible on any other workstation.

Different applications might need the same type of status information, although their global tasks are different. For example, a load management system can consider I/O statistics of parallel applications and maps application processes to workstations with high I/O bandwidth. The same status information can be used by a visualization tool to observe the I/O usage statistics in a workstation cluster. The following section presents the reference model of NSR before we list the design goals in detail.

3.1 The Reference Model

The reference model of NSR is illustrated in Fig. 1. The NSR builds a layer between measurement applications and a cluster of heterogeneous workstations. It supports on-line tools for parallel applications like e.g. load management systems, performance analysis tools, or visualization tools. However, any type of measurement application can utilize the NSR. Applications communicate with the NSR via the `Tool/NSR Interface`. Using this interface, they can manage and control the behavior of NSR. Recall that the NSR is designed as a reactive system on behalf of the measurement applications. The applications are responsible to configure the NSR with respect to their measurement requirements, i.e., the type of status information and the time at which measurements have to be initiated. The interface provides uniform access to information of any workstation that is observed by NSR. If necessary, the NSR distributes measured values between the workstations with respect to the applications' location.

The `NSR/System Interface` connects the NSR to a cluster of heterogeneous workstations and hides the details of workstation heterogeneity. The main purpose of this interface is to connect the NSR and measurement routines for various architectures. Furthermore, the interface provides a uniform data representation with unity scale. Developers are free to implement their own data measurement and data collection routines.

3.2 Design Goals

The following major objectives have to be pursued during the design of NSR with respect to the above listed measurement requirements:

Portability: The target architecture of NSR is a cluster of heterogeneous workstations. The same set of status information has to be accessible in the same way on every architecture for which an implementation exists. New architectures should be easily integrated with existing implementations.

Uniformity: For each target architecture, the measured values have to be unified with respect to their units of measure (i.e., bytes, bytes/second, etc.). Hence, status information becomes comparable also for heterogeneous workstations.

Scalability: Tools for parallel system typically need status information from a set of workstations. An increasing number of observed workstations must not slow down these applications.

Flexibility: The NSR provides a wide variety of status information. Measurement applications can determine which subset of information has to be measured on which workstations. The frequency of measurements has to be freely configurable.

Efficiency: The NSR can simultaneously serve as a measurement component for several applications. If possible, concurrent measurements of the same data should be reduced to a single measurement.

3.3 Design Aspects

According to the goals and the requirements we consider the following aspects:

Data measurement: The measurement of a workstation's status information issues system calls to the operating system. It is one of the main tasks of the operating systems to update these values and to keep them in internal management tables. System calls are time consuming and slow down the calling application. Additionally, different architectures often require different measurement routines to access the same values. For portability reasons, data measurement is decoupled from an application's additional work and performed asynchronously to it. The set of data to be measured is freely configurable by a measurement application.

Synchronization: It is likely to happen in time-shared systems that several measurement applications execute concurrently, e.g. several on-line tools for several parallel applications execute on an overlapping set of workstations in the cluster. A severe measurement overhead on multiple observed workstations can result from an increasing frequency of periodically initiated measurements. Multiple measurements of the same status information are converted to a single measurement, thereby reducing the measurement overhead. However, synchronized data measurement demands for a mechanism to control the synchronization. We use interrupts to synchronize data measurements. In case of periodic measurements, the period of time between any two measurements is at least as long as the period of time between two interrupts. However, this limits the frequency of measurement initiations and can delay the measurement application. We implemented two concepts to reduce the delays. Firstly, if a measurement application immediately requires status information then measurements can be performed independently from the interrupt period. Secondly, we use asynchronous operations for data measurement and data access.

Asynchronous Operations: Asynchronous operations are used to hide latency and overlap data measurement and computation. An identifier is immediately returned for these operations. A measurement application can wait for the completion of an operation using this identifier. Measurements are carried out by NSR, while applications can continue their execution.

Data Distribution: Apart from data measurement, the status information is transparently transferred to the location of a measurement application. A data transfer mechanism is implemented within the NSR. Again, data transfer is executed asynchronously while applications can continue their execution.

4   Implementation Aspects

4.1   Client-Server Architecture

The architecture of NSR follows the client-server paradigm. The client functionality is implemented by the *NSR library*. A process that has linked the NSR library and invokes library functions is called *NSR client* (or simply *client*). An NSR client can request service from the *NSR server* (*server*) by using these library functions. The NSR server can be seen as a global administration instance that performs data measurement and data distribution on behalf of the clients. It is implemented by a *set of replicated NSR server processes* that cooperate to serve the client requests. The server processes are intended to execute in the background as other UNIX processes. To clarify our notation we call them *NSR daemon processes* (or simply *daemons*). A single NSR daemon process resides on each workstation in the cluster for which an implementation exists.

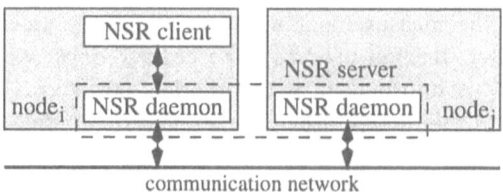

Fig. 2. The Client-Server Architecture of NSR.

An exemplary scenario of this architecture is illustrated in Fig. 2. One NSR daemon process resides on workstation node$_i$, while a second daemon resides on workstation node$_j$. The client on node$_i$ can only communicate with the local NSR daemon process. Applications can not utilize the NSR server if a local NSR daemon process is not available. The **Tool/NSR Interface** is implemented on top of this interconnection.

The local NSR daemon process controls the execution of a client request. From the viewpoint of a client, its local NSR daemon process is called *NSR master daemon*. If necessary, the NSR master daemon forwards a request to all daemons that participate in serving a client's request. The latter daemons are also called *NSR slave daemon processes*. A daemon is identified by the name of the workstation on which it resides. Note, the attributes *master* and *slave* are used to distinguish the roles each daemon plays to serve a client request. However, the attributes are not reflected in the implementation. An NSR daemon process can be both master and slave because several clients can request service from the NSR server concurrently. The main difference between master and slave daemons is that an NSR master daemon keeps additional information about a requesting client. Slave daemons do not know anything about the client that performed the request. The **NSR/System Interface** connects the workstation specific measurement routines with the NSR daemon processes. Daemons exchange messages via the communication network.

All information needed by the NSR server to serve the client requests are stored by the NSR daemon processes. Contrarily, a client is stateless with respect to data measurement. The result of a previously initiated library call does not influence the execution of future library calls. Hence, no information has to be stored between any two calls.

4.2 Status Information

The data structure **nsr_info** is the container for the status information of workstations and is communicated between the NSR components.

```
struct nsr_info { long            mtag, tick;
                  static_hostinfo       stthi;
                  dynamic_hostinfo      dynhi; };
```

The first entry **mtag** is a sequence number for the status information. Applications can identify old and new values by comparing their **mtag** values. The second entry **tick** is a timestamp for the information. It indicates the interrupt number at which the measurement was carried out. In case of periodic measurements, a master daemon uses `tick` to control data assignment if several clients request status information with different periods of time. Information about resources and interactive usage are split into two data structures. Static information is stored in `static_hostinfo` and describes a workstations e.g. in terms of type of operating system, number of CPUs, size of main memory, etc. Benchmark routines are used to determine performance indices of workstation resources. Dynamic information is stored in `dynamic_hostinfo` and represents the utilization values of workstation resources. For detailed information, we refer to the information service **node_get_info** in the OMIS document[6] [8, pp. 69 ff.]. The definition of this service matches the definition of these two data structures.

4.3 NSR Library Functions

The NSR library functions are split into two major groups. The first group covers *administrative functions* to manage the execution behavior of the NSR server. The second group includes *data access routines* to retrieve status information that has been measured by the daemons. Additional functions primarily ease the usage of the NSR.

Administrative functions are needed to configure the NSR server. They are available as *blocking* or *non-blocking* calls. The following example illustrates the difference between these two types. Initially, a client attaches itself to a set of NSR daemon processes with `nsr_iattach()`, that is a non-blocking call. The local daemon analyzes the parameters of the request, performs authentication checks for the client, and prepares one entry of its internal client management table. From the client request, the names of workstations are extracted from which status information should be retrieved. Subsequently, the master daemon forwards the request to all requested daemons which themselves store the location of the master daemon. As a result, the master daemon returns the identifier `waitID` for the request. The client can continue with its work. Later on, the client invokes `nsr_wait()` with `waitID` to finish the non-blocking call. This call checks whether all requested daemons participate to serve the client's request. The return value to the client is the identifier `nsrID`, that has to be passed with every subsequent library call. The master daemon uses `nsrID` to identify the requesting client. In any case, a non-blocking call has to be finished by `nsr_wait()` before a new call can be invoked by a client. A blocking library call is implemented by its respective non-blocking counterpart immediately followed by `nsr_wait()`.

Three additional administrative functions are provided by the NSR library. Firstly, a client can program each daemon by sending a measurement configuration message. A measurement flag determines the set of status information that should be measured by each daemon. In case of periodic measurements, a time value determines the period of time between any two measurements. A new configuration message reconfigures the execution behavior of all attached daemons. Secondly, the set of observed workstations can be dynamically changed during

---

[6] http://wwwbode.informatik.tu-muenchen.de/~omis

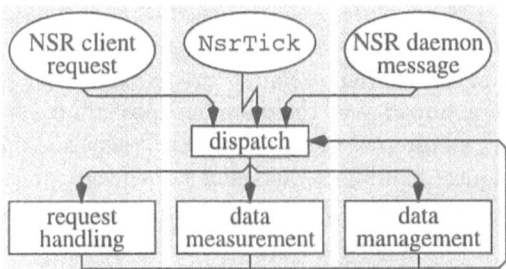

Fig. 3. Functionality of an NSR Daemon

the client's execution time. Finally, a client detaches itself from the NSR server as soon as it does not need service any more.

The second group of NSR library functions provides access to the measured data. A request for immediate measurement blocks the caller until the status information of the requested workstation is available. In case of periodic measurements, the local daemon stores the status information of every node into a local memory segment. A library call reads the memory location to access the information. A client can read the status information of all observed workstations at once. Additionally, it can request the status information of one specific workstation. The NSR master daemon stores status information independently from the execution behavior of requesting clients. Again, it is the client's responsibility to periodically read the data.

4.4 NSR Daemon Functionality

At startup time, an NSR daemon process measures a default set of status data and multicasts the values to all daemons in the cluster. During this phase, it initializes an internal timer that periodically interrupts the execution of the NSR daemon process at fixed time intervals. The period of time between two successive interrupts is called NsrTick. Any time interval used by the NSR implementation is a multiplier of this basic NsrTick. As shown in Fig. 3, three main tasks can be executed by a single daemon: *client request handling*, *data measurement*, and *data management*.

A central dispatching routine determines the control flow of the NSR daemon processes. It invokes one of the tasks depending on one of the following asynchronous events: 1) the expiration of NsrTick, 2) the arrival of an NSR client request, or 3) the arrival of another daemon message.

In the first case, a default set of status information is updated every $n \times$ NsrTick with a predefined number n as long as no client request has to be handled. This concept of *alive-messages* ensures that all daemons know about all other daemons at any time. Note, a daemon tries to propagate a *breakdown-message* in case of a failure. If client requests indicate periodic measurements then NsrTick is used to check for the expiration of a requested period of time. If necessary, the daemon measures exactly that status information requested at this point of time. The set of data to be measured is calculated as a superset of the status information currently requested by all clients.

The arrival of an NSR client request causes the dispatcher to invoke the request handling functionality of the daemon. Depending on the request type, the daemon updates an internal management table that keeps information about its local clients. Request handling also includes the forwarding of a client request to all daemons that participate to serve the client request. A result is sent to the client in response to each request.

A daemon can receive different types of messages from other daemons, i.e., alive-messages, breakdown-messages, reply-messages and update-messages. Reply-messages are sent by an NSR slave daemon in response to a forwarded client request. The result of a reply is stored in the client management table. Later on, the results of all slave daemons are checked if the daemon handles the previously mentioned nsr_wait() call. Finally, the update-messages indicate that a daemon sends updated status information in case of periodic measurements. New status information of a workstation invalidates its old values. Therefore, the daemon overwrites old workstation entries in the memory segment.

Each daemon in the cluster can perform the same functionality. Typically, several daemons cooperate to serve a client's request. Without going into further details, we state that the NSR server can handle requests from multiple clients. On the one side, applications can be implemented as a parallel program and each process can be a client that observes a subset of workstations. On the other side, several applications from different users can utilize the NSR server concurrently. Additionally, limited error recovery is provided. If necessary, a daemon that recovers from a previous failure is automatically configured to participate in serving the client requests again.

4.5 Basic Communication Mechanisms

Two types of interaction are distinguished to implement the NSR concept: the *client-server interaction* and the *server integrated interaction*, i.e., the interaction between the NSR daemons. Their implementations are based on different communication protocols.

Client-Daemon Communication:: The communication between a client and its local NSR daemon process is performed *directly* or *indirectly*. Direct communication is used for the administrative functions and involves both the NSR client and the NSR master daemon simultaneously. A TCP socket connection is established over which a client sends its requests and receives the server replies. The asynchronous communication uses a two phase protocol based on direct communication (see description above).

Indirect interaction between a client and the local daemon is used to implement data exchange in case of periodic measurements. In analogy to a mailbox data exchange scheme, the daemon writes status information to a memory mapped file independently of the clients execution state. Similarly, any client

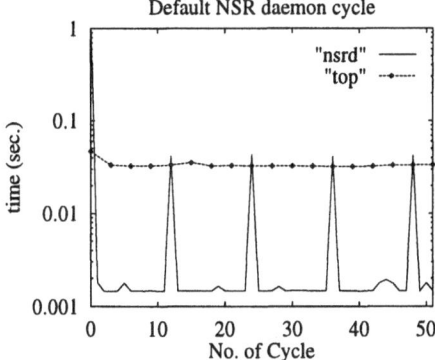

Fig. 4. Measurement duration of `top` and `nsrd`.

located on the workstation can read this data independently of the daemon's status.

Daemon-Daemon Communication: The NSR server implementation is based on the IP multicast concept. The main advantage of IP multicast over a UDP broadcast mechanism is that the dimension of a cluster is not limited to a LAN environment. A multicast environment consists of senders and receivers. Receiving processes are part of a multicast group. In our context, the NSR server is a multicast group consisting of its daemon processes. Communicated data are converted into XDR to cover the heterogeneity of different architectures. In response to a forwarded client request, an NSR slave daemon sends its reply to the master daemon via an UDP socket connection.

5 Evaluation

We will now evaluate the concept and implementation of NSR. As a point of reference, we compare the public domain software product `top` with the execution of `nsrd`, i.e., the NSR daemon process. Two experiments have been performed on a SPARC Station 10 running the SUN Solaris 2.5 operating system. Firstly, we compared the measurement overhead of `top` and `nsrd`. Secondly, we compared the average CPU utilization of both tools to determine their impact on the observed workstation.

A comparison between the measurement overhead of `top` and `nsrd` is shown in Fig. 4. It illustrates the cyclic execution behavior of `nsrd`. Both tools have been instrumented with timing routines to measure the duration of a single measurement.

The NSR daemon process measures its default set of values every $n \times$`NsrTick`, with `NsrTick` $= 5$ (seconds) and $n = 12$, while `top` updates its values every $m \times$ `NsrTick`, with $m = 3$. The set of measured information covers CPU and memory usage statistics. The set of values is approximately the same for both tools. For i mod $12 \neq 0$, the `NsrTick` just interrupts the execution of `nsrd` that subsequently performs some internal management operations. The time values of Fig. 4 indicate that the measurement routines of `top` and `nsrd` are similar with respect to their measurement overhead. The reason for this similarity is that the measurement technique is almost identically implemented in both tools.

However, the `nsrd` outperforms `top` in terms of CPU percentage that is a measure for the CPU utilization of a process. The CPU percentage is defined as the ratio of CPU time used recently to CPU time available in the same period. It can be used to determine the influence on the observed workstation. The following lines show a shortened textual output of `top` when both tools execute concurrently.

```
PID USERNAME ··· SIZE ···    CPU COMMAND
3317 roeder   ··· 1908K ··· 1.10% top
3302 root     ··· 2008K ··· 0.59% nsrd
```

The above snapshot was taken when both tools are executing on their regular level. It combines both the time each tool needs to measure its dynamic values and the time to further process these data. The `top` utility prints an list of processes that is ordered by decreasing CPU usage of the processes. The `nsrd` writes the data to the memory mapped file, converts it into the XDR format, and multicasts it to all other daemons. From this global point of view, `nsrd` outperforms `top` by a factor of 2. The overhead of `top` mainly results from its output functionality. In summary, the NSR introduces less load to the observed system than the `top` program.

6 Summary and Future Work

We presented the design and implementation of NSR that is a general-purpose measurement component to observe the status of a cluster of heterogeneous workstations. The NSR provides transparent access to static and dynamic information for any requesting measurement application. A key feature of NSR is its flexibility with respect to measurement requirements of various applications. If necessary, applications can control and observe different subsets of workstations independently. From the viewpoint of NSR, several applications can be served concurrently.

Currently, we implemented a prototype of NSR for SUN Solaris 2.5 and Hewlett Packards HP-UX 10.20 operating systems. In the future we will enhance the implementation of NSR. By that, measurement routines for various architectures like IRIX 6.2, AIX 3.x, and LINUX 2.0.x will be implemented. Additionally, we will optimize the performance of NSR to reduce the measurement overhead. Regarding performance evaluation, we will evaluate the NSR library calls and the data distribution inside the NSR server. In spring 1998, we will release the NSR software package under the GNU general public license. On some architectures, measurements are performed by reading kernel data structures. In this case, read permission to access the kernel data has to be provided for an NSR daemon process. Security aspects are not violated because the NSR does not manipulate any data. Measurement applications can be implemented without special permission flags. In production mode, a NSR daemon process can automatically start up during the boot sequence of a workstation's operating system.

The NSR already serves as the load measurement component within the LoMan project [11]. In the future, the NSR will be integrated into an OMIS compliant monitoring system to observe the execution nodes of a parallel application.

References

1. A. Beguelin. Xab: A Tool for Monitoring PVM Programs. In *Workshop on Heterogeneous Processing*, pp. 92–97, Los Alamitos, CA, Apr. 1993.
2. J.D. Case, M. Fedor, M.L. Schoffstall, and C. Davin. Simple Network Management Protocol (SNMP). Internet Activities Board – RFC 1157, May 1990.
3. D.G. Feitelson, L. Rudolph, U. Schweigelshohn, K.C. Sevcik, and P. Wong. Theory and Practice in Parallel Job Scheduling. In *IPPS'97 Workshop on Job Scheduling for Parallel Processing*, pp. 1–25, Geneva, Switzerland, Apr. 1997. University of Geneva.
4. K. Geihs and C. Gebauer. Load Monitor $\mathcal{LM}$ – Ein CORBA-basiertes Werkzeug zur Lastbestimmung in heterogenen verteilten Systemen. In *9. ITG/GI Fachtagung MMB'97*, pp. 1–17. VDE-Verlag, Sep. 1997.
5. G.A. Geist, J.A. Kohl, R.J. Manchek, and P.M. Papadopoulos. New Features of PVM 3.4 and Beyond. In *Parallel Virtual Machine — EuroPVM'95: Second European PVM User's Group Meeting, Lyon, France*, pp. 1–10, Paris, Sep. 1995. Hermes.
6. M. Hilbig. Design and Implementation of a Node Status Reporter in Networks of Heterogeneous Workstations. Fortgeschrittenenpraktikum at LRR-TUM, Munich, Dec. 1996.
7. D.J. Jackson and C.W. Humphres. A simple yet effective load balancing extension to the PVM software system. *Parallel Computing*, 22:1647–1660, Jun. 1997.
8. T. Ludwig, R. Wismüller, V. Sunderam, and A. Bode. OMIS — On-line Monitoring Interface Specification. TR. TUM-I9733, SFB-Bericht Nr. 342/22/97 A, Technische Universität München, Munich, Germany, Jul. 1997.
9. B.P. Miller, J.K. Hollingsworth, and M.D. Callaghan. The Paradyn Parallel Performance Tools and PVM. Tr., Department of Computer Sciences, University of Wisconsin, 1994.
10. C. Röder. Classification of Load Models. In T. Schnekenburger and G. Stellner (eds.), *Dynamic Load Distribution for Parallel Applications*, pp. 34–47. Teubner, Germany, 1997.
11. C. Röder and G. Stellner. Design of Load Management for Parallel Applications in Networks of Heterogeneous Workstations. TR. TUM-I9727, SFB 342/18/97 A, SFB 0342, Technische Universität München, 80290 München, Germany, May 1997.
12. C. Scheidler and L. Schäfers. TRAPPER: A Graphical Programming Environment for Industrial High-Performance Applications. In *Proc. PARLE'93, Parallel Architectures and Languages Europe*, vol. 694 of *LNCS*, pp. 403–413, Berlin, Jun. 1993. Springer.
13. D.W. Walker. The Design of a Standard Message Passing Interface for Distributed Memory Concurrent Computers. *Parallel Computing*, 20(4):657–673, Apr. 1994.
14. X/Open Group. *Systems Management: Universal Measurement Architecture Guide*, vol. G414. X/Open Company Limited, Reading, UK, Mar. 1995.

Exploiting Spatial and Temporal Locality of Accesses: A New Hardware-Based Monitoring Approach for DSM Systems

Robert Hockauf, Wolfgang Karl, Markus Leberecht, Michael Oberhuber, and Michael Wagner

Lehrstuhl fr Rechnertechnik und Rechnerorganisation (LRR-TUM)
Institut fr Informatik der Technischen Universitt Mnchen
Arcisstr. 21, D-80290 Mnchen, Germany
{karlw, leberech, oberhube}@informatik.tu-muenchen.de

Abstract. The performance of a parallel system with NUMA characteristics depends on the efficient use of local memory accesses. Programming and tool environments for such DSM systems should enable and exploit data locality.

In this paper we present an event-driven hybrid monitoring concept for the SMiLE SCI-based PC cluster. The central part of the hardware monitor consists of a content-addressable counter array managing a small working set of the most recently referenced memory regions. We show that this approach allows to provide detailed run-time information which can be exploited by performance evaluation and debugging tools.

1 Introduction

The SMiLE[1] project at LRR-TUM investigates high-performance cluster computing based on Scalable Coherent Interface SCI as interconnection technology [5]. Within this project, an SCI-based PC cluster with distributed shared memory (DSM) is being built. In order to be able to set up such a PC cluster we have developed a PCI-SCI adapter targeted to plug into the PCI bus of a PC. Pentium-II PCs equipped with these PCI-SCI adapters are connected to a cluster of computing nodes with NUMA characteristics (non-uniform memory access).

The SMiLE PC cluster as well as other SCI-based parallel systems available and accessible at LRR-TUM serve as platforms for the software developments within the SMiLE project, studying how to map parallel programming models efficiently onto the SCI hardware.

The performance of such a parallel system with DSM depends on the efficient use of local memory accesses through a parallel program. Although remote memory accesses via hardware-supported DSM deliver high communication performance, they are still an order of magnitude more expensive than local ones. Therefore, programming systems and tools for such parallel systems with NUMA

---

[1] SMiLE: Shared Memory in a LAN-like Environment

characteristics should enable and exploit data locality. However, these tools require detailed information about the dynamic behaviour of the running system.

Monitoring the dynamic behaviour of a compute cluster with hardware-supported DSM like the SMiLE PC cluster is very exacting because communication might occur implicitly on any read or write. This fact implies that monitoring must be very fine-grained too, making it almost impossible to avoid significiant probe overhead with software instrumentation.

In this paper we present a powerful and flexible event-driven hybrid monitoring system for the SMiLE SCI-based PC cluster with DSM. Our PCI-SCI adapter architecture allows the attachment of a hardware monitor which is able to deliver detailed information about the run-time and communication behaviour to tools for performance evaluation and debugging. In order to be able to record all interesting measurable entities with only limited hardware resources, the monitor exploits the spatial and temporal locality of data and instruction accesses in a similar way as cache memories in high-performance computer systems do.

Simulations of our monitoring approach helps us to make some initial design decisions. Additionally, we give an overview of the use of the hardware monitor in a hybrid performance evaluation system.

2 The SMiLE PC Cluster Architecture

The Scalable Coherent Interface SCI (IEEE Std 1596-1992) has been chosen as network fabric for the SMiLE PC cluster.

The SCI standard [6] specifies the hardware interconnect and protocols allowing to connect up to 64 K SCI nodes (processors, workstations, PCs, bus bridges, switches) in a high-speed network. A 64-Bit global address space across SCI nodes is defined as well as a set of read-, write-, and synchronization transactions enabling SCI-based systems to provide hardware-supported DSM with low-latency remote memory accesses. In addition to communication via DSM, SCI also facilitates fast message-passing. SCI nodes are interconnected via point-to-point links in ring-like arrangements or are attached to switches. The logical layer of the SCI specification defines packet-switched communication protocols. An SCI split transaction requires a request packet to be sent from one SCI node to another node with a response packet in reply to it. This enables every SCI node to overlap several transactions and allows for latencies of accesses to remote memory to be hidden.

The lack of commercially available SCI interface hardware for PCs during the initiation period of the SMiLE project led to the development of our custom PCI-SCI hardware. Its primary goal is to serve as the basis of the SMiLE cluster by bridging the PCs I/O bus with the SCI virtual bus.

Fig. 1 shows a high-level block diagram of the PCI-SCI adapter, which is described in detail in [1]. The PCI-SCI adapter is divided into three logical parts: the PCI unit, the Dual-Ported RAM (DPR), and the SCI unit.

The PCI unit interfaces to the PCI local bus. A 64 MByte address window on the PCI bus allows to intercept processor-memory transactions which are then translated into SCI transactions. For the interface to the PCI bus, the PCI9060 PCI bus master chip form PLX Technology is used. Also, packets arriving from the SCI network, have to be transformed into PCI operations.

The packets to be sent via SCI are buffered within the Dual-Ported RAM. It contains frames for outgoing packets allowing one outstanding read transaction, 16 outstanding write transactions, 16 outstanding messaging transactions using a special DMA engine for long data transfers between nodes, and one read-modify-write transaction which can be used for synchronization. Additionally, 64 incoming packets can be buffered.

The SCI unit interfaces to the SCI network and performs the SCI protocol processing for packets in both directions. This interface is implemented with the LinkController LC-1 from Dolphin Interconnect Solutions, which realizes the physical layer and part of the logical layer of the SCI specification. The SCI unit is connected to the DPR via the B-Link, the non-SCI link side of the LC-1. Control information is passed between the PCI unit and the SCI unit via a handshake bus. Two additional FPGAs, responsible for both the PCI-related and B-Link-related processing complete the design.

3 The SMiLE Hardware Monitor Extension Card

3.1 Overview

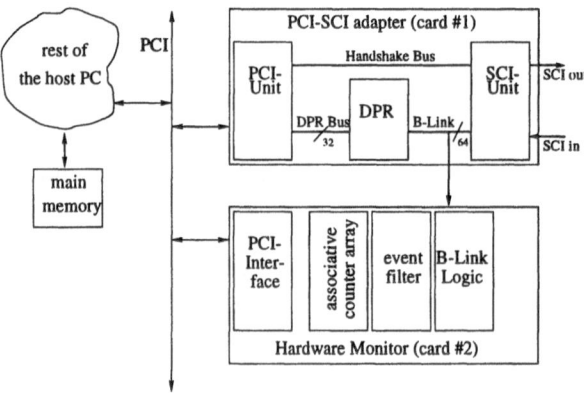

Fig. 1. The SMiLE PCI-SCI adapter and the monitor card installed in a PC

The fine-grain nature of SCI's remote memory transactions and their level of integration in the hardware makes it hard to trace and measure them in a manner necessary to do performance analysis and debugging.

As mentioned in section 2, the B-Link carries all outgoing and incoming packets to the physical SCI interface, Dolphin's LinkController LC-1. It is therefore a central sequentialization spot on which all remote memory traffic can be sensed. The SMiLE SCI hardware monitor thus is an additional PCI card attached to the original SCI adapter as shown in Fig. 1. Data that can be gathered from the B-Link includes the transaction command, the target and source node IDs, the node-internal address offset, and the packet data.

However, the design is naturally limited to remote memory traffic: a conventional plug-in PCI card does not enable us to monitor internal memory accesses. Nonetheless, remote memory accesses remain the most severe as mainly read accesses across the SCI network still account for more than one order of magnitude higher latencies than local accesses.

3.2 Working Principle

When memory access latency becomes an increasing problem during execution of machine instruction streams, and speeding up the whole main memory of a computer system appears not to be economically feasible, performance enhancements are usually achieved through the use of cache memories.

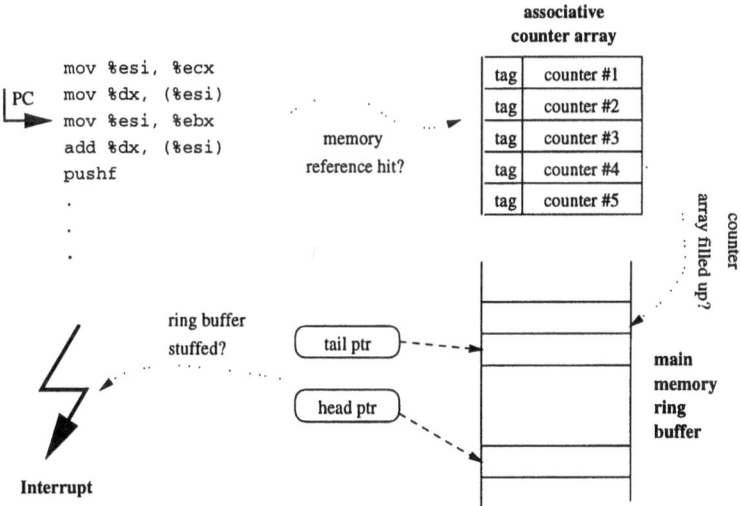

Fig. 2. The hardware monitor detects remote accesses. In order to allocate new counters, old counter value/tag pairs are flushed to the main memory ring buffer. Should it overflow, an interrupt is generated

As data accesses and instruction fetches in typical programs exhibit a great amount of temporal and spatial locality, cache memories only attempt to store a small *working set* of memory pages (cache lines). A prerequisite for this, however, is the use of a content-addressable memory as the covered address space usually dramatically exceeds the cache's capacity.

The same property holds for remote memory references in SCI-supported distributed shared memory execution of parallel programs: It is highly likely that a successive memory reference will access the same data item as its predecessor or some data close by. A hardware monitor the duty of which is to count memory accesses to specified memory regions can profit from this. The most recently addressed memory references and their associated counter can be stored in a register array with limited size. The detection of address proximity allows to combine counter values, as the accesses to neighboring data items may actually represent a single larger data object. E.g. it might be more interesting to count the accesses to an array rather than the references to its elements. Only recently unused memory references have to be flushed into external buffers, by this reducing the amount of data that has to be transported.

Drawn from these principles, the SMiLE hardware monitor uses the concept described in Fig. 2:

1. SCI remote memory accesses are sensed via the B-Link.
2. If the memory reference matches a tag in a counter array, the associated counter is incremented. If no reference is found, a new counter-tag pair is allocated and initialized to 1.
3. If no more space is available in the counter array, a counter-tag pair is flushed to a larger ring buffer in main memory. This buffer is supposed to be emptied by system software in a more or less cyclic fashion.
4. If a buffer overflow occurs nevertheless, a signal is sent to the software process utilizing the monitor in order to force the retrieval of the ring buffer's data.

Apart from a partition of counters working according to this principle (termed *dynamic* counters), the monitor also contains a set of counters that are statically preprogrammable (thus *static* counters) and are not covered in this work.

This addition to the monitor design was chosen in order to allow the user to include pre-known memory areas that are supposed to be monitored anyway, much like in conventional histogram monitors [10].

3.3 The Dynamic Coverage LRU Method

As one wants to reduce the amount of flushing and re-allocating counter-tag pairs, it makes sense to integrate the strategy of counter coverage adaption into the cache replacement mechanism. Under the prerequisite that counter tags can name not only single addresses but also continuous areas of memory for which the counter is to represent the accesses, a simple least-recently-used replacement algorithm can be adapted to this special task. The maximum address range, however, has to be predefined by the user. The Dynamic Coverage LRU method is shown in Fig. 3.

3.4 The Hardware Design

The resource saving principle of the monitor will hopefully allow us to construct the additional plug-in adapter with a small number of high-density FPGAs. The

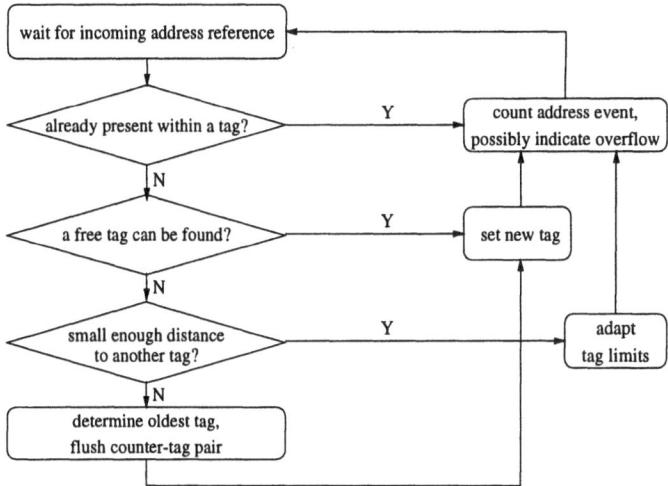

Fig. 3. The Dynamic Coverage LRU mechanism

same commercial PCI master circuit already used with the SCI bridge will serve again as its PCI interface.

The ring buffer will reside in the PCs main memory while its pointer registers will be located on the monitor hardware. Access to the monitor and its registers will be memory mapped in order to speed up accessing the device from user space. As SCI does not restrict the use of read and write transactions to main memory accesses, synchronization and configuration of a whole monitoring system can also be performed via SCI: remote monitor cards can be mapped into global shared memory in the same way that application memory is made accessible to other nodes.

4 Experimental Evaluation and Design Decisions

For initial performance assessments, a software simulator of the proposed hardware monitor served as the basis of a number of experiments. Traces were gathered by executing a number of the SPLASH-2 benchmark programs with the multiprocessor memory system simulator Limes [9]. For this, a shared memory system was assumed with the following properties:

local read/write access latency of 1 cycles, assuming they can be seen as normally cached on a usual system,

remote write latency of 20 cycles, corresponding to approx. 100 ns on a 200 MHz system (accounting for host bridge latencies when accessing the PCI bus),

remote read latency of 1000 cycles, corresponding to approx. 5 μs on a 200 MHz system [7], and

remote lock latency of 1000 cycles, as the processor gets stalled when waiting for the result of the remote atomic read-modify-write operation.

The global shared memory in the SPLASH-2 applications was distributed over the nodes in pages of 4 Kbytes in a round-robin fashion. Due to this the physical memory on a single node no longer represented a contiguous but rather interleaved area of the global shared memory. The simulated DSM system consisted of four nodes of x86 processors clocked at 200MHz.

Figures 4 and 5 represent the memory access histograms of the FFT and the radix sort programs in the SPLASH-2 suite generated by this configuration.

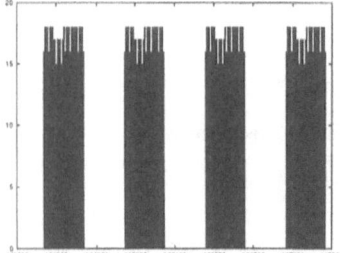

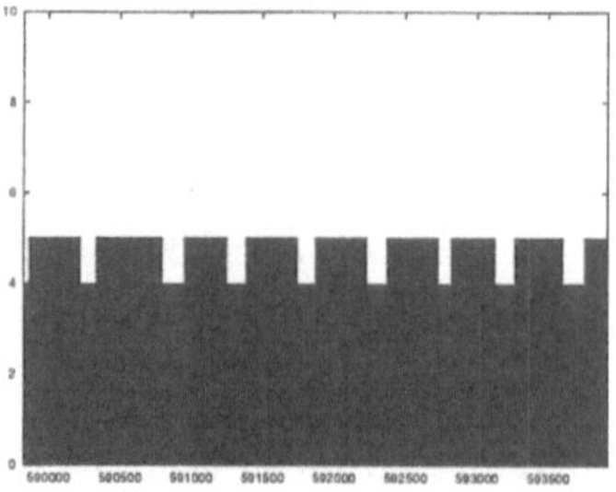

Fig. 4. Left: Memory access histogram for page 40 on node #0 of a 16K points FFT program in the SPLASH-2 suite run on a simulated 4-node SCI configuration

Fig. 5. Right: Memory access histogram for page 144 on node #0 of a 64K keys RADIX program in the SPLASH-2 suite run on a simulated 4-node SCI configuration

Both traces were run through the simulator while the number of dynamic counters was varied. This served to help us find the optimal hardware size of a physical implementation: constraints are an as lean as possible hardware realization which should still ensure low loads for buses and high-level tools.

Figures 6 and 7 display the results of these experiments.

The results follow intuition: the larger the number of counters is, the less flushes to the ring buffer occur. The same holds when the area that a counter can cover is increased: more and more adjacent acesses will be counted in a single register. Naturally, there is an optimum size for the maximum coverable range depending on the program under test: for the FFT increasing this maximal range to more than 16 bytes yields less counter range extensions. The reason for this lies in the FFT's mostly neighboring accesses to complex numbers, reflecting exactly a 16 byte granularity (two `double` values in C).

The RADIX results represent a different case: while for the overall work of the monitor the same holds as for the FFT (a larger number of counters and a bigger covered area account for less flushes), the number of counter range extensions only decreases for increased areas when a larger number of counters is working. This can, however, attributed to the linear access pattern in RADIX:

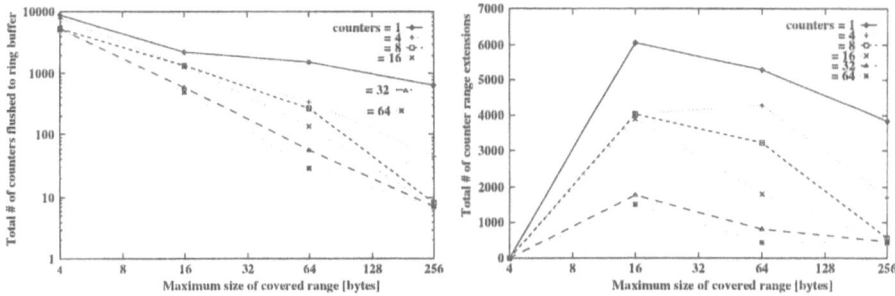

Fig. 6. Number of flushes to the ring buffer and counter coverage extensions for the FFT

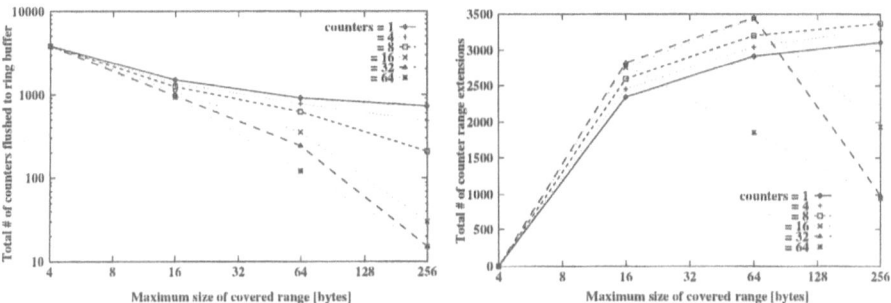

Fig. 7. Number of flushes to the ring buffer and counter coverage extensions for RADIX

successive reads and writes can always be put into the same counter until the maximum range has been reached. As soon as the counters are able to cover the whole 4K page (e. g. 256 byte range × 16 counters = 4096 bytes), the number of extensions decreases drastically.

A number of 16 or 32 counters appears to be a good compromise in a practical implementation of this monitor. Already in this size the number of accesses to main memory can be reduced significantly, even with relatively small ranges covered by the counters.

As the current monitor design always writes a packet of 16 bytes to the external ring buffer, this means that for 16 counters the amount of flushed data sums up to 84 KBytes for the FFT and 60 KBytes for RADIX. Given a simulated runtime for the FFT of 0.127 sec and 1.142 sec for RADIX, this yields a top average data rate of 661 KBytes/s. Roughly the same numbers hold for 32 counters. When the the Dynamic Coverage LRU is switched on with a maximum range of larger than 1 word, this rate instantly drops to 280 KBytes/s.

With typical PC I/O bandwidths of around 80 MB/s (PCI's nominal bandwidth of 133 MByte/s can only be reached with infinite bursts), monitoring represents a mere 0.8 % of the system's bus load and shouldn't influence program execution too much.

Switching off all monitor optimizations (range of 4 byte, only one dynamic counter) converts our hardware into a trace-writing monitor, creating trace data at a maximum rate of 1.06 MB/s.

An external ring buffer has to be sized according to these figures and the ability of a possible on-line monitoring software to make use of the sampled data. Derived from rules of thumb, ring buffers in the 1 MByte range should suffice to allow a reasonably relaxed access to flushed data without disturbing a running application.

It has to be noted that we are currently restricted to page-sized areas that can be covered by the dynamic counters of the monitor as PCI addresses are physical addresses. This means that consecutive virtual addresses may be located on different pages.

One solution to this problem would be to duplicate parts of the processor's MMU mechanisms tables on the monitoring hardware in order to be able to retranslate PCI into virtual addresses. The obvious issue of hardware complexity has so far excluded this possibility.

5 Related Work

The parallel programming in environments with distributed memory is supported by a number of tools so that there are strong efforts in a standardization of the monitoring interface [3].

Tool environments for DSM-oriented systems are less wide-spread. Mainly for software-based DSM systems performance debugging tools have been developed, which analyse traces for data locality [4] [2].

Over the last years, monitoring support has become increasingly available on research as well on commercial machines. For the CC-NUMA FLASH multiprocessor system the hardware-implemented cache coherence mechanism is complemented by components for the monitoring of fine-grained performance data (number and duration of misses, invalidations etc.) [11]. Modern CPU chips incorporate hardware counters which collect information about data accesses, cache misses, TLB misses etc. For some multiprocessor systems these information is exploited by performance analysis tools [12]

Martonosi et.al. propose a multi-dimensional histogram performance monitor for the SHRIMP multiprocessor [10].

6 Conclusion

In the paper we presented the rationale and the architecture of the distributed hardware monitor for the SMiLE DSM PC cluster. In order to be able to use the information provided by these hardware monitors a performance evaluation system needs a software infrastructure consisting of a local monitor library, a communication and managing layer providing a global monitor abstraction.

This items will be covered in an OMIS [8] compliant implementation of a prototypical DSM monitoring infrastructure.

References

1. G. Acher, H. Hellwagner, W. Karl, and M. Leberecht. A PCI-SCI Bridge for Building a PC-Cluster with Distributed Shared Memory. In *Proceedings The Sixth International Workshop on SCI-based High-Performance Low-Cost Computing*, pages 1–8, Santa Clara, CA, Sept. 1996. SCIzzL.

2. D. Badouel, T. Priol, and L. Renambot. SVMview: A Performance Tuning Tool for DSM-Based Parallel Computers. In L. Boug, P. Fraigniaud, A. Mignotte, and Y. Robert, editors, *EuroPar'96 – Parallel Processing*, number 1123 in LNCS, pages 98–105, Lyon, France, Aug. 1996. Springer Verlag.

3. A. Bode. Run-Time Oriented Design Tools: A Contribution to the Standardization of the Development Environments for Parallel and Distributed Programs. In F. Hofeld, E. Maehle, and E. W. Mayr, editors, *Proceedings of the 4th Workshop PASA'96 Parallel Systems & Algorithms*, pages 1–12. World Scientific, 1997.

4. M. Gerndt. Performance Analysis Environment dor SVM-Fortran Programs. Technical Report IB-9417, Research Centre Jlich (KFA), Central Institute for Applied Mathematics, Jlich, Germany, 1994.

5. H. Hellwagner, W. Karl, and M. Leberecht. Enabling a PC Cluster for High-Performance Computing. *SPEEDUP Journal*, 11(1), June 1997.

6. IEEE Standard for the Scalable Coherent Interface (SCI). IEEE Std 1596-1992, 1993. IEEE 345 East 47th Street, New York, NY 10017-2394, USA.

7. M. Leberecht. A Concept for a Multithreaded Scheduling Environment. In F. Hofeld, E. Maehle, and E. W. Mayr, editors, *Proceedings of the 4th Workshop on PASA'96 Parallel Systems & Algorithms*, pages 161–175. World Scientific, 1996.

8. T. Ludwig, R. Wismüller, V. Sunderam, and A. Bode. OMIS — On-line Monitoring Interface Specification (Version 2.0). TUM-I9733, SFB-Bericht Nr. 342/22/97 A, Technische Universität München, Munich, Germany, July 1997.

9. D. Magdic. *Limes: An Execution-driven Multiprocessor Simulation Tool for the i486+-based PCs*. School of Electrical Engineering, Department of Computer Engineering, University of Belgrade, POB 816 11000 Belgrade, Serbia, Yugoslavia, 1997.

10. M. Martonosi, D. W. Clark, and M. Mesarina. The SHRIMP Performance Monitor: Design and Applications. In *Proceeding 1996 SIGMETRICS Symnposium on Parallel and Distributed Tools (SPDT'96)*, pages 61–69, Philadelphia, PA, USA, May 1996. ACM.

11. M. Martonosi, D. Ofelt, and M. Heinrich. Integrating Performance Monitoring and Communication in Parallel Computers. In *SIGMETRICS'96 1996 ACM Sigmetrics Conference on Measurement and Modeling of Computer Systems*, 1996.

12. M. Zagha, B. Larson, S. Turner, and M. Itzkowitz. Performance Analysis Using the MIPS R10000 Performance Counters. In *Supercomputing SC'96*, 1996.

On the Self-Similar Nature of Workstations and WWW Servers Workload

Olivier Richard and Franck Cappello

LRI, Université Paris-Sud, 91405
Orsay, France
(auguste,fci)@lri.fr.

Abstract. This paper presents a workload characterization for workstations, servers and WWW servers. Twelve data sets built from standard UNIX tools and from access.log files are analyzed with three different time scales. We demonstrate that the workload of these resources is statistically self-similar in the periods of irregular activity.

1 Introduction

The resource sharing in a network of workstations and in Internet is the aim of a large number of research projects. Fundamental to this goal is to understand the initial workload characteristics of each resource.

Most of the mechanisms (CONDOR [1] and GLUNIX [2]) for workload distribution try to detect the periods when the user does not use its workstation. Nevertheless, even when the user uses its workstation, the initial workload may stay very low. This under-exploitation provides the potential to gather unexploited resources for more agressive users. This goal requires to study carefully the initial workload of a workstation during the user activity. The workload characterization of WWW servers is more recent. In [3], the self-similarity of WWW server workload is investigated. The automatic workload distribution of the WWW requests on mirror sites is an open issue.

We use a global approach to characterize the complexity of the workload. This approach is close to the one used for network traffic analysis [4] [5].

2 Method for the Workload Study

More than a hundred of resources are connected in our network of workstations. The machines are used for numerical processing, some heavy simulations and student and researcher works (compiler, mailer, browser, etc.). The highly computing applications saturate the computer and lead to a typical flat plot for the workload. The server workload is typically much more irregular. The workstation workload is a composition of the previous ones: null during the night, sometimes irregular and sometimes saturated during the day.

We have also analyzed two WWW servers: our lab WWW server, called LRI and the Ethernet backbone of the computing research department of the

Virginia University of technology (USA). The trace for the LRI server starts at October 31, 1995 and finishes at August 4, 1997. For the second WWW server, only the external requests at *.cs.vt.edu* domain have been recorded during a 38 days period. We analyze the hits and the transfered bytes per time.

We choose to use three time scales of analysis because the event granularities are different in a workstation, a server and a WWW server. The statistical analysis, requires about 100.000 events to give results with an acceptable quality. This number corresponds to 50 days for the minute time scale (WWW server workload analysis). It represents a daily period for the 1 second time scale (workstation workload analysis). To understand the workstation and server micro-workloads we have used a 5 milliseconds time scale (100.000 events represent 8 minutes).

The workload measurement of the servers and workstations has been done with the *vmstat* UNIX command (average percentage of the CPU availability). We have used a method based on a snoopy process with low priority to measure the workload with the 5 millisecond resolution. The accesses to the WWW servers have been recorded in the *access.log* file.

3   Self-similarity

Intuitively, a self-similar signal may be observed graphically as presenting invariant features for differents time scales. As in [4], we consider the workload signal as a wide-sense stationary stochastic process (stationary up to order 2). In such process, the mean (μ) and the variance (σ^2) stay constant over time.

Let $X(t)$ be such a stochastic process. $X(t)$ has an auto-covariance function, $R(\tau)$, and an autocorrelation function $\rho(\tau)$ of the form $R(\tau) = E[(X(t) - \mu)(X(t+\tau) - \mu)]$ and $\rho(\tau) = R(\tau)/R(0) = R(\tau)/\sigma^2$. We assume that $X(t)$ is of the form $\rho(\tau) \to \tau^{-\beta}L(\tau)$, when $\tau \to \infty$ (1), where $L(\tau)$ is slowly varying at infinity. (examples of such function are: $L(\tau) = const$, $L(\tau) = log(\tau)$).

Let $X^{(m)}$ denotes a new time series obtained by averaging the original series X in non-overlapping blocks of size m. That is: $X^{(m)} = (1/m)(X_{tm-m+1} + X_{tm-m+2} + \ldots + X_{tm})$. Let $\rho^{(m)}(\tau)$ denotes the autocorrelation function of $X^{(m)}$. If the aggregated process $X^{(m)}$ has the same autocorrelation structure as X, the process X is called exactly second order self-similar with self-similarity parameter $H = 1 - \beta/2$, i.e., $\rho^{(m)}(\tau) = \rho(\tau)$. X is called asymptotical second order self-similar with self-similarity parameter $H = 1 - \beta/2$, if we assume $\rho^{(m)}(\tau) \to \rho(\tau)$ for large m and τ.

In the next section we use three methods to test and characterize the self-similarity of a stochastic process: the variance-time graphic of the $X^{(m)}$ processes, the R/S analysis [4] and Whittle estimator [6]. All methods give an estimate of H (the self-similarity parameter). H larger than 1/2 and lower that 1 suggests the self-similarity of the signal. A rigorous introduction to the self-similar phenomenon can be found in [4].

4   Results

The data sets size is about 70.000 measurements. Eight machines have been closely analyzed in the network of workstations. The table 4 gathers the estimates of H with the three methods for the observed machines in the network of workstations. For all machines, H is larger than 1/2 and lower than 1 and β is between 0 and 1. So, together the 3 methods suggest that self-similar stochastic processes may be used to represent closely these machine workloads.

Table 1. H estimates for the CPU workload signals (1 second and 5 ms time scales), variance analysis of the $X^{(m)}$ processes, R/S analysis and Whittle estimates

	1 second time scale				5 ms time scale			
	$\beta(VAR)$	$H(VAR)$	$H(R/S)$	$H(W)$	$\beta(VAR)$	$H(VAR)$	$H(R/S)$	$H(W)$
Sun1	0.41	0.79	0.69	0.92	-	-	-	-
Sun2	0.29	0.85	0.64	-	0.31	0.84	0.78	-
Sun3	0.21	0.89	0.57	0.86	0.49	0.75	0.66	0.84
Sun4	0.48	0.75	0.64	0.99	-	-	-	-
Sun5	-	-	-	-	0.68	0.65	0.89	0.91
Sun6	-	-	-	-	0.34	0.82	0.84	0.81
HP1	0.30	0.84	0.79	0.86	-	-	-	-
HP2	0.37	0.81	0.80	0.79	-	-	-	-

The results of the estimates for the H parameter for the WWW servers with the various methods are presented in table 4. The first three raws present the estimates for H and β for three different segments of the LRI trace. The raw labeled BR gives the results for the Ethernet backbone of the computer science department (Virginia University). The values of H and *beta* for these traces indicates that both signals (hits per minute and transmitted bytes per minute) might be closely represented by self-similar processes.

Table 2. H estimates of hits per second and bytes per second for the WWW servers. The method used for each estimate is indicated between parentheses.

	Hits per second				Bytes per second			
	$\beta(VAR)$	$H(VAR)$	$H(R/S)$	$H(W)$	$\beta(VAR)$	$H(VAR)$	$H(R/S)$	$H(W)$
hit09	0.50	0.74	0.71	0.86	0.64	0.68	0.71	0.74
hit28	0.62	0.68	0.74	0.78	0.62	0.68	0.76	0.72
hit31	0.64	0.67	0.78	0.75	0.75	0.62	0.72	0.61
BR	0.43	0.78	0.88	-	0.63	0.68	0.83	-

5 Analysis of Results and Conclusion

We have shown that self-similar stochastic processes correspond to the workload of these resources (in the periods of irregular activity).

A more accurate confidence interval for the estimates of H may be obtained with the Whittle estimator applied to larger data sets or with other methods such as the wavelet methods. [7] has recently shown that some network traffics were multi-fractal instead of mono-fractal (H is varying with time).

Our result about the self-similar nature of the WWW server workload is contrary to the one given in [3]. In their data set, there are several defects : zero hit for several hours, or very punctual high volume transfers. Removing these punctual defects changes the conclusions and suggest that the self-similarity is not as rare as the authors have suggested and that high perturbations may alterate the results of the self-similarity tests.

The knowledges about self-similarity in WWW servers are getting wider. In [8], the authors propose an explanation for the nature of client site WWW traffic. The article [9] provides an interesting result: superposition of self-similar processes yields to a self-similar process with fractional Gaussian noises. [4] has presented some consequences of traffic self-similarity on communication network engineering. There are drastic implications on modeling the individual source, on the notion of "burstiness" and on the congestion management.

The methods and the results obtained in the field of communication networks may be used as the bases for further works in the workload management area. For example, the estimate of H may be used to generate synthetic traces from the fractional Gaussian noise model [10] or from the fractional ARIMA (p, d, q) processes [11]. The synthetic traces would represent the initial workload of a workstation, a server or a WWW server.

References

1. M. J. Litzkow, M. Livny, and M. W. Mutka. Condor : A hunter of idle workstations. In *8th International Conference on Distributed Computing Systems*, pages 104–111, Washington, D.C., USA, June 1988. IEEE Computer Society Press.

2. Amin M. Vahdat, Douglas P. Ghormley, and Thomas E. Anderson. Efficient, portable, and robust extension of operating system functionality. Technical Report CSD-94-842, University of California, Berkeley, December 1994.

3. Martin Arlitt and Carey L. Williamson. Web server workload characterization: The search for invariants. In *Proceedings of the ACM SIGMETRICS International Conference on Measurement and Modeling of Computer SYstems*, volume 24,1 of *ACM SIGMETRICS Performance Evaluation Review*, pages 126–137, New York, May23–26 1996. ACM Press.

4. Will E. Leland, Murad S. Taqqu, Walter Willinger, and Daniel V. Wilson. *On the Self-Similar Nature of Ethernet Traffic (Extended Version). IEEE/ACM Transactions on Networking*, 1994.

5. Vern Paxson and Sally Floyd. Wide-area traffic: the failure of Poisson modeling. *ACM SIGCOMM 94*, pages 257–268, August 1994.

6. Jan Beran. *Statistic for Long-Memory Processes*. Chapman and Hall, 1994.

7. R.H. Riedi and J. Lvy Vhel. Multifractal properties of tcp traffic: a numerical study. Technical Report 3129, INRIA, 1997.

8. Mark Crovella and Azer Bestavros. Self-similarity in World Wide Web traffic: Evidence and causes. In *Proceedings of the ACM SIGMETRICS International Conference on Measurement and Modeling of Computer SYstems*, volume 24,1 of *ACM SIGMETRICS Performance Evaluation Review*, pages 160–169, New York, May23–26 1996. ACM Press.

9. K. Krishnan. A new class of performance results for a fractional brownian traffic model. *Queueing System*, 22:277–285, 1996.

10. B. B. Mandelbrot and J. W. Van Ness. Fractional brownian motion, fractional noises and applications. *SIAM Review*, 10(4):422–437, October 1968.

11. J. Granger and R. Joyeux. An introduction to long-memory time series models and fractional differencing. *Time Series Analysis*, 1, 1980.

White-Box Benchmarking

Emilio Hernández[1] and Tony Hey[2]

[1] Departamento de Computación,
Universidad Simón Bolívar, Apartado 89000, Caracas, Venezuela,
emilio@usb.ve
[2] Department of Electronics and Computer Science,
University of Southampton, Southampton SO17 1BJ, UK
ajgh@ecs.soton.ac.uk

Abstract. Structural performance analysis of the NAS parallel benchmarks is used to time code sections and specific classes of activity, such as communication or data movements. This technique is called *white-box benchmarking* because, similarly to white-box methodologies used in program testing, the programs are not treated as black boxes. The timing methodology is portable, which is indispensable to make comparative benchmarking across different computer systems. A combination of conditional compilation and code instrumentation is used to measure execution time related to different aspects of application performance. This benchmarking methodology is proposed to help understand parallel application behaviour on distributed-memory parallel platforms.

1 Introduction

Computer evaluation methodologies based on multi-layered benchmark suites, like Genesis [1], EuroBen [2] and Parkbench [3] have been proposed. The layered approach is used for characterising the performance of complex benchmarks based on the performance models inferred from the execution of low-level benchmarks. However, it is not straightforward to relate the performance of the low-level benchmarks with the performance of more complex benchmarks.

The study of the relationship between the performance of a whole program and the performance of its components may help establish the relevance of using low-level benchmarks to characterise the performance of more complex benchmarks. For this reason we propose the structural performance analysis of complex benchmarks. A benchmarking methodology, called *white-box benchmarking*, is proposed to help understand parallel application performance. The proposed methodology differs from standard profiling in that it is not procedure oriented. Partial execution times are not only associated to code sections but also to activity classes, such as communication or data movements. These execution times may be compared to the results of simpler benchmarks in order to assess their predictive properties. The proposed methodology is portable. It only relies on MPI_WTIME (the MPI [4] timing function) and the availability of a source code preprocessor for conditional compilation, for instance, a standard C preprocessor.

Timing Method. The proposed timing method is simple enough to be portable. It is based on the combination of two basic techniques: *incremental conditional compilation* and *section timing*. Incremental conditional compilation consists of selecting code fragments from the original benchmark to form several kernels of the original benchmark. A basic kernel of a parallel benchmark can be built by selecting the communication skeleton. A second communication kernel can contain the communication skeleton plus data movements related to communication (e.g. data transfers to communication buffers). By measuring the elapsed time of both kernels, we know the total communication time (the time measured for the first kernel) and the time spent in data movements (the difference in the execution time of both benchmarks). The net computation time can be obtained by subtracting the execution time of the second kernel from the execution time of the complete benchmark. Not every program is suitable for this type of code isolation, see [5] for a more detailed discussion. Section timing is used on code fragments that take a relatively long time to execute, for example, the subroutines at the higher level of the call tree. Three executable files may be produced, the communication kernel, the communication plus data movements kernel and the whole benchmark. Optionally, if information by code section is required, an additional compile-time constant has to be set to include the timing functions that obtain the partial times. The use of MPI_WTIME allows us to achieve code portability. Two out of these three benchmark executions are usually fast because they will not execute "real" computation, but only communication and data movements. Section 2 presents an example of the use of the methodology with the NAS parallel Benchmarks [6]. In section 3 we present our conclusions.

2   Case Study with NAS Parallel Benchmarks

The NAS Parallel Benchmarks [6] are a widely recognized suite of benchmarks derived from important classes of aerophysics applications. In this work we used the application benchmarks (LU, BT and SP) and focused on the communication kernels extracted from these benchmarks. The main visible difference between these communication kernels comes from the fact that LU sends a larger number of short messages, while SP and BT send fewer and longer messages. This means that the LU communication kernel should benefit from low latency interconnect subsystems, while BT and SP communication kernels would execute faster on networks with a greater bandwidth.

2.1   Experiments

Several experiments with the instrumented version of the NAS benchmarks were conducted on a Cray T3D and a IBM SP2. For a description of the hardware and software configurations see [5]. Several experiments were conducted using the white-box methodology described here. These experiments are also described in detail in [5].

NAS Application Benchmarks (Class A size)

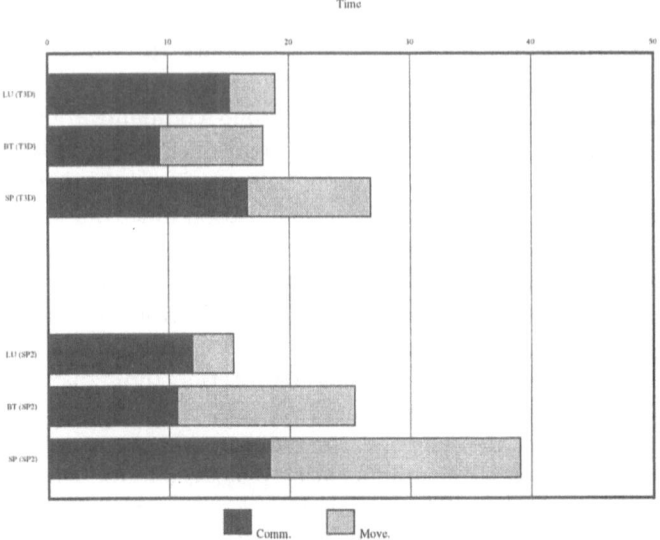

Fig. 1. Communication time and communication-related data movement time on T3D and SP2 (16 processors).

Figure 1 compares communication overhead in the T3D and the SP2 for LU, BT and SP. The communication kernel of LU runs marginally faster on the SP2 than on the T3D, while SP and BT communication kernels execute faster on the T3D. The execution of the communication kernels indicates that communication performance is not substantially better in the SP2 or the T3D. The main difference, in favour of the T3D, is the time spent in data movements related to communication, rather than communication itself.

Measurements made with COMMS1 (the Parkbench ping-pong benchmark), slightly modified for transmitting double precision elements, show that the T3D has a startup time equal to $104.359 \mu sec$ and a bandwidth equal to $3.709\ Mdp/sec$, while the SP2 has a startup time equal to $209.458 \mu sec$ and a bandwidth equal to $4.210\ Mdp/sec$, where Mdb means "millions of double precision elements".

As mentioned above, the LU communication kernel should run faster on low latency networks, while BT and SP communication kernels would execute faster on networks with a greater bandwidth. The observed behaviour of LU, BT and SP, seems to contradict the expected behaviour of these benchmarks, based on the COMMS1 results. Apart from bandwidth and startup time, many other factors may be playing an important role in communication performance, which are not measured by COMMS1. Some of these factors are network contention, the presence of collective communication functions, the fact that messages are sent from different memory locations, etc. In other words, it is clear from these experiments that communication performance may not be easily characterized by low level communication parameters (latency and bandwidth) obtained from simpler benchmarks.

3 Conclusions

White-box benchmarking is a portable and comparatively effortless performance evaluation methodology. A few executions of each benchmark are necessary to get the information presented in this article, one for the complete benchmark and the rest for the extracted kernels. A benchmark visualisation interface like GBIS [7] may easily be enhanced to incorporate query and visualisation mechanisms to facilitate the presentation of results related to this methodology.

Useful information has been extracted from the NAS parallel benchmarks case study using white-box benchmarking. Communication-computation profiles of the NAS parallel benchmarks may easily indicate the balance between communication and computation time. Communication kernels obtained by isolating the communication skeleton of selected applications may give us a better idea about the strength of the communication subsystem. Additionally, some behaviour characteristics can be exposed using white-box benchmarking, like load balance and a basic execution profile.

This methodology may be used in benchmark design rather than in application development. Programs specifically developed as benchmarks may incorporate code for partial measurements and kernel selection. A description of the instrumented code sections may also be useful and, consequently, could be provided with the benchmarks. The diagnostic information provided by white-box benchmarking is useful to help understand the variable performance of parallel applications on different platforms.

References

1. C. A. Addison, V.S. Getov, A.J.G. Hey, R.W. Hockney, and I.C. Wolton. The genesis distributed-memory benchmarks. In J. Dongarra and W. Gentzsch, editors, *Computer Benchmarks*, pages 257–271. North-Holland, 1993.
2. A. van der Steen. Status and Direction of the EuroBen Benchmark. *Supercomputer*, 11(4):4–18, 1995.
3. J. Dongarra and T. Hey. The PARKBENCH Benchmark Collection. *Supercomputer*, 11(2-3):94–114, 1995.
4. Message Passing Interface Forum. The message passing interface standard. Technical report, Univeristy of Tennessee, Knoxville, USA, April 1994.
5. E. Hernández and T. Hey. White-Box Benchmarking (longer version). Available at http://www.usb.ve/ emilio/WhiteBox.ps, 1998.
6. D. Bailey *et. al.* The NAS Parallel Benchmarks. Technical Report RNR-94-007, NASA Ames Research Center, USA, March 1994.
7. M. Papiani, A.J.G. Hey, and R.W. Hockney. The Graphical Benchmark Information Service. *Scientific Programming*, 4(4), 1995.

Cache Misses Prediction for High Performance Sparse Algorithms*

Basilio B. Fraguela[1], Ramón Doallo[1], and Emilio L. Zapata[2]

[1] Dept. Electrónica e Sistemas. Univ. da Coruña,
Campus de Elviña s/n, A Coruña, 15071 Spain
{basilio,doallo}@udc.es
[2] Dept. de Arquitectura de Computadores. Univ. de Málaga, Complejo Tecnológico
Campus de Teatinos, Málaga, 29080 Spain
ezapata@ac.uma.es

Abstract. Many scientific applications handle compressed sparse matrices. Cache behavior during the execution of codes with irregular access patterns, such as those generated by this type of matrices, has not been widely studied. In this work a probabilistic model for the prediction of the number of misses on a direct mapped cache memory considering sparse matrices with an uniform distribution is presented. As an example of the potential usability of such types of models, and taking into account the state of the art with respect to high performance superscalar and/or superpipelined CPUs with a multilevel memory hierarchy, we have modeled the cache behavior of an optimized sparse matrix-dense matrix product algorithm including blocking at the memory and register levels.

1 Introduction

Nowadays superscalar and/or superpipelined CPUs provide high speed processing, but global performances are constrained due to memory latencies even despite the fact that a multilevel memory organization is usual. Performances are further reduced in many numerical applications due to the indirect accesses that arise in the processing of sparse matrices, because of their compressed storage format [1].

Several software techniques for improving memory performance have been proposed, such as blocking, loop unrolling, loop interchanging or software pipelining. Navarro et al. [5] have studied some of these techniques for the sparse matrix-dense matrix product algorithm. They have proposed an optimized version of this algorithm as a result of a series of simulations on a DEC Alpha processor. The traditional approach for cache performance evaluation has been the software simulation of the cache effect for every memory access [7]. Different approaches consist in providing performance monitoring tools (built-in counters in current microprocessors), or the design of analytical models.

---

* This work was supported by the Ministry of Education and Science (CICYT) of Spain under project TIC96-1125-C03, Xunta de Galicia under Project XUGA20605B96, and E.U. Brite-Euram Project BE95-1564

This paper chooses this last approach and addresses the problem of the estimation of the total number of misses produced in a direct-mapped cache using a probabilistic model. In a previous work, Temam and Jalby [6] model the cache behavior for the sparse matrix-vector product of uniform banded sparse matrices, although they do not consider cross interferences. Simpler sparse linear algebra kernels than the one this paper is devoted to have been also modeled in [2], being the purpose of this work to demonstrate that it is feasible to extend this type of models to more complex algorithms, such as the one mentioned above. This fact can make possible further improvements for these codes by making easier the study of the effect of these techniques on the memory hierarchy. An extension of the proposed model for K-way associative caches has been introduced in [3].

The remainder of the paper is organized as follows: Section 2 describes the algorithm to be modeled. Basic model aparameters and concepts are introduced in Sect. 3 together with a brief explanation, due to space limitations, of the modeling process. In Sect. 4 the model is validated and used to study the cache behavior of the algorithm as a function of the block dimensions and the cache main parameters. Section 5 concludes the paper.

2 Modeled Algorithm

The optimized sparse matrix-dense matrix product code proposed in [5] is shown in Fig. 1. The sparse matrix is stored using the Compressed Row Storage (CRS) format [1]. This format uses three vectors: vector A contains the sparse matrix entries, vector C stores the column of each entry, and vector R indicates in which point of A and C a new row of the sparse matrix starts and permits knowing the number of entries per row. As a result the sparse matrix must be accessed row-wise. The dense matrix is stored in B, while D is the product matrix.

The loops are built so that variable I always refers to the rows of the sparse matrix and matrix D, J refers to the columns of matrices B and D, and K traverses the dimension common to the sparse matrix and the dense matrix. Our code is based on a IKJ order of the loops (from the outermost to the innermost loop).

This code uses one level blocking at the cache level selecting a block of matrix B with BK rows and BJ columns. The accesses to the block in the inner loop are row-wise, so a copy by rows to a temporal storage WB is desired in order to avoid self interferences and achieve a good exploitation of the spatial locality.

The performance has been further improved by applying blocking at the register level. This code transformation consists in the application of strip mining to one or more loops, loop interchanging, full loop unrolling and the elimination of redundant load and store operations. In our case, inner loop J has been completely unrolled and the load and stores for D have been taken out of loop K. This modification requires the existence of BJ registers (d1, d2, ... , dbj in the figure) to store these values, as Fig. 1 shows. The resulting algorithm has a bidimensional blocking for registers, resulting in fewer loads and stores per arithmetic operation. Besides the number of independent floating point operations in the loop body is increased.

```
1  DO J2=1, H, BJ                        19        DO WHILE (K<LK AND C(K)<LIMK)
2     LIMJ=J2+MIN(BJ, H-J2+1)-1          20           a=A(K)
                                         21           ind=C(K)
3     DO I=1, M+1                        22           d1=d1+a*WB(1,ind-j2+1)
4        R2(I)=R(I)                      23           d2=d2+a*WB(2,ind-j2+1)
5     ENDDO                                           ...
                                         24           dbj=dbj+a*W(BJ, ind-j2+1)
6     DO K2=1, N, BK                     25           K=K+1
7        LIMK=K2+MIN(BK, N-K2+1)         26        ENDDO
8        DO J=1, LIMJ-J2+1              27        D(I,J2)=d1
9           DO K=1, LIMK-K2             28        D(I,J2+1)=d2
10             WB(J,K)=B(K2+K-1,J2+J-1)              ...
11          ENDDO                        29        D(I,J2+BJ-1)=dbj
12       ENDDO
                                         30        R2(I)=K
13       DO I=1, M                       31     ENDDO
14          K=R2(I)                      32  ENDDO
15          LK=R(I+1)                    33 ENDDO

16          d1=D(I,J2)

17          d2=D(I,J2+1)

              ...
18          dbj=D(I,J2+BJ-1)
                  ↳
```

Fig. 1. Sparse matrix-dense matrix product with IKJ ordering and blocking at the memory and register levels

3 Probabilistic Model

We call intrinsic miss the one that takes place the first time a given memory block is accessed. The accesses to the block from that moment on result in hits unless the block is replaced. This happens in a direct mapped cache if and only if another memory block mapped to the same cache line is accessed. If this block belongs to the same vector as the replaced line it is called a self interference, whereas if it belongs to another vector it is called a cross-interference.

The replacement probability grows with the cache area (number of lines) affected by the accesses between two consecutive references to the considered block. This area depends on the memory location of the vectors to be accessed.

In most cases, two consecutive references to a given block are separated by accesses to several vectors that cover different areas which are measured as ratios, as they correspond to line replacement probabilities. The total area covered by these accesses in the cache is calculated adding these areas as independent probabilities[1], operation that we express with symbol $\cup$. This way our probabilistic model does not make any assumptions on the relative location of these vectors in memory.

The main parameters our model employs are depicted in Table 1. By word we mean the logical access unit, this is to say, the size of a real or an integer. We have chosen the size of a real, but the model is totally scalable. Integers are considered through the use of parameter r.

---

[1] Given the independent probabilities p_1 and p_2 of events A_1 and A_2 respectively, the probability of A_1 or A_2 is obtained as $p_1 \cup p_2 = p_1 + p_2 - p_1 p_2$

Table 1. Notation used

C_s	Cache size in words
L_s	Line size in words
N_c	Number of cache lines (C_s/L_s)
M	Number of rows of the sparse matrix
N	Number of columns of the sparse matrix
H	Number of columns of the dense matrix
BJ	Block size in the J dimension
BK	Block size in the K dimension
N_{BJ}	Number of blocks in the J dimension (H/BJ)
N_{BK}	Number of blocks in the K dimension (N/BK)
N_{nz}	Number of entries of the sparse matrix
p_n	Probability that a position in the sparse matrix contains an entry $\left(\frac{N_{nz}}{M \cdot N}\right)$
r	$\frac{\text{size of an integer}}{\text{size of a real}}$

A main issue of the proposed model we want to point out is the way the cache area affected by the accesses to a given vector is calculated. This area depends on the vector access pattern and the cache parameters. For example, a sequential access to x words loads $(x + L_s - 1)/L_s$ lines on average. The cache area they cover is

$$S_s(x) = \min\{1, (x + L_s - 1)/C_s\} \tag{1}$$

In a similar way different expressions or iterative methods have been developed to estimate the average cache area affected by the different types of accesses to a given vector. Combinig these expressions and adding the areas covered by the accesses to the different vectors the average cross interference probability is estimated. Also the self interference probability may be calculated as a function of the vector size, the cache parameters, and the vector access pattern. Miss rate calculation is directly obtained from the previous probabilities. These are the access patterns found in this code:

1. Sequential accesseses on vectors **A** and **C** within groups of entries (not positions) located in the same row in a set of BK consecutive columns in lines 19-21. There is a hit in reuse probability in the four loops (with index variables **K**, **I**, **K2** and **J2** from the inner to the outer) with decreasing weight. These accesses are slightly irregular because of the different number of entries in each considered subrow and the offset inside the vectors between the data corresponding to entries in consecutive subrows. The difference between these vectors is that **C** is always accessed at least once in each iteration of loop 13-31 in line 19, and it is always accessed once more than vector **A**, as it is used to detect the end of the subrow.
2. Access to matrix **B** in groups of BJ consecutive subcolumns of BK elements each in line 10. Recall that the access is by columns, and FORTRAN stores

matrices by columns, so it is completely sequential in each subcolumn. There is only a real reuse probability in loop 9-10, as each element is only accessed once.

3. Subrows of BJ consecutive positions of matrix D read in lines 16-80 and written in lines 27-29. The read operation has a hit in reuse probability in the loop indexed by I, and when the cache is large, also in the loop on K2. The write operations will result in hits providing they are not affected by the self interferences in the accesses to a row and the cross interferences due to the accesses in lines 19-26.

4. Vectors R and R2 have almost the same access pattern: they both are sequentially accessed in loop 3-5 and also inside the loop indexed by I (lines 14, 15 and 30: this last line gives the difference between these vectors). In the first case, only an access to the other vector takes place between two consecutive accesses to the same line, while in the second the cross interference probability is much higher. There is a hit in reuse probability in the first access to each line in the two loops when the cache is large with respect to the block size.

5. Matrix WB, which will be studied below, is accessed by rows in loops 8-11 to be filled with the values of the block to process, and by columns -thus, sequentially- in the inner loop, being the column chosen as a function of the column of the considered entry. This last fact makes the selection irregular. When this matrix fits in the cache, there is some hit in reuse probability in the first access to each line due to the previous access in loops 8-11. Conversely, the accesses in line 10 have a hit in reuse probability derived from the accesses in lines 22-24.

As an example, and due to space limitations, we shall only explain the way the number of misses on matrix WB has been modeled, as it is usually responsible for most of the misses and it has the more complex access pattern. This matrix contains a transposed copy of the block of matrix B that is being processed, so it has BJ rows and BK columns. We calculate first the number of misses for WB in the inner loop and then the number of misses during the copying process of the block of matrix B.

3.1 Misses on Matrix WB in the Inner Loop

The hit probability for a given line of matrix WB during the processing of the j-th row is calculated as

$$P_{\text{hit WB}}(j) = \sum_{i=1}^{j-1} P(1-P)^{i-1}(1-P)^{i \cdot n_1}(1 - P_{\text{cross WB}}(i)) \\ + (1-P)^{j-1}(1-P)^{j \cdot n_1}(1 - P_{\text{cross WB}}(j))P_{\text{surv WB}} \tag{2}$$

where $n_1 = \max\{0, S_s(BJ \cdot BK) - 1\}$ is the average number of lines of WB that compete with another line of this matrix for a given cache line and $P = 1 - (1 - p_n)^{Av}$ is the probability that a line of matrix WB is accessed during the processing of a subrow of the sparse matrix, being Av the average of different columns of the matrix that have elements in the same line.

In this way, $P(1-P)^{i-1}$ is the probability that the last access to the considered line has taken place i iterations before of the loop on variable I. Function $P_{\text{cross WB}}(j)$ gives the cross interference probability generated by the accesses to other matrices and vectors during the processing of j subrows of the sparse matrix. The value is calculated as

$$P_{\text{cross WB}}(j) = S_D(j) \cup S_A(j) \cup S_C(j) \cup S_s(j \cdot r) \cup S_s(j \cdot r) \tag{3}$$

where the last two terms stand for the area covered by the accesses to R and R2. The areas corresponding to the references to matrix D and vectors A and C are calculated using a deterministic algorithm further developed in [2].

The last addend in expression (2) stands for the probability of the reuse of a line that has not been referenced yet in loop I (probability $(1-P)^{j-1}$) but that has remained in the cache since the copying process. This probability of hit in the reuse means excluding both self (probability $(1-P)^{j \cdot n_1}$) and cross interferences $(1-P_{\text{cross WB}}(j))$ as well as being in cache when the copying has just finished ($P_{\text{surv WB}}$, whose calculation is not included here due to space limitations).

The number of misses on matrix WB in the inner loop is calculated multiplying the average miss probability by the number of accesses to the first element of a line in this loop, as only in the access to the beginning of a line can a miss take place:

$$F_{\text{inner WB}} = \frac{\left(1 - \frac{\sum_{j=1}^{m} P_{\text{hit WB}}(j)}{M}\right)\left(\frac{BJ+L_s-1}{L_s}\right)}{\left(\frac{N_{nz} \cdot BK}{N}\right) N_{BJ} \cdot N_{BK}} \tag{4}$$

3.2 Misses on Matrix WB in the Copying

The number of misses during the copying is obtained multiplying the average number of misses per copy process by the number of blocks, $N_{BJ} N_{BK}$. This value is estimated as

$$F_{\text{copy WB}} = \sum_{j=1}^{BJ} \sum_{i=1}^{\frac{BJ \cdot BK}{L_s}} A \cdot S_s(1) + \left(\frac{L_s}{BJ} - A\right) P_{\text{miss first}}(i,j) \tag{5}$$

where $A = \max\{0, \frac{L_s}{BJ} - 1\}$ is the average number of accesses after the first one to a given line of matrix WB during an iteration of the loop on line 8. As the equation shows, the miss probability for these accesses is the associated to an access to a line of B. As for $P_{\text{miss first}}(i,j)$, it is the probability of a miss for the first access to line i during iteration j of the loop, which is calculated as

$$P_{\text{miss first}}(i,j) = P_{\text{acc}}(j-1)(S_{\text{self}}(BJ, BK) \cup S_s(BK)) \\ +(1 - P_{\text{acc}}(j-1))((1 - P_{\text{hit WB}}(M)) \cup P_{\text{exp WB}}(i,j)) \tag{6}$$

where $P_{\text{acc}}(j)$ if the probability of the line having been accessed in the j previous iterations, which is $\min\{1, (L_s + j - 2)/BJ\}$ but for $j = 0$, for which the probability is null. The first access to a line results in a miss if it is not in the cache when the copying begins (probability $1 - P_{\text{hit WB}}(M)$) or if it has been replaced during the copy of the elements previously accessed, which is $P_{\text{exp WB}}(i, j)$. On the other hand, if the line has been accessed in the previous iteration, only the accesses to BK elements of matrix B located in two consecutive columns (whose area we approach by a completely sequential access) and the accesses to a row of matrix WB can have replaced the considered line. This last probability is calculated using the deterministic algorithm we have mentioned above.

4 Cache Behavior of the Algorithm

The model was validated with simulations on synthetic matrices made by replacing the references to memory by functions that calculate the position to be accessed and feed it to the dineroIII cache simulator, belonging to the WARTS tools [4]. Table 2 shows the model accuracy for some combinations of the input parameters using synthetic matrices. The average error obtained in the trial set was under 5%.

Table 2. Predicted and measured misses and deviation of model for optimized sparse matrix-dense matrix product with an uniform entries distribution. Order $(M = N)$, N_{nz} and H in thousands, C_s in Kwords and number of misses in millions

Order	N_{nz}	p_n	H	BJ	BK	C_s	L_s	predicted misses	measured misses	Dev.
2	20	0.005	0.2	25	500	8	4	1.524	1.516	-0.55%
2	20	0.005	0.2	25	500	16	4	1.162	1.163	0.10%
2	20	0.005	0.2	25	500	32	4	0.962	0.972	1.08%
2	20	0.005	0.2	25	500	8	8	0.887	0.872	-1.70%
2	20	0.005	0.2	25	500	16	8	0.667	0.660	-0.99%
2	20	0.005	0.2	25	500	32	8	0.552	0.553	0.17%
10	100	0.001	10	20	1000	8	4	646	672	4.05%
10	100	0.001	10	20	1000	16	4	608	612	0.72%
10	100	0.001	10	20	1000	32	4	549	546	-0.45%
10	100	0.001	10	20	1000	64	4	490	492	0.41%
10	100	0.001	10	20	1000	8	8	401	419	4.40%
10	100	0.001	10	20	1000	16	8	373	378	1.35%
10	100	0.001	10	20	1000	32	8	321	323	0.65%
10	100	0.001	10	20	1000	64	8	287	290	0.86%

As a first result of our modeling, a linear increase of the number of misses with respect to N and M due to blocking technique is shown. The exception are the accesses to matrix D, as they have an access stride M, which makes them increase noticeably when M is a divisor or multiple of C_s. Figure 2 illustrates the evolution of the number of misses with respect to the block size for a given matrix. We have supposed a value of BJ between 8 and 30, this is, that there

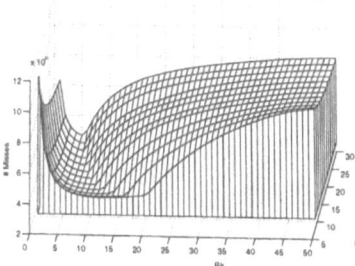
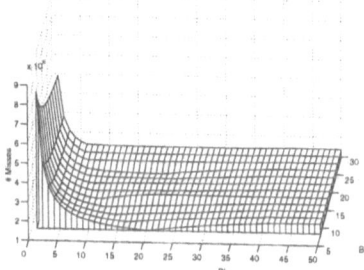

Fig. 2. Number of misses on a 16Kw cache with $L_s = 8$ for a $M = N = 5000$ sparse matrix with $p_n = 0.02$ and $H = 100$ as a function of the block size (BK in hundreds)

Fig. 3. Number of misses on a 16Kw cache with $L_s = 8$ for a $M = N = 5000$ sparse matrix with $p_n = 0.002$ and $H = 100$ as a function of the block size (BK in hundreds)

are up to 30 registers available for the blocking at the register level. The number of misses grows in the two directions of the axis tending asymptotically to a maximum when $BJ \times BK > C_s$. The minimum number of misses is obtained in a case in which the block fits completely in the cache ($BJ = 22$, $BK = 700$). Besides, among all of the possible block sizes the best has been the one with the greatest value of BJ that multiplied by a multiple of 100 (the variation of the BK value in the figure is 100 and that of BJ is 2) gives the greatest value under C_s. The reason is that the greater BJ is, lesser blocks there are in the J direction, reducing the number of accesses. In addition, accesses to matrix **WB** in the inner loop take place sequentially in the J direction, whose components are stored in consecutive memory positions. As a result exploitation of the spatial locality is improved with the block size increase in this direction. Simulations showed that the actual optimum block was ($BJ = 20$, $BK = 800$), with 5.6% less misses, which is a slight difference due to the small errors of the model.

Although the evolution of the number of misses usually follows the previous structure, we can obtain important variations depending on several factors. For example, Fig. 3 shows a similar plot for a matrix with lesser p_n. Here the optimum block size is ($BJ = 24$, $BK = 2000$), which is much greater than C_s (almost three times). This is because p_n is much smaller in this matrix, reducing remarkably the replacement probability due to self and cross interferences. In this way much greater blocks may be used, reducing the number of accesses and misses. On the other hand, although there are little variations, the number of misses is stabilized when the block size is similar to or greater than C_s because the interference increase is balanced with the lesser number of blocks to be used. The optimum block size was checked with simulations which indicated the same value for BJ and a value of 2100 for BK, with 0.04% less misses.

Finally, in Fig. 4 we study the cache behavior as a function of the cache parameters. *Log Cs* axis stands for the logarithm base two of C_s measured in Kw. There is an increase of the number of misses when the block size is greater

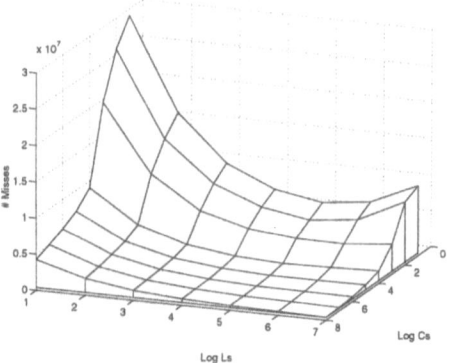

Fig. 4. Number of misses for a $M = N = 5000$ sparse matrix with $p_n = 0.02$ and $H = 100$ using a block $BJ = 22$, $BK = 700$ as a function of C_s and L_s

than C_s. Besides this value is always much greater for small line sizes ($L_s \leq 8$ words). The optimum line size turns out to be 32 when the cache size is greater than or similar to that of the block. This means a balance between the spatial locality exploitation in the access to **WB** in the inner loop of the algorithm and he self interference probability. As C_s grows, so does the optimum line size due the reduction in the self and cross interference probabilities.

5 Conclusions

The proposed model may be parameterized and considers matrices with an uniform distribution. It significantly extends the previous models in the literature as it incorporates the three possible types of misses (intrinsic misses, self and cross interferences) in the cache and it takes into account the propagation of data between successive loops of the code (reusability). Besides the modeled code size is noticeably greater than those used in most previous works and their predictions are highly reliable. An effort has been done to structure the model in algorithms and formulae related to typical access patterns so that they can be reused when studying other codes.

The time required by our probabilistic model to get the miss predictions is much smaller than that of the simulations, being this difference larger the larger the dimensions and/or sparsity degree of the considered sparse matrix. For example, the time required to generate the trace corresponding to the product of a matrix with $M = N = 1500$ and 125K entries with a block size ($BJ = 10$, $BK = 500$) by a matrix of the same size (almost 246M references) and process it with dineroIII simulating a 16Kw cache with a line size of 8 words was around 20 minutes on a 433 MHz DEC Alpha processor, while our model required between 0.1 and 0.2 seconds.

As for the time required to obtain the best block size for a given matrix and cache, the time required to do one simulation for each possible pair (BJ, BK)

in Fig. 2 would be around 54.5 hours on this machine. Anyway, as different locations of the vectors in memory generate different numbers of misses, some other simulations would be needed to get an average for each pair. On the other hand our model required only 350.5 seconds.

As shown in Sect. 4, it is possible to analyze the behavior of the number of misses with respect to the basic characteristics of the cache (its size and its line size), the block dimensions and the features of the matrix. Besides the behavior of each vector may be studied in a separate manner.

References

1. Barrett, R., Berry, M., Chan, T., Demmel, J., Donato, J., Dongarra, J., Eijkhout, V., Pozo, R., Romine, C., H. van der Vorst, H.: Templates for the Solution of Linear Systems: Building Blocks for Iterative Methods. SIAM Press (1994).

2. Fraguela, B.B.: Cache Misses Prediction in Sparse Matrices Computations. Technical Report UDC-DES-1997/1. Departamento de Electrónica e Sistemas da Univerdade da Coruña (1997).

3. Fraguela, B.B, Doallo, R., Zapata, E.L.: Modeling Set Associative Caches Behavior for Irregular Computations. To appear in Proc. ACM Sigmetrics/Performance Joint Int'l. Conf. on Measurement and Modeling of Computer Systems, Madison, Wisconsin, (1998).

4. Lebeck, A.R., Wood, D.A.: Cache Profiling and the SPEC Benchmarks: A Case Study. IEEE Computer, **27(10)** (1994) 15–26.

5. Navarro, J.J., García, E., Larriba-Pey, J.L., Juan, T.: Block Algorithms for Sparse Matrix Computations on High Performance Workstations. Proc. ACM Int'l. Conf. on Supercomputing (ICS'96) (1996) 301–309.

6. Temam, O., Jalby, W.: Characterizing the Behaviour of Sparse Algorithms on Caches. Proc. IEEE Int'l. Conf. on Supercomputing (ICS'92) (1992) 578–587.

7. Uhlig, R.A, Mudge, T.N.: Trace-Driven Memory Simulation: A Survey. ACM Computing Surveys **29** (1997) 128–170.

h-Relation Models for Current Standard Parallel Platforms

Rodríguez C., Roda J.L., Morales D.G., Almeida F.

Dpto. E.I.O. y Computación
Universidad de La Laguna - Tenerife - Spain

Abstract. This paper studies the validity of the BSP h-relation hypothesis on four current standard parallel platforms. The error introduced by the influence of the number of processors is measured on five communication patterns. We also measure the influence of the communication patterns on the time invested in an h-relation. The asynchronous nature of many current standard message passing programs do not easily fits inside the BSP model. Often this has been criticized as the most serious drawback of BSP. Based in the h-relation hypothesis we propose an extension to BSP model valid for standard message passing parallel programs. The use and accuracy of h-relation models on standard message passing programs are illustrated using a parallel algorithm to compute the Discrete Fast Fourier Transform.

1 Introduction

The development of a reasonable abstraction of parallel machines is a formidable challenge. A simple and precise model of parallel computation is necessary to guide the design and analysis of parallel algorithms. Among the plethora of solutions proposed, PRAM, Networks, LogP and BSP models are the most popular.

Section 2 introduces the BSP model. Sections 3 presents the four representative platforms used in our study: a 10 Mbits Coaxial Ethernet Local Area Network, an UTP Ethernet LAN, an IBM SP2 and a Silicon ORIGIN 2000. The IBM Scalable POWERparallel SP2 used in the experiments is a distributed memory parallel computer with 44 processors. Processors or nodes are interconnected through a High Performance Switch (HPS). The HPS is a bi-directional multistage interconnection network. The computing nodes are Thin2, each powered by a 66MHz Power2 RISC System/6000 processor. All the algorithms were implemented in PVMe [3], the improved version of the message-passing software PVM [2]. The Origin 2000 system we have used has 64 R10000 processors (196 MHz). Origin systems use distributed shared memory (DSM). To a processor, main memory appears as a single addressable space. A four-port crossbar switch interconnects the processor, the main memory, the router board and the I/O subsystem. The router board interfaces the node with the CrayLink Interconnect fabric. Both the 10 Mb/sec. Coaxial LAN and the UTP LAN are composed of Sun Sparc workstations at 70MHz running Solaris 2.4. The UTP LAN uses a FORE-SYSTEMS ES-3810 Ethernet Workgroup Switch that interconnects all the computers of the LAN. A theoretical 10 point-to-point Mb/sec is guaranteed. As for the Origin 2000, the PVM version used in both LANs was 3.3.11.

Instead of following the common classical approach of using a library adapted to the proposed model, like Active Messages for the LogP [1] or the Oxford BSP library for BSP [5], we tried to check the validity of the models in current Standard Message Passing libraries. The influence of the number of processors in the time spent in an h-relation is measured for the most common communication patterns in section 3. Both the LogP and the BSP models disregard the effects of the pattern. Most algorithms can be built around a small set of communication primitives such as one to all, all to one, all to all, permutations and reductions if an appropriate data layaout is used. Is it reasonable to treat everything as the general case? Section 4 measures the impact of communication patterns in the h-relation hypothesis and estimates the BSP-like gaps and latencies for the considered architectures. The asynchronous nature of many PVM/MPI programs do not easily fits inside the BSP model. Often this has been criticized as the most serious drawback of BSP. In section 5 we propose an extension to BSP, the EBSP model, for current standard message passing parallel programs. The use of the BSP and EBSP models in conventional PVM/MPI programs will be illustrated in this section using the Fast Fourier Transform. Conclusions are presented in Section 6.

2 The Bulk Synchronous Parallel Model (BSP)

The BSP model [Val90] views a parallel machine as a set of p processor-memory pairs, with a global communication network and a mechanism for synchronizing all the processors. A BSP calculation consists of a sequence of supersteps. The cost of a BSP program can be calculated simply by summing the cost of each separated superstep executed by the program; in turn, for each superstep the cost can be decomposed into: (i) local computation; (ii) global exchange of data and (iii) barrier synchronization. The communication pattern performed on a given superstep is called an h-relation if the maximum number of packets a processor communicates in the superstep is h.

$$h = max \{ in_i @ out_i / i \epsilon \{0, ..., p-1\} \}$$

Where @ is a binary operator, usually the maximum or the sum. Values in_i and out_i respectively denote the number of packets entering and leaving processor i. Both the particular operation @ and the size of the unit packet depend on the particular architecture. The operation @ is closer to the maximum ($@ = max$) for machines allowing a complete parallel processing of incoming/outgoing messages. The correct interpretation for each case depends on the number of input/output ports of the network interface. The results showed on this paper correspond to take as operator @ the sum.

The two basic BSP parameters that model a parallel machine are: the gap g, which reflects per-processor network bandwidth, and the minimum duration of a superstep L, which reflects the latency to send a packet through the network as well as the overhead to perform a global synchronization. The fundamental of the BSP model lays on the h-relation hypothesis. It states that the communication time spent on an h-relation is given by gh. Let denote by W the maximum time spent in local computation by any processor during the superstep. The BSP model guess that the running time of a superstep is bounded by the formula:

$$Time\ Superstep = W + gh + L$$

3 The Influence of the Number of Processors in the Validity of the h-relation Hypothesis for Five Different Patterns in Four Different Architectures

In the following paragraphs we will study the variation of the h-relation time on the four architectures with the number of processors under five different communication patterns: Exchange (abbreviated E), PingPong (PP), OneToAll (OA), AllToOne (AO) and AllToAll (AA). Five hundred experiments were carried out for each architecture, each pattern and each number of processors. The h-relation size varies between 6720 and 1720320 bytes. In all the tables presented in this work, time is given in seconds.

The rest of the section is dedicated to the different communication patterns Π in $\{E, PP, OA, AO, AA\}$. For each pattern Π, two tables denoted $\Pi.1$ and $\Pi.2$ are presented. Sub-column labeled $Time$ in tables $\Pi.1$ gives the times for 8 processors using the patterns Π. It is observed that the time $T_\Pi(h)$ spent on an h-relation size, follows a linear equation $T_\Pi(h) = L_\Pi + g_\Pi h$. To obtain the general linear approach to the h-relation time we have computed the least square fit of the average times for the different number of intervening processors. Tables $\Pi.2$ present the values of L_Π and g_Π for the four architectures. Columns labeled $MaxErr$ in tables $\Pi.1$ contain the maximum percentage of error computed according to the formula:

$$MaxErr = 100 * (max_i\{|Time_{\Pi,i}(h) - (L_\Pi + g_\Pi h)| / \ i \in H_\Pi\} / (min_i\{Time_{\Pi,i}(h)\})$$

Where $Time_{\Pi,i}(h)$ denotes the time spent when i processors communicate according to pattern Π. Index i varies in $H_\Pi = \{2, 4, 6, 8\}$ processors for the Exchange and PingPong patterns, and i is in $H_\Pi = \{4, 6, 8\}$ for the OneToAll, AllToOne and AllToAll patterns. An injection pattern is characterized by the existence of a subset S of processors of the total set of available processors H, such that each processor p of S sends its message to a different processor $d(p)$ of H. We say that an injection pattern is an Exchange when: $d(d(p)) = p$ for any processor p in S. An injection pattern is called a PingPong if and only if $S \cap d(S) = \emptyset$. The PingPong pattern gives place to an $h = m$-relation, where m is the message size. For the Exchange, the h-relation is $h = 2 * m$. Table E.1 shows a factor of almost 3 between the times of the Origin and the IBM SP2. The same factor appears between the UTP LAN and the COA LAN. However, the values in column $MaxErr$ prove that the IBM SP2 is among the 4 architectures the most invariant in the number of processors. It is remarkably the technological advance observed in the low value of MaxErr for the UTP LAN when compared with the

Table 1. The Exchange Pattern (E.1)

Exchange	ORIGIN 2000		IBM SP2		UTP LAN		COA LAN	
h-relation	Time	MaxErr	Time	MaxErr	Time	MaxErr	Time	MaxErr
6720	0.000080	26.43	0.000310	16.78	0.013278	26.25	0.033285	160.92
26880	0.000227	17.78	0.000839	7.48	0.046496	14.18	0.132507	133.94
107520	0.001058	4.71	0.003103	4.04	0.173081	7.57	0.573673	172.24
430080	0.004230	4.32	0.012738	0.48	0.674435	5.03	2.323600	179.46
1720320	0.017021	4.25	0.050868	0.69	2.745955	7.02	9.226720	177.65

Table 2. Values of g_E and L_E (E.2)

	ORIGIN 2000	IBM SP2	UTP LAN	COA LAN
L_E	-1.1759E-05	6.1566E-05	-9.3670E-04	-3.2527E-04
g_E	9.8942E-09	2.9675E-08	1.5923E-06	5.3772E-06

178% of the COA LAN. The negative values of the Exchange latency L_E in table E.2 are due to the unsuitability of taking the byte as packet size unit. However, this choice was determined by the goal of comparing different architectures. The PingPong times appear in Table PP.1. As for the Exchange, there is an almost perfect invariance of the IBM SP2 communication time in the number of intervening couples. For this pattern $h = m$, opposite to what occurs for the Exchange, there is no parallelism between inputs and outputs. The time ratio between the IBM SP2 and the Origin diminishes from three to a factor of two. The same decreasing is observed for the UTP LAN and COA LAN. There is a decreasing both in the time and in the MaxErr columns for the COA LAN architecture. This is due to the smaller number of collisions.

The Personalized One to All communication pattern measures the outbound node performance. From Table OA.1 follows that the OneToAll communication time is better aproached by a linear by pieces function, but, fore sake of simplicity, we approach its behavior by a single linear function. Although it no appears in the tables, for small values of h, the time slowly grows with the number of processors. This is due to the heavier influence of the latency. For larger values of h, time slightly decreases with the number of processors, due to the increasing parallelism introduced in the communications. This phenomenon is specially outstanding for the COA LAN [7]. Observe the decreasing in the g value of the COA LAN to $1.072E - 06$ from $4.08E - 06$ for the PingPong pattern. For that reason the OneToAll is the fastest pattern for the COA LAN architecture.

The Personalized All to One Communication Pattern measures the inbound node performance. Each processor sends a distinct message of size m to the receiver processor. This experiment was implemented by making the receiver processor call the routine $pvm\_recv()$ $p - 1$ times, and the senders making a $pvm\_send()$. As for the OneToAll, we will approach the AllToOne time by a linear function $L_{AO} + g_{AO} * m * (p - 1)$ in the h-relation size: $h = m * (p - 1)$.

Table 3. The PingPong Pattern (PP.1)

PingPong	ORIGIN 2000		IBM SP2		UTP LAN		COA LAN	
h-relation	Time	MaxErr	Time	MaxErr	Time	MaxErr	Time	MaxErr
6720	0.000107	54.26	0.000386	8.58	0.015953	20.08	0.027919	107.54
26880	0.000454	7.53	0.001197	10.87	0.056722	6.97	0.112579	88.11
107520	0.001945	2.29	0.004811	0.87	0.220280	3.59	0.443787	80.56
430080	0.008132	2.95	0.018752	0.24	0.850265	2.37	1.773765	80.13
1720320	0.032797	4.16	0.074292	0.52	3.376577	6.48	7.023510	80.97

Table 4. Values of g_{PP} and L_{PP} (PP.2)

	ORIGIN 2000	IBM SP2	UTP LAN	COA LAN
L_{PP}	-7.5956E-05	1.5691E-04	6.2546E-03	5.6284E-03
g_{PP}	1.9071E-08	4.3467E-08	1.9609E-06	4.0818E-06

Table 5. The OneToAll Pattern (OA.1)

OneToAll	ORIGIN 2000		IBM SP2		UTP LAN		COA LAN	
h-relation	Time	MaxErr	Time	MaxErr	Time	MaxErr	Time	MaxErr
6720	0.000248	141.77	0.000383	47.26	0.022926	39.91	0.018732	25.57
26880	0.000354	20.09	0.001022	9.73	0.046733	11.31	0.033738	13.01
107520	0.000962	37.79	0.003597	4.37	0.170056	17.36	0.118535	10.19
430080	0.005203	8.60	0.014700	2.46	0.744115	23.66	0.471055	10.21
1720320	0.020398	9.06	0.059125	2.73	2.818493	17.40	1.846068	7.28

Table 6. Values of g_{OA} and L_{OA} (OA.2)

	ORIGIN 2000	IBM SP2	UTP LAN	COA LAN
L_{OA}	-3.4072E-05	2.0023E-05	1.0634E-02	7.6290E-03
g_{OA}	1.1704E-08	3.4320E-08	1.6493E-06	1.0725E-06

Table 7. The AllToOne Pattern (AO.1)

AllToOne	ORIGIN 2000		IBM SP2		UTP LAN		COA LAN	
h-relation	Time	MaxErr	Time	MaxErr	Time	MaxErr	Time	MaxErr
6720	0.000232	142.86	0.000449	26.87	0.015997	74.77	0.021609	65.20
26880	0.000281	37.37	0.001133	14.50	0.027941	12.98	0.031973	13.64
107520	0.000745	19.97	0.003719	5.63	0.114966	5.81	0.125891	8.10
430080	0.003445	6.19	0.014139	3.21	0.469320	5.78	0.512352	6.11
1720320	0.014098	9.78	0.057669	1.40	1.814855	7.00	2.006803	4.93

Table 8. Values of g_{AO} and L_{AO} (AO.2)

	ORIGIN 2000	IBM SP2	UTP LAN	COA LAN
L_{AO}	4.0212E-05	9.1160E-05	3.5069E-03	6.9769E-03
g_{AO}	8.1586E-09	3.3386E-08	1.0521E-06	1.1619E-06

Table 9. The AllToAll Pattern (AA.1)

AllToAll	ORIGIN 2000		IBM SP2		UTP LAN		COA LAN	
h-relation	Time	MaxErr	Time	MaxErr	Time	MaxErr	Time	MaxErr
6720	0.000411	181.56	0.000694	46.40	0.040892	104.27	0.046908	286.30
26880	0.000529	76.85	0.001287	20.63	0.059168	12.76	0.221194	120.10
107520	0.001091	16.81	0.003605	5.68	0.170425	7.96	0.655296	93.53
430080	0.004102	16.36	0.014036	2.87	0.700719	6.14	2.348122	78.44
1720320	0.019053	3.70	0.055054	3.39	2.749624	4.26	9.041233	73.94

Unless for the COA LAN, for all the architectures, the inbound bandwidth g_{AO} has decreased compared with the outbound bandwidth g_{OA}. This results to be the fastest pattern for the Origin and UTP LAN architectures.

In the Personalized All to All Communication Pattern each processor has to send $p-1$ different messages of length m to the other $p-1$ processors. Processor i cyclically sends the messages, to processors $i+1, i+2, ..., i-1$. The AllToAll time can be modeled by a linear function: $L_{AA} + g_{AA}2m(p-1)$. The IBM SP2 achieves its better performance for this pattern. While the performance of the UTP LAN only doubles the COA LAN performance for a pattern free of collisions like the PingPong, it is more than three times faster for this pattern. This is a consequence of the improving achieved in the bisection bandwidth.

4 The Influence of the Communication Pattern in the h-relation

BSP states that the actual times spent on these five patterns for the same h-relation have to be similar. The influence of the communication pattern in the time spent in an h-relation is shown in Table 11. To obtain the general linear approach to the h-relation time, we have computed the least square fit of the average times $T_{average}$ of the set Π of patterns and the different number of processors:

$$T_{average}(h) = \Sigma_{k\epsilon\Pi}(\Sigma_{i\epsilon H_\Pi}Time_{\Pi,i}(h)/|H|)/|\Pi|$$

These values of L and g appear in Table 12. There is a factor of ten between the BSP-PVM values for the IBM SP2 $g = 3.44*E-08$ and the corresponding Oxford BSP library values: $g' = 35 * E - 8$, $L' = 4.62 * E - 4$ for the same machine

Table 10. Values of g_{AA} and L_{AA} (AA.2)

	ORIGIN 2000	IBM SP2	UTP LAN	COA LAN
L_{AA}	2.3919E-05	3.9166E-04	1.2658E-02	8.7903E-02
g_{AA}	1.0984E-08	3.1578E-08	1.5957E-06	5.2131E-06

Table 11. AvErr and MaxErr for all the architectures

	ORIGIN 2000		IBM SP2		UTP LAN		COA LAN	
h-relation	AvErr	MaxErr	AvErr	MaxErr	AvErr	MaxErr	AvErr	MaxErr
6720	59.82	304.00	20.58	70.37	29.82	141.41	42.59	10.13
26880	25.75	103.00	8.24	19.34	15.81	20.07	42.30	86.66
107520	26.11	108.63	12.59	33.01	13.67	5.04	43.15	49.51
430080	25.97	120.06	10.76	30.38	13.41	19.12	42.96	57.48
1720320	24.79	126.06	10.81	29.61	13.83	81.88	42.81	228.00

[4]. Columns labeled *AvErr* and *MaxErr* respectively show the average and maximum errors defined as:

$$AvErr = 100 * ((\Sigma_{i \epsilon \Pi} |T_i(h) - (g * h + L)|/|\Pi|)/(\Sigma_{i \epsilon \Pi} T_i(h)/|\Pi|))$$
$$MaxErr = 100 * ((max_{i \epsilon \Pi} |T_i(h) - (g * h + L)|)/(min_{j \epsilon \Pi} T_j(h)))$$

The table proves that the time invested in moving data in the IBM SP2 and the UTP LAN is more independent of the specific communication pattern than in the other two architectures.

5 Extending the BSP Model to Current Standard Message Passing Libraries: EBSP

The barrier synchronization after each step imposed by BSP, does not completely agree with the way PVM and MPI programs are written. Still, PVM and MPI programs can be divided in "Message steps" that we will call "M-steps" in roughly the same sense than BSP "supersteps". In a "M-step" a processor $i = 0, ..., p-1$, performs some local computation, send the data needed by other processors and receives the data it needs for the next M-step. Processors may be in different M-steps at a given time, since no global barrier synchronization is used. However, as in pure BSP we assume that the total number of M-steps R, performed by all the p processors is the same, and communications always occur among processors in adjacent steps $k - 1$ and k (computation on any processor can be arbitrarily divided to achieve this goal). The time $t_{s,i}$ when processor i finishes its step s is bounded by the "BSP-like-time" T_s given by:

$$T_1 = max\{w_{1,i}\} + g * max\{in_{1,i}@out_{1,i}\} + L$$
$$T_s = T_{s-1} + max\{w_{s,i}\} + g * max\{in_{s,i}@out_{s,i}\} + L \tag{1}$$

Table 12. The values of g and L for all the architectures

	ORIGIN 2000	IBM SP2	UTP LAN	COA LAN
L	-1.5327E-05	8.0835E-05	3.8229E-03	1.3220E-02
g	1.2192E-08	3.4454E-08	1.5107E-06	2.4318E-06

where $i = 0, 1, ..., p - 1$, $s = 2, ..., R$; @ is in $\{+, \ max\}$; $w_{s,i}$ is the time spent in computing by processor i in M-step s; $in_{s,i}$ and $out_{s,i}$ respectively denote the number of messages sent and received by processor i in the step s. Gap and Latency values g and L can be computed as proposed in the former paragraph. Thus the "BSP-like-time", T_R gave us an upper bound approximation to the execution time of a PVM/MPI program.

A closer bound to the actual PVM/MPI time $t_{s,i}$ when processor i finishes its s-th M-step is the value $\Phi_{s,i}$ given by the **Extended BSP model (EBSP)** we propose here. Let us define for a given step s and processor i the set $\Omega_{s,i}$ of **in-partners** of i in step s as $\Omega_{s,i} = \{ j \ / $ processor j sends a message to processor i in step $s \ \} \cup \{ i \}$. The **EBSP** time of an PVM/MPI program is given by the formulas:

$$
\begin{aligned}
\Phi_{1,i} &= max\{w_{1,j} \ / \ j \epsilon \ \Omega_{1,i}\} + g * h_{1,i} + L \\
h_{1,i} &= max\{in_{1,j} + out_{1,j}/j \ \epsilon \ \Omega_{1,i}\} && i = 0, 1, ..., p-1, \\
\Phi_{s,i} &= max\{\Phi_{s-1,j} + w_{s,j}/j \ \epsilon \ \Omega_{s,i}\} + g * h_{s,i} + L && and \\
h_{s,i} &= max\{in_{s,j} + out_{s,j}/j \ \epsilon \ \Omega_{s,i}\} && s = 2, ..., R
\end{aligned} \quad (2)
$$

The total time of a PVM/MPI program in the **ESBP** model is given by $\Psi = max\{\Phi_{R,j}/j \ \epsilon \ \{0, ..., p-1\}\}$ where R is the total number of steps. Instead a global barrier synchronization, the EBSP model implies a synchronization among partners. Formula (2) becomes the BSP-like time of formula (1) when the family of sets $\Omega_{s,i}$ is the whole set of processors $\{0, ..., p-1\}$. (This is the case, if the BSP barrier synchronization implies $\Omega_{s,i} = \{0, ..., p-1\}$ for all i.

5.1 Example: The Fast Fourier Transform

We will illustrate the BSP and EBSP models using the parallel algorithm to compute the Fast Fourier Transform described in [6]. The total time predicted by the model for this algorithm is:

$$
T_{log(p)+1} = \Psi = D(N/p) + FN/p \ log(N/p) + \sum_{i=0}^{log(p-1)} (L + g(N2^{i/p}) + VN2^{i/p}) \quad (3)
$$

Table 13 contains the values of the three computational constants D, F and V. The good accuracy of the h-relation model for a 524.288 floats FFT example is shown in Table IV. Columns labeled MODEL in Table 14 present the time computed according to formula (3). Under the REAL label are the actual times.

Table 13. Division (D), sequential FFT (F) and combinations (V) constants

Constants	D	F	R
ORIGIN 2000	2.3928E-07	2.8453E-07	4.8103E-07
IBM SP2	5.5161E-07	5.8598E-08	8.6916E-07
UTP-COA LAN	6.0400E-07	1.4400E-06	2.6500E-06

Table 14. FFT real times versus h-relation model time

Times	2	4	8
ORIGIN 2000	1.61	0.96	0.85
MODEL	1.55	0.86	0.56
ERROR	3.68	10.85	34.10
IBM SP2	3.36	1.90	1.27
MODEL	3.21	1.83	1.18
ERROR	1.59	3.85	6.82
UTP LAN	12.36	11.31	11.03
MODEL	11.00	8.41	7.27
ERROR	11.01	25.65	34.13
COA LAN	11.91	10.80	10.94
MODEL	12.76	12.01	11.73
ERROR	7.14	11.17	7.22

Entries in columns ERROR give the error percentage computed by the formula $ERROR = 100 * (REAL - MODEL)/REAL$.

The accuracy of the model is specially good for the IBM SP2 and acceptable for the ORIGIN 2000 and UTP LAN. The error grows with the number of processors. The curious exception is the Coaxial LAN. On this architecture, the time of the model for the two first PingPong communications ($i = 0$ and $i = 1$) is over the actual time while it is under the actual time of the third communication ($i = 2$) producing a compensation that leads to a false accuracy. In fact, this is the only architecture where the model fails its prediction of a decreasing behavior in the actual time (see columns 4 and 8 in row COA LAN).

6   Conclusions

We have studied the validity of the h-relation hypothesis on four representative different platforms: a 10 Mbits Coaxial Ethernet Local Area Network, an UTP Ethernet LAN, an IBM SP2 and a Silicon ORIGIN 2000. Current Standard Message Passing libraries have been used instead of specific BSP environments. The maximum and average errors introduced by the influence of the pattern and the number of processors, has been measured. The collective computation provided by MPI and the new version of PVM 3.4 makes feasible the use of a methodology compatible with the Bulk Synchronous Programming Model.

However, the more relaxed nature of many PVM/MPI programs does not easily fit inside the BSP model. We have proposed a new model, EBSP, that extends the BSP model to current standard message passing parallel programs. The use and accuracy of the BSP and EBSP model has been illustrated using a Fast Fourier Transform example.

Acknowledgments Part of this research has been done at C4 (CESCA-CEPBA) and at IAC in Tenerife.

References

1. Culler, D., Karp, R., Patterson, D., Sahay, A., Schauser, K.E., Santos, E., Subramonian, R., Eicken, T. LogP: Towards a Realistic Model of Parallel Computation. Proceedings of the 4th ACM SIGPLAN, Sym. Principles and Practice of Parallel Programming. May 1993.
2. Geist, A., Beguelin, A., Dongarra, J., Jiang, W., Mancheck, R., Sunderam, V. PVM: Parallel Virtual Machine - A Users Guide and Tutorial for Network Parallel Computing. MIT Press. 1994.
3. IBM PVMe for AIX User's Guide and Subroutine Reference Version 2, Release 1. Document number GC23-3884- 00. IBM Corp. 1995.
4. Marin, J., Martinez, A. Testing PVM versus BSP Programming, VIII Jornadas de Paralelismo, Sept 1997. pp.153-160.
5. Miller, R., Reed, J.L. The Oxford BSP Library Users' Guide. Technical Report, Programming Research Group, University of Oxford. 1993
6. Roda, J., Rodriguez, C., Almeida, F., Morales, D.G. Predicting the Performance of Injection Communication Patterns on PVM. 4th European PVM/MPI Users' Group Meeting. Springer-Verlag LNCS. pp. 33-40. Cracow. Nov-97.
7. Roda, J., Rodriguez, C., Almeida, F., Morales, D.G. Prediction of Parallel Algorithms Performance On Bus-Based Networks using PVM. 6th EuroMicro Workshop on Parallel and Distributed Processing. pp. 57-63. Madrid. Jan-98
8. Valiant, L.G. A Bridging Model for Parallel Computation. Communications of the ACM, 33(8): 103-111, August 1990.

Practical Simulation of Large-Scale Parallel Programs and Its Performance Analysis of the NAS Parallel Benchmarks

Kazuto Kubota[1] Ken'ichi Itakura[2] Mitsuhisa Sato[1] and Taisuke Boku[2]

[1] Real World Computing Partnership, Tsukuba, Japan
[2] University of Tsukuba, Tsukuba, Japan

Abstract. A simulation technique for very large-scale data parallel programs is proposed. In our simulation method, a data parallel program is divided into computation and communication sections. When the control flow of the parallel program does not depend on the contents of network messages, the computation time on each processor is calculated independently. An instrumentation tool called EXCIT is used to calculate the execution time on the target architecture and generate message traces. The communication time is calculated on the message traces by using a network simulator, which is generated by a network simulator generating system INSPIRE. With our tool set, the behavior of parallel programs on thousands processors can be estimated within a practical time span. We demonstrate our method to analyze the class B problems of LU and MG programs of the NAS Parallel Benchmarks with various parameters such as cache size and network bandwidth examined. We found that communication overhead affects the total execution time considerably, while cache effect is small.

1 Introduction

Recent massively parallel processors (MPP) architectures are scalable to very large numbers of processors. Some huge scale MPP systems with thousands of processors, such as CP-PACS[1] and ASCI Red[2], have actually been built.

In this paper, we describe a practical simulation technique for very large-scale data parallel programs. While the exact simulation of a parallel system is an effective technique to analyze the performance of such large-scale systems in designing a new large-scale systems, it is difficult to simulate its behavior in practical time, especially in case of large-scale parallel systems. In several approaches[3][4], analytical modeling is used to understand the scalability of systems. However, it is difficult to examine the user program's behavior in detail because many factors such as program implementation, compiler optimization and network collision are not taken into account. For the system with only a small number of Processing Units (PU), there is another approach that the entire system is actually simulated by both instruction simulator and network simulator[5]. Due to the limitation of the simulation time and the memory size on a simulation platform, it is difficult to apply this approach to large-scale parallel programs.

Our simulation approach is based on code instrumentatio technique and network simulation. At the first step, we execute the instrumented version of the original parallel program independently to calculate the execution time of the computation part of the parallel program on the target architecture and to get its message trace files. We assume that the control flow of the parallel program does not depend on the contents of network messages. Then, we compute the communication time in the target network using the message trace files. We use the instrumentation tool EXCIT[6] to instrument codes to estimate the computation time in the target processor. INSPIRE[7] is a network simulator generating system which generates the network simulator given by a target network description. Our simulation technique has the following advantages:

- Up to thousands of processors can be simulated in a reasonable time.
- Real size of programs can be analyzed.
- Computation and communication overlap can be estimated precisely.

To demonstrate the effectiveness of our technique, we apply it to analyze the class B size MG kernel and the LU application in the NAS Parallel Benchmarks[8]. We analyze the performance of these programs, varying the number of processors up to 1024 processors, some parallel processor parameters such as cache size and network parameters including network topology, bandwidth, and communication overhead. The simulation results help us to determine which hardware parameters are suitable for parallel programs. This information is also effective in selecting an appropriate platform and number of processors and program tuning.

2 Simulation Method of Large-Scale Parallel Programs

Our goal in this study is to analyze the performance of a large-scale parallel system. Data parallel program is one of important models in large-scale scientific applications. We assume that our simulated program is a data parallel program and the control flow of the parallel program does not depend on the contents of network messages.

2.1 Structure of Data Parallel Program

The typical structure of a data parallel program using four processors is shown in Figure 1. The data parallel program is decomposed into the following sections:

- **Computation section**: The section in which only computation is executed (**Thick** lines in PU0). The execution time of the computation section is denoted by *Tcomp*.

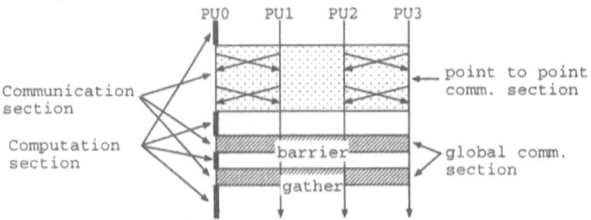

Fig. 1. Structure of data parallel program.

- **Communication section**: The section in which computation and communication are executed (Thin lines in PU0). The execution time of the computation section is denoted by *Tcomm*. The communication section is divided into further two sections.
  - **Point to point communication section**: The section which contains **send** /**recv** and the computations surrounding **send** / **recv** . The start and end points of this section is specified by the user. The execution time of this section(Tpp) consists of computation time(Tpp_{comp}) and communication time(Tpp_{comm}).
  - **Global communication section**: The section of global communications such as **barrier** and **gather** . The execution time of the global communication section(Tg) consists of computation time(Tg_{comp}) and communication time(Tg_{comm}).

2.2 Simulation Steps

Figure 2 illustrates the overview of our simulation system. The simulation steps are as follows:

- The original parallel programs (prog#0~prog#3) are instrumented to generate the instrumented version of codes (prog'#0~prog'#3, prog"#0).
- prog'#0~prog'#3 are executed independently to generate the message trace. The computation time on the target architecture is also calculated by another instrumented code(prog"#0).
- Network simulation on these message trace is performed to compute the communication time.

The instrumentation tool, called EXCIT, is used to instrument the original code to compute *Tcomp* on the target architecture and generate the message trace into the files. Message passing functions in prog"#0 are replaced by dummy functions. *Tcomp* of PU0 on the target architecture is calculated by executing prog"#0. The execution time of the computation section is about the same as the time when that program is executed as a parallel program, because we assume that the control flow of the parallel program does not depend on the contents of messages. In most data parallel programs, the computation time of

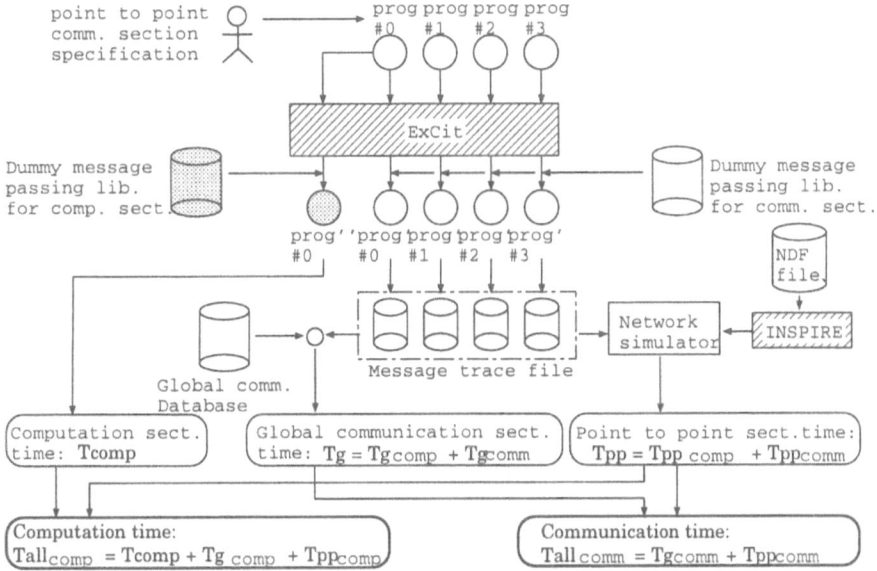

Fig. 2. Overview of Data parallel program simulation.

Fig. 3. Message trace file.

each PU program is approximately the same. Therefore, we just only examined the *Tcomp* of the PU0 program.

Message traces are generated by executing prog'#0~prog'#3. Communication functions in prog'#0~prog'#3 are replaced by trace functions which output communication function information such as source PU, destination PU, tags etc. Figure 3 shows examples of the message trace files. Each PU has its trace file. In the examples, "recv" shows a block receive. Source and destination and message tag are described, but message data is not recorded. "exec" shows the consumed cycles between communication functions. Computation and communication overlap can be calculated precisely, because communications are simulated with their preceding and succeeding computations.

By using a network simulator generated by INSPIRE, Tpp is calculated on the message trace files. Tpp_{comp} is calculated by summing the "exec" entry of each message trace file, and taking the maximum result of all files. Tpp_{comm} is

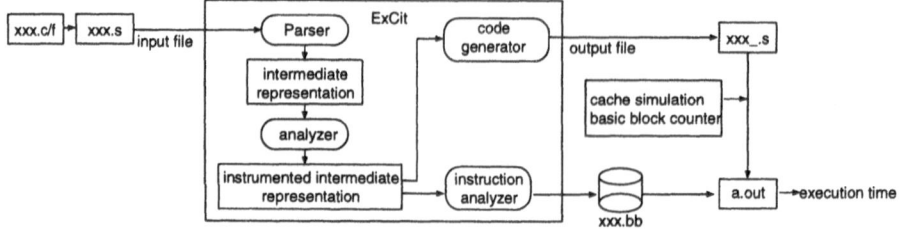

Fig. 4. Execution cycle calculation by EXCIT.

calculated by subtracting Tpp_{comp} from Tpp. The number of global communications are extracted from the message trace files and Tg_{comp} and Tg_{comm} are obtained by referring to the database file. The computation and communication time of global communications appeared in the parallel program are calculated in advance and stored in the database file.

Finally, the computation time($Tall_{comp}$) and the communication time($Tall_{comm}$) of the parallel program are obtained by the following equation.

$$Tall_{comp} = Tcomp + Tpp_{comp} + Tg_{comp}, \quad Tall_{comm} = Tpp_{comm} + Tg_{comm}$$

2.3 EXCIT and Execution Cycle Counting

EXCIT is an assembler level instrumentation tool to insert additional user code into the original program. In our simulation system, the instrumented code calculates the execution cycle of the program at execution time Figure 4 shows its overview. In EXCIT, the parser translates the **xxx.s** input file into an intermediate representation. This intermediate representation is then analyzed by the analyzer and information is inserted into the intermediate representation as follows.

- The source code is divided into basic blocks. Hook functions to call a basic block counter is inserted at the top of each of the basic blocks.
- The load and store instructions are extracted. Hook functions are inserted to call a cache simulator before the load and store instructions.

The code generator generates an instrumented code **xxx_.s** from the instrumented intermediate representation. This code is linked with the basic the block counter and the cache simulator, and an object code(**a.out**) is generated.

Instruction analyzer calculates the typical execution cycle of each basic block and generates a table file **xxx.bb**; the typical execution cycle of each basic block is recorded in this file. The typical execution cycle is the execution cycle of each basic block under the following conditions.

- The instruction cache and data cache are hit.
- The floating point operation latency is the typical cycle.
- No exceptions happen.

The typical execution cycle is calculated by simulating the instructions in all basic blocks one time.

An execution cycle is obtained by summing the typical execution cycles of the basic blocks and cache miss penalties. Assume a user specifies a section where the execution cycle is to be calculated. When the program enters the section, the cycle counter is cleared to zero. Every time each basic block is passed, its typical execution cycle is added to the cycle counter. Concurrently, a cache simulation is performed when a load or a store instruction is executed. When the section is finished, the cache miss penalties are added to the counter, and the execution cycle of the current section is obtained. It should be noted that the execution cycle changes according to the cache performance. We applied this execution cycle calculation method to the SPECfp92[9] benchmark programs. The error ratio is within 20% when compared with the real execution time. In our experiments, a MicroSparc-II CPU model was used.

2.4 Network Simulator Generating System INSPIRE

INSPIRE generates a clock level network simulator from NDF (Network Description FILE). NDF describes the hardware resources of a network, a network connection and a routing algorithm. Many kinds of network topology descriptions, including 2D mesh and Hyper-cube, are provided as libraries. Message trace files are taken as input into the network simulator and a simulation is performed. The user can specify several network parameters such as bandwidth and overheads in the network simulation.

3 Performance Analysis of the NAS Parallel Benchmarks

In this section, using our simulation technique, we analyze the class B problem size MG kernel and the LU application in the NAS Parallel Benchmark suite (ver.2.3). In this simulation, we examine the effect of cache size, network topology, communication overhead(μ_0), bandwidth(b/w), and the number of PUs.

3.1 Simulation Setup

In the simulation, we used 100 MHz, 2 integer pipes and 1 floating pipe CPU model. It has four way set associative, logical mapped L1 cache and its size varied 8192, 16384 and 65536 Byte. L1 miss penalty is 5 cycles. L2 cache size is 1 MByte and its miss penalty is 20 cycles. The number of PUs varied 16, 64, 128, 512 and 1024 in the MG, and 16, 32, 64, 128, 256 in the LU.

First, we show the effect of cache size on computation time. **L1 cache** sizes used here are **8192, 16384** and **65536** Bytes. Next, we examine the effect of network topology on communication time. A complete network(**complete**), a two dimensional mesh network(**2d mesh**), Hyper Crossbar Network(**HXB**)[1] and a ring network(**ring**) are compared. In the HXB and the two dimensional mesh network, we choose the nearest cubic shape and the nearest square shape

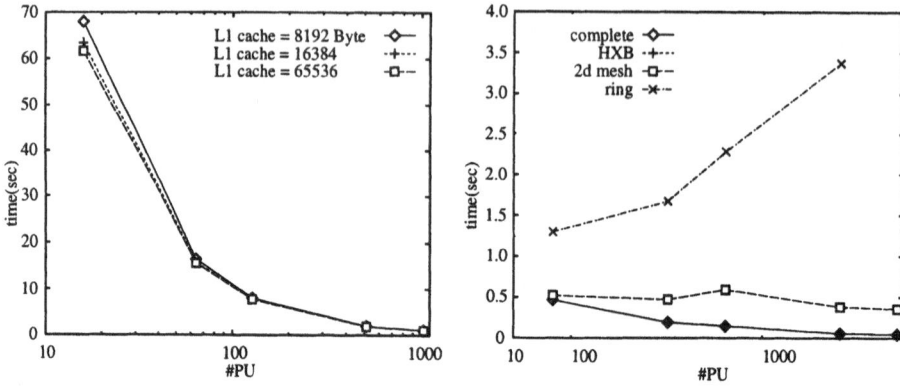

Fig. 5. MG: Cache size effect on computation time (HXB, b/w=100MB/sec, μ_0 =0).

Fig. 6. MG: Topology vs. communication time (b/w= 100MB/sec, $\mu_0 = 0$).

respectively. Then, we fixed the network topology to the HXB, and examined the effect of the μ_0 and the **bandwidth**. In our network model, when communication functions such as `mpi_send()` are called, the message goes to the network with the μ_0 overhead.

3.2 Kernel Benchmark: MG

The MG kernel is a program which solves three dimensional discrete Poisson equations by multi-grid method. The class B problem size is $256 \times 256 \times 256$. The message pattern of the MG is a hierarchy communication pattern. A hierarchy gird is used in the MG, and adjacent PUs on the hierarchy gird communicate with each other. In Figure 5, cache size doesn't greatly affects the computation times. In case of small number of PUs, cache miss rate should be high, because the working set size per a PU is large. But actually, when 16 PUs were used, the hit rate of a level one cache was more than 90%. The data using locality seems to be high in the MG. So, we fixed the cache size at 16 Kbytes in the following network simulation. Figure 6 shows the effect of network topology on the communication time. The complete and the HXB networks show almost the same performance. The two dimensional mesh network shows some overhead while the ring network shows very high overhead. Figure 7 shows the effect of bandwidth on the communication time when μ_0 is 0. The communication time is large when the number of PUs are small, because the message size is larger when the number of PUs are small, and larger size message takes a long communication time when bandwidth is narrow. When compared with the computation time of the MG, the communication time is less than 7%. Therefore we can conclude that the influence of bandwidth in the MG is small. Figure 8 shows the effect of the μ_0 on the communication time. The influence of μ_0 appears at any number of PUs. Figure 9 shows the relationship between the efficiency and the number of PUs. We calculated efficiency using the time of 16 PUs. When the number of PUs reaches 1024, and if μ_0 is 500 μsec, the efficiency declines to 40%.

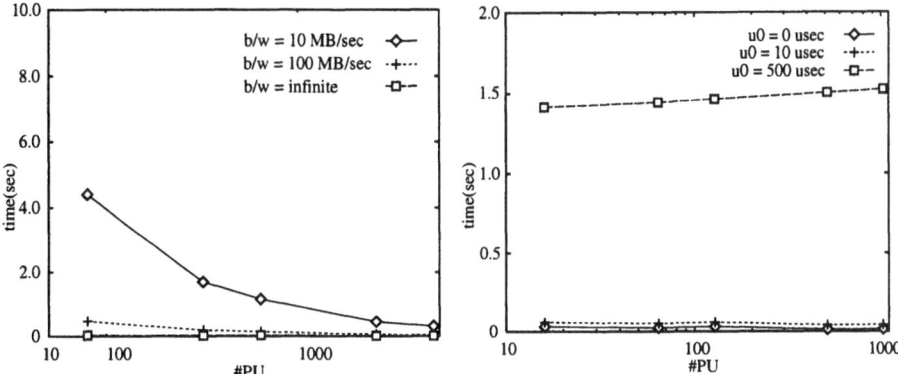

Fig. 7. MG: Bandwidth vs. communication time (HXB, $\mu_0 = 0$).

Fig. 8. MG: μ_0 vs. communication time (HXB, b/w= infinite).

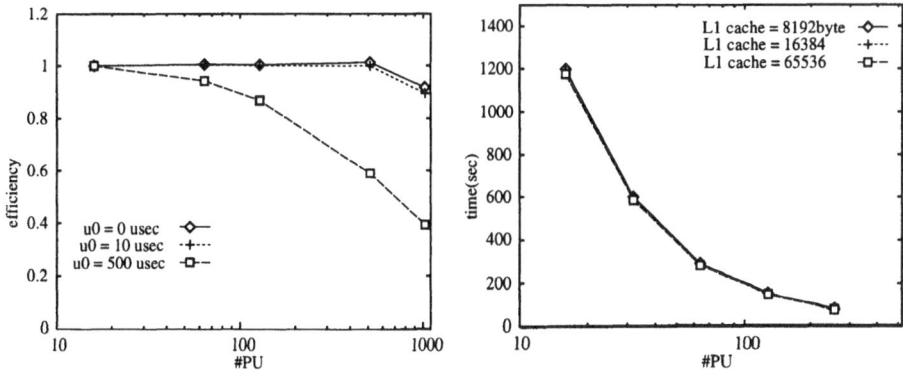

Fig. 9. MG: μ_0 vs. efficiency (HXB, b/w= infinite).

Fig. 10. LU:Cache size effect on computation time(HXB,b/w=100MB/sec,μ_0 =0).

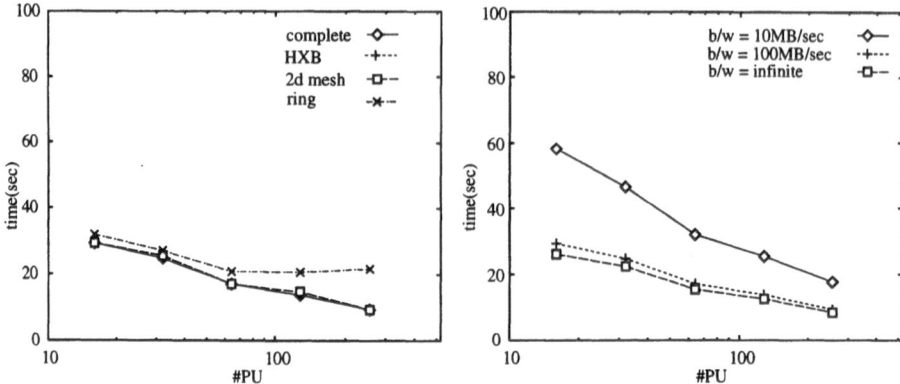

Fig. 11. LU: Topology vs. communication time (b/w= 100MB/sec, $\mu_0 = 0\mu$sec). **Fig. 12.** LU: Bandwidth vs. communication time (HXB, $\mu_0 = 0$).

3.3 Application Benchmark: LU

The LU application is a Navier-Stokes equation solver based on the Symmetric Successive Over-Relaxation (SSOR) method. Here the class B problem size is $102\times102\times102$. The message pattern of the LU is adjacent communication. Similar to the MG, computation time is less dependent on the cache size(Figure 10). In the following network simulation, we fixed the cache size at 16 Kbytes similar to the MG experiment. In Figure 11, the complete network and the HXB shows almost the same performance. However, in contrast to the MG case, the two dimensional mesh network shows high performance, because the LU message pattern is adjacent communication. In the ring network, communication overhead rose when the number of PUs were larger than 128. Figure 12 shows the effect of the bandwidth on the communication time. By the similar reason to the case of the MG, the influence of the bandwidth is small. Figure 13 shows the effect of the μ_0 on the communication time when the bandwidth is infinite. The influence of μ_0 is large, because the influence of μ_0 appears when the number of PUs are not only small but also large. Figure 14 shows the relationship between μ_0 and efficiency. When the number of PUs becomes larger than 64, efficiency begins to decrease. For example, when there are 256 PUs and μ_0 is 500 μsec , efficiency declines to 50%.

3.4 Scalability Analysis Using Simulation Results

Using the simulation results; we can find the balance point between CPU performance and the network performance. Assume 80% efficiency is required. In figure 14, the bandwidth is already infinite. Therefore it is impossible to adjust 80% efficiency by increasing the bandwidth when μ_0 is 500 μsec . Therefore in case of 128 PUs, we must decrease the μ_0 to less than 232 μsec to achieve 80%

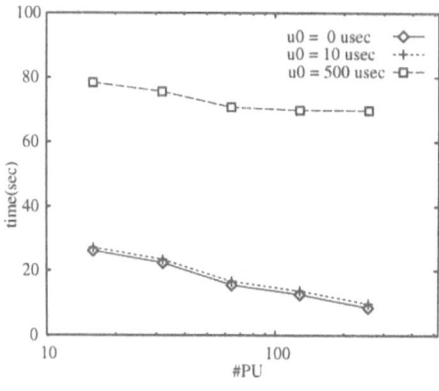 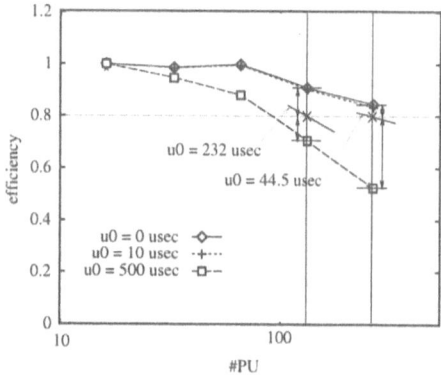

Fig. 13. LU: μ_0 vs. communication time (HXB, b/w= infinite).

Fig. 14. LU: μ_0 vs. efficiency (HXB, b/w= infinite). This graph also shows the balance point of CPU performance and Network performance when efficiency is 80%. This topic is described in section 3.4.

efficiency. And for 256 PUs, we should decrease the μ_0 to less than 44.5 μsec to achieve the same efficiency. These μ_0 are obtained by interpolation.

4 Conclusions

In this paper, we have described a simulation technique for very large-scale data parallel programs. Many factors such as program implementation, complier optimization can be estimated accurately by the simulation based on instrumentation tool. Computation and communication overlap can be calculated precisely, because communications are simulated with its neighboring computations. The class B problems in the NAS Parallel Benchmarks can be examined within a practicable time. From the simulation results, we found that in the MG and the LU, communication overhead can greatly affect the communication time, while cache size doesn't cost much execution time.

We are going on the performance analysis of the rest of the NAS Parallel Benchmarks. We believe that the results will provide an effective guideline for both architect and application users.

References

1. A. Ukawa, "Status of the CP-PACS Project," International Symposium on Lattice Field Theory, 1994.
2. T.G. Mattson, D. Scott, and S.R. Wheat, "A TeraFLOP Supercomputer in 1996: the ASCI TFLOPS System," Proceedings of the International Parallel Processing Symposium, 1996.

3. Maurice Yarrow and Rob Van der Wijngaart, "Communication Improvement for the LU NAS Parallel Benchmark: A Model for Efficient Parallel Relaxation Schemes," NAS Technical Report NAS-97-032, 1997

4. Edward Rothberg, Jaswinder Pal Singh, and Anoop Guputa, "Working Sets, Cache Sizes, and Node Granularity Issues for Large-Scale Multiprocessors," Proc. of ISCA'93, pp.14-25, 1993.

5. Taisuke Boku, Masahiro Mishima, Ken'ichi Itakura, Hiroshi Nakamura, and Kisaburo Nakazawa, "VIPPES: A performance preevaluation system for parallel processors," HPCN Europe'96, 1996.

6. Kazuto Kubota, Ken'ichi Itakura, Mitsuhisa Sato, and Taisuke Boku, "Accurate performance analysis based on code instrumentation," IPSJ SIG Report, 97ARC123-12, pp.67-72, 1997(In Japanese).

7. Taisuke Boku, Tomoaki Harada, Takashi Sone, Hiroshi Nakamura, and Kisaburo Nakazawa, "INSPIRE: A general purpose network simulator generating system for massively parallel processors," Proc. of PERMEAN'95, pp.24-33, 1995.

8. David Bailey, Tim Harris, William Saphir, Rob van der Wijngaart, Alex Woo, and Maurice Yarrow, "The NAS Parallel Benchmarks 2.0," NASA Ames Research Center Report, NAS-05-020, 1995.

9. http://www.spec.org

Assessing LogP Model Parameters for the IBM-SP

Iskander Kort and Denis Trystram

LMC-IMAG
BP 53 Domaine Universitaire
38041, Grenoble Cedex 9,France
{kort,trystram}@imag.fr

Abstract. This paper deals with LogP-model parameters assessment for the IBM-SP machine . An extension of LogP has been used. This extension allows the communication of arbitrary sized messages. Moreover, a distinction has been made between the sender and the receiver overheads. We present a methodology to assess the overhead parameter under the MPI communication library. Then, we use a microbenchmark to assess the L and g parameters.

1 Introduction

During the last decade many parallel computers have been devized. Most of them consist of a set of processors which communicate by sending messages via an interconnection network. The major advantage of these computers is *scalability*, which corresponds to the ability to integrate a large number of processors while sustaining good performance. However, writing efficient programs for such computers is still a hard work. In fact, the performance can decrease drastically due to large communication costs. In order to overcome this difficulty it is essential to use a relatively precise model describing the target computer for program design and analysis.

Since the interconnection network is the most important bottleneck in nowadays computers, many researchers have assessed some parameters related to this component, mainly latency and bandwidth. In [12], models for the Intel Paragon and a transputer-based parallel machine have been proposed. In [8] measures of latency and bandwidth of the Thinking Machine $CM5$ have been provided. A more general study dealing with a variety of current parallel computers can also be found in [4].

Another important issue in parallel programming is portability. Indeed, a program that runs efficiently on a given computer may show poor performance on another one. In order to get efficient and portable parallel programs one should use generic computational models. Such models describe a parallel computer using a set of parameters. These latters reflect the most pertinent features of current computers and are intended to be instantiated when dealing with a specific machine. Many generic models are presented in the literature such as Phase-PRAM/APRAM/BSP/LogP [1] [7].

This paper deals with the assessment of LogP model parameters for an IBM-SP multicomputer [9–11]. Previous work showed that this model is suitable for many parallel architectures. In [2] the model parameters were assessed for a CM5 multicomputer. In [3] assessments for the Intel Paragon, the Meiko CS2 and a workstation cluster are provided. In [6], the model parameters were assessed for some local area networks.

We present LogP in next section, then we show in Sect.3 how to assess the model parameters.

2 The LogP Model

The LogP model was introduced in [2]. It presents a good trade-off between realism and ease of use. Indeed, it describes a wide range of current parallel architectures with a few simply defined parameters. These parameters are:

L: The network *latency*, that is the time needed to transmit a message of one or a few words.

o: The communication *overhead* , that is the time it takes the sender or the receiver to process a message.

g: The *gap*, that is the minimum delay between two successive sends or two successive receives. In other words, $1/g$ is the maximum bandwidth per processor.

P: The processors' number.

We will rather use an extension of LogP where messages may have arbitrary sizes. So, the model parameters become functions of message size. Moreover, we split the o parameter into two new ones: o_s and o_r. The o_s (resp. o_r) parameter is the time needed by the sender (resp. receiver) to manage an outgoing (resp. incoming) message.

3 Assessing LogP Parameters

We describe in this section the experiments for assessing the LogP parameters. In these experiments, we used the MPI communication library[1] because of its wide use [13][5]. In order to design the right experiments and to be able to interpret the obtained results, the communication subsystem behavior should be studied at first.

3.1 The Communication Subsystem Behavior

We check in this section if the communication overhead is equal to the send primitive delay. This is equivalent to check if the sender processes entirely the outgoing messages within the send primitive or if it does some extra operations

---

[1] More specifically, the IBM implementation was used (MPI-F version 1.43 under the Aix 3.2 operating system).

after the primitive completion. To answer this question, we performed the experiment shown by Fig. 1. In this experiment, a processor P_0 sends a message to a processor P_1 and then performs some computations. We measure the MPI-Recv delay for various computation loads.

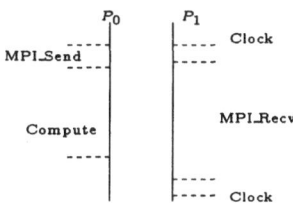

Fig. 1. The communication subsystem behavior.

The experiment shows that the MPI_Recv delay does not depend on the computation load when the message size is less than 1840 bytes. However, when the message size is greater than this value, the MPI-Recv delay increases along with the computation load. Therefore, the messages are sent by blocks. Each block is of 1840 bytes, that is eight packets. These blocks are due to the *sliding window* protocol used in the IBM-SP1-2 to acknowlege packets [9].

Since the communication overhead is greater than the send primitive delay, it is essential to find out a new methodology to assess this parameter.

3.2 Assessing the Overhead

Assessing the o_s Parameter. As shown in the previous section, the outgoing messages are not entirely processed within the send primitive. So, in order to assess o_s we should force the sender to finish the message processing. This can be done using the MPI_Test primitive. This primitive is intended to check the non blocking communications completion. However, it can be called with dummy arguments to insure communications progress. We define the following methodology to assess o_s:

- Estimate the MPI_Test delay when no communication is pending ($TestDelay$).
- Given an n bytes message, estimate the number of MPI_Test calls necessary to complete the communication ($TestCount$). To do so, we perform an experiment similar to the one in Fig.1. However, in this case, P_0 calls $TestCalls$ times MPI_Test before the computation phase. Each time MPI_Test is called, it processes pending communications, whenever this is possible. We repeat the experiment with increasing values of $TestCalls$. We stop when the MPI_Recv delay does not increase any longer with the computation load.
- Finally we measure o_s using the program below.

This methodology has led to the results in Fig.2. In this figure, we denote by $o_s(asynchronous)$ (resp. $o_s(synchronous)$) the sender overhead when the

Program 1 o_s assessment program.

```
P0 code
MPI_Barrier(MPI_COMM_WORLD);
Os = MPI_Wtime();
MPI_Bsend(m,n,MPI_CHAR,1,TAG,MPI_COMM_WORLD);
for (i=0; i < TestCount;i++) do
  MPI_Test(MPI_REQUEST_NULL,&flag,&status);
Os = MPI_Wtime() - Os - TestCount * TestDelay
P1 Code
MPI_Barrier(MPI_COMM_WORLD);
MPI_Recv(m,n,MPI_CHAR,0,TAG,MPI_COMM_WORLD,&status);
```

MPI_Bsend (resp. MPI_Ssend) primitive is used. Notice that the MPI_Bsend delay is smaller than the overhead except when n is less than 1840 bytes. More specifically, the ratio MPI_Bsend-Delay/Overhead is of about 35% for long messages. Moreover, the overhead increases linearly with the message size. In order to get the best fit, we split the measures into five intervals and we established an equation for each interval (see Table 3.2).

The comparison of the overheads in the synchronous and asynchronous cases shows that:

- when $n \leq 4KB, o_s(synchronous) > o_s(asynchronous)$, this is due to the *handshaking* operation undertaken by the sender in the synchronous mode.
- when $n > 4KB, o_s(asynchronous) > o_s(synchronous)$. Moreover, the difference $o_s(asynchronous) - o_s(synchronous)$ increases along with the message size. Indeed, MPI_Bsend sends the message according to the rendez-vous protocol[2]. On the other hand this primitive is intended to be local[3]. To cope with these two contradictory properties, the message is buffered at the sender processor which explains the above mentionned diffrence.

Table 1. The sender overhead in the asynchronous and synchronous modes: n is the message size in bytes, delays are expressed in μs.

	MPI_Bsend	MPI_Ssend
$0 \leq n \leq 4KB$	$o_s = 0.040 * n + 30$	$o_s = 0.040 * n + 110$
$4KB < n \leq 16KB$	$o_s = 0.081 * n - 90$	$o_s = 0.053 * n + 20$
$16KB < n \leq 32KB$	$o_s = 0.073 * n + 70$	$o_s = 0.054 * n + 175$
$32KB < n \leq 48KB$	$o_s = 0.058 * n + 530$	$o_s = 0.037 * n + 700$
$n > 48KB$	$o_s = 0.050 * n + 870$	$o_s = 0.034 * n + 890$

---

[2] In this protocol a message is not sent unless a receive request is posted.

[3] A send primitive is said to be local if its completion is not conditioned by the receiver state.

Assessing the o_r Parameter. In order to assess the receiver communication overhead, Prog. 2 is used. In this program both processors first synchronize. Then, processor P_0 sends a message m to P_1. Processor P_1 reads the *IBM-SP global clock*, posts a receive request and finally calls MPI_Test so many times till all the packets of m are received (*flag* becomes *true*). The *loopDelay*() function calculates the delay of *call* iterations when no communication is pending. This delay is given by the equation:

$$delay = 4.48 * call - 10$$

The obtained results are illustrated in Fig.2. Parameter o_r increases linearly along with message size according to the equations:

$$for\ 0 < n \leq 16KB,\ o_r = 0.059 * n + 10$$
$$for\ n > 16K,\qquad o_r = 0.035 * n + 420$$

To end this section, we compare the o_s and o_r parameters. In the asynchronous case, o_s is equal to o_r when $n \leq 4KB$ and $o_s > o_r$ otherwise. This is due to message buffering at the sender. In the synchronous case, o_s can be considered to be equal to o_r when $n \leq 16KB$. However, $o_s > o_r$ when $n > 16KB$.

Program 2 o_r assessment program.

```
p0 code
MPI_Barrier(MPI_COMM_WORLD);
MPI_Bsend(m,n,MPI_CHAR,1,TAG,MPI_COMM_WORLD);
p1 code
MPI_Barrier(MPI_COMM_WORLD);
call = 0;
Or = MPI_Wtime();
MPI_Irecv(m,n,MPI_CHAR,0,TAG,MPI_COMM_WORLD,&request);
repeat
    MPI_Test(&request,&flag,&status);
    call + +;
until (flag == true);
Or = MPI_Wtime() - Or - loopDelay(call);
```

3.3 Assessing the L and g Parameters

The assessment of g and L is more difficult than that of o, so only small messages will be considered in this section. In order to assess g we used the *microbenchmark* presented in [3]. The microbenchmark measures the delay of issuing M requests (denoted by $Delay(M)$), as shown in Fig. 3(a). A request is a light-weight asynchronous remote procedure call. Figure 3(b) shows the average request delay (ie $Delay(M)/M$) under LogP. Notice that there are three regimes: send-only, transition and steady-state. In the send-only regime, no reply arrives during the

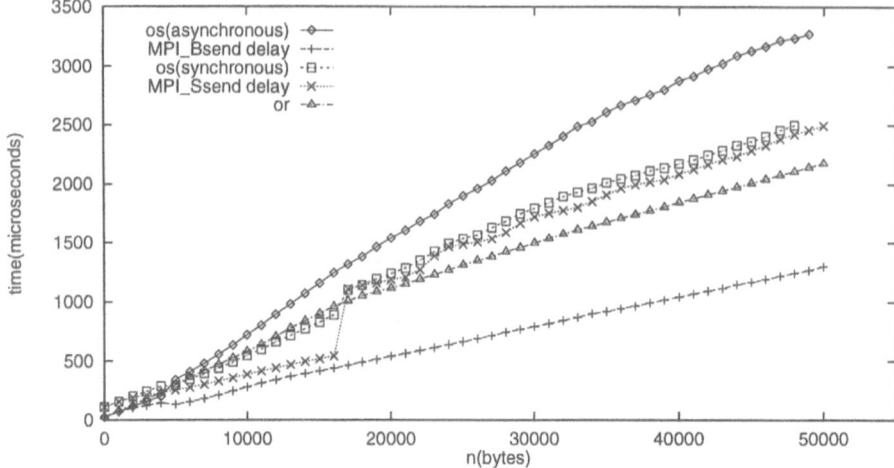

Fig. 2. Comparison of communication overhead and the send primitives delay.

issue phase. So, the average request delay is equal to o_s. For greater values of M, some replies arrive during the issue phase. This corresponds to the transition regime. Beyond a certain value of M, the capacity limit is reached. This corresponds to the steady state. If the communication bottleneck is the network then the average request delay equals g. On the other hand, if the communication bottleneck is the processor, then the average request delay equals $o_s + o_r$. To distinguish these two cases, the experiments are done with different values of Δ (computation time). If the curve limit increases even with small values of Δ then we are in the second case. Otherwise, the network is the communication bottleneck.

We implemented the microbenchmark using MPI immediate communication primitives. Figure 4 shows the microbenchmark signature for a 1000-bytes message. Experiments were performed with $\Delta = 0\mu s$ and $\Delta = 20\mu s$. Since the two curves converge to different values we deduce that the overhead is the performance bottleneck in the IBM-SP. The microbenchmark gave the following values: $o_s^{micro} = 76\mu s$, $o_r^{micro} = 73\mu s$. These values are close to those found when we applied our methodology: $o_s = 72\mu s$, $o_r = 82\mu s$.

Notice that the transition regime starts for a small value of M (for $M = 5$). This means that the interconnection network latency may be neglected.

4 Concluding Remarks

In this paper we presented a methodology to assess the LogP overhead parameter at the sender and receiver processors. The obtained results show that the sender overhead is more important when using asynchronous communications. This is due to message buffering.

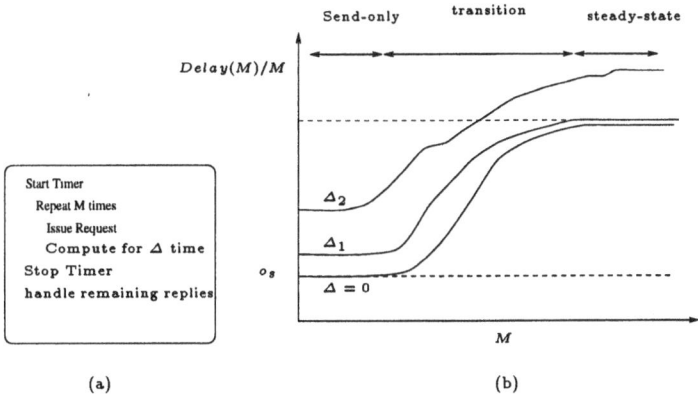

Fig. 3. (a) The microbenchmark pseudo-code. (b) The microbenchmark signature.

Moreover, the microbenchmark showed that the overhead is the performance bottleneck in the IBM-SP. The microbenchmark gave results which are close to those obtained using our methodology. However, this microbenchmark is not very convenient when one wants to find out the $o(n)$ function. Indeed, in this case a signature must be established for each message size. Also, the microbenchmark underestimates the overhead when message size is greater than 1840 bytes.

We are planning to use this model in predicting some communication patterns delays such as broadcast and scatter. The model will also be used in some scheduling algorithms and heuristics for LogP.

Acknowledgments The authors are grateful to the anonymous referees for their valuable suggestions.

References

1. C. Boeres. *Versatile Communication Cost Modelling for Multicomputer Task Scheduling Heuristics.* PhD thesis, University of Edinburgh, 1996.
2. D. E. Culler, R. M. Karp, D. Patterson, A. Sahay, K. E. Schauser, E. Santos, R. Subramonian, and T. Von Eicken. LogP: Towards a Realistic Model of Parallel Computation. In *Fourth ACM SIGPLAN Symposium on principles and practice of parallel programming*, 1993.
3. D. E. Culler, L. T. Liu, R. P. Martin, and C. Yoshikawa. LogP Performance Assessment of Fast Network Interfaces. *IEEE Micro*, February 1996.
4. J. J. Dongarra and T. Dunigan. Message Passing Performance of Various Computers. Technical report, UT-CS-95-299, Computer science department, University of Tennesee,Knoxville, Tenessee, 1995.
5. H. Franke, E. Wu, P. Pattnaik, and M. Snir. MPI Programming Environment for IBM SP1/SP2. In *International Conference on Distributed Computing Systems IEEE Computer Society Press Los Alamitos California*, 1995.
6. K. K. Keeton, T. E. Anderson, and D. A. Patterson. Logp Quantified: The Case for Low-overhead Local Area Networks. In *Hot Interconnects III: A Symposium on*

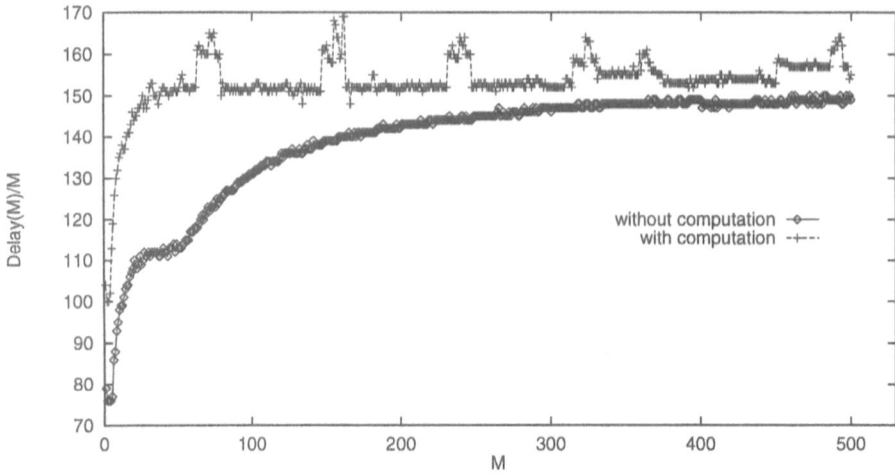

Fig. 4. The microbenchmark signature for a 1000-bytes message under the IBM-SP.

High Performance Interconnects, Stanford University, Stanford California, August 10-12 1995.

7. Z. Li, Mills P., and J. Reif. Models and Resource Metrics for Parallel and Distributed Computation. In *Annual Hawaii International Conference on System Sciences, Parallel Algorithms Software Technology Track, IEEE Press*, 1995.

8. L. Mengjou, T. Rose, et al. Performance Evaluation of the CM-5 Interconnection Network. Technical report, University of Minnesota, 1993.

9. M. Snir et al. The Communication Software and Parallel Environment of the IBM-SP2. *IBM systems journal*, 34(2), 1995.

10. B. Stunkel et al. The SP2 High Performance Switch. *IBM systems journal*, 34(2), 1995.

11. G. Tengwall. SP2 Architecture and Performance. *springer verlag*, 1994.

12. C. Tron. *Modèles Quantitatifs de Machines Parallèles: Les réseaux d'interconnection*. PhD thesis, Institut National Polytechnique de Grenoble, France, 1994.

13. University of Tenessee. *MPI Forum: A Message Passing Interface Standard*, 1994.

Communication Pre-evaluation in HPF

Pierre Boulet[1] and Xavier Redon[2]

[1] LIP, École Normale Supérieure de Lyon, 46, allée d'Italie, F-69364 Lyon cedex 07, France
[2] LIFL, Univeristé des Sciences et Technologies de Lille, Bâtiment M3, Cité Scientifique, F-59655 Villeneuve d'Ascq cedex, France

Abstract. Parallel computers are difficult to program efficiently. We believe that a good way to help programmers write efficient programs is to provide them with tools that show them how their programs behave on a parallel computer. Data distribution is the major performance factor of data-parallel programs and so automatic data layout for High Performance Fortran programs has been studied by many researchers recently. The communication volume induced by a data distribution is a good estimator of the efficiency of this data distribution.

We present here a symbolic method to compute the communication volume generated by a given data distribution during the program writing phase (before compilation). We stay machine-independent to assure portability. Our goal is to help the programmer understand the data movements its program generates and thus find a good data distribution. Our method is based on parametric polyhedral computations. It can be applied to a large class of regular codes.

1 Introduction

Parallel computing has become the solution of choice for heavy scientific computing programs. Unfortunately parallel computers require considerable knowledge and programming skills to exploit their full potential. A major mean to reduce the programming complexity is to use high-level languages. High Performance Fortran (HPF) is such a language. It follows the data-parallel programming paradigm where the computations are directed by the data. Indeed, the computations occur on the processor that "owns" the data being written. So the data distribution is a very important efficiency factor when programming with a data-parallel language.

Our goal is to help the programmer find a good data-distribution. We want a tool that is executed when the program is being written. Work on such a tool has started at the LIFL with HPF-builder [6], a tool that interactively displays the arrays as they are distributed and aligned by the HPF directives. This tool also allows to change the data distribution graphically.

We propose here a new step in the development of a tool that helps the programmer understand how data move in its program. Given a data-distribution, we are able to compute the volume of the communications generated by a program. We use symbolic computation tools to stay free of the problem size. All this is done at the language level, thus retaining portability and machine independence. This tool is intended as a mean to write a reasonably efficient program that can be tuned for a particular parallel machine with profiling systems later.

The sequel of this paper is organized as follows. In Sect. 2 we briefly review related works, then in Sect. 3 we describe the problem we consider and present its modelization in Sect. 4. We then detail the tools used to solve our problem along with an example in Sect. 5. And we finally conclude in Sect. 6.

2   Related Work

Many researchers [1, 9, 10, 13–15] have studied automatic data distribution. Estimating communication costs has been the key factor to determine the quality of a data distribution. Most of the previous works [12, 7, 11] have studied compile-time estimation of these communication costs. Indeed, they use the fact that most program parameters are known at that time and in many cases, these studies also use machine (and compiler) dependent data. Our work differs from previous work by the techniques used and the stage of program development we focus on: the program writing phase. We also use exact parameterized methods and we stay compiler and machine independent, we work at the language level.

Description of the Problem

3.1   General Remarks

The problem we study in this paper is the evaluation of the communication volume in a HPF program before compilation. The goal is to help the programmer understand the communications generated by his program and find a good data distribution (or a more efficient way to code his algorithm).

The communications we consider are at the PROCESSORS level: as in HPF, we use an abstract target machine. We stay at the language level, thus allowing to find a data distribution that is well suited to the problem, and thus retaining portability. Our aim is not to find the best data distribution for a given machine, but a good one for any machine (and compiler).

In the future, if some compiler optimization techniques such as overlap areas to vectorize communications are used by most compilers, we could adapt our evaluation techniques to these optimizations. In a first step, though, we remain at the language level to validate our approach. We believe indeed that a good data distribution at the language level should not be a bad one for any compiler, regardless of the compilation techniques used.

3.2   Modelization

We consider that the only communication generation statements are storage statements, we do not take into account I/O statements. For each of these storage statements, the surrounding loops define an iteration domain that "shapes" the communication pattern.

Given the mathematical tools we use in the following, we have to restrict ourselves to loop bounds that define a polyhedron, that is loop bounds defined as extrema of affine functions of the surrounding loop indices and parameters. For the same reason, we restrict the array access functions to affine functions. This class of loops contains most of the linear algebra routines.

To simplify the following discussion, we make the hypotheses below without loss of generality:

- Scalar values are described as arrays with one element.
- Storage statements read only one array reference that may not be aligned with the result value. Indeed, any more complex statement can be decomposed in a sequence of such statements with temporary storage [3, 2].
- All the arrays of the considered storage statement are aligned with respect to the same template T. This restriction makes sense since the distributions of computation related arrays should be related in some way. Actually most HPF programs contain only one template with all the arrays aligned onto. The domain of this template T is noted $\mathcal{D}_T$

Given the previous restrictions, a storage statement S is represented by:

$$W_S(\phi_{W_S}(I)) = f_S(R_S(\phi_{R_S}(I))) \tag{1}$$

where the iteration vector I has value in the iteration domain $\mathcal{D}_S$ defined by the surrounding loops. W_S and R_S are arrays and ϕ_{W_S} and ϕ_{R_S} are affine access functions.

We call *operation* an instance $S(I)$ of a statement S for a given value I of the iteration vector.

The number of communications generated by a given statement is the number of array elements that need to be communicated for some operation generated by this statement. There is a communication when the array elements being read are not on the same virtual processor than the one the written element is. As array elements can be replicated, we will focus on the template elements.

4 Algebraic Method to Evaluate HPF Communications

As said in the previous section, we evaluate the communication cost in a HPF program by counting the number of communications between template elements.

4.1 Formula for Communication Cost Evaluation

There is a communication between two template elements if they are not distributed onto the same virtual processor and if the computation of a value to be stored in the first template element uses a value stored in the second one.

To be more precise, let us consider a storage statement S (as defined in the previous section) and a template T. We also need to define some functions:

- Function $\mathcal{O}_T^{\mathcal{S}}$ gives, for an element J of $\mathcal{D}_T$, the set of storage operations from $\mathcal{S}$ which compute values to be stored in $T(J)$;
- Function τ gives, for an operation o from $\mathcal{S}$, the set of template subscripts from where o may read its data.
- Last, function π_T is the distribution function which maps the template T on the virtual processors.

The number of template communications generated by statement $\mathcal{S}$ is equal to the number of elements in the union of the sets $\mathcal{C}_J(\mathcal{S})$:

$$\mathcal{C}(\mathcal{S}) = \bigcup_{J \in \mathcal{D}_T} \mathcal{C}_J(\mathcal{S}) = \bigcup_{J \in \mathcal{D}_T} \{(J, o) \mid o \in \mathcal{O}_T^{\mathcal{S}}(J),\ \pi_T(J) \notin \pi_T(\tau(o))\} \ . \tag{2}$$

Note that $\mathcal{C}(\mathcal{S})$ is a set of couples and not of mere operations. Indeed we may have to count several times the same operation, so we added the template subscript to distinguish different occurrences of the same operation. Computing the set $\mathcal{C}(\mathcal{S})$ implies the application of a change of basis on the program loop nests. Indeed, in the original program, the loops are used to enumerate the operations in the program execution order and the set $\mathcal{C}(\mathcal{S})$ enumerates them following the template iteration space.

4.2 Formal Representation of HPF Alignments

To perform this change of basis the only informations needed are the HPF ALIGN directives. Basically HPF alignments are a sub-set of linear alignments but a replication symbol * has been added. Because of the replication symbol, a HPF alignment cannot be defined by a linear transformation from the array space to the template space.

A convenient way to represent a HPF alignment α is to use two linear transformations γ and δ. Transformation δ defines the replication part of the alignment. Let I be a subscript of an array A aligned on a template T using α. The set of the subscripts of T on which the data $A(I)$ is stored is:

$$\{J \mid J \in \mathcal{D}_T,\ \gamma(I) = \delta(J)\} = \delta^{-1}(\gamma(I)) \ .$$

Let us consider the alignment:

!HPF$ ALIGN A(i) WITH T(i,*)

The first transformation from the array space to an intermediate template space is obtained by removing the replication symbols in the directive: $\gamma\ :\ i \mapsto i$. The second transformation is the projection from the template space to the intermediate template space: $\delta\ :\ \begin{pmatrix} i \\ j \end{pmatrix} \mapsto i$.

With this representation of a HPF alignment one can explicit the function τ of Sect. (4.1). Let us consider a statement $\mathcal{S}$ as defined in (1) and an alignment $\tau_{R_{\mathcal{S}}}$ for array $R_{\mathcal{S}}$ on the template T defined by $(\gamma_{R_{\mathcal{S}}}, \delta_{R_{\mathcal{S}}})$. In this context, the following holds:

$$\forall I \in \mathcal{D}_T,\ \tau(\mathcal{S}(I)) = \tau_{R_{\mathcal{S}}}(\phi_{R_{\mathcal{S}}}(I)) = \delta_{R_{\mathcal{S}}}^{-1}(\gamma_{R_{\mathcal{S}}}(\phi_{R_{\mathcal{S}}}(I))) \ .$$

4.3 Enumeration of Operations in the Template Iteration Space

The main problem of enumerating operations in the template space is that there may be several operations corresponding to a template element $T(J)$. I.e. more than one operation may produce a value to be stored in $T(J)$. In fact, the solution lies again in the formal representation of HPF alignments. Let us denote by τ_{W_S} the alignment for array W_S. According to Sect. 4.2, the alignment can be written as $\tau_{W_S} = \gamma_{W_S} \circ \delta_{W_S}^{-1}$. Therefore, the set $\mathcal{O}_T^S(J)$ introduced in the beginning of this section is defined by:

$$\mathcal{O}_T^S(J) = \{\mathcal{S}(I) \mid I \in \mathcal{D}_S,\ \phi_{W_S}(I) \in \gamma_{W_S}^{-1}(\delta_{W_S}(J))\} \ .$$

Let us consider the code fragment below:

```
!HPF$ ALIGN M(i,j) WITH T(i,*)
DO i=1,n
   DO j=1,m
      M(i,j) = ...           (s1)
   END DO
END DO
```

The aligment of M on T is defined by: $\left(\gamma \ : \ \begin{pmatrix} i \\ j \end{pmatrix} \mapsto i,\ \delta \ : \ \begin{pmatrix} i \\ j \end{pmatrix} \mapsto i \right)$. Since the subscript function for M is the identity, the set $\mathcal{O}_T^{s1}$ is such that:

$$\mathcal{O}_T^{s1}\begin{pmatrix} i' \\ j' \end{pmatrix} = \left\{ s1 \begin{pmatrix} i \\ j \end{pmatrix} \mid \gamma \begin{pmatrix} i \\ j \end{pmatrix} = \delta \begin{pmatrix} i' \\ j' \end{pmatrix} \right\} = \left\{ s1 \begin{pmatrix} i' \\ j \end{pmatrix} \mid j \in [1, m] \right\} \ .$$

This means that a template element T(i,j) owns the full row of rank i of array A.

In conclusion, the set $\mathcal{C}(\mathcal{S})$ may be computed using the subscript functions in $\mathcal{S}$ for W_S and R_S, the alignments of W_S and R_S on the template T and the distribution of T on the virtual processors:

$$\mathcal{C}(\mathcal{S}) = \bigcup_{J \in \mathcal{D}_T} \{(J, \mathcal{S}(I)) \mid I \in \Phi_{W_S}^{-1}(J),\ \pi_T(J) \notin \pi_T(\Phi_{R_S}(I))\} \tag{3}$$

with the functions Φ_{W_S} and Φ_{R_S} defined as follows:

$$\Phi_{W_S} = \tau_{W_S} \circ \phi_{W_S} = \delta_{W_S}^{-1} \circ \gamma_{W_S} \circ \phi_{W_S} \ ,$$
$$\Phi_{R_S} = \tau_{R_S} \circ \phi_{R_S} = \delta_{R_S}^{-1} \circ \gamma_{R_S} \circ \phi_{R_S} \ .$$

4.4 Formal Representation of HPF Distributions

A HPF distribution directive for a template T can be represented using a projection ρ_T and an integer vector κ_T of size the dimension of the virtual processors grid P. Projection ρ_T selects the dimensions of T to be distributed on P. The dimension of T projected on the ith dimension of P is distributed according to the pattern CYCLIC($(\kappa_T)_i$). We denote by $A^{\min}$ (respectively by $A^{\max}$) the vector of lower bounds (respectively the vector of upper bounds) for array A. In this context the distribution π_T can be explicited:

$$\pi_T(J) = P^{\min} + (\rho_T(J - T^{\min}) \div \kappa_T)\%(P^{\max} - P^{\min} + \mathbb{1}) . \qquad (4)$$

In the previous expression, the operator $\div$ (respectively the operator $\%$) represents an element-wise integer division (respectively modulo) on vectors. One may remark that a BLOCK distribution can be achieved on the ith dimension of P using a relevant $(\kappa_T)_i$ value:

$$(\kappa_T)_i = \left\lceil \frac{T_i^{\max} - T_i^{\min} + 1}{P_i^{\max} - P_i^{\min} + 1} \right\rceil .$$

5   Tools for the Evaluation of HPF Communications

In the previous section we have reduced the problem of evaluating HPF communications of a statement $\mathcal{S}$ to counting the elements of a set $\mathcal{C}(\mathcal{S})$. This section presents the tools used to automatically build $\mathcal{C}(\mathcal{S})$ and to count its elements. We illustrate the use of these tools on the following HPF program:

```
Program MatInit
!HPF$ PROCESSORS P(8,8)                         do i=1,n
!HPF$ TEMPLATE T(n,m)                             do j=1,m
!HPF$ DISTRIBUTE T(CYCLIC,CYCLIC) ONTO P              A(i,j)=B(i)  (S1)
  real A(n,m), B(n)                               end do
!HPF$ ALIGN A(i,j) WITH T(i,*)                  end do
!HPF$ ALIGN B(i) WITH T(i,1)                    end
```

We have integrated the different tools into an interactive program called CIPOL. CIPOL provides a lisp-like textual interface to the tools, and pretty-prints their results.

5.1   Manipulation of Polyhedra

Our main tool is the polyhedral library developed by Wilde [16]. Most of the operations on polyhedra (union, intersection, image, etc.) are implemented in this library which allows to define a polyhedron by a set of constraints or a set of rays. For our application, we only need the first definition scheme.

The generic sets $\Phi_{W_S}^{-1}(J)$ and $\Phi_{R_S}(I)$ described in the previous section are computed using the *image* and *pre-image* functions of the polyhedral library.

The generic sets for our example are:

$$\Phi_A^{-1}(j_1, j_2, n, m) = \left\{ \begin{matrix} 1 \leq j_1 \leq n \wedge i_1 = j_1 \\ 1 \leq i_2 \leq m \end{matrix} \right. , \Phi_B(i_1, i_2, n, m) = \left\{ \begin{matrix} 1 \leq i_1 \leq n \wedge j_1' = i_1 \\ j_2' = 1 \end{matrix} \right. .$$

The first set means that the ith row of array A is duplicated on each element of the ith row of template T. The second set shows that array B is aligned with the first column of template T.

5.2 Using the PIP Software to Compute Inclusion

The evaluation of the non inclusion in definition (2) is not easy to implement since it implies that a generic condition must be verified for *each* value from a given set. It is simpler to verify an inclusion condition. In this case we just have to verify that a condition is verified for *one* value from a given set. Hence an inclusion condition can be modelized by an integer programming problem and can be resolved by a software such as PIP (see [8]). So, in place of counting the number of elements in $\mathcal{C}(\mathcal{S})$ we compute the number of elements in the set $\overline{\mathcal{C}(\mathcal{S})}$:

$$\overline{\mathcal{C}(\mathcal{S})} = \bigcup_{J \in \mathcal{D}_T} \{(J, \mathcal{S}(I)) \mid I \in \Phi_{W_\mathcal{S}}^{-1}(J),\ \pi_T(J) \in \pi_T(\Phi_{R_\mathcal{S}}(I))\} \ . \tag{5}$$

The final result is obtained using the following relation:

$$\mathrm{Card}(\mathcal{C}(\mathcal{S})) = \mathrm{Card}(\bigcup_{J \in \mathcal{D}_T} \{(J, \mathcal{S}(I)) \mid I \in \Phi_{W_\mathcal{S}}^{-1}(J)\}) - \mathrm{Card}\left(\overline{\mathcal{C}(\mathcal{S})}\right) \ . \tag{6}$$

The PIP software is able to find the lexicographical minimum of a parametric set of integer vectors defined by a set of linear constraints $S(P)$. It may also take into account linear constraints on the parameters, this other set of constraints C is called the context. We denote by $\mathrm{lexmin}(C, S(P))$ the result computed by PIP. Since the initial set of integer vectors is parametric, PIP does not return an unique vector but a *quast* (Quasi-Affine Selection Tree). Indeed, the minimum depends on the values of the parameters, hence PIP splits the domain of the parameters in sub-domains on which the minimum can be expressed in a parametric way. If there is no solution for a sub-domain of the parameter space (because for these values of the parameters $S(P)$ is void), PIP denotes by $\perp$ the lack of solution.

Let us solve the following integer program:

$$\mathrm{lexmin}(C, S(I, J)),\ C = \begin{cases} J \in \mathcal{D}_T \\ I \in \Phi_{W_\mathcal{S}}^{-1}(J) \end{cases},\ S(I, J) = \begin{cases} J' \in \Phi_{R_\mathcal{S}}(I) \\ \pi_T(J) = \pi_T(J') \end{cases} . \tag{7}$$

Consider now the sub-domains of the parameter space for which PIP gives a solution other than $\perp$. It is easy to deduce, from what we said about PIP, that the number of elements in these sub-domains is equal to the number of elements in $\overline{\mathcal{C}(\mathcal{S})}$.

Remember that PIP only deals with linear constraints with respect to the parameters and the variables of the problem. Function π_T involves euclidian divisions but there is a well known method to linearize π_T that may be found in [5]. We just have to introduce three new integer vectors to replace the initial definition (4) of the distribution function by a definition which gives $\pi_T(J)$ as the solution of the following system (the $*$ operator is an element-wise vector multiplication):

$$\begin{cases} \rho_T(J - T^{\min}) = N * \kappa_T + R \wedge N = Q * (P^{\max} - P^{\min} + 1\!\!1) + \pi_T(J) - P^{\min} \\ P^{\min} \leq \pi_T(J) \leq P^{\max} \wedge N \geq 0 \wedge Q \geq 0 \wedge 0 \leq R < \kappa_T \end{cases} .$$

One may note that N_i gives the block number for $T(J)$ with respect to the ith dimension. The previous system is effectively linear only if κ_T and $P^{\max} - P^{\min}$ are constant vectors. Computation of a parametric solution in other cases is left for future work but it is possible to obtain a result by asking the user to provide the values of some key parameters.

Integer program (7) may so be rewritten with only linear constraints.

For our program example MatInit, the template is distributed on the processor grid using two CYCLIC patterns, hence the distribution is defined by:

$$\rho_T \begin{pmatrix} j_1 \\ j_2 \end{pmatrix} = \begin{pmatrix} j_1 \\ j_2 \end{pmatrix}, \quad \kappa_T = \begin{pmatrix} 1 \\ 1 \end{pmatrix} .$$

The integer problem to solve is $\mathrm{lexmin}(C(n,m), S(i_1, i_2, j_1, j_2))$ with

$$C(n,m) = \{ 1 \leq j_1 \leq n \wedge 1 \leq j_2 \leq m \wedge i_1 = j_1 \wedge 1 \leq i_2 \leq m ,$$

$$S(i_1, i_2, j_1, j_2) = \begin{cases} j_1' = i_1 \wedge j_2' = 1 \wedge 1 \leq p_1 \leq 8 \wedge 1 \leq p_2 \leq 8 \\ j_1 - 1 = 8q_1 + p_1 - 1 \wedge j_2 - 1 = 8q_2 + p_2 - 1 \\ j_1' - 1 = 8q_1' + p_1 - 1 \wedge j_2' - 1 = 8q_2' + p_2 - 1 \\ q_1 \geq 0 \wedge q_2 \geq 0 \wedge q_1' \geq 0 \wedge q_2' \geq 0 \end{cases} .$$

One may note that, in this example, there is no variable representing the remainder in the division by κ_T (as R in (5.2)) since the block sizes are equal to 1. The result of PIP is that there exists a solution not equal to $\perp$ in the polyhedron defined by:

$$\overline{C}(n,m) = \{ 1 \leq j_1 \leq n \wedge 1 \leq j_2 \leq m \wedge i_1 = j_1 \wedge 1 \leq i_2 \leq m \wedge j_2 - 1 = 8q \wedge q \geq 0 .$$

The new parameter q is used to express that $j_2 - 1$ must be a multiple of 8.

5.3 Counting the Elements of the Communication Set

The last stage of our method consists in counting the number of integer vectors in the sub-domains computed by PIP. These parametric sub-domains $D(I, J, P)$ are defined in function of a parameter J representing the subscript of a general template element $T(J)$, in function of a parameter I which represents the subscript of an operation storing a value on $T(J)$ and in function of a vector of program parameters P. We need to compute the number of integer vectors in $D(I, J, P)$ in function of the program parameters. Fortunately, Loechner and Wilde have extended the polyhedron library to include a function able to count the number of integer vectors in a parametric polyhedron (see [4]). Like PIP, this function splits the parameter space in sub-domains on which the result can be given by a parametric expression. Hence, to apply relation (6) one has to implement an addition on Quasts.

For the example MatInit the final result (in the context $n \geq 1$ and $m \geq 1$) is

$$\mathrm{Count}(C(n,m)) - \mathrm{Count}(\overline{C}(n,m)) = n.m \left(\frac{7m}{8} - \left[0, \frac{7}{8}, \frac{3}{4}, \frac{5}{8}, \frac{1}{2}, \frac{3}{8}, \frac{1}{4}, \frac{1}{8} \right]_m \right)$$

The brackets on the previous expression denote a periodic number: if we denote by v the vector $(0, \frac{7}{8}, \frac{3}{4}, \frac{5}{8}, \frac{1}{2}, \frac{3}{8}, \frac{1}{4}, \frac{1}{8})$, the value of the periodic number is $v_{m\%8}$.

When the parameter m is a multiple of 8 we have the expected result of $\frac{7n.m^2}{8}$ atomic communications at the template level.

For a detailed description of how the pre-evaluation can be done in an automatic way take a look at the report available at the URL
ftp://ftp.lifl.fr/pub/reports/AS-publi/an98/as-182.ps.gz

6 Conclusion

Data partitioning is a major performance factor in HPF programs. To help the programmer design a good data distribution strategy, we have studied the evaluation of the communication cost of a program during the writing of this program.

We have presented here a method to compute the communication volume of a HPF program. This method is based on the polyhedral model. So, we are able to handle loop nests with affine loop bounds and affine array access functions. Our method is parameterized and machine independent. Indeed all is done at the language level. An implementation is done using the polyhedral library and the PIP software.

Ongoing work includes extending this method to a larger class of programs and adding compiler optimizations in the model. The last point is quite important since the pure counting of elements exchanged is only one of the factors in the actual communication costs. We will have to recognize special communications patterns as broadcasts which can be implemented more efficiently than general communications.

We are also integrating this method in the HPF-builder tool [6].

References

1. Jennifer M. Anderson and Monica S. Lam. Global optimizations for parallelism and locality on scalable parallel machines. *ACM Sigplan Notices*, 28(6):112–125, June 1993.
2. Vincent Bouchitte, Pierre Boulet, Alain Darte, and Yves Robert. Evaluating array expressions on massively parallel machines with communication/computation overlap. *International Journal of Supercomputer Applications and High Performance Computing*, 9(3):205–219, 1995.
3. S. Chatterjee, J. R. Gilbert, R. S. Schreiber, and S.-H. Tseng. Automatic array alignment in data-parallel programs. In ACM Press, editor, *Twentieth Annual ACM SIGPLAN-SIGACT Symposium on Principles of Programming Languages*, pages 16–28, Charleston, South Carolina, January 1993.
4. Philippe Clauss, Vincent Loechner, and Doran Wilde. Deriving formulae to count solutions to parameterized linear systems using ehrhart polynomials: Applications to the analysis of nested-loop programs. Technical Report RR 97-05, ICPS, apr 1997.
5. Fabien Coelho. *Contributions to HPF Compilation*. PhD thesis, Ecole des mines de Paris, October 1996.
6. Jean-Luc Dekeyser and Christian Lefebvre. Hpf-builder: A visual environment to transform fortran 90 codes to hpf. *International Journal of Supercomputing Applications and High Performance Computing*, 11(2):95–102, 1997.
7. Thomas Fahringer. Compile-time estimation of communication costs for data parallel programs. *Journal of Parallel and Distributed Computing*, 39(1):46–65, November 1996.
8. Paul Feautrier. Parametric integer programming. *RAIRO Recherche Opérationnelle*, 22:243–268, September 1988.

9. Paul Feautrier. Towards automatic distribution. *Parallel Processing Letters*, 4(3):233–244, 1994.

10. M. Gupta. *Automatic Data Partitioning on Distributed Memory Multicomputers.* PhD thesis, College of Engineering, University of Illinois at Urbana-Champaign, September 1992.

11. Manish Gupta and Prithviraj Banerjee. Compile-time estimation of communication costs of programs. *Journal of Programming Languages*, 2(3):191–225, September 1994.

12. Ken Kennedy and Ulrich Kremer. Automatic data layout for High Performance Fortran. In Sidney Karin, editor, *Proceedings of the 1995 ACM/IEEE Supercomputing Conference, December 3–8, 1995, San Diego Convention Center, San Diego, CA, USA*, New York, NY 10036, USA and 1109 Spring Street, Suite 300, Silver Spring, MD 20910, USA, 1995. ACM Press and IEEE Computer Society Press.

13. Kathleen Knobe, Joan D. Lukas, and Guy L. Steele. Data optimization: Allocation of arrays to reduce communication on SIMD machines. *Journal of Parallel and Distributed Computing*, 8:102–118, 1990.

14. Jingke Li and Marina Chen. Index domain alignment: Minimizing cost of cross-referencing between distributed arrays. In *Frontiers 90: The 3rd Symposium on the Frontiers of Massively Parallel Computation*, College Park, MD, October 1990.

15. S. Wholey. Automatic data mapping for distributed-memory parallel computers. In ACM, editor, *Conference proceedings / 1992 International Conference on Supercomputing, July 19–23, 1992, Washington, DC*, INTERNATIONAL CONFERENCE ON SUPERCOMPUTING 1992; 6th, pages 25–34, New York, NY 10036, USA, 1992. ACM Press.

16. Doran Wilde. A library for doing polyhedral operations. Master's thesis, Oregon State University, Corvallis, Oregon, dec 1993.

Modeling the Communication Behavior of Distributed Memory Machines by Genetic Programming*

L. Heinrich-Litan, U. Fissgus, St. Sutter, P. Molitor, and Th. Rauber

Computer Science Institute, Department for Mathematics and Computer Science,
Martin-Luther-Universität Halle-Wittenberg, D-06099 Halle (Saale), Germany
<*name*>@informatik.uni-halle.de

Abstract. Due to load imbalance and communication overhead the behavior of the runtime of distributed memory machines is very complex. The contribution of this paper is to show that runtime functions predicting the execution time of the communication operations can be generated by means of the genetic programming paradigm. The runtime functions generated dominate those presented in literature, till today.

1 Introduction

Distributed memory machines (DMMs) provide large computing power which can be exploited for solving large problems or computing solutions with high accuracy. Nevertheless, DMMs are still not broadly accepted. One of the main reasons is the costly development process for a specific parallel algorithm on a specific DMM. This is due to the fact that parallel algorithms on DMMs may show a complex runtime behavior caused by communication overhead and load imbalance. Thus, there is considerable research effort to model the performance of DMMs. This includes modeling the runtimes of communication operations with parametrized formulas [6, 5, 4, 8, 1]. The modeling of the execution time of communication operations can be used in compiler tools, in parallelizing compilers or simply as concise information of the performance behavior for the application programmer.

Genetic algorithms (GAs) are stochastic optimization techniques which simulate the natural evolutionary process of beings. They often outperform conventional optimization methods when applied to difficult problems. We refer to [2] for an overview on complex problems in the area of industrial engineering, where GAs have been successfully applied. The genetic programming approach (GP) has been introduced by Koza [7] and is a special form of a genetic algorithm.

The contribution of this paper is to show that runtime functions of high quality, which model the execution time of communication operations, can be modeled by the genetic programming approach. To illustrate the effectiveness of the approach we have chosen [8] for comparison. In [8], collective communication

---

* This work has been supported in part by DFG grant Ra 524/5 and Mo 645/5.

operations from the communication libraries PVM and MPI are investigated and compared. Their execution time is modeled by runtime functions found by curve fitting. In this paper we demonstrate the effectiveness of GPs to model the execution time of communication operations on DMMs, by example. Especially we demonstrate that GP generates runtime functions of higher quality than those published in literature till now.

The paper is structured as follows. Section 2 gives a brief overview on the investigations made by [8]. Section 3 presents how performance modeling can be attacked by GPs. Experimental results are shown and discussed in Section 4.

2 Runtime Functions Generated by Curve Functions

First, we give an overview on the communication operations which we use in our experiments. Then we present the model of the communication behavior on the IBM SP2 obtained by curve fitting.

The specific execution platform that [8] used for their experiments is an IBM SP2 with 37 nodes and 128 MByte main memory per processor from GMD St.Augustin, Germany, with IBM MPI-F, Version 1.41. They investigate and compare different communication operations. Due to space limitation, we only report on two communication operations.

Single-transfer operation: In MPI, the standard point-to-point communication operations are the blocking `MPI_Send()` and `MPI_Recv()`. For blocking send, the control does not return to the executing process before the send buffer can be reused safely. Some implementations use an intermediate buffer.

Scatter operation: A single-scatter operation is executed by the global operation `MPI_Scatter()` which must be called by all participating processes. As effect, the specified root process sends a part of its own local data to each other process.

In [8], the runtimes of MPI communication operations on the IBM SP2 are modeled by runtime formulas found by curve fitting that depend on various machine parameters including the number of processors, the bandwidth of the interconnecting networks, and the startup times of the corresponding operations. The investigations resulted in the parameterized runtime formula $f_{scatter}^{cf}(p, b) = 10.28 \cdot 10^{-6} + 91.59 \cdot 10^{-6} \cdot p + 0.030 \cdot 10^{-6} \cdot p \cdot b$ [μsec] for the scatter operation, and $f_{single}^{cf}(b) = 211.8 \cdot 10^{-6} + 0.030 \cdot 10^{-6} \cdot b$ [μsec] for the single-transfer operation. The parameters b and p are the message size in bytes and the number of processors, respectively.

3 The Genetic Programming Approach

The usual form of genetic algorithm was described by Goldberg [3]. Genetic algorithms are stochastic search techniques based on the mechanism of natural selection and natural genetics. They start with an initial population, i.e., an initial set of random solutions which are represented by chromosomes. The chromosomes evolve through successive iterations, called generations. During each generation, the solutions represented by the chromosomes of the population are evaluated using some measures of fitness. To create the next generation, new chromosomes, called offsprings, are formed. The chromosomes involved in the generation of a new population are selected according to their fitness values. Fitter chromosomes have higher probabilities of being selected. After several generations, the algorithm converges to the best chromosome which hopefully represents the optimal or suboptimal solution to the problem. GAs whose chromosomes are syntax trees of formulas, i.e., computer programs, are GPs.

In the next sections we present the GP we have applied for the problem of finding runtime functions of high quality modeling the execution time of the MPI communication operations. We use terminology given in [7].

3.1 The Chromosomes

In genetic programming, a chromosome is an element of the set of all possible combinations of functions that can be composed recursively from a set F of basic functions and a set T of terminals. Each particular function $f \in F$ takes a specified number of arguments, specifying the arity of f. The basic functions we allowed in our application are addition $(+)$ and multiplication $(\cdot)$ both of arity 2, and the operations sqr, $sqrt$, ln, and exp, all of arity 1. The set T is composed of one or two variables p and b (p specifies the number of processors and b specifies the message size), and the constants 1 to 10, 0.1, and 10^{-6}. We represent each chromosome by the syntax tree corresponding to the expression.

In the following, we identify a chromosome with the runtime function which is represented by that chromosome in order to make the diction easier.

3.2 Fitness Function

Fitness is the driving force of Darwinian natural selection and, likewise, of genetic programming. It may be measured in many different ways. The accuracy of the fitness function with respect to the application is crucial for the quality of the results produced by GP. In the application handled here experiments have shown that the average percental error of a chromosome as fitness results in best solutions. Of course, as the aim is to minimize the average percental error, a chromosome is said to be fit if its average percental error is small. Let Pos be the set of measuring points and $m(\alpha)$ be the measured data at measuring point α, then the percental error $error_f(\alpha)$ of a runtime function f at point α and the average percental error $error_f$ of f are given by

$$error_f(\alpha) = \frac{abs(m(\alpha) - f(\alpha))}{m(\alpha)} \cdot 100 \quad \text{and} \quad error_f = \frac{\sum_{\alpha \in Pos} error_f(\alpha)}{|Pos|},$$

respectively.

There are two problems we have to care about when applying this fitness function. The populations can contain chromosomes whose average percental error is greater than the maximal decimal value (which is about $1.79 \cdot 10^{308}$ in GNU g++). In this case, we have to redefine the error of these chromosomes to be some large value. The other problem is that some operations of the syntax trees are not defined, e.g., $\ln a$ with $a \leq 0$. The fitness function of our GP replaces these faulty operations by constants and adds a penalty to the error. Chromosomes with faulty operations are not taken as final result of the GP.

3.3 Crossover and Mutation Operators

The crossover operator for GP creates variation in the population by producing new offsprings that consist of parts taken from each parent. It starts with two parental syntax trees selected with error-inverse-proportionate selection, and produces two offspring syntax trees by independently selecting, using an uniform probability distribution, one random point in each parent to be the crossover point for that parent. The offsprings are produced by exchanging the crossover fragments. To direct the algorithm to generate smooth runtime functions we restricted the solution region to syntax trees of depth up to a constant c.

The mutation operator introduces random changes in structures randomly selected from the population. It begins by selecting a point within the syntax tree by random. Then, it removes the fragment which is below this mutation point and inserts a randomly generated subtree at that point. The depth of the new syntax tree is also limited by c.

3.4 Initialization and Termination

There are several methods to initialize the GP. At the beginning of our experiments we generated each chromosome of the first population recursively top down. The semantics of a node v of depth less than c is taken from $F \cup T$ by random. If a node has depth c, the semantics is randomly taken from the set T of terminals. Experiments showed, that the results are better if we generate a part of the initial chromosomes by approximation, and the rest of them as described above. These initial approximation functions represent linear dependencies between the measuring points and the measured data.

In our experiments, the GP stops after having considered a predefined number of generations.

4 Experimental Results

We begin by quantifying the quality of the runtime functions given in [8]. Then we compare these runtime functions to those generated by GP and discuss the results. We close the section by presenting a runtime formula generated by GP.

We evaluated the runtime functions given in [8] (see Section 2) by using the fitness function specified in Section 3.2. The test set of measured data contained communication times for $p = 4, 8, 16, 32$ processors and message sizes between 4 and 800 bytes with stepsize 20, between 800 and 4.000 with stepsize 400, and between 4.000 and 240.000 with stepsize 4.000.

Table 1. Deviations of the runtime functions

Operation	Approach	$error_f$	% of $\alpha \in Pos$ with $error_f(\alpha) < x$						
			< 0.01	< 0.1	< 1	< 10	< 25	< 50	< 100
scatter	curve fitting	57.36	0.0	0.6	2.6	23.8	49.0	72.5	83.4
	GP	9.68	0.0	2.3	14.2	62.9	93.7	97.7	100.0
single-transfer	curve fitting	71.90	0.0	1.2	20.9	47.4	59.4	67.4	73.4
	GP	14.64	0.3	1	17.6	59.6	75.6	93	100.0

To generate runtime functions by GP, we have to set some parameters. By setting the size of each population to 50, the maximum number of generations to 12000, the maximum depth of syntax trees to 15, and the probabilities for applying crossover, mutation, and the copy operation to 80%, 10% and 10% respectively, we obtained runtime functions by GP which dominate by far the runtime functions from [8] obtained by curve fitting.

In Table 1 we compare the deviations of the runtime functions found by curve fitting to the deviations of the best runtime functions generated after having run the GP algorithm 3 times, which takes about 6 hours on a SUN SPARC station Ultra 1, 128 MByte RAM. The first column specifies the communication operation, the second column the used approach and the third column the average percental error of the corresponding runtime function. Columns 4 to 10 give the percentage of measuring points $\alpha \in Pos$ for which the percental error $error_f(\alpha)$ is less than x for $x = 0.01, 0.1, 1, 10, 25, 50$ and 100, respectively.

Figure 1 graphically compares the deviation of the runtime function from [8] and the deviation of the runtime function generated by GP for the scatter operation with respect to some set of measured data. The curves differ very much for small values of parameter b whereby the curve generated by GP dominates by far the curve found by curve fitting. For large parameter values, both runtime functions predict the execution time of the scatter operation roughly alike. Figure 2 shows the comparison of the deviations of the runtime functions found by curve fitting and GP, respectively, for the single-transfer operation (which depends only on parameter b) with respect to several sets of measured data. Figure 2 shows only the cut-out determined by the small message sizes (up to 500 byte).

Let us discuss the results. The disadvantage of curve fitting is that usually only one simple function is selected to model a communication operation for a large range of message sizes. This may lead to poor results in some regions since often different methods are used for different message sizes. GP on the other hand allows for generating complex functions.

We close the section by presenting the runtime formula $f_{scatter}^{GP}$ predicting the execution time of the scatter operation which has been generated by GP :

$$f_{scatter}^{GP}(p, b) = 10^{-6} \cdot (b \cdot p \cdot (p^{1/2} - 1)^{1/2})^{1/2} \cdot$$
$$(b \cdot p \cdot 0.000625 + 0.00005 \cdot b + 0.00375 \cdot p^2 + 0.003 \cdot p + \ln(e^{p/2}) + 49)^{1/2}.$$

It shows that the runtime formulas generated by GPs are rather complex. To our opinion, this is the only disadvantage of the approach we presented.

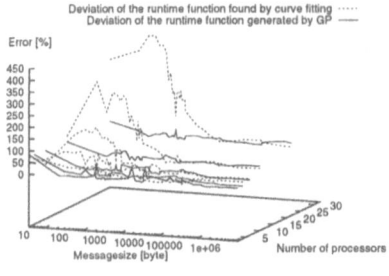
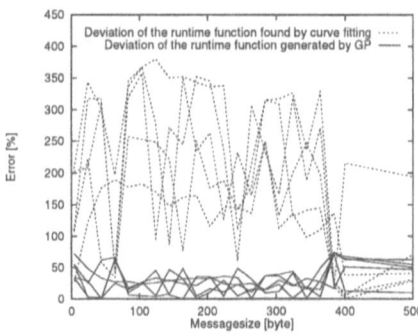

Fig. 1. Deviation of the runtime function found by curve fitting and deviation of the runtime function generated by GP for the scatter operation.

Fig. 2. Deviation of the runtime function obtained by curve fitting and deviation of the runtime function generated by GP for the single-transfer operation.

5 Conclusions

We showed by example that genetic programming provides runtime functions predicting the execution time of communication operations of DMMs which dominate the runtime functions found by curve fitting.

The next step of our investigations we will make is to automatically generate different runtime functions for different intervals of the parameters. The intervals themselves have to be determined by genetic programming, too. Using such non uniform runtime functions seems to be necessary to predict the execution functions more accurately, as DMMs use various communication protocols depending on the message size.

References

1. Foschia, R., Rauber, Th., and Rünger, G.: Prediction of the Communication Behavior of the Intel Paragon. In *Proceedings of the 1997 IEEE MASCOTS Conference*, pp.117-124, 1997.
2. Gen, M., and Cheng, R.: *Genetic Algorithms* & Engineering Design. John Wiley & Sons, Inc., New York, 1997.
3. Goldberg, D.: *Genetic Algorithms in Search, Optimization and Machine Learning*, Addison Wesley, Reading, MA, 1989.
4. Hu, Y., Emerson, D., and Blake, R.: The communication performance of the Cray T3D and its effect on iterative solvers. *Parallel Computing*, 22:829-844, 1996.
5. Hwang, K., Xu, Z., and Arakawa, M.: Benchmark Evaluation of the IBM SP2 for Parallel Signal Processing. *IEEE Transactions on Parallel and Distributed Systems*, 7(2):522-536, 1996.
6. Johnson, L.: Performance Modeling of Distributed Memory Architecture. *Journal of Parallel and Distributed Computing*, 12:300-312, 1991.
7. Koza, J.: *Genetic Programming*. The MIT Press, 1992.
8. Rauber, Th., and Rünger, G.: PVM and MPI Communication Operations on the IBM SP2: Modeling and Comparison. In Proceedings of HPCS 97, 1997.

Representing and Executing Real-Time Systems

Rafael Ramirez

National University of Singapore
Information Systems and Computer Science Department
Lower Kent Ridge Road, Singapore 119260
rafael@iscs.nus.sg

Abstract. In this paper, we describe an approach to the representation, specification and implementation of real-time systems. The approach is based on the notion of concurrent object-oriented systems where processes are represented as objects. In our approach, the behaviour of an object (its safety properties and time requirements) is declaratively stated as a set of temporal constraints among events which provides great advantages in writing concurrent real-time systems and manipulating them while preserving correctness. The temporal constraints have a procedural interpretation that allows them to be executed, also concurrently. Concurrency issues and time requirements are separated from the code, minimizing dependency between application functionality and concurrency/timing control.

1   Introduction

Parallel computers and distributed systems are becoming increasingly important. Their impressive computation to cost ratios offer a considerable higher performance than that possible with sequential machines. Yet there are few commercial applications written for them. The reason is that programming in these environments is substantially more difficult than programming for sequential machines, in respect of both correctness (to achieve correct synchronization) and efficiency (to minimize slow interprocess communication). While in traditional sequential programming the problem is reduced to make sure that the program's final result (if any) is correct and that the program terminates, in concurrent programming it is not necessary to obtain a final result but to ensure that several properties hold during program execution. These properties are classified into safety properties, those that must always be true, and progress (or liveness) properties, those that must eventually be true. Partial correctness and termination are special cases of these two properties. To make things even worse, there exists a wide variety of parallel architectures and a corresponding variety of concurrent programming paradigms. For most problems, it is not possible to envisage a general concurrent algorithm which is well suited to all parallel architectures. Real-time systems are inherently concurrent systems which, in addition to usually require synchronisation and communication with both their environment and within their own components, need to execute under timing constraints. We propose an approach which aims to reduce the inherent complexity of writing concurrent real-time systems. Our approach consists of the following:

1. In order to incorporate the benefits of the declarative approach to concurrent real-time programming, it is necessary that a program describe more than its final result. We propose a language based on classical first-order logic in which all safety properties and time requirements of programs are *declaratively* stated.

2. Programs are developed in such a way that they are not specific to any particular concurrent programming paradigm. The proposed language provides a framework in which algorithms for a variety of paradigms can be expressed, derived and compared.

3. Applications can be specified as objects, which provides encapsulation and inheritance. The language object-oriented features produce structured and understandable programs, therefore reducing the number of potential errors.

4. Concurrency issues and timing requirements are separated from the rest of the code, minimizing dependency between application functionality and concurrency control. In addition, concurrency issues and timing requirements can also be independently specified. Thus, it is in general possible to test different synchronisation schemes without modifying the timing requirements and vice versa. Also, a synchronisation/time scheme may be reused by several applications. This provides great advantages in terms of program flexibility, reuse and debugging.

2 Related Work

This work is strongly related to Tempo ([8]) and Tempo++ ([17, 16]). Tempo is a declarative concurrent programming language based on first-order logic. It is declarative in the sense that a Tempo program is both an algorithm and a specification of the safety properties of the algorithm. Tempo++ extends Tempo (as presented in [8]) by adding numbers and data structures (and operations and constraints on them), as well as supporting object-oriented programming. Both languages explicitly described processes as partially ordered set of events. Events are executed in the specified order, but their execution times are only *implicit*. Here, we extend Tempo++ to support real-time by making event execution times *explicit* as well as allowing the specification of the timing requirements in terms of relations among these execution times.

Concurrent logic programming is also an important influence for this work. This is because concurrent logic programming languages (e.g. Parlog [6], KL1 [22]) and their object-oriented extensions (e.g. Polka [4], Vulcan [10]) preserve many of the benefits of the abstract logic programming model, such as the logical reading of programs and the use of logical terms to represent data structures. However, although these languages preserve many benefits of the logic programming model, and their programs explicitly specify their final result, important program properties, namely safety and progress properties, remain implicit. These properties have to be preserved by using control features such as modes and sequencing, producing programs with little or no declarative reading [17]. In addition, traditional concurrent logic programming languages do not provide support for real-time programming and thus, are not suitable for this kind of applications.

Concurrent constraint programming languages [19] and their object-oriented and real-time extensions (e.g. [21, 20]) suffer from the same problems as concurrent logic languages. Program safety and progress properties remain implicit. These properties are preserved by checking and imposing value constraints on shared variables in the store. Also, there is no clear separation of programs concurrency control, timing requirements and application functionality. In addition, the concurrent constraint model is most naturally suited for shared memory architectures not being easily adapted to model distributed programming systems.

Our work is also related to specification languages for concurrent real-time systems. It is closer to languages based on temporal logic, e.g. Unity [3] and TLA [13], and real-time extensions of temporal logic, e.g. [12], than process algebras [9, 14] and real-time extensions of state-transition formalisms [2, 5]. Unity and TLA specifications can express the safety and progress properties of concurrent systems but they are not executable. Both formalisms model these systems by sequences of actions that modify a single shared state which might be a bottleneck if the specifications were to be executed.

Section 3 introduces the core concepts of our approach to real-time programming, namely events, precedence constraints and real-time, and outlines a methodology for the development of concurrent real-time systems (to make the paper sel-contained there is a small overlap with [8] in Sections 3.1 and 3.3). Section 4 describes how these concepts can be extended by adding practical programming features such as data structures and operations on them as well as supporting object-oriented programming. Finally, Section 5 summarizes our approach and its contributions as well as some areas of future research.

3   Events, Constraints and Real-Time

3.1   Events and Precedence

Many researchers, e.g. [11, 15], have proposed methods for reasoning about temporal phenomena using partially ordered sets of events. Our approach to the specification and implementation of real-time systems is based on the same general idea. We propose a language in which processes are explicitly described as partially ordered sets of events. The event ordering relation $X < Y$, read as "X precedes Y", is the main primitive predicate in the language (there are only two primitive predicates), its domain is the set of events, and is defined by the following axioms (the last axiom is actually a corollary of the other three):

$$\forall X \forall Y \forall Z (X < Y \land Y < Z \to X < Z)$$
$$\forall X \forall Y (time(Y, eternity) \to X < Y)$$
$$\forall X (X < X \to time(X, eternity))$$

$$\forall X \forall Y (Y < X \land time(Y, eternity) \to time(X, eternity))$$

The meaning of predicate $time(X, Value)$ is "event X is executed at time $Value$" and $eternity$ is interpreted as a time point that is later than all others. Events are atomically executed in the specified order (as soon as its predecessors have been executed) and no event is executed for a variable that has execution time $eternity$. $Long$-$lived$ processes (processes comprising a large or infinite set of events) are specified allowing an event to be associated with one or more other events: its offsprings. The offsprings of an event E are named $E + 1$, $E + 2$, etc., and are implicitly preceded by E, i.e. $E < E + N$, for all N. The first offspring $E + 1$ of E is referred to by $E+$. Syntactically, offsprings are allowed in queries and bodies of constraint definitions, but not in their heads.

Long-lived processes may not terminate because body events are repeatedly introduced. Interestingly, an infinitely-defined constraint need not cause non-termination: a constraint, all whose arguments have time value $eternity$ will not be expanded. The time value of an event E can be bound to $eternity$ (as a consequence of the axioms defining "$<$") by enforcing the constraint $E < E$. No offsprings of E will be executed. Their time values are known to be bound to $eternity$ since they are implicitly preceded by E and the time value of E is $eternity$.

In the language, disjunction is specified by the disjunction operator ';' (which has lower priority than ',' but higher than '$\leftarrow$'). The clause $H \leftarrow Cs1; \ldots ; Csn$ abbreviates the set of clauses $H \leftarrow Cs1, \ldots, H \leftarrow Csn$. In the absence of disjunction, a query determines a unique set of constraints. An interpreter produces any execution sequence satisfying those constraints, it does not matter which one. With disjunction, a single set of constraints must be chosen from among many possible sets, i.e., a single alternative must be selected from each disjunction.

In the language described above, processes are explicitly described as partially ordered set of events. Their behavior is specified as logical formulas which define temporal constraints among a set of events. This logical specification of a process has a well defined and understood semantics and allows for the possibility of employing both specification and verification techniques based on formal logic in the development of concurrent systems.

3.2 Real-Time

Time requirements in the language may be specified by using the primitive predicate $time(X, Value)$. This constrains the execution time of event X by forcing X to be executed at time $Value$. In this way, quantitative temporal requirements (e.g. maximal, minimal and exact distance between events) can be expressed in the language. For Instance, Maximal distance between two events E and F may be specified by the constraint $max(E, F, N)$, meaning "event E is followed by event F within N time units", and defined by

$$max(E, F, N) \leftarrow E < F, time(E, Et), time(F, Ft), Ft \prec Et + N.$$

where $\prec$ is the usual *less than* arithmetic relationship among real numbers. Thus, maximal distance between families of events, i.e. events and their off-springs, can be specified by the constraint $max*(E, F, N)$, meaning "occurrences of events E, $E+$, $E++,\ldots$ are respectively followed by occurrences of events F, $F+$, $F++$, ... within N time units", and defined by

$$max* (E, F, N) \leftarrow max(E, F, N), max* (E+, F+, N)$$

Minimal and exact distance between two events, as well as other common quantitative temporal requirements, may be similarly specified.

Example 1. Consider an extension to the traditional producer and consumer system where each product cannot be held inside the buffer longer than time t. The system may be represented by the timed Petri net in Figure 1.

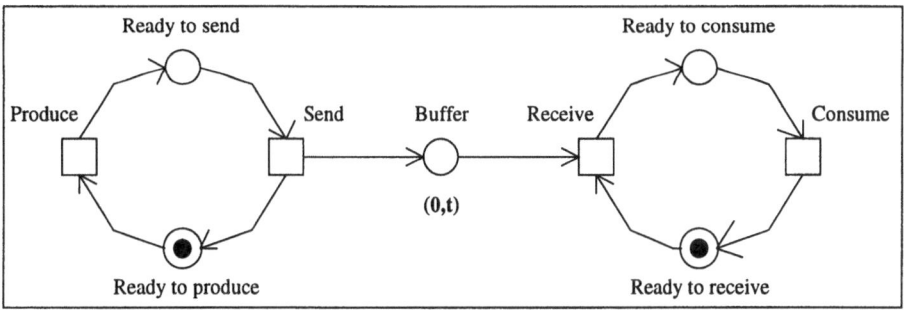

Fig. 1. A producer-consumer system with timed buffer

The producer and consumer components of the system may be respectively specified by $P <*S$, $S <*P+$ and $R <*C$, $C <*R+$, where P and S represent occurrences of events *produce* and *send* and R and C represent occurrences of events *receive* and *consume* ($P+$, $S+$, $R+$, $C+$, ... represent later occurrences of these events). The behaviour of the buffer may be specified by

$$tbuf(S, R, T) \leftarrow max(S, R, T), tbuf(S+, R+, T); tbuf(S+, R, T).$$

where T is the longest time the buffer can hold a product.

3.3 Development of Concurrent Real-Time Systems

In our approach, the specification of the behaviour of the processes in the system is a program, i.e. it is possible to directly execute the specification. Thus, a program P may be transformed into a program that logically implies P (in [7] some transformations rules that can be applied to programs are presented). The derived program is guaranteed to have the same safety properties as the original one, though its progress properties may differ, e.g. one may terminate and the other not. The program may be incrementally strengthened by introducing timing constraints to specify the system time requirements. Finally, the program can be turned into a concurrent one by grouping constraints into processes. This final step affects neither the safety nor the progress properties of the algorithm, provided that some restrictions are observed.

4   Object-Oriented Programming

Our approach to the specification and implementation of real-time systems is based on an extension to the logic presented in the previous section. The logic is extended by adding data structures and operations on them, by allowing values to be assigned to events for inter process communication and by supporting object-oriented programming. A detailed discussion of these ideas can be found in [17].

Our language supports object-oriented programming by allowing a class to encapsulate a set of constraints, specified by a constraint query, together with the related constraint definitions, specified by a set of clauses, in such a way that it describes a set of potential run-time objects. The constraint query defines a partial order among a set of events and the timing requirements on their execution times, and the constraint definitions provide meaning to the user-defined constraints in the query. Both the query and definitions are local to the class. Each of the class run-time objects corresponds to a concurrent process. The name of a class may include variable arguments in order to distinguish different instances of the same class. Events appearing in the constraint query of an object implicitly belong to that object. If an event is shared between several objects (it belongs to two or more objects), it cannot be executed until it has been enabled by all objects that share it. In order to specify the object computation, actions may be associated with events. The representation and manipulation of data as well as the spawning of new objects is handled by these actions. Our current implementation of the language assumes an action to be a definite goal. In order to execute an event, the goal associated with it (if any) has to be solved first. Objects communicate via shared events' values. Shared events represent communication channels and values assigned to them represent messages.

A novel feature of the approach described here is that an object can *partially* inherit another object, i.e. an object can inherit either another object's temporal constraints or actions. Thus, inheritance of concurrency issues and inheritance of code are independently supported. This allows an object to have its synchronisation scheme inherited from another object while defining its own code, or vice versa, or even inherit its synchronisation scheme and code from two different objects.

Our language appears to add to the proliferation of concurrent programming paradigms: processes (objects) communicate via a new medium, *shared events*. However, our objective is rather to simplify matters by providing a framework in which algorithms for a variety of concurrent programming paradigms can be expressed, derived, and compared. Among the paradigms we have considered are synchronuos message passing, asynchronous message passing and shared mutable variables.

Execution: Our current implementation uses a constraint set CS containing the constraints still to be satisfied, and a try list TL containing the events that are to be tried but have not yet been executed. The interpreter constantly takes events from TL and checks if they are *enabled*, i.e. if they are not preceded by any other event (according to CS), and the timing constraints on their execution times are satisfiable, in which case they are executed. The order in which the events are tryed is determined by the timing constraints in CS.

5 Conclusions

We have described an approach to the representation, specification and implementation of concurrent real-time systems. The approach is based on the notion of concurrent object-oriented systems where processes are represented as objects. In the approach, each object is explicitly described as a partially ordered set of events and executes its events in the specified order. The partial order is defined by a set of temporal constraints and object synchronisation and communication are handled by shared events. object behaviour (safety properties and time requirements) are declaratively stated which provides great advantages in writing real-time systems and manipulating them while preserving correctness. The specification of the behaviour of an object has a procedural interpretation that allows it to be executed, also concurrently. Our approach can also be used as the basis for a development methodology for concurrent systems. First, the system can be specified in a perpicuous manner, and then this specification may be incrementally strengthened and divided into objects that communicate using the intended target paradigm. An object can *partially* inherit another object, i.e. an object can inherit either another object's temporal constraints, actions or actions definitions. Thus, inheritance of concurrency issues and inheritance of code are independently supported.

Current status. A prototype implementation of the complete language has been written in Prolog, and used to test the code of a number of applications. The discussion of these applications is out of the scope of this paper.

Future work. In the language presented, the action associated with an event can, in principle, be specified in any programming language. Thus, different types of languages, such as the imperative languages, should be considered and their interaction with the model investigated.

Events are considered atomic. Instead of being atomic, they could be treated as time intervals during which other events can occur ([1] and [11]). Such events can be further decomposed to provide an arbitrary degree of detail. This could be useful in deriving programs from specifications.

Object behaviour is specified as logical formulas which define temporal constraints among a set of events. This logical specification of an object has a well defined and understood semantics and we are planning to look carefully into the possibility of employing both specification and verification techniques based on formal logic in the development of concurrent real-time systems.

References

1. Allen, J.F. 1983. *Maintaining knowledge about temporal intervals*. Comm. ACM 26, 11, pp.832-843.
2. Alur, R., and Dill, D.L. 1990. *Automata for modeling real-time systems*, in ICALP'90: Automata, Languages and Programming, LNCS 443, pp.322-335. Springer-Verlag.
3. Chandy, K.M. and Misra, J. 1988. *Parallel Program Design*. Addison-Wesley.
4. Davison, A. 1991. *From Parlog to Polka in Two Easy Steps*, in PLILP'91: 3rd Int. Symp. on Programming Language Implementation and LP, Springer LNCS 528, pp.171-182, Passau, Germany, August.
5. Dill, D.L. 1989. *Timing assumptions and verification of finite-state concurrent systems*, in CAV'89: Automatic Verification Methods for Finite-state Systems, LNCS 407, pp. 197-212, Springer-Verlag.
6. Gregory, S. 1987. *Parallel Logic Programming in PARLOG*, Addison-Wesley.
7. Gregory, S. 1995. *Derivation of concurrent algorithms in Tempo*. In LOPSTR95: Fifth International Workshop on Logic Program Synthesis and Transformation.
8. Gregory, S. and Ramirez, R. 1995. *Tempo: a declarative concurrent programming language*. Proc.of the ICLP (Tokyo, June), MIT Press, 1995.
9. Hoare, C.A.R. 1985. *Communicating Sequential Processes*, Prentice Hall.
10. Kahn, K.M., Tribble, D., Miller, M.S., and Bobrow, D.G. 1987. *Vulcan: Logical Concurrent Objects*, In Research Directions in Object-Oriented Programming, B. Shriver, P. Wegner (eds.), MIT Press.
11. Kowalski R., and Sergot, M. 1986. *A Logic-based Calculus of Events*, New Generation Computing, 4, 1, pp.67-95.
12. Koymans, R. 1990. *Specifying real-time properties with metric temporal logic*, Real-time Systems, 2, 4, pp.255-299.
13. Lamport, L. 1994. *The temporal logic of actions*. ACM Trans. on Programming Languages and Systems, 16, 3, pp. 872-923.
14. Milner, R. 1989. *Communication and Concurrency*, Prentice Hall.
15. Pratt, V. 1986. *Modeling concurrency with partial orders*, International Journal of Parallel Programming, 1(15):33-71.
16. Ramirez, R. 1995. *Declarative concurrent object-oriented programming in Tempo++*. In Proceedings of the ICLP'95 Workshop on Parallel Logic Programming Japan, T. Chikayama, H. Nakashima and E. Tick (Ed.).
17. Ramirez, R. 1996. *A logic-based concurrent object-oriented programming language*, PhD thesis, Bristol University.
18. Ramirez, R. 1996. *Concurrent object-oriented programming in Tempo++*. In Proceedings of the Second Asian computing Science Conference (Asian'96), Singapore. LNCS 1179, pp. 244-253. Springer-Verlag
19. Saraswat V. 1993. *Concurrent constraint programming languages*, PhD thesis, Carnegie-Mellon University, 1989. Revised version appears as *Concurrent constraint programming*, MIT Press, 1993.
20. Saraswat V. 1993 et al. *Programming in timed concurrent constraint languages*. In Constraint Programming - Proceedings of the 1993 NATO ACM Symposium, pp. 461-410. Springer-Verlag.

21. Smolka, G. 1995. *The Oz programming model*, Lecture Notes in Computer Science Vol. 1000, Springer-Verlag, pp.324-243.
22. Ueda, K. and Chikayama, T. 1990. *Design of the kernel language for the parallel inference machine*. Computer Journal 33, 6, pp.494-500.

Fixed Priority Scheduling of
Age Constraint Processes

Lars Lundberg

Department of Computer Science, University of Karlskrona/Ronneby,
S-372 25 Ronneby, Sweden,
Lars.Lundberg@ide.hk-r.se

Abstract. Real-time systems often consist of a number of independent processes which operate under an age constraint. In such systems, the maximum time from the start process L_i in cycle k to the end in cycle $k+1$ must not exceed the age constraint A_i for that process. The age constraint can be met by using fixed priority scheduling and periods equal to $A_i/2$. However, this approach restricts the number of process sets which are schedulable.

In this paper, we define a method for obtaining process periods other than $A_i/2$. The periods are calculated in such a way that the age constraints are met. Our approach is better in the sense that a larger number of process sets can be scheduled compared to using periods equal to $A_i/2$.

1 Introduction

Real-time systems often consist of a number of independent periodic processes. These processes may handle external activities by monitoring sensors and then producing proper outputs within certain time intervals. A similar example is a process which continuously monitors certain variables in a database. When these variables or sensors change, the system have to produce certain outputs within certain time intervals. These outputs must be calculated from input values which are fresh, i.e. the age of the input value must not exceed certain time limits. The processes in these kinds of systems operate under the age constraint.

The age constraint defines a limit on the maximum time from the point in time when a new input value appears to the point in time when the appropriate output is produced. Figure 1 shows a scenario where a value E_i appears shortly after process L_i has started its k:th cycle (denoted L_i^k). Process L_i starts its execution by reading the sensor or variable. Consequently, E_i will not affect the output in cycle k. The output F_i corresponding to value E_i (or fresher) is produced at the end of cycle $k+1$. The age constraint A_i is defined as the maximum time between the beginning of the process' execution in cycle k to the end of the process' execution in cycle $k+1$.

A scheduling scheme can be either static or dynamic. In dynamic schemes the priority of a process is decided at run-time, e.g. the earliest deadline algorithm [3]. In static schemes, processes are assigned a fixed priority, e.g. the rate-monotone algorithm [4]. Fixed priority scheduling is relatively easy to implement and it requires less overhead than dynamic schemes.

Most studies in this area have looked at scenarios where the computation time and the period of a process are known. However, for age constraint processes the period is not known. We have instead defined a maximum time between the beginning of the process' execution in cycle k to the end of the process' execution in cycle $k+1$. We would like to translate this restriction into a period for the process thus making it possible to use fixed priority schemes.

The age constraint is met by specifying a process period $T_i' = A_i/2$ (we will use the notation T_i for other purposes), thus obtaining a set of processes with known periods T_i' and computation times C_i. For such scenarios, it is well known that rate-monotone scheduling is optimal, and a number of schedulability tests have been obtained [4]. Specifying a process period $T_i' = A_i/2$ for age constraint processes is, however, an unnecessary strong restriction, which do not allow that the start of L_i in cycle k and the end in cycle $k+2$ may be separated by a time greater than $3T_i'$, whereas the age constraint allows a separation of up to $2A_i = 4T_i'$.

In this paper we show that, by using rate-monotone scheduling, it is possible to define periods which are better than using periods $T_i' = A_i/2$. Our method is better in the sense that we will be able to schedule a larger number of process sets than using $T_i' = A_i/2$.

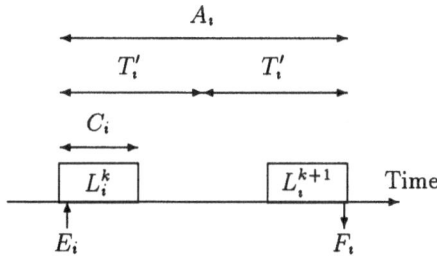

Fig. 1. The age constraint for process L_i.

2 Calculating process periods

Consider a set of n processes $\overline{L} = [L_1, L_2, ..., L_n]$, with associated age constraints A_i and computation times C_i. The priority of each process is defined by its age constraint A_i. The smaller the value A_i, the higher the priority of L_i, i.e. the priority order is the same as for rate-monotone scheduling with $T_i' = A_i/2$. We assume preemptive scheduling and we order the processes in such a way that $A_i \leq A_{i+1}$.

We start by considering L_1. This process has the highest priority, and is thus not interrupted by any other process. We want to select as long periods as possible, thus minimizing processor utilization. Obviously, the period of a process must not exceed $A_i - C_i$. Since L_1 has the highest priority, it is safe to set the period of L_1 to $A_1 - C_1$ In order to distinguish our periods from the ones which are simply equal to half the age constraint A_i, we denote our periods as T_i, i.e. $T_1 = A_1 - C_1$.

We now consider process L_2. The maximum response time R_2 of process L_2 is defined as the maximum time from the release of L_2 in cycle k to the time that L_2 completes in the same cycle. If the execution of L_2 is not interrupted by any other process, the response time is simply equal to the computation time C_2, i.e. $R_1 = C_1$. However, the execution of L_2 may be interrupted by L_1. Process L_1 may in fact interfere with as much as $\lceil R_2/T_1 \rceil C_1$ [3]. Consequently, $R_2 = C_2 + \lceil R_2/T_1 \rceil C_1$. The only unknown value in this equation is R_2. The equation is somewhat difficult to solve due to the ceiling function ($\lceil R_2/T_1 \rceil$). In general there could be many values of R_2 that solve this equation. The smallest such value represents the worst-case response time for process L_2. It has been shown that R_2 can be obtained from this equation by forming a recurrence relationship. The technique for doing this is shown in [3].

The beginning of L_2 in cycle k and the end of L_2 in cycle $k+1$ may be separated with as much as $T_2 + R_2$, where T_2 is the period that we will assign to process L_2. From the age constraint we know that $T_2 + R_2 \leq A_2$. In order to minimize processor utilization we would like to select as long a period T_2 as possible. Consequently, $T_2 = A_2 - R_2$.

In general, the maximum response time of process i can be obtained from the relation $R_i = C_i + \sum_{j=1}^{i-1} \lceil R_i/T_j \rceil C_j$ [3]. When we know R_i, the cycle time for L_i is set to $T_i = A_i - R_i$.

Figure 2 shows a set with three processes L_1, L_2 and L_3, defined by $A_1 = 8$, $C_1 = 2$, $A_2 = 10$, $C_2 = 2$, $A_3 = 12$ and $C_3 = 2$. From these values we obtain the period $T_1 = A_1 - C_1 = 8 - 2 = 6$. The maximum response time R_2 for process L_2 is obtained from the relation $R_2 = C_2 + \lceil R_2/T_1 \rceil C_1 = 2 + \lceil R_2/6 \rceil 2$. The smallest value R_2 which solves this equation is 4, i.e. $R_2 = 4$. Consequently, $T_2 = A_2 - R_2 = 10 - 4 = 6$. The maximum response time R_3 for process L_3 is obtained from the relation $R_3 = C_3 + \lceil R_3/T_1 \rceil C_1 + \lceil R_3/T_2 \rceil C_2 = 2 + \lceil R_3/6 \rceil 2 + \lceil R_3/6 \rceil 2$. The smallest value R_3 which solves this equation is 6, i.e. $R_3 = 6$. Consequently, $T_3 = A_3 - R_3 = 12 - 6 = 6$. In figure 2, the first release of L_1 is done at time 2, the first release of L_2 is done at time 1 and the first release of L_3 is done at time 0.

In the worst-case scenario, process L_i may suffer from the maximum response time R_i in two consecutive cycles, i.e. in order to meet the age constraint we know that $2R_i \leq A_i$. Consequently, there is no use in selecting a T_i smaller than R_i. However, as long as we obtain T_i which are longer than or equal to R_i, the age constraint will be met. Therefore, process L_i can be scheduled if and only if $R_i \leq A_i/2$.

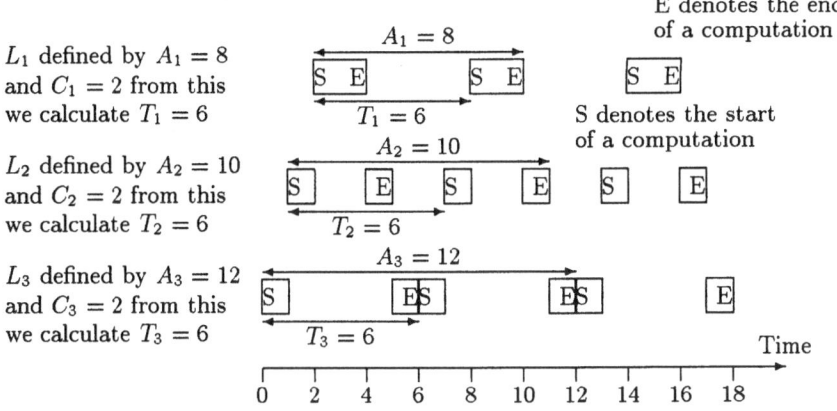

L_1 defined by $A_1 = 8$ and $C_1 = 2$ from this we calculate $T_1 = 6$

L_2 defined by $A_2 = 10$ and $C_2 = 2$ from this we calculate $T_2 = 6$

L_3 defined by $A_3 = 12$ and $C_3 = 2$ from this we calculate $T_3 = 6$

E denotes the end of a computation

S denotes the start of a computation

Fig. 2. A set with three processes L_1, L_2 and L_3, defined by $A_1 = 8, C_1 = 2, A_2 = 10, C_2 = 2, A_3 = 12$ and $C_3 = 2$.

Theorem 1. *A set of processes which is schedulable using rate-monotone priority assignment and $T'_i = A_i/2$ is also schedulable using our scheme.*

Proof. The difference between our scheme and rate-monotone with $T'_i = A_i/2$ is that we use different periods T_i. Since the process set is schedulable using the T'_i periods we know that $R'_i \leq T'_i (1 \leq i \leq n)$, where R'_i denotes the maximum response time using the T'_i periods.

By use of induction, we show that $R_i \leq R'_i (1 \leq i \leq n)$.

- $R_1 = C_1 \leq R'_1 = C_1$
- If $R_j \leq R'_j (1 \leq j \leq x < n)$, then $T'_j = A_j/2 \leq A_j - R_j = T_j$. If $T'_j \leq T_j$, then
  $$R_{x+1} = C_{x+1} + \sum_{j=1}^{x} \lceil R_{x+1}/T_j \rceil C_j \leq R'_{x+1} = C_{x+1} + \sum_{j=1}^{x} \lceil R'_{x+1}/T'_j \rceil C_j.$$

Consequently, $R_i \leq R'_i \leq T'_i = A_i/2$, i.e. $R_i \leq A_i/2$, which means that process $L_i (1 \leq i \leq n)$ can be scheduled using our scheme.

Consider the processes in figure 2. If we would have used the periods $A_i/2$, we would have got $T'_1 = 4, T'_2 = 5$ and $T'_3 = 6$. This would have resulted in a utilization of $C_1/T'_1 + C_2/T'_2 + C_3/T'_3 = 2/4 + 2/5 + 2/6 = 1.23 > 1$, i.e. the process set would not have been schedulable. Consequently, our scheme is better than rate-monotone and $T'_i = A_i/2$ in the sense that we are able to schedule a larger number of process sets.

3 Simple analysis

In the scheme that we propose, the period of a process depends on the priority of the process. The period for a process L_i gets shorter if the priority of L_i is reduced and vice versa. This property makes our scheme hard to analyze. However, for the limited case when there are two processes, a thorough analysis is possible.

Theorem 2. *The priority assignment used in our scheme is optimal for all sets containing two processes.*

Proof. Consider two processes L_1 and L_2, such that $A_1 \leq A_2$. Assume that these two processes are schedulable if the priority of L_2 is higher than the priority of L_1. We will now show that if this is the case, the two processes are also schedulable if L_1 has higher priority than L_2.

If L_1 and L_2 are schedulable when the priority of L_2 is higher than the priority of L_1, then $R_1 = C_1 + \lceil R_1/(A_2 - C_2) \rceil C_2 = C_1 + kC_2 \leq A_1/2$ (for some integer $k > 0$). Consequently, $C_2 \leq A_1/2 - C_1$.

If we consider the opposite priority assignment we know that the schedulability criterion is that $R_2 = C_2 + \lceil R_2/(A_1 - C_1) \rceil C_1 \leq A_2/2$. Since $C_2 \leq A_1/2 - C_1$, we know that the maximum interference from process L_1 on process L_2 is C_1, i.e. $R_2 = C_2 + C_1$. Since $C_2 \leq A_1/2 - C_1$ and $R_2 = C_2 + C_1$, we know that $R_2 \leq A_1/2$, and since $A_1 \leq A_2$, we know that $R_2 \leq A_2/2$, thus proving the theorem.

Theorem 3. *All sets containing two processes L_1 and L_2 for which $C_1/(A_1 - C_1) + C_2/(A_2 - C_2)$ is less than $2(\sqrt{2} - 1) = 0.83$ are schedulable using our scheme, and there are process sets containing two processes for which $C_1/(A_1 - C_1) + C_2/(A_2 - C_2) = 2(\sqrt{2} - 1) + e$ (for any $e > 0$) which are not schedulable using our scheme.*

Proof. We assume that $A_1 \leq A_2$, and that we use our scheme for calculating periods and priorities.

We want to find the minimum value $C_1/(A_1 - C_1) + C_2/(A_2 - C_2)$, such that the process set is not schedulable. We know that as long as $C_1/(A_1 - C_1) \leq 1$, process L_1 can be scheduled. This is a trivial observation.

Process L_2 can be scheduled if $R_2 \leq A_2/2$, i.e. we want to minimize $C_1/(A_1 - C_1) + C_2/(A_2 - C_2)$ under the constraint that $R_2 = C_2 + \lceil R_2/(A_1 - C_1) \rceil C_1 > A_2/2$.

If $\lceil R_2/(A_1 - C_1) \rceil = k$ (for some integer $k > 0$), then the minimum for $C_1/(A_1 - C_1) + C_2/(A_2 - C_2)$ is obtained when $R_2 = k(A_1 - C_1) = C_2 + \lceil R_2/(A_1 - C_1) \rceil C_1 = C_2 + kC_1 => C_2 = k(A_1 - 2C_1)$. Consequently, we want to find the k which minimizes $C_1/(A_1 - C_1) + k(A_1 - 2C_1)/(A_2 - k(A_1 - 2C_1))$. Since, $C_2 = k(A_1 - 2C_1) \leq A_2/2$ we see that $A_2 - k(A_1 - 2C_1) > 0$, and since $0 \leq A_1 - 2C_1$, we see that the minimum for $C_1/(A_1 - C_1) + k(A_1 - 2C_1)/(A_2 - k(A_1 - 2C_1))$ is obtained for $k = 1$. Consequently, we want to minimize $C_1/(A_1 - C_1) + (A_1 - 2C_1)/(A_2 - (A_1 - 2C_1))$, under the constraint that $R_2 = A_1 - C_1 > A_2/2$. The minimum is obviously obtained when $A_1 - C_1$ is as small as possible, i.e. when $A_1 - C_1 = A_2/2 + e$ (for some infinitely small positive number e). Consequently, we want to minimize $C_1/(A_2/2 + e) + (A_2/2 + e - C_1)/(A_2 - (A_2/2 + e - C_1))$. Without loss of generality we assume that $A_2 = 2$. In that case we obtain the following function (disregarding e)

$$f(C_1) = C_1 + (1 - C_1)/(1 + C_1)$$

From this we obtain the derivative of f:

$$f'(C_1) = 1 + ((1 - C_1) - (1 + C_1))/(1 + C_1)^2$$

By setting $f'(C_1) = 0$ we will find the values C_1 which minimizes f.

$$1 + ((1 - C_1) - (1 + C_1))/(1 + C_1)^2 = 0 => 1 + 2C_1 + C_1^2 + 1 - 2C_1 = 0 => C_1 = \sqrt{2} - 1.$$

From this we that min $f(C_1) = f(\sqrt{2} - 1) = 2(\sqrt{2} - 1) = 0.83$.

It is interesting to note that the schedulability bound for $C_1/(A_1 - C_1) + C_2/(A_2 - C_2)$ is the same as the schedulability bound for rate-monotone scheduling and two processors. However, in that case we have $C_1/T_1' + C_2/T_2' = 2C_1/A_1 + 2C_2/A_2 = 0.83$. At this point we do not know if it is a coincident that the values are the same or not. It would be interesting to examine the case with three processes and see if the schedulability bound for $C_1/(A_1 - C_1) + C_2/(A_2 - C_2) + C_3/(A_3 - C_3) = 3(\sqrt[3]{2} - 1) = 0.78$, which is the bound for rate-monotone scheduling and three processes.

4 Improving the scheme

In the previous sections we assumed that the interference from a higher priority process L_j affected the maximum response time R_{j+x} for a process L_{j+x} according to the formula $R_{j+x} = C_{j+x} + \cdots + \lceil R_{j+x}/T_j \rceil C_j$. However, if $T_{j+x} = kT_j$ (for some integer $k > 0$), we can adjust the phasing of L_j and L_{j+x} in such a way that the interference of L_j on L_{j+x} is limited to $(\lceil R_{j+x}/T_j \rceil - 1)C_j$ (see figure 3). The phasing is adjusted in such a way that a release of process L_{j+x} always occurs at exactly the same time as a release of process L_j. Consequently, if $T_{j+x} < kT_j \leq T_{j+x} - C_j$ we can extend the period of L_{j+x} to kT_j. In order to distinguish the periods obtained when using the optimized version of the scheme from the ones obtained using the unoptimized version, we denote the optimized period for process L_j as t_j.

In order to obtain the optimized periods t_j, for a set containing n processes, we start with the periods T_j and we then use the following algorithm:

```
t₁ = T₁
for y = 2 to n loop
    t_y = T_y
    for j = 1 to y − 1 loop
        if t_y < kt_j ≤ T_y − C_j then t_y = kt_j
    end loop
    y = y + 1
end loop
```

The execution of the process set is started by releasing all processes at the same time.

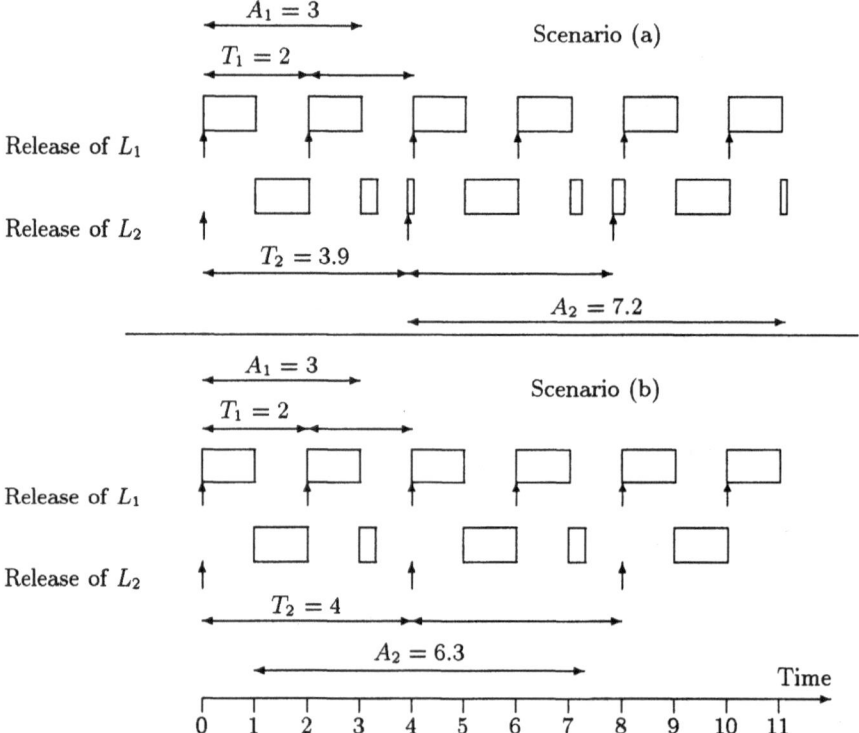

Fig. 3. In scenario (a) there are two processes L_1 and L_2. Process L_1 has a period $T_1 = 2$ and a computation time $C_1 = 1$. Process L_2 has a period $T_2 = 3.9$ and a computation time $C_2 = 1.3$. In this scenario we are able to meet the age constraints $A_1 = 3$ and $A_2 = 7.2$, i.e. $A_2 = T_2 + R_2 = T_2 + C_2 + 2C_1 = 3.9 + 1.3 + 2 = 7.2$. In scenario (b) we consider the same processes, with the exception that the period of L_2 has been extended, i.e. $T_2 = 2T_1 = 4$. The phasing of L_1 and L_2 has also been adjusted such that a release of L_2 always coincides with a release of L_1. In this scenario we are able to meet the age constraints $A_1 = 3$ and $A_2 = 6.3$, i.e. $A_2 = T_2 + R_2 = T_2 + C_2 + C_1 = 4 + 1.3 + 1 = 6.3$. Consequently, by extending the period of L_2 we were able to meet tougher age constraints.

Theorem 4. *A set of processes which is schedulable using the unoptimized periods T_i and an arbitrary phasing of processes is also schedulable using the optimized periods t_i, provided that we are able to adjust the phasing of the processes.*

Proof. Let r_i denote the maximum response time for process L_i using the periods t_i and the optimized phasing. Obviously, $T_i \leq t_i$. From this we conclude that $r_i = C_i + \sum_{j=1}^{i-1} \lceil r_i/t_j \rceil C_j \leq R_i = C_i + \sum_{j=1}^{i-1} \lceil R_i/T_j \rceil C_j$. Consequently, if $R_i \leq A_i/2$, then $r_i \leq A_i/2$.

Figure 4 shows a process set which is schedulable using the improved version of our scheme. This process set would not have been schedulable using the unoptimized version of our scheme. Consequently, the improved version of the scheme is better in the sense that we are able to schedule a larger number of process sets.

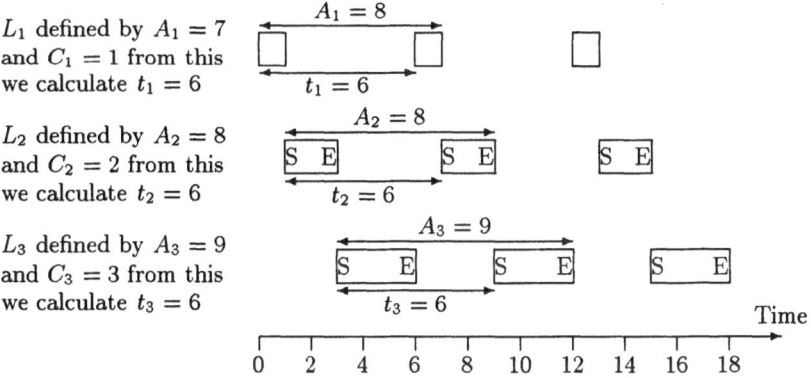

L_1 defined by $A_1 = 7$ and $C_1 = 1$ from this we calculate $t_1 = 6$

L_2 defined by $A_2 = 8$ and $C_2 = 2$ from this we calculate $t_2 = 6$

L_3 defined by $A_3 = 9$ and $C_3 = 3$ from this we calculate $t_3 = 6$

Fig. 4. Three processes which are schedulable using the improved scheme.

5 Conclusions

In this paper a method for scheduling age constraint processes has been presented. An age constraint process is defined by two values: the maximum time between the beginning of the process' execution in cycle k to the end of the process' execution in cycle $k+1$ (the age constraint), and the maximum computation time in each cycle, i.e. the period of each process is not explicitly defined. We present an algorithm for calculating process periods. Once the periods have been calculated the process set can be executed using preemption and fixed priority scheduling. The priorities are defined by the rate-monotone algorithm.

Trivially, the age constraint can be met by using periods equal to half the age constraint. However, the periods obtained from our method are better in the sense that they make it possible to schedule a larger number of process sets than we would have been able to do if we had used periods equal to half the age constraint. All process sets which are schedulable using periods equal to half the age constraint are also schedulable using our method.

A simple analysis of our method shows that the rate-monotone priority assignment algorithm is optimal for all process sets containing two processes, using our process periods. We also show that all process sets with two processes, for which $C_1/(A_1 - C_1) + C_2/(A_2 - C_2)$ is less than $2(\sqrt{2} - 1) = 0.83$, are schedulable using our scheme. There are, however, process sets containing two processes for which $C_1/(A_1 - C_1) + C_2/(A_2 - C_2) = 2(\sqrt{2} - 1) + e$ (for any $e > 0$) which are not schedulable using our scheme, i.e. we provide a simple schedulability test for process sets containing two processes.

We also define an improved version of our method. The improved version capitalizes on the fact that there is room for optimization when the period of one process is an integer multiple of the period of another process. The improved version of the method is better than the original version in the sense that we are able to schedule a larger number of process sets. All process sets which are schedulable using the original version are also schedulable using the improved version.

Previous work on age constraint process have concentrated on creating cyclic interleavings of processes [1]. Other studies have looked at age constraint processes which communicate [5]. One such scenario is scheduling of age constraint ocesses in the context of hard real-time database systems [2].

References

1. W. Albrecht, and R. Wisser, "Schedulers for Age Constraint Tasks and Their Performance Evaluation", In Proceedings of the Third International Euro-Par Conference, Passau, Germany, August 1997.
2. N. C. Audsley, A. Burns, M.F. Richardson, and A.J. Wellings, "Absolute and Relative Temporal Constraints in Hard Real-Time Databases", In Proceedings of IEEE Euromicro Workshop on Real-Time Systems, Los Alamitos, California, February 1992.
3. A. Burns, and A. Wellings, "Real-Time Systems and Programming Languages", Addison-Wesley, 1996.
4. J.P. Lehoczky, L. Sha, and Y. Ding, "The rate monotone scheduling algorithm: exact characterization and average case behavior", In Proceedings of IEEE 10th Real-Time Systems Symposium, December 1989.
5. X. Song, and J. Liu, "How Well can Data Temporal Consistency be Maintained", In Proceedings of the 1992 IEEE Symposium on Computer Aided Control Systems design, Napa, California, USA, March 1992.

Workshop 03: Scheduling and Load Balancing

Susan Flynn Hummel, Graham Riley and Rizos Sakellariou

Co-chairmen

Introduction

Mapping a parallel computation onto a parallel computer system is one of the most important questions for the design of efficient parallel algorithms. In the case of irregular data structures, the problem of initially distributing and then maintaining an even workload as computation progresses, becomes very complex. This workshop will discuss the state-of-the-art in this area.

Besides the discussion of novel techniques for mapping and scheduling irregular dynamic or static computations onto a processor architecture, a number of open problems are of special interest for this workshop: One is the development of partitioning algorithms used in numerical simulation, taking into account not only the cutsize but also the special characteristics of the application and their impact on the partitioning. Another topic of special relevance for this workshop concerns re-partitioning algorithms for irregular and adaptive finite element computations that minimize the movement of vertices in addition to balancing the load and minimising the cut of the resulting new partition. A third topic is the development of dynamic load balancing algorithms that adapt themselves to the special characteristic of the underlying parallel computer, easing the development of parable applications.

Other topics, in this uncomplete list of open problems, relate to application-specific graph partitioning, adaptable load balancing algorithms, scheduling algorithms, novel applications of scheduling and load balancing, load balancing on workstation clusters, parallel graph partitioning algorithms, etc.

The Papers

The papers selected for presentation in the workshop were split into four sessions each of which consists of work which covers a wide spectrum of different topics. We would like to thank sincerely the more than 20 referees that assisted us in the reviewing process.

Rudolf Berrendorf considers the problem of minimising load imbalance along with communication on distributed shared memory systems; a graph-based technique using data gathered from execution times of different program tasks and memory access behaviour is described. An approach for repartitioning and data remapping for numerical computations with highly irregular workloads on distributed memory machines is presented by Leonid Oliker, Rupak Biswas, and

Harold Gabow. Sajal Das and Azzedine Boukerche experiment with a load balancing scheme for the parallelisation of discrete event simulation applications. The problem of job scheduling on a parallel machine is tackled by Christophe Rapine, Isaac Scherson and Denis Trystram, who analyse an on-line time/space scheduling scheme.

The second session starts with a paper by Wolf Zimmermann, Martin Middendorf and Welf Loewe who consider k-linear scheduling algorithms on the LogP machine. Cristina Boeres, Vinod Rebello and David Skillicorn consider also a LogP-type model and investigate the effect of communication overheads. The problem of mesh partitioning and a measure for efficient heuristics is analysed by Ralf Diekmann, Robert Preis, Frank Schlimbach and Chris Walshaw. Konstantinos Antonis, John Garofalakis and Paul Spirakis analyse a strategy for balancing the load on a two-server distributed system.

The third session starts with Salvatore Orlando and Raffaele Perego who present a method for load balancing of statically predicted stencil computations in heterogeneous environments. Fabricio Alves Barbosa da Silva, Luis Miguel Campos and Isaac Scherson consider the problem of parallel job scheduling again. A general architectural framework for building parallel schedulers is described by Gerson Cavalheiro, Yves Denneulin and Jean-Louis Roch. A dynamic loop scheduling algorithm which makes use of information from previous executions of the loop is presented by Mark Bull. A divide-and-conquer strategy motivated by problems arising in finite element simulations is analysed by Stefan Bischof, Ralf Ebner and Thomas Erlebach.

The fourth session starts with a paper by Dingchao Li, Akira Mizuno, Yuji Iwahori and Naohiro Ishii that proposes an algorithm for scheduling a task graph on an heterogeneous parallel architecture. Mario Dantas considers a scheduling mechanism within an MPI implementation. An approach of using alternate schedules for achieving fault tolerance on a network of workstations is described by Dibyendu Das. An upper bound for the load distribution by a randomized algorithm for embedding bounded degree trees in arbitrary networks is computed by Jaafar Gaber and Bernard Toursel.

Optimizing Load Balance and Communication on Parallel Computers with Distributed Shared Memory

Rudolf Berrendorf

Central Institute for Applied Mathematics
Research Centre Jülich
D-52425 Jülich, Germany
r.berrendorf@fz-juelich.de

Abstract. To optimize programs for parallel computers with distributed shared memory two main problems need to be solved: load balance between the processors and minimization of interprocessor communication. This article describes a new technique called data-driven scheduling which can be used on sequentially iterated program regions on parallel computers with a distributed shared memory. During the first execution of the program region, statistical data on execution times of tasks and memory access behaviour are gathered. Based on this data, a special graph is generated to which graph partitioning techniques are applied. The resulting partitioning is stored in a template which is used in subsequent executions of the program region to efficiently schedule the parallel tasks of that region. Data-driven scheduling is integrated into the SVM-Fortran compiler. Performance results are shown for the Intel Paragon XP/S with the DSM-extension ASVM and for the SGI Origin2000.

1 Introduction

Parallel computers with a global address space share an important abstraction appreciated by programmers as well as compiler writers: the global, linear address space seen by all processors. To build such a computer in a scalable and economical way, such systems usually distribute the memory with the processors. Parallel computers with physically distributed memory but a global address space are termed distributed shared memory machines (DSM). Examples are SGI Origin2000, KSR-1, and Intel Paragon XP/S with ASVM [3]). To implement the global address space on top of a distributed memory, techniques for multi-cache systems are used which distinguish between read and write operations. If processors read from a memory location, the data is copied to the local memory of that processor where it is cached. On a write operation of a processor, this processor gets exclusive ownership of this location and all read copies get invalidated. The unit of coherence (and therefore the unit of communication) is a cache line or a page of the virtual memory system. For this reason, care has to be taken to avoid false sharing (independent data objects are mapped to the same page).

There are two main problems to be solved for parallel computers with a distributed shared memory and a large number of processors: load balancing and minimization of interprocessor communication. Data-driven scheduling is a new approach to solve both problems. The compiler modifies the code such that at run-time data on task times and memory access behaviour of the tasks is gathered. With this data a special graph is generated and partitioned for communication minimization and load balance. The partitioning result is stored in a template and it is used in subsequent executions of the program region to efficiently schedule the parallel tasks to the processors.

The paper is organized as follows. After giving an overview of related work in section 2, section 3 gives an introduction to SVM-Fortran. In section 4 the concept of data-driven scheduling is discussed and in section 5 performance results are shown for an application executed on two different machines. Section 6 concludes and gives a short outlook of further work.

2   Related Work

There are a number of techniques known for parallel machines which try to balance the load, or to minimize the interprocessor communication, or both.

Dynamic scheduling methods are well known as an attempt to balance the load on parallel computers, usually for parallel computers with a physically shared memory. The most rigid approach is self scheduling [10] where each idle processor requests only one task to be executed. With Factoring [7]) each idle processor requests at the beginning of the scheduling process larger chunks of tasks to reduce the synchronization overhead. All of these dynamic scheduling techniques take in some sense a greedy approach and therefore they have problems if the tasks at the end of the scheduling process have significantly larger execution times than tasks scheduled earlier. Another disadvantage is the local scheduling aspect with respect to one parallel loop only and the fact that the data locality aspect is not taken into account.

There are several scheduling methods known which have a main objective in generating data locality. Execute-on-Home [6] uses the information of a data distribution to execute tasks on that processor to which the accessed data is assigned (i.e. the processor which owns the data). An efficient implementation of the execute-on-home scheduling based on data distributions is often difficult if the data is accessed in an indirect way. In that case, run-time support is necessary. CHAOS/PARTI [13] (and in a similar manner RAPID [4]) is an approach to handle indirect accesses as it is for example common in sparse matrix problems. In an inspector phase the indices for indirection are examined and a graph is generated which includes through the edges the relationship between the data. Then the graph is partitioned and the data is redistributed according to the partitioning.

Many problems in the field of technical computing are modeled by the technique of finite elements. The original (physical) domain is partitioned into discrete elements connected through nodes. A usual approach of mapping such problems to parallel computers is the partitioning (e.g. [8]) of the resulting grid such that equally sized subgrids are mapped to the processors. For regular grids, partitioning can be done as a geometric partitioning. Non-uniform computation times involved with each node and data that need to be communicated between the processors have to be taken into account in the partitioning step. The whole problem of mapping such a finite element graph onto a parallel computer can be formulated as a graph partitioning problem where the nodes of the graph are the tasks associated which each node of the finite element grid and the edges represent communication demands. There are several software libraries available which can be used to partition such graphs (e.g. Metis [8], Chaco [5], Party [9], JOSTLE [12]).

Getting realistic node costs (i.e. task costs) is often difficult for non-uniform task execution times. Also, graph partitioners take no page boundaries into account when they map finite element nodes onto the memory of a DSM-computer and thus they ignore the false sharing problem. [11] discusses a technique to reorder the nodes after the partitioning phase with the aim to minimize communication.

3   SVM-Fortran

SVM-Fortran [2] is a shared memory parallel Fortran77 extension targeted mainly towards data parallel applications on DSM-systems. SVM-Fortran supports coarse-grained functional parallelism where a parallel task itself can be data parallel. A compiler and run-time system is implemented on several parallel machines such as Intel Paragon XP/S with ASVM, SGI Origin2000, and SUN and DEC multiprocessor machines.

SVM-Fortran provides standard features of shared memory parallel Fortran languages as well as specific features for DSM-computers. In SVM-Fortran the main concept to generate data locality and to balance the load is the dedicated assignment of parallel work to processors, e.g. the distribution of iterations of a parallel loop to processors.

Data locality is not a problem to be solved on the level of individual loops but it is a global problem. SVM-Fortran uses the concept of processor arrangements and templates as a tool to specify scheduling decisions globally via template distributions. Loop iterations are assigned to processors according to the distribution of the appropriate template element. Therefore, in SVM-Fortran templates are used to distribute the work rather than used to distribute the data as it is done in HPF. Different to HPF, it is not necessary for the SVM-Fortran-compiler to know the distribution of a template at compile time.

4   Data-Driven Scheduling

User-directed scheduling, where the user specifies the distribution of work to processors (e.g. specifying a block distribution for a template), makes sense and is efficient if the user has reliable information on the access behaviour (interprocessor communication) and execution times (load balance) of the parallel tasks in the application. But the prerequisite for this is often a detailed understanding of the program, all data structures, and their content at run-time. If the effort to gain this information is too high or if the interrelations are too complex (e.g. indirect addressing on several levels, unpredictable task execution times), the help of a tool or an automated scheduling is desirable.

Data-driven scheduling tries to help the programmer in solving the load balancing problem as well as the communication problem. Due to the way data-driven scheduling works, the new technique can be applied to program regions which are iterated several times (sequentially iterated program regions; see Fig. 1 for an example). This type of program region is common in iterative algorithms (e.g. iterative solvers), nested loops where the outer loop has not enough parallelism, or nested loops where data dependencies prohibit the parallel execution of the outer loop. The basic idea behind data-driven scheduling is to gather run-time data on task execution times and memory access behaviour on the first execution of the program region, use this information to find a good schedule which optimizes communication and load balance, and use the computed schedule in subsequent executions of the program region.

For simplicity, we restrict the description to schedule regions with one parallel loop inside, although more complex structures can be handled. Fig. 1 shows a small example program. This program is an artificial program to explain the basic idea. In a real program, the false sharing problem introduced with variable a could be easily solved with the introduction of a second array. In this example, the strategy to optimize for data locality and therefore how to distribute the iterations of the parallel loop to the processors might differ dependent on the content of the index field ix. On the other side, the execution time for each of the iterations might differ significantly based on the call to work(i) and therefore load balancing might be a challenging task. Data-driven scheduling tries to model both optimization aspects with the help of a weighted graph. The compiler generates two different codes (code-1, code-2) for the program region.

Code-1 is executed on the first execution of the program region[1]. Beside the normal code generation (e.g. handling of PDOs), the code is further instrumented to gather statistics. At run-time, execution times of tasks and accesses to shared variables (marked with the VARS-attribute in the directive) are stored in parallel in distributed data structures. Reaching the end of the schedule region, the gathered data is processed. A graph is generated by the run-time system where each node of the graph represents a task and the node weights are the task's execution time. To get a realistic value, measured execution times have to be subtracted by overhead times (measurement times, page miss times etc.) and synchronization times.

To explain the generation of edges in the graph, have a look at Fig. 2(a) which shows the assignments done in the example program. Task 1 and 3 access page 1,

---

[1] The user can reset a schedule region through a directive in which case code-1 is executed again, for example if the content of an index field has changed.

```
C       --- simple, artificial example to explain the basic idea
C       --- a memory page shall consist of 2 words
        PARAMETER (n=4)
        REAL,SHARED,ALIGNED a(2,n)        ! shared variable
        INTEGER ix(n)                     ! index field
        DATA /ix/1,2,1,2
C       --- declare proc. field, template, and distribute template ---
CSVM$ PROCESSORS:: proc(numproc())
CSVM$ TEMPLATE templ(n)
CSVM$ DISTRIBUTE (BLOCK) ONTO proc:: templ
        ...
C       --- sequentially iterated program region ---
        DO iter=1,niter
C           --- parallel loop enclosed in schedule region ---
CSVM$ SCHEDULE_REGION(ID(1),VARS(a))
CSVM$ PDO (STRATEGY (ON_HOME (templ(i))))
            DO i=1,n
                a(1,i) = a(2,ix(i))       ! possible communication
                CALL work(i)              ! possible load imbalance
            ENDDO
CSVM$ SCHEDULE_REGION_END
        ENDDO
```

Fig. 1. (Artificial) Example of a sequentially iterated program region.

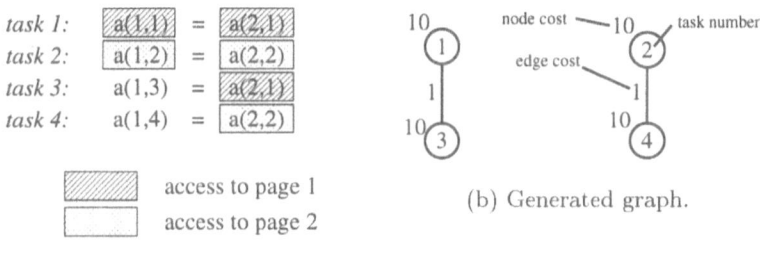

(a) Executed assignments.

(b) Generated graph.

Fig. 2. Memory accesses and resulting graph for example program.

| | task | | | | #tasks |
	1	2	3	4	
page 1	w/r		r		2
page 2		w/r		r	2
page 3			w		1
page 4				w	1
task time	10	10	10	10	

Fig. 3. Page accesses and task times (w=write access, r=read access).

task 2 and 4 access page 2 either with a read access or with a write access. Page j (j=3,4) is accessed only within task j. Table 4 shows the corresponding access table as it is generated internally as a distributed sparse data structure at run-time. The most interesting column is labeled #tasks and gives the number of different tasks accessing a page. If this entry is 1, no conflicts exist for this page and therefore no communication between processors is necessary for this page, independent of the schedule of tasks to processors. If two or more tasks access a page (#tasks > 1) of which at least one access is a write access, communication might be necessary due to the multi-cache protocol if these tasks are not assigned to the same processor. This is modeled in the graph with an edge between the two tasks/nodes with edge costs of one page miss. Fig. 2(b) shows the graph corresponding to the access table in table 4 (for this simple example, task times are assumed 10 units and a page miss counts 1 unit).

The handling of compulsory page misses is difficult as this depends on previous data accesses. If the distribution of pages to processors can be determined at run time (e.g. in ASVM[3]) and if this distribution is similar for all iterations, then these misses can be modeled by artificial nodes in the graph which are pre-assigned to processors.

In a next step, this graph is partitioned. We found that multilevel algorithms as found in most partitioning libraries give good results over a large variety of graphs. The partitioning result (i.e. the mapping of task i to processor j) is stored in the template given in the PDO-directive. The data gathering phase is done in parallel, but the graph generation and partitioning phase is done sequentially on one processor. We investigate to use parallel partitioners in the future.

Code-2 is executed at the second and all subsequent executions of the program region and this code is optimized code without any instrumentation. The iterations of the parallel loop are distributed according to the template distribution which is the schedule found in the graph partitioning step.

5 Results

We show performance results for the ASEMBL-program which is the assembly of a finite element matrix done in a sequential loop modeling time steps as it is used in many areas of technical computing. The results were obtained on an Intel Paragon XP/S with the extension ASVM [3] implementing SVM in software (embedded in the MACH3 micro kernel) and on a SGI Origin2000 with a fast DSM-implementation in hardware. A multilevel algorithm with Kernighan-Lin refinement of the Metis library [8] was used for partitioning the graph.

For the Paragon XP/S, a data set with 900 elements was taken, and for the Origin2000 a data set with 8325 elements was chosen. Due to a search loop in every parallel task, task times differ. The sparse matrix to be assembled is allocated in the shared memory region; the variable accesses are protected by page locks where only one processor can access a page but accesses to different pages can be done in parallel. The bandwidth of the sparse matrix is small, therefore strategies which partition the iteration space in clusters will generate better data locality.

Fig. 4 shows speedup values for the third sequential time step. Because of the relative small size of the shared array and the large page size on Paragon XP/S (8 KB), the performance improvements are limited to a small number of processors.

On both systems, the performance with the Factoring strategy is limited because data locality (and communication) is of no concern with this strategy. Particularly on the Paragon system with high page miss times, this strategy shows no real improvements. Although the shared matrix has a low bandwidth and therefore the Block strategy is in favor, differences in task times cause load imbalances with the Block strategy. Data-driven Scheduling (DDS) performs better than the other strategies on both systems.

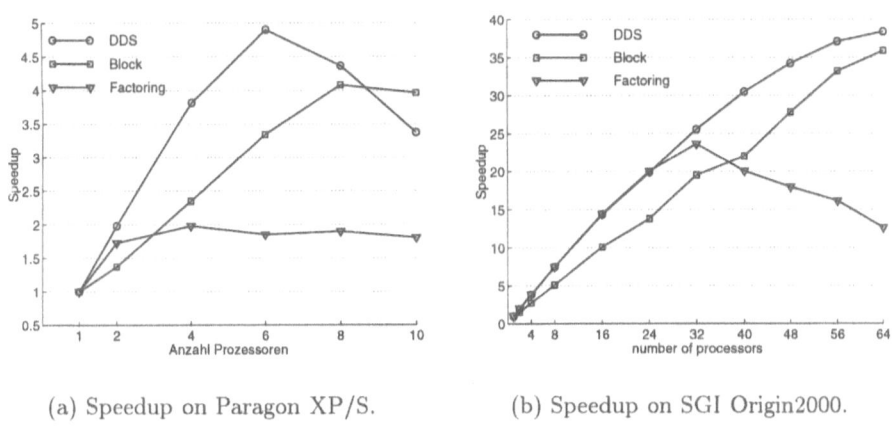

(a) Speedup on Paragon XP/S.　　　(b) Speedup on SGI Origin2000.

Fig. 4. Results for ASEMBL.

6　Summary

We have introduced data-driven scheduling to optimize interprocessor communication and load balance on parallel computers with a distributed shared memory. With the new technique, statistics on task execution times and data sharing are gathered at run-time. With this data, a special graph is generated and graph partitioning techniques are applied. The resulting partitioning is stored in a template and used in subsequent executions of that program region to efficiently schedule the parallel tasks to the processors. We have compared this method with two loop scheduling methods on two DSM-computers where data-driven scheduling performs better than the other two methods.

Currently, we use deterministic values for all model parameters. To refine our model, we plan to investigate into non-deterministic values, e.g. number of page misses, page miss times.

7 Acknowledgments

Reiner Vogelsang (SGI/Cray) and Oscar Plata (University of Malaga) gave me access to SGI Origin2000 machines. Heinz Bast (Intel SSD) supported me on the Intel Paragon. I would like to thank the developers of the graph partitioning libraries I used in my work, namely: George Karypis (Metis), Chris Walshaw (Jostle), Bruce A. Hendrickson (Chaco), and Robert Preis (Party).

References

1. R. Berrendorf, M. Gerndt. Compiling SVM-Fortran for the Intel Paragon XP/S. *Proc. Working Conference on Massively Parallel Programming Models (MPPM'95)*, pages 52–59, Berlin, October 1995. IEEE Society Press.
2. R. Berrendorf, M. Gerndt. SVM Fortran reference manual version 1.4. Technical Report KFA-ZAM-IB-9510, Research Centre Jülich, April 1995.
3. R. Berrendorf, M. Gerndt, M. Mairandres, S. Zeisset. A programming environment for shared virtual memory on the Intel Paragon supercomputer. In *Proc. Intel User Group Meeting*, Albuquerque, NM, June 1995. http://www.cs.sandia.gov/ISUG/ps/pesvm.ps.
4. C. Fu, T. Yang. Run-time compilation for parallel sparse matrix computations. In *Proc. ACM Int'l Conf. Supercomputing*, pages 237–244, 1996.
5. B. Hendrickson, R. Leland. The Chaco user's guide, version 2.0. Technical Report SAND95-2344, Sandia National Lab., Albuquerque, NM, July 1995.
6. High Performance Fortran Forum. *High Performance Fortran Language Specification*, 2.0 edition, January 1997.
7. S. F. Hummel, E. Schonberg, L. E. Flynn. Factoring - a method for scheduling parallel loops. *Comm. ACM*, 35(8):90–101, August 1992.
8. G. Karypis, V. Kumar. Analysis of multilevel graph partitioning. Technical Report 95-037, Univ. Minnesota, Department of Computer Science, 1995.
9. R. Preis, R. Dieckmann. *The PARTY Partitioning-Library, User Guide, Version 1.1*. Univ. Paderborn, September 1996.
10. P. Tang, P.-C. Yew. Processor self-scheduling for multiple nested parallel loops. In *Proc. IEEE Int'l Conf. Parallel Processing*, pages 528–535, August 1986.
11. K. A. Tomko, S. G. Abraham. Data and program restructuring of irregular applications for cache-coherent multiprocessors. In *Proc. ACM Int'l Conf. Supercomputing*, pages 214–225, July 1994.
12. C. Walshaw, M. Cross, M.G. Everett, S. Johnson, K. McManus. Partitioning & mapping of unstructured meshed to parallel machine topologies. In *Proc. Irregular 95: Parallel Algorithms for Irregularly Structured Problems*, volume 980 of *LNCS*, pages 121–126. Springer, 1995.
13. J. Wu, R. Das, J. Saltz, H. Berryman, S. Hiranandani. Distributed memory compiler design for sparse problems. *IEEE Trans. Computers*, 44(6), 1995.

Performance Analysis and Portability of the PLUM Load Balancing System

Leonid Oliker[1], Rupak Biswas[2], and Harold N. Gabow[3]

[1] RIACS, NASA Ames Research Center, Moffett Field, CA 94035, USA
[2] MRJ, NASA Ames Research Center, Moffett Field, CA 94035, USA
[3] CS Department, University of Colorado, Boulder, CO 80309, USA

Abstract. The ability to dynamically adapt an unstructured mesh is a powerful tool for solving computational problems with evolving physical features; however, an efficient parallel implementation is rather difficult. To address this problem, we have developed PLUM, an automatic portable framework for performing adaptive numerical computations in a message-passing environment. PLUM requires that all data be globally redistributed after each mesh adaption to achieve load balance. We present an algorithm for minimizing this remapping overhead by guaranteeing an optimal processor reassignment. We also show that the data redistribution cost can be significantly reduced by applying our heuristic processor reassignment algorithm to the default mapping of the parallel partitioner. Portability is examined by comparing performance on a SP2, an Origin2000, and a T3E. Results show that PLUM can be successfully ported to different platforms without any code modifications.

1 Introduction

The ability to dynamically adapt an unstructured mesh is a powerful tool for efficiently solving computational problems with evolving physical features. Standard fixed-mesh numerical methods can be made more cost-effective by locally refining and coarsening the mesh to capture these phenomena of interest. Unfortunately, an efficient parallelization of these adaptive methods is rather difficult, primarily due to the load imbalance created by the dynamically-changing nonuniform grid. Nonetheless, it is generally thought that unstructured adaptive-grid techniques will constitute a significant fraction of future high-performance supercomputing.

With this goal in mind, we have developed a novel method, called PLUM [7], that dynamically balances processor workloads with a global view when performing adaptive numerical calculations in a parallel message-passing environment. The mesh is first partitioned and mapped among the available processors. Once an acceptable numerical solution is obtained, the mesh adaption procedure [8] is invoked. Mesh edges are targeted for coarsening or refinement based on an error indicator computed from the solution. The old mesh is then coarsened, resulting in a smaller grid. Since edges have already been marked for refinement, the new mesh can be exactly predicted before actually performing the refinement step. Program control is thus passed to the load balancer at this time.

If the current partitions will become load imbalanced after adaption, a repartitioner is used to divide the new mesh into subgrids. The new partitions are then reassigned among the processors in a way that minimizes the cost of data movement. If the remapping cost is compensated by the computational gain that would be achieved with balanced partitions, all necessary data is appropriately redistributed. Otherwise, the new partitioning is discarded. The computational mesh is then refined and the numerical calculation is restarted.

2 Dynamic Load Balancing

2.1 Repartitioning the Initial Mesh Dual Graph

Repeatedly using the dual of the initial computational mesh for dynamic load balancing is one of the key features of PLUM [7]. Each dual graph vertex has a computational weight, w_{comp}, and a remapping weight, w_{remap}. These weights model the processing workload and the cost of moving the corresponding element from one processor to another. Every dual graph edge also has a weight, w_{comm}, that models the runtime communication. New computational grids obtained by adaption are represented by modifying these three weights. If the dual graph with a new set of w_{comp} is deemed unbalanced, the mesh is repartitioned.

2.2 Processor Reassignment

New partitions generated by a partitioner are mapped to processors such that the data redistribution cost is minimized. In general, the number of new partitions is an integer multiple F of the number of processors, and each processor is assigned F partitions. Allowing multiple partitions per processor reduces the volume of data movement but increases the partitioning and reassignment times [7].

We first generate a similarity measure M that indicates how the remapping weights w_{remap} of the new partitions are distributed over the processors. It is represented as a matrix where entry M_{ij} is the sum of the w_{remap} values of all the dual graph vertices in new partition j that already reside on processor i. Various cost functions are usually needed to solve the processor reassignment problem using M for different machine architectures. We present three general metrics: TotalV, MaxV, and MaxSR, which model the remapping cost on most multiprocessor systems. TotalV minimizes the total volume of data moved among all the processors, MaxV minimizes the maximum flow of data to *or* from any single processor, while MaxSR minimizes the sum of the maximum flow of data to *and* from any processor. Experimental results [2] have indicated the usefulness of these metrics in predicting the actual remapping costs. A greedy heuristic algorithm to minimize the remapping overhead is also presented.

TotalV Metric. The TotalV metric assumes that by reducing network contention and the total number of elements moved, the remapping time will be reduced. In general, each processor cannot be assigned F unique partitions corresponding to their F largest weights. To minimize `TotalV`, each processor i must be assigned F partitions j_{i_f}, $f = 1, 2, \ldots, F$, such that the objective function

$$\mathcal{F} = \sum_{i=1}^{P} \sum_{f=1}^{F} M_{ij_{i\_f}}$$

is maximized subject to the constraint

$$j_{i\_r} \neq j_{k\_s}, \text{ for } i \neq k \text{ or } r \neq s; \quad i, k = 1, 2, \ldots, P; \quad r, s = 1, 2, \ldots, F.$$

We can optimally solve this by mapping it to a network flow optimization problem described as follows. Let $G = (V, E)$ be an undirected graph. G is *bipartite* if V can be partitioned into two sets A and B such that every edge has one vertex in A and the other vertex in B. A *matching* is a subset of edges, no two of which share a common vertex. A *maximum-cardinality matching* is one that contains as many edges as possible. If G has a real-valued cost on each edge, we can consider the problem of finding a maximum-cardinality matching whose total edge cost is maximized. We refer to this as the *maximally weighted bipartite graph* (MWBG) problem (also known as the *assignment* problem).

When $F = 1$, optimally solving the `TotalV` metric trivially reduces to MWBG, where V consists of P processors and P partitions in each set. An edge of weight M_{ij} exists between vertex i of the first set and vertex j of the second set. If $F > 1$, the processor reassignment problem can be reduced to MWBG by duplicating each processor and all of its incident edges F times. Each set of the bipartite graph then has $P \times F$ vertices. After the optimal solution is obtained, the solutions for all F copies of a processor are combined to form a one-to-F mapping between the processors and the partitions. The optimal solution for the `TotalV` metric and the corresponding processor assignment of an example similarity matrix is shown in Fig. 1(a).

The fastest MWBG algorithm can compute a matching in $O(|V|^2 \log |V| + |V||E|)$ time [3], or in $O(|V|^{1/2}|E| \log(|V|C))$ time if all edge costs are integers of absolute value at most C [5]. We have implemented the optimal algorithm with a runtime of $O(|V|^3)$. Since M is generally dense, $|E| \approx |V|^2$, implying that we should not see a dramatic performance gain from a faster implementation.

MaxV Metric. The metric `MaxV`, unlike `TotalV`, considers data redistribution in terms of solving a load imbalance problem, where it is more important to minimize the workload of the most heavily-weighted processor than to minimize the sum of all the loads. During the process of remapping, each processor must pack and unpack send and receive buffers, incur remote-memory latency time, and perform the computational overhead of rebuilding internal and shared data structures. By minimizing $\max(\alpha \times \max(\texttt{ElemsSent}), \beta \times \max(\texttt{ElemsRecd}))$, where α and β are machine-specific parameters, `MaxV` attempts to reduce the total remapping time by minimizing the execution time of the most heavily-loaded processor. We can solve this optimally by considering the problem of finding a maximum-cardinality matching whose maximum edge cost is minimum. We refer to this as the *bottleneck maximum cardinality matching* (BMCM) problem.

To find the BMCM of the graph G corresponding to the similarity matrix, we first need to transform M into a new matrix M'. Each entry M'_{ij} represents

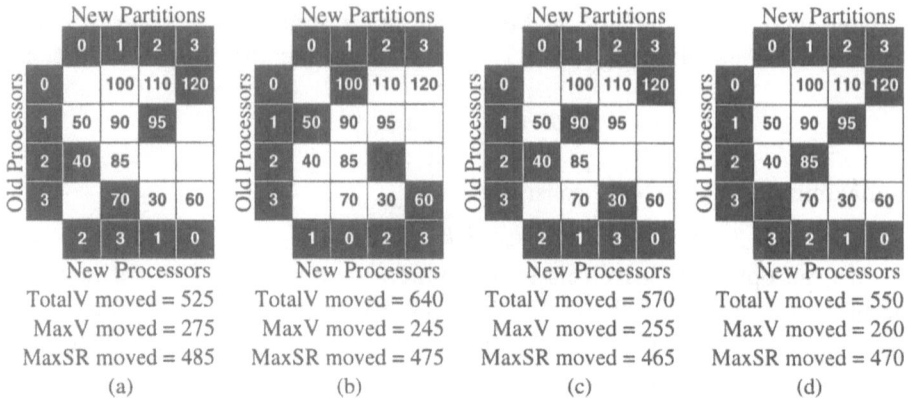

Fig. 1. Various cost metrics of a similarity matrix M for $P = 4$ and $F = 1$ using (a) the optimal MWBG, (b) the optimal BMCM, (c) the optimal DBMCM, and (d) our heuristic algorithms

the maximum cost of sending data to or receiving data from processor i and partition j:

$$M'_{ij} = \max\left(\left(\alpha \sum_{y=1}^{P} M_{iy}, y \neq j\right), \left(\beta \sum_{x=1}^{P} M_{xj}, x \neq i\right)\right).$$

Currently, our framework for the MaxV metric is restricted to $F = 1$.

We have implemented the BMCM algorithm of Bhat [1] which combines a maximum cardinality matching algorithm with a binary search, and runs in $O(|V|^{1/2}|E| \log |V|)$. The fastest known BMCM algorithm, proposed by Gabow and Tarjan [4], has a runtime of $O((|V| \log |V|)^{1/2}|E|)$.

The new processor assignment for the similarity matrix in Fig. 1 using this approach with $\alpha = \beta = 1$ is shown in Fig. 1(b). Notice that the total number of elements moved in Fig. 1(b) is larger than the corresponding value in Fig. 1(a); however, the maximum number of elements moved is smaller.

MaxSR Metric. Our third metric, MaxSR, is similar to MaxV in the sense that the overhead of the bottleneck processor is minimized during the remapping phase. MaxSR differs, however, in that it minimizes the sum of the heaviest data flow *from* any processor and *to* any processor, expressed as $\left(\alpha \times \max(\texttt{ElemsSent}) + \beta \times \max(\texttt{ElemsRecd})\right)$. We refer to this as the *double* bottleneck maximum cardinality matching (DBMCM) problem. The MaxSR formulation allows us to capture the computational overhead of packing and unpacking data, when these two phases are separated by a barrier synchronization. Additionally, the MaxSR metric may also approximate the many-to-many communication pattern of our remapping phase. Since a processor can either be sending or receiving data, the overhead of these two phases should be modeled as a sum of costs.

We have developed an algorithm for computing the minimum MaxSR of the graph G corresponding to our similarity matrix. We first transform M to a new matrix M''. Each entry M''_{ij} contains a pair of values (*Send, Receive*) corresponding to the total cost of sending and receiving data, when partition j is mapped to processor i:

$$M''_{ij} = \left\{ S_{ij} = \left(\alpha \sum_{y=1}^{P} M_{iy}, y \neq j \right), R_{ij} = \left(\beta \sum_{x=1}^{P} M_{xj}, x \neq i \right) \right\}.$$

Currently, our algorithm for the MaxSR metric is restricted to $F = 1$.

Let $\sigma_1, \sigma_2, \ldots, \sigma_k$ be the distinct *Send* values appearing in M'', sorted in increasing order. Thus, $\sigma_i < \sigma_{i+1}$ and $k \leq P^2$. Form the bipartite graph $G_i = (V, E_i)$, where V consists of processor vertices $u = 1, 2, \ldots, P$ and partition vertices $v = 1, 2, \ldots, P$, and E_i contains edge (u, v) if $S_{uv} \leq \sigma_i$; furthermore, edge (u, v) has weight R_{uv} if it is in E_i.

For small values of i, graph G_i may not have a perfect matching. Let $i_{\min}$ be the smallest index such that $G_{i_{\min}}$ has a perfect matching. Obviously, G_i has a perfect matching for all $i \geq i_{\min}$. Solving the BMCM problem of G_i gives a matching that minimizes the maximum *Receive* edge weight. It gives a matching with MaxSR value at most $\sigma_i + \text{MaxV}(G_i)$. Defining

$$\text{MaxSR}(i) = \min_{i_{\min} \leq j \leq i} (\sigma_j + \text{MaxV}(G_j)),$$

it is easy to see that MaxSR(k) equals the correct value of MaxSR. Thus, our algorithm computes MaxSR by solving k BMCM problems on the graphs G_i and computing the minimum value MaxSR(k). However, we can prematurely terminate the algorithm if there exists an $i_{\max}$ such that $\sigma_{i_{\max}+1} \geq \text{MaxSR}(i_{\max})$, since it is then guaranteed that the MaxSR solution is MaxSR($i_{\max}$).

Our implementation has a runtime of $O(|V|^{1/2}|E|^2 \log |V|)$ since the BMCM algorithm is called $|E|$ times in the worst case; however, it can be decreased to $O(|E|^2)$. The following is a brief sketch of this more efficient implementation.

Suppose we have constructed a matching $\mathcal{M}$ that solves the BMCM problem of G_i for $i \geq i_{\min}$. We solve the BMCM problem of G_{i+1} as follows. Initialize a working graph G to be G_{i+1} with all edges of weight greater than MaxV(G_i) deleted. Take the matching $\mathcal{M}$ on G, and delete all unmatched edges of weight MaxV(G_i). Choose an edge (u, v) of maximum weight in $\mathcal{M}$. Remove edge (u, v) from $\mathcal{M}$ and G, and search for an augmenting path from u to v in G. If no such path exists, we know that MaxV(G_i) =MaxV(G_{i+1}). If an augmenting path is found, repeat this procedure by choosing a new edge (u', v') of maximum weight in the matching and searching for an augmenting path. After some number of repetitions of this procedure, the maximum weight of a matched edge will have decreased to the desired value MaxV(G_{i+1}). At this point our algorithm to solve the BMCM problem of G_{i+1} will stop, since no augmenting path will be found.

To see that this algorithm runs in $O(|E|^2)$, note that each search for an augmenting path uses time $O(|E|)$ and that there are $O(|E|)$ such searches. A successful search for an augmenting path for edge (u, v) permanently eliminates it from all future graphs, so there are at most $|E|$ successful searches. Furthermore, there are at most $|E|$ unsuccessful searches, one for each value of i.

The new processor assignment for the similarity matrix in Fig. 1 using the DBMCM algorithm with $\alpha = \beta = 1$ is shown in Fig. 1(c). Notice that the MaxSR solution is minimized; however, the number of `TotalV` elements moved is larger than the corresponding value in Fig. 1(a), and more `MaxV` elements are moved than in Fig. 1(b). Also note that the optimal similarity matrix solution for `MaxSR` is provably no more than twice that of `MaxV`.

Heuristic Algorithm. We have developed a heuristic greedy algorithm that gives a suboptimal solution to the `TotalV` metric in $O(|E|)$ steps [7]. All partitions are initially flagged as unassigned and each processor has a counter set to F that indicates the remaining number of partitions it needs. The non-zero entries of the similarity matrix M are then sorted in descending order. Starting from the largest entry, partitions are assigned to processors that have less than F partitions until done. If necessary, the zero entries in M are also used. Oliker and Biswas [7] proved that a processor assignment obtained using the heuristic algorithm can never result in a data movement cost that is more than twice that of the optimal `TotalV` assignment. In addition, experimental results in Sec. 3.1 demonstrate that our heuristic quickly finds high quality solutions for all three metrics. Applying this heuristic algorithm to the similarity matrix in Fig. 1 generates the new processor assignment shown in Fig. 1(d).

2.3 Remapping Cost Model

Once the reassignment problem is solved, a model is needed to quickly predict the expected redistribution cost for a given architecture. Our redistribution algorithm consists of three major steps: first, the data objects moving out of a partition are stripped out and placed in a buffer; next, a collective communication distributes the data to its destination; and finally, the received data is integrated into each partition and the boundary information is consistently updated. This remapping procedure closely follows the superstep model of BSP [9].

The expected time for the redistribution procedure on bandwidth-rich systems can be expressed as $\gamma \times$ `MaxSR` $+ O$, where `MaxSR` $=$ max(`ElemsSent`) $+$ max(`ElemsRecd`), γ is the total computation and communication cost to process each redistributed element, and O is the sum of all constant overheads [7]. This formulation demonstrates the need to model the `MaxSR` metric when performing processor reassignment. By minimizing `MaxSR`, we can guarantee a reduction in the computational overhead of our remapping algorithm. To compute γ and O, a simple least squares fit through several data points for various redistribution patterns and their corresponding runtimes can be used. This procedure needs to be performed only once for each architecture, and the values of γ and O can then be used in actual computations to estimate the redistribution cost.

3 Experimental Results

The 3D_TAG parallel mesh adaption procedure [8] and the PLUM global load balancing strategy [7] have been implemented in C and C++, with the parallel activities in MPI for portability. All experiments were performed on the wide-node SP2 at NASA Ames, the Origin2000 at NCSA, and the T3E at NASA Goddard, without any machine-specific optimizations.

The computational mesh used in this paper is one used to simulate an acoustics wind-tunnel experiment of a UH-1H helicopter rotor blade [7]. Three different cases are studied, with varying fractions of the domain targeted for refinement based on an error indicator calculated directly from the flow solution. The strategies, called Real_1, Real_2, and Real_3, subdivided 5%, 33%, and 60% of the 78,343 edges of the initial mesh. This increased the number of mesh elements from 60,968 to 82,489, 201,780, and 321,841, respectively.

3.1 Comparison of Reassignment Algorithms

Table 1 presents a comparison of our five different processor reassignment algorithms in terms of the reassignment time (in secs) and the amount of data movement. Results are shown for the Real_2 strategy on the SP2 with $F = 1$. The PMeTiS [6] case does not require any explicit processor reassignment since we choose the default partition-to-processor mapping given by the partitioner. The poor performance for all three metrics is expected since PMeTiS is a global partitioner that does not attempt to minimize the remapping overhead. Previous work [2] compared the performance of PMeTiS with other partitioners.

Table 1. Comparison of reassignment algorithms for Real_2 on the SP2 with $F = 1$

| Algthm. | P = 32 | | | | P = 64 | | | |
	TotalV Metric	MaxV Metric	MaxSR Metric	Reass. Time	TotalV Metric	MaxV Metric	MaxSR Metric	Reass. Time
PMeTiS	58297	5067	7467	0.0000	67439	2667	4452	0.0000
MWBG	34738	4410	5822	0.0177	38059	2261	3142	0.0650
BMCM	49611	4410	5944	0.0323	52837	2261	3282	0.1327
DBMCM	50270	4414	5733	0.0921	54896	2261	3121	1.2515
Heuristic	35032	4410	5809	0.0017	38283	2261	3123	0.0088

The execution times of the other four algorithms increase with the number of processors because of the growth in the size of the similarity matrix; however, the heuristic time for 64 processors is still very small and acceptable. The total volume of data movement is obviously the smallest for the MWBG algorithm since it optimally solves for the TotalV metric. In the optimal BMCM method, the maximum of the number of elements sent or received is explicitly minimized, but all the other algorithms give almost identical results for the MaxV metric. In our helicopter rotor experiment, only a few localized regions of the domain incur a dramatic increase in the number of grid points between refinement levels. These newly-refined regions must shift a large number of elements onto other processors in order to achieve a balanced load distribution. Therefore, a similar MaxV solution should be obtained by any reasonable reassignment algorithm.

The DBMCM algorithm optimally reduces MaxSR, but achieves no more than a 5% improvement over the other algorithms. Nonetheless, since we believe that the MaxSR metric can closely approximate the remapping cost on many architectures, computing its optimal solution can provide useful information. Notice that the minimum `TotalV` increases slightly as P grows from 32 to 64, while `MaxSR` is dramatically reduced by over 45%. This trend continues as the number of processors increases, and indicates that `PLUM` will remain viable on a large number of processors, since the per processor workload decreases as P increases.

Finally, observe that the heuristic algorithm does an excellent job in minimizing all three cost metrics, in a trivial amount of time. Although theoretical bounds have only been established for the `TotalV` metric, empirical evidence indicates that the heuristic algorithm closely approximates both `MaxV` and `MaxSR`. Similar results were obtained for the other edge-marking strategies.

3.2 Portability Analysis

The top three plots in Fig. 2 illustrate parallel speedup for the three edge-marking strategies on the SP2, Origin2000, and T3E. Two sets of results are presented for each machine: one when data remapping is performed after mesh refinement, and the other when remapping is done before refinement. The speedup numbers are almost identical on all three machines. The **Real_3** case shows the best speedup values because it is the most computation intensive. Remapping data before refinement has the largest relative effect for **Real_1**, because it has the smallest refinement region and predictively load balancing the refined mesh returns the biggest benefit. The best results are for **Real_3** with remapping before refinement, showing an efficiency greater than 87% on 32 processors.

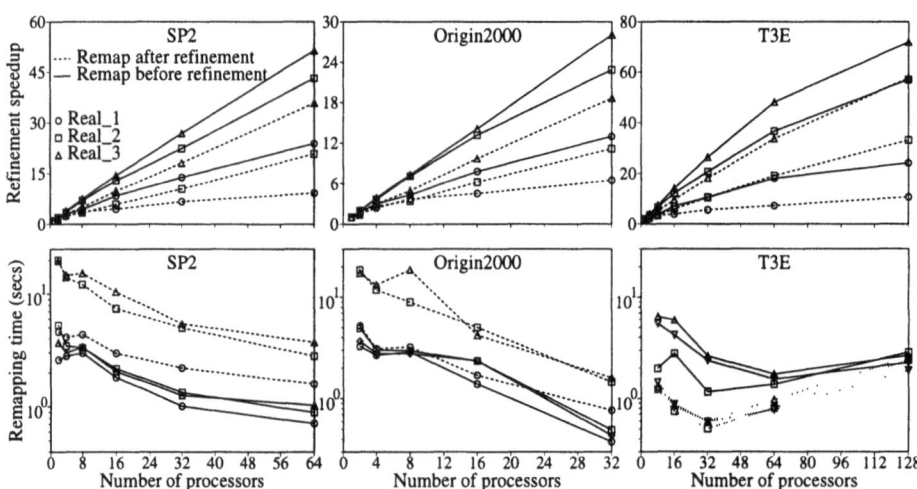

Fig. 2. Refinement speedup (top) and remapping time (bottom) within PLUM on the SP2, Origin2000, and T3E, when data is redistributed after or before mesh refinement

To compare the performance on the SP2, Origin2000, and T3E more critically, one needs to look at the actual times rather than the speedup values. Table 2 shows how the execution time (in secs) is spent during the refinement and subsequent load balancing phases for the **Real_2** case when data is remapped

Table 2. Anatomy of execution times for Real_2 on the Origin2000, SP2, and T3E

| | Adaption Time | | | Remapping Time | | | Partitioning Time | | |
P	O2000	SP2	T3E	O2000	SP2	T3E	O2000	SP2	T3E
2	5.261	12.06	3.455	3.005	3.440	2.648	0.628	0.815	0.701
4	2.880	6.734	1.956	3.005	3.440	1.501	0.584	0.537	0.477
8	1.470	3.434	1.034	2.963	3.321	1.449	0.522	0.424	0.359
16	0.794	1.846	0.568	2.346	2.173	0.880	0.396	0.377	0.301
32	0.458	1.061	0.333	0.491	1.338	0.592	0.389	0.429	0.302
64		0.550	0.188		0.890	0.778		0.574	0.425
128			0.121			1.894			0.599

before the subdivision phase. The processor reassignment times are not presented since they are negligible compared to other times, as is evident from Table 1. Notice that the T3E adaption times are consistently more than 1.4 times faster than the Origin2000 and three times faster than the SP2. One reason for this performance difference is the disparity in the clock speeds of the three machines. Another reason is that the mesh adaption code does not use the floating-point units on the SP2, thereby adversely affecting its overall performance.

The bottom three plots in Fig. 2 show the remapping time for each of the three cases on the SP2, Origin2000, and T3E. In almost every case, a significant reduction in remapping time is observed when the adapted mesh is load balanced by performing data movement prior to refinement. This is because the mesh grows in size only after the data has been redistributed. In general, the remapping times also decrease as the number of processors is increased. This is because even though the total volume of data movement increases with the number of processors, there are actually more processors to share the work. The remapping times when data is moved before mesh refinement are reproduced for the Real_2 case in Table 2 since the exact values are difficult to read off the log-scale.

Perhaps the most remarkable feature of these results is the peculiar behavior of the T3E when $P \geq 64$. When using up to 32 processors, the remapping performance of the T3E is very similar to that of the SP2 and Origin2000. It closely follows the redistribution cost model given in Sec. 2.3, and achieves a significant runtime improvement when remapping is performed prior to refinement. However, for 64 and 128 processors, the remapping overhead on the T3E begins to increase and violates our cost model. The runtime difference when data is remapped before and after refinement is dramatically diminished; in fact, all the remapping times begin to converge to a single value! This indicates that the remapping time is no longer affected only by the volume of data redistributed but also by the interprocessor communication pattern. One way of potentially improving these results is to take advantage of the T3E's ability to efficiently perform one-sided communication.

Another surprising result is the dramatic reduction in remapping times when using 32 processors on the Origin2000. This is probably because network contention with other jobs is essentially removed when using the entire machine. When using up to 16 processors, the remapping times on the SP2 and the Origin2000 are comparable, while the T3E is about twice as fast. Recall that the remapping phase within PLUM consists of both communication and computation. Since the results in Table 2 indicate that computation is faster on the Origin2000, it is reasonable to infer that bulk communication is faster on the SP2. These results generally demonstrate that our methodology within PLUM is effective in significantly reducing the data remapping time and improving the parallel performance of mesh refinement.

Table 2 also presents the PMeTiS partitioning times for Real_2 on all three systems; the results for Real_1 and Real_3 are almost identical because the time to repartition mostly depends on the initial problem size. There is, however, some dependence on the number of processors used. When there are too few processors, repartitioning takes more time because each processor has a bigger share of the total work. When there are too many processors, an increase in the communication cost slows down the repartitioner. Table 2 demonstrates that PMeTiS is fast enough to be effectively used within our framework, and that PLUM can be successfully ported to different platforms without any code modifications.

4 Conclusions

In this paper, we verified the effectiveness of our PLUM load balancer for adaptive unstructured meshes on a helicopter acoustics problem. We developed three generic metrics to model the remapping cost on most multiprocessor systems. Optimal solutions for these metrics, as well as a heuristic approach were implemented. We showed that the data redistribution overhead can be significantly reduced by applying our heuristic processor reassignment algorithm to the default mapping given by the global partitioner. Portability was demonstrated by presenting results on the three vastly different architectures of the SP2, Origin2000, and T3E, without the need for any code modifications. Results showed that, in general, PLUM will remain viable on large numbers of processors. However, our redistribution cost model was violated on the T3E when 64 or more processors were used. Future research will address the improvement of these results, and the development of a more comprehensive remapping cost model.

References

1. Bhat, K.: An $O(n^{2.5} \log_2 n)$ time algorithm for the bottleneck assignment problems. AT&T Bell Laboratories Unpublished Report (1984)
2. Biswas, R., Oliker, L.: Experiments with repartitioning and load balancing adaptive meshes. NASA Ames Research Center Technical Report NAS-97-021 (1997)
3. Fredman, M., Tarjan, R.: Fibonacci heaps and their uses in improved network optimization algorithms. J. ACM **34** (1987) 596–615
4. Gabow, H., Tarjan, R.: Algorithms for two bottleneck optimization problems. J. of Alg. **9** (1988) 411–417
5. Gabow, H., Tarjan, R.: Faster scaling algorithms for network problems. SIAM J. on Comput. **18** (1989) 1013–1036
6. Karypis, G., Kumar, V.: Parallel multilevel k-way partitioning scheme for irregular graphs. University of Minnesota Technical Report 96-036 (1996)

7. Oliker, L., Biswas, R.: PLUM: Parallel load balancing for adaptive unstructured meshes. NASA Ames Research Center Technical Report NAS-97-020 (1997)
8. Oliker, L., Biswas, R., Strawn, R.: Parallel implementation of an adaptive scheme for 3D unstructured grids on the SP2. Springer-Verlag LNCS **1117** (1996) 35–47
9. Valiant, L.: A bridging model for parallel computation. Comm. ACM **33** (1990) 103–111

Experimental Studies in Load Balancing*

Azzedine Boukerche and Sajal K. Das

Department of Computer Sciences, University of North Texas, Denton, TX. USA

Abstract. This paper takes an experimental approach to the load balancing problem for parallel simulation applications. In particular, it focuses upon a conservative synchronization protocol, by making use of an optimized version of Chandy&Misra null message method, and propose a dynamic load balancing algorithm which assumes no compile time knowledge about the workload parameters. The proposed scheme is also implemented on an Intel Paragon A4 multicomputer, and the performance results for several simulation models are reported.

1 Introduction

A considerable number of research projects on load balancing in parallel and distributed systems in general has been carried out in the literature due to the potential performance gain from this service. However, all of these algorithms are not suitable for parallel simulation since the synchronization constraints exacerbate the dependencies between the LPs. In this paper, we consider the load balancing problem associated with the conservative synchronization protocol that makes use of an optimized version of Chandy-Misra Null messages protocol [3]. Our primary goal is to minimize the synchronization overhead in conservative parallel simulation, and significantly reduce the total execution time of the simulation by evenly distributing the workload among processors [1, 2].

2 Load Balancing Algorithm

Before the load balancing algorithm can determine the new assignment of processes, it must receive information from the processes. We propose to implement the load balancing facility as two kinds of processes: *load balancer* and *process migration* processes. A load balancer makes decision on *when* to move *which* process to *where*, while migration process carries out the decision made by the load balancer to move processes between processors.

To prevent the bottleneck and reduce the message traffic over a fully distributed approach. we propose a Centralized approach (CL), and a Multi-Level (ML) hierarchical scheme; where processors are grouped and work loads of processors are balanced hierarchically through multiple levels. At the present time,

* Part of this research is supported by Texas Advanced Technology Program grant TATP-003594031

we settle with a Two-Level scheme. In level 1, we use a centralized approach that makes use of a *(Global) Load Balancing Facitilty, (GLBF)*, and the global state consisting of process/processors mapping table. The motivation for this is that many contemporary parallel computers are usually equipped with a front-end host processor (e.g., hypercube multiprocessor machines). The load balancer $(GLBF)$ sends periodically a message $< Request\_Load >$ to a specific processor, called $first\_proc$, within each clusters requesting the average load of each processor within each groups.

In level 2, the processors are partitioned into clusters, and the processors within each group are structured as a virtual ring, and operate in parallel to collect the work loads of processors. A virtual ring is designed to be traversed by a token which originates at a particular processor, called *first_proc*, passes through intermediate processors, and ends it traversal at a preidentified processor called *last_proc*. Each of the ring will have its own circulating token, so that information (i.e., work loads) gathering within the rings is concurrent. As the token travels through the processors of the ring, it accumulates the information (the load of each processors and *id* of the processors that contain the highest/lowest loads), so that when it arrives at the *last_proc*, information have been gathering from all processors of the ring. When all processors have responded, the load balancer $(GLBF)$ computes the new load balance. Once the load balancer makes the decision on *when to move which LP to where*, it sends $Migrate_{Request}$ to the migration process which in turns initiates the migration mechanism. Then, the migration process sends the (heavily overloaded) selected process to the lighted underloaded neighbor processor.

The computational load in our parallel simulation model consists of executing the null and the real messages. We wish to distribute the null messages and the real messages among all available processors. We define the (normalized) Load at each processor (Pr_k) as follows $Load_k = \mathcal{F}(R_{avg}^k, N_{avg}^k) = \alpha R_{avg}^k / R_{avg} + (1 - \alpha)N_{avg}^k / N_{avg}$; where R_{avg}^k, and N_{avg}^k are respectively the average CPU-queue length for real messages and null messages at each processor Pr_k; and α and $(1 - \alpha)$ are respectively the relative weights of the corresponding parameters. The value of α was determined empirically.

3   Simulation Experiments

The experiments have been conducted on a 72 nodes Intel Paragon, and we examined a general distributed communication model (DCM), see Fig. 1, modeling a national network consisting of four regions where the subnetworks are a toroid, a circular loops, a fully connected, and a pipeline networks. These regions are connected through four centrally located delays. A node in the network is represented by a process. Final message destinations are not known when a message is createad. Hence, no routing algorithm is simulated. Instead one third of arrival messages are forwarded to the neighboring nodes. A uniform distribution is employed to select which neigbor receives a message. Consequently the volumes of messages between any two nodes is inversely proportional to their hop distance.

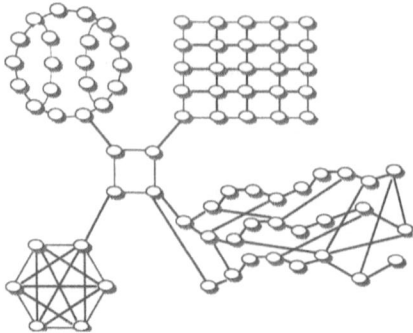

Fig. 1. Distributed Communication Model

Messages may flow between any two nodes and typically several paths may be employed. Nodes receive messages at varying rates that are dependant on traffic patterns. Since there are numerous deadlocks in this model, it provides a stress test for any conservative synchronization mechanisms. In fact, deadlocks occurs so frequently and null messages are generated so often that the load balancing strategy is almost a necessity in this model.

Various simulation conditions were created by mimicking the daily load fluctuation found in large communication network operating across various time zones in the nation [1]. In the pipline region, for instance, we arranged the sub-region into stages, and all processes in the same stages perform the same normal distribution with a standard deviation 25%. The time required to execute a message is significantly larger than the time to raise an event message in the next stage (message passing delay).

The experimental data was obtained by averaging several trial runs. The execution time (in seconds) as function of the number of processors are presented in the form of graphs. We also report the *synchronization overhead* in the form of null message ratio (NMR) which is defined as the number of null messages processed by the simulation using Chandy-Misra null-message approach divided by the number of real messages processed. the performance of our dynamic load balancing algorithms were compared with a static partitioning.

Figure 2 depicts the results obtained for the DCM model. We observe that the load balancing algorithm improves the running time. In other words, the running time for both network models, ie., Fully Connected and Distributed Communication, decreases as we increase the number of processors. The results in Fig. 6 indicate a reduction of 50% in the running time using the CL dynamic load balancing strategy compared with the static one when the number of processors is increased from 32 to 64, and about 55-60% reduction when we use the ML dynamic load balancing algorithm.

Figure 3 displays NMR as a function of the number of processors employed in the network models. We observe a significant reduction of null-messages for all load balancing schemes. The results show that NMR increases as the number of processors increases for both population levels. For instance, If we confine ourselves to less than 4 processors, we see approximately 30-35% reduction of

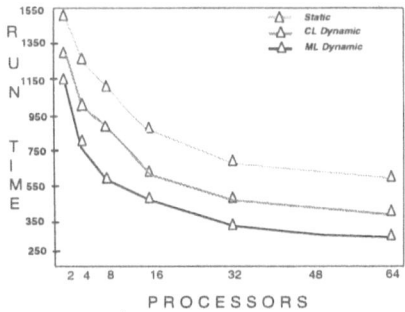

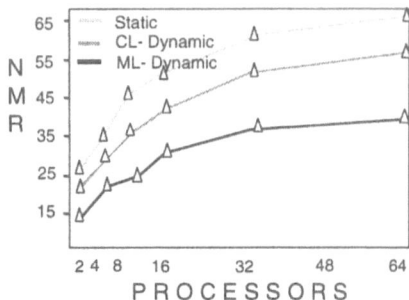

Fig. 2. Run Time Vs. Nbr of Processors   **Fig. 3.** NMR Vs. Nbr of Processors

the synchronization overhead using the CL dynamic load balancing algorithm over the static one. Increasing the number of processors from 4 to 16, we observe about 35-40% reduction of NMR. We also observe a 45% reduction using CL strategy over the static one, when 64 processors. are used, and a reduction of more than 50% using ML scheme over the static one.

These results conclude that the multi-level load balancing strategy significantly reduces the synchronization overhead when compared to a centralized method. In other words, a careful dynamic load balancing improves the performance of a conservative parallel simulation.

4 Conclusion and Future Work

We have described a dynamic load balancing algorithm based upon a notion of CPU-queue length utilization, and in which process migration takes place between physical processors. The synchronization protocol makes use of Chandy-Misra null message approach. Our results indicate that careful load balancing can be used to further improve the performance of Chandy-Misra's approach. We note that a Multi-Level approach seems to be a promising solution in reducing further the execution time of the simulation. An improvement between 30% and 50% in the event throughput were also noted which resulted in a significant reduction in the running time of the simulation models.

References

1. Boukerche, A., Das, S. K.: Dynamic Load Balancing Strategies For Parallel Simulations. TR-98. University of North Texas.
2. Boukerche, A., and Tropper C., "A Static Partitioning and Mapping Algorithm for Conservative Parallel Simulations", IEEE/ACM PADS'94, 1994, 164–172.
3. Misra, J., "Distributed Discrete-Event Simulation", ACM *Computing Surveys*, Vol. 18, No. 1, 1986, 39–65.

On-Line Scheduling of Parallelizable Jobs

Christophe Rapine[1], Isaac D. Scherson[2] and Denis Trystram[1]

[1] LMC – IMAG, Domaine Universitaire, BP 53, 38041 GRENOBLE cedex 9, France
[2] University of California, Irvine, Department of Information and Computer Science,
Irvine, CA 92717-3425, U.S.A.

Abstract. We consider the problem of efficiently executing a set of parallel jobs on a parallel machine by effectively scheduling the jobs on the computer's resources. This problem is one of optimization of resource utilization by parallel computing programs and/or the management of multi-users requests on a distributed system. We assume that each job is parallelizable and can be executed on any number of processors. Various on-line scheduling strategies of time/space sharing are presented here. The goal is to assign jobs to processors in space and time such that the total execution time is optimized.

1 Introduction

1.1 Description of the Computational Model

A model of computation is a description of a computer together with its program execution rules. The interest of a model is to provide a high level and abstract vision of the real computer world such that the design and the theoretical development and analysis of algorithms and of their complexity are possible. The most commonly adopted parallel computer model is the PRAM model [4]. A PRAM architecture is a *synchronous* system consisting of a global shared-memory and an unbounded number of processors. Each processor owns a small local memory and is able to execute any basic instruction in one unit of time. After each basic computational step a synchronization is done and each processor can access a global variable (for reading or writing) in one time unit. This model provides a powerful basis for the theoretical analysis of algorithms. It allows in particular the classification of problems in term of their intrinsic parallelism. Basically, the parallel characterization of a problem $\mathcal{P}$ is based on the order of magnitude of the time $T_\infty(n)$ of the best known algorithm to solve an instance of size n. $T_\infty(n)$ provides a lower bound for the time to solve $\mathcal{P}$ in parallel. However, in addition to answer the question *how fast the problem can be solved ?*, one important characteristic is *how efficient is the parallelization ?*, i.e. what is the order of magnitude of the number of processors $P_\infty(n)$ needed to achieve $T_\infty(n)$. This is an essential parameter to deal with in order to implement an algorithm on an actual parallel system with limited resources. This is best represented by the ratio μ_∞ between the *work* of the PRAM algorithm, $W_\infty(n) = T_\infty(n).P_\infty(n)$ and the total number of instructions in the algorithm W_{tot}. In this paper we consider a physically-distributed logically-shared memory machine composed of m

identical processors linked by an interconnection network. In the light of modern fast processor technology, compared to the PRAM model, communications and synchronizations become the most expensive operations and must be considered as important as computations.

1.2 Model of Jobs

For reasons of efficiency, algorithms must be considered at a high level of *granularity*, i.e. elementary operations are to be grouped into *tasks*, linked by precedence constraints to form programs. A general approach leads to write a parallel algorithm for v virtual processors with $m \ll v \ll P_\infty$. The execution of the program needs a scheduling policy a, static and/or dynamic, which directly influences the execution time T_m of the program. We define the *work* of the application on m processors as the space-time product $W_m = m \times T_m$. Ideally speaking, one may hope to obtain a parallel time equal to T_{seq}/m, where T_{seq} denotes the sequential execution time of the program. This would imply a conservation of the work. $W_m = W_{tot}$. Such a linear speed-up is hard to reach even with an optimal scheduling policy, due mainly to communication delays and the intrinsic non-parallelism of the algorithm (even in the PRAM model we may have $\mu_\infty \gg 1$). We introduce the ratio μ_m to represent the "penalty" of a parallel execution with respect to a sequential one and define it as:

$$\mu_m = \frac{W_m}{W_{tot}} = \frac{m.T_m}{T_{seq}}$$

1.3 Discussion on Inefficiency

The *Inefficiency* factor (μ_m) is an experimental parameter which reflects the quality of the parallel execution, even though it can be computed theoretically for an application and a scheduling. Let us remark that the parameter μ_m depends theoretically on n and on the algorithm itself. Practically speaking, most existing implementations of large industrial and/or academic scientific codes show a small constant upper bound. However, we can assume some general properties about the inefficiency that seem realistic. First, it is reasonable to assume that the work W_p is an increasing function of p. Hence inefficiency is also an increasing function of p : $\mu_p \le \mu_{p+1}$ for any p. Moreover, we can also assume that the parallel time T_p is a non-increasing function of p, all the less for "reasonable" number of processors. Hence, $(p+1)T_{p+1} \le (1+\frac{1}{p})p\, T_p \Rightarrow \mu_{p+1} \le (1+\frac{1}{p})\mu_p$. Developing the inequality, for any number of processors $p, q \ge 1$ we get $\mu_{p+q} \le \frac{p+q}{p}\mu_p$. In other words, $\frac{\mu_p}{p}$ is a non-increasing function of p. As a consequence, and as μ_1 is 1 by definition, μ_p is bounded by p, which intuitively means that in the worst case the execution of the application on p processors leads in fact to a sequential execution, i.e. the application is not parallel.

1.4 Competitivity

The factor μ_p is a performance characteristic of the parallel execution of a program. It can be easily computed *a posteriori* after an execution in the same way than speedup or efficiency. A more theoretical and precise performance guarantee is the *competitive ratio*: an algorithm is said to be $\rho(m)$ competitive if for any instance $\mathcal{I}$ to be scheduled on m processors, $T_m(\mathcal{I}) \leq \rho(m).T_m^*(\mathcal{I})$, where $T_m(\mathcal{I})$ denotes the time of the execution produced by the algorithm and $T_m^*(\mathcal{I})$ is the optimal time on m processors. Our goal is, given multiprocessor tasks individually characterized by an inefficiency μ_p when executed on p processors, to determine the competitive ratio for the whole system.

2 Scheduling Independent Jobs

We define an application as a set of jobs linked by precedence constraints. Most parallel programming paradigms assume that a parallel program produces independent subproblems, to be treated concurrently. Consider the following problem: *Given a m-processor parallel machine and a set of N independent parallelizable jobs* $(J_i)_{1,N}$ *whose duration is not known until they terminate, what competitive ratio can we guarantee for the execution?* We assume that each job J_i has a certain inefficiency μ_p^i when executed on p processors and that it can be preempted. The question can be restated as: *Given a set of inefficiencies , what is the competitive ratio we may hope for from an on-line scheduling strategy?* In the following, we denote by $T^*(\mathcal{I})$ the optimal execution time and by $T_a(\mathcal{I})$ the one of the scheduling strategy a. The competitive ratio ρ_a^* of a is defined as the maximum on all the instances $\mathcal{I}$ of the fraction $T_a(\mathcal{I})/T^*(\mathcal{I})$. Another interesting performance guarantee is the comparison between $T_a(\mathcal{I})$ and the optimal execution time $T_{\mathcal{S}}^*(\mathcal{I})$ when the parallelization of the jobs is not allowed, i.e. jobs are computed one at a time on single processors. We denote by ρ_a^{seq} the maximum on $\mathcal{I}$ of $T_a(\mathcal{I})/T_{\mathcal{S}}^*(\mathcal{I})$. This competitive ratio will highlight the (possible) gain to allow the multiprocessing of jobs compared to the classical approach we present below where jobs are purely sequential.

2.1 Two Extreme Strategies: Graham and Gang Scheduling

The scheduling of multiprocessor jobs has recently become a matter of much research and study. In off-line scheduling theory, several articles have been published by J.Błażewicz and al. [2, 1], considering multiprocessor tasks but each one requiring a *fixed* number of processors for execution. If we look at the field of on-line scheduling, the most important contributions focused on the the case of one-processor tasks. In this classical approach any task requires only one processor for execution. The most famous algorithm is due to Graham [3] and consists simply in a greedy affectation of the tasks to the processors such that a processor cannot be idle if there is a remaining unscheduled job. Its competitive ratio is $2 - \frac{1}{m}$, which is the best possible ratio for any deterministic scheduling strategy when the computation costs of the tasks are not known till their completion. At the other end of the spectrum, the Gang scheduling analyzed recently in Scherson [5] assumes multiprocessor jobs and schedule them allowing the whole machine, m processors, to all the tasks. For the *Happiness* function defined by Scherson, the Gang strategy is proved to be the best one for preemptive multiprocessor job scheduling. Certainly, if we are concerned with the competitive ratio, the best strategy will be a compromise between allowing one processor per job (Graham) and the whole machine to any job (Gang). In particular we hope that multiprocessing of the job may improve the competitive ratio $(2 - \frac{1}{m})$ of Graham. In the following, let denote $\mu_p = \max_{1,N} \mu_p^i$.

Lemma 1. *The competitive ratio of Gang scheduling is equal to μ_m.*

Proof. Consider any instance $\mathcal{I}$. Let a_i be the work of job J_i, and $\mathcal{W}_{tot}$ the total amount of work of the program. We have $T_{Gang} = \sum_{i=1}^{N} \frac{\mu_m^i}{m} a_i \leq \frac{\mu_m}{m} \mathcal{W}_{tot}$ Noticing that $\mathcal{W}_{tot}/m$ is always a lower bound of T^*, it follows that $\rho_{Gang}^* \leq \mu_m$. Consider now a particular instance consisting of m identical jobs of work a. Gang scheduling produces an execution time of $\mu_m a$, while assigning one processor per job produces a schedule of length exactly a. As this schedule is one-processor job, it proves that $\rho_{Gang}^{seq} \geq \mu_m$, which gives the result. $\square$

Lemma 2. *For Graham scheduling, $\rho_{Graham}^{seq} = 2 - \frac{1}{m}$ and $\rho_{Graham}^* \geq \frac{m}{\mu_m}$*

Proof. Let consider the scheduling of a single job of size a. Any one-processor scheduling delivers an execution time of a, while the optimal one is reached giving the whole machine to the job. Thus $T^* = \frac{\mu_m}{m} a$ and so $\rho_{Graham}^* \geq \frac{m}{\mu_m}$. $\square$

3 SCHEduling Multiprocessors Efficiently: SCHEME

A Graham scheduling breaks into two phases: in the first one all processors are busy, getting an efficiency of one. Inefficiency happens when less than m tasks remain. Our idea is to preempt these tasks and to schedule them with a Gang policy in order to avoid idle time.

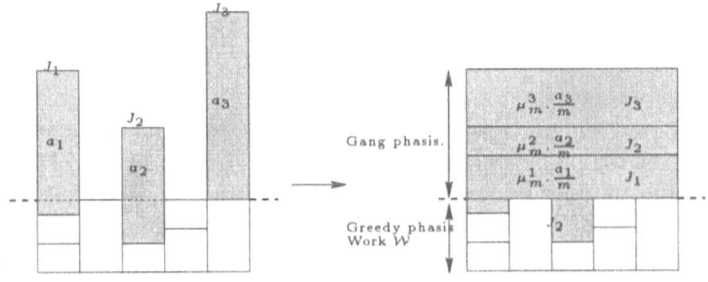

Fig. 1. *Principle of a SCHEME scheduling*

3.1 Analysis of the Sequential Competitivity

Let's consider first time t in the Graham scheduling when less than m jobs are not completed. Let $J_1, \ldots, J_k$, $k \leq m-1$, be these unfinished jobs, and denote by $a_1, \ldots, a_k$ their remaining amount of work. Let $W = m.t$ be the work executed between time 0 and t. We have the majoration $T_{sc}(m) \leq \frac{W}{m} + \frac{\mu_m}{m} \sum_{i=1}^{i=k} a_i$. Introducing the total amount of work of the jobs, $W_{tot} = W + \sum_{i=1}^{i=k} a_i$, and the maximum amount of work remaining for a job, $a = max\{a_1, \ldots, a_k\}$, it follows that $T_{sc}(m) \leq W_{tot}/m + \frac{\mu_m - 1}{m} \sum_{i=1}^{i=k} a_i \leq W_{tot}/m + \frac{m-1}{m}(\mu_m - 1)a$

To obtain an expression of sequential competitivity, we need to minorate the one-processor optimal time. A lower bound is always W_{tot}/m. As we consider a one-processor job scheduling, T_S^* is greater than the sequential time of any task. Hence we have $T_S^* > a$. We get the following lemma:

Lemma 3. *SCHEME has a sequential competitivity of* $(1 - \frac{1}{m})\mu_m + \frac{1}{m}$

Proof. It is sufficient to prove that the bound is tight. Consider the following instance composed of $m - 1$ jobs of size m and m jobs of size 1. The optimal schedule allocate one processor per tasks of size m and assign all the small jobs to the last processor. It realizes an execution time $m = W_{tot}/m$. A possible execution of SCHEME scheduling allocates first one small jobs per processor and then uses the gang strategy for the $m - 1$ large jobs. It conducts to an execution time $T_{sc}(m) = 1 + (m - 1)\mu_m$, which realizes the bound. □

3.2 Analysis of the Competitive Ratio

We determine here the competitive ratio of the SCHEME scheduling compared to an optimal multiprocessor jobs scheduling. Let consider an instance of the problem on m processors, and S^* an optimal schedule. We decompose S^* into temporal area slices $(A_i^*)_i$, each slice corresponding to the jobs compute between time t_{i-1} and t_i defined recursively as follow: $t_0 = 0$, and t_{i+1} is the maximal date such that in the time interval $[t_i, t_{i+1}[$ no job is preempted or completed. Notice that in the slice A_i^*, either all the jobs are computed on one processor, the slice is said "one-processored", or at least one job is executed on $p \geq 2$ processors. Let denote by A_i the area needed in the SCHEME schedule to compute the instructions of A_i^*. Notice that any piece of job in the SCHEME scheduling is either process on one or on m processors.

• Let A_i^* be a one-processored slice. Let denote by G_i the amount of instructions of A_i^* executed on m processors in A_i and by W_i the amount of instructions of A_i^* executed on one processor in A_i. By definition $A_i^* = G_i + W_i$. At least one of the jobs is entirely computed on one processor in the SCHEME schedule, hence $W_i \geq A_i^*/m$. In A_i, the area to compute the instruction of G_i is increased by a factor μ_m. Thus $A_i = \mu_m G_i + W_i = \mu_m A_i^* - (\mu_m - 1)W_i \leq \mu_m A_i^* - (\mu_m - 1)\frac{A_i^*}{m} \leq ((1 - \frac{1}{m})\mu_m + \frac{1}{m})A_i^*$. It follows that $A_i \leq \rho_{sc}^{seq}(m)A_i^*$.

• Let A_i^* be a multi-processored slice. There exists a job J computes on $p \geq 2$ processors in the slice. Let a be its amount of instruction and denote by

$B = A_i^* - \mu_p a$ the remaining area less J. In the SCHEME schedule, the area A_i needed to execute the instructions of A_i^* is at most $\mu_m(a + B) = \mu_m(A_i^* - (\mu_p - 1)a)$. Moreover $A_i^* \leq \frac{\mu_p}{p}a \times m$. Hence: $A_i \leq \mu_m(1 - \frac{\mu_p-1}{\mu_p}\frac{p}{m})A_i^*$ The maximum ratio appears minimizing the term $\frac{\mu_p-1}{\mu_p}p$. Noticing that p/μ_p and μ_p are increasing functions, maximal value is reached for $p = 2$. Hence: $A_i \leq (1 - \frac{2}{m} + \frac{2}{m\mu_2})\mu_m A_i^*$. We call the ratio factor $\rho_{sc}^{par}(m)$. For any slice A_i^* of S^*, we have the corresponding area in S bounded by $\max\{\rho_{sc}^{seq}(m),\ \rho_{sc}^{par}(m)\}.A_i^*$. Noticing that the SCHEME schedule contains no idle time, we get the following lemma:

Lemma 4. *The competitive ratio of SCHEME scheduling is equal to*

$$\rho_{sc}^*(m) = (1 - \frac{1}{m})\mu_m + \frac{1}{m}\max\{1, \frac{2-\mu_2}{\mu_2}\mu_m\}$$

Proof. To prove that the ratio is tight, consider the instance composed of $m - 2$ jobs of size 1 and one job of size $(2/\mu_2)$. A possible schedule allocating one processor per unit job and two for the last job, completes at time 1. The SCHEME scheduling leads to time execution $\frac{\mu_m}{m}((m - 2) + \frac{2}{\mu_2})$. □

4   Concluding Remarks

We are currently experimenting the mixed strategy SCHEME shown to be (theoretically) promising. Other more sophisticated scheduling strategies are also under investigation.

References

1. J. Błażewicz, P. Dell'Olmo, M. Drozdowski, and M.G. Speranza. Scheduling multiprocessors tasks on three dedicated processors. *Information Processing Letters*, 41:275–280, April 1992.
2. J. Błażewicz, M. Drabowski, and J. Węglarz. Scheduling multiprocessor tasks to minimize schedule length. *IEEE Transactions on Computers*, 35(5):389–393, 1986.
3. R.L. Graham. Bounds on multiprocessing timing anomalies. *SIAM Journal on Applied Mathematics*, 17(2):416–429, March 1969.
4. R.M. Karp. On the computational complexity of combinatorial problems. *Networks*, 5:45–68, 1975.
5. I. D. Scherson, R. Subramanian, V. L. M. Reis, and L. M. Campos. Scheduling computationally intensive data parallel programs. In *Placement dynamique et répartition de charge : application aux systèmes parallèles et répartis (École Française de Parallélisme, Réseaux et Système)*, pages 107–129. Inria, July 1996.

On Optimal k-linear Scheduling
of Tree-Like Task Graphs for LogP-Machines

Wolf Zimmermann[1], Martin Middendorf[2], and Welf Löwe[1]

[1] Institut für Programmstrukturen und Datenorganisation, Universität Karlsruhe,
76128 Karlsruhe, Germany, E-Mail:{zimmer|loewe}@ipd.info.uni-karlsruhe.de
[2] Institut für Angewandte Informatik und Formale Beschreibungsverfahren,
Universität Karlsruhe, 76128 Karlsruhe, Germany, E-Mail mmi@aifb.uni-karlsruhe.de

Abstract. A k-linear schedule may map up to k directed paths of a task graph onto one processor. We consider k-linear scheduling algorithms for the communication cost model of the LogP-machine, i.e. without assumption on processor bounds. The main result of this paper is that optimal k-linear schedules of trees and tree-like task graphs G with n tasks can be computed in time $O(n^{k+2} \log n)$ and $O(n^{k+3} \log n)$, respectively, if $o \geq g$. These schedules satisfy a capacity constraint, i.e., there are at most $\lceil L/g \rceil$ messages in transit from any or to any processor at any time.

1 Introduction

The LogP-machine [3,4] assumes a cost model reflecting latency of point-to-point-communication in the network, overhead of communication on processors themselves, and the network bandwidth. These communication costs are modeled with parameters Latency, $o$verhead, and $g$ap. The gap is the inverse bandwidth of the network per processor. In addition to L, o, and g, parameter P describes the number of processors. Furthermore, the network has finite capacity, i.e. there are at most $\lceil L/g \rceil$ messages in transit from any or to any processor at any time. If a processor would attempt to send a message that exceeds this capacity, then it stalls until the message can be sent without violation of the capacity constraint. The LogP-parameters have been determined for several parallel computers [3, 4, 6, 8]. These works confirmed all LogP-based runtime predictions.

Much research has been done in recent years on scheduling task graphs with communication delay [9–11,14,21,23,25], i.e. $o = g = 0$. In the following, we discuss those works assuming an unrestricted number of processors. Papadimitriou and Yannakakis [21] have shown that given a task graph G with unit computation times for each task, communication latency $L \geq 1$, and integer $T_{\max}$, it is NP-complete to decide whether G can be scheduled in time $\leq T_{\max}$. Their results hold no matter whether redundant computations of tasks are allowed or not. Finding optimal time schedules remains NP-complete for DAGs obtained by the concatenation of a join and a fork (if redundant computations are allowed) and for fine grained trees (no matter if redundant computations are allowed or not). A task graph is fine grained if the granularity γ which is a constant closely related to the ratio of computation and communication is < 1 [22]. Gerasoulis and Yang [9] find schedules for task-graphs guaranteeing the factor $1 + \frac{1}{\gamma}$ of the optimal time if recomputation is not allowed. For some special classes of task graphs, such as join, fork, coarse grained (inverse) trees, an optimal schedule can be computed in polynomial time [2, 9, 15, 25]. If recomputation is allowed some coarse grained DAGs can also be scheduled optimally in polynomial time [1, 5, 15]. The problem to find optimal schedules having a small amount of redundant computations has been addressed in [23].

Recently, there are some works investigating scheduling task graphs for the LogP-machine without limitations on the number of processors [12, 16–18, 20, 24, 26] and with limitations on the number of processors [19, 7]. Most of these works discuss approximation algorithms. In particular, [18, 26] generalizes the result of Gerasoulis and Yang [9] to LogP-machines using linear schedules. [20] shows that optimal linear schedules for trees and tree-like task graphs can be computed in polynomial time if the cost of each task is at least $g - o$. However, linear schedules for trees often have bad performance. Unfortunately, if we drop the linearity, even scheduling trees becomes NP-complete [20]. [24] shows that the computation of a schedule of length at most B is NP-complete even for fork and join trees and $o = g$. They discuss several approximation algorithms for fork and join trees. [12] discusses an optimal scheduling algorithm for some special cases of fork graphs. In this paper, we generalize the linearity constraint and allow instead to map up to k directed paths on one processor. We show that it is possible to compute optimal k-linear schedules on trees and tree-like task graphs in polynomial time if $o = g$. In contrast to most other works on scheduling task graphs for the LogP-machine, we take the capacity constraint into account.

Section 2 gives the basic definitions used in this paper. Section 3 discusses some normalization properties for k-linear schedules on trees. Section 4 presents the scheduling algorithm with an example that shows the improvement of k-linearity w.r.t. linear schedules.

2 Definitions

A *task graph* is a directed acyclic graph $G = (V, E, \tau)$, where the vertices are sequential tasks, the edges are data dependencies between them, and τ_v is the computation cost of task $v \in V$. The set of *direct predecessors* $PRED_v$ of a task v in G is defined by $PRED_v = \{u \in V \mid (u, v) \in E\}$. The set of *direct successors* $SUCC_v$ is defined analogously. *Leaves* are tasks without predecessors, *roots* are tasks without successors. A task v is *ancestor* (*descendant*) of a task u iff there is a path from v to u (u to v). The set of ancestors (descendants) of u is denoted by ANC_u ($DESC_u$). A set of tasks $U \subseteq V$ is *independent* iff for all $u, v \in V$, $u \neq v$, neither $u \in ANC_v$ nor $v \in ANC_u$. A set of tasks $U \subseteq V$ is *k-independent* iff there is an independent subset $W \subseteq U$ of size k. G is an *inverse tree* iff $|SUCC_v| \leq 1$ and there is exactly one root. In this case, $SUCC_v$

denotes the unique successor of non-root tasks v. G is *tree-like* iff for each root v, the subgraph of G induced by the ancestors of v is a tree.

A *schedule* for G specifies for each processor P_i and each time step t the operation performed by processor P_i at time t, provided there starts one. The operations are the execution of a task $v \in V$ (denoted by the task), sending a message containing the result of a task v to another processor P (denoted by $send(v, P)$), and receiving a message containing the result of a task[1] v from processor P (denoted by $recv(v, P)$). A schedule is therefore a partial mapping $s : \mathbb{P} \times \mathbb{N} \rightarrow OP$, where $\mathbb{P}$ denotes an infinite set of processors and OP denotes the set of operations. s must satisfy some restrictions:

(i) Two operations on the same processor must not overlap in time. The computation of a task v requires time τ_v. Sending from v or receiving a message sent by v requires time o.

(ii) If a processor computes a task $v \in V$ then any task $w \in PRED_v$ must be computed or received before on the same processor.

(iii) If a processor sends a message of a task, it must be computed or received before on the same processor.

(iv) Between two consecutive send (receive) operations there must be at least time g, where v is the message sent earlier (received later).

(v) For any send operation there is a corresponding receive operation, and vice versa. Between the end of a send operation and the beginning of the corresponding receive operation must be at least time L.

(vi) Any $v \in V$ is computed on some processor.

(vii) The capacity constraints are satisfied by s.

A *trace of s induced by processor P* is the sequence of operations performed by P together with their starting time. $tasks(P)$ denotes the set of tasks computed by processor P. A schedule s for G is *k-linear*, $k \in \mathbb{N}$, iff $tasks(P)$ is not $(k + 1)$-independent for every processor P, i.e. at most k directed paths of G are computed by P. The *execution time $TIME(s)$* of a schedule s is the time when the last operation is finished. A *sub-schedule s_v induced by a task v* computes just the ancestors of v and performs all send and receive operations of the ancestors except v on the same processors at the same time as s, i.e.

$$s_v(P, t) = \begin{cases} s(P, t) & \text{if } s(P, t) \in ANC_v, \\ & s(P, t) = recv(w, P') \text{ for a processor } P' \text{ and } v \neq w \in ANC_v, \\ & s(P, t) = send(w, P') \text{ for a processor } P' \text{ and } v \neq w \in ANC_v, \\ \text{undefined} & \text{otherwise} \end{cases}$$

It is easy to see that s_v is a schedule of the subgraph of G induced by ANC_v.

A schedule s for an inverse tree $G = (V, E, \tau)$ is *non-redundant* if each task of G is computed exactly once, for each $v \in V$ there exists at most one send operation, and there exists no send operation if v is computed on the same processor as its successor.

---

[1] For simplicity, we say sending and receiving the task v, respectively.

3 Properties of k-linear Schedules

In this section we prove some general properties of k-linear schedules on trees. These properties are used to restrict the search space for optimal schedules.

Lemma 1. *Let $G = (V, E, \tau)$ be an inverse tree. For every schedule s for G there exists a non-redundant schedule s' with $TIME(s') \leq TIME(s)$. If s is k-linear, then also s'.*

Proof: Let s be a schedule for G. First we construct a sub-schedule of s where each node is computed only once. Assume that a task is computed more than once in s. If the root is computed more than once just omit all but one computation of the root. Otherwise let v be computed more than once such that $SUCC_v$ is computed only once. Let $SUCC_v$ be computed on processor P. Then there exists a computation of v on a processor P_0 and a (possibly empty) sequence $send(v, P_1)$, $recv(v, P_0)$, $send(v, P_2), \ldots,$ $send(v, P)$, $recv(v, P_k)$ of corresponding send and receive operations of v such that v is received on P before the computation of $SUCC_v$ starts. Now, all other computations of v are omitted and the above sequence of send and receive operations is replaced by $send(v, P); recv(v, P_0)$, and remove all other operations sending or receiving v, provided $P_0 \neq P$. If $P_0 = P$, then all operations sending or receiving v are remove. This process is repeated until each task is computed exactly once. It is clear that the obtained schedule s' is non-redundant, $TIME(s') \leq TIME(s)$, and the above transformations preserve k-linearity. ∎

Lemma 2. *Let $G = (V, E, \tau)$ be an inverse tree. For every non-redundant k-linear schedule for G there is a non-redundant k-linear schedule s' satisfying $TIME(s') \leq TIME(s)$ such that for every processor P, the subgraph of G induced by $tasks(P)$ is an inverse tree with at most k leaves.*

Proof: Suppose there is a processor P such that the subgraph G_P of G induced by $tasks(P)$ is not an inverse tree. Since every subgraph of G is a forest consisting of inverse trees, G_P must have more than one connected component. Since s is k-linear, each connected component contains at most k leaves. Let s'' be the schedule obtained from s by computing each of these connected components on separate (new) processors, and updating accordingly the operations that send tasks to processor P. It is easy to see that this transformation neither affects the execution time nor the k-linearity. The claim follows by induction. ∎

Corollary 1. *Let $G = (V, E, \tau)$ be an inverse tree. Then there exists an optimal k-linear schedule s such that for every $v \in V$ the sub-schedule s_v induced by v is an optimal k-linear schedule for G_v which is non-redundant and for each processor P the subgraph induced by the tasks $tasks(P)$ is an inverse tree with at most k leaves.*

Proof: Let s be an optimal k-linear schedule for G that is non-redundant. If for a $v \in V$ the sub-schedule s_v is not optimal replace it by an optimal k-linear sub-schedule that is non-redundant. ∎

Thus, for obtaining optimal k-linear schedules, it is sufficient to consider non-redundant schedules that satisfy Corollary 1.

4 The Scheduling Algorithm

In this section, we present the algorithm computing optimal k-linear schedules for task graphs, if $o = g$. We first reduce the scheduling problem to a special case of the 1-machine scheduling problem with precedence constraints and release dates, i.e. a task-graph $G = (V, E, \tau)$ is to be scheduled on one processor where for every task there is a time r_v (*release date*) such that v cannot be scheduled before r_v:

Lemma 3. *Let $G = (V, E, \tau)$ be an inverse tree. Let s be a schedule for G that satisfies the property of Corollary 1. For a $v \in V$ consider the sub-schedule s_v and let P be the processor computing v. Let tr be the trace of P in s_v where receive operations are replaced by computations of length o computing the received tasks, and G'_v obtained from the subgraph of G_v induced by the tasks$(P) \cup S$ of s_v. Then tr is an optimal 1-machine schedule for G' with release dates $TIME(s_v) + L + o$ for $v \in S$.*

Proof: No $v \in S$ can be received before time $t_v = TIME(s_v)$, because $send(v, P)$ cannot be scheduled earlier than t_v. Since every reception of a task costs time o, tr is a 1-machine schedule for G' with release dates $t_v + L + o$ for $v \in S$. Since s is optimal tr is also optimal. ∎

ALGORITHM 1
(1) **for** $v \in V$ according to a topological order of G **do**
(2) $t_v := \infty;\ s_v := \emptyset;$ – – try to compute optimal k-linear schedule
(3)   **for** every independent $U \subseteq ANC_v,\ 0 < |U| \leq k$ **do**
(4) $s'_v := optimal\_trace(G, v, U, t)$
(5)   **if** $TIME(s_v) < t_v$ **then** – – a better schedule is found
(6) $t_v := TIME(s'_v);$
(7) $s_v := s'_v \uplus \biguplus_{u \in PRED(U)} s_u \uplus \{s_v(P_u, t') = send(u, P_v)\};$
          where $s'_v(t' + L + o, P_v) = recv(u, P_u)$
(8)   **end;** – – **if**
(9)   **end;** – – **for**
(10) **end;** – – **for**

Algorithm 1 computes (in topological order) for each task $v \in V$ an optimal schedule. In the iteration step, for every possible set R for the subtree rooted at v as defined in Lemma 3, the optimal trace for the processor computing v is derived (function *optimal_trace*). According to Lemma 3, one of these schedules is optimal for the subtree rooted at v. The algorithm uses the fact that R can is uniquely defined by its leaves.

Theorem 1 (Correctness of Algorithm 1). *Let $G = (V, E, \tau)$ be an inverse tree with root r. The schedule s_r computed by Algorithm 1 is an optimal k-linear schedule for G.*

Proof: We prove by induction on $|ANC_v|$ that for every $v \in V$ s_v is an optimal k-linear schedule. If $|ANC_v| = 0$, then v is a leaf and the claim is trivial. If v is not a leaf, the claim follows directly by induction hypothesis and Lemma 3. By line (7) every send operation is scheduled $L + o$ time steps earlier than the corresponding receive operations. Hence, a processor can perform at most $\lceil L/o \rceil$ receive operations for every time interval of length L. Since the corresponding send operations start at the latest possible time, the capacity constraint is satisfied. ∎

Theorem 2. *Let $G = (V, E, \tau)$ be a tree. Algorithm 1 computes in time $O(|V|^{k+2} \log |V|)$ an optimal k-linear schedule for G.*

Proof: For every $v \in V$, loop (3)–(8) is executed at most $O(|ANC_v|^k)$ times, because the number of subsets of size at most k is $O(n^k)$. Thus, if *optimal_trace* can be implemented by a polynomial algorithm, Algorithm 1 is polynomial. It is known that a simple greedy algorithm with time complexity $O(|V| \log |V|)$ computes an optimal schedule a 1-machine scheduling problem [13], i.e. *optimal_trace* can be executed in time $O(|ANC_v| \cdot \log |ANC_v|)$. Hence, the time complexity of loop (3)–(8) is $O(|ANC_v|^{k+1} \log |ANC_v|)$ by the above discussion. Summing over all $v \in V$ yields the claimed time complexity of Algorithm 1 ∎

Example: (Execution of Algorithm 1) We compute optimal 1- and 2-linear schedules for the inverse tree $G = (V, E, \tau)$ shown in Fig. 1(a) (where $\tau_v = 1$ for all $v \in V$) with the parameters $L = 2$ and $o = 1$ (cf. Fig. 1(b) and (c)). Fig. 1(d) shows an optimal schedule. The schedules are visualized by Gantt-charts. The x-axis corresponds to time, the y-axis to y. We place a box at (t, i), if processor P_i starts at time t an operation. The horizontal length of the box corresponds to the cost (i.e. τ_v if v is computed and o if it is send or receive operation). Send and receive operations are drawn as dark boxes. White boxes denote the

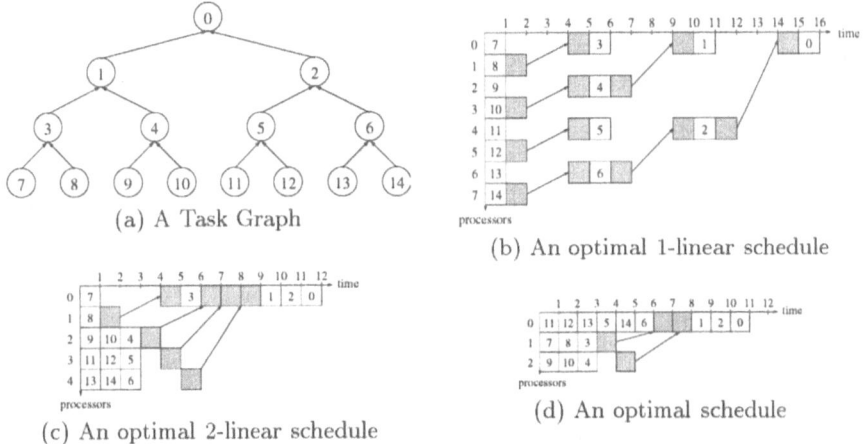

(a) A Task Graph

(b) An optimal 1-linear schedule

(c) An optimal 2-linear schedule

(d) An optimal schedule

Fig. 1. A Task Graph with optimal 1- and 2-linear Schedules

Table 1. Execution Times of Optimal k-linear Schedules s_v

	$v \in \{7,\ldots,14\}$	$v \in \{3,\ldots,6\}$	$v \in \{1,2\}$	$v = 0$
1-linear	$t_v = 1$	$t_v = 6$	$t_v = 11$	$t_v = 16$
2-linear	$t_v = 1$	$t_v = 3$	$t_v = 8$	$t_v = 12$
3-linear	$t_v = 1$	$t_v = 3$	$t_v = 7$	$t_v = 11$
8-linear	$t_v = 1$	$t_v = 3$	$t_v = 7$	$t_v = 11$

computation of a task. Table 1 shows the executions times of the optimal k-linear schedules for the tasks $v \in V$. All tasks on the same level of the tree have the same schedules up to renaming. The table also shows that optimal schedules for G have execution time 11. The optimal schedule for tasks 7–14 is obviously the computation of the task. The optimal 1-linear schedule for task 3 computes tasks 7 and 3 on one processor and task 8 on another processor. The optimal 2-linear schedule computes the tasks 7, 8 and 3 sequentially. For 1-linear schedules, an optimal trace for task 1 and task 0 starts at task 7 (time 11 and 16, respectively). For 2-linear schedules, an optimal trace for task 1 starts with task 7 (time 8), and an optimal trace for task 0 starts with task 7 and 2 (time 12). ∎

Obviously, an optimal n-linear schedule for an inverse tree G is optimal among all schedules. Thus, Algorithm 1 can be used to compute an optimal schedule. Although this problem is NP-hard and the algorithm is exponential, many subsets of V can be ignored, because they are not independent. As our example shows, using symmetry may reduce the number of subsets of V even further. E.g. if all tasks of the task graph in Fig. 1(a) have equal weights if they are the same level, then only 66 subsets instead of 2^{15} need to be examined. Although there are still $2^{O(|V|)}$ possibilities to be examined, the constant in the exponent is much smaller than 1, which makes Algorithm 1 more feasible than a general brute-force search.

The generalization of Algorithm 1 to tree-like task graphs is straightforward: Let $G = (V, E, \tau)$ be a tree-like task graph and O be the set of its roots. Then an optimal schedule for G is obtained by scheduling separately the subgraph induced by ANC_o (which is a tree by definition of tree-like) for every $o \in O$ according to Algorithm 1. Hence, Theorem 2 implies directly the

Corollary 2. Let $G = (V, E, \tau)$ be a tree-like task graph. Then, an optimal k-linear schedule for G can be computed in time $O(|V|^{k+3} \log |V|)$.

5 Conclusion

We have shown that optimal k-linear schedules for trees and tree-like task-graphs can be computed in polynomial time on the LogP-machine with an unbounded number of processors, if $o = g$ and $k = O(1)$. For every task graph G that is an inverse tree, Algorithm 1 constructs optimal k-linear schedules for the subtrees G_v of G rooted at v, where the tasks are processed from the leaves to the root.

The main step is to construct an optimal k-linear schedule for G_v, when for all proper ancestors u of v the optimal k-linear schedules for G_u are already known. The subgraph of G_v induced by the tasks computed by the processor computing v is always a tree with at most k leaves. Algorithm 1 is polynomial because there is only a polynomial number of independent sets over the ancestors of v of size at most k and for a given independent set, an optimal trace can be computed in polynomial time. Thus, it is possible to consider all possibilities in polynomial time.

The generalization of the approach for $g > o$ is not straightforward. The principal approach remains the same, but computing the optimal traces is more difficult. For this, more normalization properties are required. [20] discusses some of those for 1-linear schedules. However, it is not clear whether these apply to k-linear schedules in general, and whether other normalization properties are required.

[18, 26] show that for every task graph G 1-linear schedules guaranteeing the factor $1+1/\gamma$ of the optimal time can be computed in polynomial time, where γ is the granularity of G. Obviously, with increasing k the execution time of optimal k-linear schedules does not increase. In fact, the example in Section 4 shows that it may decrease drastically (in our example by 25%). The question raises whether it is possible to compute k-linear schedules with better performance guarantees for increasing k.

References

1. F.D. Anger, J. Hwang, and Y. Chow. Scheduling with sufficient loosely coupled processors. *Journal of Parallel and Distributed Computing*, 9:87–92, 1990.
2. P. Chretienne. A polynomial algorithm to optimally schedule tasks over an ideal distributed system under tree-like precedence constraints. *European Journal of Operations Research*, 2:225–230, 1989.
3. D. Culler, R. Karp, D. Patterson, A. Sahay, K. E. Schauser, E. Santos, R. Subramonian, and T. von Eicken. LogP: Towards a realistic model of parallel computation. In *4th ACM SIGPLAN Symposium on Principles and Practice of Parallel Programming (PPOPP 93)*, pp. 1–12, 1993. published in: SIGPLAN Notices (28) 7.
4. D. Culler, R. Karp, D. Patterson, A. Sahay, K. E. Schauser, E. Santos, R. Subramonian, and T. von Eicken. LogP: A practical model of parallel computation. *Communications of the ACM*, 39(11):78–85, 1996.
5. S. Darbha and D.P. Agrawal. SBDS: A task duplication based optimal algorithm. In *Scalable High Performance Conference*, 1994.
6. B. Di Martino and G. Ianello. Parallelization of non-simultaneous iterative methods for systems of linear equations. In *Parallel Processing: CONPAR 94 - VAPP VI*, volume 854 of *Lecture Notes in Computer Science*, pp. 253–264. Springer, 1994.
7. J. Eisenbiegler, W. Löwe, and W. Zimmermann. Optimizing parallel programs on machines with expensive communication. In *Europar' 96 Parallel Processing Vol. 2*, volume 1124 of *Lecture Notes in Computer Science*, pp. 602–610. Springer, 1996.
8. Jrn Eisenbiegler, Welf Löwe, and Andreas Wehrenpfennig. On the optimization by redundancy using an extended LogP model. In *International Conference on Advances in Parallel and Distributed Computing (APDC'97)*, pp. 149–155. IEEE Computer Society Press, 1997.

9. A. Gerasoulis and T. Yang. On the granularity and clustering of directed acyclic task graphs. *IEEE Transactions on Parallel and Distributed Systems*, 4:686–701, June 1993.

10. J.A. Hoogreven, J.K. Lenstra, and B. Veltmann. Three, four, five, six or the complexity of scheduling with communication delays. *Operations Research Letters*, 16:129–137, 1994.

11. H. Jung, L. M. Kirousis, and P. Spirakis. Lower bounds and efficient algorithms for multiprocessor scheduling of directed acyclic graphs with communication delays. *Information and Computation*, 105:94–104, 1993.

12. I. Kort and D. Trystram. Scheduling fork graphs under LogP with an unbounded number of processors. Submitted to *Europar '98*, 1998.

13. E.L. Lawler. Optimal sequencing of a single machine subject to precedence constraints. *Management Science*, 19:544–546, 1973.

14. J.K. Lenstra, M. Veldhorst, and B. Veltmann. The complexity of scheduling trees with communication delays. *Journal of Algorithms*, 20:157–173, 1996.

15. W. Löwe and W. Zimmermann. On finding optimal clusterings in task graphs. In N. Mirenkov, editor, *Parallel Algorithms/Architecture Synthesis pAs '95*, pp. 241–247. IEEE, 1995.

16. W. Löwe and W. Zimmermann. Programming data-parallel – executing process parallel. In P. Fritzson and L. Finmo, editors, *Parallel Programming and Applications*, pp. 50–64. IOS Press, 1995.

17. W. Lwe and W. Zimmermann. Upper time bounds for executing PRAM-programs on the LogP-machine. In M. Wolfe, editor, *Proceedings of the 9th ACM International Conference on Supercomputing*, pp. 41–50. ACM, 1995.

18. Welf Löwe, Wolf Zimmermann, and Jörn Eisenbiegler. On linear schedules for task graphs for generalized LogP-machines. In *Europar '97: Parallel Processing*, volume 1300 of *Lecture Notes in Computer Science*, pp. 895–904, 1997.

19. W. Lwe, J. Eisenbiegler, and W. Zimmermann. Optimizing parallel programs on machines with fast communication. In *9. International Conference on Parallel and Distributed Computing Systems*, pp. 100–103, 1996.

20. M. Middendorf, W. Löwe, and W. Zimmermann. Scheduling inverse trees under the communication model of the LogP-machine. Accepted for publication in *Theoretical Computer Science*.

21. C.H. Papadimitriou and M. Yannakakis. Towards an architecture-independent analysis of parallel algorithms. *SIAM Journal on Computing*, 19(2):322 – 328, 1990.

22. C. Picouleau. New complexity results on the uet-uct scheduling algorithm. In *Proc. Summer School on Scheduling Theory and its Applications*, pp. 187–201, 1992.

23. J. Siddhiwala and L.-F. Cha. Path-based task replication for scheduling with communication cost. In *Proceedings of the International Conference on Parallel Processing*, volume II, pp. 186–190, 1995.

24. J. Verriet. Scheduling tree-structured programs in the LogP-model. Technical Report UU-CS-1997-18, Dept. of Computer Science, Utrecht University, 1997.

25. T. Yang and A. Gerasoulis. DSC: Scheduling parallel tasks on an unbounded number of processors. *IEEE Transactions on Parallel and Distributed Systems*, 5(9):951–967, 1994.

26. W. Zimmermann and W. Löwe. An approach to machine-independent parallel programming. In *Parallel Processing: CONPAR 94 - VAPP VI*, volume 854 of *Lecture Notes in Computer Science*, pp. 277–288. Springer, 1994.

Static Scheduling Using Task Replication for LogP and BSP Models

Cristina Boeres[1], Vinod E. F. Rebello[1] and David B. Skillicorn[2]

[1] Departamento de Ciência da Computação
Universidade Federal Fluminense (UFF), Niterói, RJ, Brazil
{boeres,vefr}@pgcc.uff.br
[2] Computing and Information Science
Queen's University, Kingston, Canada
skill@qucis.queensu.ca

Abstract. This paper investigates the effect of communication overhead on the scheduling problem. We present a scheduling algorithm, based on *LogP*-type models, for allocating task graphs to networks of processors. The makespans of schedules produced by our *multi-stage scheduling approach* (MSA) are compared with other well-known scheduling heuristics. The results indicate that new classes of scheduling heuristics are required to generate efficient schedules for realistic abstractions of today's parallel computers. The scheduling strategy of MSA can also be used to generate BSP-structured programs from more abstract representations. The performance of such programs are compared with "conventional" versions.

1 Introduction

Programmers face daunting problems when attempting to achieve efficient execution of parallel programs on parallel computers. There is a considerable range in communication performance on the platforms currently in use. More importantly, the sources of performance penalties are varied, so that the best implementation of a program on two different architectures may differ widely in its form. Understandably, programmers wish to build programs at a level of abstraction where these variations can be ignored, and have the program automatically transformed to an appropriate form for each target.

One common abstraction is a directed acyclic graph whose nodes denote *tasks*, and whose edges denote data dependencies (and hence communication, if the tasks are mapped to distinct processors). There are two related issues in transforming this DAG: clustering the tasks that will share each processor, and arranging the communication actions of the program to optimally use the communication substrate of the target architecture.

Clustering tasks reduces communication costs, but may also reduce the available parallelism. The objective, therefore, is to find the optimal task granularity with respect to the communication characteristics of the target machine. In task clustering without replication, tasks are partitioned into disjoint sets or clusters and exactly one copy of each task is scheduled. In task clustering with duplication, a task may have several copies belonging to different clusters, each of which is scheduled independently. In architectures with high communication costs, this often produces schedules with smaller total execution times [12]. The task scheduling (or clustering) problem, with or without task duplication, has been well studied and is known to be NP-hard [15].

A more complicated issue is the choice of a model of the target's communication system. The standard model in the scheduling community is the *delay* model, where the sole architectural parameter for the communication system is the latency, that is the transit time for each message [15]. A scheduling technique based on latency is unlikely to produce good schedules on most of today's popular parallel computers because it assumes a fully-connected network. In practice, latency depends heavily on congestion in the shared communication paths. Furthermore, the dominant cost of communication in today's architectures is that of crossing the network boundary, a cost that arises partly from the protocol stack and partly from interaction with other messages leaving and entering the processor at about the same time. In practice, the latter dominates the communication cost [10] and cannot be modelled as a latency.

The delay model also assumes that communication and computation can be completely overlapped. This may not be possible because: the CPU may have to manage, or at least initiate communication (even when the details are handled by a network interface card); the CPU may have to copy data from user space to kernel space; and it may be desirable to postpone some communication to allow messages to be packed together, or routed in a congestion-free way [10].

These architectural properties have motivated new parallel programming models such as LogP [6] and BSP [16], both of which have accurate cost models. Given a program, it is possible to accurately predict its runtime given a few program and architecture parameters. However, both models require writing programs in a slightly awkward way, and so it is useful to consider using a scheduling technique to map more general programs into the required structured form.

The *LogP model* [6] is an MIMD message-passing model with four architectural parameters: the latency, L; the overhead, o, the CPU penalty incurred by a communication action; the gap, g, a lower bound on the interval between two successive communication actions, designed to prevent overrun of the communication capacity of the network (and hence related to the inverse of the network's bisection bandwidth); and the number of processors, P. Because the parameter g depends on the target architecture, it is hard to write $LogP$ programs in a general and architecture-independent way. Nevertheless, $LogP$-based predictions of runtimes have been confirmed in practice [6,7].

BSP is similar, except that it takes a less prescriptive view of communication actions. Programs are divided into supersteps which consist of three sequential phases: local computation, communication between processors, and a barrier synchronisation, after which communicated data becomes visible on destination processors. The architecture parameters are p, the number of processors, g the inverse of the effective network bandwidth, and l the time taken for a barrier synchronisation. The g parameter is significantly different from that of the $LogP$ model despite their superficial similarity. BSP does not limit the rate at which data are inserted into the network. Instead, g is measured by inserting traffic to uniformly-random destinations as fast as possible and measuring how long it takes to be delivered in the steady state (note that this includes an accounting for latency). In a superstep where each processor spends time w_i in computation, and sends and receives some quantity of data h_i, the BSP cost model charges

$$MAXw_i + MAXh_ig + l$$

for the superstep. The units of g are instruction execution times per word, and of l are instruction execution times, so that the sum is sensible. This cost model has been shown to be extremely accurate [8,9]. Note that the second term accounts for the fact that the cost of communication is dominated by fan-in and fan-out, while the inability of many architectures to overlap computation and communication is taken into account by summing the first two terms, rather than taking their maximum.

To the best of our knowledge, few if any, scheduling algorithms exist for communication models which consider such parameters as those discussed above. Löwe *et al.* [13] proposed a clustering algorithm with task replication for the $LogP$ model, but it only operates on restricted types of algorithms. ETF [11] and DSC [17] which do not use task replication, and PY [15], PLW [14] which do, are well-known and effective scheduling algorithms based on the delay model. When applied to parallel computers, delay model approaches implicitly assume two properties: the ability to *multicast*, i.e. to send a number of different messages to different destinations simultaneously; and that the communication overhead and task computation on a processor can completely overlap, i.e. communication does not require computation time. Neither assumption is particularly realistic, as discussed earlier.

In this paper, we briefly present a versatile *multi-stage scheduling approach* (MSA) for general DAGs on $LogP$-type machines (with bounded numbers of processors) which can also generate BSP-structured schedules. MSA considers the following parameters, adopted by the *Cost and Latency Augmented DAG* (CLAUD) model [5] (developed independently of $LogP$). In the CLAUD model, a DAG G represents the application and $\mathcal{P}$ is a set of m identical processors. The overhead associated with the *sending* and *receiving* of a message is denoted by λ_s and λ_r, respectively, and the communication latency or delay τ. The CLAUD model very naturally models $LogP$, given the following parameter relationships: $L = \tau$ ($LogP$ only considers fully connected networks), $o = \frac{\lambda_s + \lambda_r}{2}$, $g = \lambda_s$ and $P = m$, and BSP with $h = \lambda_s + \lambda_r$. (Note the delay model is an instance of these models).

2 A Multi-stage Scheduling Approach

The Multi-stage Scheduling Approach (MSA) currently consists of two main stages: the first transforms the input program DAG into one whose characteristics (e.g. granularity, message lengths) are better suited to the target architecture; and the second stage maps this resulting DAG onto the target architecture. Each stage consists of two algorithms which are described in greater detail in [1, 3].

A parallel application is represented by a directed acyclic graph (DAG) $G = (V, E, \varepsilon, \omega)$, where V is the set of tasks, E is the precedence relation among them, $\varepsilon(v_i)$ is the execution cost of task $v_i \in V$ and $\omega(v_i, v_j)$ is the weight associated to the edge $(v_i, v_j) \in E$ representing the amount of data units transmitted from v_i to v_j. Presently, MSA only considers input DAGs whose vertices have unit execution costs and edges have unit weight. For the duration of each communication overhead (known here as a *communication event*), the processor is effectively *blocked* unable to execute other tasks in V or even to send or receive other messages. Therefore, these sending and receiving overheads can also be viewed as "tasks" executed by processors.

In parallel architectures with high overheads, the number of communication events can be reduced by bundling many small messages into fewer, larger ones. This can be achieved by clustering together tasks with common ancestors or successors and restricting the necessary *communication events* of the cluster to take place only before and after the execution of all of the tasks belonging to that cluster. This allows messages for the same destination processor, from various tasks within the cluster, to be combined and sent as one. In contrast, other clustering techniques (for example [17]) allow communication to occur as soon as each task within the cluster finishes its execution. This leads to a subtle difference between these clustering approaches: the latter only reduces the total communication cost through the elimination of messages, by allocating the source and destination tasks to the same processor; the former, tries to minimise the cost of sending the remaining messages as well.

The First Stage. This stage attempts to transform the input DAG G into one which will have better communication behaviour on the target architecture. The formation of *clusters*, each containing a task v_i of G and *replicated* copies of some of its ancestors, minimises communication in 3 ways: no cost is incurred for communication within the cluster; the bundling of messages to a common destination reduces the number of times that overheads are incurred; and the execution of the replicated tasks within the cluster can hide some of the communication cost between v_i and ancestors outside the cluster. The degree of replication, determined by characteristics of both the input DAG and the target architecture, is represented by a single parameter γ, the *clustering factor*. Determining the best value of γ, i.e. the value that leads to the smallest makespan is discussed elsewhere [1,2]. This clustering process is based on the *replication principle* (also adopted by PY [15]) and on the *latest-finish time* of a task i.e. the latest time that a cluster can finish and still have the schedule achieve the minimum makespan assuming all communication costs are γ.

The result of this *replication* algorithm is a set of n clusters of tasks (n being the number of tasks in G). Every cluster contains $\gamma + 1$ tasks except those in the first band which may have fewer. Furthermore, the latest finish time of every cluster is a multiple of $\gamma + 1$. Therefore, groups of parallel clusters exist in non-overlapping *bands* or *computation intervals*. Later, gaps will be created, known as *communication intervals*, between adjacent bands in which only communication event tasks will be scheduled.

For target architectures where the *overheads* of communication are much higher than the delay, minimising the in- and out-degrees of the clusters is essential for effective scheduling. When overheads are negligible and communication cost is dominated by the delay, it is better to minimise the length of messages. A second algorithm is now applied to specify the appropriate dependencies between required clusters taking into consideration the communication characteristics of the target architecture.

The problem of reducing the communication costs of the schedule has now been simplified to iteratively considering the communication costs between at most three adjacent computation intervals [1]. For each computation band, this involves finding the combination of immediate ancestor clusters (containing the required immediate ancestor tasks of all tasks in the the current band), which will incur smallest communication cost. The exact solution is computationally expensive and instead a heuristic approach has been proposed [1, 3].

The Second Stage. This stage performs a *cluster merging* step to map each cluster's distinct virtual processor to a physical one, minimising communication costs by applying Brent's Principle [4]. Communication is not permitted during execution of a cluster. When a relatively small overhead cost is associated with sending and receiving events, the bundling of messages can actually increase the makespan. In such cases, the edges of the final *DAG* can be *unbundled* and the clusters *opened* to allow messages to be sent after each task is executed.

Once the supertasks have been allocated to the physical processors, excess copies of tasks allocated to the same processor are removed and the final schedule is defined for the new *DAG* based on the parameters λ_s, λ_r and τ. The required communication event tasks can now be allocated and the start times of all tasks estimated.

3 The Structure of BSP Programs

To write efficient BSP programs a good strategy is to balance the computation across the processes; balance the communication fan-in and fan-out; and minimise the number of supersteps, since this determines the number of times l is incurred. This problem is similar to the problem of clustering for task scheduling in the sense that the number of supersteps can be reduced by increasing the amount of local computation carried out in a process. However, this also reduces the amount of parallelism and thus, just as the scheduling problem tries to balance parallelism against communication, here parallelism has to be balanced against communication and synchronisation.

From the previous discussion of MSA, it is clear that there are similarities in the structure of efficient programs for both BSP and *LogP*-based models. BSP's notion of local computation is most accurately captured by the cluster definition of MSA since communication only occurs after all of the tasks within the cluster have been executed.

The cost of the superstep computation, w_i, in BSP is equivalent to the computation band of constant cost $\gamma + 1$ in MSA. Since each cluster in the band has the same size, balanced computation is achieved through task replication – a novel feature for BSP programs. The communication and synchronisation costs are represented by the communication interval in MSA. The exploitation of these similarities suggests the following. An approach similar to that adopted by MSA can be used to generate efficient BSP versions of MIMD programs automatically. The efficient execution of programs depends on their schedule which, in turn, is influenced by architectural parameters. Any change of target architecture is likely to require a change to the program structure i.e. the superstep size.

BSP emphasises architecture-independence and portability of programs across architectures. If MSA is able to generate effective program schedules whose costs are known to programmers, many of the benefits of the BSP approach can be maintained in a more general programming setting. Further, if MSA generates schedules with good execution times for general-purpose programs on general parallel machines then this implies that the superstep discipline is not too restrictive. This might be expected to increase the use and acceptance of BSP, particularly for irregular problems where superstep structure is sometimes seen as overly restrictive.

4   The Influence of the Overheads on Task Scheduling

The impact of $LogP$-type parameters on the scheduling of tasks can be analysed by comparing of makespans generated by various scheduling approaches for a suite of (in this case, unit execution task unit data communication) DAGs over a range of values for communication overheads and latency.

Table 1 presents three scheduling approaches with a selection of DAGs which include: an irregular DAG based on a molecular physics application (Ir_{41}); two random graphs with 80 and 186 tasks (Ra_{80}, Ra_{186}); a binary in- and out-tree of 511 tasks (It_{511}, Ot_{511}); and a diamond DAG (Di_{400}). The application of MSA to a larger variety of DAGs with differing sizes and characteristics (including: course- and fine-grain graphs; regular DAGs such as binary trees and diamond DAGs; and irregular and random graphs) is presented in [1–3]. Table 1 also presents the makespans of MSA schedules generated under the BSP model in order to investigate the additional costs incurred by the superstep discipline.

For the $LogP$-type architectures, MSA is compared with two recently published scheduling algorithms (DSC [17] and PLW [14]) both of which have been proved to generate schedules within a factor of two of the optimal under the delay model. While the delay model-based PLW ignores the overhead parameters (o or λ_s and λ_r), DSC is in fact based on an extension of this model (the linear model) and treats the overhead as an additional latency parameter.

One objective of this paper is to show that for architectures with *LogP*-type characteristics, even scheduling strategies which simply add the overhead parameter to the latency do not perform well. Furthermore, in practice, the execution time of the schedule produced can be much *worse* than predicted.

The actual execution times (ATs) of the schedules produced by PLW, DSC and MSA were measured on a discrete event driven simulation of a parallel machine. The main reason for adopting a simulator is to facilitate the study of the relationship between the various parameters and the effects of hardware and operating system techniques to improve message-passing performance.

Table 1. A comparison of predicted and simulated makespans for the DSC, PLW and MSA scheduling heuristics. λ_s and λ_r are the sending and receiving overheads, respectively, and τ is the latency. The number of processors required by the schedule is represented by Prs, l is the hardware cost of barrier synchronisation and γ the clustering factor.

DAG	λ_s	λ_r	τ	DSC Prs	PT	AT	PLW Prs	PT	AT	MSA γ	Prs	PT	AT	$MSA_{BSP}(l=0)$ γ	Prs	PT	AT	$MSA_{BSP}(l=2)$ γ	Prs	PT	AT
Ir_{41}	0	0	1	13	16	16	22	16	16	3	10	16	16	3	10	21	21	3	10	21	27
Ir_{41}	0	0	2	9	21	21	20	22	22	3	10	19	18	2	8	24	26	6	10	27	31
Ir_{41}	0	0	4	8	31	31	19	27	27	6	10	24	23	8	10	31	31	15	7	35	33
Ir_{41}	0	0	8	3	44	42	19	39	39	12	5	28	27	15	7	39	34	15	7	39	36
Ir_{41}	1	1	1	7	26	44	22	16	36	2	8	30	30	6	10	33	31	12	5	33	33
Ir_{41}	1	1	2	5	31	41	20	22	43	6	10	33	32	6	10	39	32	12	5	36	35
Ir_{41}	1	1	4	8	40	52	19	27	40	2	3	37	33	8	10	41	37	14	5	42	39
Ir_{41}	1	1	8	4	52	57	19	39	60	12	5	36	36	15	7	45	39	15	7	45	41
Ir_{41}	2	0	1	7	26	55	22	16	44	8	10	25	25	6	10	29	28	6	10	29	32
Ir_{41}	2	0	2	5	31	43	20	22	47	11	6	25	25	8	10	31	31	15	7	33	33
Ir_{41}	2	0	4	8	40	62	19	27	39	11	6	27	26	15	7	35	32	15	7	35	34
Ir_{41}	2	0	8	4	52	63	19	39	64	11	6	31	30	15	7	39	36	15	7	39	38
Ir_{41}	2	1	1	5	31	47	22	16	49	6	10	33	33	12	5	34	32	12	5	34	34
Ir_{41}	2	1	2	5	31	50	20	22	56	8	10	34	34	12	4	37	34	12	4	37	36
Ir_{41}	2	1	4	5	38	57	19	27	45	11	6	34	34	15	7	41	37	15	7	41	39
Ir_{41}	2	1	8	2	43	58	19	39	72	11	6	38	37	15	7	45	40	15	7	45	42
Ir_{41}	2	2	1	5	31	55	22	16	55	11	6	43	35	12	5	38	35	12	5	38	37
Ir_{41}	2	2	2	8	40	67	20	22	66	11	6	43	36	12	5	41	37	12	5	41	39
Ir_{41}	4	0	1	5	31	73	22	16	72	11	6	28	27	15	7	32	31	15	7	32	33
Ir_{41}	4	0	2	8	40	90	20	22	77	11	6	29	28	15	7	33	33	15	7	33	34
Ir_{41}	4	0	4	3	44	68	19	27	58	11	6	31	30	15	7	35	34	15	7	35	35
Ir_{41}	4	0	8	1	41	41	19	39	90	11	6	35	34	15	7	39	38	15	7	39	40
Ir_{41}	4	1	1	8	40	91	22	16	77	11	6	34	34	12	5	36	34	12	5	36	36
Ir_{41}	4	1	2	3	38	85	20	22	83	11	6	34	34	12	5	39	36	12	5	39	38
Ir_{41}	4	1	4	5	43	75	19	27	60	12	5	35	35	15	7	41	38	15	7	41	40
Ir_{41}	4	4	1	5	43	82	22	16	93	1	1	41	41	1	1	41	41	1	1	41	41
Ra_{80}	0	0	4	27	25	25	41	31	31	6	48	23	22	6	48	45	45	6	48	49	49
Ra_{80}	4	2	2	27	40	193	49	21	145	14	25	76	56	14	25	76	56	14	25	76	59
Ra_{186}	0	0	4	66	20	22	87	19	19	4	52	21	19	2	90	28	32	2	90	34	38
Ra_{186}	4	2	2	64	31	112	111	17	93	11	78	69	47	11	78	69	48	11	78	69	50
It_{511}	0	0	4	98	37	39	167	27	27	1	85	25	25	2	64	28	28	6	64	33	33
It_{511}	4	2	2	95	61	151	341	21	67	1	85	45	45	2	64	52	52	2	64	60	60
Ot_{511}	0	0	4	146	25	27	264	13	13	8	256	9	9	8	256	9	9	8	256	9	9
Ot_{511}	4	2	2	121	43	171	308	13	81	8	256	9	9	8	256	9	9	8	256	9	9
Di_{400}	0	0	4	27	146	146	165	140	140	6	37	121	116	2	29	136	136	2	29	174	174
Di_{400}	4	2	2	24	236	317	101	97	397	29	67	338	260	28	32	341	277	23	34	314	296

Under the delay model (where $\lambda_s = \lambda_r = 0$), the performance of both DSC and PLW for the *DAG* Ir_{41} progressively worsens in comparison with MSA as the *DAG* effectively becomes more fine-grained due to the increasing latency value. Considering the other graphs under this model, one can argue that the improvements in the makespans for PLW over DSC are due mostly to the benefits of task replication which very much depend on the graph structure (e.g. compare Ra_{80} and Ot_{511}). Although MSA also employs replication, it generally utilses fewer processors to greater effect. As expected under this model, the three approaches predict the execution times of their schedules quite accurately.

Under *LogP*-type models (i.e. when the overheads are nonzero), DSC and PLW performs poorly on two counts: the ATs of the generated makespans in many cases are worse than the single processor execution time; and the predictions are not very accurate. In the case of DSC, this is partly because the overhead cost is added to the latency, thus specifying that communication events can be overlapped with task computation. PLW produces the same schedule under both the *LogP* and delay models, and thus even poorer predictions (PTs), because it ignores overheads all together. While adopting the linear model for PLW may improve matters with respect to the ATs, both PLW and DSC still apply the fundamental assumptions of the delay model which no longer hold on *LogP*-type machines, e.g. because of the multicasting assumption they tend to create clusters with large out-degrees.

The predicted makespans (PTs) of the schedules produced by MSA are, on the whole, quite reasonable although there is scope for improvement. Compared to the predictions for the BSP model, those for the *LogP* model are in general less accurate. The reason is that it is difficult to predict the communication behaviour since, for example, the order in which messages are sent from a cluster is not defined, or atleast not under the control of the schedule. This effect, hidden by the barrier synchronisations, is more pronounced when strict banding is relaxed.

Under BSP, a barrier synchronisation occurs after each pair of computation and communication intervals. In the simulation, the parameter l only models the "system" cost of signalling that all processors have reached the barrier. For different graphs, the reason why the ATs of the *LogP* schedules are in some cases much better than those of the BSP ones (especially under delay model conditions) is because MSA considered it better to uncluster the supertasks. For the BSP model this is not possible, so MSA attempts to counter this effect by usually increasing γ (compare the values used by MSA with those of the two MSA$_{BSP}$ schedules for the graph Ir_{41}). A larger γ may sometimes lead to fewer bands with fewer, but larger, smaller in- or out-degree clusters. In the entries where γ and the number of processors are the same for the MSA and MSA$_{BSP}$ schedules, the difference in ATs represents the total cost for the barriers. On the whole, it seems that the MSA scheduling approach does manage to produce reasonable *DAG* transformations (and thus schedules) for the BSP model particularly under *LogP* model conditions. Note also that the ATs of the BSP schedules are better than those of DSC and PLW when the delay is large and particularly when the overheads are nonzero.

5 Conclusions

A scheduling strategy has been proposed to produce good schedules for a range of parallel computers, including those in which the overheads due to sending and receiving are relatively high compared to the latency. When these *communication events* (the sending and receiving overheads) are considered, communication is not totally overlapped with computation of tasks as assumed in a number of scheduling heuristics. In the case where overheads are significant, task replication is exploited to decrease the number of communication events. In an attempt to decrease these further, messages are bundled into fewer, larger ones. MSA effectively transforms the structure of the input *DAG* into one which is more amenable to allocation on the target architecture. For realistic models (*LogP* and *BSP*), initial results indicate that MSA generates better schedules than some well known conventional delay model-based approaches. This suggests that a new class of scheduling heuristics are required to generate efficient schedules for today's parallel computers.

The definition of *computational* and *communication* intervals facilitates both a more precise prediction of the number of *communication events*, especially when the bundling of messages is necessary, and the exploitation of parallelism. Note that the main objective when defining such intervals is not to specify synchronisation points but rather to easily identify potential parallel clusters for efficient allocation onto physical processors and for the scheduling of communication event tasks. In the final *LogP* schedule, these clearly defined intervals may no longer exist since unnecessary idle times will be removed. For the BSP model, however, these interval are maintained in the schedule. Under this model, although barrier synchronisation is an additional cost to be incurred by programs, MSA is able to reduce its impact by appropriately determining the sizes and the number of supersteps. Given that, for *LogP*-type machines, the makespans of BSP-structured schedules are not significantly worse than conventional ones and that their predictions relatively accurate, it seems that the superstep discipline is not too restrictive and that an approach to generating BSP-structured programs using a scheduling strategy such as MSA could be advantageous to programmers.

Acknowledgements

The authors would like thank Tao Yang and Jing-Chiou Liou for providing the DSC and PLW algorithms, respectively, and for their help and advice.

References

1. C. Boeres. *Versatile Communication Cost Modelling for Multicomputer Task Scheduling Heuristics*. PhD thesis, Department of Computer Science, University of Edinburgh, May 1997.

2. C. Boeres and V. E. F. Rebello. Versatile task scheduling of binary trees for realistic machines. In M. Griebl C. Lengauer and S. Gorlatch, editors, *Proceedings of the 3rd International Euro-Par Conference on Parallel Processing (Euro-Par'97)*, LNCS 1300, pages 913–921, Passau, Germany, August 1997. Springer-Verlag.

3. C. Boeres and V. E. F. Rebello. A versatile cost modelling approach for multicomputer task scheduling. Accepted for publication in *Parallel Computing*, 1998.

4. R. P. Brent. The parallel evaluation of general arithmetic expressions. *J. ACM*, 21:201–206, 1974.

5. G. Chochia, C. Boeres, M. Norman, and P. Thanisch. Analysis of multicomputer schedules in a cost and latency model of communication. In *Proceedings of the 3rd Workshop on Abstract Machine Models for Parallel and Distributed Computing*, Leeds, UK., April 1996. IOS press.

6. D. Culler *et al.* LogP: Towards a realistic model of parallel computation. In *Proceedings of the Fourth ACM SIGPLAN Symposium on Principles and Practice of Parallel Programming*, San Diego, CA, May 1993.

7. B. Di Martino and G. Ianello. Parallelization on non-simultaneous iterative methods for systems of linear equations. In *Parallel Processing (CONPAR'94-VAPP VI)*, LNCS 854, pages 254–264. Springer-Verlag, 1994.

8. J. M. D. Hill, P. I. Crumpton, and D. A. Burgess. The theory, practice, and a tool for BSP performance prediction. In *Proceedings of the 2nd International Euro-Par Conference on Parallel Processing (Euro-Par'96)*, volume 1124 of *LNCS*, pages 697–705. Springer-Verlag, August 1996.

9. J. M. D. Hill, S. Donaldson, and D. B. Skillicorn. Stability of communication performance in practice: from the Cray T3E to networks of workstations. Technical Report PRG-TR-33-97, Oxford University Computing Laboratory, October 1997.

10. J. M. D. Hill and D. B. Skillicorn. Lessons learned from implementing BSP. In *High-Performance Computing and Networks*, Springer Lecture Notes in Computer Science Vol. 1225, pages 762–771, April 1997.

11. J-J. Hwang, Y-C. Chow, F. D. Anger, and C-Y. Lee. Scheduling precedence graphs in systems with interprocessor communication times. *SIAM Journal of Computing*, 18(2):244–257, 1989.

12. H. Jung, L. Kirousis, and P. Spirakis. Lower bounds and efficient algorithms for multiprocessor scheduling of DAGs with communication delays. In *Proc. ACM Symposium on Parallel Algorithms and Architectures*, pages 254–264, 1989.

13. W. Lowe, W. Zimmermann, and J. Eisenbiergler. Optimization of parallel programs on machines with expensive communication. In L. Bouge, P. Fraigniaud, A. Mignotte, and Y. Robert, editors, *Proceedings of the 2nd International Euro-Par Conference on Parallel Processing (Euro-Par'96)*, LNCS 1124, pages 602–610, Lyon, France, August 1996. Springer-Verlag.

14. M. A. Palis, J-C. Liou, and D. S. L. Wei. Task clustering and scheduling for distributed memory parallel architectures. *IEEE Transactions on Parallel and Distributed Systems*, 7(1):46–55, January 1996.

15. C. H. Papadimitriou and M. Yannakakis. Towards an architecture-independent analysis of parallel algorithms. *SIAM Journal of Computing*, 19:322–328, 1990.

16. D. B. Skillicorn, J. M. D. Hill, and W. F. McColl. Questions and answers about BSP. *Scientific Programming*, 6(3):249–274, 1997.

17. T. Yang and A. Gerasoulis. DSC: Scheduling parallel tasks on an unbounded number of processors. *IEEE Transactions on Parallel and Distributed Systems*, 5(9):951–967, 1994.

Aspect Ratio for Mesh Partitioning

Ralf Diekmann[1], Robert Preis[1], Frank Schlimbach[2], and Chris Walshaw[2]

[1] University of Paderborn, Germany,
{diek@,preis@hni.}uni-paderborn.de
[2] University of Greenwich, London, UK,
{sf634,C.Walshaw}@gre.ac.uk

Abstract. This paper deals with the measure of *Aspect Ratio* for mesh partitioning and gives hints why, for certain solvers, the *Aspect Ratio* of partitions plays an important role. We define and rate different kinds of *Aspect Ratio*, present a new center-based partitioning method which optimizes this measure implicitly and rate several existing partitioning methods and tools under the criterion of *Aspect Ratio*.

1 Introduction

Almost all numerical scientific simulation codes belong to the class of *data-parallel applications*: their parallelization executes the same kind of operations on every processor but on different parts of the data. This requires the partitioning of the mesh into equal-sized subdomains as preprocessing step. Together with the additional constraint of minimizing the number of cut edges (i.e. minimizing the total interface length in the case of FE-mesh partitioning), the problem is NP-complete. Thus, a number of graph (mesh) partitioning heuristics have been developed in the past (e.g. [8]), and used in practical applications.

Most of the existing graph partitioning algorithms optimize the *balance* of subdomains plus the number of cut edges, the *cut size*. This is sufficient for many applications, because an equal balance is necessary for a good utilization of the processors of a parallel system and the *cut size* indicates the amount of data to be transfered between different steps of the algorithms. The efficiency of parallel versions of global iterative methods like (relaxed) Jacobi, Conjugate Gradient (CG) or Multigrid is (mainly) determined by these two measures. But if the decomposition is used to construct pre-conditioners (DD-PCG) [1, 2], cut and balance may not longer be the only efficiency determining factors. The *shape* of subdomains heavily influences the quality of pre-conditioning and, thus, the overall execution time [10]. First attempts at optimizing the *Aspect Ratio* of subdomains weigh elements depending on their distance from a subdomain's center (e.g. [5]) and include this weight into the cost function of local iterative search algorithms like the Kernighan-Lin heuristic [6].

In the following section we will define the *Aspect Ratio* of subdomains and give hints why it can improve the overall execution time of domain decomposition methods. Section 3 introduces a new mesh partitioning heuristic which implicitly optimizes the *Aspect Ratio* and Section 4 finishes with experimental results. Further information on this topic can be found in [4].

2 Aspect Ratios

An example that *cut size* is not always the right measure in mesh partitioning can be found in Fig. 1. The sample mesh is partitioned into two parts with different AR's and different cuts. A Poisson problem with homogeneous Dirichlet-0 boundary conditions is solved with DD-PCG. It can be observed that the number of iterations is determined by the Aspect Ratio and that the cut size (number of neighboring elements) would be the wrong measure.

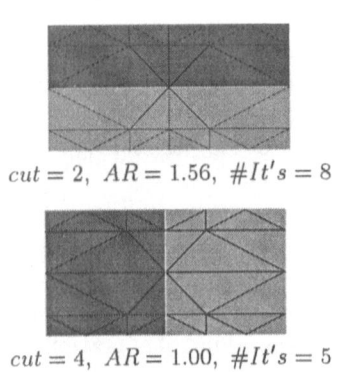

$cut = 2, \quad AR = 1.56, \quad \#It's = 8$

$cut = 4, \quad AR = 1.00, \quad \#It's = 5$

Fig. 1. *Aspect Ratio vs. Cut Size.*

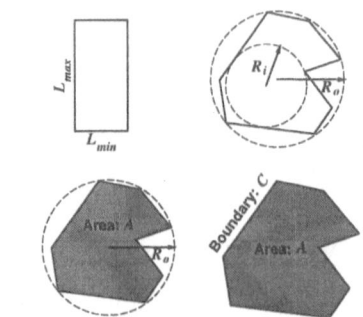

Fig. 2. Different definitions of *Aspect Ratio AR:* $\frac{L_{max}}{L_{min}}$, $\frac{R_o^2}{R_i^2}$, $\frac{R_o^2}{A}$ and $\frac{C^2}{16A}$.

Possible definitions of AR can be found in Fig. 2. The first two are motivated by common measures in triangular mesh generation where the quality of triangles are expressed in either $\frac{L_{max}}{L_{min}}$ (longest to shortest boundary edge) or $\frac{R_o^2}{R_i^2}$ (area of smallest circle containing the domain to area of largest inscribed circle). The definition of $AR = \frac{R_o^2}{R_i^2}$ expresses the fact that circles are perfect shapes. Unfortunately, the circles are quite expensive to find for arbitrary polygons: using Voronoi-diagrams, both can be determined in $O(2n \log n)$ steps where n is number of nodes of the polygon. $AR = \frac{R_o^2}{A}$ (A is the area of the domain) is another measure favoring circle-like shapes. It still requires the determination of the smallest outer circle but turns out to be better in practice. We can do a further step and approximate R_o by the length C of the boundary of the domain (which can be determined fast and updated incrementally in $O(1)$). The definition of $AR = \frac{C^2}{16A}$ additionally assumes that squares are perfect domains. Circles offer a better circumfence/area ratio but force neighboring domains to become non-convex (Fig. 3). For a sub-domain with area A and circumfence C $AR = \frac{C^2}{16A}$ is the ratio between the area of a square with circumfence C and A.

For irregular meshes and partitions, the first definition $AR = \frac{L_{max}}{L_{min}}$ does not express the *shape* properly. Fig. 4 shows an example. P_1 is perfectly shaped, but as the boundary towards P_4 is very short, $\frac{L_{max}}{L_{min}}$ is large. The circle-based

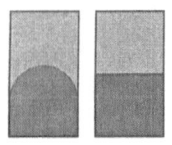

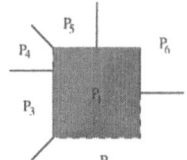

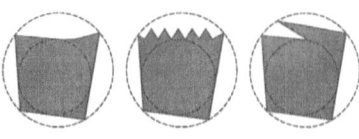

Fig. 3. Circles as perfect shapes?

Fig. 4. $\frac{L_{\max}}{L_{\min}}$?

Fig. 5. Problems of $AR := \frac{R_\circ^-}{R_i^2}$.

measures usually fail to rate complex boundaries like "zig-zags" or inscribed corners. Fig. 5 shows examples which have the same AR each but are very different in shape.

3 Mesh Partitioning

The task of mesh partitioning is to divide the set of elements of a mesh into a given number of parts. The usual optimization criteria are the *balance* and the *cut*, which is the number of crossing edges of the *element graph* (the element graph expresses the data-dependencies between elements in the mesh). The calculation of a balanced partition with minimum cut is NP-complete. We will consider methods included in the tools **JOSTLE** [11], **METIS** [7] and **PARTY** [9].

The *coordinate partitioning* (**COO**) method cuts the mesh by a straight line perpendicular to the axis of the longest elongation such that both parts have an equal number of elements, resulting in a stripe-wise partition. If used recursively (**COO_R**), it usually results in a more box-wise partition. There are many examples for *greedy* methods , in which one element after another is added to a subdomain. We will consider the variations in which the next element to be added is either chosen in a breadth-first manner (**BFS**) or as the one which increases the cut least of all (**CFS**).

The new method *Bubble* (**BUB**) is designed to create compact domains and the idea is to grow subdomains simultaneously from different *seed* elements. A partition is represented by a set of (initial random) seeds, one for each part. Starting from the seeds, the parts are grown in a breadth-first manner until all elements are assigned to a subdomain. Each subdomain then computes its center based on the graph distance measure, i.e. it determines the element which is "nearest" to all others in the subdomain. These center-elements are taken as new seeds and the process is started again. The iteration stops if the seeds do not move any more, or if an already evaluated configuration is visited again. In general, the method stops after few iterations and produces connected and very compact parts, i.e. the shape of the parts is naturally smooth. As drawback, the parts do not have to contain the same number of elements. Therefore we integrated a global load balancing step based on the diffusion method [3] as post-processing trying to optimize either the cut (**BUB+CUT**) or the Aspect Ratio (**BUB+AR**).

We also implemented the meta-heuristic *Simulated Annealing* (**SA**) for the shape optimization and used $AR = \frac{C^2}{16A}$ as optimization function. If the parameters are chosen carefully, SA is able to find very good solutions. Unfortunately, it also requires a very large number of steps.

4 Results

We compare the described approaches with respect to the number of global iterations, the cut size and the Aspect Ratio in Fig. 6 and 7. Methods 1-7 are included in PARTY and 8-11 are default settings of the tools. The partitions are listed with the method numbers with increasing number of iterations.

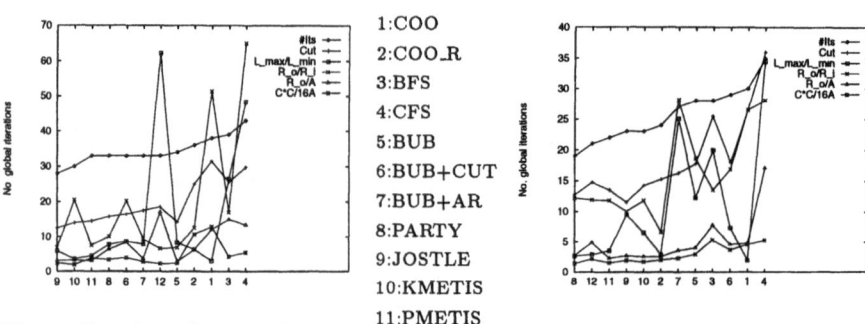

1:COO
2:COO_R
3:BFS
4:CFS
5:BUB
6:BUB+CUT
7:BUB+AR
8:PARTY
9:JOSTLE
10:KMETIS
11:PMETIS
12:SA

Fig. 6. Results of example *turm* (531 elements) into 8 parts.

Fig. 7. Results of example *cooler* (749 elements) into 8 parts.

The Aspect Ratios $\frac{L_{max}}{L_{min}}$ and $\frac{R_o}{R_i}$ do not, whereas the values of $\frac{R_o}{A}$ and $\frac{C^2}{16A}$ roughly follow the number of iterations. A low *cut* also leads to a fairly low number of iterations. The coordinate and greedy methods usually result in very high, whereas the bubble variations in adequate iteration numbers. Very low iteration numbers can be observed for the tools and Simulated Annealing.

References

1. S. Blazy, W. Borchers, and U. Dralle. Paralleliziation methods for a characteristic's pressure correction scheme. In *Flow Simulation with High-Perf. Comp. II*, 1995.
2. J.H. Bramble, J.E. Pasciac, and A.H. Schatz. The construction of preconditioners for elliptic problems by substructering i.+ii. *Math. Comp.*, 47+49, 1986+87.
3. G. Cybenko. Load balancing for distributed memory multiprocessors. *J. Par. Distr. Comp.*, 7:279–301, 1989.
4. R. Diekmann, F. Schlimbach, and C. Walshaw. Quality balancing for parallel adaptive fem. In *IRREGULAR*, LNCS. Springer, 1998.
5. N. Chrisochoides et.al. Automatic load balanced partitioning strategies for pde computations. In *Int. Conf. on Supercomp.*, pages 99–107, 1989.
6. C. Farhat, N. Maman, and G. Brown. Mesh partitioning for implicit computations via iterative domain decomposition. *Int. J. Num. Meth. Engrg.*, 38:989–1000, 1995.

7. G. Karypis and V. Kumar. A fast and high quality multilevel scheme for partitioning irregular graphs. Technical Report 95-035, University of Minnesota, 1995.

8. B.W. Kernighan and S. Lin. An effective heuristic procedure for partitioning graphs. *The Bell Systems Technical Journal*, pages 291–308, Feb 1970.

9. R. Preis and R. Diekmann. The party partitioning-library, user guide, version 1.1. Technical Report TR-RSFB-96-024, Universität–GH Paderborn, Sep 1996.

10. D. Vanderstraeten, R. Keunings, and C. Farhat. Beyond conventional mesh partitioning algorithms... In *SIAM Conf. on Par. Proc.*, pages 611–614, 1995.

11. C. Walshaw, M. Cross, and M.G. Everett. A localised algorithm for optimising unstructured mesh partitions. *Int. J. Supercomputer Appl.*, 9(4):280–295, 1995.

A Competitive Symmetrical Transfer Policy for Load Sharing*

Konstantinos Antonis, John Garofalakis, Paul Spirakis

[1] Computer Technology Institute, P.O. Box 1122, 26110 Patras, Greece
Phone: (+30)61-273496, Fax: (+30)61-222086
{antonis , garofala , spirakis}@cti.gr
[2] University of Patras, Department of Computer Engineering and Informatics
26500 Rion, Patras, Greece.

Abstract. Load Sharing is a policy used to improve the performance of distributed systems by transferring workload from heavily loaded nodes to lightly loaded ones in the system. In this paper, we propose a dynamic and symmetrical technique for a two-server system, called *Difference-Initiated (DI)*, in which transferring decisions are based on the difference between the populations of the two servers. In order to measure the performance of this policy, we apply the SSP analytical approximation technique proposed in [3]. Finally, we compare the theoretically derived results of the DI technique with two of the most commonly used dynamic techniques: the *Sender-Initiated (SI)*, and the *Receiver-Initiated (RI)* which were simulated.

1 Introduction

Transferring of jobs between two nodes lies in the heart of any distributed load sharing algorithm. Especially in the cases of *dynamic* or *adaptive* load sharing techniques (in which scheduling decisions are based on the current state of the system), the transferring of jobs from a heavily loaded node to a lightly loaded node, requires in most cases the mutual knowledge of the current state of the two nodes (server and receiver). Such dynamic load sharing techniques are the nearest neighbor approach, the sender-initiated, the receiver-initiated, etc.

In this work, we propose a *symmetrical transfer policy* for load sharing between two nodes, that uses a threshold on the difference of the populations of the respective queues, in order to make decisions for remote processing. We evaluate the performance of this policy, by applying the State Space Partition (SSP) supplementary queueing approximation technique [3], and also simulations. In the simulation model that we developed, we incorporated two of the most widely used dynamic load sharing policies, the *sender-initiated (SI)* and *receiver-initiated (RI)* [1], in order to compare their performance with our policy, which we call *difference-initiated (DI)*.

* This work was partially supported by ESPRIT LTR Project no. 20244 - ALCOM-IT basic research program of the E.U. and the Operational Program for Research and Development 2 No. 3.3 453 sponsored by the General Secreteriat for Research and Technology.

2 The Difference-Initiated Technique

2.1 The Model

We consider two serving queues as the processing nodes, which are subject to the following simple load sharing mechanism: when the difference of their queue lengths is greater than or equal to a threshold c, then jobs are transferred from the heavier loaded queue to the lightly loaded queue at a constant rate β (i.e. the transfer period is exponential of mean length $1/\beta$). The above algorithm behaves symmetrically, either as server-intitiated or receiver-initiated, depending on whether the difference between the two queue lengths exceeds or is under the corresponding threshold c. A server probes for the queue length of the other server, every time a job arrives to or departs from its own queue.

2.2 The State Space Partition (SSP) Technique

It is easy to observe that the behaviour of our system as time progresses, balances between 3 elementary queueing systems, QS_1, QS_2, and QS_3, when $|n_1 - n_2| < c$, $n_1 - n_2 \geq c$, and $n_2 - n_1 \geq c$, respectively. This technique computes the steady-state probability $P(n_1, n_2)$ for finding n_1 and n_2 customers at $queue_1$ and $queue_2$ respectively. The approximation technique conjectures that:

$$P(n_1, n_2) = A\ P_1(n_1, n_2) + B\ P_2(n_1, n_2) + C\ P_3(n_1, n_2) \tag{1}$$

where: $P_1(n_1, n_2)$, $P_2(n_1, n_2)$, $P_3(n_1, n_2)$ denote the steady-state probabilities for finding n_1 and n_2 customers at $queue_1$ and $queue_2$, for the three elementary, easily solvable systems above, with no constraints for $(n_1 - n_2)$, and

- $A = Prob\{\epsilon_1\}$, where $\epsilon_1 = \{(n_1, n_2) : |n_1 - n_2| < c\}$,
- $B = Prob\{\epsilon_2\}$, where $\epsilon_2 = \{(n_1, n_2) : n_1 - n_2 \geq c\}$,
- $C = Prob\{\epsilon_3\}$, where $\epsilon_3 = \{(n_1, n_2) : n_2 - n_1 \geq c\}$.

In applying equation 1 we are confronted with the problem of estimating A, B, and C. So, we use the following iteration technique:

Iteration Algorithm

- Solve QS_1, QS_2, QS_3 by any existing technique for Product Form Queueing Networks (Mean Value Analysis, LBANC, etc.) and get $P_i(n_1, n_2)$, $i = 1, 2, 3$.
- Assume initial values for A, B, C, namely $A^{(0)}$, $B^{(0)}$, $C^{(0)}$. Here we select: $A^{(0)} = P_1(\epsilon_1)$, $B^{(0)} = P_1(\epsilon_2)$, $C^{(0)} = P_1(\epsilon_3)$.
- $i = 1$
- $(\star)$ Compute the ith iteration value of $P(n_1, n_2)$ by equation 1 using $A^{(i-1)}$, $B^{(i-1)}$, $C^{(i-1)}$. Call it $P^{(i)}(n_1, n_2)$
- Find new values $A^{(i)}$, $B^{(i)}$, $C^{(i)}$ by using the following equation:

$$A^{(i)} = A^{(i-1)}P_1(\epsilon_1) + B^{(i-1)}P_2(\epsilon_1) + C^{(i-1)}P_3(\epsilon_1) \tag{2}$$

Similarly, for $B^{(i)}$ and $C^{(i)}$, summing over ϵ_2 and ϵ_3, respectively.
- $i = i + 1$
- If a convergence criterion for A, B, C is not satisfied goto $(\star)$.
- Use $P_{(i+1)}(n_1, n_2)$ as an approximation to $P(n_1, n_2)$.

3 Simulation Results and Comparisons

Three different transfer policies for load sharing between two FIFO servers have been analysed or simulated and compared: the difference-initiated (DI), the sender-initiated (SI), and the receiver-initiated (RI) technique. The analytical model of DI was described in section 2, while the models for the other two simulated techniques can be found in [1]. We mention here, that only one job can be in transmission any time for each direction. We assume Poisson arrival times, and exponential service and transfer times for all the above policies. The parameters used are the following: T_s = the threshold adopted for SI, T_r = the threshold adopted for RI, c = the threshold adopted for DI, λ_1, λ_2 = the arrival rates for the two servers, μ_1, μ_2 = the service rates for the two servers, β = the transfer rate.

In order to have comparative systems, corresponding to the three policies, we choose to have $T_s = c$ and $T_r = 0$ in all cases. We compare the three above transfer policies concerning the average response time for a job completion (the total time spent in the two server system by a single job). We have taken analytical (SSP method) or simulation results for the above performance measures for DI including: homogeneous and heterogeneous arrival times and variable β, for heavier and lighter workloads. The same results for SI and RI were taken by our simulation model.

3.1 Comparative Results for Homogeneous Arrival Streams

First, we consider the lightly loaded conditions. Our results have shown that RI has the worst average response time. On the ohter hand, DI behaves better, especially for large β (which means a low transfer cost, see fig. 1). As the arrival rate grows, the differences between the performance measures of the three techniques are decreased, but DI continues to behave a little better. The results for SI and RI are likely to be almost identical under heavily loaded conditions. We think that under more heavier conditions, RI should present better results compared to SI, confirming the results of [2]. DI performs better than the other two because of its symmetrical behaviour.

3.2 Comparative Results for Heterogeneous Arrival Streams

Considering the extremal case of fig. 2, we can see that DI continues to perform better than the other two techniques. DI balances the load between the two servers very well, and its results are very close to the results of the SI policy. On the other hand, RI presents very poor results, because of its nature, since it can not move a big number of jobs from the heavier to the lighter server. The explaination is that the other two policies can move jobs even when the lightly loaded server is not idle. According to the definition of RI and the threshold used in this case, a server has first to be idle, in order to receive a job from another server.

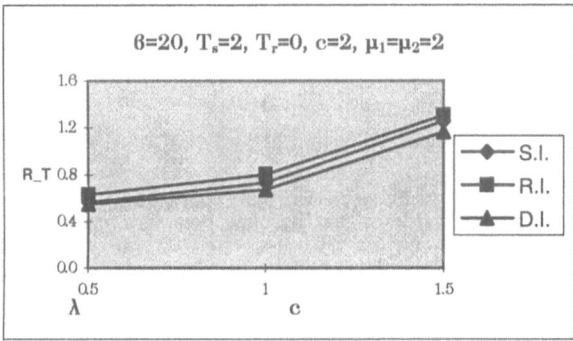

Fig. 1. Comparative results for the av. res. times (homogeneous arrival times, $\beta=20$).

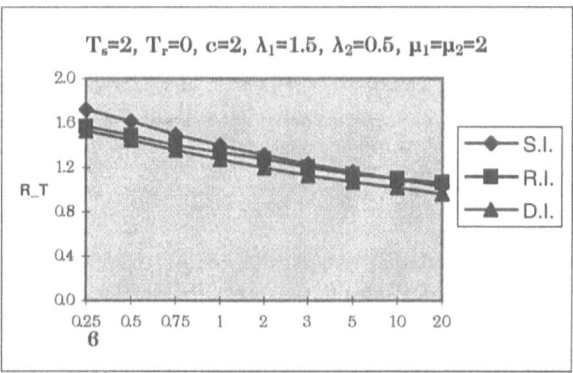

Fig. 2. Comparative results for the av. res. time (heterogeneous arrival times).

References

1. S. P. Dandamudi, "The Effect of Scheduling Discipline on Sender-Initiated and Receiver-Initiated Adaptive Load Sharing in Homogeneous Distributed Systems", *Technical Report, School of Computer Science, Carleton University*, TR-95-25 1995.
2. D. L. Eager, E. D. Lazowska, and J. Zahorjan, "A Comparison of Receiver-Initiated and Sender-Initiated Adaptive Load Sharing", *Performance Evaluation*, Vol. 6, March 1986, pp. 53-68.
3. J. Garofalakis, and P. Spirakis, "State Space Partition Modeling . A New Approximation Technique", *Computer Technology Institute, Technical Report*, TR. 88.09.56, Patras, 1988.

Scheduling Data–Parallel Computations on Heterogeneous and Time–Shared Environments

Salvatore Orlando[1] and Raffaele Perego[2]

[1] Dip. di Matematica Appl. ed Informatica, Università Ca' Foscari di Venezia, Italy
[2] Istituto CNUCE, Consiglio Nazionale delle Ricerche (CNR), Pisa, Italy

Abstract. This paper addresses the problem of load balancing data–parallel computations on heterogeneous and time-shared parallel computing environments, where load imbalance may be introduced by the different capacities of processors populating a computer, or by the sharing of the same computational resources among several users. To solve this problem we propose a run–time support for parallel loops based upon a hybrid (static + dynamic) scheduling strategy. The main features of our technique are the absence of centralization and synchronization points, the prefetching of work toward slower processors, and the overlapping of communication latencies with useful computation.

1 Introduction

It is widely held that distributed and parallel computing disciplines are converging. Advances in network technology have in fact strongly increased the network bandwidth of state–of–the–art distributed systems. A gap still exists with respect to communication latencies, but this difference too is now much less dramatic. Furthermore, most Massively Parallel Systems (MPSs) are now built around the same off–the–shelf superscalar processors that are used in high performance workstations, so that fine–grain parallelism is now exploited intra–processor rather than inter–processor. The convergence of parallel and distributed computing disciplines is also more evident if we consider the programming models and environments which dominate current parallel programming: MPI, PVM and High Performance Fortran (HPF) are now available for both MPSs and NOWs.

This trend has a direct impact on the research carried out on the two converging disciplines. For example, parallel computing research has to deal with problems introduced by heterogeneity in parallel systems. This is a typical feature of distributed systems, but nowadays heterogeneous NOWs are increasingly used as parallel computing resources [2], while MPSs may be populated with different off–the–shelf processors (e.g. a SGI/Cray T3E system at the time of this writing may be populated with 300 or 450 MHz DEC Alpha processors). Furthermore, some MPSs, which were normally used as batch machines in *space sharing*, may now be concurrently used by different users as *time shared* multiprogrammed environments (e.g. an IBM SP2 system can be configured in a way that allows users to specify whether the processors must be acquired as shared or dedicated resources). This paper focuses on one of the main problems programmers thus have to deal with on both parallel and distributed systems: the problem of "system" load imbalance due to heterogeneity and time sharing of resources. Here we restrict our view to data–parallel computations expressed by means of parallel loops.

In previous works we considered imbalances introduced by non–uniform data–parallel computations to be run on homogeneous, distributed memory, MIMD parallel systems [11,9,10]. We devised a novel compiling technique for parallel loops and a related run-time support (SUPPLE). This paper shows how SUPPLE can also be utilized to implement loops in all the cases where load imbalance is not a characteristic of the user code, but is caused by variations in capacities of processing nodes. Note that, much other research has been conducted in the field of run-time supports and compilation methods for irregular problems [13,12,7]. In our opinion, besides SUPPLE, many of these techniques can be also adopted when load imbalance derives from the use of a time–shared or heterogeneous parallel system. These techiques should also be compared with those specifically devised to face load imbalance in NOW environments [4, 8].

SUPPLE is based upon a hybrid scheduling strategy, which dynamically adjusts the workloads in the presence of variations of processor capacities. The main features of SUPPLE are the efficient implementation of regular stencil communications, the hybrid (static + dynamic) scheduling of chunks of iterations, and the exploitation of aggressive chunk prefetching to reduce waiting times by overlapping communication with useful computation. We report performance results of many experiments carried out on an SGI/Cray T3E and an IBM SP2 system. The synthetic benchmarks used for the experiments allowed us to model different situations by varying a few important parameters such as the computational grain and the capacity of each processor. The results obtained suggest that, in the absence of a priori knowledge about the relative capacities of the processors that will actually execute the program, the hybrid strategy adopted in SUPPLE yields very good performance.

The paper is organized as follows. Section 2 presents the synthetic benchmarks and the machines used for the experiments. Section 3 describes our run–time support and its load balancing strategy. The experimental results are reported and discussed in Section 4, and, finally, the conclusions are drawn in Section 5.

2 Benchmarks

We adopted a very simple benchmark program that resembles a very common pattern of parallelism (e.g. solvers for differential equations, simulations of physical phenomena, and image processing applications). The pattern is *data-parallel* and "regular", and thus considered easy to implement on homogeneous and dedicated parallel systems. In the benchmark a bidimensional array is updated iteratively on the basis of the old values of its elements, while array data referenced are modeled by a *five-point stencil*. The simple HPF code illustrating the benchmark is shown below.

```
      REAL A(N1,N2), B(N1,N2)
!HPF$ TEMPLATE D(N1,N2)
!HPF$ DISTRIBUTE D(BLOCK,BLOCK)
!HPF$ ALIGN A(i,j), B(i,j) WITH D(i,j)
      .........
      DO k= 1,N_ITER
!HPF$   INDEPENDENT
```

```
FORALL (i= 2:N1-1,  j= 2:N2-1)
  B(i,j) = Comp(A(i,j), A(i-1,j), A(i,j-1), A(i+1,j), A(i,j+1))
END FORALL
A = B
END DO
.........
```

Note the BLOCK distribution to exploit data locality by reducing off local-memory data references. The actual computation performed by the benchmark above is hidden by the function Comp(). We thus prepared several versions of the same benchmark where Comp() was replaced with a dummy computation characterized by known and fixed costs. Moreover, since it is important to observe the performance of our loop support when we increase the number of processors, for each different grid P of processors we modified the dimensions of the data set to keep the size of the block of data allocated to each processor constant. Finally, another feature that we changed during the tests is N_ITER, the number of iterations of the external sequential loop. This was done to simulate the behavior of real applications such as solvers of differential equations, which require the same parallel loops to be executed many times, and image filtering applications, which usually perform the update of the input image in just one step.

3   The SUPPLE Approach

SUPPLE (SUPport for Parallel Loop Execution) is a portable run-time support for parallel loops [9, 11, 10]. It is written in C with calls to the MPI library.

The main innovative feature of SUPPLE [9] is its ability to allow data and computation to be dynamically migrated, without losing the ability to exploit all the static optimizations that can be adopted to efficiently implement stencil data references.

Stencil implementation is straightforward, due to the regularity of the *blocking* data layout adopted. For each array involved SUPPLE allocates to each processor enough memory to host the block partition, logically subdivided into an *inner* and a *perimeter* region, and a surrounding *ghost* region. The *ghost* region is used to buffer the parts of the *perimeter* regions of the adjacent partitions that are owned by neighboring processors, and are accessed through non local references. The *inner* region contains data elements that can be computed without fetching external data, while to compute data elements belonging to the *perimeter* region these external data have to be waited for. Loop iterations are statically assigned to each processor according to the *owner computes rule*, but, to overlap communications with useful computations, iterations are reordered [3]: the iterations that assign data items belonging to the *inner* region (which refer local data only) are scheduled between the asynchronous sending of the *perimeter* region to neighboring processors and the receiving of the corresponding data into the *ghost* region. This static scheduling may be changed at run-time by migrating iterations, but, in order to avoid the introduction of irregularities in the implementation of stencil data reference, only iterations updating the *inner* region can be migrated. We group these iterations into *chunks* of fixed size g, by statically *tiling* the inner region. SUPPLE migrates chunks and associated data tiles instead of single iterations.

At the beginning, to reduce overheads, each processor statically executes its chunks, which are stored in a queue Q, hereafter *local* queue. Once a processor understands that its local queue is *becoming empty*, it autonomously decides to start the dynamic part of its scheduling policy. It tries to balance the processor workloads by asking overloaded partners for migrating both chunks and corresponding data tiles. Note that, due to stencil data references, to allow the remote execution of a chunk, the associated tile must be accompanied by a surrounding area, whose size depends on the specific stencil features. Migrated chunks and data tiles are stored by each receiving processor in a queue *RQ*, called *remote*. Since load balancing is started by underloaded processors, our technique can be classified as *receiver initiated* [6, 14]. In the following we detail our hybrid scheduling algorithm.

During the initial *static* phase, each processor only executes local chunks in *Q* and measures their computational cost. Note that, since the possible load imbalance may only derive from different "speeds" of the processors involved, chunks that will possibly be migrated and stored in *RQ* will be considered as having the same cost as the ones stored in *Q*. Thus, on the basis of the knowledge of the chunk costs, each processor estimates its *current load* by simply inspecting the size of its queues *Q* and *RQ*.
When the estimated local load becomes lower than a machine-dependent *Threshold*, each processor autonomously starts the *dynamic* part of the scheduling technique and starts asking other processors for remote chunks. Correspondingly, a processor p_j, which receives a migration request from a processor p_i, will grant the request by moving some of its workload to p_j only if its *current load* is higher than *Threshold*. To reduce the overheads which might derive from requests for remote chunks which cannot be served, each processor, when its *current load* becomes lower than *Threshold*, broadcasts a so-called *termination message*. Therefore the *round-robin* strategy used by underloaded processors to choose a partner to be asked for further work skips terminated processors. Once an overloaded processor decides to grant a migration request, it must choose the most appropriate number of chunks to be migrated. To this end, SUPPLE uses a modified *Factoring* scheme [5], which is a *Self Scheduling* heuristics formerly proposed to address the efficient implementation of parallel loops on shared–memory multiprocessors
Finally, the policies exploited by SUPPLE to manage data coherence and termination detection are also fully distributed and asynchronous. A *full/empty-like* technique [1] is used to asynchronously manage the coherence of migrated data tiles. When processor p_i sends a chunk b to p_j, it sets a flag marking the data tiles associated with b as invalid. The next time p_i needs to access the same tiles, p_i checks the flag and, if the flag is still set, waits for the updated data tiles from node p_j. As far as termination detection is concerned, the role of a processor in the parallel loop execution finishes when it has already received a termination message from all the other processors, and both its queues *Q* and *RQ* are empty.

In summary, unlike other proposals [12, 4], the dynamic scheduling policy of SUPPLE is fully distributed and based upon local knowledge about the local workload, and thus there is no need to synchronize the processors in order to exchange updated information about the global workload. Moreover, SUPPLE may also be employed for applications composed of a single parallel loop, such as filters for image processing. Unlike other proposals [13, 8], it does not exploit past knowledge about the work-

360

load distribution at previous loop iterations, since dynamic scheduling decisions are asynchronously taken concurrently with useful computation.

4 Experimental Results

All experiments were conducted on an IBM SP2 system and an SGI/Cray T3E. Note that both machines might be heterogeneous, since both can be equipped with processors of different capacities. The T3E can in fact be populated with DEC Alpha processors of different generations[1], while IBM enables choices from three distinct types of nodes – high, wide or thin – where the differences are in the number of per–node processors, the type of processors, the clock rates, the type and size of caches, and the size of the main memory. However, we used the SP2 (a 16 node system equipped with 66 MHz POWER 2 wide processors) as a *homogeneous time-shared environment*. To simulate load imbalance we simply launched some compute-bound processes on a subset of nodes. On the other hand, we used the T3E (a system composed of 64 300 MHz - DEC Alpha 21164 processors) as a space-shared heterogeneous system. Since all the nodes of our system are identical, we had to simulate the presence of processors of different speeds by introducing an extra cost in the computation performed by those processors considered "slow". Thus, if the granularity of Comp() is μ μsec (including the time to read and write the data) on the "fast" processors, and F is the *factor of slowdown* of the "slow" ones, the granularity of Comp() on the "slow" processors is $(F \cdot \mu)$ μsec. To prove the effectiveness of SUPPLE, we compared each SUPPLE implementation of e benchmark with an equivalent static and optimized implementation, like the one ploited by a very efficient HPF compiler.

We present several curves, all plotting an *execution time ratio* (ETR), i.e. the ratio of the time taken by the static scheduling version of the benchmark *over* the time taken by the SUPPLE hybrid scheduling version. Hence, a ratio greater than one corresponds to an improvement in the total execution time obtained by adopting SUPPLE with respect to the static version. Each curve is relative to a distinct granularity μ of Comp(), and plots the ETRs as a function of the number of processors employed. The size of the data set has been modified according to the number of processors to keep the size of the sub-block statically assigned to each processor constant.

4.1 Time–shared Environments

First of all, we show the results obtained on the SP2. Due to the difficulty in running probing experiments because of the unpredictability of the workloads, as well as for the exclusive use of some resources (in particular, resources as the SP2 high–performance switch and/or the processors themselves) by other users, the results in this case are not so exhaustive as those reported for the tests run on the T3E. As previously mentioned, on the SP2 we ran our experiments after the introduction of a synthetic load on a subset of the nodes used. In particular, we launched 4 compute–bound processes on

[1] The Pittsburgh Supercomputing Center has a T3E system with 512 application processors of which half runs at 300 MHz and half at 450 MHz.

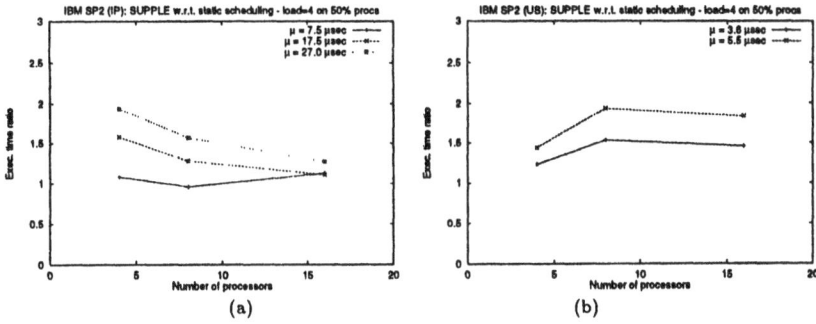

Fig. 1. SP2 results: ETR results for experiments exploiting (a) the Ethernet, and (b) the high-performance switch

50% of the processing nodes employed, while the loads were close to zero on the rest of the nodes. This corresponds to "slow" processors characterized by a value of F (*factor of slowdown*) equal to 5. The two plots in Figure 1 show the ETRs obtained for various μ and numbers of processors. The size of the sub-block owned by each processor was kept constant and equal to 512×512, while N_ITER was equal to 5. As regards the SUPPLE parameters, the *Threshold* value was set to 0.02 msec, and the chunk size g to 32 iterations.

Figure 1.(a) shows the SP2 results obtained by using the Ethernet network and the IP protocol. Note that, even in this case as in the following ones, the performance improvement obtained with SUPPLE increases in proportion to the size of the grain μ. For $\mu = 27$ μsec, and 4 processors, we obtained the best result: SUPPLE implementation is about 100% faster than the static counterpart. For smaller values of μ, due to the overheads to migrate data, the load imbalance paid by the static version of the benchmark is not large enough to justify the adoption of a dynamic scheduling technique.

Figure 1.(b) shows, on the other hand, the results obtained by running the parallel benchmarks in time–sharing on the SP2 by exploiting the US high–performance switch. Due to the better communication framework, in this case the ETR is favorable to SUPPLE even for smaller values of μ.

4.2 Heterogeneous Environments

All the experiments regarding heterogeneous environments were conducted on the T3E. Heterogeneity was simulated, so that we were able to perform a lot of tests, also experimenting different factors F of slowdown.

Iterative benchmark Figure 2 reports the results for the iterative benchmark, where the external sequential loop is repeated 20 times (N_ITER = 20). For all the tests we kept fixed the size of the sub-block assigned to each processor (512×512). Figures 2.(a) and 2.(b) are relative to an environment where only 25% of all processors are "slow".

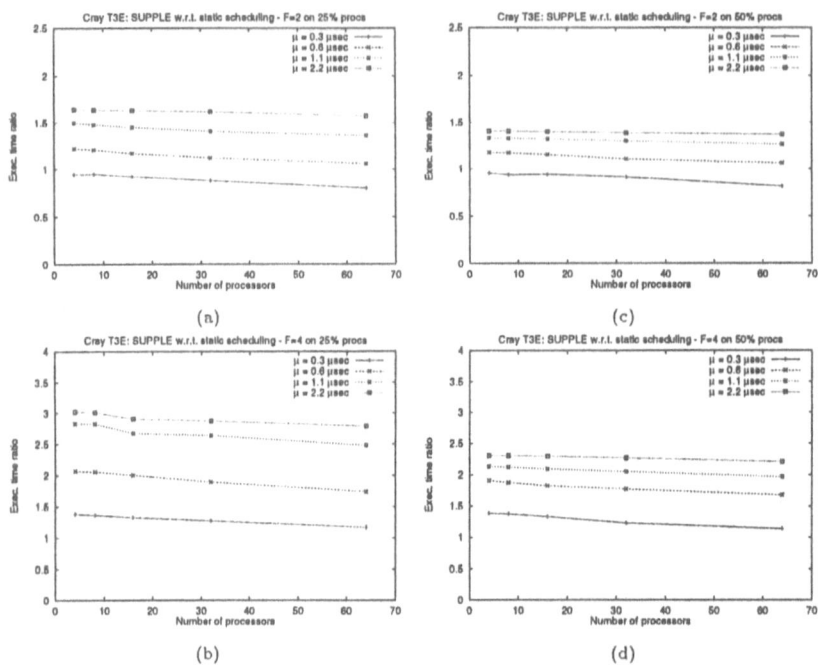

Fig. 2. T3E results: ETR results for various values of F and μ

The factors F of slowdown of these processors are 2 and 4, respectively. Figures 2.(c) and 2.(d) refer to an environment where half of the processors employed are "slow".

Each curve in all these figures plots the ETR for a benchmark characterized by a distinct μ. With respect to the SP2 tests, in this case we were able to reduce μ (μ now ranges between 0.3 and 2.2 μsec), always obtaining an encouraging performance with our SUPPLE support. Note that such grains are actually very small: for μ equal to 0.3μsec, about 85% of the total time is spent on memory accesses (to compute addresses and access data element covering the five-point stencil), and only 15% on arithmetic operations. We believe that the encouraging results obtained are due to the smaller overheads and latencies of the T3E network, as well as to the absence of time–sharing of the processor, thus speeding up the responsiveness of "slow" processors to requests coming from "fast" ones. The reductions in granularity made it possible to enlarge the chunk size g ($g = 128$) without losing the effectiveness of the dynamic scheduling algorithm. The *Threshold* ranged between 0.02 and 0.06 msec, where larger values were used for larger μ.

Looking at Figure 2, we can see that better results (SUPPLE execution times up to 3 times lower than those obtained with static scheduling) were obtained when the "slow" processors were only 25% of the total number. The reason for this behavior is clear: in this case, we have more "fast" processors to which the extra workload previously assigned to the "slow" processors can be dynamically distributed. The execution times obtained with the static implementation are, on the other hand, almost independent of the percentage of "slow" processors. In fact, even if only one of the processors was "slow", its execution time would dominate the overall completion time.

We also tested SUPPLE on a homogeneous system (i.e. a balanced one) in order to evaluate its overhead w.r.t. a static implementation, which, in this case, is optimal. The overhead is almost constant for data sets of the same sizes subdivided into a given number of chunks, but its influence becomes larger for smaller granularities because of the shorter execution time and the limited possibility of hiding communication latencies with computations. Thus, for $\mu = 2.2\,\mu\text{sec}$ the two execution times are almost comparable, while for $\mu = 0.3\mu\text{sec}$, the static version of the benchmark becomes 60% faster than SUPPLE. We verified that the overhead introduced by SUPPLE is due to some undesired migration of chunks, and to the dynamic scheduling technique which entails polling the network interface to check for incoming messages (even if these messages do not actually arrive).

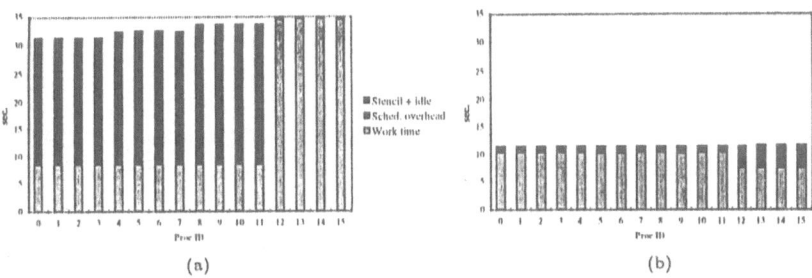

Fig. 3. Work time and overheads for a static (a) and a SUPPLE (b) version of the benchmark

Finally, we instrumented the static and the SUPPLE versions of a specific benchmark to evaluate the ratio between the execution time spent on useful computations and on dynamic scheduling overheads. The features of the benchmark used in this case are the following: a 2048 × 2048 data set distributed over a grid of 4 × 4 T3E processors, and an external sequential loop iterated for 20 times. The simulated unbalanced environment was characterized by 25% of "slow" processors, with a slowdown factor of 4. Figure 3.(a) show the results obtained by running the static implementation of this benchmark. It is worth noting the work time on the "slow" processors, which is 4 times the work time on the "fast" ones. The black portions of the bars show idle times on "fast" processors while waiting for border data. These idle times are thus due to communications implementing stencil data references, where corresponding send/receive on fast/slow processors are not synchronized due to their different capacities.

Figure 3.(b), on the other hand, shows the SUPPLE results on the same benchmark. Note the redistribution of workloads from "slow" to "fast" processors. Thanks to dynamic load balancing, idle times disappeared, but we have larger scheduling overheads due to the communications used to dynamically migrate chunks. These overheads are clearly larger on "slow" processors, which spend a substantial part of their execution time on giving away work and on receiving the results of migrated iterations.

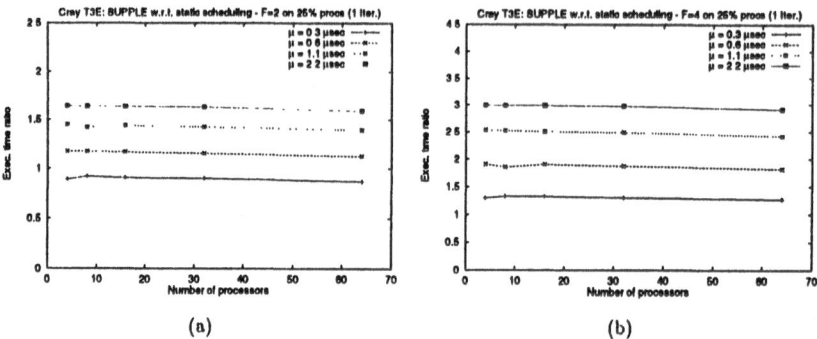

Fig. 4. T3E results: ETR results obtained running a *single iteration* benchmark for various values of F and μ

Single iteration benchmark As explained above, one of the advantages of SUPPLE is that it can also be used for balancing parallel loops that have only to be executed once. In this case some overheads cannot be overlapped, and at the end of loop execution "slow" processors have to wait for the results of iterations executed remotely. Figure 4 shows the encouraging ETRs obtained by our SUPPLE implementation w.r.t. the static one. Note that all the results plotted in the figure refer to unbalanced environments where only 25% of the processors are "slow". Moreover, due to the larger data sets used for these tests, the ETRs are in some cases even more favorable for SUPPLE than the ones for iterative benchmarks.

5 Conclusions

We have discussed the implementation on heterogeneous and/or timed-shared parallel machines of regular and uniform parallel loop kernels, with statically predictable stencil data references. We have assumed that no information about the capacities of the processors involved is available until run-time, and that, in time-shared environments, these capacities may change during run-time. To implement the kernel benchmark we employed SUPPLE, a run-time support that we had previously introduced to compile non-uniform loops.

The tests were conducted on an SGI/Cray T3E and an IBM SP2. We compared the SUPPLE results with a static implementation of the benchmark, where data and computations are evenly distributed to the various processing nodes. The SP2, a parallel system that can be used as a time-shared NOWs, was loaded with artificial compute-bound processes before running the tests. On the other hand, we needed to simulate a heterogeneous SGI/Cray T3E, i.e. a machine whose nodes may be equipped with different off-the-shelf processors and/or memory organization. The performance results were very encouraging. On the SP2, where half of the processors were loaded with 4 compute-bound processes, the SUPPLE version of the benchmark resulted at most 100% faster than the static one. On the T3E, depending on the amount of "slow" processors, on the number of processors employed and the granularity of loop iterations, the SUPPLE version reached percentages of performance improvement ranging between 20% and 270%.

Further work has to be done to compare our solution with other dynamic scheduling strategies proposed elsewhere. More exhaustive experiments with different benchmarks and dynamic variations of the system loads are also required to fully evaluate the proposal. However, we believe that hybrid strategies like the one adopted by SUPPLE can be profitably exploited in many cases where locality exploitation and load balance must be solved at the same time. Moreover, our strategy can be easily integrated in the compilation model of a high level data parallel language.

References

1. R. Alverson et al. The Tera computer system. In *Proc. of the 1990 ACM Int. Conf. on Supercomputing*, pages 1–6, 1990.
2. A. Anurag, G. Edjlali, and J. Saltz. The Utility of Exploiting Idle Workstations for Parallel Computation. In *Proc. of the 1997 ACM SIGMETRICS*, July 1997.
3. S. Hiranandani, K. Kennedy, and C. Tseng. Evaluating Compiler Optimizations for Fortran D. *J. of Parallel and Distr. Comp.*, 21(1):27–45, April 1994.
4. S. Flynn Hummel, J. Schmidt, R. N. Uma, and J. Wein. Load-Sharing in Heterogeneous Systems via Weighted Factoring. In *Proc. of the 8th SPAA*, July 1997.
5. S.F. Hummel, E. Schonberg, and L.E. Flynn. Factoring: A Method for Scheduling Parallel Loops. *Comm. of the ACM*, 35(8):90–101, Aug. 1992.
6. V. Kumar, A.Y. Grama, and N. Rao Vempaty. Scalable Load Balancing Techniques for Parallel Computers. *J. of Parallel and Distr. Comp.*, 22:60–79, 1994.
7. J. Liu and V. A. Saletore. Self-Scheduling on Distributed-Memory Machines. In *Proc. of Supercomputing '93*, pages 814–823, 1993.
8. M. Colajanni M. Cermele and G. Necci. Dynamic Load Balancing of Distributed SPMD Computations with Explicit Message-Passing. In *Proc. of the IEEE Workshop on Heterogeneous Computing*, pages 2–16, 1997.
9. S. Orlando and R. Perego. SUPPLE: an Efficient Run-Time Support for Non-Uniform Parallel Loops. Technical Report TR-17/96, Dipartimento di Mat. Appl. ed Informatica, Università di Venezia, Dec. 1996. To appear on J. of System Architecture.
10. S. Orlando and R. Perego. A Comparison of Implementation Strategies for Non-Uniform Data Parallel Computations. Technical Report TR-9/97, Dipartimento di Mat. Appl. ed Informatica, Università di Venezia, April 1997. Under revision for publication on the J. of Parallel and Distr. Comp.
11. S. Orlando and R. Perego. A Support for Non-Uniform Parallel Loops and its Application to a Flame Simulation Code. In *Proc. of the 4th Int. Symposium, IRREGULAR '97*, pages 186–197, Paderborn, Germany, June 1997. LNCS 1253, Spinger-Verlag.
12. O. Plata and F. F. Rivera. Combining static and dynamic scheduling on distributed-memory multiprocessors. In *Proc. of the 1994 ACM Int. Conf. on Supercomputing*, pages 186–195, 1994.

13. J. Saltz et al. Runtime and Language Support for Compiling Adaptive Irregular Programs on Distributed Memory Machines. *Software Practice and Experience*, 25(6):597–621, June 1995.
14. M.H. Willebeek-LeMair and A.P. Reeves. Strategies for Dynamic Load Balancing on Highly Parallel Computers. *IEEE Trans. on Parallel and Distr. Systems*, 4(9):979–993, Sept. 1993.

A Lower Bound for Dynamic Scheduling of Data Parallel Programs

Fabricio Alves Barbosa da Silva[2], Luis Miguel Campos[1], Isaac D. Scherson[1,2]

[1] Information and Comp. Science, University of California, Irvine, CA 92697 U.S.A.
{isaac,lcampos}@ics.uci.edu[***]
[2] Université Pierre et Marie Curie, Laboratoire ASIM, LIP6, Paris, France.[†]
fabricio.silva@asim.lip6.fr[‡]

Abstract. Instruction Balanced Time Slicing (IBTS) allows multiple parallel jobs to be scheduled in a manner akin to the well-known gang scheduling scheme in parallel computers. IBTS however allows for time slices to change dynamically and, under dynamically changing workload conditions is a good non-clairvoyant scheduling technique when the parallel computer is time sliced one job at a time. IBTS-parallel is proposed here as a dynamic scheduling paradigm which improves on IBTS by allowing also dynamically changing space sharing of the computer's processors. IBTS-parallel is also non-clairvoyant and it is characterized under the competitive ratio metric. A lower bound on its performance is also derived.

1 Introduction

A solution to the dynamic parallel job scheduling problem is proposed together with its complexity analysis. The problem is one of defining how to share, in an optimal manner, a parallel machine among several parallel jobs. A job is defined as a set of data parallel threads. One important characteristic that makes the problem dynamic, as defined in [3], is the possibility of arrival of new jobs at arbitrary times. In the static case, the set of jobs to be executed is already defined when scheduling decisions are made, and arbitrary job arrivals are not permitted.

In this paper we propose a new scheduling algorithm dubbed *IBTS-Parallel*, which is derived from the IBTS algorithm. IBTS stands for instruction balanced time slicing, and was originally defined in [6]. IBTS is a non-clairvoyant [3] scheduling algorithm designed to optimize the competitive ratio (CR) metric, also defined in [6][4].

In addition to the theoretical analysis of IBTS-Parallel, we present an experimental analysis performed using a general purpose event driven simulator developed by our research group.

[***] Supported in part by the Irvine Research Unit in Advanced Computing and NASA under grant #NAG5-3692.
[†] Professor Alain Greiner is gratefully acknowledged.
[‡] Supported by Capes, Brazilian Government, grant number 1897/95-11.

2   Previous Work

Various classifications for the parallel job scheduling problem have been suggested in the literature. The classification used here is based on the way in which computing resources are shared : temporal sharing, also known as time slicing or preemption; and space slicing, also called partitioning. These two classifications are orthogonal, and may lead to a taxonomy based on the possible options. Table 1 shows the scheduling policies adopted by commercial and research systems, and was borrowed from [2]. It is worth noting that the lack of consensus on which scheduling policy is best among those showed in table 1 is total. The problem is that the assumptions leading to and justifying the different schemes are usually quite different, which makes difficult the comparison between different solutions.

			time slicing		
			yes		no
			independent PEs	gang scheduling	
			global queue	local queue	
space slicing: yes	flexible		Mach	Paragon/service Meiko/timeshare KSR/interactive transputers Tera/streams Chrysalis	Medusa Butterfly@LLNL Cray T3E Meiko/gang Paragon/gang SGI/gang Tera/PB MAXI/gang	IBM SP2, Victor Meiko/batch Paragon/slice KSR/bath 2-level/bottom TRAC, MICROS Amoeba
	structured			NX/2 on iPSC/2 nCUBE	CM-5 Cedar DHC on SP2 DQT on RWC-1	Cray T3D CM-2 PASM hypercubes
	no		IRIX on SGI NYU Ultra Dynix 2-level/top Hydra/C.mmp	Star OS Psyche Elxsi AP1000	MasPar MP2 Alliant FX/8 Chagori on K2	Illiac IV MPP GF11 Warp

Fig. 1. Scheduling policies followed in current commercial and research systems

We address the scheduling problem by using the same assumptions (i.e. model and metrics) described by Subramaniam and Scherson in [6][4]. In [6], Subramaniam and Scherson studied a scheduling policy named Instruction Balanced Time Slicing (IBTS), which performs well under the competitive ratio metrics [4]. Each job is allocated on all P processing elements (PE) of the machine for a finite quantum (time-slice). All time-quanta devoted to the various jobs are equal as measured in machine instructions (that is, all jobs are preempted after equal number of machine instructions).

Note that time quanta may vary in absolute value while still being equal when measured in number of instructions. A good analysis of gang scheduling, which is similar to IBTS but imposes an invariant constant time slice can be found in [5].

In [4] the programming model used is the data parallel (V-RAM) model[1]. The scheduling problem was defined as a optimization problem, and the *compet-*

itive ratio was the metric used as the objective function. The competitive ratio (CR) was defined in function of the non clairvoyance of the scheduler.

The competitive ratio (CR) is based on the *happiness* [4] concept. The happiness metric attempts to capture the satisfaction of a job as a function of the scheduling decisions made by a scheduler. The CR is then the ratio between the happiness achieved by a knowledgeable *malicious adversary*, and the that achieved by a partially ignorant scheduler. IBTS is hence a non-clairvoyant algorithm, and the malicious adversary always has more information that the scheduler itself. It is that hidden information that is used by the adversary to keep the happiness as low as possible. By minimizing the CR the scheduler approaches the result that would be obtained by an adversary with global knowledge.

$$CR = \overset{max}{y'} \frac{Happiness(x, y')}{Happiness(x, y)} \tag{1}$$

In equation 1, x represents the input to the algorithm, y is the result obtained by the scheduler and y' is the result obtained by the adversary.

Also:

$$Happiness(x, y) = \frac{\int_0^T min_j \wp^j(\tau) d\tau}{T} \tag{2}$$

Where $\wp^j$ is the power delivered to job j, and T is the time at which the last job completes. In analogy with physics terminology, the power delivered to a job is defined as:

$$\wp = \frac{W}{\Gamma} \tag{3}$$

Γ is the running time of a job as a function of the number of statements and the mean time completion of each statement (the mean time completion is computed according to the type of the statement: local statements, remote access statements, etc, and is machine architecture dependent). W is the *work*, or the processor-time product in the ideal world, i.e. it is the product of the ideal number of processors for execution of the job and the running time assuming all statements as local statements.

$\wp$ is valid for the single job case. For multiple jobs we have:

$$W^j = \int_0^{T_j} \wp^j(\tau) d\tau \tag{4}$$

Another related definition is the inefficiency of running the job on a machine:

$$\eta = \frac{P'\Gamma}{W}, \eta > 1 \tag{5}$$

Γ is the running time of a job, W is the work and P' is the number of processors actually allocated to the job.

The most important result contained in [4][6] is that, of all non-clairvoyant schedulers, instruction-balanced time-slicing has the least CR, with the corresponding proof. As a consequence of the proof it was verified that the least possible CR is equal to η_{max}, which is the maximum inefficiency among all jobs that are running in a machine at a given moment.

3 IBTS-parallel

In IBTS, each job is allocated to the whole machine and executes a predefined number of instructions before being preempted. However, not all jobs necessarily use all processors of the machine at all times. In order to further optimize the space sharing in parallel scheduling policies should allocate simultaneously multiple jobs. We use the fact that IBTS is the non-clairvoyant algorithm with the least CR to propose a modified version of IBTS that also permits a better spatial allocation of the machine.

Let us start by considering a machine with N processors in a MIMD architectural model as described in [4]. In our modified version of IBTS the machine is shared by more than one job at any given time. Jobs are preempted after all threads, running in parallel, execute a fixed number I of instructions. The resulting scheduling algorithm is dubbed IBTS-parallel. The use of many different spatial scheduling strategies are possible with IBTS-Parallel (first-fit, best-fit, etc.

IBTS-Parallel has at least the same performance than IBTS under the CR metric, as stated in the lemma below.

Lemma 1. *The CR of IBTS-parallel is* $\leq \eta_{max}$

Proof. It follows from [4][6] by verifying that the two properties of IBTS are valid for IBTS-parallel :

1. At any time, an equal amount of power is supplied to all running jobs
2. The job that finishes last incurs the maximum amount of work

Using the lemma above, we can state the main result of this section.

Theorem 1. $\eta_{max}^{IBTS\text{-}par} \leq \eta_{max}^{IBTS}$

In other words, IBTS represents the worst case of IBTS-parallel

Proof. As we are running multiple jobs in parallel, the execution time of each job will necessarily be smaller than if we run one job after another, as is the case in IBTS since one job allocates all processing elements of a machine. From the definition of inefficiency :

$$\eta = \frac{P'\Gamma}{W} \tag{6}$$

P' and W do not change from IBTS to IBTS-parallel. The time Γ is smaller or equal for all jobs under IBTS-parallel as compared to pure IBTS. So, all jobs have smaller or equal inefficiencies under IBTS-parallel as compared to pure IBTS, which makes the maximum inefficiency in IBTS-parallel smaller than or equal to the corresponding quantity in IBTS.

4   Simulation and Verification

To verify the results above, we used a general purpose event driven simulator, developed by our research group for studying a variety of related problems (e.g., dynamic scheduling, load balancing, etc). The simulator accepts two different formats for describing jobs. The first is a fully qualified DAG. The second is a set of parameters used to describe the job characteristics such as computation/communication ratio. When the second form is used the actual communication type, timing and pattern are left unspecified and it is up to the simulator to convert this user specification into a DAG, using probabilistic distributions, provided by the user, for each of the parameters. Other parameters include the spawning factor for each thread, a thread life span, synchronization pattern, degree of parallelism (maximum number of thread that can be executed at any given time), depth of critical path, etc. Even-though probabilistic distributions are used to generate the DAG, the DAG itself behaves in a completely deterministic way.

Once the input is in the form of a DAG, and the module responsible for implementing a particular scheduling heuristics is plugged into the simulator, several experiments can be performed using the same input by changing some of the parameters of the simulation such as the number of processing elements available or the topology of the network, among others. The outputs can be recorded in a variety of formats for later visualization.

For this study we grouped parallel jobs in classes where each class represents a particular degree of parallelism (maximum number of threads that can be executed at any given time). We divided the workload into ten different classes with each class containing 50 different jobs. The arrival time of a job is described by a Poisson random variable with an average rate of two job arrivals per time slice. The actual job selection is done in a round robin fashion by picking one job per class. This way we guarantee the interleaving of heavily parallel jobs with shorter ones.

We distinguish the classes of computation and of communication instructions in the various threads that compose a job. A communication forces the thread to be suspended until the communication is concluded. If the communication is concluded during the currently assigned time-slice the thread resumes execution. All threads are preempted only after executing 100 instructions (duration of a time slice), with the caveat that any thread that executed one or more communication instruction will be preempted at the end of the time-slice regardless of how many instructions it was able to execute in the current time-slice. We used a factor of 0.001 communications per computation instructions.

The practical implementation of IBTS-parallel was based on a greedy algorithm applied at the beginning of each workload change.During the workload change interval, called cycle, the workload is obviously assumed constant. Thus, the eligible threads of queued jobs are allocated to processors using the first fit strategy for each time slice. Clearly, after all eligible threads are scheduled on a processor for some time slice (slot), the temporal sequence is repeated periodically until a workload change again occurs. We considered in simulations a machine with 1024 processors.

Preliminary results are shown in table 1. They have been normalized to show the total execution time of IBTS-parallel to be 100%.

Table 1. Preliminary experimental results

IBTS		IBTS-parallel	
Total Running Time (%)	Total Idle Time (%)	Total Running Time (%)	Total Idle Time (%)
123.6	41.9	100	28.2

It is important to dissect the value obtained for *idle time*. This value is the result of following three factors:

1. Communications
2. Absence of ready threads
3. Lack of job virtualization

The first is a natural consequence of threads communicating among themselves.. The second reflects the fact that jobs arrive and finish at random times and at any given instance there might not be any job ready to be scheduled. The last is a result of not allowing individual threads of a same job to be scheduled at different time-slices. This factor is, we believe, the chief reason behind the high idle time measured and should be greatly reduced by extending IBTS-parallel with job virtualization [4] techniques.

References

1. Blelloch,G. E.: Vector Models for Data-parallel Computing *MIT Press*,Cambridge, MA (1990)
2. Feitelson, D.: Job Scheduling in Multiprogrammed Parallel Systems *IBM Research Report RC 19970*, Second Revision (1997)
3. Motwani, R., Phillips, S., Torng, E., Non-clairvoyant scheduling, *Theoretical Computer Science*, (1994) 130(1):17- 47
4. Scherson, I. D., Subramaniam R., Reis, V. L. M., Campos, L. M.. Scheduling Computationally Intensive Data Parallel Programs. *Ecole Placement Dynamique et Re'partition de charge* (1996) 39–61
5. Squillante, M. S., Wang, F., Papaefthymiou, M., An Analysis of Gang Scheduling for Multiprogrammed Parallel Computing Environments, *Proceedings of the 8th Annual ACM Symposium on Parallel Algorithms and Architectures*, (1996) 89-98
6. Subramaniam, R.: A Framework for Parallel Job Scheduling *PhD Thesis*, University of California at Irvine (1995)

A General Modular Specification for Distributed Schedulers

Gerson G. H. Cavalheiro[†1], Yves Denneulin[2] and Jean-Louis Roch[1]

[1] Projet Apache[‡]
Laboratoire de Modlisation et Calcul, BP 53
F-38041 Grenoble Cedex 9, France
[2] Laboratoire d'Informatique Fondamentale de Lille
Cit Scientifique
F-59655 Villeneuve d'Ascq Cedex, France

Abstract. In parallel computing, performance is related both to algorithmic design choices at the application level and to the scheduling strategy. Concerning dynamic scheduling, general classifications have been proposed. They outline two fundamental units, related to control and information. In this paper, we propose a generic modular specification, based not on two but on four components. They and the interactions between them are precisely described. This specification has been used to implement various scheduling algorithms in two different parallel programming environments: PM^2 (ESPACE) and Athapascan (APACHE).

1 Introduction

From various general scheduling systems classifications for parallel architectures [2,6], it is possible to extract two fundamental elements: a *control unit* to compute the distribution of work in a parallel machine and an *information unit* to evaluate the load of the machine. These units can implement different load balancing mechanisms, privileging either a (class of) machine architecture or an (class of) application structure.

A dynamic scheduling system (DSS for short) is particularly interesting in case of an unpredictable behavior at execution time. Irregular applications and networks of workstations (NOW) are typical cases where the execution behavior is unknown *a priori*. Usually a DSS for a tuple {*application, machine*} is specified by an entry in a scheduling classification, and is implemented as a black-box: it is difficult to tune it or reuse it for another application or machine without rewriting totally or partially its code. Some works in the literature discuss this problem and present possible solutions, like [4] where scheduling services are grouped in families and presented as functions prototypes.

In this paper we discuss the gap between the scheduling classifications and the implementation of DSSs. We present a modular structure for general scheduling

---

[†] CAPES–COFECUB Brazilian scholarship
[‡] Supported by CNRS, INPG, INRIA et UJF

systems. This general architecture, presented as a framework, aims at avoiding the introduction of additional constraints to implement the different classes of scheduling. This framework introduces two new functional units: a *work creation unit* and an *executor unit* and specifies an interface between the components.

2 General Organization of a Dynamic Scheduling System

In this work a DSS is implemented at software level to control the execution of an application on a parallel architecture. It was conceived to be independent from both application and hardware and to support various classes of scheduling algorithms. The DSS runs on a parallel machine composed of several computing nodes, each one having a local memory and at least one real processor. There is also a support for communications among them.

The notion of job. Units of work are produced at the application level; from them, *jobs* are built and inserted in a graph of precedence (DAG), denoted G. G evolves dynamically; it represents at each instant of execution the current state of the application jobs. Each node in G represents a job; each edge, a precedence constraint between two jobs. So, a job is the basic element to be scheduled, it corresponds to a finite sequence of instructions, and has properties to allow its manipulation by the scheduler. At a given instant of time, a job is in one of the following states: *waiting*, the job has precedence constraints in G; *ready*, the job can be executed; *preempted*, the execution was interrupted; *running*, the job's instructions are executed; or *completed*, the precedence constraints in G imposed by the job are relaxed.

Correction of a scheduling algorithm. The scheduling manages the execution of jobs on the machine resources. For this, any DSS has to respect the following properties, that define its semantics:
(P_1) for the application: only jobs in the *ready* state can be put in the *running* state by the DSS;
(P_2) for the computing resources: if some resources of the machine (nodes) are idle while there are jobs in the *ready* state, at least one job will be put by the DSS in the *running* state after a finite delay. This delay, which may sometimes be bounded, defines the *liveness* of the DSS.
A DSS verifying those properties is said to be *correct*.

3 A Modular Architecture for a Parallel Scheduling

The DSS is implemented as a framework based on four modules (Fig. 1): a *Job Builder*, to construct and manage a graph of jobs; a *Scheduling Policy*, where the scheduling algorithm is implemented; a *Load Level*, to evaluate the load of the machine; and an *Executor*, which implements a pool of threads used to execute the jobs. Each module defines an interface to its services. An access to one of them represents a *read* or a *write* operation, both accomplished in a non-blocking fashion. A read operation is a request for data; it allows a module to get a data handled by another one. Write operations also provide resources for data exchange, but in the opposite direction: a module decides to send a data to another module (e.g. a data or status changes). As a reaction to the use of a service, an internal activity can be triggered and executed concurrently with the caller before the service ends. A short description of each module follows.

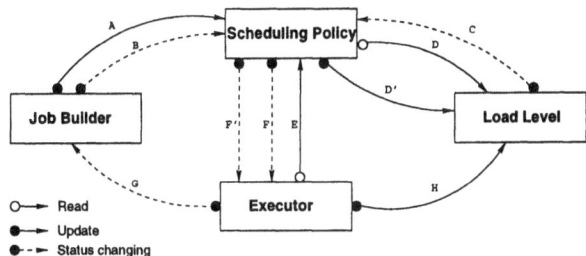

Fig. 1. Internal communication scheme between modules on the framework.

Job Builder unit. Its role is to build *jobs* from the tasks defined by the application, eventually giving access to their dependencies. So, it manages the evolutions of G. This module is attached to the application to receive and handle the tasks creation requests. The Job Builder unit must be informed when a job terminates to update the states of the jobs in G. In Fig. 1, arrow G shows the task Executor sending a job termination event to the Job Builder.

Scheduling Policy unit This is the core of the scheduling framework; it takes decisions about the placement of the jobs. In order to know how the application evolves, the Scheduling Policy receives from the Job Builder every change in G, e.g. when a job is created (arrow A in Fig. 1) or when its state changed (arrow B). To place the jobs to respect the load balancing scheme, the Scheduling Policy may consider the load on the parallel machine (arrow D). Notice that after taking the decisions, the load information must be updated (arrow D'). The Scheduling Policy must also react to two signals: an irregular load distribution detected by the Load Level (arrow C) and the requests of jobs to execute (arrow E) from the Executor. It can also create a new virtual processor (arrow F) or destruct one (arrow F').

Executor unit. The Executor is implemented over a support that provides fine-grained activities; it furnishes virtual processors, called threads, to process the execution of the application jobs. Each thread executes an infinite loop of two actions: 1) send a request (arrow E) of a job to the Scheduling Policy; and 2) execute sequentially the job's instructions. The load information is updated after and before the execution of each job (arrow H).

Load Level unit. The Load Level unit collects the load informations in the computing system. These data are updated after the requests for write operations from the Scheduling Policy and Executor units. If an irregular distribution of the load in the machine is detected, this unit sends a message to the Scheduling Policy unit (arrow C). The definition of the load and the notion of irregular distribution depends on the Scheduling Policy unit.

4   Implemented Environments

This general specification has been used to build Athapascan-GS and GTLB, both DSSs for two distinct parallel programming environments. Athapascan-GS [5] was implemented on top of Athapascan-0 [1] for the Athapascan-1 macro data-flow language (INRIA APACHE Project, LMC laboratory) to support static and dynamic scheduling algorithms. Athapascan-GS can choose in the set of ready jobs the one that will be triggered in order to optimize a specific index of performance, like memory space, execution time or communication. The Generic Threads Load Balancer (GTLB) [3] defines a generic scheduler for highly irregular applications of tree search that belongs to the Branch&Bound family; it was implemented on the top of PM^2 (ESPACE project, LIFL laboratory) to support schedulers based on algorithms of mapping with migration. On both implementations, this design provides reusability opportunity in the development of specific scheduling algorithms.

5   Conclusion

In this paper, we have proposed a specification design to implement dynamic scheduling systems. This work completes classical classifications which analyze only two units of scheduling algorithms, dedicated respectively to the control and the load information. We claim that such a distinction is not precise enough; the major argument is that it does not take into account the relationships with the application (that produces tasks) and the execution (that consumes tasks). Two new modules are proposed: one dedicated to the work creation and another to the execution support, a protocol that specifies their interactions was also presented.

Two examples of DSSs implementing this general structure were presented: Athapascan-GS and GTLB, both supporting various scheduling algorithms.

References

1. J. Briat, I. Ginzburg, M. Pasin and B. Plateau. Athapascan Runtime: Efficiency for Irregular Problems. In *Proc. of the 3th Euro-Par Conference*. Passau, Aug. 1997.
2. T.L. Casavant and J.G. Kuhl. A Taxonomy of Scheduling in General-Purpose Distributed Computing Systems. *IEEE Trans. Soft. Eng.*. V. 14(2): 141-154, Fev. 1988.
3. Y. Denneulin. *Conception et ordonnancement des applications hautement irrgulires dans un contexte de paralllisme grain fin*. PhD Thesis, Universit de Lille, Jan. 1998.
4. C. Jacqmot. *Load Management in Distributed Computing Systems*: Towards Adaptative Strategies. DII, Universit Catholique de Louvain, PhD Thesis, Louvain-la-Neuve, Jan. 1996.
5. J.-L. Roch et all. *Athapascan-1*. Apache Project, Grenoble, http://www-apache.imag.fr Oct. 1997.
6. M.H. Willebeek-LeMair and A.P. Reeves. Strategies for Dynamic Load Balancing on Highly Parallel Computers. *IEEE Trans. Par. and Dist. Syst.* V. 4(9): 979-993, Sept. 1993.

Feedback Guided Dynamic Loop Scheduling: Algorithms and Experiments

J. Mark Bull

Centre for Novel Computing, Dept. of Computer Science,
University of Manchester, M13 9PL, U.K.
markb@cs.man.ac.uk

Abstract. Dynamic loop scheduling algorithms can suffer from overheads due to synchronisation, loss of locality and small iteration counts. We observe that timing information from previous executions of the loop can be utilised to reduce these overheads. We introduce two new algorithms for dynamic loop scheduling which implement this type of feedback guidance, and report experimental results on a distributed shared memory architecture. Under appropriate circumstances, these algorithms are observed to give significant performance gains over existing loop scheduling techniques.

1 Introduction

Minimising load imbalance is a key activity in producing efficient implementations of applications on parallel architectures. Since loops are the most significant source of parallelism in many applications, the scheduling of loop iterations to processors can be an important factor in determining performance. Most of the existing techniques for dynamic loop scheduling on shared memory machines are variants of, or are derived from, *guided self-scheduling* (GSS) [6]. These techniques share some common characteristics—they assume no prior knowledge of the workload associated with the iterations of the loop, and they all divide the loop iterations into a set of *chunks*, where there are more chunks than processors. The key observations that motivate the new algorithms presented here are the following:

- Some prior knowledge of the workload associated with the iterations of the loop may be available, particularly if we can assume that the workload has not changed too much since the previous execution of the loop. This is often the case in simulation of physical systems where the parallel loop is over a spatial domain and is executed at every time-step.
- Dividing the loop iterations into more chunks than processors can hurt performance, as it may incur overheads associated with additional synchronisation, loss of both inter- and intra-processor locality, and reduced efficiency of loop unrolling or pipelining.

Our algorithms are designed to utilise knowledge about the workload derived from the measured execution time of previous occurrences of the loop with the aim of reducing the incurred overheads.

2   Feedback-Guided Scheduling Algorithms

2.1   Self-scheduling Algorithms

The simplest scheduling algorithm of all is *block* or *static* scheduling, which divides the number of the loop iterations by the number of processors as equally as possible. No attempt is made to dynamically balance the load, but overheads are kept to a minimum: no synchronisation is required and the chunk size is as large as possible. Of dynamic algorithms, the simplest is *self-scheduling* [8] in which a central queue of iterations is maintained. As soon as a processor finishes an iteration, it removes the next available iteration from the queue and executes it. *Chunk self-scheduling* [3] allows each processor to remove a fixed number k of iterations from the queue instead of one. This reduces the overheads at the risk of poorer load balance.

Guided self-scheduling (GSS) [6] attempts to overcome the difficulty of choosing a chunk size by dynamically varying the chunk size as the execution of the loop progresses, starting with large chunks and making them progressively smaller. There is also the possibility that the load may not be well balanced— for example if the loop has N iterations and the first N/p iterations contain more than $1/p$th of the workload. A number of algorithms have been proposed which are essentially similar to GSS, and differ mainly in the manner in which the chunk size is decreased as the algorithm progresses. These include *adaptive guided self-scheduling* [1], *factoring* [2], *tapering* [4] and *trapezoid self-scheduling* [9].

Affinity scheduling [5] uses per-processor work queues in order to reduce contention between processors. Each processor initially has on its queue the iterations it would have been assigned under block scheduling. Another benefit of affinity scheduling is that it allows temporal locality of data across multiple executions of the parallel loop to be exploited. Variants of affinity scheduling to reduce synchronisation costs have been proposed in [7] and [10].

2.2   New Algorithms

In order to exploit timing information, we require access to a fast, accurate hardware clock. Since our objective is to keep the chunk size a large as possible, we only permit measurement of the execution time of whole chunks, and not of iterations within a chunk.

The first algorithm we describe is a feedback guided version of block scheduling. We divide the loop iterations into p chunks, not necessarily of equal size, where p is the number of processors. On each execution of the parallel loop we measure the total execution time for the chunk of iterations assigned to each processor. Dividing this time by the number of iterations on each processor gives us a mean load per iteration for that chunk, and hence a piecewise constant approximation to the true workload. We then choose new bounds for each processor to use for the next execution of the loop, based on an equipartition of the area under this piecewise constant function. By keeping a running total of the areas under each constant piece, this can be achieved in $O(p)$ time. This could be reduced to $O(\log_2(p))$, using parallel prefix operations, but this would only be worthwhile for large values of p.

The second algorithm is a feedback guided version of affinity scheduling. On each execution of the parallel loop we measure the total execution time for the chunk of iterations initially assigned to *and subsequently executed by* each processor. We derive a mean load per iteration, and hence a piecewise constant approximation to the true workload in the same way as for the feedback guided block algorithm. Equipartitioning this approximation to the workload gives us the initial assignment of iterations for the next execution of the loop. This is similar in spirit to the *dynamic affinity* algorithm of [7], which initialises each processor with the same number of iterations as it executed on the previous execution of the loop.

3   Experimental Evaluation

3.1   Test Problems

To test our algorithms we use a workload consisting of a Gaussian distribution whose midpoint oscillates sinusoidally, defined at the ith iteration and the tth loop execution by

$$w_i^t = \exp\left\{-\left(\frac{i - (N/2 + N\sin(2\pi t/\tau)/4)}{N/8}\right)^2\right\}$$

where N is the number of iterations, and τ is the period of the oscillation. By varying τ we can control how rapidly the load distribution changes from one execution of the loop to the next—a large value gives a nearly static load, a small value gives a rapidly varying one. We execute the parallel loop 1000 times with values of the period τ varying between 10 and 1000.

We use three different loop bodies. Loop Body 1 simply causes the processor to spin for a time Dw_i^t where D is a fixed (wall clock) time period:

```
for (i=0; i<N; i++){
    interval = D * load(i);
    spin(interval);
}
```

Loop Body 2 increments the ith element of an array a number of times proportional to w_i^t:

```
for (i=0; i<N; i++){
    khi = (N/2) * load(i);
    for (k=0; k<khi; k++) a[i]++;
}
```

For small chunk sizes, more than one processor at a time may be updating the same cache line of a, resulting in a significant number of coherency misses in a cache coherent system. In Loop Body 3 each iteration updates elements in a column of a two-dimensional array using the corresponding element of a second array and its four nearest neighbours, where the number of elements updated is proportional to w_i^t.

```
for (i=1; i<=N; i++){
   khi = N * load(i);
   for (k=1; k=<khi; k++){
      a[i][k] = 0.125 * (4.0 * b[i][k] + b[i+1][k] + b[i-1][k]
                                       + b[i][k+1] + b[i][k-1] );
   }
}
```

On alternate executions of the loop the roles of the two arrays are reversed: the elements of a are used to update b. This loop body is designed to test the algorithms' ability to exploit data affinity between subsequent executions. We can now define the four test problems used:

Problem 1 Loop body 1, $N = 1000$, $D = 10^{-4}$ seconds.
Problem 2 Loop body 1, $N = 1000$, $D = 10^{-3}$ seconds.
Problem 3 Loop body 2, $N = 10000$.
Problem 4 Loop body 3, $N = 1000$.

3.2 Results and Discussion

We ran each of the four test problems on 16 processors of a Silicon Graphics Origin 2000 system (with 195 MHz R10000 processors), using the following scheduling algorithms: block, feedback guided (FG) block, affinity, feedback guided affinity, simple self-scheduling (SS), guided self-scheduling (GSS) and trapezoid self-scheduling (TSS), recording the execution time T_p. We also ran the block scheduling algorithm on one processor to give a sequential reference time T_s. The computational cost of feedback guidance is low: each call to the hardware clock requires approximately $0.35\mu s$ and the computation of the new partition for 16 processors approximately $45\mu s$. Figures 1 to 4 show the efficiency (computed as T_s/pT_p) as a function of the period τ of the oscillating workload.

For Problem 1 the workload on each iteration is sufficiently small for the overheads of synchronised access to work queues to dominate the efficiency of the self-scheduling algorithms. The result of this is that simple self-scheduling (which requires $O(N)$ accesses to the central queue), and affinity scheduling (which requires $O(p\log(N/p^2))$ accesses to each of the p local queues) perform no better than block scheduling. GSS with $O(p\log N)$ accesses and TSS with $4p$ accesses to the central queue respectively, are more efficient. Adding feedback guidance to affinity scheduling does nothing to reduce the total number of queue accesses, even though there are fewer accesses to queues of other processors, and the additional overhead of timing each chunk of iterations is incurred. FG block

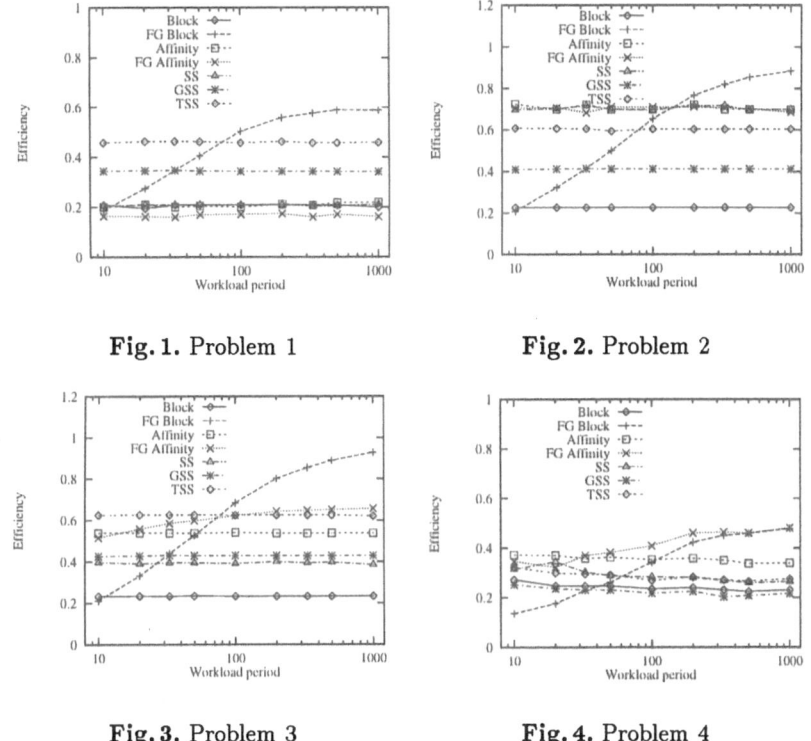

Fig. 1. Problem 1

Fig. 2. Problem 2

Fig. 3. Problem 3

Fig. 4. Problem 4

scheduling avoids all synchronisation costs, and for sufficiently slowly changing workload ($\tau \geq 100$) it is the most efficient of the algorithms tested.

For Problem 2 the workload on each iteration is larger, and synchronisation overhead is less important. SS and affinity scheduling successfully balance the load, whereas GSS and TSS (with larger chunks) do not. Feedback guidance does not improve affinity scheduling, but again for sufficiently slowly changing workload ($\tau \geq 200$), FG block scheduling shows the best performance.

For Problem 3, false-sharing induced communication is the most significant overhead. Both SS and GSS produce many small chunks, with consecutive chunks likely to be executed on different processors. Affinity scheduling also produces many small chunks, but there is a higher likelihood of consecutive chunks being executed on the same processor. TSS produces only $4p$ chunks, and thus it performs well even though it may not successfully balance the load. Apart from the fastest moving workload ($\tau = 10$), FG affinity scheduling is more efficient than affinity scheduling as it further increases the likelihood of consecutive chunks being executed on the same processor. FG block scheduling is again poor for rapidly evolving workload, as it does not balance the load well, but as τ increases, the advantage of having only p chunks becomes significant. For these cases, it is by far the most efficient algorithm.

For Problem 4, both synchronisation and communication overheads are important. None of the central queue algorithms perform well, as they do not exploit data affinity between executions of the parallel loop. For rapidly changing workload ($\tau \leq 20$), affinity scheduling is the most efficient, but as the workload becomes more predictable, FG affinity scheduling gives the best performance. For small τ, FG block scheduling is very poor, but as τ increases, its performance approaches that of feedback guided affinity scheduling.

4 Conclusions

We have described two algorithms which make use of feedback information to reduce the overheads associated with scheduling a parallel loop in a shared variable programming model. Our experiments show that under certain conditions, when there is sufficient correlation between the workload on successive executions of the parallel loop, these new algorithms will outperform traditional loop self-scheduling methods, sometimes by a considerable margin.

References

1. Eager, D.L. and Zahorjan, J. (1992) *Adaptive Guided Self-Scheduling*, Technical Report 92-01-01, Department of Computer Science and Engineering, University of Washington.
2. Hummel, S.F., Schonberg, E. and Flynn, L.E. (1992) *Factoring: A Practical and Robust Method for Scheduling Parallel Loops*, Communications of the ACM, vol. 35, no. 8, pp. 90–101.
3. Kruskal, C.P. and Weiss, A. (1985) *Allocating Independent Subtasks on Parallel Processors*, IEEE Trans. on Software Engineering, vol. 11, no. 10, pp. 1001–1016.
4. Lucco, S. (1992) *A Dynamic Scheduling Method for Irregular Parallel Programs*, in Proceedings of ACM SIGPLAN '92 Conference on Programming Language Design and Implementation, San Francisco, CA, June 1992, pp. 200–211.
5. Markatos, E.P. and LeBlanc, T.J. (1994) *Using Processor Affinity in in Loop Scheduling on Shared Memory Multiprocessors*, IEEE Transactions on Parallel and Distributed Systems, vol. 5, no. 4, pp. 379–400.
6. Polychronopoulos, C. D. and Kuck, D. J. (1987) *Guided Self-Scheduling: A Practical Scheduling Scheme for Parallel Supercomputers*, IEEE Transactions on Computers, C-36(12), pp. 1425–1439.
7. Subramaniam, S. and D.L. Eager (1994) *Affinity Scheduling of Unbalanced Workloads*, in Proceedings of Supercomputing '94, IEEE Comp. Soc. Press, pp. 214–226.
8. Tang, P. and Tew, P-C. (1986) *Processor Self-Scheduling for Multiple Nested Parallel Loops*, in Proceedings of 1986 Int. Conf. on Parallel Processing, pp. 528–535, St. Charles, IL.
9. Tzen, T.H. and Ni, L.M., (1993) *Trapezoid Self-Scheduling Scheme for Parallel Computers*, IEEE Trans. on Parallel and Distributed Systems, vol. 4, no. 1, pp. 87–98.
10. Yan, Y., C. Jin and X. Zhang (1997) *Adaptively Scheduling Parallel Loops in Distributed Shared-Memory Systems* IEEE Trans. on Par. and Dist. Systems, vol. 8, no. 1, pp. 70–81.

Load Balancing for Problems with Good Bisectors, and Applications in Finite Element Simulations

Stefan Bischof, Ralf Ebner, and Thomas Erlebach

Institut für Informatik
Technische Universität München
D-80290 München, Germany
{bischof|ebner|erlebach}@in.tum.de
http://www{mayr|zenger}.in.tum.de/

Abstract. This paper studies load balancing issues for classes of problems with certain bisection properties. A class of problems has α-bisectors if every problem in the class can be subdivided into two subproblems whose weight (i.e. workload) is not smaller than an α-fraction of the original problem. It is shown that the maximum weight of a subproblem produced by Algorithm HF, which partitions a given problem into N subproblems by always subdividing the problem with maximum weight, is at most a factor of $\lfloor 1/\alpha \rfloor \cdot (1 - \alpha)^{\lfloor 1/\alpha \rfloor - 2}$ greater than the theoretical optimum (uniform partition). This bound is proved to be asymptotically tight. Two strategies to use Algorithm HF for load balancing distributed hierarchical finite element simulations are presented. For this purpose, a certain class of weighted binary trees representing the load of such applications is shown to have $1/4$-bisectors. The maximum resulting load is at most a factor of $9/4$ larger than in a perfectly uniform distribution in this case.

1 Introduction

Load balancing is one of the major research issues in the context of parallel computing. Irregular problems are often difficult to tackle in parallel because they tend to overload some processors while leaving other processors nearly idle. For these applications it is very important to find methods for obtaining a balanced distribution of load. Usually, the load is created by processes that are part of a (parallel) application program. During the run of the application, these processes perform certain calculations independently but have to communicate intermediate results or other data using message passing. One is usually interested in achieving a balanced load distribution in order to minimize the execution time of the application or to maximize system throughput.

Load balancing problems have been studied for a huge variety of models, and many different solutions regarding strategies and implementation mechanisms have been proposed. A good overview of recent work can be obtained from [5], for example. [6] reviews ongoing research on dynamic load balancing, emphasizing the presentation of models and strategies within the framework of general classification schemes.

In this paper we study load balancing for a very general class of problems. The only assumption we make is that all problems in the class have a certain bisection property. Such classes of problems arise, for example, in the context of distributed hierarchical finite element simulations. We show how our general results can be applied to numerical applications in several ways.

2 Using Bisectors for Load Balancing

In many applications a computational problem cannot be divided into many small problems as required for an efficient parallel solution directly. Instead, a strategy similar to *divide and conquer* is used repeatedly to divide problems into smaller subproblems. We refer to the division of a problem into two smaller subproblems as *bisection*. Assuming a weight function w that measures the resource demand, a problem p cannot always be bisected into two subproblems p_1 and p_2 of equal weight $w(p)/2$. For many classes of problems, however, there is a bisection method that guarantees that the weights of the two obtained subproblems do not differ too much. The following definition captures this concept more precisely.

Definition 1. *Let* $0 < \alpha \leq \frac{1}{2}$. *A class* $\mathcal{P}$ *of problems with weight function* $w : \mathcal{P} \to \mathbb{R}^+$ *has* α-*bisectors if every problem* $p \in \mathcal{P}$ *can be efficiently divided into two problems* $p_1 \in \mathcal{P}$ *and* $p_2 \in \mathcal{P}$ *with* $w(p_1) + w(p_2) = w(p)$ *and* $w(p_1), w(p_2) \in [\alpha w(p); (1 - \alpha)w(p)]$.

Note that this definition requires for the sake of simplicity that all problems in $\mathcal{P}$ can be bisected, whereas in practice this is not the case for problems whose weight is below a certain threshold. We assume, however, that the problem to be divided among the processors is big enough to allow further bisections until the number of subproblems is equal to the number of processors. This is a reasonable assumption for most relevant parallel applications.

Figure 1 shows Algorithm HF, which receives a problem p and a number N of processors as input and divides p into N subproblems by repeated application of α-bisectors to the heaviest remaining subproblem. A perfectly balanced load distribution on N processors would be achieved if a problem p of weight $w(p)$ was divided into N subproblems of weight exactly $w(p)/N$ each. The following theorem gives a worst-case bound on the ratio between the maximum weight among the N subproblems produced by Algorithm HF and this ideal weight $w(p)/N$. All proofs are omitted in this paper due to space constraints but can be found in [1].

Theorem 2. *Let* $\mathcal{P}$ *be a class of problems with weight function* $w : \mathcal{P} \to \mathbb{R}^+$ *that has* α-*bisectors. Given a problem* $p \in \mathcal{P}$ *and a positive integer* N, *Algorithm HF uses* $N - 1$ *bisections to partition* p *into* N *subproblems* $p_1, \ldots, p_N$ *such that*

$$\max_{1 \leq i \leq N} w(p_i) \leq \frac{w(p)}{N} \cdot \left\lfloor \frac{1}{\alpha} \right\rfloor \cdot (1 - \alpha)^{\lfloor \frac{1}{\alpha} \rfloor - 2}.$$

```
Input: problem p, positive integer N
begin
    P ← {p};
    while |P| < N do
        begin
            q ← a problem in P with maximum weight;
            bisect q into q₁ and q₂;
            P ← (P ∪ {q₁, q₂}) \ {q};
        end;
        output P;
end.
```

Fig. 1. Algorithm HF (Heaviest Problem First)

It is not difficult to see that the bound given in Theorem 2 is asymptotically tight. This means that for sufficiently large N there exists a sequence of bisections such that the maximum load generated by Algorithm HF is arbitrarily close to this bound. For this purpose, the original problem is first split into a number of subproblems $p_1, \ldots, p_n$ of equal weight. Each of $p_1, \ldots, p_{n-1}$ is then split into $\lfloor 1/\alpha \rfloor$ subproblems using a sequence of $\lfloor 1/\alpha \rfloor - 1$ exact α-bisections, whereas p_n is split similarly but without doing the final bisection step.

The construction outlined above suggests that the bound from Theorem 2 might not be tight for small values of N, however. Indeed, we can prove that for all $\alpha \leq 1/5$ and for all $N \leq 1/\alpha$, Algorithm HF partitions a problem from a class of problems with α-bisectors into N subproblems such that the weight of the heaviest subproblem is at most $w(p)(1 - \alpha)^{N-1}$. Hence, the ratio between the maximum weight and the ideal average weight is bounded by $N(1 - \alpha)^{N-1}$, which is smaller than $\lfloor 1/\alpha \rfloor (1-\alpha)^{\lfloor 1/\alpha \rfloor - 2}$ for the range where this bound applies.

3 Application of Algorithm HF to Distributed Finite Element Simulations

In this section, we present the application of Algorithm HF for load balancing in the field of distributed numerical simulations with the finite element (FE) method [2]. The FE method is used in statics analysis, for example, to calculate the response of objects under external load. In this paper, we consider a short cantilever under plane stress conditions, a problem from plane elasticity. The quadratic cantilever is uniformly loaded on its upper side. Its left side is fixed as shown in Figure 2, top left. The definition of the object's shape, its structural properties, and the boundary conditions lead to a system of elliptic partial differential equations.

We substructure the physical domain of the cantilever recursively. With the method of [4], a tree data structure is built reflecting the hierarchy of the substructured domain (see Figure 2). This data structure contains a system of linear

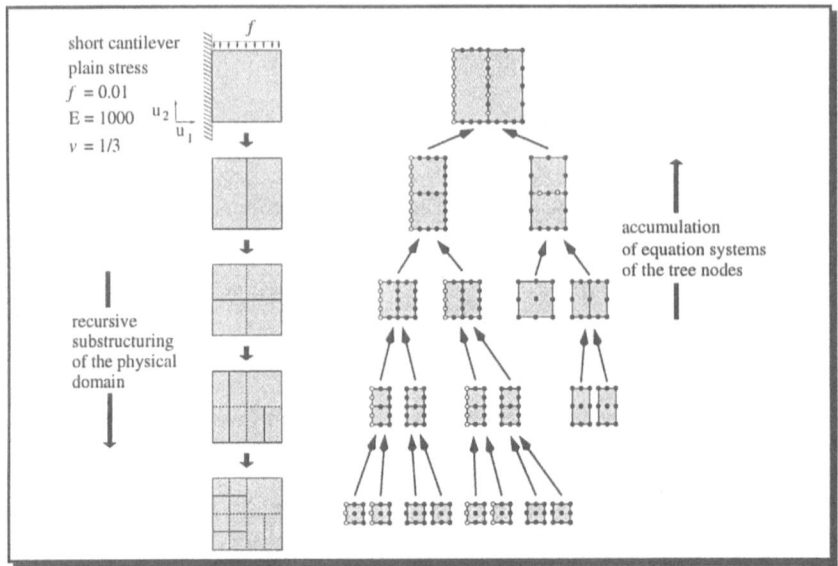

Fig. 2. The example application

equations $Su = f$ for each tree node. Its *stiffness matrix S* determines the unknown displacement values in the vector u at the black points shown in Figure 2
(right), dependent on the external and reaction forces f. In the leaves, the system of equations is constructed by standard FE discretization and the so-called
virtual displacement method. In an internal node, the system of linear equations
is assembled out of the equation systems of its children.

For the solution of all those equation systems, we use an iterative solver which
traverses the tree several times, promoting displacements in top-down direction
and reaction forces in bottom-up direction. Since the adaptive structure of the
tree is not known *a priori*, a good load balancing strategy is essential before the
parallel execution of the solving iterations.

We assign a load value $\ell(p) = C_b n_b(p) + C_s n_s(p)$ to each tree node p, with
$n_b(p)$ black points on the border without boundary conditions, and $n_s(p)$ black
points (not ghost points) belonging to the separator line of node p. The load
value $\ell(p)$ models the computing time of node p, where the constants C_s and C_b
are independent of node p, and $C_s \approx 6C_b$. These values can be easily collected
during the tree construction phase.

In order to distribute the work load among the processors, we need to know
the computing time of whole subtrees of the FE tree. So for each node p, we
accumulate the load in the subtree with root p to get the weight value $\mathsf{w}(p)$:

$$\mathsf{w}(p) = \begin{cases} \ell(p) & \text{if } p \text{ is leaf} \\ \ell(p) + \mathsf{w}(c_1) + \mathsf{w}(c_2) & \text{if } p \text{ is root or internal node} \end{cases}$$

where c_1 and c_2 are the children of p.

For a bisection step according to Algorithm HF, we remove the root node p
from the tree of weight values. This yields two subtrees rooted at the children

c_1 and c_2 of p, and the weight of the root node p is ignored for the remainder of the load balancing phase. The results of Theorem 2 are well approximated because $\ell(p)$ (work load for the one-dimensional separator) is negligible compared to $w(p)$ (work load for the two-dimensional domain) in our application with commensurately large FE trees.

Algorithm HF chops N subtrees off the FE tree, each of which can be traversed in parallel by the iterative solver. These N subtrees contain the main part of the solving work and may be distributed over the available N processors, whereas the upper $N-1$ tree nodes cannot exploit the whole number of processors, anyway.

Another strategy divides the *entire* given FE tree into N partitions of approximately equal size by cutting edges (and not root nodes of subtrees). For this purpose, we set $w(p) = \ell(p)$ in each tree node. So this application of Algorithm HF takes into account that the main memory resources of the processors become the limiting factor if very high accuracy of the simulation is required.

4 Weighted Trees with Good Bisectors

Let $\mathcal{T}$ be the set of all rooted binary trees with node weights $\ell(v)$ satisfying:

(1) $\ell(p) \le \ell(c_1) + \ell(c_2)$ for nodes p with two children c_1 and c_2
(2) $\ell(p) \ge \ell(c)$ if c is a child of p

The weight of a tree $T = (V, E)$ in $\mathcal{T}$ is defined as $w(T) = \sum_{v \in V} \ell(v)$.

This class $\mathcal{T}$ of binary trees models the load of applications in hierarchical finite element simulations, as discussed in Section 3. Recall that in these applications the domain of the computation is repeatedly subdivided into smaller subdomains. The structure of the domains and subdomains yields a binary tree in which every node has either two children or is a leaf. The resource demands (CPU and main memory) of the nodes in this FE tree are such that the resource demand at a node is at most as large as the sum of the resource demands of its two children.

Note that (1) and (2) ensure that the two subtrees obtained from a tree in $\mathcal{T}$ by removing a single edge are also members of $\mathcal{T}$. The following theorem shows that trees from the class $\mathcal{T}$ can be $\frac{1}{4}$-bisected by removal of a single edge unless the weight of the tree is concentrated in the root.

Theorem 3. *Let $T = (V, E)$ be a tree in $\mathcal{T}$, and let r be its root. If $\ell(r) \le \frac{3}{4} w(T)$, then there is an edge $e \in E$ such that the removal of e partitions T into subtrees T_1 and T_2 with $w(T_1), w(T_2) \in [\frac{1}{4} w(T); \frac{3}{4} w(T)]$.*

According to Theorem 2 a problem p from a class of problems that has $\frac{1}{4}$-bisectors can always be subdivided into N subproblems $p_1, \ldots, p_N$ such that $\max_{1 \le i \le N} w(p_i) \le \frac{w(p)}{N} \cdot \frac{9}{4}$. The following corollary gives a condition on trees in $\mathcal{T}$ that ensures that they can be subdivided into N subproblems using $\frac{1}{4}$-bisectors.

Corollary 4. *Let $T = (V, E)$ be a tree in $\mathcal{T}$, and let r be its root. Let N be a positive integer. If $w(T) \ge \frac{4}{3}(N-1)\ell(r)$, Algorithm HF partitions T into N*

subtrees by cutting exactly $N - 1$ edges such that the maximum weight of the resulting subtrees is at most $\frac{9}{4} \cdot \frac{w(T)}{N}$.

Note that an optimal min-max k-partition of a weighted tree (i.e., a partition with minimum weight of the heaviest component after removing k edges) can be computed in linear time [3]. These algorithms are preferable to our approach using Algorithm HF in the case of trees that are to be subdivided by removing a minimum number of edges. Since the heaviest subtree in the optimal solution does obviously not have a greater weight than the maximum generated by Algorithm HF, the bound from Corollary 4 still applies.

5 Conclusion

The existence of α-bisectors for a class of problems was shown to allow good load balancing for a surprisingly large range of values of α. The maximum load achieved by Algorithm HF is at most a factor of $\lfloor 1/\alpha \rfloor \cdot (1 - \alpha)^{\lfloor 1/\alpha \rfloor - 2}$ larger than the theoretical optimum (uniform distribution).

Load balancing for distributed hierarchical FE simulations was discussed, and two strategies for applying Algorithm HF were presented. The first strategy tries to make the best use of the available parallelism, but requires that the nodes of the FE tree representing the load of the computation have good separators. The second strategy tries to partition the entire FE tree into subtrees with approximately equal load. For this purpose, it was proved that a certain class of weighted trees, which include FE trees, has 1/4-bisectors. Here, the trees are bisected by removing a single edge. Partitioning the trees by removing a minimum number of edges ensures that only a minimum number of communication channels of the application must be realized by network connections. Our results provide worst-case bounds on the maximum load achieved for applications with good bisectors in general and for distributed hierarchical finite element simulations in particular. For the latter application, we showed that the maximum resulting load is at most a factor of 9/4 larger than in a perfectly uniform distribution.

At present we are implementing the load balancing methods proposed in this paper and integrating them into the existing finite element simulations software. We expect considerable improvements as compared to the static (compile-time) processor allocation currently in use. See [1] for first results.

References

1. Stefan Bischof, Ralf Ebner, and Thomas Erlebach. Load Balancing for Problems with Good Bisectors, and Applications in Finite Element Simulations: Worst-case Analysis and Practical Results. SFB-Bericht 342/05/98 A, SFB 342, Institut für Informatik, Technische Universität München, 1998. http://wwwmayr.in.tum.de/berichte/1998/TUM-I9811.ps.gz.
2. Dietrich Braess. *Finite Elemente*. Springer, Berlin, 1997. 2. überarbeitete Auflage.
3. Greg N. Frederickson. Optimal Algorithms for Tree Partitioning. In *Proceedings of the Second Annual ACM-SIAM Symposium on Discrete Algorithms SODA '91*, pages 168–177, New York, 1991. ACM Press.

4. Reiner Hüttl. *Ein iteratives Lösungsverfahren bei der Finite-Element-Methode unter Verwendung von rekursiver Substrukturierung und hierarchischen Basen.* PhD thesis, Institut für Informatik, Technische Universität München, 1996.

5. Behrooz A. Shirazi, Ali R. Hurson, and Krishna M. Kavi. *Scheduling and Load Balancing in Parallel and Distributed Systems.* IEEE Computer Society Press, Los Alamitos, CA, 1995.

6. Thomas Schnekenburger and Georg Stellner, editors. *Dynamic Load Distribution for Parallel Applications.* TEUBNER-TEXTE zur Informatik. Teubner Verlag, Stuttgart, 1997.

An Efficient Strategy for Task Duplication in Multiport Message-Passing Systems

Dingchao Li[1], Yuji Iwahori[1], Tatsuya Hayashi[2] and Naohiro Ishii[3]

[1] Educational Center for Information Processing
[2] Department of Electrical and Computer Engineering
[3] Department of Intelligence and Computer Science
Nagoya Institute of Technology
liding@center.nitech.ac.jp

Abstract. important problem in compilers for parallel machines. In this paper, we present a duplication strategy for task scheduling in multiport message-passing systems. Through a performance gain analysis, we establish a condition under which duplicating a parent task of a task is beneficial. We also show that, by incorporating this strategy into two well-known priority-based scheduling algorithms, significant reductions in the execution time can be achieved.

1 Introduction

Scheduling is a sequential optimization technique used to exploit the parallelism inherent in programs. In this paper, we consider the problem of scheduling the tasks of a given program in a multiport message-passing system with the aim of minimizing the overall execution time of the program. Researchers have investigated two versions of this problem, depending on whether or not task duplication (or recomputation) is allowed. In general, for the same program, task scheduling with duplication produces a schedule with a smaller makespan (i.e., total execution time) than when task duplication is not allowed.

Some duplication based scheduling algorithms have been introduced in [2, 4, 5, 7]. However, all of these algorithms assume the availability of unlimited number of processors. If the number of processors available is less than the number of the processors they require, there could be a problem. In this paper, we aim to develop an efficient and practical algorithm with task duplication for scheduling tasks onto a bounded number of available processors. In this area, the work described in [8] is close to ours. The algorithm determines the tasks on the critical paths of a given program before scheduling and tries to select such tasks for duplication in every scheduling step. However, a critical path may be shortened and new critical paths may be generated during the scheduling process, because assigning a task and a parent task of the task to a single processor makes the communication overhead between them become zero. Therefore, our algorithm does not attempt to identify whether a task is critical. Instead, it dynamically analyzes the increase over the program execution time, created possibly by duplicating a parent task of a task, to guide a duplication decision. We will show that incorporating this strategy with some known scheduling algorithms such as the mapping heuristic [9] and the dynamic level method [10] can improve their performance without introducing additional computational complexity.

The remainder of the paper is organized as follows: system and program models are described in Section 2 and in Section 3, we present a duplication strategy for task scheduling. The performance evaluation of the strategy is discussed in Section 4. Finally, Section 5 gives some concluding remarks.

2   System and Program Models

We consider a system consisting of m identical programmable processors $P_i(i = 1 \cdots m)$, each with a private or local memory and fully connected by a multi-port interconnection network. In this system, the tasks of a parallel program mapped to different processors communicate solely by message-passing. In every communication step, a processor can send distinct messages to other processors and simultaneously receive multiple messages from these processors. We assume that communication can be overlapped by computation and the communication time can be neglected if two tasks are mapped to a processor.

A parallel program is modeled as a weighted directed acyclic graph, or task graph. Let $G = (\Gamma, A, \mu, c)$ denote a task graph, where Γ is a finite set of tasks $T_i(i = 1, \cdots, n)$, A is a set of directed arcs representing dependence constraints among tasks, μ is an execution time function whose value $\mu(T_i)$ (time units) is the execution time of task T_i, and c is a communication cost function whose value $c(T_i, T_j)$ denotes the amount of communication from T_i to T_j. The arc (T_i, T_j) from T_i to T_j asserts that task T_j cannot start execution until the message from T_i arrives at the processor executing T_j. If $(T_i, T_j) \in A$, T_i is said to be a parent task of T_j and T_j is said to be a child task of T_i. In task graph G, tasks without parent tasks are known as entry tasks and tasks without child tasks are known as exit tasks. We assume, without loss of generality, that G has exactly one entry task and exactly one exit task.

For task T_i, let $est(T_i)$ and $lst(T_i)$ denote the earliest start and completion times, respectively. The computation of the earliest start and completion times can be determined in a top-down fashion, starting with the entry task and terminating at the exit task. Mathematically,

$$est(T_i) = \min_{(T_j, T_i) \in A} \max_{(T_k, T_i) \in A, k \neq j} \{ect(T_j), ect(T_k) + c(T_k, T_i)\}, \tag{1}$$

$$ect(T_i) = est(T_i) + \mu(T_i). \tag{2}$$

Similarly, let $lst(T_i)$ and $lct(T_i)$ denote the latest start and completion times of T_i, respectively. The latest start and completion times of T_i can be given by

$$lct(T_i) = \min_{(T_i, T_j) \in A} \{lst(T_j) - c(T_i, T_j)\}, \tag{3}$$

$$lst(T_i) = lct(T_i) - \mu(T_i). \tag{4}$$

Note that the latest start time of T_i indicates how long the start of T_i can be delayed without increasing the minimum execution time of the graph.

3 Scheduling with Duplication

For machine and program models described above, this section develops a duplication based algorithm in an attempt to find a one-to-one mapping from each task to its assigned processor such that the program execution time is minimized. Our algorithm consists of two parts. The first part decides which ready task should be assigned to an idle processor. This part is heuristic in nature and we can make the choice by applying one of the existing strategies such as HLF(Highest Level First), ETF(Earliest Task First), LP(Longest Path), LPT(Longest Processing Time), and CP(Critical Path). Given the task in the first part, the second part decides whether a parent of the task should be duplicated.

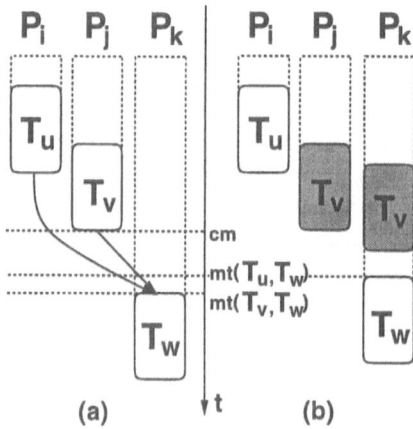

Fig. 1. The task duplication strategy. (a) The schedule before duplication. (b) The schedule after duplication.

Before discussing our task duplication strategy, we introduce cm to represent the current time of the event clock, $P(T_u)$ to represent the processor that processes task T_u and $st(P(T_u), T_u)$ to represent the time when T_u starts execution on $P(T_u)$. We also assume that processor P_k is idle at cm and task T_w is selected for scheduling, as shown in Fig. 1. The start time at which task T_w can run on processor P_k in case of no task duplication is thus given by

$$st_{nocopy}(P_k, T_w) = \max_{\substack{(T_u, T_w) \in A, P(T_u) \neq P_k}} \{cm, st(P(T_u), T_u) + \mu(T_u) + c(T_u, T_w)\}. \tag{5}$$

Consider the fact that if the start time of a task is later than the latest start time of the task, then the overall completion time of the whole graph will be increased. Thus, we can define the increase incurred due to the execution of task T_w on processor P_k as follows.

$$\delta_{nocopy}(P_k, T_w) = st_{nocopy}(P_k, T_w) - lst(T_w). \tag{6}$$

Obviously, if $\delta_{nocopy}(P_k, T_w) \leq 0$, we should stop duplicating the parent tasks of T_w and assign T_w onto processor P_k, since duplicating its parent tasks is unable to shorten the schedule length even if its start time is reduced.

Assume now that $\delta_{nocopy}(P_k, T_w) > 0$. To minimize the overall execution time, we need to consider task duplication for reducing this increase as much as possible. Let us use $mt(T_u, T_w)$ to denote the time when the message for task T_w from T_u arrives at processor $P(T_w)$; i.e., $mt(T_u, T_w) = st(P(T_u), T_u) + \mu(T_u) + c(T_u, T_w)$ if $P(T_u) \neq P(T_w)$. From Fig. 1, it is clear that we should select a task T_v, as the candidate for duplication, such that

$$mt(T_v, T_w) = \max_{(T_u, T_w) \in A, P(T_u) \neq P(T_w)} \{st(P(T_u), T_u) + \mu(T_u) + c(T_u, T_w)\}.$$

Further let $it(P_k)$ denote the time when processor P_k becomes idle. Thus, the time at which T_v can start execution on P_k can be calculated below.

$$st(P_k, T_v) = \max_{(T_s, T_v) \in A, P(T_s) \neq P_k} \{it(P_k), st(P(T_s), T_s) + \mu(T_s) + c(T_s, T_v)\}. \tag{7}$$

Consequently, in this case, the time at which task T_w can start execution on processor P_k is that

$$st_{copy}(P_k, T_w) = \max_{(T_u, T_w) \in A, P(T_u) \neq P_k, u \neq v} \{cm, st(P(T_u), T_u) + \mu(T_u) + c(T_u, T_w), st(P_k, T_v) + \mu(T_v)\}. \tag{8}$$

This produces an increase as follows.

$$\delta_{copy}(P_k, T_w) = st_{copy}(P_k, T_w) - lst(T_w). \tag{9}$$

Comparing $\delta_{copy}(P_k, T_w)$ with $\delta_{nocopy}(P_k, T_w)$, we can finally establish the following condition: duplicating a parent task of T_w to the current processor should only occur if

$$\delta_{copy}(P_k, T_w) < \delta_{nocopy}(P_k, T_w). \tag{10}$$

The above duplication strategy clearly takes $O(p)$ times in the worst case, where p is the maximum number of parent tasks of every task in G. Therefore, incorporating this strategy into some known scheduling algorithms such as the mapping heuristic [9] and the dynamic level method [10] does not introduce additional time complexity, since these algorithms generally require $O(nm)$ times for selecting a ready task and an idle processor for assignment, where n is the number of tasks and m is the number of processors.

4 Experimental Study

This section presents an experimental study of the effects of the task duplication strategy described in Section 3. By incorporating the strategy into the mapping heuristic (MH) and the dynamic level method (DL) without global clock heuristic, we implemented two scheduling algorithms on a Sun workstation using the programming language C. Whereas we compared the performance of each enhanced algorithm with that of the original algorithm under various parameter settings, due to space limitations, here we show only some of the salient results.

In our experiments, we generated 350 random task graphs and scheduled them onto a machine consisting of eight identical processors, fully interconnected through full-duplex interprocessor links. The task graphs generated randomly ranged in size from 40 to 160 nodes with increments of 20. Each graph was constructed by generating a directed acyclic graph and removing all transitive arcs. For each vertex, the execution time and the number of the child tasks were picked from an uniform distribution whose parameters are input data. The communication time of each arc was determined alike. The communication-to-computation (C/C) ratio value was settled to be 5.0, where the C/C ratio of a task graph is defined as the average communication time per arc divided by the average execution time per task.

The performance analysis was based on the average schedule length obtained by each algorithm. Tables 1 and 2 show the experimental results, where *Impr.* represents the percentage performance improvement of each enhanced algorithm compared with that of the original algorithm. It can be seen that the performance of each algorithm with duplication is consistently better than the algorithm without duplication. The performance improvement compared with the original mapping heuristic varies from 4.01% to 8.17%, and the performance improvement compared with the original dynamic level algorithm varies from 3.42% to 6.90%. These results confirm our expectation that the proposed duplication strategy is efficient and practical.

Table 1. Results from the mapping heuristics with and without duplication.

No. of graphs	No. of tasks	Avg. sche. leng. MH without dup.	Avg. sche. leng. MH with dup.	Impr. (%)
50	40	649.58	609.64	6.15
50	60	860.88	826.34	4.01
50	80	1093.38	1023.16	6.42
50	100	1310.74	1231.68	6.03
50	120	1515.46	1401.70	7.51
50	140	1711.06	1583.58	7.45
50	160	1962.92	1802.51	8.17

Table 2. Results from the dynamic level algorithms with and without duplication.

No. of graphs	No. of tasks	Avg. sche. leng. DL without dup.	Avg. sche. leng. DL with dup.	Impr. (%)
50	40	602.60	572.41	5.01
50	60	788.61	761.66	3.42
50	80	1037.65	993.82	4.22
50	100	1226.73	1172.94	4.38
50	120	1454.92	1367.88	5.98
50	140	1643.25	1536.92	6.47
50	160	1891.74	1761.24	6.90

5 Conclusions

We have presented a novel strategy for task duplication in multiport message-passing systems. The strategy is based on the analysis of the increase over the total program execution time, created possibly by duplicating a parent task of a task. Experimental results have shown that incorporating it into two well-known scheduling algorithms either produces the same assignments as the original algorithms, or it produces better assignments. In addition, the costs, both in implementation effort and compile time, are very low. Future work includes a more complete investigation of the impact of varying the communication-to-computation ratio. Experimental studies on various multiprocessor platforms are also needed.

Acknowledgment

This work was supported in part by the Ministry of Education, Science and Culture under Grant No. 09780263, and by a Grant from the Artificial Intelligence Research Promotion Foundation under contract number 9AI252-9. We would like to thank the anonymous referees for their valuable comments.

References

1. B. Kruatrachue and T.G. Lewis, Grain Size Determination for Parallel Processing, IEEE Software, pp. 23-32, Jan., 1988.
2. J.Y. Colin and P. Chritienne, C.P.M. Scheduling with Small Communication Delays and Task Duplication, Operations Research, vol. 39, no. 4, pp. 680-684, July, 1991.
3. Y.C.Chung and S.Ranka, Application and Performance Analysis of a Compile-Time Optimization Approach for List Scheduling Algorithms on Distributed-Memory Multiprocessors, Proc. of Supercomputing'92, pp. 512-521, 1992.
4. H.B. Chen, B. Shirazi, K. Kavi, and A.R. Hurson, Static Scheduling Using Linear Clustering with Task Duplication, Proc. of ISCA International Conference on Parallel and Distributed Computing and Systems, pp. 285-290, 1993.

5. J. Siddhiwala and L.F. Chao, Path-Based Task Replication for Scheduling with Communication Costs, Proc. of the 1995 International Conference on Parallel Processing, vol. II, pp. 186-190, 1995.
6. M. A. Palis, J. Liou and D. S. L. Wei, Task Clustering and Scheduling for Distributed Memory Parallel Architectures, IEEE Trans. on Parallel and Distributed Systems, vol. 7, no. 1, pp. 46-55, Jan. 1996.
7. S. Darbha and D.P. Agrawal, Optimal Scheduling Algorithm for Distributed-Memory Machines, IEEE Trans. on Parallel and Distributed Systems, vol. 9, no. 1, pp. 87-95, Jan. 1998.
8. K.K. Kwok and I. Ahmad, Exploiting Duplication to Minimize the Execution Times of Parallel Programs on Message-Passing Systems, Proc. of the sixth IEEE Symposium on Parallel and Distributed Processing, pp. 426-433, 1994.
9. H.El-Rewini and T.G.Lewis, "Scheduling Parallel Program Tasks onto Arbitrary Target Machines," J. Parallel and Distributed Computing 9, pp. 138-153, 1990.
10. G.C.Sin and E.A.Lee, "A Compile-Time Scheduling Heuristic for Interconnection-Constrained Heterogeneous Processor Architectures," IEEE Trans. on Parallel and Distributed Syst., vol. 4, no. 2, pp. 175-187, 1993.

Evaluation of Process Migration for Parallel Heterogeneous Workstation Clusters

M.A.R. Dantas

Computer Science Department,
University of Brasilia (UnB),
Brasilia, 70910-900, Brazil
mardantas@computer.org

Abstract. It is recognised that the availability of a large number of workstations connected through a network can represent an attractive option for many organisations to provide to application programmers an alternative environment for high-performance computing. The original specification of the Message-Passing Interface (MPI) standard was not designed as a comprehensive parallel programming environment and some researchers agree that the standard should be preserved as simple and clean as possible. Nevertheless, a software environment such as MPI should have somehow a scheduling mechanism for the effective submission of parallel applications on network of workstations. This paper presents the performance results and benefits of an alternative lightweight approach called *Selective - MPI* (S-MPI), which was designed to enhance the efficiency of the scheduling of applications on parallel workstation cluster environments.

1 Introduction

The standard message-passing interface known as MPI (Message Passing Interface) is a specification which was originally designed for writing applications and libraries for distributed memory environments. The advantages of a standard message-passing interface are portability, ease-of-use and providing a clearly defined set of routines to the vendors that they can design their implementations efficiently, or provide hardware and low-level system support.Computation on workstation clusters when using the MPI software environment might result in low performance, because there is no elaborate scheduling mechanism to efficiently submit processes to these environments. Consequently, although a cluster could have available resources (e.g. lightly-loaded workstations and memory), these parameters are not completely considered and the parallel execution might be inefficient.This is a potential problem interfering with the performance of parallel applications, which could be efficiently processed by other machines of the cluster.

In this paper we consider the *mpich* [2] as a case study implementation of MPI. In addition, we are using ordinary Unix tools therefore concepts employed in the proposed system can be extended for similar environments. An approach called **Selective - MPI (S-MPI)** [1], is an alternative design to provide MPI users with the same attractive features of the existing environment and covers some of the functions of a high-level task scheduler. Experimental results of demonstrator applications indicate that the proposed solution enhances successfully the performance of applications with a significant reduction on the elapsed-time. The present paper is organised as follows. In section 2, we outline some details of the design philosophy of the S-MPI. Experimental results are presented in section 3. Finally, section 4 presents conclusions.

2   S-MPI Architecture

Figure 1, illustrates the architecture of the S-MPI environment compared to the *mpich* implementation. Difference between the two environments are mainly the layer where the user interface sits and the introduction of a more elaborate scheduling mechanism.

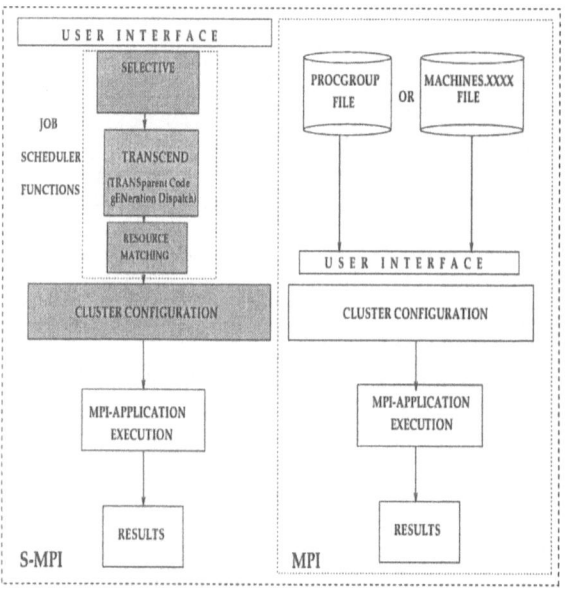

Fig. 1. Architecture differences between S-MPI and *mpich*.

The S-MPI approach covers functions of a high-level task scheduler monitoring workstations workload, automatically compiling (or recompiling) the parallel code, matching processes requirements with resources available and finally selecting dynamically an appropriate cluster configuration to execute parallel applications efficiently.

3 Experimental Results

Configurations considered in this paper use a distributed parallel configuration composed of heterogeneous clusters of SUNs and SGIs (Silicon Graphics) workstations, in both cases using the *mpich* implementation. These workstations were private machines which could have idle cycles (i.e. considering a specific threshold of the average CPU queue length load index of the workstations) and memory available. Operating systems and hardware characteristics of these workstations are illustrated by table 1.

Table 1. General charateristics of the SUN and SGI clusters.

CLUSTERS	SUN	SGI
Machine Architecture	SUN4u - Ultra	IP22
Clock(MHz)	143-167	133-250
Data/Instruction Cache (KB)	64	32-1024
Memory(MB)	32-250	32-250
Operating System	SOLARIS 5.5.1	IRIX 6.2
MPI version	*mpich* 1.1	*mpich* 1.1

We have implemented two demonstrator applications with different ratio between computation and communication, The first application, was a gaussian algorithm to solve a system of equations. The second algorithm implemented was a matrix multiplication, which is also an essential component in many applications. Our algorithm is implemented using a *master-slave* interaction.

The improved performance trend on the S-MPI environment executing the matrix multiplication is presented in figure 2. This figure demonstrates that the S-MPI has a better performance than the MPI configurations. The SUN-MPI and SGI-MPI environments represent the use of the specific machines defined in the static *machines.xxx* files. The SUN/SGI-MPI configuration shows results of the use of the *procgroup* file. This static file was formed with the address of heterogeneous machines from both SUN and SGI clusters. Finally, the results of SUN/SGI-SMPI illustrated the use of the automatic cluster configuration approach of the S-MPI.

Figure 3 shows the enhanced performance of the S-MPI approach compared to the MPI for solving a system of 1000 equations. This figure shows a downward trend of the elapsed-time for the S-MPI environment in comparison to the conventional SUN and SGI MPI environments.

4 Conclusions

In this paper we have evaluated the performance results and benefits of a lightweight model called **Selective - MPI (S-MPI)**. The system co-exists cleanly with the

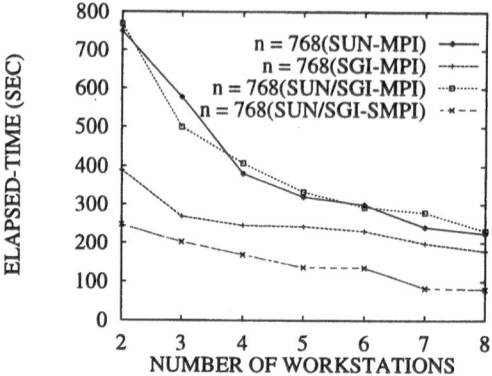

Fig. 2. Elapsed-time of the matrix multiplication on MPI and S-MPI environments.

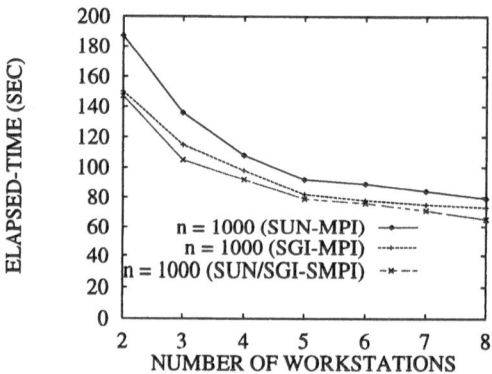

Fig. 3. Elapsed-time of the Gaussian algorithm on MPI and S-MPI environments.

existing concepts of the MPI and preserves the original characteristics of the software environment. In addition, S-MPI provides application programmers with a more transparent and enhanced environment to submit MPI parallel applications without layers of additional complexities for heterogeneous workstation clusters. Moreover, the concept clearly has advantages providing users with the same friendly interface and ensuring that all available resources will be evaluated before configuring a cluster to run MPI applications.

References

1. M.A.R. Dantas and E.J. Zaluska. Efficient Scheduling of MPI Applications on Network of Workstations. *Accepted Paper on Future Generation Computer Systems Journal*, 1997.
2. William Gropp Patrick Bridges, Nathan Doss. Users' Guide to mpich, a Portable Implementation of MPI. *Argonne National Laboratory* http://www.mcs.anl.gov, 1994.

Using Alternative Schedules for Fault Tolerance in Parallel Programs on a Network of Workstations

Dibyendu Das*

Department of Computer Science and Engineering,
Indian Institute of Technology, Kharagpur 721302, India
deedee@cse.iitkgp.ernet.in

1 Introduction

In this work a new method has been devised whereby a parallel application running on a cluster of networked workstations (say, using PVM) can adapt to machine faults or machine reclamation [1] without any program intervention in a transparent manner. Failures are considered fail-stop in nature rather than Byzantine ones [4]. Thus, when machines fail, they are effectively shut-out from the network dying an untimely death. Such machines do not contribute anything further to a computation process. For simplicity both failures as well as reclamation are referred to as faults henceforth.

Alternative schedules are used for implementing this adaptive fault tolerance mechanism. For parallel programs representable by static weighted task graphs, this framework builds a number of schedules, statically at compile time, instead of a single one. Each such schedule has a different machine requirement i.e. schedules are built for 1 machine, 2 machines and so on till a maximum of M machines. The main aim of computing alternative schedules is to have possible schedules to one of which the system can fall back upon for fault recovery during run-time. For example, if a parallel program starts off with a 3-machine schedule and one of them faults during run-time the program adapts by switching to a 2-machine schedule computed beforehand. However, the entire program is not restarted from scratch. Recomputation of specific tasks and resending of messages if required are taken care of.

2 Alternative Schedule Generation Using Clustering with Cloning

Schedules in this work are generated using the concept of task clustering with cloning [2], [5] as cloning of tasks on multiple machines have a definite advantage as far as adapting to faults is concerned. The Duplication Scheduling Heuristic by Kruatrachue et al. [2], [3] has been used to generate alternative schedules with differing machine requirements like 1,2,3, ...

* The author acknowledges financial assisstance in the form of K. S. Krishan Fellowship

3 The Schedule Switching Strategy

The major problem at run-time is to facilitate a schedule switch once some machine/s is/are thrown out of the cluster due to fault. As the run-time system may need a new schedule to be invoked at a fault juncture, such a schedule should already exist on the machines. In addition, it is not known at compile time, which cluster of the new schedule will be invoked on which machine. Once the set of completed tasks on each live machine is known the schedule switch to an alternative schedule should take place in such a manner that the resultant program completes as quickly as possible. Also, the switching algorithm itself should be simple and fast.

3.1 Algorithm SchedSwitch

The algorithm for schedule switching is based on minimal weighted bipartite graph matching. The main idea is to shift from a currently executing thread (cluster of a schedule executing on a machine) to a virtual cluster of an alternative schedule such that the overall scheduling time is reduced as much as possible. In order to do this a complete weighted bipartite graph $G = (V, E)$ is constructed where $V = M_{live} \cup V_{Cl}$. M_{live} is the set of live machines (except the faulty ones) and V_{Cl} is the set of clusters of an alternative schedule to which a schedule switch can take place. The set of edges $E = (v_1, v_2), v_1 \in M_{live}, v_2 \in V_{Cl}$. Each edge is weighted, the weight being computed via the following strategy:

The weight of the matching between cluster i in the pre-computed schedule with machine j is as follows. The not done tasks of cluster i are tried to be scheduled in the machine j as early as possible. While constructing this schedule, some of the tasks in cluster i may start late as compared to their start time in the original alternative schedule. The maximum of these delays among the tasks in cluster i is computed, and the edge e is annotated with this value. A minimum weighted bipartite matching algorithm is applied to G thus created to determine the best mapping of clusters to machines.

3.2 Static and Run-Time Strategies for Schedule Switching

One of the major requirements for an adaptive program - be it for fault adaptation or machine reclamation, is that the adaptive mechanism must be reasonably fast. Keeping this in mind two strategies have been designed for this framework - STSR (Static Time Switch Resolution) and RTSR (Run Time Switch Resolution). STSR is a static time approach whereby the possible schedule switches are computed statically given certain fault configurations - usually single machine faults. RTSR is a run-time approach whereby the schedule switching algorithm is applied during adaptation. This is a more general approach compared to STSR.

STSR is a costly approach even if applied for only single machine failures as it is not known at what point a machine may fail and hence all possibilities are enumerated and the possible schedule switches computed. The complexity of the STSR strategy is $O(C_\mu V^{C_\mu}((V)^2 + C_\mu^2 log C_\mu))$, where the maximum number of clusters formed by a schedule S in the set of schedules CS generated is denoted as C_μ. Also a maximum of $O(V)$ nodes are present in the task graph. Though this method is compute-intensive it allows a very fast run-time adaptation. In addition, this algorithm parallelizable. This can be used to reduce its running time.

A modified strategy called $STSR_{mod}$ has also been developed. In this modified strategy, even when C_μ is high all possible schedule switches need not be considered. This is based on the observation that we get identical machine to cluster matchings over certain intervals of time as the weights of the matchings change only on some discrete events, such as the completion of a task or the start of a new task on a machine. The complexity of $STSR_{mod}$ reduces to $O(C_\mu(V)^3)$.

In RTSR, the switching algorithm is applied at run-time instead of compile time. Hence, adaptation time is longer than in STSR. However, RTSR has a greater flexibility being able to handle a wider range of fault configurations. The complexity of RTSR is $O((C_\mu^2)logC_\mu + (V)^2)$.

4 Experiment

The simulation experiment uses a parallel application structured as a FFT graph each node being of weight 2 units and each edge of 25 units. The bar graphs shown in Figure 1 are the scheduling times for the number of machines provided. Here, the number of maximum machines in use is 10 while the lowest is 2. The figure depicts how the total scheduling time T_{sw} (including faults) changes as machines fail one by one. The three plots shown are the plots of T_{sw} for machine failures with inter-machine failure time set to 10, 20 and 40 time units respectively. T_{sw} includes the retransmission and recomputation costs. The lowest plot shows the change in T_{sw} with inter-machine failure of 10 units, the middle one shows the change for inter-machine failure of 20 units and the topmost one for 40 units. The upper two plots do not extend in entirety because the program gets over (i.e. finishes) before the later failures can take place. The lowest T_{sw} plot for $P = 10$ is the value of the ultimate scheduling time if the 10-machine is initially run and a fault occurs after 10 time units forcing a switch to a 9-machine schedule. With the new 9-machine schedule if the program runs for another 10 units before faulting, the T_{sw} using the 8-machine schedule is plotted next. The other plots have similar connotations but with different inter-machine failure times.

5 Conclusion

In this work we have described the use of computing alternative schedules as a new framework, which can be used easily to adapt to fluctuations which may either cripple the running of a program or substantially reduce its performance on a network of workstations. This strategy, hitherto unused, can be effectively employed to compute a set of good schedules and used later at run-time without the overhead of generating a good schedule at run-time which uses fewer

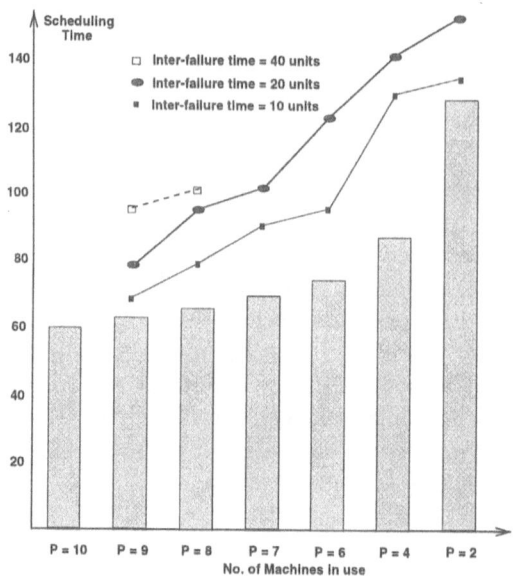

Fig. 1. Performance for multiple single machine faults for a FFT task graph

machines. Thus, our strategy is a hybrid off-line approach coupled with on-line support for effective fault tolerance or machine eviction on a machine cluster without user intervention.

References

1. A. Acharya, G. Edjlali, and J. Saltz. The utility of exploiting idle workstations for parallel computation. In *Proceedings of the ACM Sigmetrics Int'l Conf. on Measurement and Modeling of Computer Systems '97, Seattle*, June 1997.
2. B. Kruatrachue and T. Lewis. Grain size determination for parallel programs. *IEEE Software*, pages 23–32, Jan 1988.
3. H. El-Rewini, T. Lewis, and H. H. Ali. *Task Scheduling for Parallel and Distributed Systems.* Englewood Cliffs, NJ: Prentice-Hall, 1994.
4. G. Tel. *Introduction to Distributed Algorithms.* Cambridge University Press, 1994.
5. M. A. Palis, J. Liou, and D. S. L. Wei. Task Clustering and Scheduling for Distributed Memory Parallel Architectures. *IEEE Transactions on Parallel and Distributed Systems, Vol. 7, No. 1*, pages 46–55, Jan 1996.

Dynamic and Randomized Load Distribution in Arbitrary Networks

J.Gaber and B.Toursel

L.I.F.L., Université des Sciences et Technologies de Lille1
59655 Villeneuve d'Ascq cedex -France-
{gaber,toursel}@lifl.lifl.fr

Abstract. We present the analysis of a randomized load distribution algorithm that dynamically embed arbitrary trees in a distributed network with an arbitrary topology. We model a load distribution algorithm by an associeted Markov chain and we show that the randomized load distribution algorithm spreads any M-node tree in a distributed network with N vertices with load $O(\frac{M}{N} + \delta(\varepsilon))$, where $\delta(\varepsilon)$ depends on the convergence rate of the associated Markov chain. Experimental results obtained by the implementation of these load distribution algorithms, to validate our framework, on the two-dimensional mesh of the massively parallel computer MasPar MP-1 with 16,384 processors, and on a network of workstations are also given.

1 Introduction

We model the load distribution problem that dynamically maintains evolving an arbitrary tree in a distributed network as an on-line embedding problem. These algorithms are dynamic in the sense that the tree may starts as one node and grows by dynamically spawning children. The nodes are incrementally embedded as they are spawned.

Bhatt and Cai present in [1,2] a randomized algorithm that dynamically embeds an arbitrary M-node binary tree on a N processor binary hypercube with dilation $O(\log \log N)$ and load $O(M/N + 1)$. Leighton and al. present in [3,4] two randomized algorithms for the hypercube and the butterfly. The first algorithm achieves, with high probability, a load $O(M/N + \log N)$ and respectively dilation 1 for the hypercube and dilation 2 for the butterfly. The second algorithm achieves, for the hypecube, a dilation $O(1)$ and load $O(M/N+1)$ with high probability. Kequin Li presents in [5] an optimal randomized algorithm that achieves dilation 1 and load $O(M/N)$ in linear arryas and rings. The algorithm concerns a model of random trees called the *reproduction tree* model [5]. The load distribution algorithm that we describe here work for every arbitrary tree (binary or not) on a *general network*. The analysis use mathematical tools derived from both the Markov chain theory and the numerical analysis of matrix iterative schemes.

2 The Problem

Consider the following needed notations. Let P_j the jth vertex of the host graph and k a given integer. We define, for each vertex P_j, the set

$$\mathcal{V}(P_j) = \{P_{n(j,1)}, P_{n(j,2)}, ..., P_{n(j,k)}\}.$$

This set defines a *logical neighbourhood* for P_j. $n(j, \delta)$ refers to the δ-neighbour of P_j. The terms *logical neighbour* denotes that an element of $\mathcal{V}(P_j)$ is not necessarily closed to P_j in the distributede network.

Consider the behavior of the following mapping algorithm. At any instant in time, any task allocated to some vertex u that does not have k children can spawn a new child task. The newly spawned children must be placed on vertices with satisfying the following conditions as proposed by the paradigm of Bhatt and Cai: (1) without foreknow how the tree will grow in the future, and (2) without accessing any global information and (3) once a task is placed on a particular vertex, it cannot be reallocated subsequently. Hence, the process migration is disallowed and the placement decision must be implemented within the network in a distributed manner, and locally without any global information.

The mapping algorithm that we will describe is randomized and operates as follows. The children of any parent node v initially in a vertex P_j are randomly and uniformly placed on distinct vertices of the set $\mathcal{V}(P_j)$. The probability that a vertex in $\mathcal{V}(P_j)$ is chosen is $1/k$. At the start, the root is placed on an initial vertex which can be fixed or randomly choosen. The randomness is crucial to the success of the algorithm as it will be seen in its analysis.

3 Analysis

As was mentionned, as each node is spawned as a child of a node v initially in P_j, it is assigned a random vertex in the set $\mathcal{V}(P_j)$. The probability that any particular vertex is chosen is $1/k$. As our process has the property that the choice of a vertex destination at any step ℓ depends only on the father's node, i.e., depends only on the state $\ell - 1$, we have thus a Markov's chain whose state space is the set of the N vertices. We construct its transition probability matrix $A = (a_{ij})_{0 \leq i,j \leq N}$ where

$$a_{ij} = \begin{cases} \frac{1}{k} & \text{if } P_j \in \mathcal{V}(P_i) \\ 0 & \text{otherwise} \end{cases}$$

A node distribution can be considered to be a probability distribution and the mapping of the newly spawned nodes is the computation of a one-step transition probabilities. Let p_ℓ be an N vector where any entry $p_\ell(i)$ is the proportion of objects in state i at time ℓ. We denote by p_0 the initial probability distribution. At the state ℓ, we have $p_\ell = p_0 A^\ell$, $\forall \ell \geq 1$ We know that if the matrix A is regular, then the Markov chain has stationary transition probabilities since $\lim_{\ell \to \infty} A^\ell$ exists. This means that for large values of ℓ, the probability of being in state i after ℓ transitions is approximatively equal to $\frac{1}{N}$ (the ith entry of $\tilde{p}$) no matter what the initial state was. In other words, as a consequence of the asymptotic behavior of a such Markov process, the distribution converges to uniform distribution which allocates the same amount of nodes to each vertex for very large M. Let p_ℓ be the probability distribution at step ℓ. We have

$$p_\ell = p_{\ell-1}A \quad \text{et} \quad \tilde{p} = \lim_{\ell \to \infty} p_\ell = \left(\tfrac{1}{N}, \tfrac{1}{N}, ..., \tfrac{1}{N} \right)$$

Let $c(v)$ be the number of outcomes where vertex v is selected. We consider that the Markov chain reaches the stationary phase beginning from $t(\varepsilon)$ (e.g., we choose the smallest t such that each entries of p_t is equal to $1/N + \varepsilon$ where ε goes to 0).

Denote by $k^{(\ell)}$ the number of spawned nodes at any step ℓ. We have $M = \sum_{\ell=0}^{\bar{\ell}} k^{(\ell)}$, with $\bar{\ell} \in [t, \infty[$.

We have $c(v) = \sum_{\ell=0}^{\bar{\ell}} p_\ell(v) k^{(\ell)}$ which become [6] $\underbrace{\sum_{\ell=0}^{t(\varepsilon)} (p_\ell(v) - \tilde{p}_\ell(v)) k^{(\ell)}}_{\alpha(v)} + \dfrac{M}{N}$

we obtain the vector $\alpha = \sum_{\ell=0}^{t(\varepsilon)} (p_\ell - \tilde{p}) k^{(\ell)}$. We need now to bound α. As stated by

Cybenko in [7], if γ is subdominant modulus of A, then we have $\| \underbrace{p_0 A^\ell}_{p_\ell} - \tilde{p}\|^2 \leq \gamma^{2\ell} \|p_0 - \tilde{p}\|^2$

and thus $\|p_\ell - \tilde{p}\| \leq \gamma^\ell \|p_0 - \tilde{p}\|$ and $\|\alpha\| = \|\sum_{\ell=0}^{t(\varepsilon)} k^{(\ell)}(p_\ell - \tilde{p})\| \leq \sum_{\ell=0}^{t(\varepsilon)} k^{(\ell)} \|p_\ell - \tilde{p}\| \leq$

$\sum_{\ell=0}^{t(\varepsilon)} k^{(\ell)} \gamma^\ell \|p_0 - \tilde{p}\|$ thus $\|\alpha\| \leq \|p_0 - \tilde{p}\| \sum_{\ell=0}^{t(\varepsilon)} k^\ell \gamma^\ell$

Note that we have use the fact that the number of spawned nodes $k^{(\ell)}$ at any step ℓ is at most k^ℓ. We obtain $\|\alpha\| \leq \|p_0 - \tilde{p}\| \dfrac{(k\gamma)^{t(\varepsilon)+1} - 1}{k\gamma - 1}$

Recall that $\|.\|_\infty$ of a vector denotes its maximum entry. Denote by $\delta(\varepsilon) = \|\alpha\|_\infty$ the maximum entry of α. Since $p_0 - \tilde{p} = \left(1 - \tfrac{1}{N}, -\tfrac{1}{N}, ... \right)$, we have $\|p_0 - \tilde{p}\|_\infty = 1 - \tfrac{1}{N}$

and

$$\delta(\varepsilon) \leq \left(1 - \frac{1}{N}\right) \frac{1 - (k\gamma)^{t(\varepsilon)+1}}{1 - k\gamma} \tag{1}$$

In view of (1), for small γ, we obtain a small value $\delta(\varepsilon)$. Recall that $\gamma \in]-1, 1]$. For $\gamma = \tfrac{1}{k} + a$. We compute the Taylor series expansion of $\delta(\varepsilon)$ at $\tfrac{1}{k}$. This reveals that $\delta(\varepsilon) \leq t(\varepsilon) + 1$, where $t(\varepsilon)$ is the rate of convergence of A (note that $t(\varepsilon)$ is bounded).

With the above analysis we obtain the following theorem.

Theorem 1. *Given a randomized mapping algorithm of a process graph that is an arbitrary dynamic k-ary tree on any arbitrary topology. If the mapping's* *stochastic matrix associated with the logical neighbourhood is regular then the number of nodes mapped on a single vertex of the network is $O(\frac{M}{N} + \delta(\varepsilon))$, where $\delta(\varepsilon) = (1 - \frac{1}{N}) \dfrac{(k\gamma)^{t(\varepsilon)+1} - 1}{k\gamma - 1}$, γ is the subdominant eigenvalue of the mapping's stochastic matrix, $t(\varepsilon)$ the number of steps necessary to converge within ε, M is the number of nodes and N the number of vertices of the network.*

4 Experimentations

We ran a parallel implementation of an algorithm wherein an arbitrary tree grows during the course of the computation (as in branch-and-bound serach, divide-and-conquer or game tree evaluation), on a 128×128 mesh. The following table gives the experimental results in terms of the maximum load obtained when the randomized algorithm is (resp. is not) used to balance load (each line shows the result in terms of the maximum load obtained when we embed an arbitrary tree. The first column gives the total nodes number of this tree).

Number of nodes	load without randomized algorithm	load with randomized algorithm	ideal load
3123	93.00	1.17	1
55791	219.00	4.59	3.41
65135	4.00	4.00	3.98
99325	16.00	6.82	6.06
204383	64.00	13.43	12.47
731923	64.00	45.86	44.67
1640123	256.00	102.59	100.11

We ran, on a network of 8 stations with a multithreaded environement, a parallel implementation of an algorithm wherein an arbitrary binary tree grow during the course of the computation. The following table gives the effective load of each station obtained with and without the randomized algorithm. The value ToTal denotes the total nodes number of the embeded arbitrary tree.

5 Conclusion

Theorem 1 establishes that for a given mapping function, a simple randomized load distribution algorithm as described in section 2 maintains dynamically evolving an arbitrary tree on a general distributed network with a load $O(\frac{M}{N} + \delta(\varepsilon))$, where $\delta(\varepsilon)$ depends on the mapping function. This implies that we can easily compare mapping functions just by computing eigenvalues of the associated adjancy matrix.

N=8			
Numéro Station	load without randomized algorithm	load with randomized algorithm	ideal load *To Tal/N*
0	4171	16106	17271.50
1	4012	16164	17271.50
2	7302	17050	17271.50
3	7289	16012	17271.50
4	3605	16627	17271.50
5	3412	16597	17271.50
6	51722	20394	17271.50
7	56659	19222	17271.50
ToTal	138172.00		

Acknowledgment

We are grateful to F.T.Leighton for the helpful comments and K.Li for helpful discussions. We are also grateful to F.Chung[8], S.Bhatt and G.Cybenko for providing us a helpful papers and suggestions.

References

1. S.N. Bhatt and J.-Y. CAI. Talk a walk, grow a tree. *in Proc. 29th Annual IEEE Symposuim on Foundations of Computer, Science, IEEE CS, Washington, DC, pp. 469-478*, 1988.
2. S.Bhatt and J-Y.Cai. Taking random walks to grow trees in hypercubes. *Journal of the ACM*, 40(3):741–764, July 1993.
3. F.T. Leighton, M.J. NEWMAN, A.G. RANADE, and E.J. SCHWABE. Dynamic tree embeddings in butterflies and hypercubes. *Siam Journal on computing*, 21(4):639–654, August 1992.
4. F.T. Leighton. *Introduction to parallel algorithms and architectures*. Morgan Kauffmann Publishers., 1992. Traduit en Français par P.FRAIGNAUD et E.FLEURY, International Thomson publishing France, 1995.
5. K. Li. Analysis of randomized load distribution for reproduction trees in linear arrays and rings. *Proc. 11th Annual International Symposium on High Performance Computers, Winnipeg, Manitoba, Canada (10-12), July*, 1997.
6. J. Gaber. Plongement et manipulations d'arbres dans les architectures distribuées. *Thèse LIFL*, Janvier 1998.
7. G. Cybenko. Dynamic load balancing for distributed memory architecture. *Journal of Parallel and Distributed Computing*, 7:279–301, 1989.
8. Fan.R.K. Chung. *Spectral Graph Theory*. AMS., 1997.

Workshop 04
Automatic Parallelization and High-Performance Compilers

Jean-François Collard

Co-chairmen

Presentation

This workshop deals with all topics concerning automatic parallelization techniques and the construction of parallel programs using high performance compilers. Topics of interest include the traditional fields of compiler technology, but also the interplay between compiler technology and communication libraries or run-time support. Of the 16 papers submitted to this workshop, 7 were accepted as regular papers, 2 as short papers, and 7 were rejected.

Organization

The first session focuses on data placement and data access. In "Data Distribution at Run-Time: Re-Using Execution Plans," Beckmann and Kelly show how data placement optimization techniques can be made efficiently available in run-time systems by a mixed compile- and run-time technique. On the contrary, the approach by Kandemir *et al.* in "Enhancing Spatial Locality using Data Layout Optimizations" to improve cache performance in uni- and multi-processor systems is purely static. They propose an array restructuring framework based on a combination of hyper-plane theory and reuse vectors. However, when data structures are very irregular, such as meshes, the compiler alone can in general extract very little information. In "Parallelization of Unstructured Mesh Computations Using Data Structure Formalization," Koppler introduces a small description language for mesh structures which allows him to propose a special-purpose parallelizer for the class of applications he tackles. It is worth noticing the wide spectrum of techniques, ranging from completely static methods to purely run-time ones, explored in this field. This definitely illustrates the difficulty of the problem, and the three papers mentioned above make significant contributions asserted by real-life case studies.

The second session starts with "Parallel Constant Propagation," where Knoop presents an extension to parallel programs of a classical optimization of sequential programs: constant propagation. Another classical sequential optimization, extended to parallel programs, is redundancy elimination [2]. It is well known, however, that redundancies can be an asset in the parallel setting. Eisenbiegler takes benefit of this property in his paper "Optimization of SIMD Programs with Redundant Computations," and reports very encouraging execution time improvements. Finally, in "Exploiting Coarse Grain Parallelism from FORTRAN by Mapping it to IF1," Lachanas and Evripidou describe the parallelization of Fortran programs through conversion to single assignment. This work is also interesting for its smart use of two separately available tools: the front-end of Parafrase 2 and the back-end of the SISAL compiler.

In the third session, Feautrier presents in "A Parallelization Framework for Recursive Tree Programs" a novel framework to analyze dependencies in programs with recursive data. It is most noteworthy that a topically related paper has recently been published elsewhere [3], illustrating that the analysis of programs with recursive structures currently is a matter of great interest. How to extend the static scheduling techniques crafted by the author [1] to this framework is an exciting issue. Like scheduling, tiling is a well-known technique to express the parallelism in programs at compile-time. In their paper "Optimal Orthogonal Tiling," Andonov, Rajopadhye and Yanev bring a new analytical solution to the problem of determining the tile size that minimizes the total execution time.

A mixed compile- and run-time technique is addressed in the last paper. In "Enhancing the Performance of Autoscheduling in Distributed Shared Memory Multiprocessors," Nikolopoulos, Polychronopoulos and Papatheodoro present a technique to enhance the performance of autoscheduling, a parallel program compilation and execution model that combines automatic extraction of parallelism, dynamic scheduling of parallel tasks, and dynamic program adaptability to the machine resources.

Conclusion

When trying to gain retrospect on the papers presented in this workshop, one may notice that the borderline between compile-time and run-time is getting blurred in data dependence techniques, data placement, and the exploitation of parallelism. In other words, intensive research is being conducted to benefit from the best of both worlds so as to cope with real-size applications. This orientation is very encouraging and we can hope the end-user will soon benefit from the nice work proposed by the papers of this workshop.

References

1. P. Feautrier. Some efficient solutions to the affine scheduling problem, part I, one dimensional time. *Int. J. of Parallel Programming*, 21(5):313–348, October 1992.
2. J. Knoop, O. Rüthing, and B. Steffen. Optimal code motion: Theory and practice. *ACM Transactions on Programming Languages and Systems, TOPLAS*, 16:1117–1155, 1994.
3. Y. A. Liu. Dependence analysis for recursive data. In *Int. Conf. on Computer Languages*, pages 206–215, Chicago, Illinois, May 1998. IEEE.

Data Distribution at Run-Time:
Re-using Execution Plans

Olav Beckmann and Paul H J Kelly

Department of Computing, Imperial College
180 Queen's Gate, London SW7 2BZ, U.K.
{ob3,phjk}@doc.ic.ac.uk

Abstract. This paper shows how data placement optimisation techniques which are normally only found in optimising compilers can be made available efficiently in run-time systems. We study the example of a delayed evaluation, self-optimising (DESO) numerical library for a distributed-memory multicomputer. Delayed evaluation allows us to capture the control-flow of a user program from within the library at run-time, and to construct an optimised execution plan by propagating data placement constraints backwards through the DAG representing the computation to be performed. In loops, essentially identical DAGs are likely to recur. The main concern of this paper is recognising opportunities where an execution plan can be re-used. We have adapted both conventional parallelising compiler techniques and hardware dynamic branch prediction techniques in order to ensure that our run-time optimisations need not perform any more work than a parallelising compiler would have to do unless there is a prospect of better performance.

1 Introduction

Parallel libraries have two major advantages as a parallel programming model. Firstly, they are convenient because user programs simply call library operators which hide all aspects of parallelism internally. Secondly, there is ample evidence [6] that compiled programming models do not yet get close to purpose-built libraries in terms of performance. It is therefore often worthwhile to invest in a highly optimised, machine-specific library of common numerical subroutines. Hitherto, the disadvantage with such libraries has been that opportunities for optimisation across library calls have been missed.

In this paper, we study a delayed evaluation, self-optimising (DESO) vector-matrix library for a distributed-memory multicomputer. The idea of DESO is to delay actual execution of function calls for as long as possible. Evaluation is forced by the need to access array elements[1]. Delayed evaluation then provides the opportunity at run-time to construct an execution plan which minimises redistribution by propagating data placement constraints backwards through the DAG representing the computation to be performed. We will study a detailed example in Section 2.

---

[1] The most common reasons for accessing array elements are output and conditional tests. We will refer to statements where this happens as *force points*.

Key issues. The main challenge in optimising at run-time is that the optimiser itself has to be very efficient. We achieve this by

- Working from aggregate loop nests, which have been optimised in isolation and which are not re-optimised at run-time. It is precisely the point of the library approach that we have invested in an implementation of selected operators which have been pre-optimised offline.
- Using a purely mathematical formulation for data distributions, which allows us to calculate, rather than search for optimal placements. We will not expand on this aspect of our methodology here.
- Re-using execution plans for previously optimised DAGs. A value-numbering scheme is used to capture cases where this may be possible. The value numbers are used to index a cache of optimisation results, and we use a technique adapted from hardware dynamic branch prediction for deciding whether to further optimise DAGs we have previously encountered.

Context and Structure of this Paper. This paper builds on related work in the field of automatic data placement [8], run-time parallelisation [4, 9, 10] and conventional compiler and architecture technology [1, 5]. In our earlier paper [3] we described the basic idea of a lazy, self-optimising parallel vector-matrix library. In this current paper, we extend that work by presenting the techniques we use to avoid re-optimisation of previously encountered problems. Below, we begin in Section 2 by discussing two alternative strategies for run-time optimisation. Following that, Section 3 presents our techniques for avoiding re-optimisation where appropriate. Section 4 shows performance results and Section 5 concludes.

2 Issues in Run-Time Optimisation

This section discusses and compares two different basic strategies for performing run-time optimisation. We will refer to them as "Forward Propagation Only" and "Forward And Backward Propagation". We use the conjugate gradient iterative algorithm for solving linear systems $Ax - b = 0$ to illustrate both strategies. The pseudocode for this algorithm can be found in Figure 2. We use the following terminology: n is the number of operator calls in a sequence, a the maximum arity of operators, m is the maximum number of different methods per operator. If we work with a fixed set of data placements, s is the number of different placements, and in DAGs, d refers to the degree of the shared node (see [8]) with maximum degree.

Forward Propagation Only. This is the only strategy open to us if we perform run-time optimisation of a sequence of library operators under *strict evaluation*. The strategy is illustrated in Figure 1: we optimise the placement of each node based on information purely about its ancestors. The total optimisation time for a sequence of operator calls under this strategy has *linear* complexity in the number of operators. However, as we have illustrated in Figure 1, the price we pay for using such an algorithm is that it may give a significantly suboptimal

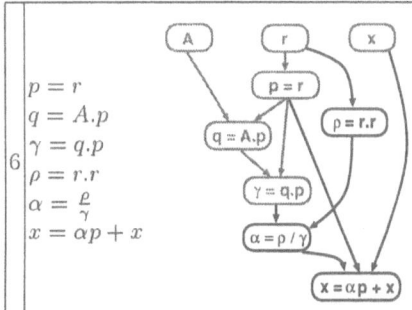

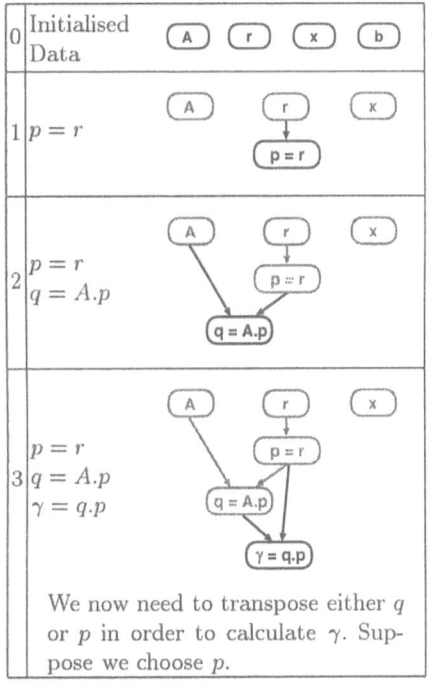

0	Initialised Data
1	$p = r$
2	$p = r$ $q = A.p$
3	$p = r$ $q = A.p$ $\gamma = q.p$

We now need to transpose either q or p in order to calculate γ. Suppose we choose p.

6	$p = r$ $q = A.p$ $\gamma = q.p$ $\rho = r.r$ $\alpha = \frac{\ell}{\gamma}$ $x = \alpha p + x$

We now require p to be aligned with x, but we have just transposed it, so we have to transpose back.

We have made optimisation very easy by deciding the placement of each new node generated based simply on the placement of its immediate ancestor nodes. However, this can result in sub-optimal placements.

Fig. 1. Run-time optimisation of the first iteration of the CG algorithm (see Figure 2) with *Forward Propagation Only*.

answer. This problem is present even for trees, but it is much worse for DAGs since shared nodes are not taken into account properly.

Forward And Backward Propagation. Delayed evaluation gives us the opportunity to propagate placement constraint information backwards through a DAG since we accumulate a full DAG before we begin to optimise. This type of optimisation is much more complex than *Forward Propagation Only*. Mace [8] has shown it to be NP-complete for general DAGs, but presents algorithms with complexity $O((m + s^2)n)$ for trees and with complexity $O(n \times s^{d+1})$ for a restricted class of DAG. The point to note here, though, is that *Forward and Backward Propagation* does give us the opportunity to find the optimal solution to a problem, provided we are prepared to spend the time required.

3 Re-using Execution Plans

The previous section has shown how delayed evaluation gives us the opportunity to derive optimal execution plans, but potentially at not insignificant cost. In real programs, essentially identical DAGs often recur. In such situations, our delayed evaluation, run-time approach is set to suffer a significant performance disadvantage over compile-time techniques unless we can reuse the results of previous optimisations we have performed.

$$r^{(0)} = b - Ax^{(0)}$$
for $i = 1, \ldots, i_{max}$
$\quad \rho_{i-1} = r^{(i-1)^T} r^{(i-1)}$
$\quad$ if $\quad i = 1$
$\quad\quad p^{(1)} = r^{(0)}$
$\quad$ else $\quad \beta_{i-1} = \rho_{i-1}/\rho_{i-2}$
$\quad\quad\quad p^{(i)} = r^{(i-1)} + \beta_{(i-1)}p^{(i-1)}$
$\quad$ endif
$\quad q^{(i)} = Ap^{(i)}$
$\quad \alpha_i = \rho_{i-1}/p^{(i)^T} q^{(i)}$
$\quad x^{(i)} = x^{(i-1)} + \alpha_i p^{(i)}$
$\quad r^{(i)} = r^{(i-1)} - \alpha_i q^{(i)}$
$\quad$ check convergence
end

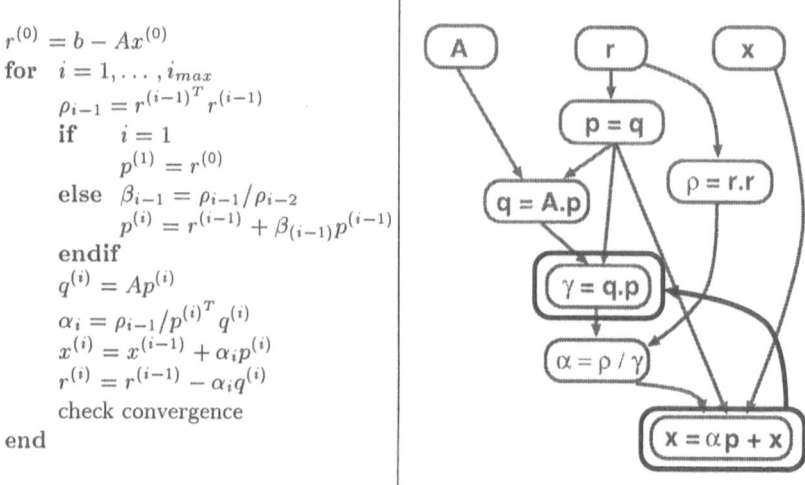

Fig. 2. Left: Pseudocode for the conjugate gradient iterative algorithm. **Right:** Optimisation with *Forward And Backward Propagation*: we take account of the use of p in the update of x in deciding the correct placement for the earlier calculation of γ.

This section shows how we can ensure that our optimiser does not have to carry out any more work than an optimising compiler would have to do, unless there is the prospect of better performance than the compiler could deliver. We discuss first the problem of how to recognise a DAG, i.e. optimisation problem, which we have encountered before and then the issue of whether to optimise further, or, re-optimise, such a DAG.

Recognising Opportunities for Reuse. The full optimisation problem is a large structure. To avoid having to traverse it for comparing with previously encountered DAGs, we derive a hashed "value number" [1,5] for each node.

- Our value numbers have to encode data placements and placement constraints, not actual data values. For nodes which are already evaluated, we simply apply a hash function to the placement descriptor of that node. For nodes which are not yet evaluated, we have to apply a hash function to the *placement constraints* on that node.
- The key observation is that by seeking to take account of all placement constraints on a node, we are in danger of deriving an algorithm for calculating value numbers which has the same O-complexity as *Forward and Backward Propagation* optimisation algorithms: each node in a DAG can potentially exert a placement constraint over every other node.
- Our algorithm for calculating value numbers is therefore based on *Forward Propagation Only*: we calculate value numbers for unevaluated nodes by applying a hash function to the placement constraints deriving from their immediate ancestors.

– Since we do not store the "full DAG" information, we cannot easily detect hash conflicts. We return to this point shortly.

When to Re-Use and When to Optimise. Because our value numbers are calculated on *Forward Propagation Only* information, we have to address the problem of how to handle those cases where nodes which have identical value numbers are used in a different context later on; in other words, how to to avoid the drawbacks of *Forward Propagation Only* optimisation. This is a branch-prediction problem, and we use a technique adapted from hardware dynamic branch prediction (see [7]) for predicting heuristically whether identical value numbers will result in identical future use of the corresponding node and hence identical optimisation problems.

Caching Execution Plans. Value numbers and 'dynamic branch prediction' together provide us with a fairly reliable mechanism for recognising the fact that we have encountered a node in the same context before. Assuming that we optimised the placement of that node when we first encountered it, our task is then simply to re-use the placement which the optimiser derived. We do this by using a "cache" of optimised placements, which is indexed by value numbers. Each cache entry has a valid-tag which is set by our branch prediction mechanism.

Competitive Optimisation. As we showed in Section 2, full optimisation based on *Forward And Backward Propagation* can be very expensive. Each time we invoke the optimiser on a DAG, we therefore only spend a limited time optimising that DAG. For a DAG which we encounter only once, this means that we only spend very little time trying to eliminate the worst redistributions. For DAGs which recur, our strategy is to gradually improve the execution plan used until our optimisation algorithm can find no further improvements.

The final point we need to address is how to handle hash conflicts. The result of a hash conflict on a value number will be that we use a sub-optimal placement for a node which we had previously optimised. In order to detect this, our system has been instrumented to record the communication cost of executing a DAG under an optimised execution plan. An increase in this cost on a "re-use" of that plan indicates a hash conflict.

Summary. We use the full optimisation information, i.e. *Forward and Backward Propagation*, to optimise. We obtain access to this information by delayed evaluation. We use a scheme based on *Forward Propagation Only*, with linear complexity in program length, to ensure that we re-use the results of previous optimisations.

4   Performance

In this Section, we show performance figures for our library on the Fujitsu AP3000 multicomputer here at Imperial College. As a benchmark we used the

Table 1. Time in milliseconds for 20 iterations of Conjugate Gradient, with a convergence test every 10 iterations, on the AP3000, with 300MHz UltraSparc-2 nodes (average figures, outlying points were omitted). N denotes timings without any optimisation, O timings with optimisation but no caching, and C timings with optimisation and caching of optimisation results. *Memory* shows time spent in `malloc()` and `free()`, *Overhead* the cost of maintaining data distribution descriptors and suspended library calls at run-time. *Speedup* is the speedup due to optimisation and execution plan re-use.

	P	N	Comp.	Memory	Overh.	Comms.	Opt.	Total	Speedup
N	1	256	51.14	0.81	3.97	0.00	0.00	55.93	·
O	1	256	51.22	0.81	3.88	0.00	3.59	59.50	0.94
C	1	256	51.13	0.79	3.25	0.00	0.30	55.47	1.01
N	4	512	51.30	1.03	3.47	30.51	0.00	86.30	·
O	4	512	51.79	0.86	3.06	22.35	3.73	81.79	1.06
C	4	512	51.69	0.94	2.45	21.94	0.31	77.32	1.12
N	9	768	51.72	1.12	3.46	34.78	0.00	91.10	·
O	9	768	51.96	0.91	3.08	26.06	3.74	85.75	1.06
C	9	768	51.90	0.98	2.46	25.91	0.31	81.55	1.12
N	16	1024	52.05	1.03	3.48	45.37	0.00	101.93	·
O	16	1024	52.29	0.88	3.16	35.99	3.82	96.14	1.06
C	16	1024	52.13	0.95	2.49	35.65	0.31	91.53	1.11

conjugate gradient iterative algorithm. The pseudo-code for the algorithm and the source code when implemented using our library were shown in Figure 2 and the timing data are in Table 1.

- Our optimiser avoids two out of three vector transpose operations per iteration. This can not be seen from the data in Table 1, it was determined analytically and by tracing communication.
- Optimisation achieves a reduction in communication time of between 20% and 30%. We do not achieve more because a significant proportion of the communication in this algorithm is due to reduce-operations which are unaffected by our current optimisations.
- Run-time overhead and optimisation time are independent of the amount of parallelism used. We suspect that the slight difference is due to cache effects.
- The reason why run-time overhead is reduced by optimisation is that performing fewer redistributions also results in spending less time inspecting data placement descriptors. Caching achieves a further reduction in overhead; this is because it is cheaper to read placement descriptors from cache than to generate them by function calls.
- Without caching of optimisation results, we achieve an overall speedup of around 6%. On platforms which have less powerful processors than the 300MHz UltraSparc-2 nodes we use here, the cost of optimising afresh each time can easily outweigh the benefit of reduced communication.
- With caching of optimisation results, the time we spend optimising is negligible, and we achieve overall speedups of around 12%.

5 Conclusions

We have presented a technique for interprocedural data placement optimisation which exploits run-time control-flow information and is applicable in contexts where the calling program cannot be analysed statically. We present preliminary experimental evidence that the benefits can easily outweigh the run-time costs.

Related Work. There is a huge body of work on data mappings for regular problems, [2] is but one example. Our work relies on this in producing optimised implementations for library operators. However, the problem we seek to address in this paper is different — how to perform interprocedural optimisation over a sequence of such pre-optimised operators.

Saltz *et al.* [10] address the basic problem of how to parallelise loops where the dependence structure is not known statically. Loops are translated into an *inspector* loop which determines the dependencies at run-time and constructs a schedule, and an *executor* loop which carries out the calculations planned by the inspector. Saltz *et al.* discuss the possibility of reusing a previously constructed schedule, but rely on user annotations for doing so. Ponnusamy *et al.* [9] propose a simple conservative model which avoids the user having to indicate to the compiler when a schedule may be reused. Benkner *et al.* [4] describe the reuse of parallel schedules via explicit directives in HPF+: REUSE directives for indicating that the schedule computed for a certain loop can be reused and SCHEDULE variables which allow a schedule to be saved and reused in other code sections.

Value numbering schemes were pioneered by Ershov [5], who proposed the use of "convention numbers" for denoting the results of computations and avoid having to recompute them. More recent work on this subject has been done, e.g., by Alpern *et al.* [1].

Run-Time vs. Compile-Time Optimisation. The particular example studied in this paper would have been amenable to compile-time analysis. We refer back to Section 1 and the arguments of convenience and efficiency we outlined there for why we parallelise this application via a parallel library. using a library then forces us to optimise at run-time. Further, we have shown that a runtime optimiser need not actually perform any more work than a compiler.

To compare the quality of run-time and compile-time schedules, consider the loop opposite, assuming that there are no force-points inside the loop and that the loop is encountered a number of times, evaluation being forced after the loop each time.

```
for(i = 0; i < N; ++i) {
  if <unknown condition>
    <do A>
  else
    <do B>
}
```

This loop can potentially have 2^N control-paths. A compile-time optimiser would have to find one compromise execution plan for all invocations of the loop. With our approach, we optimise the actual DAG which is generated on each occasion. If the number of different DAGs is high, compile-time methods would probably have the edge over ours, since we cannot reuse execution plans. If, however, the number of different DAGs is small, our execution plans for the *actual* DAGs will be superior to compile-time compromise solutions, and by reusing them, we limit the time spent optimising.

Future work. The most exciting next step is to store cached execution plans persistently, so that they can be reused subsequently for this or similar applications. Although we can derive some benefit from exploiting run-time control-flow information, we have the opportunity to make run-time optimisation decisions based on run-time properties of data; we plan to extend this work to address sparse matrices shortly. The run-time system has to make on-the-fly data placement decisions. An intriguing question raised by this work is to compare this with an optimal off-line schedule.

Acknowledgements. This work was partially supported by the EPSRC, under the Futurespace and CRAMP projects (refs. GR/J 87015 and GR/J 99117). We thank the Imperial College Parallel Computing Centre for the use of their AP3000 machine.

References

1. B. Alpern, M. N. Wegman, and F. K. Zadeck. Detecting equalities of variables in programs. In *15th Annual ACM Symposium on Principles of Programming Languages*, pages 1–11, San Diego, California, Jan. 1988.
2. J. M. Anderson, S. P. Amarasinghe, and M. S. Lam. Data and computation transformations for multiprocessors. *SIGPLAN Notices*, 30(8):166–178, Aug. 1995.
3. O. Beckmann and P. H. J. Kelly. Runtime interprocedural data placement optimisation for lazy parallel libraries (extended abstract). In Lengauer *et al.*, editor, *Proceedings of Euro-Par '97, Passau, Germany*, number 1300 in LNCS, pages 306–309. Springer Verlag, Aug. 1997.
4. S. Benkner, P. Mehrotra, J. V. Rosendale, Zima, and Hans. High-level management of communication schedules in HPF-like languages. Technical Report TR-97-46, Institute for Computer Applications in Science and Engineering, NASA Langley Research Center, Hampton, VA 23681, USA, Sept. 1997.
5. A. P. Ershov. On programming of arithmetic operations. *Communications of the ACM*, 1(8):3–6, 1958. Three figures from this article are in CACM 1(9):16.
6. W. D. Gropp. Performance driven programming models. In *MPPM'97, Proceedings of the 3rd International Working Conference on Massively Parallel Programming Models*, London, U.K., Nov. 1997. To appear.
7. J. L. Hennessy and D. A. Patterson. *Computer Architecture A Quantative Approach*. Morgan Kaufman, San Mateo, California, 1st edition, 1990.
8. M. E. Mace. *Storage Patterns in Parallel Processing*. Kluwer, 1987.
9. R. Ponnusamy, J. Saltz, and A. Choudhary. Runtime compilation techniques for data partitioning and communication schedule reuse. In *Proceedings of Supercomputing '93: Portland, Oregon, November 15–19, 1993*, pages 361–370, New York, NY 10036, USA, Nov. 1993. ACM Press.

10. J. H. Saltz, R. Mirchandaney, and K. Crowley. Run-time parallelization and scheduling of loops. *IEEE Transactions on Computers*, 40(5):603–612, May 1991.

Enhancing Spatial Locality via Data Layout Optimizations

M. Kandemir[1]   A. Choudhary[2]   J. Ramanujam[3]   N. Shenoy[2]   P. Banerjee[2]

[1] EECS Dept., Syracuse University, Syracuse, NY 13244 (mtk@ece.nwu.edu)
[2] ECE Dept., Northwestern University, Evanston, IL 60208
({choudhar,nagaraj,banerjee}@ece.nwu.edu)
[3] ECE Dept., Louisiana State University, Baton Rouge, LA 70803 (jxr@ee.lsu.edu)

Abstract. This paper aims to improve locality of references by suitably choosing array layouts. We use a new definition of spatial reuse vectors that takes into account memory layout of arrays. This capability creates two opportunities. First, it allows us to develop an array restructuring framework based on a combination of hyperplane theory and reuse vectors. Second, it allows us to observe the effect of different array layout optimizations on spatial reuse vectors. Since the iteration space based locality optimizations also change the spatial reuse vectors, our approach allows us to compare the iteration-space based and data-space based approaches in terms of their effects on spatial reuse vectors. We illustrate the effectiveness of our technique using an example from the BLAS library on the SGI Origin distributed shared-memory machine.

1 Introduction

In most computer systems, exploiting locality of reference is key to high levels of performance. It is well-known that caching of data whether it is private or shared improves memory latency as well as processor utilization. In fact, increasing the cache hit rates is one of the most important factors in reducing the average memory latency. Although cache hit rates in uniprocessors can be improved by optimizing the organization of the cache such as carefully selecting the cache size, line size and associativity, there are still tasks to be done from software. The programmers and compiler writers often attempt to modify the access patterns of a program so that the majority of accesses are made to the nearby memory. Several efforts have been aimed at iteration space transformations and scheduling techniques to improve locality [10, 9, 13]; these techniques improve data locality *indirectly* as a result of modifying the iteration space traversal order.

In this paper, we focus on an alternative approach to the data locality optimization problem. Unlike traditional compiler techniques, we focus directly on the data space, and attempt to transform data layouts so that better locality is obtained. We present a data restructuring technique which can improve cache performance in uniprocessors as well as multiprocessor systems. Our technique is based on the concept of reuse vectors [9, 13] and uses linear algebra techniques to help a compiler translate potential reuse in a given program into locality. In our technique, we use data transformations expressed by linear full-rank transformation matrices. We also compare data transformations with linear loop transformations that can be expressed by non-singular transformation matrices. We are particularly interested in optimizing spatial locality, that in turn causes better utilization of the long cache lines which represents the current trend in cache architecture.

This paper is organized as follows. In the next section, we outline the background and motivations for our work. Section 3 presents a memory layout representation based on hyperplane theory from linear algebra. In Section 4, we give a method to compute a reuse vector, for a given layout expressed by hyperplanes. We also illustrate how this new definition of reuse vector can be employed to guide data layout transformations. In Section 5, we present our preliminary results on the **syr2k** example from the BLAS library. In Section 6 we review the related work and conclude the paper in Section 7.

2 Background

We represent each iteration of a loop nest of depth n by a loop iteration vector $I = (i_1, i_2, ..., i_n)$ where i_k is the value of the kth loop from the outermost. Each reference to an m–dimensional array U in such a loop nest is assumed to be an affine function of the iteration represented by I, i.e., I is mapped onto data element $A_U I + o_U$. Here A_U is an $m \times n$ matrix called the *access matrix* and the m-vector o_U is called the *offset vector* [13, 9]. Consider the following example.

Example 1: do $i = li, \ ui$
        do $j = lj, \ uj$
            $U(j+1,i-1) = V(i-1,j+1) + W(j-1,i+j+1) + X(j) + Y(i)$
        enddo
        enddo

Here, for array U, $A_U = \begin{pmatrix} 0 & 1 \\ 1 & 0 \end{pmatrix}$, and $o_U = \begin{pmatrix} 1 \\ -1 \end{pmatrix}$. Two references to an array are said to belong to a uniformly generated reference (UGR) set if they have the same access matrix [3].

We say that reuse exists during the execution of a loop nest if the same data element or nearby data elements (e.g., the elements stored consecutively in a column for a column-major array) are accessed more than once. Given a loop nest, it is important to know where data reuse exists. A simple way of detecting the loops which carry reuse is to derive a set of vectors called reuse vectors [13, 9]. The objective here is first to use vectors to represent directions in which reuse occurs, and then to transform the loop nest such that these vectors will have some desired properties, i.e., forms.

Consider a reference with an access matrix A_U and offset vector o_U. Two iterations I and J reference the same data element if $A_U I + o_U = A_U J + o_U \Rightarrow A_U(J - I) = 0 \Rightarrow A_U r_{Ut} = 0 \Rightarrow r_{Ut} \in Ker\{A_U\}$ where $r_{Ut} = J - I$. In this case we say that reference A_U exhibits temporal reuse; that is, the same data element is used by more than one iteration; r_{U_t} is termed as temporal reuse vector and $Ker\{A_U\}$ is referred as temporal reuse vector space [13]. As an example, consider the reference $X(j)$ in Example 1. The access matrix of this reference is $(0, 1)$. Therefore, $r_{X_t} \in Ker\{(0, 1)\}$; that is, $r_{X_t} \in span\{(1, 0)^T\}$. This means that there is a temporal reuse carried by the i loop. That is, for a fixed j, successive iterations of the i loop access the same element of array X. For concreteness, we say that $(1, 0)^T$ is the temporal reuse vector for that reference. In general, the loop corresponding to the first non-zero element in a reuse vector is said to carry the associated reuse. It should be noted that all references that belong to a single UGR set have the same access matrix; therefore, they have the same temporal reuse vector. The reuse vector for the reference $Y(i)$ on the other hand is $(0, 1)^T$. Intuitively, a reuse vector such as $(0, 1)^T$ means that the successive iterations of the innermost loop will access the same data element. In a sense, this is a desired reuse vector form; because, the reuse is exploited in the innermost loop where it is most useful. On the other hand, a reuse vector such as $(1, 0)^T$ implies that the same data item is accessed in the successive iterations of the outer loop. If the trip count of the innermost loop is very large then it might be the case that between two uses the data item is flushed from cache. In that case we say that reuse has not converted into locality. The main objective of the compiler-based data reuse analysis can be cast as the problem of transforming ʼrograms such that as many reuses as possible will be converted into locality. In ιr terms, this means to transform programs such that the resulting codes will ₊ve desired reuse vector forms.

In addition to temporal reuse, spatial reuse is very important in scientific codes. Assuming a column-major memory layout for arrays, a spatial reuse occurs when the same column of an array is accessed more than once. Consider Example 1 again. Assuming column-major memory layouts for all arrays, we note that successive iterations of j loop will access elements in the same column of array U. In that case, we say that array U has spatial reuse with respect to j loop. Similarly, successive values of i loop access elements of the same column of array V; that is, array V has spatial reuse with respect to loop i. It should be noted that spatial reuse is defined with respect to a given memory layout, which is column-major in our case. The spatial reuse exhibited by array W however is slightly more complicated to analyze. For this array, elements of a given column c are accessed if and only if $i + j = c - 1$. In mathematical terms, in order for two iteration vectors I and J access elements residing in the same column, they should satisfy the condition $A_{U_s}I = A_{U_s}J$ where A_{U_s} is the matrix A_U (for an array U) with the first row deleted [13]. From this condition, since A_{U_s} represents a linear mapping, $A_{U_s}(J - I) = 0 \Rightarrow J - I \in Ker\{A_{U_s}\} \Rightarrow r_{U_s} \in Ker\{A_{U_s}\}$ where $r_{U_s} = J - I$.

In this case, r_{U_s} is termed as the spatial reuse vector and $Ker\{A_{U_s}\}$ is referred as spatial reuse vector space [13]. When there is no confusion, we use r_U instead of r_{U_s} for the sake of clarity. The loop which corresponds to the first non-zero element of the generator vector of $Ker\{A_{U_s}\}$ is said to carry spatial reuse. In Example 1, for array U, the spatial reuse is carried by j loop. In this example, there exists a spatial reuse vector for each reference and these are $r_U = (0,1)^T$, $r_V = (1,0)^T$, and $r_W = (1,-1)^T$. Recall that all these reuse vectors are with respect to column-major memory layout. It is possible that a reference may have more than one spatial reuse vector. In that case, these individual reuse vectors constitute a matrix called spatial reuse matrix, which will be denoted in this paper by R_U for an array U or for a reference to an array U when there is no confusion. Individual spatial reuse matrices from individual references to possibly different arrays constitute a combined reuse matrix [9]. The order of the vectors in the combined reuse matrix might be important. Like Li [9], we assume that they are prioritized from left to right according to the frequency of occurrence.

An important attribute of a reuse vector is its height which is defined as the number of dimensions from the first non-zero entry to the last entry. One way of increasing cache locality is to introduce more leading zeros in the reuse vector, called *height reduction* [9]. In this context, the best possible spatial reuse vector is $r = (0, 0, \cdots, 0, 1)^T$ meaning that the spatial reuse will be carried by the innermost loop in the nest.

We emphasize the fact that temporal reuse can only be manipulated by iteration space (loop) transformations. Data space transformations cannot directly affect the temporal reuse exhibited by a reference. Since, in this paper we are interested only in data layout transformations, we consider only spatial reuse vectors. In addition, we only focus on self-spatial reuses [13].

3 Hyperplanes: An Abstraction for Array Layouts

This section reviews our work [6] on representing memory layouts of multi-dimensional arrays using hyperplanes. In an m-dimensional data space, a *hyperplane* can be defined as a set of tuples $\{(a_1, a_2, ..., a_m) \mid g_1 a_1 + g_2 a_2 + ... + g_m a_m = c\}$ where g_1, g_2,...,g_m are rational numbers called hyperplane coefficients and c is a rational number called hyperplane constant [12]. For convenience, we use a row vector $g^T = (g_1, g_2, ..., g_m)$ to denote an hyperplane family (with different values of c) whereas g corresponds to the column vector representation.

In a two-dimensional data space, we can think of a hyperplane family as parallel lines for a fixed coefficient set and different values of c. An important property of the hyperplanes is that two data points (in our case, array elements) d_1 and d_2 belong to the same hyperplane g if

$$g^T d_1 = g^T d_2. \tag{1}$$

For example, a hyperplane vector such as $(0,1)^T$ indicates that two array elements belong to the same hyperplane as long as they have the same value for the column index (i.e. the second dimension); the value for the row index does not matter.

We note that a hyperplane family can be used to partially define memory layout of an array. For example, in two-dimensional array case, a hyperplane family represented by $(0,1)$ indicates that the array in question is divided into columns such that each column corresponds to a hyperplane with a different c value (hyperplane constant). The data elements that have the same c value (that is, the elements which make up a column) are stored in memory consecutively (ordered by their column indices). We assume in a broader sense that two data elements which lie along an hyperplane with a specific c value have spatial locality. Another way of saying this is that two elements d_1 and d_2 have spatial locality if they satisfy equation (1). Notice that this definition of spatial locality is coarse and does not hold in array boundaries; but it is suitable for our purposes.

In higher dimensions, we need more than one hyperplane. We refer the interested reader to [6] for an in-depth discussion of hyperplane based layout representation. In the remainder of this paper, for the sake of simplicity we mainly focus on two-dimensional arrays. Our results easily extend to higher dimensional arrays.

4 Reuse Vectors under Different Layouts

4.1 Determining Spatial Reuse Vectors

The definition of the spatial reuse vector presented earlier is based on the assumption that the memory layout for all arrays is column-major. In this section, we give a definition of spatial reuse vector under a given memory layout. We start with the following theorem. The reader is referred to [7] for the proofs of all the theorems in this paper.

Theorem 1. *Let g represent a memory layout for an array U, and r_U the spatial reuse vector associated with a reference represented by the access matrix A_U. Then, the following equality holds between g, r_U and A_U:*

$$g^T A_U r_U = 0 \qquad (2)$$

If the memory layout of an array is represented by a matrix L (as in three- or higher dimensional cases), then the equation (2) should be satisfied for each row of L. This theorem gives us the relation between memory layouts and spatial reuse vectors and is very important. Notice that for a given g^T, in general, from $g^T A_U r_U = 0$, the spatial reuse vector r_U can always be found as

$$r_U \in Ker\{g^T A_U\} \qquad (3)$$

Let us consider the following example assuming row-major memory layout for all arrays:

Example 2: do $i = li,\ ui$
    do $j = lj,\ uj$
      do $k = lk,\ uk$
        $U(i+k,j+k) = V(i+j,i+k,j+k) + W(k,j,i)$
      enddo
    enddo
  enddo

The access matrices for this loop nest are $A_U = \begin{pmatrix} 1 & 0 & 1 \\ 0 & 1 & 1 \end{pmatrix}$, $A_V = \begin{pmatrix} 1 & 1 & 0 \\ 1 & 0 & 1 \\ 0 & 1 & 1 \end{pmatrix}$, and

$A_W = \begin{pmatrix} 0 & 0 & 1 \\ 0 & 1 & 0 \\ 1 & 0 & 0 \end{pmatrix}$. The row-major layout for 2-dimensional arrays is represented

by $g^T = (1, 0)$ and by $L = \begin{pmatrix} 1 & 0 & 0 \\ 0 & 1 & 0 \end{pmatrix}$ for 3-dimensional arrays. Using relation (3),

we find the following reuse matrices: $R_U = \begin{pmatrix} 0 & 1 \\ 1 & 0 \\ 0 & -1 \end{pmatrix}$, $R_V = \begin{pmatrix} 1 \\ -1 \\ -1 \end{pmatrix}$, and $R_W =$

$\begin{pmatrix} 1 \\ 0 \\ 0 \end{pmatrix}$. Most of the previous research has concentrated on optimizing r_U given

a fixed memory layout. For instance, if an array is stored in column-major order in memory, i.e., $g^T = (0, 1)$, from equation (2) $g^T A_U r_U = 0 \Rightarrow A_{U,} r_U = 0$; this is the definition of spatial reuse vector used previously [9, 13].

4.2 Reproducing the Effects of Iteration Space Transformations

In this subsection, we show how to reproduce the effect of a linear loop transformation by using linear data transformations instead. Li [9] shows that an iteration space transformation T transforms a reuse vector r_U into $r_U' = T r_U$. In this paper, we treat temporal locality as a special case of spatial locality. That is, accessing the same element can be considered accessing the same column (for a column-major layout).

Theorem 2. *Given a 'single' loop nest, if a reference is optimized for spatial locality using an iteration space transformation matrix T, the same effect can be obtained by a corresponding data transformation matrix M.*

Consider the following matrix-multiplication example assuming column-major layout:

Example 3: do $i = li,\ ui$
　　　　do $j = lj,\ uj$
　　　　　do $k = lk,\ uk$
　　　　　　$C(i,j) = C(i,j) + A(i,k) * B(k,j)$
　　　　　enddo
　　　　enddo
　　　　enddo

The spatial reuse matrices are $R_C = \begin{pmatrix} 1 & 0 \\ 0 & 0 \\ 0 & 1 \end{pmatrix}$, $R_A = \begin{pmatrix} 1 & 0 \\ 0 & 1 \\ 0 & 0 \end{pmatrix}$, and $R_B =$

$\begin{pmatrix} 1 & 0 \\ 0 & 0 \\ 0 & 1 \end{pmatrix}$. By taking into account the frequency of occurrence of each vector, we can write a combined reuse matrix as $R = \begin{pmatrix} 1 & 0 & 0 \\ 0 & 0 & 1 \\ 0 & 1 & 0 \end{pmatrix}$. Here the leftmost column has the highest priority whereas the rightmost column has the lowest. In this case, Li's algorithm [9] finds the following matrix to optimize the locality $T = \begin{pmatrix} 0 & 1 & 0 \\ 0 & 0 & 1 \\ 1 & 0 & 0 \end{pmatrix}$. The new reuse matrix is $R' = TR = \begin{pmatrix} 0 & 0 & 1 \\ 0 & 1 & 0 \\ 1 & 0 & 0 \end{pmatrix}$. Notice that the reuse vector of highest priority is optimized very well, i.e., its height is reduced to 1. The transformed program is as follows:

```
do u = lu, uu
  do v = lv, uv
    do w = lw, uw
      C(w,u) = C(w,u) + A(w,v) * B(v,u)
    enddo
  enddo
enddo
```

In the optimized program, the references $C(w, u)$ and $A(w, v)$ have spatial locality in the innermost loop (w–loop) whereas the reference $B(v, u)$ has temporal locality in the innermost loop. Since, in this paper, we treat temporal locality in a loop as a special case of spatial locality, we conclude that in this program all three references have spatial locality in the innermost loop.

Next we show how to obtain the effect of this transformation with data layout transformations. Notice that after the transformation T, the final spatial reuse matrices for individual references are $R_C' = \begin{pmatrix} 0 & 0 \\ 0 & 1 \\ 1 & 0 \end{pmatrix}$, $R_A' = \begin{pmatrix} 0 & 1 \\ 0 & 0 \\ 1 & 0 \end{pmatrix}$, and

$R_B' = \begin{pmatrix} 0 & 0 \\ 0 & 1 \\ 1 & 0 \end{pmatrix}$. Now, from each reuse matrix, we choose the most frequently used reuse vector (considering all reuse vectors in all reuse matrices), and use it as a target reuse vector for the associated array. In this example, the target reuse vector happens to be $(0, 0, 1)^T$ for all references.

For array C: $g^T A_C r_C' = 0 \Rightarrow (g_1, g_2) \begin{pmatrix} 1 & 0 & 0 \\ 0 & 1 & 0 \end{pmatrix} \begin{pmatrix} 0 \\ 0 \\ 1 \end{pmatrix} = 0 \Rightarrow (g_1, g_2)^T \in Ker\{(0, 0)\}$

For array A: $g^T A_A r_A' = 0 \Rightarrow (g_1, g_2) \begin{pmatrix} 1 & 0 & 0 \\ 0 & 0 & 1 \end{pmatrix} \begin{pmatrix} 0 \\ 0 \\ 1 \end{pmatrix} = 0 \Rightarrow (g_1, g_2)^T \in Ker\{(0, 1)\}$

For array B: $g^T A_B r_B' = 0 \Rightarrow (g_1, g_2) \begin{pmatrix} 0 & 0 & 1 \\ 0 & 1 & 0 \end{pmatrix} \begin{pmatrix} 0 \\ 0 \\ 1 \end{pmatrix} = 0 \Rightarrow (g_1, g_2)^T \in Ker\{(1, 0)\}$

Thus, the array C can have any layout (say row-major), the array A should be row-major, and the array B should be column-major. Notice that if we assign these layouts, then we have spatial locality for all the references with respect to the innermost loop, a result that has also been obtained by the loop transformation T given earlier.

From the previous discussion, we can conclude that given a single loop nest where each array has one UGR set [3], it is always possible to obtain the locality effect of a linear loop transformation by corresponding linear data transformations.

4.3 Optimizing for the Best Locality

So far, we have shown that under certain conditions, it is possible to reproduce the effect of a loop transformation by data transformations. By studying the effect of a loop transformation on a spatial reuse vector, we can find a corresponding data transformation which can generate the same effect. Now, we go one step further, and prove a stronger result.

Theorem 3. *Given a loop nest where each array has one UGR set, then it is always possible to optimize the array layouts such that each reference will have spatial locality with respect to the innermost loop.*

Consider the following example with a column-major layout for all arrays.

Example 4: do $i = li, \ ui$
$\qquad$ **do** $j = lj, \ uj$
$\qquad\qquad$ **do** $k = lk, \ uk$
$\qquad\qquad\qquad$ $U(i,j,k) = U(i\text{-}1,j,k\text{+}1) + U(i,j,k\text{-}1) + V(j\text{+}1,k\text{+}1,i\text{-}1)$
$\qquad\qquad$ **enddo**
$\qquad$ **enddo**
$\qquad$ **enddo**

Ignoring the temporal reuses, the spatial reuse matrices for individual arrays are $R_U = (1,0,0)^T$ and $R_V = (0,1,0)^T$. Therefore, the combined reuse matrix is $R = \begin{pmatrix} 1 & 0 \\ 0 & 1 \\ 0 & 0 \end{pmatrix}$. Notice that the first column occurs more frequently (three times to be exact). The best iteration space transformation from the locality point of view (using Li's approach [9]) is $T = \begin{pmatrix} 0 & 0 & 1 \\ 0 & 1 & 0 \\ 1 & 0 & 0 \end{pmatrix}$ which would generate $R' = TR = \begin{pmatrix} 0 & 0 \\ 0 & 1 \\ 1 & 0 \end{pmatrix}$. Unfortunately, due to the data dependence involving array U, this transformation is not legal. In search for a second best transformation, we have two options:

For the first option, $T_1 = \begin{pmatrix} 0 & 1 & 0 \\ 1 & 0 & 0 \\ 0 & 0 & 1 \end{pmatrix}$ which gives $R' = T_1 R = \begin{pmatrix} 0 & 1 \\ 1 & 0 \\ 0 & 0 \end{pmatrix}$. In the transformed nest, the spatial locality for array U is exploited in the second

for each array $U \in \mathcal{U}$
    compute the access matrix A_U representing U
    $i = n$
    while $(p(i) = 1)$
      $i = i - 1$
    end while
    $r = e_i$
    compute $\mathcal{N} = Ker\{(A_U r)^T\}$
    use the elements of the basis set of $\mathcal{N}$ as rows of the layout matrix for U
end for each

Fig. 1. Algorithm for optimizing spatial locality.

innermost loop (v–loop); and the spatial locality for array V is exploited in the outermost loop (u–loop).

For the second option, $T_2 = \begin{pmatrix} 1 & 0 & 0 \\ 0 & 0 & 1 \\ 0 & 1 & 0 \end{pmatrix}$ which gives $R' = T_2 R = \begin{pmatrix} 1 & 0 \\ 0 & 0 \\ 0 & 1 \end{pmatrix}$. In the transformed nest, the spatial locality for array U is exploited in the outermost loop (u–loop); and the spatial locality for array V is exploited in the innermost loop (w–loop). We note that while T_1 improves the spatial locality of U slightly, T_2 improves the spatial locality for V.

It is easy to see that data layout transformations can easily reproduce the effects of T_1 and T_2 (see [7]). However, neither T_1 nor T_2 (nor their data transformation counterparts) is able to optimize the spatial locality for both arrays in the innermost loop. The reason for the failure of the iteration space transformations is the data dependences which prevent the most desired loop permutation.

We now show that how a different data layout transformation can optimize the same nest. We use the best possible spatial reuse vector as the desired (or target) vector for both the arrays. In other words, $r_U' = r_V' = (0, 0, 1)^T$. For array U, $g^T A_U r_U' = 0 \Rightarrow (g_1, g_2, g_3) \begin{pmatrix} 1 & 0 & 0 \\ 0 & 1 & 0 \\ 0 & 0 & 1 \end{pmatrix} \begin{pmatrix} 0 \\ 0 \\ 1 \end{pmatrix} = 0$; therefore, $(g_1, g_2, g_3)^T \in Ker\{(0, 0, 1)\}$ meaning that the array U should have a memory layout such that the last dimension should be the fastest changing dimension. The row-major layout is such a memory layout with the layout matrix $L_U = \begin{pmatrix} 1 & 0 & 0 \\ 0 & 1 & 0 \end{pmatrix}$. For array V, $g^T A_V r_V' = 0 \Rightarrow (g_1, g_2, g_3) \begin{pmatrix} 0 & 1 & 0 \\ 0 & 0 & 1 \\ 1 & 0 & 0 \end{pmatrix} \begin{pmatrix} 0 \\ 0 \\ 1 \end{pmatrix} = 0$; thus, $(g_1, g_2, g_3)^T \in Ker\{(0, 1, 0)\}$ meaning that the memory layout of array V must be such that the second dimension is the fastest changing dimension. An example layout matrix which satisfies that is $L_V = \begin{pmatrix} 1 & 0 & 0 \\ 0 & 0 & 1 \end{pmatrix}$.

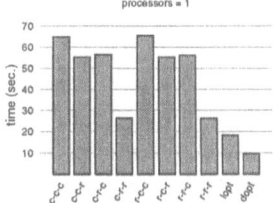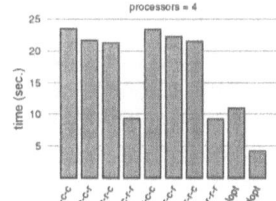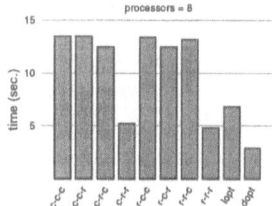

Fig. 2. Execution times (in sec.) of syr2k with different layouts on SGI Origin.

In this case, for all the references the spatial reuse will be exploited in the innermost loop. This example clearly shows that the data transformations may be effective in some cases where the iteration space transformations fail.

4.4 Optimizing for Shared-Memory Multiprocessors

In a shared-memory multiprocessor case, we should take into account the issues related to parallelism as well. As a rule, to prevent a common form of false sharing, a loop which carries spatial reuse should not be parallelized [9]. False sharing occurs when two processors access the same coherence unit (at least one of them writes) without sharing a data element [4].

We assume that the parallelism decisions are made by a previous pass in the compilation process, and the information for a loop nest is available to our algorithm as a form of vector p where $p(i)$ (the ith element) is one if the loop i is parallelized otherwise it is zero. Our memory layout determination algorithm is given in Figure 1. In the figure, U is the set of all arrays referenced in the nest. The symbol n denotes the number of loops in the nest, and e_i is a vector will all zero entries except for the ith entry which is 1. The algorithm assumes that there is no conflict (in terms of layout requirements) between references to a particular array. If there is a conflict, then a conflict resolution scheme should be applied [6]. For each array, a subset of all possible spatial reuse vectors starting with the best possible vector are tried. The sequence of trials correspond to $e_n,...,e_2,e_1$. A vector is rejected as target spatial reuse vector if the associated loop is parallelized. Otherwise e_i with the largest i is selected as target reuse vector. Once a spatial reuse vector is chosen, the remainder of the algorithm involves only computation of a null set and a set of basis vectors for it. Since optimizing compilers for shared-memory multiprocessors are quite successful in parallelizing the outermost loops, the algorithm terminates quickly; and the spatial locality in the innermost loops is exploited without incurring severe false sharing.

5 Experimental Results

We conducted experiments on an SGI Origin, which is a distributed-shared-memory machine that uses 195MHz R10000 processors, and has 32KB L1 data cache and 4MB L2 unified cache. We used the `syr2k` code (in C) from BLAS to show that our data layout transformation framework can handle complicated access patterns. Assuming a column-major memory layout, all the arrays have poor locality. The version that is optimized by loop transformations (from [9], assuming column-major layouts) optimizes spatial locality for all the arrays, but has two drawbacks: First, the loop bounds and array subscript expressions are very complicated, and second, the temporal locality for array C in the original program has been converted to spatial locality. Our framework, on the other hand, decides suitable memory layouts. Figure 2 shows the performances for eight possible (permutation-based) layouts on an SGI Origin using 1024×1024 double matrices. The legend x-y-z means that the memory layouts for C, A and B are x, y and z respectively, where c means column-major and r means row-major. The layout optimized version –obtained by our framework– (marked `dopt`) and the loop optimized version (marked `lopt`) are also shown as the last two bars. Notice that our framework optimizes the performance substantially, and no dimension re-indexing (permutation-based layout transformation) can obtain this performance.

6 Related Work

The compiler work in optimizing loop nests for locality can be divided into two groups: (1) iteration space based optimizations and (2) data layout optimizations.

In the first group, Wolf and Lam [13] present definitions of different types of reuses and propose an algorithm to maximize locality by evaluating a subset of legal loop transformations. In contrast, Li [9] uses the concept of 'reuse distance'. His algorithm constructs a combined reuse matrix and tries to reduce its 'height' using techniques from linear algebra. McKinley et al. [10] offer a unified optimization technique consisting of loop permutation, loop fusion and loop distribution. Our work on locality is different from those mentioned. First, we focus on data space transformations instead of iteration space transformations. Second, we use a new definition of spatial reuse vector whereas the reuse vectors used in [9] and [13] are oriented for a uniform memory layout.

In the second group, O'Boyle and Knijnenburg [11] focus on restructuring the code given a data transformation matrix and show its usefulness in optimizing spatial locality; in contrast, we concentrate more on the problem of determining suitable layouts by taking into account false sharing as well. Anderson et al. [1] propose a data transformation technique—comprised of permutations and strip-mining—that restructures the data in the shared memory space such that the data for each processor are stored in nearby locations. In contrast, we focus on a larger space of data transformations. Cierniak and Li [2] and Kandemir et al. [5] propose optimization techniques that combine loop and data transformations in a unified framework, but restrict the transformation space. Even in the restricted space, they perform sort of exhaustive search. Jeremiassen and Eggers [4] use data transformations to reduce false sharing. Our framework also considers reducing a common form of false sharing as well. Leung and Zahorjan [8] present an array restructuring framework to optimize locality in uniprocessors. Our work differs from theirs in several points: (1) our technique is based on explicit representation of memory layouts which is central to a unified loop and data transformation framework such as ours; (2) our technique finds optimal memory layouts in a single step rather than first determining a transformation matrix and then refining it for minimizing memory space using Fourier-Motzkin elimination; and (3) we explicitly take false sharing into account for multiprocessors.

7 Conclusions

In this paper, we have presented a definition of reuse vector under a given memory layout. For this, we have represented the memory layout of a multidimensional array using hyperplane families. Then we have presented a relation between layout representation and spatial reuse vector. We have shown that this relation can be exploited at least in two ways. For a given layout and reference, we can find the spatial reuse vector *or* for a given desired spatial reuse vector, we can determine the target layout. This second usage allows us to develop a data layout reorganization framework based on the existing compiler technology. Given a loop nest, our framework can optimize the memory layouts of arrays by considering the best possible reuse vectors.

References

1. J. Anderson, S. Amarasinghe, and M. Lam. Data and computation transformations for multiprocessors. *Proc. 5th SIGPLAN Symp. Prin. & Prac. Para. Prog.*, Jul. 1995.
2. M. Cierniak, and W. Li. Unifying data and control transformations for distributed shared memory machines. *Proc. SIGPLAN '95 Conf. Prog. Lang. Des.& Impl.*, Jun. 1995.
3. D. Gannon, W. Jalby, and K. Gallivan. Strategies for cache and local memory management by global program transformations, *J. Para. & Dist. Comp.*, 5:587–616, 1988.
4. T. Jeremiassen, and S. Eggers. Reducing false sharing on shared memory multiprocessors through compile time data transformations. *Proc. 5th SIGPLAN Symp. Prin. & Prac. Para. Prog.*, Jul. 1995.
5. M. Kandemir, J. Ramanujam, and A. Choudhary. Compiler algorithms for optimizing locality and parallelism on shared and distributed memory machines. *Proc. 1997 International Conference on Parallel Architectures and Compilation Techniques*, pp. 236–247, Nov. 1997.
6. M. Kandemir, A. Choudhary, N. Shenoy, P. Banerjee, and J. Ramanujam. A data layout optimization technique based on hyperplanes. TR CPDC-TR-97-04, Northwestern Univ.
7. M. Kandemir, A. Choudhary, J. Ramanujam, and P. Banerjee. Optimizing spatial locality in loop nests using linear algebra. *Proc. 7th Workshop Compilers for Parallel Computers*, 1998.

434

8. S-T. Leung, and J. Zahorjan. Optimizing data locality by array restructuring. Technical Report, CSE Dept., University of Washington, TR 95-09-01, Sep. 1995.

9. W. Li. Compiling for NUMA parallel machines. Ph.D. dissertation, Cornell University, 1993.

10. K. McKinley, S. Carr, and C.W. Tseng. Improving data locality with loop transformations. *ACM Transactions on Programming Languages and Systems,* 1996.

11. M. 'Boyle, and P. Knijnenburg. Non-singular data transformations: Definition, validity, applications. *Proc. 6th Workshop on Compilers for Parallel Computers,* pp. 287–297, 1996.

12. J. Ramanujam, and P. Sadayappan. Compile-time techniques for data distribution in distributed memory machines. In *IEEE Trans. Para. & Dist. Sys.,* 2(4):472–482, Oct. 1991.

13. M. Wolf, and M. Lam. A data locality optimizing algorithm. In *Proc. ACM SIG-PLAN 91 Conf. Programming Language Design and Implementation,* pp. 30–44, June 1991.

Parallelization of Unstructured Mesh Computations Using Data Structure Formalization

Rainer Koppler

Department of Computer Graphics and Parallel Processing (GUP)
Johannes Kepler University, A-4040 Linz, Austria/Europe
koppler@gup.uni-linz.ac.at

Abstract. This paper introduces a concept for semi-automatic parallelization of unstructured mesh computations called *data structure formalization*. Unlike existing concepts it does not expect knowledge about parallelism but just enough knowledge about the application semantics such that a formal description of the data structure implementation can be given. The parallelization tool *Parlamat* uses this description for deduction of additional information about arrays and loops such that more efficient parallelization can be achieved than with general tools. We give a brief overview of our data structure modelling language and first experiences with *Parlamat*'s capabilities by means of the translation of some real-size applications from Fortran 77 to HPF.

1 Introduction

Unstructured meshes are used in a large number of scientific codes, for example, in computational fluid dynamics and structural analysis codes. Unfortunately, these codes are hard to parallelize efficiently. Automatic parallelization cannot satisfy today's expectations due to restrictions of compile-time techniques. On the other hand, languages such as High Performance Fortran (HPF) [9] require considerable skill and efforts from programmers, in particular with real-size unstructured mesh codes.

In this paper we narrow the gap between the above two approaches using a novel concept called *data structure formalization*. Unlike existing semi-automatic approaches our concept does not confront users with parallelism but just expects knowledge about the sequential application such that the user can provide a brief and formal description of the mesh data structure. We show that such a description provides enough information in order to let a software tool perform automatic array distribution and detect parallelism in loops containing indirect array accesses. We implemented a tool with such capabilities that utilizes the results for translation of the Fortran 77 code to HPF. The tool's back-end automatically inserts directives for array and loop distribution according to known HPF parallelization techniques. First experiences with the application of our concept to real-size codes gave encouraging results and justify development of special-purpose parallelization systems for particular application classes.

2   Parallelization of Mesh Computations

The main reason why today's parallelizing compilers fail to produce efficient results is the complexity of unstructured mesh implementations. For example, finite element meshes consist of several arrays, where some are used as index arrays and others store application data. First, complexity inhibits automatic distribution of arrays in a block or cyclic fashion because each mesh partition yields a set of irregular array distributions. Run-time partitioning techniques [18] address this problem but they are expensive and require parallel loops to be known. Secondly, automatic detection of parallel loops in unstructured mesh codes is unfeasible using known techniques. Such codes are dominated by *indirect* array accesses, which prevent application of compile-time dependence tests. Consequently, most parallelizers leave loops containing indirect array accesses sequential. Some approaches allow speculative parallel execution [2] but require additional run-time analysis, which is costly in terms of run time and storage overhead.

Therefore, researchers became aware of the fact that knowledge from the programmer must be specified to the compiler in order to achieve efficient parallelization. They designed extensions to sequential languages, where HPF is the most popular result. Further, several parallelization strategies for unstructured mesh codes using the family of HPF languages have been developed [5, 16]. However, these strategies still keep programmers far away from automatic parallelization. It has hardly been considered that they have been applied only to small and simple kernels and manual application to large-scale codes is time-consuming and error-prone [8]. Further, in most cases the HPF programmer and the code author are not the same person.

The semi-automatic concept we describe in this paper provides a basis for automatic application of parallelization strategies for unstructured mesh codes. In our tool we use a parallelization strategy that makes use of facilities provided by HPF. We adopted basic ideas from existing work on this subject and defined a general framework for automatic application of these ideas and our own additions. Our strategy consists of *code parallelization* and *mesh data preparation*. Details on the two steps are beyond the scope of this paper and can be found elsewhere. In the first step directives for array and loop distribution are inserted [5,16]. The second step consists of mesh partitioning and renumbering [16,13].

3   Data Structure Formalization

As mentioned earlier automated compiler techniques cannot parallelize unstructured mesh computations efficiently due to lack of information that is accessible to a compiler. An obvious solution to this problem is specification of missing information to the compiler in a systematic way, for example, specification of the mesh data structure. McManus [13] considers systematic description of data structures impractical because in theory there are unlimited possibilities of implementing an unstructured mesh.

Although in theory there may be an unlimited variety of mesh implementations, in practice programmers prefer implementation variants that provide a good compromise between low storage overhead and simplicity and efficiency of algorithms. A comprehensive collection of different unstructured mesh codes showed us that in fact there exists *one* particular variant that is preferably used by Fortran 77 programmers. The main reasons for the dominance of this variant are legibility and efficiency of algorithms that are built up on it. A mesh implementation that follows this variant is shown in Fig. 1.

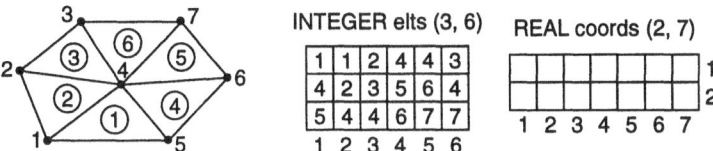

Fig. 1. Implementation of a triangular mesh

Thereupon we generalized the implemention variant for application to arbitrary mesh structures that can be embedded into the Entity-Relationship (ER) diagram shown in Fig. 2a. Further we developed a tiny modelling language that provides description of mesh structures and mapping of a mesh structure to variables of the application code. Whenever the data structure of an application can be described through this language, semantic knowledge about it can be specified to the compiler in machine-readable form.

A remarkable property of our modelling language is that it does not confront the user with parallelism but just expects enough knowledge about the application semantics such that a "formal documentation" of parts of the sequential code can be given. Hence, the language is accessible to a broad range of users, even to those who are inexperienced in parallelization or who do not want to get in touch with parallelism.

The rest of this section gives a brief overview of our data structure modelling language. First we describe some basic concepts and terms, which emerged during generalization of the implementation variant mentioned above. Then we show applications of the language by means of different mesh structures taken from real applications and benchmarks.

A mesh consists of a number of entities – so-called *objects*, which belong to different *object types*, for example, vertex, edge, triangle, or tetrahedron. Mesh applications define a hierarchy between object types, that is, objects may consist of several *subobjects* with lower dimensionality. For example, a hexahedron can be viewed as a composition of eight vertices, twelve edges, or six quadratic faces. Each application defines a particular hierarchy, which reflects a subdiagram of the ER diagram shown in Fig. 2a.

Mesh applications associate object types with specific properties – so-called *attributes*. For example, fluid dynamics applications usually associate vertices with density, velocity, and energy [14], or a solid mechanics application may associate hexahedrons with normal stresses and shear stresses [12]. We call application-specific properties *data attributes*. Besides those, each object type that refers to one or several subobjects has a *structure attribute*. For example, the structure attribute of the triangle type may be a triplet of vertex numbers.

The modelling language covers two major issues: (1) definition of application-specific object types with their attributes, which yields a subdiagram of the ER diagram shown in Fig. 2a, and (2) mapping of attributes to array variables of the application code.

Application-specific object types are defined through derivation of pre-defined *base types*, which are enclosed by rectangles in Fig. 2a. A definition specifies the name of the derived type, a base type, possibly a subobject type, and a list of data attributes. Figure 3a shows a simplified mesh specification taken from a fluid dynamics application and Fig. 2b depicts its corresponding ER diagram. The pre-defined type *Vertex* provides the pre-defined data attributes x and y for the description of vertex coordinates.

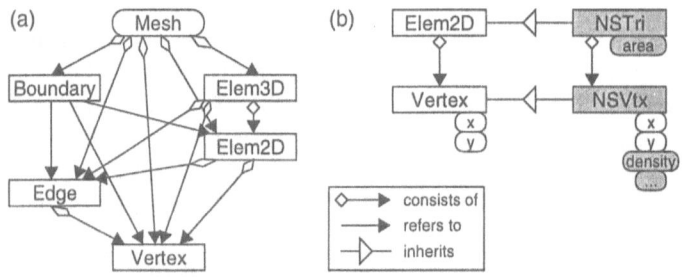

Fig. 2. (a) Universal ER diagram (b) Application-specific ER diagram

Mapping of attributes to variables of the application code assumes a mesh implementation as it is shown in Fig. 1. Each attribute of object type T is implemented by an array, which is indexed by numbers of T objects. Figure 3b shows how attributes of the mesh specified in Fig. 3a can be mapped to the arrays given in Fig. 1.

(a) object NSVtx
    is Vertex
    with density, velocity, energy.

    object NSTri
    is Elem2D(NSVtx)
    with area.

(b) NSVtx(i).x -> coords(1,i).
    NSVtx(i).y -> coords(2,i).
    NSVtx(i).density -> d(i).
    ...
    NSTri(i) -> elts(:, i).
    NSTri(i).area -> a(i).

Fig. 3. Specification of a triangular mesh: (a) structure (b) attribute mapping

Some applications implement special treatment of objects that lie on the boundary of a mesh. Such objects may require additional attributes and may be collected in additional object lists. Our modelling language allows definition of boundary object types through derivation of the pre-defined base type *Boundary*. Figure 4a shows an example of a boundary vertex type. The array **bndlst** stores indices of all *NSVtx* objects that belong to the boundary vertex type *SlippingVtx*.

Finally the language also provides support for applications that use hybrid meshes. The number of subobjects of a hybrid object type is not constant but must be specified for each object separately, for example, using an additional array or array section. Figure 4b shows the definition and mapping of a hybrid object type we found in a Poisson solver. The code uses mixed meshes consisting of tetrahedra and hexahedra.

```
(a)   object SlippingVtx           (b)   object FEMSolid is Elem3D(FEMNode).
         is Boundary(NSVtx)               object FEMNode is Vertex.
         with info.
                                          FEMSolid(i).# -> elts(1,i).
      SlippingVtx(i) -> bndlst(i).        ! 1st value = number of FEMNode indices
      ! bndlst contains NSVtx indices
                                          FEMSolid(i) -> elts(2:9,i).
      SlippingVtx(i).info -> bndinf(i).   ! room for up to eight FEMNode indices
```

Fig. 4. (a) Boundary object type (b) Hybrid object type

In this paper we describe our modelling language using simple examples. However, we have also studied mesh implementation strategies designed for multigrid methods [7] and dynamic mesh refinement [10] and observed that these strategies do not necessitate changes to the current language design.

4   The Parallelization System

Usually today's parallelizers are applicable to *every* program written in a given language and their analysis capabilities are restricted to the language level. By way of contrast, very few parallelizers have been developed that focus on special classes of applications, for example, unstructured mesh computations. Such systems receive more knowledge about the application than general-purpose parallelizers, thus they require almost no user intervention and can produce more efficient results.

Parlamat (Parallelization Tool for Mesh Applications) is a special-purpose parallelizer, which uses our modelling language for specification of knowledge about the implementation of mesh data structures. It consists of a compiler that translates Fortran 77 codes to HPF and a tool for automatic mesh data preparation. The compiler uses the specification for deduction of knowledge for array distribution and dependence analysis and for recognition of particular programming idioms.

4.1 Additional Array Semantics

After *Parlamat* has carried out basic control and data flow analysis, it performs a special analysis that determines additional semantics for arrays of the application code. It examines the use of arrays in loops over mesh objects, in particular, array indexing and assignments to INTEGER arrays, and extends symbol table information by the following properties:

- *Index type*: Arrays that are indexed using a loop variable receive the object type associated with the loop variable as index type. Other arrays, for example, coords in Fig. 3b, receive this type using the mesh specification.
- *Contents type*: Index arrays that store object indices, which can be detected through analysis of assignments, receive the type of these objects as contents type. Other arrays, for example, elts in in Fig. 3b receive this type using the mesh specification.
- *Uniqueness*: Index arrays that do not contain repeated values are called unique. This property is set using knowledge from the mesh specification or through analysis of index array initializations in the application code. For example, elts, which is the structure array of an *Elem2D* type, is not unique by definition, which can be seen from Fig. 1.

The first two properties play an important role for array distribution. Arrays that show the same index type become aligned to each other. The distribution of object types is determined by the mesh partition. The third property is used by the dependence test.

Before special analysis starts, the mesh specification is evaluated in order to determine the above properties for arrays occuring in the specification. After this, the analysis proceeds in an interprocedural manner, where properties are propagated from actual to dummy arguments and vice versa.

4.2 Dependence Analysis

Dependence analysis [1] is essential in order to classify loops as sequential or parallel. For each array that is written at least once in a loop a dependence test checks if two iterations refer to the same array element. *Parlamat* performs dependence analysis for loops over mesh objects. Analysis consists of a flow-sensitive scalar test, which detects privatizeable scalars and scalar reductions, and a flow-insensitive test for index-typed arrays.

Direct array accesses, as they appear in unstructured mesh codes, can be handled sufficiently by conventional tests. On the contrary, indirect array accesses must be passed on to a special test. The core of our test is a procedure that checks if two indirect accesses may refer to the same array element. The procedure performs a pairwise comparison of the *dependence strings* of both accesses. Loosely speaking, a dependence string describes the index chain of an indirect access as a sequence of contents-typed accesses that end up with the loop variable. Figure 5 shows two loops containing simple indirect accesses and

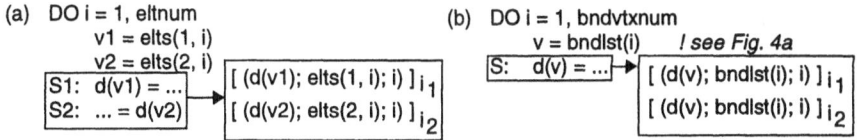

Fig. 5. Loop-carried dependences: (a) present, (b) absent

comparisons of their dependence strings. The notation $[\mathbf{x}]_i$ stands for "value of expression $\mathbf{x}$ in iteration i".

The procedure compares strings from the right to the left. In order to detect loop-carried dependences it assumes that $i_1 \neq i_2$. Hence, in Fig. 5a $[\mathbf{v1}]_{i_1} = [\mathbf{v2}]_{i_2}$ is true iff $\mathtt{elts}(1, i_1) = \mathtt{elts}(2, i_2)$. At this point the procedure requires information about the *uniqueness* of $\mathtt{elts}$. As mentioned above, $\mathtt{elts}$ is not unique by definition. Hence, the condition is true and the accesses $\mathtt{d(v1)}$ and $\mathtt{d(v2)}$ will definitely overlap. For example, with respect to the mesh shown in Fig. 1 this will occur for $i_1 = 3$ and $i_2 = 2$. Unless both accesses appear in reduction operations, the true dependence $S_1 \delta^t S_2$ disallows parallelization.

Figure 5b shows how *Parlamat* can disprove dependence using uniqueness. The array $\mathtt{bndlst}$ is the structure array of a boundary object type and is unique by definition as a list of boundary objects does not contain duplicates. Hence, the loop can be parallelized. With respect to the code, where the loop was taken from, this is an important result, because the loop is enclosed by a time step loop and thus executed many times.

5 Experiences

We consider special-purpose systems such as *Parlamat* primarily useful for parallelization of real-size application codes, where it helps to save much work. On the other hand, even a special-purpose solution should be applicable to a broad range of applications. In order to keep this goal in mind we based the development of our solution on a set of various unstructured mesh codes, which comprises real-size codes as well as benchmarks. In this paper we report on experiences with three codes taken from this set:

1. The benchmark *euler* [11] is the kernel of a 2D fluid dynamics computation. We mention it here because at the moment its HPF version is the *only* example of a parallel unstructured mesh code that is accepted by one of today's HPF compilers.
2. The structural mechanics package *FEM3D* [12] solves linear elasticity problems in 3D using meshes consisting of linear brick elements and an element-wise PCG method.
3. The fluid dynamics package *NSC2KE* [14], which has been used in several examples throughout this paper, solves the Euler and Navier-Stokes equations in 2D with consideration of turbulences. It uses meshes consisting of triangles.

We regard the codes FEM3D and NSC2KE as real-size applications for two reasons. First, they are self-contained packages that solve real-world problems. Secondly, they consist of 1100 and 3000 lines of code, respectively. Table 1 gives a brief characterization of the three codes. The lengths of the data structure specifications show that specifications – even if they belong to large applications – usually fit into one screen page.

Table 1. Characterization of three unstructured mesh applications

application	data structure specification				DO loops			
	object types	boundary object types	attribute mappings	length in lines	parallel	sequential I/O	others	total
euler	2	–	7	11	5 (3)	2	1	8 (3)
FEM3D	2	–	10	14	25 (5)	6	29	60 (5)
NSC2KE	3	6	38	56	66 (17)	15	22	103 (21)

The table also reports on parallelization of DO loops. It shows how much loops of each application can be marked as parallel through safe use of the INDEPENDENT directive, sometimes after minor automatic restructuring. Some of the remaining loops cannot be parallelized because they contain I/O statements that refer to one and the same file. Numbers given in parentheses refer to loops that are executed many times, for example, within a time step loop. The statistics indicate that the knowledge accessible to *Parlamat* helps to achieve a high number of parallel loops for all three applications. Furthermore nearly all of the frequently executed loops can be parallelized, which should lead to good parallel speed-ups.

Our experiences with the HPF versions of the above three codes are very limited due to several restrictions of today's HPF compilers. In order to handle arbitrary parallel loops containing indirect array accesses, an HPF compiler must support the *inspector-executor* technique [15]. Our attempts to compile the HPF version of euler using ADAPTOR 4.0 [4] and *pghpf* 2.2 [17] gave evidence that these compilers do not support this technique. Thereupon we switched to a pre-release of the HPF+ compiler *vfc* [3]. The HPF+ version could be compiled with *vfc* and executed on a Meiko CS-2 without problems. On the other hand, the HPF+ versions of FEM3D and NSC2KE were not accepted by *vfc* due to preliminary compiler restrictions.

In spite of problems with HPF compilers we could verify the correctness of the transformations applied to FEM3D and NSC2KE. Due to the nature of the language HPF programs can be compiled also with a Fortran 9X compiler and executed on a sequential computer. We tested *Parlamat*'s code transformation and mesh data preparation facilities thoroughly on a single-processor workstation through comparison of the results produced by the HPF version and the original version. The differences between the results are hardly noticeable and stem from the fact that renumbering of mesh object changes the order of arithmetic operations with real numbers, which causes different rounding.

6 Related Work

McManus [13] and Hascoet [8] investigated automatization concepts for another parallelization strategy for unstructured mesh computations. Unlike the HPF strategy described here they focus on execution of the original code as node code on a distributed-memory computer after some changes such as communication insertion or adjustment of array and loop bounds have been made. It is obvious that this strategy is less portable than HPF.

McManus considers some steps of his five-step strategy automateable within the framework of the *CAPTools* environment. However, such a general-purpose system requires extensive knowledge from the user in order to achieve efficient results. Hascoet has designed a special-purpose tool that provides semi-automatic identification of program locations where communication operations must be inserted. The tool requests information including the loops to be parallelized, association of loops with mesh entities, and a so-called *overlap automaton*.

7 Conclusion

In this paper we introduced a concept for semi-automatic parallelization of unstructured mesh computations. The core of this concept is a modelling language that allows users to describe the implementation of mesh data structures in a formal way. The language is simple and general enough to cover a broad variety of data structure implementations. We also showed that data structure specifications provide the potential for automatization of a strategy that translates Fortran 77 codes to HPF. In particular, we outlined their usefulness for efficient loop parallelization by means of enhanced dependence testing.

Finally we reported on application of our concept to a set of unstructured mesh codes, with two real-size applications among them. Our experiences show that data structure formalization yields significant savings of work and provides enough knowledge in order to transform applications with static meshes into highly parallel HPF equivalents. Hence, they justify and recommend the development of special-purpose parallelizers such as *Parlamat*.

We believe that the potential of data structure formalization for code parallelization has not been exploited yet by far. The current version of *Parlamat* utilizes semantic knowledge just for most obvious transformations. In particular, applications using dynamic mesh refinement provide several opportunities for exploitation of semantic knowledge for recognition of coarse-grain parallelism. We regard development of respective HPF parallelization strategies and concepts for automatic translation as interesting challenges for the future.

444

Acknowledgement

The author thanks Siegfried Benkner for access to the HPF+ compiler *vfc*.

References

1. Banerjee, U.: Dependence Analysis for Supercomputing. Reading, Kluwer Academic Publishers, Boston (1988)
2. Blume, W., Eigenmann, R., Hoeflinger, J., Padua, D., Petersen, P., Rauchwerger, L., Tu, P.: Automatic Detection of Parallelism. IEEE Parallel and Distributed Technology **2** (1994) 37–47
3. Benkner, S., Pantano, M., Sanjari, K., Sipkova, V., Velkow, B., Wender, B.: VFC - Compilation System for HPF+ – Release Notes 0.91. University of Vienna (1997)
4. Brandes, T.: ADAPTOR Programmer's Guide Version 4.0. GMD Technical Report, German National Research Center for Information Technology (1996)
5. Chapman, B., Zima, H., Mehrotra, P.: Extending HPF for Advanced Data-Parallel Applications. IEEE Parallel And Distributed Technology **2** (1994) 59–70
6. Cross, M., Ierotheou, C., Johnson, S., Leggett, P.: CAPTools – semiautomatic parallelisation of mesh based computational mechanics codes. Proc. HPCN '94 **2** (1994) 241–246
7. Das, R., Mavriplis, D., Saltz, J., Gupta, S., Ponnusamy, R.: The Design and Implementation of a Parallel Unstructured Euler Solver Using Software Primitives. AIAA Journal **32** (1994) 489–496
8. Hascoet, L.: Automatic Placement of Communications in Mesh-Partitioning Parallelization. ACM SIGPLAN Notices **32** (1997) 136–144
9. High-Performance Fortran Forum: High Performance Fortran Language Specification – Version 2.0. Technical Report, Rice University, TX (1997)
10. Kallinderis, Y., Vijayan, P.: Adaptive Refinement-Coarsening Scheme for Three-Dimensional Unstructured Meshes. AIAA Journal **31** (1993) 1140–1447
11. Mavriplis, D.J.: Three-Dimensional Multigrid for the Euler Equations. AIAA Paper 91-1549CP, American Institute of Aeronautics and Astronautics (1991) 824–831
12. Mathur, K.: Unstructured Three Dimensional Finite Element Simulations on Data Parallel Architectures. In: Mehrotra, P., Saltz, J., Voigt, R. (eds.): Unstructured Scientific Computation on Scalable Multiprocessors. MIT Press (1992) 65–79
13. McManus, K.: A Strategy for Mapping Unstructured Mesh Computational Mechanics Programs onto Distributed Memory Parallel Architectures. Ph.D. thesis,University of Greenwich (1996)
14. Mohammadi, B.: Fluid Dynamics Computations with NSC2KE – A User Guide. Technical Report RT-0164, INRIA (1994)
15. Mirchandaney, R., Saltz, J., Smith, R., Nicol, D., Crowley, K.: Principles of run-time support for parallel processors. Proc. 1988 ACM International Conference on Supercomputing (1988) 140–152
16. Ponnusamy, R., Hwang, Y.-S., Das, R., Saltz, J., Choudhary, A., Fox, G.: Supporting Irregular Distributions Using Data-Parallel Languages. IEEE Parallel and Distributed Technology **3** (1995) 12–14
17. The Portland Group, Inc.: pghpf User's Guide. Wilsonville, OR (1992)
18. Ponnusamy, R., Saltz, J., Das, R., Koelbel, Ch., Choudhary, A.: Embedding Data Mappers with Distributed Memory Machine Compilers ACM SIGPLAN Notices **28** (1993) 52–55

Parallel Constant Propagation

Jens Knoop

Universität Passau, D-94030 Passau, Germany
knoop@fmi.uni-passau.de
phone: ++49-851-509-3090 fax: ++49-...-3092

Abstract. *Constant propagation* (*CP*) is a powerful, practically relevant optimization of sequential programs. However, systematic adaptions to the parallel setting are still missing. In fact, because of the computational complexity paraphrased by the catch-phrase "state explosion problem", the successful transfer of sequential techniques is currently restricted to bitvector-based optimizations, which because of their structural simplicity can be enhanced to parallel programs at almost no costs on the implementation and computation side. CP, however, is beyond this class. Here, we show how to enhance the framework underlying the transfer of bitvector problems obtaining the basis for developing a powerful algorithm for *parallel constant propagation* (*PCP*). This algorithm can be implemented as *easily* and as *efficiently* as its sequential counterpart for *simple constants* computed by state-of-the-art sequential optimizers.

1 Motivation

In comparison to automatic parallelization (cf. [15, 16]), optimization of *parallel* programs attracted so far only little attention. An important reason may be that straightforward adaptions of sequential techniques typically fail (cf. [10]), and that the costs of rigorous, correct adaptions are unacceptably high because of the combinatorial explosion of the number of interleavings manifesting the possible executions of a parallel program. However, the example of unidirectional *bitvector* (UBv) problems (cf. [2]) shows that there are exceptions: UBv-problems can be solved as *easily* and as *efficiently* as for sequential programs (cf. [9]). This opened the way for the successful transfer of the classical bitvector-based optimizations to the parallel setting at almost no costs on the implementation and computation side. In [6] and [8] this has been demonstrated for *partial dead-code elimination* (cf. [7]), and *partial redundancy elimination* (cf. [11]).

  Constant propagation (*CP*) (cf. [4, 3, 12]), however, the application considered in this article, is beyond bitvector-based optimization. CP is a powerful and widely used optimization of sequential programs. It improves performance by replacing terms, which at compile-time can be determined to yield a unique constant value at run-time by that value. Enhancing the framework of [9], we develop in this article an algorithm for *parallel constant propagation* (*PCP*). Like the bitvector-based optimizations it can be implemented as *easily* and as *efficiently* as its sequential counterpart for *simple constants* (*SCs*) (cf. [4, 3, 12]), which are computed by state-of-the-art optimizers of sequential programs.

The power of this algorithm is demonstrated in Figure 1. It detects that y has the value 15 before entering the parallel statement, and that this value is maintained by it. Similarly, it detects that b, c, and e have the constant values 3, 5, and 2 before and after leaving the parallel assignment. Moreover, it detects that independently of the interleaving sequence taken at run-time d and x are assigned the constant values 3 and 17 inside the parallel assignment. Together this allows our algorithm not only to detect the constancy of the right-hand side terms at the edges **48**, **56**, **57**, **60**, and **61** after the parallel statement, but also the constancy of the right-hand side terms at the edges **22**, **37**, and **45** inside the parallel statement. We do not know of any other algorithm achieving this.

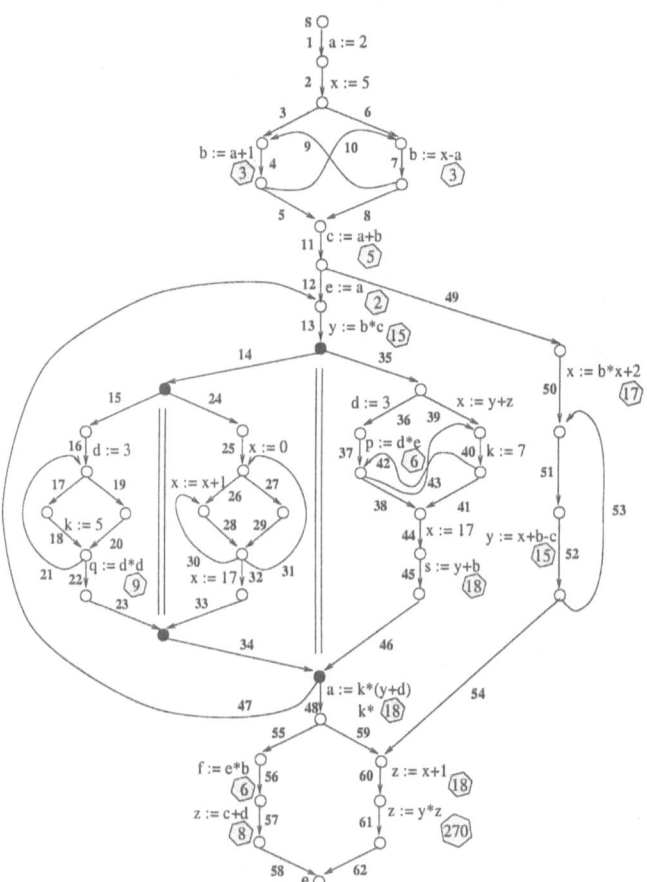

Fig. 1. The power of *parallel constant propagation*.

Importantly, this result is obtained without any performance penalty in comparison to sequential CP. Fundamental for achieving this is to conceptually decompose PCP to behave differently on sequential and parallel program parts. Intuitively, on sequential program parts our PCP-algorithm coincides with its sequential counterpart for SCs, on parallel program parts it computes a safe approximation of SCs, which is tailored with respect to side-conditions allowing us as for unidirectional bitvector problems to capture the phenomena of *inter-ference* and *synchronization* of parallel components completely without having to consider interleavings at all. Summarizing, the central contributions of this article are as follows: (1) Extending the framework of [9] to a specific class of non-bitvector problems. (2) Developing on this basis a PCP-algorithm, which works for reducible and irreducible control flow, and is as efficient and can as easily be implemented as its sequential counterpart for SCs. An extended presentation can be found in [5].

2 Preliminaries

This section sketches our parallel setup, which has been presented in detail in [9]. We consider parallel imperative programs with shared memory and interleaving semantics. Parallelism is syntactically expressed by means of a **par** statement. As usual, we assume that there are neither jumps leading into a component of a parallel statement from outside nor vice versa. As shown in Figure 1, we represent a parallel program by an edge-labelled *parallel flow graph* G with node set N, and edge set E. Moreover, we consider terms $t \in \mathbf{T}$, which are inductively built from variables $v \in \mathbf{V}$, constants $c \in \mathbf{C}$, and operators $op \in \mathbf{Op}$ of arity $r \geq 1$. Their *semantics* is induced by an *interpretation* $I = (\mathbf{D}' \cup \{\bot, \top\}, I_0)$, where $\mathbf{D}'$ denotes a non-empty data domain, $\bot$ and $\top$ two new data not in $\mathbf{D}'$, and I_0 a function mapping every constant $c \in \mathbf{C}$ to a datum $I_0(c) \in \mathbf{D}'$, and every r-ary operator $op \in \mathbf{Op}$ to a total, strict function $I_0(op) : \mathbf{D}^r \to \mathbf{D}$, $\mathbf{D} =_{df} \mathbf{D}' \cup \{\bot, \top\}$ (i.e., $I_0(op)(d_1, \ldots, d_r) = \bot$, whenever there is a j, $1 \leq j \leq r$, with $d_j = \bot$). $\Sigma = \{\sigma \mid \sigma : \mathbf{V} \to \mathbf{D}\}$ denotes the set of *states*, and $\sigma_\bot$ the distinct *start state* assigning $\bot$ to all variables $v \in \mathbf{V}$ (this reflects that we do not assume anything about the context of a program being optimized). The *semantics* of terms $t \in \mathbf{T}$ is then given by the evaluation function $\mathcal{E} : \mathbf{T} \to (\Sigma \to \mathbf{D})$. It is inductively defined by: $\forall t \in \mathbf{T} \, \forall \sigma \in \Sigma$.

$$\mathcal{E}(t)(\sigma) =_{df} \begin{cases} \sigma(x) & \text{if } t = x \in \mathbf{V} \\ I_0(c) & \text{if } t = c \in \mathbf{C} \\ I_0(op)(\mathcal{E}(t_1)(\sigma), \ldots, \mathcal{E}(t_r)(\sigma)) & \text{if } t = op(t_1, \ldots, t_r) \end{cases}$$

In the following we assume $\mathbf{D}' \subseteq \mathbf{T}$, i.e., the set of data $\mathbf{D}'$ is identified with the set of constants $\mathbf{C}$. We introduce the *state transformation function* $\theta_\iota : \Sigma \to \Sigma$, $\iota \equiv x := t$, where $\theta_\iota(\sigma)(y)$ is defined by $\mathcal{E}(t)(\sigma)$, if $y = x$, and by $\sigma(y)$ otherwise. Denoting the set of all *parallel program paths*, i.e., the set of interleaving sequences, from the unique start node $\mathbf{s}$ to a program point n by $\mathbf{PP}[\mathbf{s}, n]$, the set of states Σ_n being possible at n is given by $\Sigma_n =_{df} \{\theta_p(\sigma_\bot) \mid p \in \mathbf{PP}[\mathbf{s}, n]\}$. Here, θ_p denotes the straightforward extension of θ to parallel paths. Based on Σ_n we can now define the set $\mathcal{C}_n$ of all terms yielding a unique constant value at program point $n \in N$ at run-time:

$$\mathcal{C}_n =_{df} \bigcup \{\mathcal{C}_n^d \mid d \in \mathbf{D}'\}$$

with $C_n^d =_{df} \{t \in \mathbf{T} \mid \forall \sigma \in \Sigma_n. \; \mathcal{E}(t)(\sigma) = d\}$ for all $d \in \mathbf{D}'$. Unfortunately, even for sequential programs C_n is in general not decidable (cf. [12]). Simple constants recalled next are an efficiently decidable subset of C_n, which is computed by state-of-the-art optimizers of sequential programs.

3 Simple Constants

Intuitively, a term is a (sequential) *simple constant* (*SC*) (cf. [4]), if it is a program constant or all its operands are simple constants (in Figure 2, a, b, c and d are SCs, whereas e, though it is a constant of value 5, and f, which in fact is not constant, are not). Data flow analysis provides the means for computing SCs.

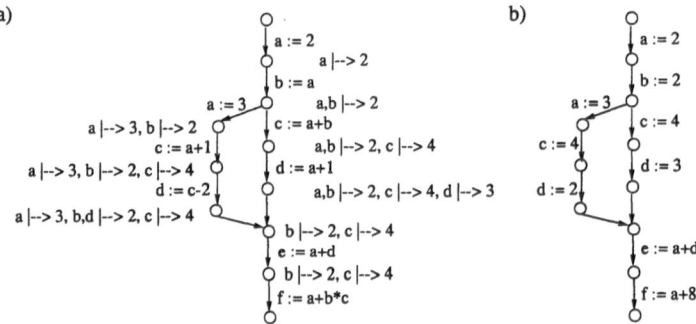

Fig. 2. Simple constants.

Data Flow Analysis. In essence, *data flow analysis* (*DFA*) provides information about the program states, which may occur at a specific program point at run-time. Theoretically well-founded are DFAs based on *abstract interpretation* (cf. [1]). Usually, the abstract semantics, which is tailored for a specific problem, is specified by a *local semantic functional* $[\![\ ]\!] : E \to (\mathcal{L} \to \mathcal{L})$ giving abstract meaning to every program statement (here: every edge $e \in E$ of a sequential or parallel flow graph G) in terms of a transformation function on a complete lattice $(\mathcal{L}, \sqcap, \sqsubseteq, \bot, \top)$. Its elements express the DFA-information of interest.

Local semantic functionals can easily be extended to capture finite paths. This is the key to the so-called *meet over all paths* (*MOP*) approach. It yields the intuitively desired solution of a DFA-problem as it directly mimics possible program executions: it "meets" all informations belonging to a program path leading from **s** to a program point $n \in N$.

The *MOP-Solution:* $\forall l_0 \in \mathcal{L}. \; MOP_{l_0}(n) = \sqcap \{ [\![p]\!](l_0) \mid p \in \mathbf{P}[s, n] \}$

Unfortunately, this is in general not effective, which leads us to the *maximal fixed point* (*MFP*) approach. Intuitively, it approximates the *greatest* solution of a system of equations imposing consistency constraints on an annotation of the program with DFA-information with respect to a start information $l_0 \in \mathcal{L}$:

$$\mathbf{mfp}(n) = \begin{cases} l_0 & \text{if } n = s \\ \sqcap \{ [\![(m,n)]\!](\mathbf{mfp}(m)) \mid (m,n) \in E \} & \text{otherwise} \end{cases}$$

The greatest solution of this equation system, denoted by $\mathtt{mfp}_{l_0}$, can effectively be computed, if the semantic functions $[\![e]\!]$, $e \in E$, are monotonic, and $\mathcal{L}$ is of finite height. The solution of the *MFP*-approach is then defined as follows:

The *MFP*-Solution: $\forall l_0 \in \mathcal{L} \; \forall n \in N. \; MFP_{l_0}(n) =_{df} \mathtt{mfp}_{l_0}(n)$

Central is now the following theorem relating both solutions. It gives sufficient conditions for the correctness and even precision of the *MFP*-solution (cf. [3, 4]):

Theorem 1 (Safety and Coincidence).

1. Safety: $MFP(n) \sqsubseteq MOP(n)$, *if all $[\![e]\!]$, $e \in E$, are monotonic.*
2. Coincidence: $MFP(n) = MOP(n)$, *if all $[\![e]\!]$, $e \in E$, are distributive.*

Computing Simple Constants. Considering $\mathbf{D}$ a flat lattice as illustrated in Figure 3(a), the set of states Σ together with the pointwise ordering is a complete lattice, too. The computation of SCs relies then on the local semantic functional $[\![\;]\!]_{sc} : E \to (\Sigma \to \Sigma)$ defined by $[\![e]\!]_{sc} =_{df} \theta_e$ for all $e \in E$, with respect to the start state $\sigma_\perp$. Because the number of variables occurring in a program is finite, we have (cf. [3]):

Lemma 1. Σ *is of finite height, and all functions $[\![e]\!]_{sc}$, $e \in E$, are monotonic.*

Hence, the *MFP*-solution $MFP_{\sigma_\perp}^{sc}$ of the SC-problem can effectively be computed inducing the formal definition of (sequential) *simple constants*.

$$C_n^{sc} =_{df} C_n^{sc,mfp} =_{df} \{ t \in \mathbf{T} \mid \exists d \in \mathbf{D}'. \; \mathcal{E}(t)(MFP_{\sigma_\perp}^{sc}(n)) = d \}$$

Defining dually the set $C_n^{sc,mop} =_{df} \{ t \in \mathbf{T} \mid \exists d \in \mathbf{D}'. \; \mathcal{E}(t)(MOP_{\sigma_\perp}^{sc}(n)) = d \}$ induced by the $MOP_{\sigma_\perp}^{sc}$-solution, we obtain by means of Theorem 1(1) (cf. [3]):

Theorem 2 (SC-Correctness). $\forall n \in N. \; C_n \supseteq C_n^{sc,mop} \supseteq C_n^{sc,mfp} = C_n^{sc}$.

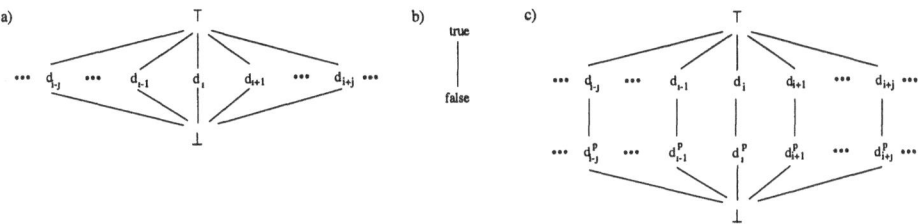

Fig. 3. a), b) Flat (data) lattices.   c) Extended lattice.

4   Parallel Constant Propagation

In a parallel program, the validity of the SC-property for terms occurring in a parallel statement depends mutually on the validity of this property for other terms; verifying it cannot be separated from the interleaving sequences. The costs of investigating them explicitly, however, are prohibitive because of the well-known *state explosion problem*. We therefore introduce a safe approximation of SCs, called *strong constants (StCs)*. Like the solution of a unidirectional bitvector (UBv) problem they can be computed without having to consider interleavings at all. The computation of StCs is based on the local semantic functional $[\![\ ]\!]_{stc}$: $E \to (\Sigma \to \Sigma)$ defined by $[\![e]\!]_{stc}(\sigma) =_{df} \theta_e^{stc}(\sigma)$ for all $e \in E$ and $\sigma \in \Sigma$, where θ_e^{stc} denotes the state transformation function of Section 2, where, however, the evaluation function $\mathcal{E}$ is replaced by the simpler $\mathcal{E}_{stc}$, where $\mathcal{E}_{stc}(t)(\sigma)$ is defined by $I_0(c)$, if $t = c \in \mathbf{C}$, and $\bot$ otherwise. The set of strong constants at n, denoted by $\mathcal{C}_n^{stc}$, is then analogously defined to the set of SCs at n.

Denoting by $\mathcal{F}_{\mathbf{D}} =_{df} \{ Cst_d \mid d \in \mathbf{D} \} \cup \{ Id_{\mathbf{D}} \}$ the set of constant functions Cst_d on $\mathbf{D}$, $d \in \mathbf{D}$, enlarged by the identity $Id_{\mathbf{D}}$ on $\mathbf{D}$, we obtain:

Lemma 2. $\forall e \in E \ \forall v \in \mathbf{V}.\ [\![e]\!]_{stc}|_v \in \mathcal{F}_{\mathbf{D}} \backslash \{ Cst_{\top} \}$, where $|_v$ denotes the restriction of $[\![e]\!]_{stc}$ to v.

Intuitively, Lemma 2 means that (1) all variables are "decoupled": each variable in the domain of a state $\sigma \in \Sigma$ can be considered like an (independent) slot of a bitvector. (2), the transformation function relevant for a variable is either a constant function or the identity on $\mathbf{D}$. This is quite similar to UBv-problems. There, for every bit (slot) of the bitvector the set of data is given by the flat lattice $\mathcal{B}$ of Boolean truth values as illustrated in Figure 3(b), and the transformation functions relevant for a slot are the two constant functions and the identity on $\mathcal{B}$. However, there is also an important difference. $\mathcal{F}_{\mathbf{D}}$ is not closed under pointwise meet: consider the pointwise meet of $Id_{\mathbf{D}}$ and Cst_d for some $d \in \mathbf{D}$; it is given by the "peak"-function p_d with $p_d(x) = d$, if $x = d$, and $\bot$ otherwise. These two facts together with an extension allowing us to mimic the effect of pointwise meet in a setting with only constant functions and the identity are the key for extending the framework of [9] focusing on the scenario of bitvector problems to the scenario of StC-like problems.

4.1   Parallel Data Flow Analysis

As in the sequential case, the abstract semantics of a parallel program is specified by a local semantic functional $[\![\ ]\!] : E \to (\mathcal{L} \to \mathcal{L})$ giving meaning to its statements in terms of functions on an (arbitrary) complete lattice $\mathcal{L}$. In analogy to their sequential counterparts, they can be extended to capture finite parallel paths. Hence, the definition of the parallel variant of the *MOP*-approach, the *PMOP*-approach and its solution, is obvious:

The *PMOP*-**Solution:** $\forall l_0 \in \mathcal{L}.\ PMOP_{l_0}(n) = \bigsqcap \{ [\![p]\!](l_0) \mid p \in \mathbf{PP}[s, n] \}$

The point of this section now is to show that as for UBv-problems, also for the structurally more complex StC-like problems the *PMOP*-solution can efficiently be computed by means of a fixed point computation.

StC-like problems. StC-like problems are characterized as UBv-problems by (the simplicity of) their local semantic functional $[\![\ ]\!] : E \to (D \to D)$. It specifies the effect of an edge e on a single component of the problem under consideration (considering StCs, for each variable $v \in V$), where D is a flat lattice of some underlying (finite) set D' enlarged by two new elements $\bot$ and $\top$ (cf. Figure 3(a)), and where every function $[\![\ e\ ]\!]$, $e \in E$, is an element of $\mathcal{F}_D =_{df} \{\ Cst_d \mid d \in D\} \cup \{Id_D\}$. As shown before, $\mathcal{F}_D$ is not closed under pointwise meet. The possibly resulting peak-functions, however, are here essential in order to always be able to model the effect when control flows together. Fortunately, this can be mimiced in a setting, where all semantic functions are constant ones (or the identity). To this end we extend D to D_X by inserting a second layer as shown in Figure 3(c), considering then the set of functions $\mathcal{F}_{D_X} =_{df} (\mathcal{F}_D \backslash \{Id_D\}) \cup \{Id_{D_X}\} \cup \{\ Cst_{d^p} \mid d \in D\}$. The functions Cst_{d^p}, $d \in D'$, are constant functions like their respective counterparts Cst_d, however, they play the role of the peak-functions p_d, and in fact, after the termination of the analysis, they will be interpreted as peak-functions.

Intuitively, considering StCs and synchronization, the "doubling" of functions allows us to distinguish between components of a parallel statement assigning a variable under consideration a unique constant value not modified afterwards along *all* paths (Cst_d), and those doing this along *some* paths (Cst_{d^p}). While for the former components the variable is a total StC (or shorter: an StC) after their termination (as the property holds for all paths), it is a *partial* StC for the latter ones (as the property holds for some paths only). Keeping this separately in our data domain, gives us the handle for treating synchronization and interference precisely. E.g., in the program of Figure 1, d is a total StC for the parallel component starting with edge **15**, and a partial one for the component starting with edge **35**. Hence, d is an StC of value 3 after termination of the complete parallel statement as none of the parallel "relatives" destroys this property established by the left-most component because it is a partial StC of value 3, too, for the right-most component, and it is transparent for the middle one. On the other hand, d would not be an StC after the termination of the parallel statement, if, let's say, d would be a partial StC of value 5 of the right-most component.

Obviously, all functions of $\mathcal{F}_{D_X}$ are distributive. Moreover, together with the orderings $\sqsubseteq_{seq}$ and $\sqsubseteq_{par}$ displayed in Figure 4, they form two complete lattices with least element $Cst_\bot$ and greatest element $Cst_\top$, and least element $Cst_\bot$ and greatest element Id_{D_X}, respectively. Moreover, $\mathcal{F}_{D_X}$ is closed under function composition (for both orderings). Intuitively, the "meet" operation with respect to $\sqsubseteq_{par}$ models the merge of information in "parallel" join nodes, i.e., end nodes of parallel statements, which requires that all their parallel components terminated. Conversely, the "meet" operation modelling the merge of information at points, where sequential control flows together, which actually would be given by the pointwise (indicated below by the index "*pw*") meet-operation on

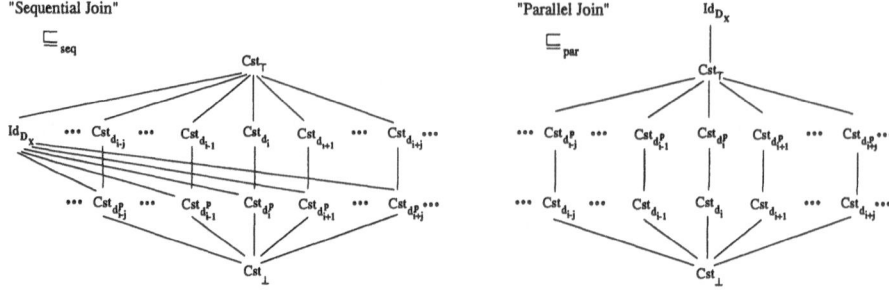

Fig. 4. The function lattices for "sequential" and "parallel join".

functions of $\mathcal{F}_{I\!D}$, is here equivalently modelled by the "meet" operation with respect to $\sqsubseteq_{seq}$. In the following we drop this index. Hence, $\sqcap$ and $\sqsubseteq$ expand to $\sqcap_{seq}$ and $\sqsubseteq_{seq}$, respectively. Recalling Lemma 2 and identifying the functions Cst_{d^p}, $d \in I\!D'$, with the peak-functions p_d, we obtain for every sequential flow graph G and start information out of $\mathcal{F}$:

Lemma 3. $\forall n \in N.\ MOP_{pw}(n) = MOP(n) = MFP(n)$

Lemma 3 together with Lemma 4, of which in [9] its variant for bitvector problems was considered, are the key for the efficient computation of the effects of "interference" and "synchronization" for StC-like problems.

Lemma 4 (Main-Lemma). *Let $f_i : \mathcal{F}_{I\!D_X} \to \mathcal{F}_{I\!D_X}$, $1 \leq i \leq q$, $q \in I\!N$, be functions of $\mathcal{F}_{I\!D_X}$. Then:* $\exists k \in \{1,...,q\}.\ f_q \circ ... \circ f_2 \circ f_1 = f_k \ \wedge \ \forall j \in \{k + 1,...,q\}.\ f_j = Id_{I\!D_X}$.

Interference. As for UBv-problems, also for StC-like problems the relevance of Main Lemma 4 is that each possible interference at a program point n is due to a single statement in a parallel component, i.e., due to a statement whose execution can be interleaved with the statement of n's incoming edge(s), denoted as n's interleaving predecessors $IntPred(n) \subseteq E$. This is implied by the fact that for each $e \in IntPred(n)$, there is a parallel path leading to n whose last step requires the execution of e. Together with the obvious existence of a path to n that does not require the execution of any statement of $IntPred(n)$, this implies that the only effect of interference is "destruction". In the bitvector situation considered in [9], this observation is boiled down to a predicate *NonDestruct* inducing the constant function "true" or "false" indicating the presence or absence of an interleaving predecessor destroying the property under consideration. Similarly, we here define for every node $n \in N$ the function

$$InterferenceEff(n) =_{df} \sqcap\{[\![e]\!] \mid e \in IntPred(n) \wedge [\![e]\!] \neq Id_{I\!D}\}$$

These (precomputable) constant functions *InterferenceEff*(n), $n \in N$, suffice for modelling interference.

Synchronization. In order to leave a parallel statement, all parallel components are required to terminate. As for UBv-problems, the information required to model this effect can be hierarchically computed by an algorithm, which only considers purely sequential programs. The central idea coincides with that of interprocedural DFA (cf. [14]): one needs to compute the effect of complete subgraphs, in this case of complete parallel components. This information is computed in an "innermost" fashion and then propagated to the next surrounding parallel statement. In essence, this three-step procedure $\mathcal{A}$ is a hierarchical adaptation of the functional version of the *MFP*-approach to the parallel setting. Here we only consider the second step realizing the synchronization at end nodes of parallel statements in more detail. This step can essentially be reduced to the case of parallel statements G with purely sequential components $G_1, \ldots, G_k$. Thus, the global semantics $[\![G_i]\!]^*$ of the component graphs G_i can be computed as in the sequential case. Afterwards, the global semantics $[\![G]\!]^*$ of G is given by:

$$[\![G]\!]^* = \sqcap_{par} \{ [\![G_i]\!]^* \mid i \in \{1, \ldots, k\} \} \quad (\textit{Synchronization})$$

Central for proving the correctness of this step, i.e., $[\![G]\!]^*$ coincides with the *PMOP*-solution, is again the fact that a single statement is responsible for the entire effect of a path. Thus, it is already given by the projection of this path onto the parallel component containing the vital statement. This is exploited in the synchronization step above. Formally, it can be proved by Main Lemma 4 together with Lemma 2 and Lemma 3, and the identification of the functions Cst_{d^p} with the peak-function p_d. The correctness of the complete hierarchical process then follows by a hierarchical coincidence theorem generalizing the one of [9]. Subsequently, the information on StCs on parallel program parts can be fed into the analysis for SCs on sequential program parts. In spirit, this follows the lines of [9], however, the process is refined here in order to take the specific meaning of peak-functions represented by Cst_{d^p} into account.

4.2 The Application: Parallel Constant Propagation

We now present our algorithm for PCP. It is based on the semantic functional $[\![\ ]\!]_{pcp} : E \to (\Sigma \to \Sigma)$, where $[\![e]\!]$ is defined by $[\![e]\!]_{stc}$, if e belongs to a parallel statement, and by $[\![e]\!]_{sc}$ otherwise. Thus, for sequential program parts, $[\![\ ]\!]_{pcp}$ coincides with the functional for SCs, for parallel program parts with the functional for StCs. The computation of parallel constants proceeds then essentially in three steps. (1) Hierarchically computing (cf. procedure $\mathcal{A}$) the semantics of parallel statements according to Section 4.1. (2) Computing the data flow information valid at program points of sequential program parts. (3) Propagating the information valid at entry nodes of parallel statements into their components.

The first step has been considered in the previous section. The second step proceeds almost as in the sequential case because at this level parallel statements are considered "super-edges", whose semantics ($[\![\ ]\!]^*$) has been computed in the first step. The third step, finally, can similarly be organized as the corresponding step of the algorithm of [9]. In fact, the complete three-step procedure evolves as an adaption of this algorithm. Denoting the final annotation computed by the PCP-algorithm by $Ann^{pcp} : N \to \Sigma$, the set of constants detected, called *parallel simple constants (PSCs)*, is given by

$$\mathcal{C}_n^{psc} =_{df} \{t \in \mathbf{T} \mid \exists d \in \mathbf{D}'. \; \mathcal{E}(t)(Ann_n^{pcp}) = d\}$$

The correctness of this approach, whose power has been demonstrated in the example of Figure 1, is a consequence of Theorem 3:

Theorem 3 (PSC-Correctness). $\forall n \in N. \; \mathcal{C}_n \supseteq \mathcal{C}_n^{psc}$.

Aggressive PCP. PCP can detect the constancy of complex terms inside parallel statements as e.g. of $y + b$ at edge **45** in Figure 1. In general, though not in this example, this is the source of *second-order effects* (cf. [13]). They can be captured by incrementally reanalyzing affected program parts starting with the enclosing parallel statement. In effect, this leads to an even more powerful, aggressive variant of PCP. Similar this holds for extensions to subscripted variables. We here concentrated on scalar variables, however, along the lines of related sequential algorithms, our approach can be extended to subscripted variables, too.

5 Conclusions

Extending the framework of [9], we developed a PCP-algorithm, which can be implemented as easily and as efficiently as its sequential counterpart for SCs. The key for achieving this was to decompose the algorithm to behave differently on sequential and parallel program parts. On sequential parts this allows us to be as precise as the algorithm for SCs; on parallel parts it allows us to capture the phenomena of interference and synchronization without having to consider any interleaving. Together with the successful earlier transfer of bitvector-based optimizations to the parallel setting, complemented here by the successful transfer of CP, a problem beyond this class, we hope that compiler writers are encouraged to integrate classical sequential optimizations into parallel compilers.

References

1. P. Cousot and R. Cousot. Abstract interpretation: A unified lattice model for static analysis of programs by construction or approximation of fixpoints. In *Conf. Rec. 4th Symp. Principles of Prog. Lang. (POPL'77)*, pages 238 – 252. ACM, NY, 1977.
2. M. S. Hecht. *Flow Analysis of Computer Programs*. Elsevier, North-Holland, 1977.
3. J. B. Kam and J. D. Ullman. Monotone data flow analysis frameworks. *Acta Informatica*, 7:309 – 317, 1977.
4. G. A. Kildall. A unified approach to global program optimization. In *Conf. Rec. 1st Symp. Principles of Prog. Lang. (POPL'73)*, pages 194 – 206. ACM, NY, 1973.
5. J. Knoop. Constant propagation in explicitly parallel programs. Technical report, Fak. f. Math. u. Inf., Univ. Passau, Germany, 1998.
6. J. Knoop. Eliminating partially dead code in explicitly parallel programs. *TCS*, 196(1-2):365 – 393, 1998. (Special issue devoted to *Euro-Par'96*).

7. J. Knoop, O. Rüthing, and B. Steffen. Partial dead code elimination. In *Proc. ACM SIGPLAN Conf. on Prog. Lang. Design and Impl. (PLDI'94)*, volume *29*,6 of *ACM SIGPLAN Not.*, pages 147 – 158, 1994.

8. J. Knoop, B. Steffen, and J. Vollmer. Code motion for parallel programs. In *Proc. of the Poster Session of the 6th Int. Conf. on Compiler Construction (CC'96)*, pages 81 – 88. TR LiTH-IDA-R-96-12, 1996.

9. J. Knoop, B. Steffen, and J. Vollmer. Parallelism for free: Efficient and optimal bitvector analyses for parallel programs. *ACM Trans. Prog. Lang. Syst.*, 18(3):268 – 299, 1996.

10. S. P. Midkiff and D. A. Padua. Issues in the optimization of parallel programs. In *Proc. Int. Conf. on Parallel Processing, Volume II*, pages 105 – 113, 1990.

11. E. Morel and C. Renvoise. Global optimization by suppression of partial redundancies. *Comm. ACM*, 22(2):96 – 103, 1979.

12. J. H. Reif and R. Lewis. Symbolic evaluation and the global value graph. In *Conf. Rec. 4th Symp. Principles of Prog. Lang. (POPL'77)*, pages 104 – 118. ACM, NY, 1977.

13. B. K. Rosen, M. N. Wegman, and F. K. Zadeck. Global value numbers and redundant computations. In *Conf. Rec. 15th Symp. Principles of Prog. Lang. (POPL'88)*, pages 2 – 27. ACM, NY, 1988.

14. M. Sharir and A. Pnueli. Two approaches to interprocedural data flow analysis. In S. S. Muchnick and N. D. Jones, editors, *Program Flow Analysis: Theory and Applications*, chapter 7, pages 189 – 233. Prentice Hall, Englewood Cliffs, NJ, 1981.

15. M. Wolfe. *High performance compilers for parallel computing.* Addison-Wesley, NY, 1996.

16. H. Zima and B. Chapman. *Supercompilers for parallel and vector computers.* Addison-Wesley, NY, 1991.

Optimization of SIMD Programs with Redundant Computations

Jörn Eisenbiegler

Institut für Programmstrukturen und Datenorganisation,
Universität Karlsruhe, 76128 Karlsruhe, Germany
eisen@ipd.info.uni-karlsruhe.de

Abstract. This article introduces a method for using redundant computations in automatic compilation and optimization of SIMD programs for distributed memory machines. This method is based on a generalized definition of parameterized data distributions, which allows an efficient and flexible use of redundancies. An example demonstrates the benefits of this optimization method in practice.

1 Introduction

edundant computations, i.e. computations that are done on more than one pro-ssor, can improve the performance of parallel programs on distributed memory machines [2]. In the examples given in section 5, the execution time is improved by nearly one magnitude. Optimization methods (e.g. [5, 7]) which use redundant computations have two major disadvantages: First the task graph of the program has to be investigated node per node. Second the communication costs have to be estimated without knowledge about the final data distribution.

To avoid these disadvantages, other methods as for example proposed in [3, 4, 8] use parameterized data distributions. But, the data distributions given there do not allow redundant computations in a more flexible way than "all or nothing". Therefore, this article introduces a scheme for defining parameterized data distributions, which allow the efficient and flexible use of redundant computations.

Based on these definitions, this article describes a method for optimizing SIMD programs for distributed memory systems. This method is divided into three steps. As in [1], the first step uses heuristics (as described in [3, 4, 8]) to propose locally good data distributions. However, these distributions are not used on the level of loops but on the level of parallel assignments. In the second step, additional data distributions (which lead to redundant computations) are calculated which connect subsequent parallel assignments in a way such that no communication is needed between them. The last step selects a globally optimal proposition for each parallel assignment. In contrast to [1], an efficient algorithm is described instead of 0–1 integer programming. The benefits of this optimization method are shown by the example of numerically solving of a differential equation.

2 Parameterized Data Distributions

In order to define parameterized data distributions, a definition for data distributions is required. Generally, a data element can exist on any number of processors and a processor can hold any number of data elements. Therefore, distributions have to be defined as *relations* between elements and processors:

Definition 1 (data distribution). *A data distribution is a set* $D \subseteq N_0 \times N_0$ *of tuples.* $(p, i) \in D$ *iff element i is valid on processor p, i. e. element i has the correct value on processor p.*

As an example, assume a machine with $P = 4$ processors and an array a with 15 elements. The set $D = \{$ $(0,0)$, $(0,1)$, $(0,2)$, $(0,10)$, $(0,11)$, $(0,12)$, $(0,13)$, $(0,14)$, $(1,1)$, $(1,2)$, $(1,3)$, $(1,4)$, $(1,5)$, $(1,13)$, $(1,14)$, $(2,4)$, $(2,5)$, $(2,6)$, $(2,7)$, $(2,8)$, $(3,7)$, $(3,8)$, $(3,9)$, $(3,10)$, $(3,11)\}$ is a data distribution where the elements $1, 2, 4, 5, 7, 8, 10, 11, 13, 14$ are valid on exactly two out of four processors.

For practical purposes in a compiler or for hand optimization of non–trivial programs, such data structures are too large. Therefore, parameterized data distributions are used. Each combination of parameters represents a certain data distribution. The correlation between the parameters and the data distributions they represent is the *interpretation*, i.e. an interpretation of a parameterized data distribution maps parameters to a data distribution as defined in Definition 1:

Definition 2 (parameterized data distribution).
A parameterized data distribution space is a pair (S, I), *where S is an arbitrary set and* $I : S \rightarrow 2^{N_0 \times N_0}$ *is interpretation function. Every* $P \in S$ *is a parameterized data distribution, i.e.* $(p, i) \in I(P)$ *iff element i is valid on processor p.*

Usually, the set S (the set of parameters) is a set of tuples whose elements describe the number of processors, block length etc. for each array dimension. In order to carry on with the example, regard the data distribution space (N_0^3, I) with interpretation I:

$$I(b, a, r) := \{(p, i) | a + kbP + pb - r \leq i < a + kbP + (p+1)b + r, \ k \in Z\}$$

The parameter b is the block size, the parameter a the alignment, and the parameter r describes redundancy. In this data distribution space, the data distribution D described above is parameterized with $(3, -1, 1)$, because $I(3, -1, 1) = D$. Note, that the correlation between processors and data elements is still relational, even if the interpretation I of the parameters is a function. Therefore definition 2 is a generalization of former definitions (e.g. [3,8]).

The subset relation defines a semi–ordering for data distributions. This semi–ordering can be transferred to parameterized data distributions:

Definition 3 (semi–ordering). *Let* (S, I) *be a parameterized data distribution space,* $P_1, P_2 \in S$: $P_1 \leq P_2$, *iff* $I(P_1) \subseteq I(P_2)$.

Theorem 1. *Let* (S, I) *be a parameterized data distribution space,* $P_1, P_2 \in S$. *There are no communication costs for changing the distribution of an array from* P_1 *to* P_2, *iff* $P_1 \geq P_2$.

Proof. Let $P_1 \geq P_2$. Then, every data element which is valid on a processor in P_1 is also valid their in P_2, since $I(P_1) \supseteq I(P_2)$. Therefore, no data elements have to be transferred and no communication costs occur. If $P_1 \not\geq P_2$, there is at least one data element valid on a processor in $I(P_2)$ which is not valid their in $I(P_1)$. This data element has to be transferred generating communication costs.

The data distribution state of a program is described by the data distribution of each array in the program. This state is defined by a data distribution function, which maps arrays to parameterized data distributions.

Definition 4 (data distribution functions). *Let ID be the name space of a program and* (S, I) *be a parameterized data distribution space. Then, a function* $f : ID \rightarrow S$ *is called a data distribution function. Let* f_1 *and* f_2 *be two data distribution functions.* $f_1 \leq f_2$ *iff* $\forall x \in ID : f_1(x) \leq f_2(x)$.

3 Parallel Array Assignments

With the definitions given in section 2, it is possible to distinguish between the input and output data distribution of a parallel assignment. As an example, regard the data distribution space described above and the parallel assignment

> **par** i **in** $[1, n-1]$ $x[i] := x[i-1] + x[i] + x[i+1]$;

If the output data distribution of x is to be $(3, -1, 0)$, the input data distribution for x must be at least $(3, -1, 1)$. This distinction is not possible with former definitions of parameterized data distributions, since it is not possible to describe the input data distribution there.

In order to optimize data distributions for arrays, the execution of an SIMD–program can now be seen as a sequence of array assignments. Every assignment has an input and an output data distribution function. Between two assignments the data distribution state has to be changed from the output data distribution of the first to a data distribution which is larger or equal to the input data distribution of the second assignment. Note that there is no difference between the remapping of data and getting the necessary input for an assignment. In this article, the overlapping of communication and computation is not examined. However, this might be used for further optimization.

4 Optimization

4.1 Proposition

In the first step of the optimization, heuristics propose locally good data distributions for each parallel assignment. This paper does not go into detail about searching locally good data distributions, since this problem is well–investigated (e.g. [3, 8]). It is allowed to propose more than one data distribution for one parallel assignment using more than one proposition method to exploit the benefits of different heuristics.

The methods mentioned above only compute the output data distributions for the array written in the assignment. Therefore the corresponding input and output data distribution functions have to be calculated. If there are no data elements of an array needed in a data distribution function, this array is declared as undefined. As the result of this step, each parallel assignment is applied with several proposals of pairs of input and output data distribution functions.

4.2 Redundancy

In the second step of optimization, additional propositions are introduced which connect subsequent parallel assignments in a way such that no communication is necessary. Let a_1 and a_2 be two subsequent parallel assignments and (i_1, o_1) and (i_2, o_2) pairs of input and output distribution functions proposed for a_1 and a_2, respectively. Let n be the array written in a_1 and $d_1 \cup d_2$ be a parameterized data distribution which is larger than d_1 and d_2. First, (i'_1, o'_1) is set to the pair of distribution functions corresponding to the proposal $i_2(n) \cup o_2(n)$ for n in a_1 (as above). Hence, no communication is needed to redistribute n from $o'_1(n)$ to $i_2(n)$. Second, to avoid communication for other arrays, (i^*_1, o^*_1) is set to

$$o^*_1(x) = \begin{cases} o'_1(x) \cup i_2(x) & x \neq n \\ o'_1(x) & x = n \end{cases} \qquad i^*_1(x) = \begin{cases} i'_1(x) \cup i_2(x) & x \neq n \\ i'_1(x) & x = n \end{cases}$$

As a result of this calculation, o^*_1 is larger than i_2, and therefore no communication is needed to redistribute from o^*_1 to i_2. However, in most cases some data elements have to be calculated redundantly on more than one processor for a_1.

The method can be applied to all combinations of proposals for a_1 and a_2. Especially, it can be applied to cost tuples for a_2 which were themselves proposed this way. This leads to (potential) sequences in the program where no communication is needed. At the end of this step, again all parallel assignments are applied with several pairs of input and output data distribution functions.

4.3 Configuration

The configuration step tries to select the globally optimal data distribution for each parallel assignment from the local proposals. If the program is not oblivious (see [5]), it can not be guaranteed that the optimal selection is found due to the lack of compile time knowledge.

In order to find an optimal solution for arbitrary programs, the configuration method by Moldenhauer [6] is adopted. The proposals for the parallel assignments are converted to cost tuples of the form (i, c, o) where i and o are the proposed input and output distribution functions and c are the costs for computing this parallel assignment with the proposed distributions. To find an optimal solution for the whole program, these sets of cost tuples are combined to sets of cost tuples for larger program blocks.

Let a_1 and a_2 be two subsequent program blocks and M_1 and M_2 be the sets of cost tuples proposed or calculated for these blocks. If a_1 is executed with (i_1, c_1, o_1) and a_2 is executed with (i_2, c_2, o_2) then $a_1; a_2$ is executed with $(i, c, o) = (i_1, c_1, o_1) \circ_t (i_2, c_2, o_2)$ where

$$i(x) = \begin{cases} i_1(x) \text{ if } i_1(x) \text{ or } o_1(x) \text{ is defined} \\ i_2(x) \text{ else if } i_2(x) \text{ is defined} \end{cases}$$

$$c = c_1 + c_2 + \text{redistribution\_costs}(o_1, i_2)$$

$$o(x) = \begin{cases} o_2(x) \text{ if } o_2(x) \text{ is defined} \\ o_1(x) \text{ else} \end{cases}$$

For further combination with other program blocks, the cost tuples with minimal costs for all combinations of input and output data distribution functions which can occur using the proposals are calculated, i.e. the set of cost tuples $M :=$ $M_1 \circ M_2$ for the sequence $a_1; a_2$ is

$$M := M_1 \circ M_2 := \{(i, c, o) | (i, c, o) \text{ is cost–minimal in}$$
$$\{(i, c^*, o) = (i_1, c_1, o_1) \circ_t (i_2, c_2, o_2) | (i_k, c_k, o_k) \in M_k\}\}$$

Rules for other control constructs can be given analogously and are omitted in this article. Using these rules, the set of cost tuples for the whole program can be calculated. In the set for the whole program the element with minimal costs is selected. For oblivious programs it represents the optimal selection from the proposals. For non–oblivious programs heuristics have to be used for the combination of the sets. To obtain the optimal proposals for each parallel assignment, the origin of the tuples has to be remembered during the combination step.

This configuration problem is not in APX, i.e. no constant approximation can be found. Accordingly, the algorithm described above is exponential. But in contrast to the solution proposed in [1], it is only exponential in the number of array variables simultaneously alive. Since this number is small in real applications the method is feasible.

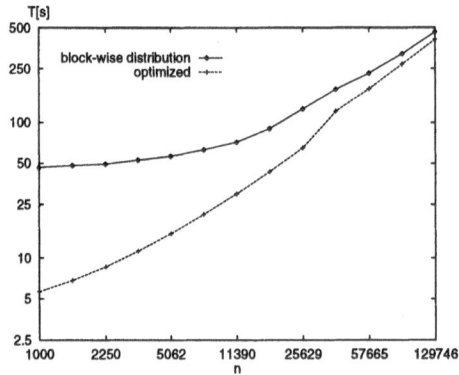

Fig. 1. Solving a differential equation

5 Practical Results

The optimization method described above is implemented in a prototype compiler. This section shows the optimization results compared to results that can be achieved if no redundancy is used and data distribution is not done for parallel assignments but for loops only. All measurements were done on an 8 processor Parsytec PowerXplorer.

Figure 1 shows the runtimes for solving a differential equation in n points. If no redundant computations are allowed, a blockwise distribution is optimal. Therefore the optimized program is compared to a version using blockwise distribution. Especially for small input sizes, the optimized program is nearly one magnitude faster than the blockwise distributed version.

6 Conclusion and Further Work

This article introduces a formal definition for parameterized data distributions that allow redundancies. This makes it possible to transfer the concept of redundant computations to the world of parameterized data distributions such that compilers can use redundant computations efficiently.

Based on these definitions, a method for optimizing SIMD programs on distributed systems is given. This method allows combining several local optimization methods for computing local proposals. Furthermore, redundant computations are generated to save communication. For oblivious programs the method selects the optimal proposals for each array assignment. The practical results show a large gain in performance.

In the future different data distribution spaces, i.e. parameters and their interpretations, have to be examined to select suitable ones. Additionally, optimal redistribution schemes for these data distribution spaces should be selected.

References

1. R. Bixby, K. Kennedy, and U. Kramer. Automatic data layout using 0–1 integer programming. In M. Cosnard, G. R. Gao, and G. M. Silberman, editors, *Parallel Architecture and Compilation Techniques (A-50)*. IFIP, Elsevier Science B.V. (North–Holland), 1994.
2. J. Eisenbiegler, W. Löwe, and A. Wehrenpfennig. On the optimization by redundancy using an extended LogP model. In *International Conference on Advances in Parallel and Distributed Computing (APDC'97)*, pages 149–155. IEEE Computer Society Press, March 1997.
3. M. Gupta. *Automatic Data Partitioning on Distributed Memory Multicomputers*. PhD thesis, Department of Computer Science, University of Illinois at Urbana-Champaign, 1992.
4. P. Z. Lee. Efficient algorithms for data distribution on distributed memory parallel computers. *IEEE Transactions on Parallel and Distributed Systems*, 8(8):825 – 839, August 1997.

5. W. Löwe, W. Zimmermann, and J. Eisenbiegler. Optimization of parallel programs with expensive communication. In *EUROPAR '96, Parallel Processing*, volume 1124 of *Lecture Notes in Computer Science*, pages 602–610, 1996.

6. H. Moldenhauer. *Kostenbasierte Konfigurierung für Programme und SW–Architekturen*. Logos Verlag, Berlin, 1998. Dissertation, Universität Karlsruhe.

7. C. Papadimitriou and M. Yannakakis. Towards an architecture-independent analysis of parallel algorithms. *SIAM Journal on Computing*, 19(2):322 – 328, 1990.

8. M. Philippsen. *Optimierungstechniken zur Übersetzung paralleler Programmiersprachen*. Number 292 in VDI Fortschritt-Berichte, Reihe 10: Informatik. VDI-Verlag GmbH, Düsseldorf, 1994. Dissertation, Universiät Karlsruhe.

Exploiting Course Grain Parallelism from FORTRAN by Mapping it to IF1

Adrianos Lachanas and Paraskevas Evripidou

Department of Computer Science,University of Cyprus
P.O.Box 537,CY-1687 Nicosia, Cyprus
Tel:+357-2-338705 (FAX 339062), `skevos@turing.cs.ucy.ac.cy`

Abstract. FORTRAN, a classical imperative language is mapped into IF1, a machine-independent dataflow graph description language with Single Assingment Semantics (SAS). Parafrase 2 (P2) is used as the front-end of our system. It parses the source code, generates an intermediate representation and performs several types of analysis. Our system extends the internal representation of P2 with two data structures: the Variable to Edge Translation Table and the Function to Graph Translation Table. It then proceeds to map the internal representation of P2 into IF1 code. The generated IF1 is then processed by the back-end of the Optimizing SISAL Compiler that generates parallel executables on multiple target platforms. We have tested the correctness and the performance of our system with several benchmarks. The results show that even on a single processor there is no performance degradation from the translation to SAS. Furthermore, it is shown that on a range of processors, reasonable speedup can be achieved.

1 Introduction

The translation of imperative languages to Single Assingment Semantics was thought to be a hard, if not an impossible task. In our work we tackled this problem by mapping Fortran to a Single Assingment Intermediate Form, IF1. Mapping Fortran to SAS is a worthwhile task by itself. However, our work has also targeted the automatic and efficient extraction of parallelism. In this paper we present the compiler named **Venus** that transforms a classical imperative language (Fortran, C) into an inherently parallel applicative format (IF1).The Venus system uses a multi-language precompiler (Parafrase 2 [5]) and a multi-platform post compiler (Optimizing SISAL Compiler [4]) in order to simplify the translation process and to provide multi-platform support [1].

The P2 intermediate representation of the source code is processed by Venus to generate a dataflow description of the source program. The IF1 dataflow graph description is then passed to the last stage, the Optimizing SISAL Compiler (OSC), which performs optimizations and produces parallel executables. Since IF1 is machine independent and OSC has been implemented successfully on many major parallel computers, the final stage can generate executables for a wide range of computer platforms [4].

Our systems maps the entire Fortran 77 semantics into IF1. The only construct that is not yet supported is I/O. We have tested our system with a variety of benchmarks. Our analysis have shown that there is no performance degradation from the mapping to SAS. Furthermore, the SAS allows for more efficient exploitation of parallelism. In this paper we concentrate on the description of the mechanism for mapping coarse grain Fortran constructs into SAS.

2 Compilation Environment

The Target Language (IF1): IF1 is a machine independent dataflow graph description language that adheres to single assignment semantics. It was developed as the intermediate form for the compilation of the SISAL language. As a language, IF1 is based on hierarchical acyclic graphs with four basic components: nodes, edges, types, and graph boundaries. The **IF1 Graph Component** is an important tool in IF1 for procedural abstraction. Graph boundaries are used to encapsulate related operations. In the case of compound nodes, the graph serves to separate the operations of the various functional parts. Graph boundaries can also delimit functions. The **IF1 Edge Component** describes the flow of data between nodes and graphs, representing explicit data dependencies. The **IF1 Node Component** represents operations in the IF1 dataflow graph. Nodes are referenced using their *label* which must be unique within the graph. All nodes are functions, and have both input and output ports.

The **front-end**, Parafrase 2 [5], is a source to source multilingual restructuring compiler that translates Fortran to Cedar Fortran. The P2 invokes different *passes* to transform the program to an Internal Representation: in the form of a parse tree, symbol table, call graph and dependency graph [5]. P2 does dependency analysis, variable alias analysis, and loop parallelization analysis.

Venus starts by traversing the Parse Tree of P2. The basic module of the parse tree is the function and the subroutine. Each such module contains a parameter list, a declarations list and a linked list of statments/expressions. For each module we generate an FGTT and update it with the paramemter and declarations lists. Then we proceed to generate a new scope for the VETT. We then continue to update the VETT with all variables in the delacration list. For each statment in the statment/expression list we generate a new IF1 node and its corresponding edges and update the VETT accordingly.

The **back-end**, the Optimizing SISAL Compiler [2] is used to produce executable code from the IF1 programs. OSC was chosen for this project since it has been successfully implemented for performance machines like Sequent, SGI, HP, Cray, Convex and most UNIX based workstations [2].

3 Fine and Medium Grain Parallelism

The problem when generating code on the basic (expression) level is the single assignment restriction. This is a concern when Fortran variables are translated into IF1 edges [1]. In Fortran a variable is a memory location that can have many values stored in it over time whereas IF1 edges carry single values. An IF1 edge is a single assignment. Our system handles this problem on the basic block level using a structure called **Variable to Edge Translating Table** (VETT) [3]. The VETT always allows the translator to be able to identify the node and port whose output represent the current value of a particular variable.

When a variable is referenced in an expression, VETT is accessed to obtain the current node and port number of the variable. The VETT fields do not need to change since the value of the variable is not changed. When a variable is referenced on the left hand side (LHS) of an assignment, the node and port entry in VETT will change to represent the new node and port that assigned/carried the new value. An UPDATED flag is set to VETT entry to show that the variable has changed value within the scope. A variable write consists only of updating the VETT entries (SN, SP) of that variable.

In IF1, arrays can be accessed sequentially or in parallel. Whether an array is accessed sequentially or in parallel is decided by Parafrase 2. During the **Sequential LHS Access** the dereferencing nodes that expose the location of the old element, the nodes that provide the value for the new element, and the reconstruction nodes that combine the deconstructed pieces of the array together with the new element value, build a new array. The mechanism used in **Parallel RHS Access** is similar to that used in sequential array access. Multiple values of the array elements are generated and are forwarded to IF1 nodes within the compound node. **Parallel LHS access** is done using a loop reduction node. Values for the array elements are generated by the parallel loop iterations and are gathered into a new array.

4   Coarse Grain Parallelism

Fortran limits the amount of parallelism we can exploit because of its parameter interface. In Fortran, all input parameters of a module are passed by reference. This allows the body of the function to have access on the input parameter and change its value.

4.1   Mapping of Functions and Subroutines into IF1 Graphs

Fortran Modules are interrelated parts of code that have common functionality. Translating modules into IF1 graphs requires additional information about the module like: the type of the module (function or subroutine), the input parameters, the output parameters and the source-destination node-port of each parameter. The translator handle this problem using a structure named **Function to Graph Translating Table** (FGTT).

The FGTT is generated in two passes. During the **First Pass** the translator builds the FGTT and stores all the information about a module such as: the name of the module, the type of module (function or subroutine) and the common variables that the module has access (see Figure 1). In IF1 there are not any global variables. The translator removes all common statements and replaces

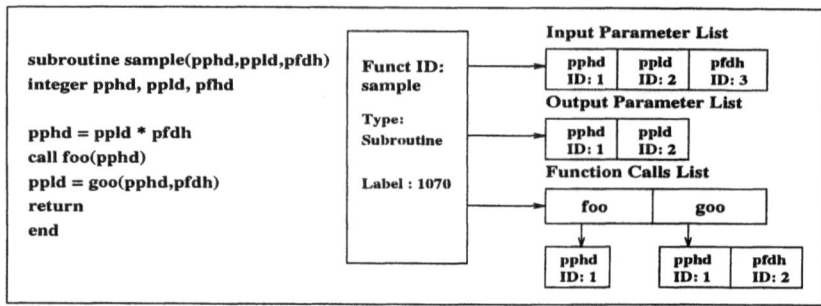

Fig. 1. FGTT functionality at coarse grain parallelism

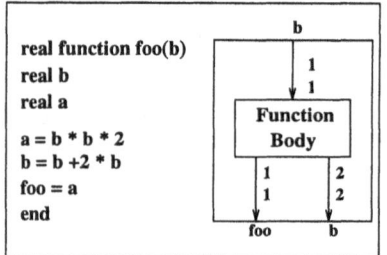 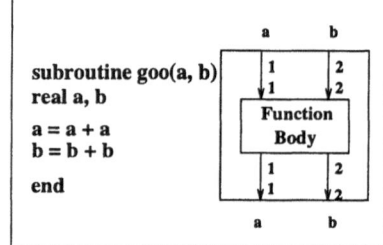

Fig. 2. (a) The IF1 graph of a Fortran function. Output port 1 is reserved for the function return value.(b)The IF1 graph of a Fortran subrouti ne.

them with input and output parameters. The values from a Fortran module are returned explicitly or implicitly. The function return values are exported at the first port of the IF1 graph (Figure 2a) whereas the other values will be returned on greater ports. At the first pass the parameters that are *assigned* at the function body are flagged as updated.

At the **Second Pass** FGTT is processed. A function's parameter can also be changed by a function call that has as input that parameter. If the parameter has changed by the function call, it should be returned by the IF1 graph. For each function, the translator accesses the function call list with their input parameters, and detect which parameters are changed by each function call.

The parameters returned by a module are only those that have been changed at the function body. This reduces the data dependencies in the dataflow model because when a function parameter does not change then its value can be fetched by the nodes ascending the function call node.

Fine and Medium Grain Parallelism at function body are translated with the aid of the VETT as described in Section 3. To create output edges from the graph, the translator traverses the FGTT output parameter list. For each output parameter the translator finds its corresponding VETT and creates an output edge with source node/port values, the corresponding VETT's source node/port field values (SN, SP).

4.2   Structured Control Flow: Nested IFs

Structured control flow of Fortran is implemented using the IF1 compound node *"Select"*. The Select node consist of at least three subgraphs: a selector subgraph to determine which code segment to execute, and at least two alternative subgraph branches (corresponding to THEN and ELSE block) containing each code branch.

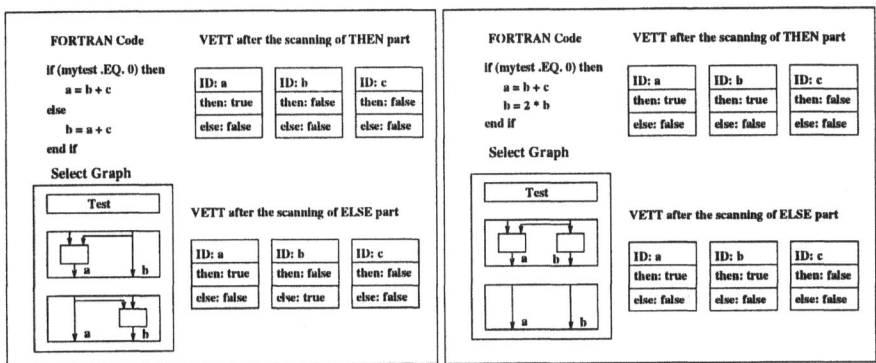

Fig. 3. Select Node handling (a) else statement is present, (b) else statement is missing.

A compound node constitutes a new IF1 scope. The code within the Fortran IF statement is scanned twice by Venus. During the first pass, the translator builds IF1 code for each alternative branch, and updates the *"then"* or *"else"* flag of VETT (see Figure 3 a, b) of variables being updated. For example, if a variable is updated on the THEN branch, then its *"then"* flag in VETT is set to 1 (True). For variables updated on ELSE branch, their *"else"* flag at VETT is set to 1.

At second pass the translator rebuilds the IF1 code for each branch and generates output edges. A variable within compound node scope that has the *"then"* or *"else"* flag in VETT set to 1, is returned by the compound node. Variables that were assigned as updated have their VETT entries changed in order to reflect the new values exported by the compound node.

5   Performance Analysis

Our benchmark suite consists of a number of Livermore Loops and several benchmarks from weather applications. In Table 1 we present the execution time on a single processor of the Livermore Loops written in Fortran, Sisal and IF1 generated from the Venus translator. As shown in the table the loops execute in time similar to their Sisal equivalent and Fortran. This shows that there is no performance degradation when we map Fortran to SAS. Furthermore the results shows that Venus can produce as efficient code from Fortran as OSC does from SISAL.

The results obtained from a multiprocessor Sequent Symmetry S81 and a HP Convex system are summarized in Table 2. (*12L/20L* is a collection of 12 and 20 Livermore Loop **subroutines**. A,E,F,G are symbols that corespond to *vtadv, cloudy, hzadv, depos* modules from Eta-dust code for regional weather prediction and dust accumulation. Eta-dust is a product of National Meteorological Center and University of Belgrade. B,C,D are symbols that corespond to modules *simple, ricard, yale* which are benchmarks taken from weather prediction fortran codes) The results are summarized in terms of speedup gained in a number of processors (ranged from 2 to 6) over 1 processor. It shows that in all cases the Venus was able to produce IF1 code with respectable speedups.

Table 1. Execution time of Livermore Loops on a Sun OS/5.

Livermore Loop	Serial Execution Time in *sec*								
	Fortran to IF1			SISAL			Fortran		
	Number of Iterations								
Loop #	300	600	1000	300	600	1000	300	600	1000
1	72.3	118.5	179.9	70.1	115.3	162.8	70.4	112.3	177.5
2	30.1	42.2	58.7	30.9	42.1	60.4	30.4	40.8	55.7
3	57.3	96.6	149.1	60.2	92.9	144.7	54.8	93.2	148.1
4	369.7	738.0	1228.2	550.7	742.1	1220.7	361.2	735.4	1199.9
5	54.6	83.6	122.2	55.7	83.9	130.4	51.7	83.1	118.3
6	26.2	35.4	57.3	26.5	33.3	59.1	24.9	32.5	58.7
7	53.8	71.7	95.9	54.4	70.2	99.3	50.9	70.5	99.6
8	119.7	134.5	154.7	123.4	154.2	162.1	102.4	127.9	145.2
9	42.2	74.5	117.5	42.3	74.9	121.4	38.7	72.9	114.7
10	32.6	51.4	74.3	31.4	52.3	72.6	30.5	50.4	72.8
11	50.9	83.6	127.5	52.3	82.6	125.3	47.4	85.7	131.4
12	33.2	52.1	77.3	33.7	53.7	74.2	32.5	52.6	74.6
12L	784.8						731.4		
20L	1746.1						1672.8		

6　Conclusions

In this paper we presented, Venus, a system thats maps Fortran programs into Single Assignment Semantics with special emphasis in the extraction of parallelism. Venus is able to process Fortran 77 with the exception of the I/O constructs. We have developed the necessary frame work for the mapping and developed two special data structures for implementing the mapping. The overall system is robust and has been thoroughly tested both on a uniprocessor and on

Table 2. Speedup gained from Fortran-to-IF1 programs over a range of processors

CPUs	Livermore Loop #														Weather Benchmarks						
	2	6	7	8	10	14	15	16	17	19	20	21	12L	20L	A	B	C	D	E	F	G
2	2.1	1.8	2.1	1.9	1.6	2.0	1.8	1.3	1.4	1.9	1.2	1.9	1.8	1.6	1.4	1.7	1.6	1.7	2.0	1.5	2.4
3	3.0	2.1	2.9	2.6	2.3	2.9	2.7	1.6	1.8	2.7	1.4	3.1	2.7	2.5	1.7	2.5	2.3	2.3	2.6	1.9	3.0
4	4.0	2.5	3.9	3.7	3.2	4.0	3.5	2.1	2.2	3.4	1.5	3.8	3.4	3.2	2.2	3.1	3.2	2.9	3.4	2.2	4.1
5	4.2	2.8	4.9	4.2	3.9	4.9	4.7	2.6	2.2	4.0	1.7	4.9	4.0	3.7							
6	5.9	3.2	5.9	4.7	4.5	5.7	5.3	4.2	3.5	4.6	2.1	5.7	4.7	4.3							

a multiprocessor. The results have shown that there is no performance penalty when we map from Fortran to SAS. Furthermore we can exploit the same level of parallelism as we do from equivalent programs written in a functional language. This paper has presented the case for combining the benefits of imperative syntax with functional semantics by transforming structured sequential Fortran code to the single assignment intermediate format, IF1. Furthermore, this project couples OSC with Fortran 77. Thus, it has the potential of exposing a much larger segment of the scientific community to portable parallel programming with OSC.

References

1. Ackerman, W. B., "Dataflow languages." In S. S. Thakkar, editor, *Selected Reprints on Dataflow and Reduction Architectures*, Computer Society Press, Washington D.C., 1987.
2. Cann, C. D., "The Optimizing SISAL Compiler." Lawrence Livermore National Laboratory, Livermore, California.
3. Evripidou P. and Robert, J. B., "Extracting Parallelism in FORTRAN by Translation to a Single Assignment Intermediate Form." *Proceedings of the 8th IEEE International Parallel Processing Symposium*, April 1994.
4. Feo, J. T., and Cann, D. C. "A report on the Sisal language project." *Journal of Parallel and Distributed Computing*, 10, (1990) 349-366.
5. Polychronopoulos, C. D., "Parafrase 2: an environment for parallelizing, partitioning, synchronizing and scheduling programs on multiprocessors." *Proceedings of the 1989 International Conference on Parallel Processing*, 1989.
6. Polychronopoulos, C. D., and Girkar, M. B., "Parafrase-2 Programmer's Manual", February 1995.
7. Skedzielewski, S, "SISAL." In B. K. Szymanski, editor, *Parallel Functional Languages and Compilers*, Addison-Wesley, Menlo Park, CA, 1991.
8. Skedzielewski, S, Glauert, J, "IF1: An intermediate form for Applicative languages." Technical Report TR M-170, University of California - Lawrence Livermore National Laboratory, August 1985.

A Parallelization Framework for Recursive Tree Programs

Paul Feautrier*

Laboratoire PRiSM,
Université de Versailles St-Quentin
45 Avenue des Etats-Unis
78035 VERSAILLES CEDEX FRANCE

Abstract. The automatic parallelization of "regular" programs has encountered a fair amount of success due to the use of the polytope model. However, since most programs are not regular, or are regular only in parts, there is a need for a parallelization theory for other kinds of programs. This paper explore the suggestion that *some* "irregular" programs are in fact regular on other data and control structures. An example of this phenomenon is the set of recursive tree programs, which have a well defined parallelization model and dependence test.

1 A Model for Recursive Tree Programs

The polytope model [7, 3] has been found a powerful tool for the parallelization of array programs. This model applies to program that use only **DO** loops and arrays with affine subscripts. The relevant entities of such programs (iteration space, data space, execution order, dependences) can be modeled as Z-polytopes, i.e. as sets of integral points belonging to bounded polyhedra. Finding parallelism depends on our ability to answer questions about the associated Z-polytopes, for which task one can use well known results from mathematics and operation research.

The aim of this paper is to investigate whether there exists other program models for which one can devise a similar automatic parallelization theory. The answer is yes, and I give as an example the *recursive tree programs*, which are defined in sections 1.4 and 1.5. The relevant parallelization framework is presented in section 2. In the conclusion, I point to unsolved problems for the recursive tree model, and suggest a search for other examples of parallelization frameworks.

1.1 An Assessment of the Polytope Model

The main lesson of the polytope model is that the suitable level of abstraction for discussing parallelization is the *operation*, i.e. an execution of a statement.

---

* e-mail : Paul.Feautrier@prism.uvsq.fr

In the case of DO loop programs, operations are created by loop iterations. Operations can be named by giving the values of the surrounding loop counters, arranged as an *iteration vector*. These values must be within the loop bounds. If these bounds are affine forms in the surrounding loop counters and constant *structure parameters*, then the iteration vector scans the integer points of a polytope, hence the name of the model.

To achieve parallelization, one has to find subsets of independant operations. Two operations are dependent if they access the same memory cell, one at least of the two accesses being a write. This definition is useful only if operations can be related to the memory cells they access. When the data structures are arrays, this is possible if subscripts are affine functions of iteration vectors. The dependence condition translates into a system of linear equations and inequations, whose unknowns are the surrounding loop counters. There is a dependence iff this system has a solution in integers.

These observations can be summarized as a set of requirements for a parallelization framework:

1. We must be able to describe, in finite terms, the set of operations of a program. This set will be called the *control domain* in what follows. The control domain must be ordered.
2. Similarly, we must be able to describe a data structure as a set of *locations*, and a fonction from locations to values.
3. We must be able to associate sets of locations to operations through the use of *address functions*.

The aim of this paper is to apply these prescriptions to the design of a parallelization framework for recursive tree programs.

1.2 Related Work

This section follows the discussion in [5]. The analysis of programs with dynamic data structures has been carried mainly in the context of pointer languages like C. The first step is the identification of the type of data structures in the program, i.e. the classification of the pointer graph. The main types are trees (including lists), DAG and general graphs. This can be done by static analysis at compile time [4], or by asking the programmer for the information. This paper uses the second solution, and the data structures are restricted to trees.

The next step is to collect information on the possible values of pointers. This is done in a static way in the following sense: the sets of possible pointer values are associated not to an operation but to a statement. These sets will be called *regions* here, by analogy to the array regions [9] in the polytope model. Regions are usually represented as *path expressions*, which are regular expressions on the names of structure attributes [8].

Now, a necessary (but not sufficient) condition for two statements to be in dependence is that two of their respective regions intersect. It is easy to see that this method incurs a loss of information which may forsake parallelisation in important cases. The main contribution of this paper is to improve the precision of the analysis for a restricted category of recursive tree programs.

1.3 Basic Concepts and Notations

The main tool in this paper is the elementary theory of finite state automata and rational transductions. A more detailed treatment can be found in Berstel's book [1].

In this paper, the basic alphabet is $\mathbb{N}$ (the set of non negative integers). ϵ is the zero-length word and the point (.) denotes concatenation. In practical applications, the alphabet is always some finite subset of $\mathbb{N}$.

A finite state automaton (fsa) is defined in the usual way by states and labelled transitions. One obtains a word of the language generated (or accepted) by an automaton by concatenating the labels on a path from an initial state to a terminal state. A rational transduction is a relation on $\mathbb{N}^* \times \mathbb{N}^*$ which is defined by a *generalized sequential automaton* (gsa): an fsa whose edges are labelled by input and output words. Each time an edge is traversed, its input word is removed at the begining of the input, and its output word is added at the end of the ouput. Gsa are also known as Moore machines. The family of rational transductions is closed by inversion (simply reverse the elements of each digram), concatenation and composition [1].

A regular language can also be represented as a regular expression: an expression built from the letters and ϵ by the operations of concatenation (.), union (+) and Kleene star. This is also true for gsa, with letters replaced by digrams.

The domain of a rational transduction is a regular language whose fsa is obtained by deleting the second letter of each edge label. There is a similar construction for the range of a rational transduction.

From one fsa c, one may generate many others by changing the set of initial or terminal states. $c(s;)$ is deduced from c by using s as the unique initial state. $c(;t)$ has t as its unique terminal state. In $c(s;t)$, both the initial and terminal states have been changed. Since rational transductions are defined by automata, similar operations can be defined for them.

1.4 The Control Domain of Recursive Programs

The following example is written in a C-like language which will be explained presently:

```
BOOLEAN tree leaf;                    void main(void) {
int tree value;                       5 : sum([]);}
void sum(address I) {                 }
1 : if(! leaf[I]) {
2 :     sum(I.1);
3 :     sum(I.2);
4 :     value[I] = value[I.1] + value[I.2]
        }
}
```

`value` is a tree whose nodes hold integers, and `leaf` is a Boolean tree. The problem is to sum up all integers on the leaves, the final result being found at the root of the tree. Addresses will be discussed in Sect. 1.5.

Let us consider statement 4. Execution of this statement results from a succession of recursive calls to `sum`, the statement itself being executed as a part of the body of the last call. Naming these operations is achieved by recording where each call to `sum` comes from: either line 2 or 3. The required call strings can be generated by a *control automaton* whose states are the functions names and the basic statements of the program. The state associated to `main` is initial and the states associated to basic statements are terminal. There is a transition from state p to state q if p is a function and q is a statement occuring in p body. The transition is labelled by the label of q in the body of p.

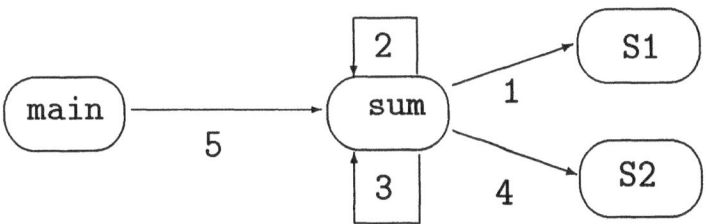

Fig. 1. The Control Automaton of `sum`

As the example shows, if the statements in each function body are labeled by ascending numbers, the execution order is exactly lexicographic ordering on the call strings.

1.5 Addressing in Trees

Remark first that most tree algorithms in the literature are expressed recursively. Observe also that in the polytope model, the same mathematical object is used as the control space and the set of locations of the data space. Hence, it seems quite natural to use trees as the prefered data structure for recursive programs.

In a tree, let us number the edges coming out of a node from left to right by consecutive integers. The name of node n is then simply the string of edge numbers which are encountered on the unique path from the root of the tree to n. The name of the root is the zero-length string, ϵ. This scheme dates back at least to Dewey Decimal Notation as used by librarians the world over.

The set of locations of a tree structure is thus $\mathbb{N}^*$, and a tree object is a partial function from $\mathbb{N}^*$ to some set of values, as for instance the integers, the floating point numbers or the characters.

Address functions map operations to locations, i.e. integer strings to integer strings. The natural choice for them is the family of *rational transductions* [1]. Consider again the above example. Notice the global declaration for the trees **value** and **leaf**. **address** is the type of integer strings. In line 4, such addresses are used to access **value**. The second address, for instance, is built by postfixing the integer 1 to the value of the address variable I. This variable is initialized to ϵ at line 5 of **main**. If the call at line 2 (resp. 3) of **sum** is executed, then a 1 (resp. 2) is postfixed to I.

The outcome of this discussion is that at entry into function **sum**, I comes either from lines 2 or 3 or 5, hence the regular equation:

$$I = \langle 5, \epsilon \rangle + I.\langle 2, 1 \rangle + I.\langle 3, 2 \rangle,$$

whose solution is the regular expression:

$$I = \langle 5, \epsilon \rangle.(\langle 1, 2 \rangle + \langle 3, 2 \rangle)^*.$$

Similarly, the second address in line 4 is given by the following rational transduction : $\langle 4, \epsilon \rangle.(\langle 2, 1 \rangle + \langle 3, 2 \rangle)^*.\langle 4, 1 \rangle$.

I conjecture that the reasoning that has been used to find the above address functions can be automated, but the details have not been worked out yet. It seems probable that this analysis will succeed only if the operations on adresses are suitably limited. Here, the only allowed operator is postfixing by a word of $\mathbb{N}^*$. This observation leads to the definition of a toy language, $\mathcal{T}$, which is similar to C, with the following restrictions:

- No pointers are allowed. They are replaced by addresses.
- The only data structures are scalars (integers, floats and so on) and trees thereof. Trees are always global variables. Addresses can only be used as local variables or functions parameters. No function may return an address.
- The only control structures are the conditionals and the function calls, possibly recursive. No loops or **goto** are allowed.

2 Dependence Analysis of $\mathcal{T}$

2.1 Parallelization Model

When parallelizing static control programs, one has first to decide the shape of the parallel version. One usually distinguishes between *control parallelism*, where operations executed in parallel are instances of different statements, and *data parallelism*, where parallelism is found among iterations of the same statement. In recursive programs, repetition of a statement is obtained by enclosing it in the body of a recursive function, as for example in the program of Fig. 2.

Suppose it is possible to decide that the operations associated to S and all operations generated by the call to **foo** are independent. The parallel version in Fig. 3 (where {^ ... ^} is used as the parallel version of { ... } [6]) is equivalent to the sequential original. The degree of parallelism of this program

```
                                          void foo(x) {
                                            {^
            void foo(x) {                    S;
              S;                              if(p) foo(y);
              if (p) foo(y);                  ^}
            }                               }
```

Fig. 2. Sequential version   **Fig. 3.** Parallel version

is of the order of the number of recursive calls to foo, which probably depends on the data set size. This is data paralellism expressed as control parallelism. A possible formalization is the following.

Let us consider a function foo and the statements $\{S_1, \ldots, S_n\}$ of its body. The statements are numbered in textual order, and i is the label of S_i. Tests in conditional statements are to be considered as elementary, and must be numbered as they occur in the program text.

Let us construct a synthetic dependence graph (SDG) for foo. The vertices of the SDG are the statements of foo. There is a dependence edge from S_i to $S_j, i < j$ iff there exists three iteration words u, v, w such that:

- u is an iteration of foo.
- Both $u.i.v$ and $u.j.w$ are iterations of some terminal statements S_k and S_l.
- $u.i.v$ and $u.j.w$ are in dependence.

Observe that u is an iteration word for foo, hence belongs to the language generated by $c(; \texttt{foo})$. Similarly, v is an iteration of S_k relative to an iteration of S_i, hence belongs to $c(S_i; S_k)$, and $w \in c(S_j; S_l)$. As a consequence, the pair $\langle u.i.v, u.j.w \rangle$ belongs to the following rational transduction:

$$h = c(; \texttt{foo})^= . \langle i, j \rangle . c(S_i; S_k) . c(S_j; S_l)^{-1},$$

in which if a is an automaton, then $a^=$ is the transduction obtained by setting each output word equal to the corresponding input word. This formula also uses an automaton as a transduction whose output words have zero length. Similarly, the inverse of an automaton is used as a transduction whose input words have zero length.

S_i and S_j may also access local scalar variables, for which the dependence calculation is trivial. Besides, one must remember to add control dependences, from the test of each conditional to all statements in its branches. Lastly, dependences between stqtements belonging to opposite arms of a conditional are to be ommited.

Once the SDG is computed, a parallel program can be constructed in several well known ways. Here, the program is put in serie/parallel form by topological sorting of the SDG. As above, I will use the C-EARTH version of the fork ... join construct, {^ ... ^}. The run time exploitation of this kind of parallelism is a well known problem [6].

2.2   The Dependence Test

Computing dependences for the body of function **foo** involves two distinct algorithms. The first, (or outermost) one enumerates all pairs of references which are to be checked for dependence. This is a purely combinatorial algorithm, of polynomial complexity, which can be easily reconstructed by the reader.

The inner algorithm has to decide whether there exists three strings x, y, w such that:

$$\langle x, y \rangle \in h, \quad \langle x, w \rangle \in f, \quad \langle y, w \rangle \in g.$$

where the first term expresses the fact that x and y are iterations of S_k and S_l which are generated by one and the same call to **foo**, the second and third ones expressing the fact that both x and y access location w. f ans g are the address transductions of S_k and S_l. The first step is to eliminate w, giving $\langle x, y \rangle \in k = g^{-1} \circ f$. k is a rational transduction by Elgot and Mezei theorem [2]. Hence, the pair $\langle x, y \rangle$ belongs to the intersection of the two transductions h and k Deciding whether $h \cap k$ is empty is clearly equivalent to deciding whether $\ell \cap =$ is empty where $=$ is the equality relation and $\ell = k^{-1} \circ h$.

Deciding whether the intersection of two transductions is empty is a well known undecidable problem [1]. It is possible however to solve it by a semi-algorithm, wich is best presented as a (one person) game. A position in the game is a triple $\langle u, v, p \rangle$ where u and v are words and p is a state of ℓ. The initial state is $\langle \epsilon, \epsilon, p_0 \rangle$, where p_0 is the initial state of ℓ. A position is a win if $u = v = \epsilon$ and if p is terminal. A move in the game consists in selecting a transition from p to q in ℓ with label $\langle x, y \rangle$. The outcome is a new position $\langle u', v', q \rangle$ where u' and v' are obtained from $u.x$ and $v.y$ by deleting their common prefix. A position is a loss if u and v begin by distinct letters: in such a case, no amount of postfixing can complete u and v to equal strings. There remains positions in which either u or v or both are ϵ. Suppose $u = \epsilon$. Then, for success, v must be the prefix of a string in the domain of ℓ when starting from p. This can be tested easily, and, if the check fails, then the position again is a loss. The situation is symmetrical if $v = \epsilon$.

This game may have three outcomes: if a win can be reached, then by restoring the deleted common prefixes, one reconstructs a word u such that $\langle u, u \rangle \in \ell$, hence a solution to the dependence problem. If all possible moves have been explored without reaching a win, then the problem has no solution. Lastly, the game can continue for ever. One possibility is to put an upper bound to the number of moves. If this bound is reached, one decides that, in the absence of a proof to the contrary, a dependence exists.

The following algorithm explores the game tree in breadth-first fashion.

Algorithm D.

1. Set $D = \emptyset$ and $L = \{ \langle \epsilon, \epsilon, p_0 \rangle \}$ where p_0 is the initial node of ℓ.
2. If $L = \emptyset$, stop. There is no dependence.
3. Extract the leftmost element of L, $\langle u, v, p \rangle$.
4. If $\langle u, v, p \rangle \in D$, restart at step 2.

5. If $u = v = \epsilon$ and if p is terminal, stop. There is a dependence.
6. If both $u \neq \epsilon$ and $v \neq \epsilon$, the position is a loss. Restart at step 2.
7. If $u = \epsilon$ and if v is not a prefix of a word in the domain of $\ell(p;)$, restart at step 2.
8. If $v = \epsilon$ and if u is not a prefix of a word in the range of $\ell(p;)$, restart at step 2.
9. Add $\langle u, v, p \rangle$ to D. Construct all the positions which can be reached in one move from $\langle u, v, p \rangle$ and add them on the right of L. Restart at step 2.

Since the exploration is breadth-first, it is easy to prove that if there is a dependence, then the algorithm will find it.

This algorithm has been implemented as a stand alone program in Objective Caml. The user has to supply the results of the analysis of the input program, including the control automaton, the address transductions, and the list of statements with their accesses. The program then computes the SDG. All examples in this paper have been processed by this pilot implementation. As far as my experience goes, the case where algorithm D does not terminate has never been encountered.

2.3 sum Revisited

Consider the problem of parallelizing the body of sum. There are already control dependences from statement 1 to 2, 3, and 4. The crucial point is to prove that there are no dependences from 2 to 3. One has to test one output dependence from value[I] to itself, two flow dependences from value[I] to value[I.1] and value[I.2], and two symmetrical anti-dependences.

Let us consider for instance the problem of the flow dependence from value[I] to value[I.1]. The ℓ transduction begins in the following way:

$$\ell = (\langle 2, 2 \rangle + \langle 3, 3 \rangle)^* . \langle 3, 2 \rangle . \ldots$$

Algorithm D finds that there is no way of crossing the $\langle 3, 2 \rangle$ edge without generating distinct strings. Hence, there is no dependence.

On the other hand, algorithm D readily finds a dependence from 2 to 4. All in all, the SDG of sum is given by Fig. 4, to which corresponds the parallel program in Fig. 5 — a typical case of parallel divide-and-conquer.

3 Conclusion and Future Work

I have presented here a new framework in which to analyze recursive tree programs. The main differences between the present method and the more usual pointer alias analysis are:

- Data structures are restricted to trees, while in alias analysis, one has to determine the *shape* of the structures. This is a weakness of the present approach.

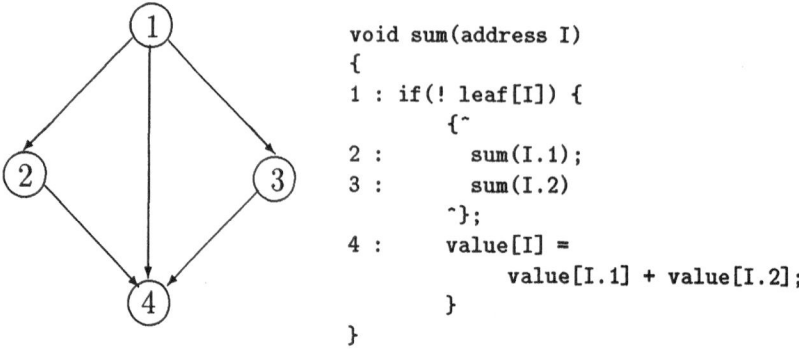

```
void sum(address I)
{
1 : if(! leaf[I]) {
        {^
2 :        sum(I.1);
3 :        sum(I.2)
        ^};
4 :        value[I] =
               value[I.1] + value[I.2];
        }
}
```

Fig. 4. The SDG of sum   **Fig. 5.** The parallel version of sum

- In $\mathcal{T}$, the operations on addresses are limited to postfixing, which, translated in the language of pointers, correspond to the usual pointer chasing.
- The analysis is operation oriented, meaning that addresses are associated to operations, not to statements. This allows to get more precise results when computing dependences.

Pointer alias analysis can be transcribed in the present formalism in the following way. Observe that, in the notations of Sect. 2.2, the iteration word x belongs to Domain(h), which is a regular language, hence w belongs to $f(\text{Domain}(h))$, which is also regular. This is the *region* associated to S_k. Similarly, w is in the region $g(\text{Range}(h))$. If the intersection of these two regions is empty, there is no dependence. When compared to the present approach, the advantage of alias analysis is that the regions can be computed or approximated without restrictions on the source program.

It is easy to prove that when $f(\text{Domain}(h)) \cap g(\text{Range}(h)) = \emptyset$ then ℓ is empty. It is a simple matter to test whether this is the case. The present implementation reports the number of cases which can be decided by testing for the emptiness of ℓ, and the number of cases where Algorithm D has to be used.

In a $\mathcal{T}$ implementation of the merge sort algorithm, there were 208 dependence tests. Of these, 24 were found to be actual dependences, 34 where solved by region intersection, and 150 required the use of algorithm D. While extrapolating from this example alone would be jumping at conclusions, it gives at least an indication of the relative power of the region intersection and of algorithm D. Incidentaly, the SDG of merge sort was found to be of the same shape as the SDG of sum, thus leading to another example of divide-and-conquer parallelism.

The $\mathcal{T}$ language as it stands clearly needs a sequential compiler, and a tool for the automatic construction of address relations. Some of the petty restrictions of Sect. 1.5 can probably be removed without endangering dependence analysis. For instance, having trees of structures or structure of trees poses no difficulty. Allowing trees and subtrees as arguments to functions would pose the usual aliasing problems. A most useful extension would be to allow trees of arrays, as found for instance in some versions of the adaptive multigrid method.

How is $\mathcal{T}$ to be used? Is it to be another programming language, or is it better used as an intermediate representation when parallezing pointer programs as in C or ML or Java? The latter choice would raise the question of translating C (or a subset of C) to $\mathcal{T}$, i.e. translating pointer operations to address operations. Another problem is that $\mathcal{T}$ is static with respect to the underlying set of locations. It is not possible, for instance, to insert a cell in a list, or to graft a subtree to a tree. Is there a way of allowing that kind of operations?

Lastly, trees are only a subset of the data structures one encounter in practice. I envision two ways of dealing, e.g., with DAGs and cyclic graphs. Adding new address operators, for instance a prefix operator:

$$\pi(a_1.\ldots.a_n) = (a_1.\ldots.a_{n-1})$$

allows one to handle doubly linked lists and trees with an upward pointer. The other possibility is to use other mathematical structures as a substrate. Finitely presented monoids or groups come immediately to mind, but there might be others.

References

1. Jean Berstel. *Transductions and Context-Free Languages*. Teubner, Stuttgart, 1979.
2. C.C. Elgot and J.E. Mezei. On relations defined by generalized finite automata. *IBM J. of Research and Development*, 47–68, 1965.
3. Paul Feautrier. Automatic parallelization in the polytope model. In Guy-René Perrin and Alain Darte, editors, *The Data-Parallel Programming Model*, pages 79–103, Springer, 1996.
4. Rakesh Ghiya and Laurie J. Hendren. Is it a tree, a dag or a cyclic graph? In *PoPL'96*, pages 1–15, ACM, 1996.
5. Joseph Hummel, Laurie J. Hendren, and Alexandru Nicolau. A general dependence test for dynamic, pointer based data structures. In *PLDI'94*, pages 218–229, ACM Sigplan, 1994.
6. Laurie J. Hendren, Xinan Tang, Yingchun Zhu, Shereen Ghobrial, Guang R. Gao, Xun Xue, Haiying Cai, and Pierre Ouellet. Compiling C for the EARTH multithreaded architecture. *Int. J. of Parallel Programming*, 25(4):305–338, August 1997.
7. Christian Lengauer. Loop parallelization in the polytope model. In Eike Best, editor, *CONCUR'93, LNCS 715*, pages 398–416, Springer-Verlag, 1993.
8. James R. Larus and Paul N. Hilfinger. Detecting conflicts between structure accesses. In *PLDI'88*, pages 31–34, ACM Sigplan, 1988.
9. Rémi Triolet, François Irigoin, and Paul Feautrier. Automatic parallelization of FORTRAN programs in the presence of procedure calls. In Bernard Robinet and R. Wilhelm, editors, *ESOP 1986, LNCS 213*, Springer-Verlag, 1986.

Optimal Orthogonal Tiling

Rumen Andonov[1], Sanjay Rajopadhye[2], and Nicola Yanev[3]

[1] LIMAV, University of Valenciennes, France. andonov@univ-valenciennes.fr
[2] Irisa, Rennes, France. Sanjay.Rajopadhye@irisa.fr
[3] University of Sofia, Bulgaria. choby@math.acad.bg

Abstract. Iteration space tiling is a common strategy used by parallelizing compilers and in performance tuning of parallel codes. We address the problem of determining the tile size that minimizes the total execution time. We restrict our attention to *orthogonal tiling*—uniform dependency programs with (hyper) parallelepiped shaped iteration domains which can be tiled with hyperplanes parallel to the domain boundaries. Our formulation includes many machine and program models used in the literature, notably the BSP programming model. We resolve the optimization problem analytically, yielding a closed form solution.

1 Introduction

Tiling the iteration space [17, 7, 14] is a common method for improving the performance of loop programs on distributed memory machines. It may be used as a technique in parallelizing compilers as well as in performance tuning of parallel codes by hand (see also [6, 13, 10, 15]). A *tile* in the iteration space is a (hyper) parallelepiped shaped collection of iterations to be executed as a single unit, with the communication and synchronization being done only once per tile. Typically, communication are performed by send/receive calls and also serve as synchronization points. The code for the body contains no communication calls.

The *tiling problem* can be broadly defined as the problem of choosing the tile parameters (notably the shape and size) in an optimal manner. It may be decomposed into two subproblems: *tile shape optimization* [5], and *tile size optimization* [2, 6, 9, 13] (some authors also attempt to resolve both problems under some simplifying assumptions [14, 15]). In this paper, we address the tile size problem, which, for a given tile shape, seeks to choose the size (length along each dimension of the hyper parallelepiped) so as to minimize the total execution time. We assume that the dependencies are uniform, the iteration space (domain) is an n-dimensional hyper-rectangle, and the tile boundaries are parallel to the domain boundary (this is called *orthogonal tiling*). A sufficient condition for this that in all dependence vectors, all non-zero terms have the same sign. Whenever orthogonal tiling is possible, it leads to the simplest form of code; indeed most compilers do not even implement any other tiling strategy. We show here that when orthogonal tiling is possible the tile sizing problem is easily solvable even in its most general, n-dimensional case.

Our approach is based on the two step model proposed by Andonov and Rajopadhye for the 2-dimensional case in [2], and which was later extended to 3 dimensions in [3]. In this model we first abstract each tile by two simple parameters: tile period, $\mathcal{P}$ and inter-tile latency, $\mathcal{L}$. We then "instantiate" the abstract model by accounting for specific architectural and program features, and analytically solve the resulting non-linear optimization problem, yielding the desired tile size. In this paper we first extend and generalize our model to the case where an n-dimensional iteration space is implemented on a k-dimensional (hyper) toroid (for any $1 \leq k \leq n - 1$). We also consider a more general form of the specific functions for $\mathcal{L}$ and $\mathcal{P}$. These functions are general enough to not only include a wide variety of machine and program models, but also be used with the BSP model [11, 16], which is gaining wide acceptance as a well founded theoretical model for developing architecture independent parallel programs.

Being an extension of previous results, the emphasis of this paper is on the new points in the mathematical framework and on the relations of our model to others proposed in the literature. More details about the definitions, notations and their interpretations, as well about some practical aspects of the experimental validations are available elsewhere [1–3].

The remainder of this paper is organized as follows. In the following section we develop the model and formulate the abstract optimization problem. In Section 3 we instantiate the model for specific machine and program parameters, and show how our model subsumes most of those used in the literature. In Section 4 we resolve the problem for the simpler HKT model which assumes that the communication cost is constant, independent of the message volume (and also a sub-case of the BSP model, where the network bandwidth is very high). Next, in Section 5 we resolve the more general optimization problem. We present our conclusions in Section 6.

2   Abstract Model Building

We first develop an analytical performance model for the running time of the tiled program. We introduce the notation required as we go along. The original iteration space is an $N_1 \times N_2 \times \ldots \times N_n$ hyper-rectangle, and it has (at least n linearly independent) dependency vectors, $d_1, \ldots d_m$. The nonzero elements of the dependency vectors all have the same sign (say positive). Hence, orthogonal tiling is possible (i.e., does not induce any cyclic dependencies between the tiles). Let the tiles be $x_1 \times x_2 \times \ldots \times x_n$ hyper-rectangles, and let $q_i = \frac{N_i}{x_i}$ be the number of tiles in the i-th dimension. The *tile graph* is the graph where each node represents a tile and each arc represents a dependency between tiles, and can be modeled by a uniform recurrence equation [8] over an $q_1 \times q_2 \times \ldots \times q_n$ hyper-rectangle. It is well known that if the x_i's are large as compared to the elements of the dependency vectors[1], then the dependencies between the tiles are *unit vectors* (or binary linear combinations thereof, which can be neglected

---

[1] In general this implies that the feasible value of each x_i is bounded from below by some constant. For the sake of clarity, we assume that this is 1.

for analysis purposes without any loss of generality). A tile can be identified by an index vector $\mathbf{z} = [z_1, \ldots, z_n]$. Two tiles, $\mathbf{z}$ and $\mathbf{z}'$, are said to be *successive*, if $\mathbf{z} - \mathbf{z}'$ is a unit vector. A simple analysis [7] shows that the earliest "time instant" (counting a tile execution as one time unit) that tile $\mathbf{z}$ can be executed is $t_z = z_1 + \ldots + z_n$.

We map the tiles to $p_1 \times p_2 \times \ldots \times p_k$ processors, arranged in a k dimensional hyper-toroid[2] (for $k < n$). The mapping is by projection onto the subspace spanned by k of the n canonical axes. To visualize this mapping, first consider the case when $k = n - 1$, where tiles are allocated to processors by projection along (without loss of generality) the n-th dimension, i.e., tile $\mathbf{z}$ is executed by processor $[z_1, \ldots, z_{n-1}]$. This yields a (virtual) array of $q_1 \times q_2 \times \ldots \times q_{n-1}$ processors, each one executing all the tiles in the n-th dimension. In general $p_i \leq q_i$, so this array is emulated by our $p_1 \times p_2 \times \ldots \times p_{n-1}$ hyper-toroid by using multiple passes (there are $\frac{q_i}{p_i}$ passes in the i-th dimension). The case when $k < n - 1$ is modeled by simply letting $p_{k+1} = \ldots = p_{n-1} = 1$.

The *period*, $\mathcal{P}$ of a tile is defined as the time between executions of corresponding instructions in two successive tiles (of the same pass) mapped to the same processor. The *latency*, $\mathcal{L}$ between tiles is defined to be the time between executions of corresponding instructions in two successive tiles mapped to *different* processors. Depending on the volume of the data transmitted and the nature of the program dependencies, the latency may be different for different dimensions of the hyper-toroid, and we use $\mathcal{L}_i$ (for $i = 1 \ldots k$) to denote the latency in the i-th dimension.

We assume that by the time the first processor, $\mathbf{1} = [1 \ldots 1]$, finishes computing its macro column, at least one of the "last" processors (i.e., one of $[1, \ldots, p_i, \ldots, 1]$, for $1 \leq i \leq k$) has finished its first tile[3]. In this case, processor $\mathbf{1}$ can *immediately* start another pass. Let $W = \prod_{i=1}^n N_i$ denote the total computation volume, $v = \prod_{i=1}^n x_i$ be the tile volume and $P = \prod_{i=1}^k p_i$ be the total number of processors. The total number of tiles is $\prod_{i=1}^n q_i = \frac{W}{v}$. Let $\widetilde{p_i}$ denote $p_i - 1$, $\widetilde{P_k} = \sum_{i=1}^k \widetilde{p_i}$ and $v_{\max} = (\prod_{i=1}^k N_i)/P$.

Let us first analyze a single pass. Each processor must execute q_n tiles, and this takes $q_n \mathcal{P}$ time. However, the last processor (i.e., the one with coordinates $[p_1, p_2, \ldots p_k]$) cannot start because of the dependencies of the tile graph. Indeed, processor $[p_1, 1 \ldots, 1]$ can only start at time $(p_1 - 1)\mathcal{L}_1$, processor $[p_1, p_2 \ldots, 1]$, at time $(p_1 - 1)\mathcal{L}_1 + (p_2 - 1)\mathcal{L}_2$, and hence processor $[p_1, p_2 \ldots, p_k]$ can only start its first pass at time $\sum_{i=1}^k \widetilde{p_i} \mathcal{L}_i$.

There are $\frac{W}{v}$ tiles, of which $P q_n$ are executed in each pass, and hence there are $\frac{W/v}{P q_n}$ passes. Because there is no idle time between passes, the last pass starts

---

[2] This is just an abstract machine architecture. Our machine model is independent of the topology, since we will later assume that communication time is independent of distance and/or contention.

[3] There is no loss of generality in this assumption. Were it not true, the processors would be idle between passes, and this would lead to a sub-optimal solution. this has been proved for 2 and 3 dimensions [2, 3] and the arguments carry over.

at time $\left(\frac{W}{Pvq_n} - 1\right)Pq_n$. Thus, the *last* processor starts executing its *last* macro column at time instant $\left(\frac{W}{Pvq_n} - 1\right)Pq_n + \sum_{i=1}^{k} \tilde{p}_i \mathcal{L}_i$. It takes another Pq_n time, and hence the total running time and the corresponding optimization problem are as follows.

$$T(x_1, \ldots x_n) = \frac{WP}{Pv} + \sum_{i=1}^{k} \tilde{p}_i \mathcal{L}_i \tag{1}$$

Prob. 1: Minimize (1) in the feasible space, $\mathcal{R}$ given below (recall that the lower bounds may be other than 1, based on the dependencies of the original recurrence).

$$\mathcal{R} = \{[x_1 \ldots x_n] \in \mathcal{Z}^n \mid 1 \le x_i \le u_i\} \tag{2}$$

where $u_i = \frac{N_i}{p_i}$ for $i = 1 \ldots k$, and $u_i = N_i$ for $k < i \le n$.

3 Machine and Program Specific Model

We now "instantiate" **Prob. 1** for a specific program and machine architecture. The code executed for a tile is the standard loop:

```
repeat
    receive(v1); receive(v2), ...,receive(vk) ;
    compute(body);
    send(v1); send(v2), ..., send(vk) ;
end
```

where **vi** denotes the message transmitted in the i-th dimension. We will now determine $\mathcal{P}$ and $\mathcal{L}_i$. Our development is based on Andonov & Rajopadhye [1], and uses standard assumptions about low level behavior of the architecture and program [4]. The sole difference is that each tile now makes k systems calls to send (and receive) messages. A tile depends directly on its n neighbors in *each* dimension. The volume of data transfer along the i-th dimension is proportional to the (hyper) surface of the tile in that dimension, i.e., $\prod_{j \ne i} x_j = v/x_i$. In the first k dimensions, this corresponds to an inter-processor communication, whereas in the dimensions $k+1 \ldots n$, the transfer is achieved through local memory. Hence the period of a tile can be written as follows.

$$\mathcal{P} = k(\beta_r + \beta_s) + \left(2\tau_c v \sum_{i=1}^{k} \frac{1}{x_i}\right) + \alpha v \tag{3}$$

Here β_s (resp. β_r) is the overhead of the **send** (resp. **receive**) system call, τ_c is the time (per byte) to copy from user to system memory, α is the computation time for a single instance of the loop body (we neglect the overhead of setting up the loop for each tile, as well as cache effects). Similarly, if $1/\tau_t$ is the network bandwidth, the latency is as follows (see [2] for details).

$$\mathcal{L}_i = \mathcal{P} + \tau_t \frac{v}{x_i} - k\beta_r$$

$$= k\beta_s + \left(2\tau_c v \sum_{i=1}^{k} \frac{1}{x_i}\right) + \alpha v + \tau_t \frac{v}{x_i} \qquad (4)$$

Note that the β_r's are subtracted because the **receive** occurs on a *different* processor, and when the sender and receiver are properly synchronized, the calls are overlapped. Substituting in Eqn. (1) and simplifying, we obtain

$$T_k(\boldsymbol{x}) = \alpha v \widetilde{P_k} + 2\tau_c \widetilde{P_k} v \sum_{i=1}^{k} \frac{1}{x_i} + \tau_t v \sum_{i=1}^{k} \frac{\widetilde{p_i}}{x_i} + \frac{2\tau_c W}{P} \sum_{i=1}^{k} \frac{1}{x_i} +$$

$$+ k(\beta_r + \beta_s)\frac{W}{Pv} + k\beta_s \widetilde{P_k} + \frac{\alpha W}{P} \qquad (5)$$

3.1 Simplifying Assumptions and Particular Cases

The model of (5) is very general. One may want to specialize it for a number of reasons—say rendering the final optimization problem more tractable, or modeling a certain class of architectures or computations. It turns out that many of these simply consist of choosing the parameters appropriately in the above function.

The HKT model, first used by Hiranandani et al. [6], corresponds to setting $\beta_r = \tau_c = \tau_t = 0$ (a slightly more general version consists of letting β_r be nonzero). This model assumes that communication cost is independent of the message size, but is dominated by the startup time(s) β_s (and β_r).

At first sight this may seem an oversimplification. However, in addition to making the mathematical problem more tractable, it is not far from the truth, as corroborated by other authors [12, 13]. Indeed, experimental as well as analytic evidence [1] shows that on machines such as the Intel Paragon the more accurate models yield *no observable difference* in the predictions. With $\tau_c = \tau_t = 0$, we obtain the *HKT cost function*:

$$T_k(\boldsymbol{x}) = \frac{\alpha W}{P} + k(\beta_r + \beta_s)\frac{W}{Pv} + (\alpha v + k\beta_s)\widetilde{P_k} \qquad (6)$$

The BSP model [11, 16] has been proposed as a formal model for developing architecture independent parallel programs. It is a bridge between the PRAM model which is general but somewhat unrealistic, and machine-specific models which lead to lack of portability and predictability of performance. Essentially, the computation is described in terms of a sequence of "super-steps" executed in parallel by all the processors. A super-step consists of (i) some local computation, (ii) some communication events *launched* during the super-step, and (iii) a synchronization which ensures that the communication events are *completed*. The time for a super-step is the sum of the times for each of the above activities.

This is very similar to our tile model: indeed, if we simply set $\tau_t = \beta_r = 0$ and $\beta_s = \frac{\beta}{k}$ (β is the BSP synchronization cost) we obtain the running time of the program under the BSP model. With this simplification, we obtain the **BSP cost function** as follows.

$$T_k(x) = \frac{\alpha W}{P} + \frac{\beta W}{Pv} + (\alpha v + \beta)\widetilde{P_k} + 2\tau_c \left(\widetilde{P_k}v + \frac{W}{P}\right) \sum_{i=1}^{k} \frac{1}{x_i} \qquad (7)$$

In the BSP model, the communication startup cost is replaced by the synchronization cost. However, in our general cost function (5) we incur the startup cost k times. As a result, if we take a particular case of the BSP model where the network bandwidth is extremely high (the communication time is dominated by the synchronization cost), then this **high bandwidth BSP cost function** is given as follows (it is similar but not identical to the HKT model).

$$T_k(x) = \frac{\alpha W}{P} + \frac{\beta W}{Pv} + (\alpha v + \beta)\widetilde{P_k} \qquad (8)$$

With some other simple modifications, our cost function can also model the overlap of communication and computation, which is often used as a performance tuning strategy. This is not detailed here due to space constraints.

4 Solution for HKT and High Bandwidth BSP Models

In this section, we will focus only on the simple models (6, 8). Our main results are that the optimal tile volume and the dimension of optimal virtual architecture can be determined as a closed form solution. These results serve two important purposes, in spite of the apparent simplicity of the model. First, they are valid for a number of current machines where the communication latency and network bandwidth are both relatively high (such as the Intel Paragon, and a number of similar machines as well as networks of workstations). Second, they give a good indication of our solution method for the more general results. We first solve the problem for the HKT model, and then show how the high bandwidth BSP model follows almost directly.

4.1 Optimal Tile Volume

It is easy to see that the running time (6) depends only on the *tile volume* and is a strictly convex function of the form $T_k(v) = \frac{A}{v} + Bv + C$, which attains its optimal value of $C + 2\sqrt{AB}$ at $\tilde{v} = \sqrt{\frac{A}{B}}$. By substituting we obtain

$$\tilde{v} = \sqrt{\frac{k(\beta_r + \beta_s)W}{P\alpha\widetilde{P_k}}} \qquad (9)$$

$$\widetilde{T_k} = \frac{\alpha W}{P} + k\beta_s\widetilde{P_k} + 2\sqrt{\frac{k\alpha(\beta_r + \beta_s)W\widetilde{P_k}}{P}} \qquad (10)$$

The optimal solution will be as given above if $\tilde{v}$ is a feasible tile volume. Now, observe that each x_i is bounded from above by $\frac{N_i}{p_i}$, and hence $v \leq \frac{W}{P}$. Since W grows much faster than P, we have $1 \leq \tilde{v} \leq \frac{W}{P}$ asymptotically. Hence we have the following result.

Theorem 1. *The optimal tile volume and the corresponding running time for the* HKT *model are given by (9-10).*

4.2 Optimal Architecture

So far, we have assumed that k is a fixed constant as are each of the p_i's. In practice, we typically have P processors, and the values of each p_i are not specified, and neither is k, and we now solve this problem. Note that the dominant term in (10) is $\frac{\alpha W}{P}$, the "ideal" parallel time with no overhead. The ideal architecture will seek to minimize the overhead whose dominant term is $O(\sqrt{\frac{W}{P}})$.

From (10) it can be easily deduced that for a given k and P, the optimal architecture is the one that minimizes $\widetilde{P_k}$, i.e., the torus with $p_i = \sqrt[k]{P}$. Substituting this in (10), the coefficient of the overhead term is proportional to the square root of the function $f(k, P) = k^2 P^{\frac{1}{k}} - k^2$. Thus $\widetilde{T_k}$ is minimized for the value of k that minimizes $f(k, P)$, i.e., $k^* = 0.625 \ln P$. Since $f(k, P)$ is monotonically decreasing up to k^* (for each P) and monotonically increasing thereafter, we have the following result.

Corollary 1. *If $n \leq \lceil k^* \rceil$ the optimal architecture is a balanced $n - 1$ dimensional torus, and for $n > \lceil k^* \rceil$ it is either a balanced $\lceil k^* \rceil$ or a $\lfloor k^* \rfloor$ dimensional torus depending respectively on whether $f(\lceil k^* \rceil, P)$ is smaller than $f(\lfloor k^* \rfloor, P)$ or not.*

It is thus clear that as the number of processors grows, the optimal architecture tends towards an $n - 1$ dimensional hyper-torus. However, k^* grows logarithmically with P, and for a limited number of processors (most practical cases), the optimal may not be $n - 1$ dimensional. Indeed, the sensitivity of k^* with respect to number of processors P is illustrated by the following: for up to 25 processors, the optimal architecture is a linear array, from 25–130 it is 2-dimensional, for 130–650 processors it is a 3-dimensional, etc.

The extension of these results to the high bandwidth BSP model (Eqn 8) is straightforward, and indeed the mathematical treatment is a little simpler. The solution is the same as that given by (9-10), but with k, β_r and β_s respectively equal to 1, 0 and β. The sole subtle difference is that the function $f(k, P)$ is $k P^{1/k} - k$, and hence $k^* = +\infty$, i.e., $f(k, P)$ is monotonically decreasing. Hence, the optimal architecture is *always* a balanced $n - 1$ dimensional torus.

Finally, note that the optimization of the processor architecture yields a second order improvement—it does not affect the dominant term $\frac{\alpha W}{P}$.

5 Solution for the BSP Cost Function

We now solve our optimization problem for the BSP cost function (7) in the feasible space specified by (2). We will show that our problem can be decomposed into two special cases. The first case is very similar to the HKT model, but the second one is more complicated, for which we first solve the corresponding *unconstrained optimization problem* and then determine where the constrained solution lies. Finally, we show that the second solution is globally optimal asymptotically.

Lemma 1. *The minimum of (7) over conv($\mathcal{R}$) is attained at:*

$$\text{either} \quad x_i = \frac{N_i}{p_i}, \text{ for } i = 1 \dots k$$
$$\text{or} \quad x_i = 1, \text{ for } i = k+1 \dots n$$

Proof. Let v be an arbitrary feasible volume and let there exist two indices l, m for $l \le k$, $k+1 \le m \le n$ such that $x_l < N_l/p_l$ and $x_m > 1$. Then by increasing x_l, decreasing x_m and keeping their product constant, the function (7) strictly decreases and we obtain the needed. ∎

Based on this, we have to look for the solution in the two regions of $\mathcal{R}$ corresponding to the above two conditions. Let $\mathcal{R}_1 = \mathcal{R} \cap x_i = \frac{N_i}{p_i}$ for $i = 1 \dots k$, and $\mathcal{R}_2 = \mathcal{R} \cap x_i = 1$ for $i = k+1 \dots n$.

5.1 Case I

The cost function in region $\mathcal{R}_1$ can be simplified to

$$T_k(\boldsymbol{x}) = (\alpha + 2\tau_c \widetilde{N_k})\frac{W}{P} + \beta \widetilde{P_k} + \frac{\beta W}{Pv} + (\alpha + 2\tau_c \widetilde{N_k})\widetilde{P_k}v \qquad (11)$$

where $\widetilde{N_k} = \sum_{i=1}^{k} \frac{p_i}{N_i}$. We can now use the same reasoning as in Section 4, and obtain the optimal volume and running time as follows:

$$\tilde{v} = \sqrt{\frac{\beta W}{P\omega}} \quad \text{and} \quad \widetilde{T_k} = (\alpha + 2\tau_c \widetilde{N_k})\frac{W}{P} + \beta \widetilde{P_k} + 2\sqrt{\frac{\beta W \omega}{P}} \qquad (12)$$

where $\omega = (\alpha + 2\tau_c \widetilde{N_k})\widetilde{P_k}$. As before, appropriate values for $x_{k+1} \dots x_n$ in $\mathcal{R}_1$ that yield the optimal volume are all equivalent. Observe that in (12) we have a factor $2\tau_c \widetilde{N_k}$ in the dominant term, and this turns out to be important as we shall see later.

5.2 Case II

Let us now consider region $\mathcal{R}_2$. Here, $v = \prod_{i=1}^{k} x_i$, but the cost function remains the same as (7), except that we have only k variables to solve for. We obtain the solution in two steps.

Unconstrained Optimization We first solve the problem in the entire positive orthant, without any constraints. This can be formulated as follows.

Prob. 2: Minimize (7) in the feasible space $\mathcal{R}_+^k = \left\{ [x_1 \ldots x_k]^\mathrm{T} \mid x_i \geq 0 \right\}$.

Let H_v be the hyperboloid defined by $\{ \boldsymbol{x} \in \mathcal{R}_+^k \mid \prod_{i=1}^k x_i = v \}$. Observe that the set of families H_v is a partition of $\mathcal{R}_+^k$, i.e., $\mathcal{R}_+^k = \bigcup_v H_v$, and hence $\min_{R_+^k} T_k(\boldsymbol{x}) = \min_v \min_{H_v} T_k(\boldsymbol{x})$. Thus, we first minimize (7) for a given tile volume v (i.e., over a given hyperboloid H_v) and then choose the volume. Now, observe that for a fixed v, (7) is of the form $A + B \sum_i \frac{1}{x_i}$ whose minimum is attained at $\boldsymbol{x} \in H_v, x_1 = x_2 = \ldots = x_k$. Hence, for a given tile volume, v, the optimal tile shape is (hyper) cubic with $x_i = \sqrt[k]{v}$, for each $i = 1 \ldots k$. Thus, $\sum_{i=1}^k \frac{1}{x_i} = \frac{k}{\sqrt[k]{v}}$, and we can define,

$$f(v) = \min_{H_v} T(\boldsymbol{x}) = \frac{A}{v} + Bv + 2k\tau_c \widetilde{P_k} v^{1-\frac{1}{k}} + kD v^{-\frac{1}{k}} + \frac{\alpha W}{P} + \beta \widetilde{P_k} \qquad (13)$$

where $A = \frac{\beta W}{P}$, $B = \alpha \widetilde{P_k}$, and $D = \frac{2\tau_c W}{P}$. Now we have to determine the optimal tile volume by minimizing $f(v)$ in the feasible space, $1 \leq v \leq v_{\max}$. It ·an easily be shown that $f(v)$ is an unimodal function and attains its minimum , the root of $f'(v) = 0$. Unfortunately this is not easily obtained in closed form ·t it is quite well approximated by the root of the function $h(v) = \frac{-A}{v^2} + B + 2\tau_c(k-1)\widetilde{P_k} - \frac{D}{v^2}$, whose zero is at $\sqrt{\frac{A+D}{2\tau_c(k-1)\widetilde{P_k}+B}}$. Thus, we obtain the following (approximate) optimal solution of the unconstrained problem.

$$\tilde{v} \approx \sqrt{\frac{\gamma W}{P}} \text{ and } \widetilde{T_k} \approx \frac{\alpha W}{P} + 2k\tau_c \left(\frac{1}{\gamma} \right)^{\frac{1}{2k}} \left(\frac{W}{P} \right)^{\frac{2k-1}{2k}} + O\left(\sqrt{\frac{W}{P}} \right) \qquad (14)$$

where $\gamma = (\beta + 2\tau_c)/((2\tau_c(k-1)+\alpha)\widetilde{P_k})$. Of course we could always determine an exact solution if needed, but we have found the approximate solution to be reasonable in practice. Moreover, as we shall see later, it illustrates some interesting points. Also recall that in the problem as we have resolved so far, we assume that k and the p_i's are fixed, as also the choice of which of the N_i's to map to the processor space.

Constrained Optimization We now address the question of the restrictions on the optimal solution imposed by the feasibility constraints (2) namely $1 \leq x_i \leq u_i$. Note that the unconstrained (global) optimal solution is on the intersection of the line defined by the vector **1**, and the hyperboloid $H_{\tilde{v}}$, where $\tilde{v}$ is the optimal tile volume (14). If this intersection is outside the feasible space (2) we will need to solve the constrained optimization problem. We have observed that the optimal running time is extremely sensitive to the tile volume, and much less dependent on the particular values of x_i (indeed this is predicted by the HKT model). We have the following cases

Case A: $\tilde{v} > v_{\max}$ In this case, the optimal volume hyperboloid does not intersect the feasible space, and according to the properties of the function $f(v)$ the optimal tile size is given by $x_i = \frac{N_i}{p_i}$, for $i = 1 \ldots k$. However, note that since $\tilde{v} = O(\sqrt{W/P})$ while $v_{\max} = O(W/P)$, this case is unlikely.

Case B: $\tilde{v} \leq v_{\max}$ Now, $H_{\tilde{v}}$ has a non-empty intersection with (2) and so our heuristic is to choose a solution by moving one of the x_i's such that we move within the feasible region.

5.3 Where is the Global Optimum?

The optimal solutions for the two cases are given, respectively by (12) for region $\mathcal{R}_1$ and by (14) for region $\mathcal{R}_2$. Most authors [6, 13, 12] have only considered region $\mathcal{R}_1$ either implicitly by not posing the problem in full generality, or by erroneously claiming that the solution is always in $\mathcal{R}_1$. This is incorrect because the final solution will always asymptotically be in $\mathcal{R}_2$, due to the additional factor, $2\tau_c \widetilde{N_i}$ in the dominant term in (12) as compared to (14). The following simple example illustrates the prediction of lemma 1: for $n = 3, k = 2, p_1 = p_2 = 4, \alpha = \beta = \tau_c = 10^{-6}, N_1 = 50, N_2 = N_3 = 1000$, the optimal time in $\mathcal{R}_1$ is 1.8 while that in $\mathcal{R}_2$ is 1.5. More illustrative instances with real data can be found in [3].

6 Conclusions

We addressed the problem of finding the tile size that minimizes the running time of SPMD programs. We formulated a discrete non-linear optimization problem using first an abstract model and then specific machine model. The resulting cost function is general enough to subsume most of those in the literature, including the BSP model. We then analytically solved the resulting discrete nonlinear optimization problem, yielding the desired solution.

There are a number of open questions. The first one is the direct extension to the non orthogonal case (when the tiles boundaries cannot be parallel to the domain boundaries). We have addressed this elsewhere (for the 2-dimensional case) and formulated a non-linear optimization problem [2], but a closed form solution is not available. Finally, experimental validation on a number of target machines is the subject of our ongoing work.

References

1. R. Andonov, H. Bourzoufi, and S. Rajopadhye. Two-dimensional orthogonal tiling: from theory to practice. In *International Conference on High Performance Computing*, pages 225–231, Tiruvananthapuram, India, December 1996. IEEE.
2. R. Andonov and S. Rajopadhye. Optimal orthogonal tiling of 2-d iterations. *Journal of Parallel and Distributed Computing*, 45(2):159–165, September 1997.

3. R. Andonov, N. Yanev, and H. Bourzoufi. Three-dimensional orthogonal tile sizing problem: Mathematical programming approach. In *ASAP 97: International Conference on Application-Specific Systems, Architectures and Processors*, pages 209–218, Zurich, Switzerland, July 1997. IEEE, Computer Society Press.

4. S. Bokhari. Communication overheads on the Intel iPSC-860 Hypercube. Technical Report Interim Report 10, NASA ICASE, May 1990.

5. P. Boulet, A. Darte, T. Risset, and Y. Robert. (pen)-ultimate tiling? *INTEGRATION, the VLSI journal*, 17:33–51, Nov? 1994.

6. S. Hiranandani, K. Kennedy, and C-W. Tseng. Evaluating compiler optimizations for Fortran D. *Journal Of Parallel and Distributed Computing*, 21:27–45, 1994.

7. F. Irigoin and R. Triolet. Supernode partitioning. In *15th ACM Symposium on Principles of Programming Languages*, pages 319–328. ACM, Jan 1988.

8. R. M. Karp, R. E. Miller, and S. V. Winograd. The organization of computations for uniform recurrence equations. *JACM*, 14(3):563–590, July 1967.

9. C-T. King, W-H. Chou, and L. Ni. Pipelined data-parallel algorithms: Part I–concept and modelling. *IEEE Transactions on Parallel and Distributed Systems*, 1(4):470–485, October 1990.

10. C-T. King, W-H. Chou, and L. Ni. Pipelined data-parallel algorithms: Part II–design. *IEEE Transactions on Parallel and Distributed Systems*, 1(4):486–499, October 1990.

11. W. F. McColl. Scalable computing. In J. van Leeuwen, editor, *Computer Science Today: Recent Trends and Developments*, volume 1000, pages 46–61. Springer Verlag, 1995.

12. H. Ohta, Y. Saito, M. Kainaga, and H. Ono. Optimal tile size adjsutment in compiling general DOACROSS loop nests. In *International Conference on Supercomputing*, pages 270–279, Barcelona, Spain, July 1995. ACM.

13. D. Palermo, E. Su, J. Chandy, and P. Banerjee. Communication optimizations used in the PARADIGM compiler for distributed memory multicomputers. In *International Conference on Parallel Processing*, pages xx–yy, St. Charles, IL, August 1994. IEEE.

14. J. Ramanujam and P. Sadayappan. Tiling multidimensional iteration spaces for non shared-memory machines. In *Supercomputing 91*, pages 111–120, 1991.

15. R. Schreiber and J. Dongarra. Automatic blocking of nested loops. Technical Report 90.38, RIACS, NASA Ames Research Center, Aug 1990.

16. L. G. Valiant. A bridging model for parallel computation. *Communications of the ACM*, 33(8):103–111, August 1990.

17. M. J. Wolfe. Iteration space tiling for memory hierarchies. *Parallel Processing for Scientific Computing (SIAM)*, pages 357–361, 1987.

Enhancing the Performance of Autoscheduling in Distributed Shared Memory Multiprocessors*

Dimitrios S. Nikolopoulos, Eleftherios D. Polychronopoulos
and Theodore S. Papatheodorou

High Performance Computing Architectures Laboratory
Department of Computer Engineering and Informatics
University of Patras
Rio 26500, Patras, Greece
e-mail:{dsn,edp,tsp}@hpclab.ceid.upatras.gr

Abstract. Autoscheduling is a parallel program compilation and execution model that combines uniquely three features: Automatic extraction of loop and functional parallelism at any level of granularity, dynamic scheduling of parallel tasks, and dynamic program adaptability on multiprogrammed shared memory multiprocessors. This paper presents a technique that enhances the performance of autoscheduling in Distributed Shared Memory (DSM) multiprocessors, targetting mainly at medium and large scale systems, where poor data locality and excessive communication impose performance bottlenecks. Our technique partitions the application Hierarchical Task Graph and maps the derived partitions to clusters of processors in the DSM architecture. Autoscheduling is then applied separately for each partition to enhance data locality and reduce communication costs. Our experimental results show that partitioning achieves remarkable performance improvements compared to a standard autoscheduling environment and a commercial parallelizing compiler.

1 Introduction

Distributed Shared Memory (DSM) multiprocessors have evolved as a powerful and viable platform for parallel computing. Unfortunately, automatic parallelization for these systems becomes often an onerous duty, requiring numerous manual optimizations and substantial effort from the programmer to sustain high performance [3]. The development of programming models and compilation techniques that exploit the advantages and overcome the architectural bottlenecks of DSM systems remains a challenge.

Among several parallel compilation and execution models proposed in the literature, *autoscheduling* [13] is one of the most promising approaches. Autoscheduling provides a compilation framework which merges the control flow and data flow models to efficiently exploit multiple levels of loop and functional parallelism. The compilation framework is coupled with an execution environment

---

* This work was supported by the European Commission under ESPRIT IV project No. 21907 (NANOS)

that schedules parallel programs dynamically on a variable number of processors, by controlling the granularity of parallel tasks at runtime. Coordination between the compiler and the execution environment achieves effective integration of multiprocessing with multiprogramming.

In this paper, we identify sources of performance degradation for an autoscheduling environment running on a DSM multiprocessor. Performance bottlenecks arise as a consequence of poor caching performance and excessive communication through the interconnection network. We present a technique that attempts to resolve these problems by augmenting autoscheduling with partitioning and clustering. We partition the application task graph in subgraphs with independent dataflows and map the derived subgraphs to processor clusters. The topology of the clusters conforms to the system architecture. Partitioning and clustering are performed dynamically at runtime. After mapping partitions to clusters, autoscheduling is applied separately to each partition to extract and schedule the parallelism.

Using application and synthetic benchmarks, we demonstrate that partitioning reduces significantly the execution time of autoscheduled programs on a 32-processor SGI Origin2000. For the same set of benchmarks, partitioned autoscheduling outperforms the native SGI loop parallelizer, even in situations where standard autoscheduling fails to do so.

The rest of this paper is organized as follows: Section 2 provides some background on autoscheduling and discuss its performance limitations. Section 3 presents our technique along with some implementation issues. Section 4 describes our evaluation framework and Section 5 provides detailed experimental results. We summarize our conclusions in Section 6.

2 Autoscheduling and Performance Limitations

An autoscheduling environment consists of a parallelizing compiler, a multithreading runtime library and an enhanced operating system scheduler. The compiler applies control and data dependence analysis to produce an intermediate program representation called the Hierarchical Task Graph (HTG) [5]. Programs are represented with a hierarchical structure consisting of simple and compound nodes. Simple nodes correspond to basic blocks of sequential code. Compound nodes represent tasks amenable to parallelization, such as parallel loops and parallel sections. The levels of the hierarchy are determined by the nesting of loops and correspond to different levels of exploitable parallelism. Nodes are connected with arcs representing execution precedences. A boolean expression with a set of firing conditions is associated with each node. When this expression evaluates to TRUE, all data and control dependences of the node are satisfied and the corresponding task is enabled for execution.

The compiler generates parallel code by means of a user-level runtime system which creates and schedules lightweight execution flows called *nano-threads* [13]. The granularity of the generated parallelism is controlled dynamically at runtime. A communication mechanism between the runtime system and the operating system scheduler keeps parallel programs informed of processor availability and enables dynamic program adaptability to varying execution conditions [14].

Each compound node in the HTG is annotated as a candidate source of parallelism. The runtime system decomposes all the compound nodes which contain enough parallelism to utilize the processors available at a given execution instance. If several compound nodes coexist at the same level of the hierarchy, parallelism can be extracted from all of them simultaneously. In that case, nanothreads originating from different compound nodes are executed in an overlapped manner across the processors. As an example, consider the HTG of the multiplication of two matrices Y, Z with complex entries, calculated as:

$$X = YZ = (A + jB)(C + jD) = (AC - BD) + j(BC + AD) \tag{1}$$

The computation consists of four independent matrix by matrix multiplications, followed by an addition and a subtraction. Each of these tasks is a compound node with inherent loop parallelism. If a sufficient number of processors is available at runtime, autoscheduling may execute the matrix multiplications simultaneously and each matrix multiplication in parallel to exploit two levels of parallelism.

Overlapped execution of nano-threads originating from different compound nodes can have undesirable effects on a DSM multiprocessor. This is particularly true when each compound node individually uses all the available processors. Three performance implications may arise in this case: Increase of conflict and capacity cache misses, heavy traffic in the interconnection network, and higher runtime overhead.

Bursts of conflict and capacity misses occur when the working sets of concurrently executing compound nodes do not fit in the caches, thus forcing the caches to alternate frequently between different working sets. This happens as processors are switched between nanothreads originating from different compound nodes. Overlapped execution of compound nodes stresses the interconnection network, due to excessive data communication. The runtime system ovehead is higher, since multiple processors are forced to access heavily shared ready queues and memory pools, in order to create and schedule nano-threads. This side-effect exacerbates contention and communication traffic due to remote memory accesses [11].

3   Performance Enhancements

Our method to alleviate the problems presented in Sect. 2 while preserving the semantics of autoscheduling, is to decompose the program HTG into partitions with independent data flows and the processor space of the program into clusters, with a one-to-one mapping between HTG partitions and processor clusters. After partitioning and clustering, autoscheduling is applied separately for each partition. Parallelism from each partition is extracted to utilize the processors of the cluster to which the partition is mapped. Processor clusters serve as memory locality domains for the partitions. This means that the working set of a partition is stored in physical memory that lies within the corresponding cluster, in order to reduce memory latency during the execution of the partition. The communication required for a partition is performed within the cluster, thus putting less pressure in the interconnection network and avoiding long distance memory accesses. Furthermore, data dependent partitions use the same processors to preserve locality. If a partition $\mathcal{P}$ is data dependent on a partition $\mathcal{Q}$, our mechanism executes $\mathcal{P}$ either in a superset, or a subset of the cluster to which $\mathcal{P}$ is mapped.

The relative size of the clusters is determined with the following procedure: Compound nodes that can be simultaneously executed are detected in the HTG and groupped in sets. The computational requirements of each compound node are estimated with execution profiling and then weighed with the aggregate computational requirements of the compound nodes that belong to the same set. Processors are distributed proportionally to each compound node according to its weighed computational requirements. The actual size of the clusters is set at runtime, and depends on their precomputed relative size and processor availability. Let P be the number of processors available to the application at an execution instance, n the number of concurrently executing compound nodes at that instance and $r_i, i = 1 \ldots n$, the computational requirement of each compound node, expressed in CPU time. Then the processor set is partitioned into n clusters $c_i, i = 1 \ldots n$, and each cluster receives p_{c_i} processors where:

$$p_{c_i} = max\{1, \frac{r_i}{\sum_{i=1}^{n} r_i} P\} \tag{2}$$

If the number of available processors does not exceed the degree of functional parallelism exposed by the application —expressed as the number of concurrently executing compound nodes—, then our technique degenerates to standard autoscheduling. Otherwise, autoscheduling is applied within the bounds of each cluster by extracting the data and/or functional parallelism of the associated HTG partition. The clusters can be dynamically reconfigured at runtime. Load imbalances between clusters are not considered. We rather rely on the proportional allocation of processors to clusters to handle these cases. However, load imbalances within the clusters across parallel loops or parallel sections are handled by the user-level scheduler which employs affinity scheduling and work stealing [15].

The topology of the clusters is kept consistent with the underlying system architecture. For DSM architectures, clusters are built incrementally using the Symmetric Multiprocessing (SMP) nodes as the basic building block. The interconnection network is exploited to form clusters with SMP nodes which are physically close to each other. The clusters can shrink or grow at runtime, either by splitting, or by joining directly connected clusters.

The entire mechanism is implemented at user-level. Partitions are detected from the HTG, and communicated to the runtime system through directives. Static information about the relative processor requirements of each partition is also communicated to the runtime system. The actual cluster sizes are computed at runtime using Eq. 2. The infrastructure for mapping the partitions to processors is provided by a hierarchy of ready queues [2]. The hierarchy corresponds to a tree with a global system queue at the root, per-processor local queues at the leaves and a number of intermediate levels of ready queues. Mapping of partitions to clusters is performed with Distributed Hierarchical Control (DHC) [4]. When a compound node is assigned to a cluster of size p, the corresponding nano-thread is mapped to a ready queue at the root of a subtree that has at least p processors at the leaves. The processor that creates the nano-thread searches the hierarchy up until it finds a queue from which a tree of the appropriate size is emanated. Nested nano-threads created by the outermost nano-thread are mapped to the queues that reside at the leaves of the subtree. Processor clusters are implicitly configured using DHC, without invoking the operating system to construct the memory locality domains.

As an example, Fig. 1.(a) illustrates the task graph of complex matrix multiplication. Following our scheme, the processor set that executes the computation is initially partitioned in four clusters of equal size, and a matrix multiplication is executed separately in each cluster. Within a cluster, the matrix multiplication is autoscheduled on a number of processors equal to the size of the cluster. When the subtraction and addition tasks are activated, pairs of neighbouring clusters are joined to form two clusters of doubled size and execute the tasks. Figures 1.(b) and 1.(c) demonstrate executions of the task graph under standard autoscheduling and our scheme respectively. Circles indicate nano-threads, while their fill patterns indicate compound nodes from which the nano-threads originate. Under standard autoscheduling, the execution sequence of nano-threads from the concurrently executing compound nodes is determined from the serialization of processor accesses to the ready queues and may change between subsequent executions of the program. With partitioning, compound nodes are still executed concurrently, but the parallelism extracted from each partition is scheduled in isolation within a cluster. Figure 2 illustrates a hierarchy of ready queues for an 8-processor sytsem and the scheduling strategy for the complex matrix multiplication example.

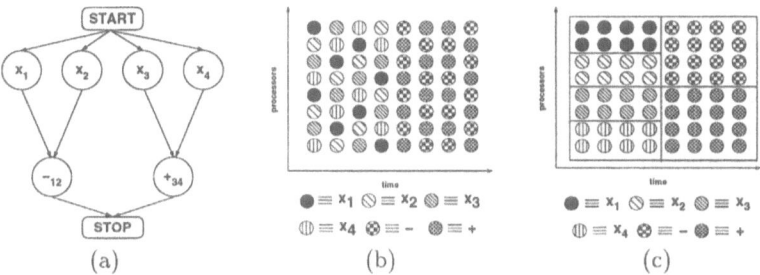

Fig. 1. Complex matrix multiplication task graph and execution with standard and partitioned autoscheduling . $\times, +, -$ indicate matrix by matrix operations.

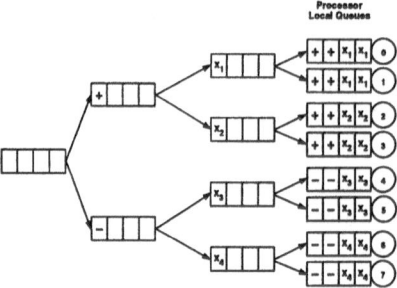

Fig. 2. Hierarchy of ready queues and enqueing of complex matrix multiplication tasks.

4 Evaluation Framework

We evaluate our technique with a number of application and synthetic benchmarks. The application benchmarks presented here were used in a previous study of autoscheduling [9]. This gives us the opportunity for a quantitative analysis of our results in direct comparison with existing work in the field. The purpose of the synthetic benchmarks is to demonstrate the performance implications of autoscheduling and the ability of our method to overcome them.

The application benchmarks include the complex matrix multiplication kernel, the kernel of a Fourier-Chebyschev spectral computational fluid dynamics code [7], and an industrial molecular dynamics code [16]. All benchmarks have two levels of exploitable parallelism, namely a constant degree of functional parallelism at the outer lelel and data parallelism at the inner level. The complex matrix multiplication and the cfd kernels use matrices of size 256×256. The molecular dynamics kernel consists of 14 parallel functions, each one executing a parallel loop with 1000 iterations. More details on the structure of these benchmarks, as well as their performance under autoscheduling on a bus-based multiprocessor can be found in [9].

Our first synthetic benchmark consists of 8 parallel dot products. Each dot product is computed with two vectors of 64 Kilobytes of double precision elements. A detailed description of the benchmark can be found in [11]. Two levels of functional and data parallelism are exploitable, similarly to the application benchmarks. The dataflow of this benchmark favours processor clustering to maintain locality. The benchmark experiences increased conflict cache misses due to the alignment of data in memory, and a low ratio of computation to overhead because of the fine thread granularity. The second synthetic benchmark consists of four parallel point LU factorizations, followed by two parallel additions of the factorized matrices. The matrix size used is 256 × 256. Autoscheduling is expected to lead to conflicts between elements of different matrices in the caches, and increased remote cache invalidations due to the all to all communication pattern of the computation. Partitioning is expected to save cache misses and localize communication.

Table 1. Maximum speedups attained for the benchmarks with the three paralellization schemes.

Benchmark	Loop Parallel		Autoscheduled		Partitioned Autosch.	
	S_{max}	p	S_{max}	p	S_{max}	p
synthetic dot product	1.42	4	1.82	4	3.52	8
synthetic LU	7.06	16	7.34	16	15.72	32
cmm kernel	7.14	16	8.51	16	10.08	16
cfd kernel	5.59	12	6.22	12	7.31	12
md kernel	1.13	14	1.03	14	1.34	28

We performed our experiments on a SGI Origin2000 [6], equipped with 32 MIPS R1000 processors. We implemented partitioning and clustering using Nth-Lib [8]. NthLib is a multithreading runtime library designed to support autoscheduling. The library is currently under development in the context of the ESPRIT project NANOS [10]. NthLib provides lightweight threads, created and scheduled entirely at user-level. The runtime system translates HTG representations into multithreading code, while attempting to preserve desirable properties such as data locality and load balancing. At the same time, NthLib communicates with the operating system to enable application adaptivity to the runtime conditions. Automatic parallelization was performed with HPF-like directives and the Parafrase-2 compiler [1,12]. Three versions of each benchmark were used in the experiments: (a) a *loop parallel* version produced by the native SGI parallelizing compiler and exploiting a single level of parallelism from loops; (b) a *autoscheduled* version, which exploits multiple levels of loop and functional parallelism; and (c) a *partitioned autoscheduled* version which uses autoscheduling enhanced with partitioning and clustering. All benchmarks were compiled with the -O3 optimization flag and executed in a dedicated environment.

5 Performance Results

In this section, we present detailed results from our experiments. Table 1 shows the maximum speedups attained for the benchmarks presented in Sect. 4 with the three parallelization schemes. The table also shows the number of processors on which the maximum speedup is attained in each case.

The obtained speedups are justified as follows: For the synthetic dot product benchmark, poor speedup is a consequence of increased runtime overhead and cache conflicts. This benchmark scales better only if the problem size and thread granularity are significantly increased. The molecular dynamics kernel exhibits poor speedups due to poor data locality and high irregularity in the native code. Our results for this code agree with those reported in [9]. The other three codes exhibit good speedups, with respect to the structure of the computations, the platform architecture and the problem size.

The primary conclusion drawn from Table 1 is that in all cases, the maximum attainable speedup is obtained from the partitioned autoscheduled versions. The partitioned autoscheduled versions of the synthetic benchmarks and the molecular dynamics kernel scale up to a higher number of processors than both the autoscheduled and the loop parallel versions. The autoscheduled versions have always better speedups than the loop parallel versions with the exception of the molecular dynamics kernel. These results are consistent with the results in [9] and strengthen the argument that autoscheduling exploits unstructured parallelism more effectively.

All benchmarks achieve better speedups and some of them do scale up to 32 processors if the problem sizes and thread granularities are adequately increased. However, we insisted in using small problem sizes to stress the ability of the runtime system to handle fine-grain parallelism, a prerequisite for the applicability of autoscheduling. We believe that parallel sections and loops with such a fine granularity are frequently met in scientific codes, and the question whether parallelizing these code fragments or not is worth the effort remains often dangling. Autoscheduling is a practical scheme for exploiting this kind of parallelism from scientific applications and our experiments intend to demonstrate this feature. A more thorough study of the codes as well as architecture-specific optimizations that could improve the speedups on our experimental platform are out of the scope of this paper.

Figure 3 illustrates the normalized execution times of the benchmarks when executed on a variable number of processors. Normalization is performed against the sequential execution time excluding the parallelization overhead. The conclusions concerning the relative performance of the three parallelization methods, agree with those derived from Table 1. The partitioned autoscheduled versions exhibit good behaviour even when executed on 32 processors, despite the fact that most benchmarks do not scale up to this point. Compared to loop parallelization and standard autoscheduling, our technique improves speedup by a factor of two and 36% respectively for the complex matrix multiplication. The corresponding numbers for the CFD kernel are 55% and 45%. For the synthetic dot product benchmark, partitioned autoscheduling is the only mechanism that achieves any speedup on 32 processors.

Two benchmarks that exhibit interesting behaviour are the synthetic LU benchmark and the molecular dynamics kernel. In the former, autoscheduling gives only marginal improvements compared to loop parallelization, while in the latter, autoscheduling results in performance degradation. In both cases, partitioning overcomes these limitations. The speedup of the synthetic LU benchmark is almost doubled with partitioning. The molecular dynamics kernel enjoys a noticeable speedup improvement of 19%.

In order to gain a better insight into our results, we performed experiments that measured memory performance under the three parallelization schemes. We used the hardware counters of MIPS R10000 and the **perfex** utility to obtain various performance statistics. Figure 4 presents a summary of the results from these experiments. The charts show the total number of cache misses and remote invalidations for each benchmark when executed on 32 processors. The results indicate clearly that partitioned autoscheduling improves memory performance

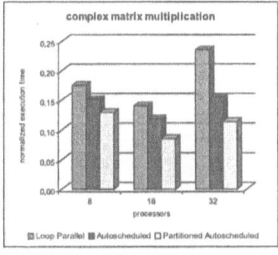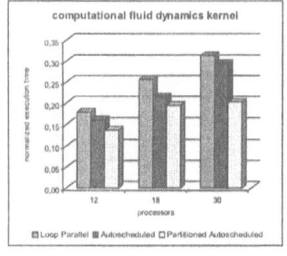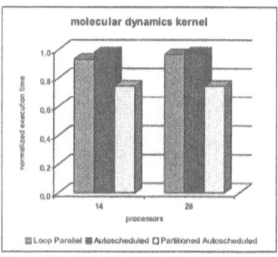

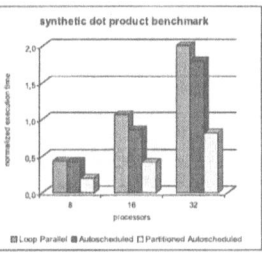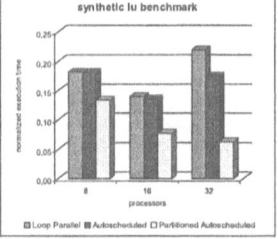

Fig. 3. Normalized execution times of the benchmarks

by reducing cache misses and interconnection network traffic. Cache misses and remote invalidations are on average halved with partitioning and clustering.

6 Conclusions

This paper demonstrated that although autoscheduling exploits potentially more parallelism, it is still amenable to enhancements in order to achieve high performance in DSM multiprocessors. We presented a simple and easy to implement technique which takes advantage of data and communication locality by applying partitioning and clustering of autoscheduled programs. The results prooved that autoscheduling remains the exploiting choice for multilevel parallelism when the mechanisms that extract and schedule the parallelism are aware of the target architecture. Several issues of partitioned autoscheduling need further investigation. Topics for further research might be the automation of the clustering procedure using program analysis in the compiler, and the evaluation of partitioned autoscheduling under multiprogramming conditions. Our current work is oriented towards these directions.

Acknowledgements

We would like to thank Constantine Polychronopoulos for his support and valuable comments, the European Center for Parallelism in Barcelona (CEPBA) for providing us access to their Origin2000, all our partners in the NANOS project and George Tsolis for his help in improving the appearance of the paper.

500

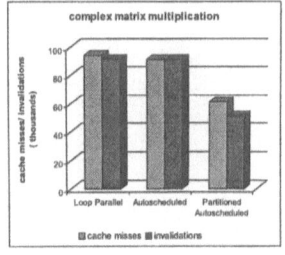
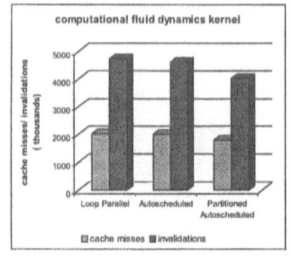
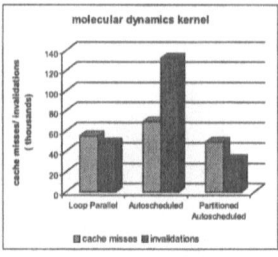

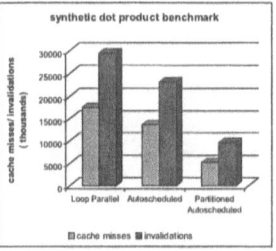
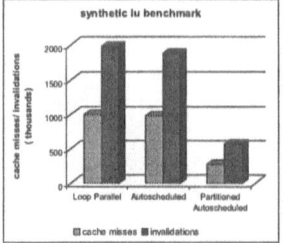

Fig. 4. Memory performance of the benchmarks

References

1. E. Ayguadé, X. Martorell, J. Labarta, M. Gonzàlez, and N. Navarro, *Exploiting Parallelism through Directives on the Nano-Threads Programming Model*, Proceedings of the 10th International Workshop on Languages and Compilers for Parallel Computing, Minneapolis, Minnesota, August 1997.
2. S. Dandamundi, *Reducing Run Queue Contention in Shared Memory Multiprocessors*, IEEE Computer, Vol. 30(3), pp. 82–89, March 1997.
3. R. Eigenmann, J. Hoeflinger and D. Padua, *On the Automatic Parallelization of Perfect Benchmarks*, IEEE Transactions on Parallel and Distributed Systems, Vol. 9(1), pp. 5–23, January 1998.
4. D. Feitelson and L. Rudolph, *Distributed Hierarchical Control for Parallel Processing*, IEEE Computer, Vol. 23(5), pp. 65–79, May 1990.
5. M. Girkar and C. Polychronopoulos, *Automatic Extraction of Functional Parallelism from Ordinary Programs*, IEEE Transactions on Parallel and Distributed Systems, vol. 3(2), pp. 166–178, March 1992.
6. J. Laudon and D. Lenoski, *The SGI Origin: A ccNUMA Highly Scalable Server*, Proceedings of the 24th Annual International Symposium on Computer Architecture, pp. 241–251, Denver, Colorado, June 1997.
7. S. Lyons, T. Hanratty and J. MacLaughlin, *Large-scale Computer Simulation of Fully Developed Channel Flow with Heat Transfer*, International Journal of Numerical Methods for Fluids, vol. 13, pp. 999–1028, 1991.
8. X. Martorell, J. Labarta, N. Navarro and E. Ayguadé, *A Library Implementation of the Nano-Threads Programming Model*, Proceedings of Euro-Par'96, pp. 644–649, Lyon, France, August 1996.
9. J. Moreira, *On the Implementation and Effectiveness of Autoscheduling for Shared-Memory Multiprocessors*, PhD Thesis, University of Illinois at Urbana-Champaign, Department of Electrical and Computer Engineering, 1995.

10. NANOS Consortium, *Nano-Threads Programming Model Specification*, ESPRIT Project No. 21907 (NANOS), Deliverable M1.D1, July 1997, also available at http://www.ac.upc.es/NANOS

11. D. Nikolopoulos, E. Polychronopoulos and T. Papatheodorou, *Efficient Runtime Thread Management for the Nano-Threads Programming Model*, Proceedings of the IPPS/SPDP'98 Workshop on Runtime Systems for Parallel Programming, LNCS vol. 1388, pp. 183–194, Orlando, Florida, March/April 1998.

12. C. Polychronopoulos, M. Girkar, M. Haghighat, C. Lee, B. Leung and D. Schouten, *Parafrase-2: An Environment for Parallelizing, Partitioning, Synchronizing and Scheduling Programs*, International Journal of High Speed Computing, Vol. 1 (1), 1989.

13. C. Polychronopoulos, N. Bitar and S. Kleiman, *Nano-Threads: A User-Level Threads Architecture*, CSRD Technical Report 1297, University of Illinois at Urbana-Champaign, 1993.

14. E. Polychronopoulos, X. Martorell, D. Nikolopoulos, J. Labarta, T. Papatheodorou and N. Navarro, *Kernel-Level Scheduling for the Nano-Threads Programming Model*, Proceedings of the 12th ACM International Conference on Supercomputing, Melbourne, Australia, July 1998.

15. E. Polychronopoulos and T. Papatheodorou, *Dynamic Bisectioning Scheduling for Scalable Shared-Memory Multiprocessors*, Technical Report LHPCA-010697, University of Patras, June 1997.

16. B. Quentrec and C. Brot, *New Method for Searching for Neighbors in Molecular Dynamics Computations*, Journal of Computational Physics, Vol. 13, pp. 430–432, 1975.

Workshop 05+15
Distributed Systems and Databases

Lionel Brunie and Ernst Mayr

Co-chairmen

Distributed Systems

The 15 papers in this workshop have been selected from the submissions to two areas: five from parallel and distributed databases, and another ten from distributed systems and algorithms.

Parallel and distributed database systems are critical for many application domains such as high-performance transaction systems, data warehousing and interoperable information systems. A number of solutions are commercially available, but many open problems remain. Future database systems must support flexible and adaptive approaches for data allocation, load balancing and parallel query processing, both at the DML and at the transaction program level. Distributed databases have to meet new challenges by the Internet and by the integration into workflow management systems. New application areas such as decision support, data mining, text handling imply new requirements with respect to functionality and performance.

Interprocessor communication has become fast enough that distributed systems can be used to solve highly parallel problems in addition to more traditional distributed problems (such as client-server applications). These distributed systems range from a local area network of homogeneous workstations to coordinated heterogenous workstations and supercomputers. Algorithmic and architectural solutions to problems from the fields of distributed and parallel processing (as well as new solutions) can often be applied or adapted to these kinds of systems. Typical examples are implementations of shared memory abstractions on top of message-passing systems, scheduling parallel applications on distributed heterogeneous systems, mechanism and abstractions for fault tolerance, and algorithms to provide elementary system functions and services.

The first session on Distributed Systems deals with networks of processors, connected via various type of networks, like hypercubic networks or tree networks. The talks are concerned with issues of load balancing, with synchronization (mutual exclusion) between neighboring processors, and with fault propagation.

The second session also addresses several areas. The first pair of talks discusses efficient protocols for inexpensive networks of workstations, the following paper presents some refinements of causality relations between non-atomic poset events in a distributed environment, and the last talk presents a new proposal to communicate channels over channels, an idea interesting for instance for mobile computing systems.

In the third session on Distributed Systems, some practical solutions to efficiency problems are presented. The first talk describes an SCI-based implementation of shared-memory and the operating system support for it, while the next paper presents a simple and efficient solution for garbage collection in an environment with task migration. Finally, a proposal based on active messages for improving communication efficiency in local networks of workstations gets discussed.

We would like to express our sincere appreciation to the other members of the Program Committee, Prof. Paul Spirakis, Prof. Friedemann Mattern, and Dr. Marios Mavronicolas, for their invaluable help in the entire selection process. We would also like to thank the numerous reviewers for their time and effort.

Parallel and Distributed Databases

The distributed and parallel databases session is composed of 5 papers. 2 papers concern parallel databases while 3 papers address distributed databases topics.

Though object oriented databases are widely accepted as a very promising paradigm, only a few papers have addressed the parallelization of object oriented queries. The first paper is one of them. Authors propose new algorithms for the paralllel processing of collection intersection join queries, i.e. queries which check for whether there is an intersection between collection attributes (e.g. sets, lists, arrays...). Performance evaluation of these algorithms is very promising.

With the huge development of the Internet, information retrieval (IR) has received a new attention. In this framework, symmetric multiprocessors, by exploiting the parallelism implicit in the queries, allow processing tremendous loads. The paper by Lu et al. investigate how to balance hardware and software resources to exploit a symmetric multiprocessor devoted to IR.

Last papers explore three of the main issues in distributed database management systems : update protocols, query optimization and concurrency control. The paper by Pedone et al. proposes to use atomic broadcast primitives to ensure query termination in replicated databases. Experiments demonstrate that this approach shows a better throughput and response time than traditional atomic commitment methods.

F. Najjar and Y. Slimani give a novel insight on distributed query optimization based on semi-joins. Using dynamic programming techniques, the optimization heuristics they propose allow exploiting all the parallelism inherent to the query.

Finally, Boukerche and al. investigate how to adapt virtual time techniques, classicaly used in distributed systems and VLSI design, to concurrency control problems in distributed databases. They show this new approach outperforms multiversion timestamp techniques as the number of conflicts or transaction size increases.

In summary, this session proposes a very exciting panel of papers which give new insights on some of the hottest issues in parallel and distributed databases.

Collection-Intersect Join Algorithms for Parallel Object-Oriented Database Systems

David Taniar[1] and J. Wenny Rahayu[2]

[1] Monash University - GSCIT, Churchill, Vic 3842, Australia
David.Taniar@fcit.monash.edu.au
[2] La Trobe University, Dept. of Computer Sc. & Comp. Eng., Bundoora, Vic 3083,
Australia
wenny@latcs1.lat.oz.au

Abstract. One of the differences between relational and object-oriented
databases (OODB) is that attributes in OODB can of a collection type
(e.g. sets, lists, arrays, bags) as well as a simple type (e.g. integer, string).
Consequently, explicit join queries in OODB may be based on collection
attributes. One form of collection join queries in OODB is collection-
intersect join queries, where the joins are based on collection attributes
and the queries check for whether there is an intersection between the
two join collection attributes We propose two algorithms for parallel processing of collection-intersect join queries. The first one is based on sort-
merge, and the second is based on hash. We also present two data partitioning methods (i.e. simple replication and "divide and partial broadcast") used in conjunction with the parallel collection-intersect join algorithms. The parallel sort-merge algorithm can only make use of the
divide and partial broadcast data partitioning, whereas the parallel hash
algorithm may have a choice which of the two data partitioning to use.

1 Introduction

In *Object-Oriented Databases* (OODB), although path expression between classes
may exist, it is sometimes necessary to perform an explicit join between two or
more classes due to the absence of pointer connections or the need for value
matching between objects. Furthermore, since objects are not in a normal form,
an attribute of a class may have a collection as a domain. Collection attributes
are often mistakenly considered merely as set-valued attributes. As the matter
of fact, *set* is just one type of collections. There are other types of collection.
The Object Database Standard *ODMG* [1] defines different kinds of collections:
particularly *set*, *list/array*, and *bag*. Consequently, object-oriented join queries
may also be based on attributes of any collection type. Such join queries are
called *collection join queries* [9]. Our previous work reported in [9] classified three
different types of collection join queries, namely: *collection-equi join*, *collection-
intersect join*, and *sub-collection join*. In this paper, we would like to focus on
collection-intersect join queries. We are particularly interested in formulating
parallel algorithms for processing such queries. The algorithms are non-trivial to

parallel object-oriented database systems, since most conventional join algorithms (e.g. hybrid hash join, sort-merge join) deal with single-valued attributes and hence most of the time they are not suitable to handle collection join queries.

Collection-intersect join queries are queries that join two classes based on an attribute of a collection type. The join predicates check for whether there is an intersection between the two collection join attributes. An *intersect* predicate can be written by applying an intersection between the two sets and comparing the intersection result with an empty set. It is normally in a form of (attr1 intersect attr2) != set(nil). Attributes attr1 and attr2 are of type set. If one or both of them are of type *bag*, they must be converted to sets. Suppose the attribute *editor-in-chief* of class *Journal* and the attribute *program-chair* of class *Proceedings* are of type sets of *Person*. An example of a collection-intersect join is to retrieve pairs of Journal and Proceedings, where the program-chairs of a conference are intersect with the editors-in-chief of a journal. The query expressed in OQL (Object Query Language) [1] can be written as follows:

```
Select A, B
From A in Journal, B in Proceedings
Where (A.editor-in-chief intersect B.program-chair) != set(nil)
```

As clearly seen that the intersection join predicates involve the creation of intermediate results through an *intersect* operator. The result of the join predicate cannot be determined without the presence of the intermediate collection result. This predicate processing is certainly not efficient. In a collection-intersect join query, the original subset predicate has to produce an intermediate set, before it can be compared with an empty set. This process checks for the smaller set twice: one for an intersection, the other for an equality comparison. Therefore, optimization algorithms for efficient processing of such queries are critical if one wants to improve query processing performance. In this paper, we present two parallel join algorithms for collection-intersection join queries. The primary intention is to solve the inefficiency imposed by the original join predicates.

An interest in *parallel OODB* among database community has been growing rapidly, following the popularity of multiprocessor servers and the maturity of OODB. The emerging between parallel technology and OODB has shown promising results [3], [5], [7], [11]. However, most research done in this area concentrated on path expression queries with pointer chasing. Explicit join processing exploiting collection attributes has not been given much attention.

The rest of this paper is organized as follows. Section 2 explains two data partitioning methods for parallel collection-intersect join algorithms. Section 3 describes a proposed parallel join algorithm for collection-intersect join queries based on a sort-merge technique. Section 4 introduces another algorithm, which is based on a hash technique. Finally, section 5 draws the conclusions and explains the future work.

2 Data Partitioning

Parallel join algorithms are normally decomposed into two steps: *data partitioning* and *local join*. Data partitioning creates parallelism, as it divides the data to multiple processors, so that the join can then be performed locally in each processor without interfering others. For collection-intersect join queries, it is not possible to have non-overlap partitions, due to the nature of collections which may be overlapped. Hence, some data needs to be replicated. Two non-disjoint partitioning methods are proposed. The first is a simple replication based on the value of the element in each collection. The second is a variant of Divide and Broadcast [4], called *"Divide and Partial Broadcast"*.

2.1 Simple Replication

Using a simple replication technique, each element in a collection is treated as a single unit, and is totally independent of other elements within the same collection. Based on the value of an element in a collection, the object is placed into a particular processor. Depending on the number of elements in a collection, the objects that own the collections may be placed into different processors. When an object is already placed at a particular processor based on the placement of an element, if another element in the same collection is also to be placed at the same place, no object replication is necessary.

As a running example, consider the data shown in Figure 1. Suppose class A and class B are *Journal* and *Proceedings*, respectively. Both classes contain a few objects shown by their OIDs (e.g., objects a to i are Journal objects and objects p to w are Proceedings objects). The join attributes are *editor-in-chief* of Journal and *program-chair* of Proceedings; and are of type collection of *Person*. The OID of each person in these attributes are shown in the brackets. For example $a(250,75)$ denotes a Journal object with OID a and the editors of this journal are Persons with OIDs 250 and 75.

Figure 2 shows an example of a simple replication technique. The **bold** printed elements are the elements which are the basis for the placement of those objects. For example, object $a(250, \mathbf{75})$ in processor 1 refers to a placement for object a in processor 1 because of the value of element 75 in the collection. And also, object $a(\mathbf{250}, 75)$ in processor 3 refers to a copy of object a in processor 3 based on the first element (i.e., element 250). It is clear that object a is replicated to processors 1 and 3. On the other hand, object $i(80, 70)$ is not replicated since both elements will place the object at the same processor, that is processor 1.

2.2 Divide and Partial Broadcast

The Divide and Partial Broadcast algorithm, shown in Figure 3, proceeds in two steps. The first step is a *divide* step, and the second step is a *partial broadcast* step. We divide class B and partial broadcast class A. The *divide* step is explained as follows. Divide class B into n number of partitions. Each partition of class B is placed in a separate processor (e.g. partition $B1$ to processor 1, partition

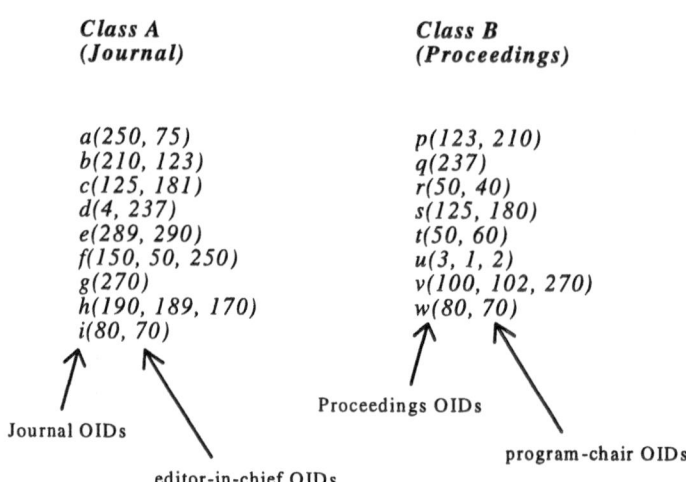

Fig. 1. Sample data

Class A	Class B	
a(250, **75**) d(**4**, 237) f(150, **50**, 250) i(**80**, 70)	r(**50**, 40) t(**50**, 60) u(**3**, 1, 2) w(**80**, 70)	Processor 1 (range 0-99)
b(210, **123**) c(**125**, 181) f(**150**, 50, 250) h(**190**, 189, 170)	p(**123**, 210) s(**125**, 180) v(**100**, 102, 270)	Processor 2 (range 100-199)
a(**250**, 75) b(**210**, 123) d(4, **237**) e(289, **290**) f(150, 50, **250**) g(**270**)	p(123, **210**) q(**237**)	Processor 3 (range 200-299)

Fig. 2. Simple replication

$B2$ to processor 2, etc). Partitions are created based on the largest element of each collection. For example, object $p(123, 210)$; the first object in class B, is partitioned based on element 210, as element 210 is the largest element in the collection. Then, object p is placed on a certain partition, depending on the partition range. For example, if the first partition is ranging from the largest element 0 to 99, the second partition is ranging from 100 to 199, and the third partition is ranging from 200 to 299, then object p is placed in partition $B3$, and subsequently in processor 3. This is repeated for all objects of class B.

```
Procedure DividePartialBroadcast
Begin
  Step 1 (divide):
    1. Divide class B based on the largest element in each collection.
    2. For each partition of B (i = 1, 2, ..., n)
         Place partition Bi to processor i
       End For

  Step 2 (partial broadcast):
    3. Divide class A based on the smallest element in each collection.
    4. For each partition of A (i = 1, 2, ..., n)
         Broadcast partition Ai to processor i to n
       End For
End Procedure
```

Fig. 3. Divide and Partial Broadcast Algorithm

The *partial broadcast* step can be described as follows. First, partition class A based on the smallest element of each collection. Clearly, this partitioning method is exactly the opposite of that in the divide step. Then for each partition Ai where $i=1$ to n, broadcast partition Ai to processors i to n. This broadcasting technique is said to be partial, since the broadcasting goes down as the partition number goes up. For example, partition $A1$ is basically replicated to all processors, partition $A2$ is broadcast to processor 2 to n only, and so on. In regard to the load of each processor, the load of the last processor may be the heaviest, as it receives a full copy of class A and a portion of class B. The load goes down as class A is divided into smaller size (e.g., processor 1). Load balanced can be achieved by applying the same algorithm to each partition but with a reverse role of A and B; that is, *divide A* and *partial broadcast B*. It is beyond the scope of this paper to evaluate the Divide and Partial Broadcast partitioning method. This has been reserved for future work. Some preliminary results have been reported in [10].

3 Parallel SORT-MERGE Join Algorithm

The partitioning strategy for the parallel sort-merge collection-intersect join query algorithm is based on the *Divide and Partial Broadcast* technique. The use of the Divide and Partial Broadcast is attractive to collection joins because of the nature of collections where disjoint partitions without replication are often not achievable. After data partitioning is completed, each processor has its own data. The join operation can then be done independently. The overall query results are the union of the results from each processor.

In the local joining process, each collection is sorted. Sorting is done within collections, not among collections. After the sorting process is completed, the merging process starts. We use a nested loop structure to compare the two collections from the two operand objects. Figure 4 shows a parallel sort-merge join algorithm for collection-intersect join queries.

```
Program Parallel-Sort-Merge-Collection-Intersect-Join
Begin
  Step 1 (data partitioning):
    Call DividePartialBroadcast

  Step 2 (local joining): In each processor
    a. Sort phase
       For each object a(c1) and b(c2) of class A and B, respectively
         Sort collection c1 and c2
       End For
    b. Merge phase
       For each object a(c1) of class A
         For each object b(c2) of class B
           Merge collection c1 and c2
           If TRUE Then
             Concatenate objects a and b into query result
           End If
         End For
       End For
End Program
```

Fig. 4. Parallel Sort-Merge Collection-Intersect Join Algorithm

4 Parallel HASH Join Algorithm

In this section we introduce a parallel join algorithm based on a *hash* method for collection-intersect join queries. Like the previous Parallel Sort-Merge algorithm, Parallel-Hash algorithm is also divided into data partitioning and local join phases. However, unlike the parallel sort-merge algorithm, data partitioning for parallel hash algorithm is available in two forms: *Divide and Partial Broadcast* and *Simple Replication*. Once data partitioning is complete, each processor has its own data, and hence local join process can proceed.

The local join process itself is divided into two steps: hash and probe. The hashing is carried out to one class, whereas the probing is performed to the other class. In the hashing part, it basically runs through all elements of each collection in a class. The probing part is done in similar way, but is applied to the other class. Figure 5 shows the pseudo-code for parallel hash join algorithm for collection-intersect join queries.

```
Program Parallel-Hash-Collection-Intersect-Join
Begin
   Step 1 (data partitioning):
     Divide and Partial Broadcast version:
       Call DivideAndPartialBroadcast partitioning
     Simple Replication version:
       Call SimpleReplication partitioning

   Step 2 (local joining): In each processor
     a. Hash
        For each object a(c1) of class A
          Hash collection c1 to a hash table
        End For
     b. Probe
        For each object b(c2) of class B
          Hash and probe collection c2 into the hash table
          If there is any match Then
            Concatenate obj b and the matched obj a into query result
          End If
        End For
End Program
```

Fig. 5. Parallel Hash Collection-Intersect Join Algorithm

5   Conclusions and Future Work

The need for join algorithms especially designed for collection-intersect join queries is clear, as collection-intersect join predicates normally require intermediate results to be generated, before the final predicate results can be determined. This is certainly not optimal. In this paper, we present two algorithms especially design for collection-intersect join queries, namely *Parallel Sort-Merge* and *Parallel Hash* algorithms. These two algorithms are designed especially for collection-intersect join queries in object-oriented databases. Data partitioning methods, which create parallelism, are based on either simple replication or "divide and partial broadcast". Parallel Sort-Merge can make use only the divide and partial broadcast method, whereas Parallel Hash can have a choice between the two partitioning methods. Once a data partitioning method is applied, local join is carried out by either a sort-merge operator (in the case of parallel sort-merge algorithm) or a hash function (in the case of parallel hash). Local join process is therefore much straightforward, as each processor performs sequential sort-merge or hash operations. Hence, the critical element is the data partitioning method.

Our future work includes evaluating the two data partitioning methods (and possibly other partitioning methods) for parallel collection-intersect join algorithms, since data partitioning plays an important role in the overall efficiency of parallel collection join algorithms.

References

1. Cattell, R.G.G. (ed.), The Object Database Standard: ODMG-93, Release 1.1, Morgan Kaufmann, 1994.
2. DeWitt, D.J. and Gray, J., "Parallel Database Systems: The Future of High Performance Database Systems", Communication of the ACM, **35**(6), pp. 85-98, 1992.
3. Kim, K-C., "Parallelism in Object-Oriented Query Processing", Proceedings of the Sixth International Conference on Data Engineering, pp. 209-217, 1990.
4. Leung, C.H.C. and Ghogomu, H.T., "A High-Performance Parallel Database Architecture", Proceedings of the Seventh ACM International Conference on Supercomputing, pp. 377-386, Tokyo, 1993.
5. Leung, C.H.C., and Taniar, D., "Parallel Query Processing in Object-Oriented Database Systems", Australian Computer Science Communications, **17**(2), pp. 119-131, 1995.
6. Mishra, P. and Eich, M.H., "Join Processing in Relational Databases", ACM Computing Surveys, **24**(1), pp. 63-113, March 1992.
7. Taniar, D., and Rahayu, W., "Parallelization and Object-Orientation: A Database Processing Point of View", Proceedings of the Twenty-Fourth International Conference on Technology of Object-Oriented Languages and Systems TOOLS ASIA'97, Beijing, China, pp. 301-310, 1997.
8. Taniar, D. and Rahayu, J.W., "Parallel Collection-Equi Join Algorithms for Object-Oriented Databases", Proceedings of International Database Engineering and Applications Symposium IDEAS'98, IEEE Computer Society Press, Cardiff, UK, July 1998.
9. Taniar, D. and Rahayu, J.W., "A Taxonomy for Object-Oriented Queries", a book chapter in Current Trends in Database Technology, Idea Group Publishing, in press (1998).
10. Taniar, D. and Rahayu, J.W., "Divide and Partial Broadcast Method for Parallel Collection Join Queries", High-Performance Computing and Networking, P.Sloot et al (eds.), Lecture Notes in Computer Science 1401, Springer-Verlag, pp. 937-939, 1998.
11. Thakore, A.K. and Su, S.Y.W., "Performance Analysis of Parallel Object-Oriented Query Processing Algorithms", Distributed and Parallel Databases **2**, pp. 59-100, 1994.

Exploiting Atomic Broadcast
in Replicated Databases*

Fernando Pedone**, Rachid Guerraoui, and André Schiper

Département d'Informatique
Ecole Polytechnique Fédérale de Lausanne
1015 Lausanne, Switzerland

Abstract. Database replication protocols have historically been built on top of distributed database systems, and have consequently been designed and implemented using distributed transactional mechanisms, such as atomic commitment. We argue in this paper that this approach is not always adequate to efficiently support database replication and that more suitable alternatives, such as atomic broadcast primitives, should be employed instead. More precisely, we show in this paper that fully replicated database systems, based on the deferred update replication model, have better throughput and response time if implemented with an atomic broadcast termination protocol than if implemented with atomic commitment.

1 Introduction

In the database context, replication techniques based on the *deferred update model* have received increasingly attention in the past years [6]. According to the deferred update model, transactions are processed locally at one server (i.e., one replica manager) and, at commit time, are forwarded for certification to the other servers (i.e., the other replica managers). Deferred update replication techniques offer many advantages over *immediate update* techniques, which synchronise every transaction operation across all servers. Among these advantages, one may cite: (a) better performance, by gathering and propagating multiple updates together, and localising the execution at a single, possibly nearby, server (thus reducing the number of messages in the network), (b) more flexibility, by propagating the updates at a convenient time (e.g., during the next dial-up connection), (c) better support for fault tolerance, by simplifying server recovery (i.e., missing updates may be demanded to other servers), and (d) lower deadlock rate, by eliminating distributed deadlocks [6].

Nevertheless, deferred update replication techniques have two limitations. Firstly, the termination protocol used to propagate the transactions to other servers for certification is usually an atomic commitment protocol (e.g., a 2PC

* Research supported by the EPFL-ETHZ DRAGON project and OFES under contract number 95.0830, as part of the ESPRIT BROADCAST-WG (number 22455).
** On leave from Colégio Técnico Industrial, University of Rio Grande, Brazil

algorithm), whose cost directly impacts transaction response time. Secondly, the certification procedure that is usually performed at transaction termination time consists in aborting all conflicting transactions.[1] Such certification procedure typically leads to a high abort rate if conflicts are frequent, and hence impacts transaction throughput.

In the context of client-server distributed systems, most replication schemes (that guarantee strong replica consistency) are based on *atomic broadcast* communication [3, 9]. An *atomic broadcast* communication primitive enables to send messages to a set of processes, with the guarantee that the processes agree on the *set* of messages delivered, and on the *order* according to which the messages are delivered [7]. With this guarantee, consistency is trivially ensured if every operation on a replicated server is distributed to all replicas using atomic broadcast. Although several authors have mentioned the possibility of using atomic broadcast to support replication schemes in a database context (e.g., [3, 11]), little work has been done in that direction.

In this paper, we show that atomic broadcast can successfully be used to improve the performance of the database deferred update replication technique. In particular, we show that, for different resilience scenarios, the deferred update replication technique based on atomic broadcast provides better transaction throughput and response time than a similar scheme based on atomic commitment.

The paper is organised as follows. In Section 2, we present the replicated database model, and we recall the principle of the deferred update replication technique. In Section 3, we describe a variation of this replicated technique based on atomic broadcast. In Section 4, we compare the transaction throughput and response time of deferred update replication based on atomic broadcast with the throughput and response time of deferred update replication based on classical atomic commitment. Section 5 summarises the main contributions of the paper and discusses some research perspectives.

2 The Deferred Update Technique

2.1 Replicated Database Model

We consider a replicated database system composed of a set of processes $\Sigma = \{p_1, p_2, \ldots, p_n\}$, each one executing in a different processor, without shared memory and access to a common clock. Each process has a replica of the database and plays the role of a replica manager. Processes may fail by crashing, and can recover after a crash. We assume that processes do not behave maliciously (i.e., we exclude Byzantine failures). If a process p is able to execute requests at a certain time τ (i.e., p did not fail or p has failed but recovered) we say that the

---

[1] Note that we do not consider here replication protocols that allow replica divergence and require reconciliations. We focus on replication schemes that guarantee *one-copy serializability* [2].

process p is *up* at time τ. Otherwise the process p is said to be *down* at time τ. We say that a process p is correct if there is a time after which p is forever up.[2]

Processes execute transactions, that are sequences of read and write operations followed by a commit or abort operation. Transactions are submitted by client processes executing in any processor (with or without a replica of the database). Our correctness criterion for transaction execution is one-copy serializability (1SR) [2].

Committed transactions are represented as T_i, T_j and T_k. The set Π_p^τ contains all committed transactions in process p at time τ. Non-committed transactions are represented as t_m and t_n. When a transaction t_m commits, it is represented as T_i, where i indicates the serial order of t_m, related to the already committed transactions in the database.

2.2 The Deferred Update Principle

In the deferred update technique, a transaction passes through some well-defined states. It starts in the *executing* state, when its read and write operations are locally executed by the process where it initiated. When the transaction requests the commit, it passes to the *committing* state and is sent to the other processes. When received by a process, the transaction is also in the committing state. A transaction in a process remains in the committing state until its fate is known by the process (i.e., *commit* or *abort*). The executing and committing states are transitory states, whereas the commit and abort states are final states.

To summarise the principle of the deferred update replication technique, we consider a particular transaction t_m executing in some process p_i. Hereafter, the *readset (RS)* and *writeset (WS)* are the sets of data items that a transaction reads and writes, respectively, during its execution.

1. Transaction t_m is initiated and executed at process p_i. Read and write operations request the appropriate locks, that are locally granted according to the strict two phase locking rule. During this phase, transaction t_m is in the executing state.

2. When transaction t_m requests the commit, t_m passes to the committing state. Its read locks are released (the write locks are maintained by p_i until t_m reaches a final state), and process p_i then triggers the *termination protocol* for t_m: the updates performed by t_m (e.g., the redo log records), as well as its readset and writeset, are sent to all the other processes.

3. As part of the termination protocol, t_m is certified by every process. The certification test guarantees one-copy serializability. The outcome of this test is the commit or abort of t_m, according to concurrent transactions that

---

[2] The notion of *forever up* is a theoretical assumption to guarantee that correct processes do useful computation. This assumption prevents cases where processes fail and recover successively without being up enough time to make the system evolve. *Forever up* means, for example, from the beginning until the end of a termination protocol.

executed in other processes. We will come back to this issue in Sections 3 and 4.

4. If t_m passes the certification test, all its write locks are requested and its updates are applied to the database. Hereafter, transaction t_m is represented as T_i. Transactions in the execution state whose locks conflict with T_i's write locks are aborted for T_i's sake.

During the time when a process p is down, p may miss some transactions by not participating in their termination protocol. However, as soon as process p is up again, p catches up with another process that has seen all transactions in the system. This recovery procedure depends on the implementation of the termination protocol (see [10] for a detailed discussion).

3 A Replication Scheme Based on Atomic Broadcast

3.1 Atomic Broadcast

An atomic broadcast primitive enables to send messages to a set of processes, with the guarantee that all processes agree on the *set* of messages delivered and the *order* according to which the messages are delivered [7]. More precisely, atomic broadcast ensures that (i) if some process delivers message m then every correct process delivers m *(uniform atomicity)*; (ii) no two processes deliver any two messages in different orders *(order)*; and (iii) if a process broadcasts message m and does not fail, then every correct process eventually delivers m *(termination)*.

It is important to notice that the properties of atomic broadcast are defined in terms of message *delivery* and not in terms of message *reception*. Typically, a process first receives a message, then performs some computation to guarantee the atomic broadcast properties, and then delivers the message. The notion of delivery captures the concept of irrevocability (i.e., a process must not forget that it has delivered a message). In the following, we use the expression *delivering a transaction t_m* to mean *delivering the message that contains transaction t_m*.

3.2 A Termination Protocol Based on Atomic Broadcast

We describe now the termination protocol for the deferred update replication technique based on atomic broadcast. On delivering a committing transaction, each process certifies the transaction and, in case of success, commits it. Once the transaction is delivered and certified successfully, it passes to the commit state and its writes are processed.

In order for a process to certify a committing transaction t_m, it has to know which transactions conflict with t_m. The notion of conflict is defined by the *precedes* relation between transactions and the operations issued by the transactions. A transaction T_j *precedes* another transaction t_m, denoted $T_j \rightarrow t_m$, if T_j committed before t_m started its execution. The relation $T_j \nrightarrow t_m$ (not $T_j \rightarrow t_m$)

means that T_j committed after t_m started its execution. Based on these definitions, we say that a transaction T_j conflicts with t_m if (1) $T_j \not\to t_m$ and (2) t_m and T_j have conflicting operations.[3]

More precisely, process p_j performs the following steps after delivering t_m.

1. *Certification test.* Process p_j aborts t_m if there is any committed transaction T_j that conflicts with t_m and that updated data items read by t_m.
2. *Commitment.* If t_m is not aborted, it passes to the commit state (hereafter t_m will be expressed as the committed transaction T_i, where i represents the sequential order of transaction t_m, and $\Pi_{p_j}^{\tau'} = \Pi_{p_j}^{\tau} \cup \{T_i\}$); process p_j tries to grant all T_i's write locks, and processes its updates. There are two cases to consider.
   (a) *There is a transaction t_n in execution at p_j whose read or write locks conflict with T_i's write locks.* In this case t_n is aborted by p_j and, therefore, all t_n's read and write locks are released.
   (b) *There is a transaction t_n, that is executed locally at p_j and requested the commit, but is delivered after T_i.* If t_n executed locally at p_j, it has all write locks on the data items it updated. If t_n commits, its writes will supersede T_i's (i.e., the ones that overlap) and, in this case, T_i need neither request these write locks nor process the updates over the database. This is similar to Thomas' Write Rule [12]. If t_n is later aborted (i.e., it does not pass the certification test), the database should be restored to a state without t_n, for instance, by applying T_i's redo log entries to the database.

Given a committing transaction t_m, the set of transactions $\{T_j \mid T_j \not\to t_m\}$ (see certification test above) is determined as follows. Given a process p, let $last_p(\tau)$ represent the last delivered and committed transaction in p at local time τ. A transaction t_m that starts its execution in process p at time τ_0 has to be checked at certification time τ_p against the transactions $\{T_{last_p(\tau_0)+1}, \ldots, T_{last_p(\tau_p)}\}$. The information $last_p(\tau_0)$ is broadcast together with the writes of T_i, and $last_p(\tau_p)$ is determined by each process p at certification time (see [10] for more details).

4 Atomic Broadcast *vs.* Atomic Commit

In this section we describe the deferred update technique based on atomic commit and compare it with the deferred update technique based on atomic broadcast using two criteria: transaction throughput and transaction response time.

4.1 The Termination Protocol Based on Atomic Commit

With an atomic commit implementation of the deferred update termination protocol, no total order knowledge is available for the processes and a conflict has to be resolved by aborting all conflicting transactions. We describe below the termination protocol based on atomic commit.

---

[3] Two operations conflict if they are issued by different transactions, access the same data item and at least one of them is a write.

1. *Certification test.* A process can make a decision on the commit or abort of a transaction t_m when it knows that all the other processes also know about t_m. This requires an interaction among processes that ensures that when a transaction is certified, all committing transaction are taken into account. On certifying transaction t_m, process p_j decides for its abort if there is a transaction t_n that does not precede t_m ($t_n \not\rightarrow t_m$) and with which t_m has conflicting operations.
2. *Commitment.* If t_m is not aborted, it passes to the commit state and p_j tries to grant all its write locks and processes its updates. Transactions that are being locally executed in p_j and have read or write locks on data items updated by t_m are aborted.

In order to certify and commit transactions, the system has to detect transactions that executed concurrently in different processes. For example, in [1] this is done associating *vector clocks* timestamps with local events (e.g., execution or commit of a transaction). These timestamps have the property of being ordered if the events are causally related. Each time a process communicates to another, it sends its vector clock.

4.2 Transaction Throughput and Response Time

The values presented in Figure 1 were obtained by means of a simulation involving a database with 10000 data items replicated in 5 database processes. Each transaction has a readset and a writeset with 10 different data items and each data item has the same probability of being accessed. These experiments consider concurrent committing transactions (i.e., transactions that have already requested the commit). The results show that total order can indeed produce better results.

In the following we compare implementations of atomic broadcast to implementations of atomic commitment. Due to lack of space, we do not describe how atomic broadcast can be implemented (see [10] for a detailed discussion). The measures shown in Figure 2 were obtained with Sparc 20 workstations (running Solaris 2.3), an Ethernet network using the TCP protocol, transactions (messages and logs) of 1024 bytes, and failure free runs (the most frequent ones in practice). The measures convey an average time to deliver a transaction message. These results show that atomic broadcast protocols have a better performance than atomic commit, when compared under the same degree of resilience.

5   Concluding Remarks

In a recent paper, Cheriton and Skeen expressed their scepticism about the adequateness of atomic broadcast to support replicated databases [4]. The reasons that were raised were (1) atomic broadcast replication schemes consider individual operations, whereas in databases, operations are gathered inside transactions, (2) atomic broadcast does not guarantee uniform atomicity (e.g., a server

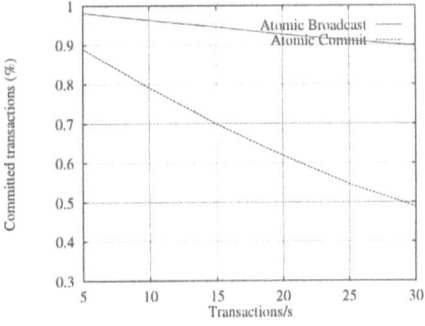

Fig. 1. Throughput

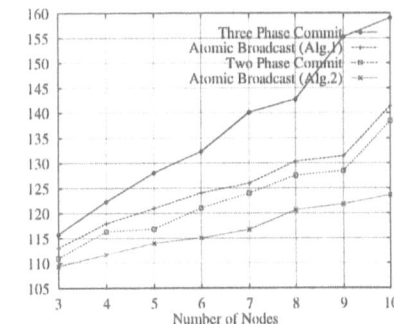

Fig. 2. Response time

could deliver a message and crash, with the other servers not even receiving it), whereas uniform atomicity is fundamental in distributed databases (all processes, even those that have later crashed, must agree to commit or not a transaction), and (3) atomic broadcast is usually considered in a crash-stop process model (i.e., a process that crashes never recovers), whereas in databases processes are supposed to recover after a crash. In this paper, we have considered an atomic broadcast primitive in a crash-recovery model that guarantees uniformity, i.e., if one process delivers a transaction, all the others also deliver it. We have shown how that primitive can efficiently be used to propagate transaction control information in a deferred update model of replication. The choice for that replication model was not casual, as some recent research has shown that immediate update models are inviable due to the nature of the applications or the side effects of synchronisation [6].

Indirectly, this paper points out the fact that existing database replication protocols are built on top of distributed database systems, using mechanisms that were developed to deal with distributed information, but not necessarily designed with replication in mind. A flagrant example of this is the atomic commitment mechanism, which we have shown can be favourably replaced by an atomic broadcast primitive, providing better throughput and response time. Our performance figures help dis-mystify a common (mis)belief that total order atomic broadcast primitives are too expensive, when compared to traditional transactional primitives, and so inappropriate for high performance systems. As already stated in [13], we also believe that replication can be successfully applied for performance, and this can be achieved without sacrificing consistency (e.g., as in [5]) or making semantic assumptions about the transactions (e.g., as in [6, 8]).

It is important however to notice that the replication scheme based on atomic broadcast presented in this paper is based on two important assumptions: (1) processes certify and commit transactions sequentially, and (2) the database is fully replicated. The first assumption can be bypassed since concurrency between transactions that come from the *Certifier* is allowed if the transactions have disjoint write sets. The second assumption about a fully replicated database is reasonable in traditional closed information systems, but is inappropriate for an open system with a large number of nodes or a large database. In such systems, replication can only be partial (i.e., processes store only subsets of the database), and the extend to which atomic broadcast can be useful in this context is an open issue.

References

1. D. Agrawal, A. El Abbadi, and R. Steinke. Epidemic algorithms in replicated databases. In *Proceedings of the Sixteenth ACM SIGACT-SIGMOD-SIGART Symposium on Principles of Database Systems*, Tucson, Arizona, 12–15 May 1997.
2. P. Bernstein, V. Hadzilacos, and N. Goodman. *Concurrency Control and Recovery in Database Systems*. Addison-Wesley, 1987.
3. K. P. Birman and R. van Renesse. *Reliable Distributed Computing with the ISIS Toolkit*. IEEE Press, 1994.
4. D. Cheriton and D. Skeen. Understanding the Limitations of Causally and Totally Ordered Communication. In *Proceedings of the 14th ACM Symposium on Operating Systems Principles*, Asheville, North Carolina, December 1993.
5. D. J. Delmolino. Strategies and techniques for using Oracle 7 replication. Technical report, Oracle Corporation, 1995.
6. J. N. Gray, P. Helland, P. O'Neil, and D. Shasha. The dangers of replication and a solution. In *Proceedings of the 1996 ACM SIGMOD International Conference on Management of Data*, Montreal, Canada, June 1996.
7. V. Hadzilacos and S. Toueg. *Distributed Systems, 2ed*, chapter 3, Fault-Tolerant Broadcasts and Related Problems. Addison Wesley, 1993.
8. H. V. Jagadish, I. S. Mumick, and M. Rabinovich. Scalable versioning in distributed databases with commuting updates. In *Proceedings of the Thirteenth International Conference on Data Engineering, April 7-11, 1997 Birmingham U.K.*, pages 520–531. IEEE Computer Society Press, April 1997.
9. M. F. Kaashoek and A. S. Tanenbaum. Group communication in the amoeba distributed operating system. In *11th International Conference on Distributed Computing Systems*, pages 222–230, Washington, D.C., USA, May 1991. IEEE Computer Society Press.
10. F. Pedone, R. Guerraoui, and A. Schiper. Exploiting atomic broadcast in replicated databases. Technical Report No 98/258, Swiss Federal Institute of Technology, Lausanne, Switzerland, 1998. Available at http://lsewww.epfl.ch/pedone.
11. A. Schiper and M. Raynal. From group communication to transaction in distributed systems. *Communications of the ACM*, 39(4):84–87, April 1996.
12. R. H. Thomas. A majority consensus approach to concurrency control for multiple copy databases. *ACM Trans. on Database Systems*, 4(2):180–209, June 1979.
13. P. Triantafillou. High availability is not enough. In J.-F. Paris and H. G. Molina, editors, *Proceedings of the Second Workshop on the Management of Replicated Data*, pages 40–43, Monterey, California, November 1992. IEEE Computer Society Press.

The Hardware/Software Balancing Act for Information Retrieval on Symmetric Multiprocessors

Zhihong Lu Kathryn S. McKinley Brendon Cahoon

Department of Computer Science
University of Massachusetts
Amherst, MA 01003
{zlu, mckinley, cahoon}@cs.umass.edu

Abstract. Web search engines, such as AltaVista and Infoseek, handle tremendous loads by exploiting the parallelism implicit in their tasks and using symmetric multiprocessors to support their services. The web searching problem that they solve is a special case of the more general *information retrieval* (IR) problem of locating documents relevant to the information need of users. In this paper, we investigate how to exploit a symmetric multiprocessor to build high performance IR servers. Although the problem can be solved by throwing lots of CPU and disk resources at it, the important questions are *how much* of *which* hardware and *what* software structure is needed to effectively exploit hardware resources. We have found, to our surprise, that in some cases adding hardware *degrades* performance rather than improves it. We show that multiple threads are needed to fully utilize hardware resources. Our investigation is based on InQuery, a state-of-the-art full-text information retrieval engine.

1 Introduction

As information explodes across the Web and elsewhere, people increasingly depend on search engines to help them to find information. Web searching is a special case of the more general *information retrieval* (IR) problem of locating documents relevant to the information need of users. In this paper, we investigate how to balance hardware and software resources to exploit a symmetric multiprocessor (SMP) architecture to build high performance IR servers. Our IR server is based on InQuery [2, 3], a state-of-the-art full-text information retrieval engine that is widely used in Web search engines, large libraries, companies, and governments such as Infoseek, Library of Congress, White House, West Publishing, and Lotus [5]. Our work is novel because it investigates a real, proven effective system under a variety of realistic workloads and hardware configurations on an SMP architecture. The previous research investigates either the IR system on massively parallel processing (MPP) architecture or it investigates only a subset of the system on SMP architecture such as the disk system or it compares the cost factors of SMP architecture with other architectures. (See [6] for a more thorough comparison with the related work). Our results provide insights for building high performance IR servers for searching the Web and other environments using a symmetric multiprocessor.

The remainder of this paper is organized as follows. The next section describes the implementation of our parallel IR server and simulator. Section 3 presents results that demonstrate the system scalability and hardware/software balancing with respect to multiple threads, CPUs, and disks. Section 4 summarizes our results and concludes.

2 A Parallel Information Retrieval Server

This section describes the implementation of our parallel IR server and simulator. We begin with a brief description of the InQuery retrieval engine [2, 3, 5]. We next present the features we model and summarize our validation of the simulator against the multithreaded implementation.

InQuery Retrieval Engine

InQuery is one of the most powerful and advanced full-text information retrieval engines in commercial or government use today [5]. It uses an inference network model, which applies Bayesian inference networks to represent documents and queries, and views information retrieval as an inference or evidential reasoning process [2, 3]. The inference networks are implemented as inverted files. In this paper, we use "collection" to refer to a set of documents, and "database" to refer to an indexed collection.

The InQuery server supports a range of IR commands such as query, document, and relevance feedback. The three basic IR commands we model are: *query, summary,* and *document.* A *query* command requests documents that match a set of terms. A query response consists of a list of top ranked document identifiers. A *summary* command consists of a set of document identifiers. A summary response includes the document titles and the first few sentences of the documents. A *document* command requests a document using its document identifier. The response includes the complete text of the document.

The Parallel IR Server

To investigate the balance between hardware and software in a IR system on a symmetric multiprocessor, we implemented a parallel IR server using InQuery as the retrieval engine. The parallel IR server exploits parallelism as follows: (1) It executes multiple IR commands in parallel by multithreading; and (2) It executes one command against multiple partitions of a collection in parallel. To expedite our investigation of possible system configurations, characteristics of IR collections, and the basic IR system performance, we implement a simulator with numerous system parameters, such as the number of CPUs, threads, disks, collection size, and query characteristics. Table 1 presents all the parameters and the values we use in our experiments.

The simulation model is driven by empirical timing measurements from the actual system. For queries, summaries, and documents, we measure CPU and

Table 1. Experimental Parameters

Parameters	Abbreviation	Values				
Query Number	QN	1000				
Terms per Query (average) shifted neg. binomial dist. (n,p,s)	TPQ	2 (4,0.8,1)				
Query Term Frequency dist. from queries	QTF	Observed Distribution				
Client Arrival Pattern Poisson process (requests per minute)	λ	6 150	30 180	60 240	90 300	120
Collection Size (GB)	CSIZE	1	2	4	8	16
Disk Number	DISK	1	2	4	8	16
CPU Number	CPU	1	2	3	4	
Thread Number	TH	1	2	4	8	16 32

disk usage for each operation, but do not measure the memory and cache effects. We model the collection and queries by obtaining document and term statistics from test collections and real query sets (See [1,6] for more details.)

We validate our simulator against the multithreaded implementation. The simulator reports response times that are 4.5% slower than the actual system on the average (See [6] for more details).

3 Experiments and Results

This section explores how software and hardware configurations affect system scalability with respect to multiple threads, CPUs, and disks.

We start with a base system that consists of one thread, CPU, and disk. This system is disk bound. We improve the performance of our IR server through better software: multithreading; and with additional hardware: CPUs and disks. We demonstrate the system scalability using two sets of experiments. In the first set of experiments, we explore the effects of threading on system scalability. In the second set of experiments, we explore system scalability by increasing the collection size from 1 GB to 16 GB. When multiple disks exist, we use a round-robin strategy to distribute the collection and its index over disks.

We assume the client arrival rate is a Poisson process. Each client issues a query and waits for response. For each query, the server performs two operations: query evaluation and retrieving the corresponding summaries. Since users typically enter short queries, we experiment with a query set that consists of 1000 short queries, with an average of 2 terms per query that mimic those found in the query set down loaded from the Web server for searching the 103rd Congressional Record [4]. All experiments measure response time, CPU and disk utilization, and determine the largest arrival rate at which the system supports a response time under 10 seconds. We chose 10 seconds arbitrarily as our cutoff point for a reasonable response time.

3.1 Threading

This section examines how the software structure, i.e., number of threads, affects system scalability. Figure 1 illustrates how the average response time and

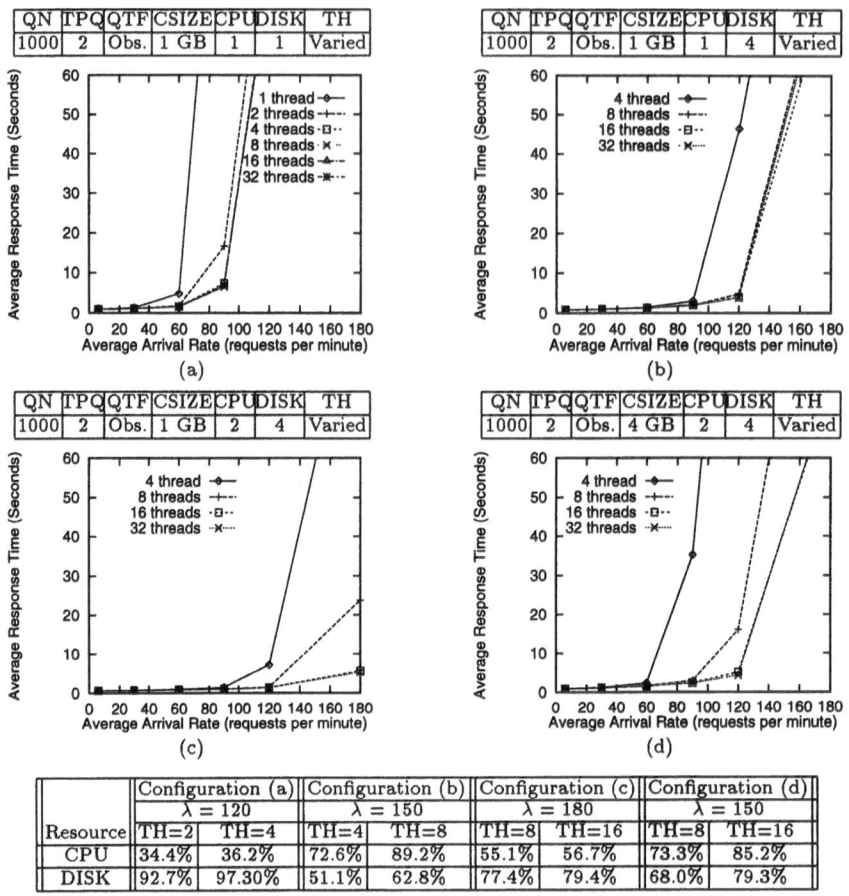

	Configuration (a)		Configuration (b)		Configuration (c)		Configuration (d)	
	$\lambda = 120$		$\lambda = 150$		$\lambda = 180$		$\lambda = 150$	
Resource	TH=2	TH=4	TH=4	TH=8	TH=8	TH=16	TH=8	TH=16
CPU	34.4%	36.2%	72.6%	89.2%	55.1%	56.7%	73.3%	85.2%
DISK	92.7%	97.30%	51.1%	62.8%	77.4%	79.4%	68.0%	79.3%

(e) hardware utilization at some interested data points

Fig. 1. Performance as the number of threads increases

resource utilization changes as the number of threads increases with varying number of CPUs and disks.

In all the configurations, the average response time improves significantly as the number of threads increases until either the disk or the CPU is overutilized (see Figure 1 (a) and (b)). Too few threads limits the system's ability to achieve its peak performance. For example in configuration (c), using 4 threads only supports 120 requests per minute for a response time under 10 seconds, while using 16 threads supports more than 180 requests per minute under the same hardware configuration. When either the CPU or the disk is a bottleneck, the system needs fewer threads to reach its peak performance. When CPUs and disks are well balanced (configuration (c) and (d)), the necessary number of threads is influenized more by the number of disks than the collection size. In both configuration (c) and configuration (d), the system achieves its peak performance using 16 threads. Additional threads do not bring further improvement.

3.2 Increasing the collection size

This section examines system scalability and hardware balancing as the collection size increases from 1 GB to 16 GB. In order to examine different hardware configurations, we consider two disk configurations as the collection size increases: fixing the number of disks, and adding disks. We vary the number of CPUs in each disk configuration.

Figure 2 illustrates the average response time and resource utilization when the collection is distributed over 16 disks which means each disk stores a database for 1/16 of the collection. Partitioning the collection over 16 disks illustrates when the system is CPU bound (see Figure 2(c)). Although performance degrades as the collection size increases, the degradation is closely related to the CPU utilization. With 1 CPU where the CPU is overutilized for 1 GB and 60 requests per minute, increasing the collection size from 1 GB to 16 GB decreases the largest arrival rate at which the system supports a response time under 10 seconds by a factor of 10 (see Figure 2(a)). With 4 CPUs where CPUs are overutilized for 1 GB and 180 requests per minute, the performance degrades much more gracefully (see Figure 2(b)). Increasing the collection size from 1 GB to 16 GB only decreases the largest arrival rate for a response time under 10 seconds by a factor of 3 for 4 CPUs.

Figure 3 illustrates the average response time and the resource utilization when the number of disks varies with the collection size and each disk stores a database for 1 GB of data. The system produces response times better than 1 GB for 2 GB using 1 CPU and 8 GB using 4 CPUs. A single CPU system thus handles a 2 GB collection faster than a 1 GB collection and a 4 CPU system handles a 2, 4, or 8 GB collection faster than a 1 GB collection in our configuration. The performance improves because work related to retrieving summaries is distributed over the disks such that each disk handles less work, relieving the disk bottleneck. By examining the utilization of CPU and disks in Figure 3(c), we see that the performance improves until the CPUs are overutilized. In the example of the single CPU system, the CPU is overutilized for a 4 GB collection. For a 2 GB collection distributed over 2 disks, the system handles 27.8% more requests than for a 1 GB collection on 1 disk.

By comparing Figure 2(c) and Figure 3(c), we find that the CPU utilization is more closely related to the number of disks rather than the collection size. We also find that adding disks degrades system performance when CPUs and disks are not well balanced. For example, for a 8 GB collection, partitioning over 8 disks using 4 CPUs results in 67.5% CPU utilization (see Figure 3(c)), while partitioning over 16 disks results in 88.0% CPU utilization (see Figure 2(c)) due to the additional overhead to access each disk. In this configuration, a system with 16 disks performs worse than 8 disks because the CPUs are a bottleneck.

4 Conclusion

In this paper, we investigate building a parallel information retrieval server using a symmetric multiprocessor to improve the system performance. Our results show that we need more than one thread to fully utilize hardware resources (4 to

526

QN	TPQ	QTF	CSIZE	CPU	DISK	TH
1000	2	Obs.	Varied	Varied	16	16

QN	TPQ	QTF	CSIZE	CPU	DISK	TH
1000	2	Obs.	Varied	Varied	Varied	16

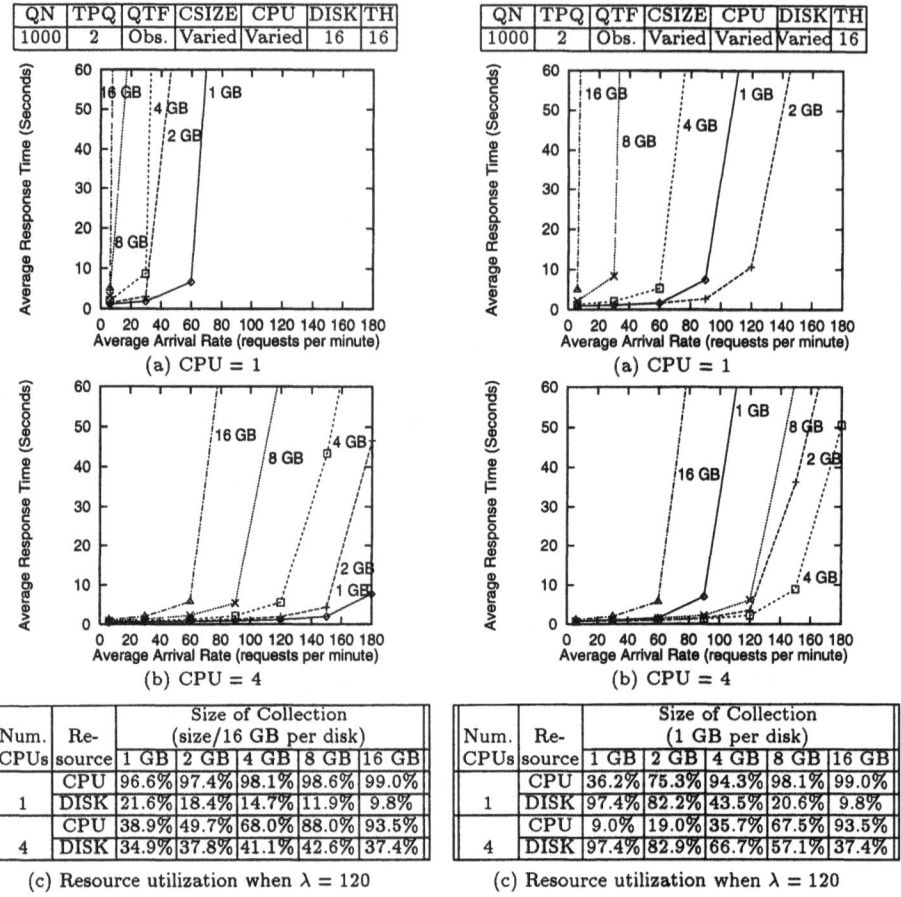

(a) CPU = 1

(b) CPU = 4

		Size of Collection				
Num. CPUs	Re-source	(size/16 GB per disk)				
		1 GB	2 GB	4 GB	8 GB	16 GB
1	CPU	96.6%	97.4%	98.1%	98.6%	99.0%
1	DISK	21.6%	18.4%	14.7%	11.9%	9.8%
4	CPU	38.9%	49.7%	68.0%	88.0%	93.5%
4	DISK	34.9%	37.8%	41.1%	42.6%	37.4%

(c) Resource utilization when $\lambda = 120$

		Size of Collection				
Num. CPUs	Re-source	(1 GB per disk)				
		1 GB	2 GB	4 GB	8 GB	16 GB
1	CPU	36.2%	75.3%	94.3%	98.1%	99.0%
1	DISK	97.4%	82.2%	43.5%	20.6%	9.8%
4	CPU	9.0%	19.0%	35.7%	67.5%	93.5%
4	DISK	97.4%	82.9%	66.7%	57.1%	37.4%

(c) Resource utilization when $\lambda = 120$

Fig. 2. Average response time and resource utilization for a collection distributed over 16 disks

Fig. 3. Average response time and resource utilization when the number of disks varies with the size of the collection

16 threads for the configurations we explored). We also show that adding hardware components can improve the performance, but these components must be well balanced since the IR workload performs significant amounts of both I/O and CPU processing. Our results show that we can search more data with no loss in performance in many instances. Although performance eventually degrades as the collection size increases, the performance degrades very gracefully if we keep the hardware utilization balanced. Our results also show that system performance for our system is more strongly related to the number of disks rather than the collection size.

Acknowledgment

This material is based on work supported in part by the National Science Foundation, Library of Congress and Department of Commerce under cooperative agreement number EEC-9209623. Kathryn S. McKinley is supported by an NSF CAREER award CCR-9624209. Any opinions, findings and conclusions or recommendations expressed in this material are the authors and do not necessarily reflect those of the sponsor.

We thank Bruce Croft, Jamie Callan, James Allan, and Chris Small for their support of this work. We also thank all the developers of InQuery without whose efforts this work would not have been possible.

References

1. B. Cahoon and K. S. McKinley. Performance evaluation of a distributed architecture for information retrieval. In *Proceedings of the Nineteenth Annual International ACM SIGIR Conference on Research and Development in Information Retrieval*, pages 110–118, Zurich, Switzerland, August 1996.
2. J. P. Callan, W. B. Croft, and J. Broglio. TREC and TIPSTER experiments with INQUERY. *Information Processing & Management*, 31(3):327–343, 1995.
3. J. P. Callan, W. B. Croft, and S. M. Harding. The INQUERY retrieval system. In *Proceedings of the 3rd International Conference on Database and Expert System Applications*, Valencia, Spain, September 1992.
4. W. B. Croft, R. Cook, and D. Wilder. Providing government information on the internet: Experiences with THOMAS. In *The Second International Conference on the Theory and Practice of Digital Libraries*, Austin, TX, June 1995.
5. InQuery. http://ciir.cs.umass.edu/info/highlights.html.
6. Zhihong Lu, Kathryn S. McKinley, and Brendon Cahoon. The hardware/software balancing act for information retrieval on symmetric multiprocessors. Technical Report TR98-25, University of Massachusetts, Amherst, 1998.

The Enhancement of Semijoin Strategies in Distributed Query Optimization

F. Najjar and Y. Slimani

Dept. Informatique - Faculté des Sciences de Tunis
Campus Universitaire - 1060 Tunis, Tunisie
yahya.slimani@fst.rnu.tn

Abstract. We investigate the problem of optimizing distributed queries by using semijoins in order to minimize the amount of data communication between sites. The problem is reduced to that of finding an optimal semijoin sequence that locally fully reduces the relations referenced in a general query graph before processing the join operations.

1 Introduction

The optimization of general queries, in a distributed database system, is an important and challenging research issue. The problem is to determine a sequence of database operations which process the query while minimizing some predetermined cost function.

Join is a frequently used database operation. It is also the most expensive, specifically in a distributed database system; it may involve large communication costs when the relations are located at different sites. Hence, instead of performing joins in one step, semijoins [1], are performed first to reduce the size of the relations so as to minimize the data transmission cost for processing queries [2]. In the next step, joins are performed on the reduced relations. The join of two relations R and S on an attribute A is denoted by $(R \bowtie_A S)$, while the semijoin from R to S on an attribute A is denoted by $S \propto_A R$. Thus, $S \propto_A R$ is defined as follows: (i) project R on the join attribute A (i.e. $R(A)$); (ii) Ship $R(A)$ to the site containing S; (iii) Join S with $R(A)$. The transmission cost of sending S to the site containing R for the join $R \bowtie_A S$ can thus be reduced.

There are two main methods to process a join operation between two relations. One is called the *nondistributed join*, where a join is performed between two unfragmented relations. The other is called the *distributed join*, where the join operation is performed between the fragments of relations. As pointed out in [5], the problem of query processing has been proved to be NP-hard. This fact justifies the necessity of resorting to heuristics.

The remaining of this paper is organized as follows: preliminaries are given in Section 2. Section 3 defines the main characteristics of two semijoin-based query optimization heuristics; then, we present and discuss the join query optimization in a fragmented database. Finally, Section 5 concludes the paper.

2 Preliminaries

A join query graph can be denoted by a graph $G = (V, E)$, where V is the set of relations and E is the set of edges. An edge $(R_i, R_j) \in E$, if there exists a join predicate on some attribute of R_i and R_j. Without loss of generality, only cyclic query graphs are considered. In addition, all attributes are renamed in such a way that two join attributes have the same name if and only if they have a join predicate between them. The relations referenced in the query are assumed to be located at different sites. The query problem is simplified to be the estimation of the data statistics and the optimization of the transmission order, so that the total data transmission is minimized.

We denote by $|S|$ the cardinality of a relation S. Let w_A be the width of an attribute A and w_{R_i} be the width of a tuple in R_i. The size of the total amount of data in R_i can then be denoted by $\|R_i\| = w_{R_i}|R_i|$. For notational simplicity, we use $|A|$ to denote the extant domain of the attribute A. $R_i(A)$ denotes the set of distinct values for the attribute A appearing in R_i. For each semijoin $R_j \propto_A R_i$, a selectivity factor, $\rho_{iA} = \frac{|R_i(A)|}{|A|}$ is used to predict the reduction effect. After the execution of $R_j \propto_A R_i$, the size of R_j becomes $\rho_{iA}\|R_i\|$. Morever, it is important to verify that a semijoin $R_j \propto_A R_i$ is *profitable*, i.e. if the cost incurred by this semijoin, $w_A|R_i(A)|$, is less than the cost of the reduction (called the benefit), which is computed in terms of avoided future transmission cost, $w_{R_i}|R_i| - \rho_{iA}|R_i|$. The profit is set to be (benefit - cost).

3 Nondistributed Join Method

In this section, we propose two heuristics. The first, namely *one-phase Parallel SemiJoins*, 1-PSJ, determines a set of parallel semijoins. The second, namely *Hybrid A* heuristic*, HA*, finds a general sequence of semijoins, which is a combination of parallel and sequential semijoins.

3.1 1-PSJ

We say that R_i is fully locally reduced if $\{j \ / \ R_i \propto_A R_j$ is feasible$\}$. We denote by $RD_i = \{j/R_i \propto R_j$ is profitable$\}$ the set of index reducers of the relation R_i. Our objective is to find the set of the most locally profitable semijoins (called applicable semijoins), $AP_i \subseteq RD_i$, such that the overall profit is maximized, and subsequently the total transmission cost (TC_i) of R_i is minimized. Furthermore, removing a profitable semijoin may increase the total profit and minimize the extra costs incurred by semijoins. Since all applicable semijoins are executed simultaneously, local optimality (with respect to R_i) can be attained. Finally, in order to reduce each relation in the query, we apply a divide-and-conquer algorithm. The total cost (TC) is minimized if all tranmission cost (TC_i) are minimized simultaneously. The details of this algorithm are given in [4].

3.2 Hybrid A*

The well known A* can be used to determine a sequence of semijoin reducers [6] for distributed query processing. The key issue of A* algorithm is to derive a heuristic function which can intelligently guide the search of a sequence of semijoins. In the A* algorithm, the search is controlled by a heuristic function f, with two arguments: the cost of reaching p from the initial node (original query graph with its corresponding profile), and the cost of reaching the goal node from p. Accordingly, $f(p) = g(p) + h(p)$, where $g(p)$ estimates the minimum cost of trajectory from the initial state to p, and $h(p)$ estimates the minimum cost from p to the goal state. The node p chosen for expansion (i.e., whose immediate successors will be generated) is the one which has the smallest value $f(p)$ among all generated nodes that have not been expanded so far.

In order to derive a general sequence of semijoins, for a node p, $g(p) = g(q) + \sum_{j \in AP_i} \cos t(R_i \propto R_j) + ||R_i'||$, where p is an immediate successor of q, and R_i' denotes the resulting relation after performing applicables semijoins to the original relation R_i. The function h is defined as the sum of the sizes of remaining relations such that the effect of the total reduction (with respect to neighboring relations) gives the best estimation. $h(p) = \sum_k (\sum_j \cos t(R_k \propto R_j) + ||R_k'||)$, where R_k is not yet reduced.

Example 1: Consider the following join query: Select A, D from R_1, R_2, R_3, R_4 where $(R_1.A = R_3.A)$ and $(R_1.B = R_2.B)$ and $(R_2.C = R_3.C)$ and $(R_3.D = R_4.D)$.

We suppose that R_1, R_2, R_3 and R_4 are located in different sites. The corresponding query graph and profile are given, respectively in Figure 1 and Table 1.

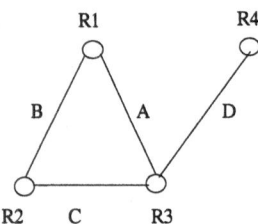

Fig. 1. Join query graph for example 1.

$1 - PSJ$ finds the set $\{R_1 \propto R_2, R_3 \propto R_1, R_3 \propto R_4\}$, with the total transmission cost to the final site R_2, $18, 370$. Whereas, $HA*$, finds the general sequence of semijoins, $R_1 \propto R_2$, $\{R_3 \propto R_1', R_3 \propto R_4\}$, with the cost of $16, 681$.

To show more insights into the performance of $1 - PSJ$ and $HA*$ heuristics, simulations were carried on different queries for n (n is being 5-12) relations

Table 1. Profile Table for example 1.

R_i	$|R_i|$	X	W_X	$\|R_i(A)\|$
R_1	1190	A	2	830
		B	1	850
R_2	3440	B	1	850
		C	3	900
R_3	2152	A	2	800
		C	3	900
		D	1	720
R_4	3100	D	1	700

involved in each query. For a comparison purpose, in addition to $1 - PSJ$ and HA^*, we also apply the original method, OM, which consists of sending all the relations directly to the final site. In Figure 2, it is apparent that as the number of relations increases ($n \geq 8$), HA^* heuristic becomes better than $1 - PSJ$. When $n \geq 9$, HA^* outperfoms the other heuristics significantly (the reduction cost is about 45%).

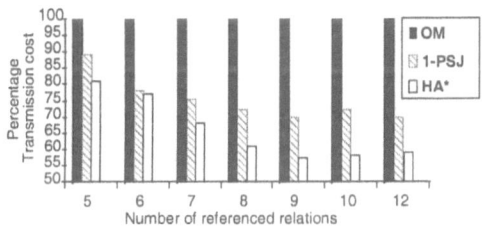

Fig. 2. Impact of the number of relations on transmission cost.

4 Distributed Join Method

A relation can be horizontally partitioned (declustered) into disjoint subsets, called *fragments,* distributed across several sites. We associate for each fragment its qualification, which is expressed by a predicate describing the common properties of the tuples in that fragment. One major issue in developing a horizontal fragmentation technique is determining the criteria to be used in guiding the fragmentation. A major difficulty is that there are no known significant criteria that can be used to partition relations horizontally.

In the context of our study, we suggest a bipartition of each relation R_i, such that, a relation is divided into mutually exclusive fragments. To represent the fragments more specifically, we propose the following formula: $|R_i| = \alpha|R_i| + (1-\alpha)|R_i| = |R_{i1}| + |R_{i2}|$, where α is a rational number ranging from 0 to 1 and R_{i1}, R_{i2} are the fragments of R_i. The above fragmentation satisfies the three conditions [2], *completeness*, *reconstruction*, and *disjointness*, which ensure a correct horizontal fragmentation. Note that bipartitioning can be applied to a relation repeatedly.

To estimate the cost of an execution plan, the database profile may have the following statistics: $|R_{ij}|$ denotes the cardinality of the fragment number j of relation R_i and $|R_{ij}(A)|$ represents the number of distinct values of attribute A in its fragment.

When semijoins are used in a fragmented database system, they have to be performed in a relation-to-fragment manner, so that they do not cause the elimination of contributive tuples. At each site containing a fragment R_{jk} to be reduced, we proceed as follows: (i) every fragment of R_i [1] must participate in reducing R_{jk}; so, find the optimal set of applicable semijoins and send values of the semijoins attributes from each fragment of R_i to R_{jk}; (ii) Merge the fragments of R_i before eliminating any tuple of R_{jk}.

Example 2: We illustrate the distributed join method (HA^* on fragmented relations) with the same previous example discussed for the nondistributed join. After partitioning, the corresponding profile is given in Table 2.

Table 2. Profile Table for Example 1.

R_{ij}	$|R_{ij}|$	X	W_X	$|R_{ij}(A)|$
R_{11}/R_{12}	119/1071	A	2	90/809
		B	1	91/818
R_{21}/R_{22}	344/3096	B	1	91/818
		C	3	97/872
R_{31}/R_{32}	215/1936	A	2	87/782
		C	3	97/872
		D	1	79/800
R_{41}/R_{42}	310/2790	D	1	76/683

The optimal general sequence is: $\{R_{12} \propto R_{22}, R_{12} \propto (R_{31} \cup R_{32})\}, \{R_{21} \propto R_{11}, R_{21} \propto (R_{31} \cup R_{32})\}, R_{42} \propto R_{31}, \{R_{32} \propto (R_{11} \cup R_{12}),$
$\{R_{32} \propto (R_{21} \cup R_{22}), R_{32} \propto R_{42}\}$; it incurs $13,959$, which is less than in nondistributed join method.

A general conclusion is that the communication cost is substantially reduced if we use a "good fragmentation". In the absence of a formal definition of a good fragmentation, we can approximate it by the α-factor. In effect, we have noted that a good choice of this criteria leads to a good fragmentation. The Fig 2 shows the effect of the α-factor on the communication cost for a given query in which the number of relations is constant and α is varied.

---

[1] R_i such that $R_j \bowtie R_i$ is applied in the query.

5 Conclusion

In this paper, we proposed two distributed query processing strategies for join queries using semijoin as a query processing tactic. For these two strategies, we present new heuristics that "intelligently" guiding the search and returning a general reducer sequence of semijoins. For the case of the distributed join strategy, we proposed a technique to bipartition each relation assuming a fixed α-factor.

References

1. P.A. Bernstein, N. Goodman, E. Wong, C. Reeve, and J.B. Rothnie. Query Processing in a System for Distributed Databases (SDD-1). *ACM TDS*, vol. 6(4), Dec. 1981, pp. 602-625.
2. S. Ceri and G. Pelagatti. *Distributed Databases: Principles and Systems*. McGraw-Hill, 1985.
3. M-S Chen and P.S Yu. Combining Join and Semijoin Operations for Distributed Query Processing. *IEEE TKDE* vol. 5(3), Jun. 1993, pp. 534-. 542.
4. F. Najjar, Y. Slimani, S. Tlili, and J. Boughizane. Heuristics to determine a general sequence of semijoins in distributed query processing. *Proc. of the 9th IASTED Int. Conf., PDCS*, Washington D. C. (USA), Oct. 1997, pp. 354-359.
5. C. Wang and M-S. Chen. On the Complexity of Distributed Query Optimization. *IEEE TKDE*, vol. 8(4), Aug. 1996, pp. 650-662.
6. H. Yoo and S. Lafortune. An Intelligent Search Method for Query Optimization by Semijoins. *IEEE TKDE*, vol. 1(2), Jun. 1989, pp. 226-237.

Virtual Time Synchronization in Distributed Database Systems Using a Cluster of Workstations[*].

Azzedine Boukerche[1] Timothy E. LeMaster[2], Sajal K. Das[1] and Ajoy Datta[2]

[1] Department of Computer Sciences, University of North Texas, Denton, TX. USA
[2] Department of Computer science, University of Nevada, Las Vegas, NV., USA

Abstract. Virtual Time is the fundamental concept in optimistic synchronization schemes. In this paper, we investigate the efficiency of a distributed database management system. synchronized by virtual time using a LAN connected collection of SPARCS. To the best of our knowledge, there is little data reporting VT performance in DDBMS. The experimental results are compared with a multiversion timestamp ordering (MVTO) concurrency control method, a widely used technique in distributed databases, and approximately 25-30% reduction in the response time is observed when using the VT synchronization scheme. Our results demonstrate that the VT synchronization approach is a viable alternative concurrency control method for distributed database systems and outpforms MVTO one.

1 Introduction

Virtual Time is the fundamental concept in optimistic synchronization schemes. Many applications for virtual time have been proposed which include parallel simulations, distributed systems, and VLSI design problems. Time Warp (TW) [4, 5] based mechanism is a technique to implement virtual time by allowing application programs to cancel previously scheduled events. While conservative synchronization methods [3] rely on blocking to avoid violation of dependency constraints, Time Warp relies on detecting synchronization errors at run time and on recovery using rollback mechanisms. The principal advantages of Time Warp over more conventional blocking-based synchronization protocols is that Time Warp offers the potential for greater exploitation of concurrency and parallelism, and more importantly greater transparency of the synchronization mechanism to the programmer. The latter is due to the fact that Time Warp is less reliant on application specific information regarding which computations

Despite the fact that research on virtual time has been ongoing for the past decade, little data reporting performance of distributed database systems

---

[*] Part of this research is supported by Texas Advanced Technology Program grant TATP-003594031

(DDBMS) synchronized by virtual time. Jefferson [4, 5] has suggested that virtual time could be used to synchronize DDBMS. Therefore, a question of considerable pragmatic interest is whether virtual time is really a viable, alternate concurrency control method for distributed database systems and whether it can provide good performance or not.

In this paper, our primary goal is to describe an implementation of a virtual time synchronization in distributed database management systems, and study its performance. This is an important step to determine the practicality of a virtual time synchronized database systems. We then compare the performance of our method with a MultiVersion Timestamp Orderings (MVTO) concurrency control method [1]. The empirical results demonstrate that the VT approach outperforms the MVTO one.

2   Virtual Time Database Synchronization

We view a database system as a collection of objects (sequential processes) that execute concurrently and communicate through timestamped messages. Each object is capable of both computation and communication, and it has an associated *virtual clock* which acts as the simulation clock for that object.

Now, let us examine the component and the structure of a database management system. There are four majors components of a DDBMS at each site: *transactions, transaction manager* (TM), *database manager* (DM), and *data*. Transactions are generated from users. The transaction manager supervises interaction between users and the DDBMS, while the database manager controls access to the data, and provides an interface between the user (via the query) and the file system. It is also responsible for controlling the simultaneous use of the database and maintaining its integrity and security.

There are no formal distinction between transaction and data objects. While data objects respond only to *Read/Write* messages, transaction objects receive messages directly from external users. Objects communicate with each other through *Request* and *Response* primitives which send messages. A *Request* either reads or writes a data value, while a *Response* returns a data value. After an object issues a *Request*, it blocks to await a response. The possibility of deadlock exists since two objects within a transaction could request data values from each other at the same time. Each would block, waiting for the other to respond to the *Request*. Virtual Time scheme prevents deadlock between transactions, since all transactions are executed in timestamps order, and all transaction timestamps are unique.

Virtual time can be viewed as a global, one-dimensional temporal coordinate system imposed on a distributed system to measure computational progress and define synchronization [4, 5]. There are two important properties of virtual time. First, it may increase or decrease with respect to real time. There is no restriction on the amount or frequency by which it changes. Second, the virtual times generated must form a total (or partial) ordering on the relation "less than."

A manager (or process) may send a message to any other manager at any time. There are no restrictions placed upon potential communication paths between processes. No fixed or static declarations are required before execution begins. The communication medium is assumed to be reliable but messages are not required to arrive in the order they are sent. Indeed, Occasionally, different transactions cause an unordered message to be received by an object. When this occur, rollback is used for synchronization. However, instead of rolling back the entire transaction, only certain actions within the transactions are rolled back. These actions are the ones that are casually connected to the errant message. This is where virtual time synchronization approach diverges from traditional one, in which the entire transaction is either aborted an retried or is blocked.

Virtual time incorporates several different data structures and control algorithms to produce optimistic processing. The most noteworthy are the Time Warp based mechanism to implement virtual time, and the use of rollback as the fundamental synchronizer in the system. The algorithms include transaction and database managers. The former is responsible for generating and submitting requests while the latter processes the requests. The rollback mechanism is used by both managers.

In this paper, we compare VT with the MVTO version based on Bernstein's algorithm [1]. A complete description of VT and MVTO can be found in [2].

3 Simulation Experiments

All experiments were conducted on a 12-processors bus based message passing Sun Microsystems SPARC Stations[1]. The SPARC stations were connected into a local area network (LAN) via ethernet. It should be pointed out that in distributed database systems, there are more structures than in distributed simulation environments, see [2] for more details.

A simple database model was used. Each database manager was responsible for a specified number of items. Whenever a transaction requested a *Read* or *Write* access to an item, it was delayed 50 milliseconds to simulate the physical I/O operations. The goals of our experiments are to study the viability of virtual time as a concurrency control method in database systems, and to compare its relative performance compared to multiversion ordering synchronization scheme. In our model, we made use of several database/transaction sizes, inter-transaction delays and varied the number of processors from 4 to 12. The experimental data which follows were obtained by averaging several trial runs. Every test run required between 8 and 50 minutes of actual time to complete all experiments. The average length of the experiments was 24 minutes.

Space limitation preclude us to include all of our experiments. Hence, we present our results below in the form of graphs of response time, and the throughput. The response time achieved is defined as the average amount of wall clock time per processor required for a transaction manager to successfully submit a

---

[1] W use a distributed memory environment.

fixed number of transactions. The throughput is defined as follow: $\frac{K*Nt}{T_{resp}}$; where T_{resp} represents the total response time of all processors; K is the number of processors used, and Nt is the number of transactions.

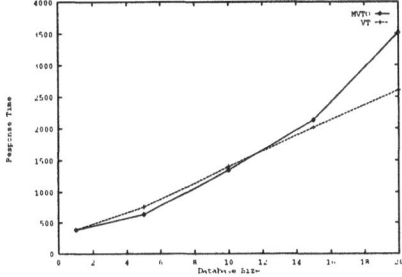 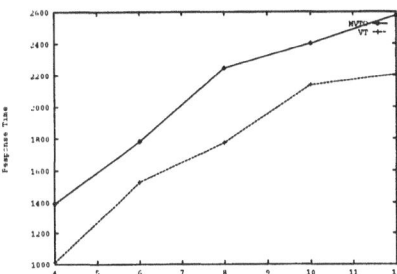

Fig. 1. Response Time Vs. Database Size

Fig. 2. Response Time Vs. Number of Processors

In our experiments, we analyzed the effect of database size on the performance (response time and throughput) of the DDBMS synchronized by virtual time. We made use of 8 processors. We varied the database size from 1 to 20. The transaction size was set to 7% of the database size. All *Read* requests were updated with a probability 25%. The Inter-transaction delay was set to zero. In Fig. 1, we present the response time as a function of database size for both synchronizations VT and MVTO. The results obtained show that MVTO and VT exhibit about the same response time when the database size is less than 12. Then, when we increase the database size from 12 to 20, VT responded faster than MVTO. We observe about 25% reduction in the response time when we use a database size of 20.

In the next set of experiments, in order to reduce the response time of DDBMS system (s), we run some preliminary tests where we varied the number of processors from 4 to 12. The results obtained for the response time are shown in Figure 2. As expected, we also observe that as we increase the number of processors, the response time increases for both synchronization schemes. Furthermore, we see that VT performs better than MVTO. For example, we see a reduction of the response time between 25-30% with VT scheme when compared to MVTO one. Similar results were obtained for the throughput metric [2].

References

1. P.A. Bernstein, V. Hadzilacos, and N. Goodman, *Concurrency Control and Recovery in Database Systems*, Addison-Wesley, 1987.
2. A. Boukerche, T. LeMaster, Sajal, K. Das, and A.K. Datta, "Virtual Time Synchronization in Distributed Systems", *Tech. Rep. TR-98*, University of North Texas.

3. A. Boukerche, C. Tropper, "Parallel Simulation on the Hypercube Multiprocessor", Distributed Computing, Spring Verglas 1993.
4. D. Jefferson and A. Motro, "The Time Warp Mechanism for Database Concurrency Control," *International Conference on Data Engineering*, 1986, 474–481.
5. D.R. Jefferson, "Virtual Time," *ACM Transactions on Programming Languages and Systems*, Vol. 7, No. 3, 1985, 404–425.

Load Balancing and Processor Assignment Statements

C. Rodríguez, F. Sande, C. León, I. Coloma, A. Delgado

Dpto. Estadística, Investigación Operativa y Computación,
Universidad de La Laguna, Tenerife, Spain

Abstract. This paper presents extensions of processor assignment statements, the language constructs to support them and the methodology for their implementation on distributed memory computers. The data exchange produced by these algorithms follows a generalization of the concept of hypercubic communication pattern. Due to the efficiency of the introduced weighted dynamic hypercubic communication patterns, the performance of programs built using the methodology are comparable to the obtained using classical message passing programs. The concept of balanced processor assignment statement eases the portability of PRAM algorithms to multicomputers.

1 Balanced Processor Assignment Statements

Algorithms for the PRAM [1] model are usually described introducing a parallel statement like:

for all x **in** $m..n$ **in parallel do** $instruction(x)$

It is essential to differentiate between the concepts of processor activation and processor assignment statements. The **for all** construction is used to denote processor assignment statements, which are usually charged with constant time, although authors avoid to explain this point.

Load balancing is a fundamental issue in parallel programming. It is straightforward to implement nested processor assignment statements using cyclic and block distribution mappings. However, the division of the processors among the tasks in groups of equal sizes, provided by cyclic and block processor distribution policies may produce, specially for irregular problems, unfair balance. Any divide and conquer algorithm can be used to illustrate this idea. Since the complexity of the resulting subproblems after the division of the problem may be considerably different, to split the current group of processors in equal sizes will likely lead to work load unbalance. One way to solve this inefficiency is to expand the language with a statement allowing the programmer to modify the processor distribution policy. This can be achieved introducing a simple variant of the **for all** statement to provide extra heuristic information $w(x)$, regarding to the suitable number of processors required for each task $instruction(x)$:

for all x **in** $a..b$ **with weight** $w(x)$ **in parallel do** $instruction(x)$

The effect of this new *"balanced parallel assignment statement"* is to adjust the number of processors according to the weights attached to each instance of *instruction(x)*.

An example of this kind of construct can be found in the *La Laguna C system* (*llc*). The *La Laguna C* tool consists in a set of C macros and functions expanding the library model for message passing (PVM, MPI, Inmos C, etc.). It provides not only load balancing but also many other features as processor virtualization, nested parallelism and support for collective operations.

Let us first consider the implementation in *llc* of the well-known quicksort algorithm. A first approach to the solution consists in using a classical processor assignment statement. In this case, the workload assigned to each processor depends of the selected pivot. The implementation of this solution in *llc* is what appears in Figure 1, substituting in line 6 the call to PARWEIGHTVIRTUAL by a call to the macro PARVIRTUAL.

```
1 void qs(int first, int last) {
2   int w1, w2, size1, size2, i, j;
3   if (first < last) {
4     partition(&i, &j, first, last);
5     size1 = w1 = (j - first + 1);   size2 = w2 = (last - i + 1);
6     PARWEIGHTVIRTUAL(w1, qs(first,j), (A+first), size1, w2, qs(i,last), (A+i), size2);
7   }
8 }
```

Fig. 1. Balancing the size of the groups according to the load

Figure 1 shows the use of one of the *llc* balanced parallel processor assignment statements, the macro PARWEIGHTVIRTUAL. At the time to execute the macro PARWEIGHTVIRTUAL at line 6, the set of available processors is divided into two subsets. One executing the first call (second parameter) and the other doing the second one (sixth parameter). The third and fourth parameters, $A + first$ and $size1$, inform that the result of the call to the function $qs(first, j)$ on the first segment is constituted by the $size1$ integers beginning at $A + first$. The two last parameters $A + i$ and $size2$ describe the fact that the result of the second call is constituted by the $size2$ integers starting in $A + i$. The first and fifth parameters of the call ($w1$ and $w2$) provide heuristic information about the number of processors required by the corresponding sorting task. In this case they are simply the sizes of the subintervals to sort. At the end of the parallel call the two sets of processors exchange their results and rejoin to form the previous set. To increase performance, instead of using virtualization, the programmer can take control of the number of available processors through the use of the *llc* variable NUMPROCESSORS.

The former example is a *Replicated problem*. We define a *Replicated* problem (or parallel algorithm) as one in which the initial input data have to be replicated in all the processors and the solution is also required in all the processors. The

```
1 void qs (int *size) {
2 int totalSize, i, j, pivot, sum, s;
3 if (NUMPROCESSORS > 1) {
4 BALANCE( *totalSize, (*size), A, decision, getWeigth);
5 if (totalSize > NUMPROCESSORS) {
6 REDUCEBYADD( sum, A[*size/2]);
7 pivot = sum / NUMPROCESSORS;
9 PAR( part( *size-1,p,&i,&j,&s), A+i, s, revPart( *size-1,p,&i,&j,&s), A+i, s);
10 *size = j + 1 + s;
11 SPLIT( qs(size), qs(size)); }
12 else qsSeq( 0, (*size-1)); }
13 else qsSeq( 0, (*size-1));
14 }
```

Fig. 2. Distributed quicksort

example in Figure 2 presents the use of processor assignment statements in a *Distributed Problem*. A *Distributed* problem is one in which both the input data and the results are distributed among the processors. The initial array is distributed among the processors and the goal is to have the items sorted according to the processor name. Through the all-to-all reduction REDUCEBYADD, the common pivot is chosen in lines 6 and 7. The current set of processors is divided in two subsets of equal size using the PAR macro in line 9: one executing the classical partition procedure *part*, and the other executes a reversed version of partition *revPart*, that stores elements greater than the pivot in the first part of the array (from 0 to j) and stores the s items smaller than the pivot between i and $size - 1$. The results are interchanged and the size of the array is updated in line 10. The recursive call to the macro SPLIT divides the group using the same processor distribution policy than the the former call to the PAR macro. The work load is equilibrated by the call to macro BALANCE at line 4. This *La Laguna C* macro generalizes the dimension exchange algorithm [4] for hypercubes. Parameters *decision* and *getWeight* are provided by the user. When called, *getWeight* returns a measure of the difficulty of the problem. From the input weights of two arrays of problems, *decision* decides the number of items in the array of problems exceeding the ideal. Furthermore, BALANCE returns in *totalSize* the sum of the sizes of the arrays in the current group.

2 Weighted Dynamic Hypercube Topologies

When nested unbalanced or balanced assignment statements are executed on a distributed system, each processor computes its *partners* in the other subsets created. Let us consider a set of nested balanced processor assignment statements:

for all x_1 **in** $1..n_1$ **with weight** $w_1(x_1)$ **in parallel do**
    for all x_2 **in** $1..n_2$ **with weight** $w_2(x_2)$ **in parallel do**
    ...

        for all x_m **in** $1..n_m$ **with weight** $w_m(x_m)$ **in parallel do**
        $instruction(x_1, x_2, ..., x_m)$

the resulting communication pattern becomes a sort of distorted dynamic hypercube. Each *nested balanced processor assignment statement* creates a "new dimension" in the current set H of available processors. In the jth nested call, the set H is partitioned in a collection of n_j subsets $\Gamma = \{H_1, ..., H_{n_j}\}$ whose cardinalitiess are distributed according to the weights of the statements. Each processor p on each of the new created sets H_t has at least one *partner* or neighbor in each of the other subsets in the collection $\Gamma - \{H_t\}$. This way a graph is associated with the nesting of balanced processor assignments in which links join *partner* processors. The way *partners* are selected in the other groups is called a *partnership policy*. The family of topologies defined this way is called *weighted dynamic hypercubes*. Although the formulas for the algorithm to support nested load balanced processor assignment statements are by far more convoluted than those for the unbalanced case, they still reduce to elementary arithmetic and logical operations. This simplicity and the good performance of current multicomputers on weighted hypercube connection patterns explain the efficiency of the implementation.

3 Computational Results

Our experiments were performed in an IBM–SP2 computer and a Silicon Graphics Origin 2000. The IBM–SP2 nodes (Power2–thin2 processors) are connected through a synchronous multistage packet switch with whormhole routing providing a real data transfer rate of 22 Mb/s [5]. The Origin 2000 is a 32 + 32 R10000 processors (196 MHz) machine with 8 GB main memory, 288 GB disk and 25.08 Gflop/s theoretical peak.

For each number of processors and for each algorithm, columns labeled S in Table 1 show the average speedup for ten experiments against the corresponding best sequential algorithm. First and fifth pairs of columns of the Table compare the results obtained for the *llc* parallel quicksort algorithm of Figure 1, (label BALVIRT) against a message passing parallel *quicksort* algorithm for a hypercube multicomputer due to professor P. B. Hansen [2] (label BH). The Hansen algorithm achieves a perfect load balance through the use of the *find* procedure due to Hoare [3]. This procedure divides the array to sort in two equal halves. On both architectures *La Laguna C* times are slightly better than those obtained by the Hansen algorithm. This advantage can be explained by the different approach taken in the implementation of nested parallelism. While nested parallelism produces a *processor assignment* in the *llc* version, the implicit solution used by the Hansen algorithm roughly corresponds to a *processor activation.* The constant complexity of the processor assignment algorithm used by the *llc* system is smaller than the logarithmic complexity of the processor activation algorithm. No inefficiency is introduced by the methodology in spite of the easiness introduced in the expression of parallel algorithms.

Table 1. Speedups and Speedup Standard Deviations for the parallel quicksort.IBM SP2 and Origin 2000, 7M integer array

	PROCS	BH S	BH SSD	UNBAL S	UNBAL SSD	MANUAL S	MANUAL SSD	BALAN S	BALAN SSD	BALVIRT S	BALVIRT SSD
SP2	2	1.49	0.05	1.27	0.17	1.53	0.05	1.23	0.20	1.15	0.18
	4	2.02	0.08	2.94	0.37	2.15	0.10	2.41	0.39	2.24	0.35
	8	2.42	0.12	2.77	0.66	2.66	0.14	3.45	0.31	3.29	0.29
	16	2.62	0.20	3.48	1.01	2.99	0.15	4.72	0.30	4.29	0.29
Origin	2	1.59	0.04	1.27	0.16	1.61	0.04	1.27	0.16	1.23	0.16
	4	2.20	0.08	1.95	0.31	2.31	0.08	2.46	0.33	2.37	0.32
	8	2.68	0.13	2.74	0.52	2.88	0.13	3.60	0.24	3.50	0.22

Since all the N elements of the array to sort have to be received by processor 0, the size of the array N constitutes a lower bound of the $Time_{par}$ of any parallel sorting algorithm. Therefore, the speedup of any replicated sorting algorithm is bounded by the logarithm of the array size.

Theorem 1. *If N is the size of the array to sort, then it holds: $SpeedUp = \frac{N \log(N)}{Time_{par}} \leq \frac{N \log(N)}{N} = \log(N)$*

This theorem explains the poor speedup in Table 1. As a consequence, the only way to increase efficiency is to consider huge arrays much larger than the ones considered in these experiences.

The results in the four rightmost speedups columns compare three different load balance approaches. The UNBAL label corresponds to the algorithm in figure 1 but substituting line 6 for the use of the unbalanced processor assignment statement PAR. The pivot is selected as the average of three randomly selected elements. The amount of work assigned to each processor depends on the selected pivot. The MANUAL label stands for a *llc* program that uses Hoare *find* procedure and unbalanced processor assignment statement. The programmer takes responsibility of load balance. The results of the algorithm of Figure 1 using the PARWEIGHT construct instead of PARWEIGHTVIRTUAL are presented under the label BALAN. Performance obtained using balanced processor assignment statements are even better than those reached using a hand-written specific load balancing algorithm designed by an experienced programmer. Label BALVIRT corresponds to the algorithm in Figure 1. Processor virtualization does not introduce appreciable inefficiency. Columns labeled SSD present for each algorithm, the Speedup Standard Deviation of the experiments. Although the average speedup of the UNBAL entries are comparable to the MANUAL entries, the standard deviation shows the unpredictability of its behavior. On the other extreme, the MANUAL case presents the smaller deviation. The BALAN and BALVIRT versions keep the standard deviation in an acceptable range. When the number of processors grows, the standard deviation keeps almost constant. The Origin 2000 scales slightly better than the IBM SP2.

The average speedup values for the distributed parallel quicksort algorithm presented in Figure 2 with eight processors are 6.66 and 6.90 for the IBM-SP2 and Silicon-Origin respectively. These values were obtained as the average of 10 experiments with 16M integer arrays randomly generated according to an uniform distribution. The speedups show a better behaviour than for the replicated problems.

4 Conclusions

A methodology to efficiently implement balanced nested processor assignment statements in constant time was presented. The algorithm to implement the nesting of such statements can be used to build tools providing the capability to mix parallelism and recursion. A non negligible advantage is that their performance can be predicted using the model presented in [5]. The computational results prove that the efficiency of this approach is comparable to the obtained using specific message passing algorithms. From an efficiency point of view, the results obtained for the different approaches considered are pretty similar, so the advantage of the proposed paradigms must not be seen from this perspective. The most important benefit resides in the fact that it provides a comfortable frame to translate PRAM algorithms to standard message–passing libraries without loss of performance.

Acknowledgements This research has been done using the resources at the Centre de Computació i Comunicacions de Catalunya (CESCA-CEPBA).

References

1. Fortune S., Wyllie, J.: Parallelism in Random Access Machines. STOC. (1978) 114–118
2. Hansen P. B.: Do hypercubes sort faster than tree machines?. Concurrency: Practice and Experience **6** (2) (1994) 143–151
3. Hoare, C. A. R.: Proof of a Program: Find. Communications of the ACM. **14** 39–45
4. Murphy, T.A., Vaughan J.: On the Relative Performance of Diffusion and Dimension Exchange Load Balancing in Hypercubes. Proceedings of the PDP'97. IEEE Computer Society Press. (1997) 29-34
5. Rodríguez, C., Roda J.L., Morales, D. G., Almeida, F.: h-relation Models for Current Standard Parallel Platforms. Proceedings of the Euro-Par'98. Springer Verlag LNCS. (1998)

Mutual Exclusion Between Neighboring Nodes in a Tree That Stabilizes Using Read/Write Atomicity

Gheorghe Antonoiu[1] and Pradip K. Srimani[1]

Department of Computer Science, Colorado State University, Ft. Collins, CO 80523

Abstract. Our purpose in this paper is to propose a new protocol that can ensure mutual exclusion between neighboring nodes in a tree structured distributed system, i.e., under the given protocol no two neighboring nodes can execute their critical sections concurrently. This protocol can be used to run a serial model self stabilizing algorithm in a distributed environment that accepts as atomic operations only send a message, receive a message an update a state. Unlike the scheme in [1], our protocol does not use time-stamps (which are basically unbounded integers); our algorithm uses only bounded integers (actually, the integers can assume values only 0, 1, 2 and 3) and can be easily implemented.

1 Introduction

Because of the popularity of the serial model and the relative ease of its use in designing new self-stabilizing algorithm, it is worthwhile to design lower level self-stabilizing protocols such that an algorithm developed for a serial model can be run in a distributed environment. This approach was used in [1] and can be compared with the layered approach use in networks protocol stacks. The advantage of such a lower level self-stabilizing protocol is that it makes the job of self-stabilizing application designer easier; one can work with the relatively easier serial model and does not have to worry about message management at lower level. Our purpose in this paper is to propose a new protocol that can be used to run a serial model self stabilizing algorithm in a distributed environment that accepts as atomic operations only send a message, receive a message an update a state. Unlike the scheme in [1], our protocol does not use time-stamps (which are basically unbounded integers); our algorithm uses only bounded integers (actually, the integers can assume values only 0, 1, 2 and 3) and can be easily implemented. Our algorithm is applicable for distributed systems whose underlying topology is a tree.

It is interesting to note that the proposed protocol can be viewed as a special class of self-stabilizing distributed mutual exclusion protocol. In traditional distributed mutual exclusion protocols, self-stabilizing or non self-stabilizing (for references in self-stabilizing distributed mutual exclusion protocols, see [2] and for non self-stabilizing distributed mutual exclusion protocols, see [3, 4]), the objective is to ensure that only one node in the system can execute its critical section at any given time (i.e., critical section execution is mutually exclusive from *all other* nodes in the system; the objective in our protocol, as in [1], is to ensure that a node executes its critical section mutually exclusive from its *neighbors* in the system graph (as opposed to all nodes in the system), i.e., multiple nodes can execute their critical sections concurrently as long as they are not neighbors to each other; in the critical section the node executes an atomic step of a serial model self-stabilizing algorithm.

2 Model

We model the distributed system, as used in this paper, by using an undirected graph $G = (V, E)$. The nodes represent the processors while the symmetric edges represent the bidirectional communication links. We assume that each processor has its unique *id*. Each node x maintains one or more local state variables and one or more local state vectors (one vector for each local variable) that are used to store copies of the local state variables of the neighboring nodes. Each node x maintains an integer variable S_x denoting its *status*; the node also maintains a local state vector LS_x where it stores the copies of the status of its neighbors (this local state vector contains d_x elements, where d_x is the degree of the node x, i.e. x has d_x many neighbors); we use the notation $LS_x(y)$ for the local state vector element of node x that keeps a copy of the local state variable S_y of neighbor node y. The *local state* of a node x is defined by its local state variables and its local state vectors. A node can both read and write its local state variables and its local state vectors; it can only read (and not write) the local state variables of its neighboring nodes and it does neither read nor write the local state vectors of other nodes. A configuration of the entire system or a system state is defined to be the vector of local states of all nodes.

Next, our model assumes read/write atomicity of [2] (as opposed to the composite read/write atomicity as in [5,6]). An atomic step (move) of a processor node consists of either reading the a local state variable of one of its neighbors (and updating the corresponding entry in the appropriate replica vector), or some internal computation, or writing of one of its local state variables; any such move is executed in finite time. Execution of a move by a processor may be interleaved with moves by other nodes – in this case the moves are concurrent. We use the notation $\mathcal{N}(x) = \{n_x[1], \ldots . n_x[d_x]\}$ to denote the set of neighbors of node x.

The process executed by each node x consists of an infinite loop of a finite sequence of Read, Conditional critical section and Write moves.

Definition 1. *An infinite execution is fair if it contains a infinite number of actions for any type.*

Remark 1. The purpose of our protocol is to ensure mutual exclusion between neighboring nodes. Each node x can execute its critical section iff the predicate Φ_x is true at node x. Thus, in a legitimate system state, mutual exclusion is ensured iff

$$\Phi_x \Rightarrow \forall y \mid y \text{ is a neighbor of } x, \quad \neg \Phi_y$$

i.e., as long as node x executes its CS, no neighbor y of node x can execute its CS ($\Phi_y \mid (y$ is a neighbor of $x)$ is false).

3 Self-Stabilizing Balance Unbalance Protocol for a Pair of Processes

We present an approach without using any shared variables unlike that in [2]. The structure of the two processes A and B is the same, i.e. infinite loop at each processor consists of an atomic read operation, critical section execution if certain predicate is true and an atomic write action. S_a and $LS_a(b)$ are two local variables maintained by process A ($LS_a(b)$ is the variable maintained at process A to store a copy of the state of its neighbor process B); process A can write on both S_a and $LS_a(b)$ and process B can only read S_a. Similarly, S_b and $LS_b(a)$ are local variables to process B: process B can write on both S_b and $LS_b(a)$ and process A can only read S_b. The difference is that the variables are now ternary, i.e., they can assume values 0, 1 or 2. The proposed algorithm is shown in Figure 1:

Process A	Process B
R_a: $LS_a(b) = S_b$;	R_b: $LS_b(a) = S_a$;
CS_a : if$(S_a = 0)$ then Execute CS	CS_b : if$(S_b = 1)$ then Execute CS
W_a: if$(LS_a(b) = S_a)$ then $S_a = S_a + 1 \bmod 3$	W_b: $S_b = LS_b(a)$;

Fig. 1. Self-Stabilizing Balance Unbalance Protocol for a Pair of Processes

Since each of the processes A and B executes an infinite loop, after a R_a action the next "A" type action is W_a, after a W_a action the next "A" type action is R_a, after a R_b action the next "B" type action is W_b, after a W_b action the next "B" type action is R_b and so on.

Remark 2. An execution of the system is an infinite execution of the processes A and B and hence an execution of the system contains an infinite number of each of the actions from the set $\{R_a, W_a, R_b, W_b\}$; thus, the execution is *fair*.

The system may start from an arbitrary initial state and the first action in the execution of the system can be any arbitrary one from the set $\{R_a, W_a, R_b, W_b\}$. Note that the global system state is defined by the variables S_a and $LS_a(b)$ in process A and the variables S_b and $LS_b(a)$ in process B.

Remark 3. When a process makes a move (the system executes an action), the system state may or may not be changed. For example, in a system state where $S_a \neq LS_a(b)$, the move W_a does not modify the system state, i.e., the system remains in the same state after the move.

Definition 2. *A move (action) that modifies the system state is called a* **modifying action**.

Our objective is to show that the system, when started from an arbitrary initial state (possibly illegitimate), converges to a legitimate state in finite time (after finitely many actions by the processes). We introduce a new binary relation.

Definition 3. *We use the notation $x \succeq y$, if $x = y$ or $x = (y+1) \bmod 3$, where $x, y \in Z_3$.*

Remark 4. The relation $\succeq$ is neither reflexive, nor transitive, nor symmetric. For example, $1 \succeq 0$, $2 \succeq 2$, $2 \not\succeq 0$, $2 \succeq 1$, $0 \not\succeq 1$, etc.

Definition 4. *Consider the ordered sequence of variable $(S_a, LS_b(a), S_b, LS_a(b))$; a system state is* **legitimate** *if (i) $S_a \succeq LS_b(a) \wedge LS_b(a) \succeq S_b \wedge S_b \succeq LS_a(b)$ and (ii) if at most one pair of successive variables in the previous sequence are unequal.*

Example 1. For example, $\{S_a = 1,\ LS_a(b) = 0,\ S_b = 0,\ LS_b(a) = 1\}$ is a legitimate state while $\{S_a = 2,\ LS_a(b) = 1,\ S_b = 1,\ LS_b(a) = 0\}$ is not.

Theorem 1. *In a legitimate state the two processes A and B execute their critical sections in mutual exclusive way, i.e., if process A is executing CS then process B cannot execute CS and vice versa, i.e. $S_a = 0 \Rightarrow S_b \neq 1$ and $S_b = 1 \Rightarrow S_a \neq 0$.*

Proof. The proof is obvious since in a legitimate state $S_a \succeq LS_b(a) \wedge LS_b(a) \succeq S_b \wedge S_b \succeq LS_a(b)$ and at most one pair of successive variables in the sequence can be unequal.

Theorem 2. *Any arbitrary move from the set $\{R_a, W_a, R_b, W_b\}$ made in a legitimate state of the system leads to a legitimate state after the move.*

Proof. Since there are only four possible moves, it is easy to check the validity of the claim. For example, consider the move R_a; the variable $LS_a(b)$ is affected; if $LS_a(b) = S_b$ before the move, this move does not change the system state; if $LS_a(b) \neq S_b$ before the move, then the system state before the move must satisfy $S_a = LS_b(a) = S_b$ (since it is legitimate) and after the move it will satisfy $S_a = LS_b(a) = S_b = LS_a(b)$ (hence, the resulting state is legitimate).

Lemma 1. *Any infinite fair execution contains infinitely many modifying actions (see Definition 2).*

Proof. By contradiction. Assume that after a finite number of moves S_a does not change its value anymore. Then after a complete loop executed by process B, $LS_b(a) = S_a$ and $S_b = S_a$. In the next loop the process A must move, which contradicts our assumption. It is easy to see that if S_a changes its value infinitely many times, any other variable changes its value infinitely many times.

Lemma 2. *For any given fair execution and for any initial state, a state such that three variables from the set $\{S_a,\ LS_b(a),\ S_b,\ LS_a(b)\}$ are equal each other is reached in a finite number of moves.*

Proof. Consider the first move that modifies the state of S_a. After this move $S_a \neq LS_a(b)$. To change again the value of S_a, the $LS_a(b)$ variable must change its value and become equal to S_a. But $LS_a(b)$ always takes the value of S_b. Since S_a change its value infinitely many times a state such that $S_a = LS_a(b)$ and $LS_a(b) = S_b$ is reached in a finite number of moves.

Lemma 3. *For any given fair execution and for any initial state, a state such that $S_a \neq LS_a(b)$, $S_a \neq S_b$, $S_a \neq LS_b(a)$, is reached in a finite number of moves.*

Proof. Since we use addition modulo 3, the variables S_a, $LS_b(a)$, S_b, $LS_a(b)$, can have values in the set $Z_3 = \{0, 1, 2\}$. When three variables from the set $\{S_a, LS_b(a), S_b, LS_a(b)\}$ are equal to each other, Lemma 2, there is a value $i \in Z_3$ such that $S_a \neq i$, $LS_a(b) \neq i$, $S_b \neq i$, $LS_b(a) \neq i$. When $LS_b(a)$, S_b, $LS_a(b)$ change their values, they only copy the value of one variable in the set $\{S_a, LS_b(a), S_b, LS_a(b)\}$. Thus, when S_a reaches for the first time the value i, the condition $S_a \neq LS_a(b)$, $S_a \neq S_b$, $S_a \neq LS_b(a)$ is met.

Theorem 3. *For any given fair execution, the system starting from any arbitrary state reaches a legitimate state in finitely many moves.*

Proof. The system reaches a state such that $S_a \neq LS_a(b)$, $S_a \neq S_b$, $S_a \neq LS_b(a)$ in a finite number of moves, Lemma 3. Let $S_a = i \in Z_3$ Since $LS_b(a) \neq i$, $S_b \neq i$ and $LS_a(b) \neq i$, S_a can not change its value until $LS_a(b)$ becomes equal to i, $LS_a(b)$ can not become equal to i until S_b becomes equal to i and S_b can not become equal to i until $LS_b(a)$ becomes equal to i. Thus, S_a can not modify its state until a legitimate state is reached.

4 Self-Stabilizing Mutual Exclusion Protocol for a Tree Network Without Shared Memory

Consider an arbitrary rooted tree; the tree is rooted at node r. We use the notation d_x for the degree of node x, $n_x[j]$ for the j-th neighbor of the node x, $\mathcal{N}(x) = \{n_x[1], \ldots n_x[d_x]\}$ for the set of neighbors of node x, $\mathcal{C}(x)$ for the set of children of x and P_x for the parent of node x; since the topology is a tree each node x knows its parent P_x and for the root node r, P_r is Null.. As before, each node x maintains a local state variable S_x (readable by its neighbors) and a local state vector LS_x used to store the copies of the states of its neighbors; we use notation $LS_x(y)$ to denote the component of the state vector LS_x that stores a copy of the state variable S_y of node y, $\forall y \in \mathcal{N}(x)$. All variables are now modulo 4 integers (we explain the reason later). We assume that each node x maintains a height variable H_x such that $H_r = 0$ for the root node and for $\forall x \neq r$, H_x is

Root node r
R_r^1: $LS_r(n_r[1]) = S_{n_r[1]}$;
$\vdots$ $\vdots$
$R_r^{d_r}$: $LS_r(n_r[d_r]) = S_{n_r[d_r]}$;
CS_r: if $S_r = 0$ then   Execute Critical Section;
W_r: if $\left(\bigwedge_{y \in C(r)} (S_r = LS_r(y)) \right)$ then $S_r = S_r + 1 \mod 4$;

Leaf node y
R_y^1: $LS_y(P_y) = S_{P_y}$;
CS_y: if $\Phi(y)$ then   Execute Critical Section;
W_y: $S_y = LS_y(P_y)$;

Internal Node x
R_x^1: $LS_x(n_x[1]) = S_{n_x[1]}$
$\vdots$ $\vdots$
$R_x^{d_x}$: $LS_x(n_x[d_x]) = S_{n_x[d_x]}$
CS_x: if $\Phi(x)$ then   Execute Critical Section
W_x: if $\left(\bigwedge_{y \in C(x)} (S_r = LS_x(y)) \right)$ then $S_x = RH_x(P_x)$;

Fig. 2. Protocol for an Arbitrary Tree

the number of edges in the unique path from node x to the root node. It is easy to see that if the root node sets $H_r = 0$ and any other node x sets its H_x to $H_{P_x} + 1$ (level of its parent plus 1), the height variables will correctly indicate the height of each node in the tree after a finite number of moves, starting from any illegitimate values of those variables. To avoid cluttering the algorithm, we do not include the rules for handling H_x variable in our algorithm specification. As before, the root node, internal nodes as well as the leaf nodes execute infinite loops of reading the states of neighbor(s), executing critical sections (if certain predicate is satisfied) and writing its new state. The protocols (algorithms) for root, internal nodes and leaf nodes are shown in Figure 2 where the predicate $\Phi(x)$ is:

$$\Phi(x) = (S_x = 0 \wedge (H_x \text{ is even})) \vee (S_x = 2 \wedge (H_x \text{ is odd}))$$

Note, as before, the state of a node x is defined by the variable S_x and the vector LS_x; the global system state is defined by the local states of all participating nodes in the tree.

Definition 5. *Consider a link or edge (x, y) such that node x is the parent of node y in the given tree. The state of a link (x, y), in a given system state, is* *defined to be the vector* $(S_x, LS_y(x), S_y, LS_x(y))$. *The state of a link is called* **legitimate** *iff* $S_x \succeq LS_y(x) \wedge LS_y(x) \succeq S_y \wedge S_y \succeq LS_x(y)$ *and at most one pair of successive variables in the vector* $(S_x, LS_y(x), S_y, LS_x(y))$ *are unequal.*

Definition 6. *The system is in a legitimate state if all links are in a legitimate state.*

Theorem 4. *For an arbitrary tree in any legitimate state, no two neighboring processes can execute their critical sections simultaneously.*

Proof. In a legitimate state (when the H variables at nodes have stabilized) for any two neighboring nodes x and y, we have (either L_x is even & L_y is odd), or (L_x is odd & L_y is even). Hence, $\Phi(x)$ and $\Phi(y)$ are simultaneously (concurrently) true iff $S_x = 2$ and $S_y = 0$ or $S_x = 0$ and $S_y = 2$. But since link (x, y) is in a legitimate state, such condition can not be met; hence, two neighboring nodes cannot execute their critical sections simultaneously.

Lemma 4. *Consider an arbitrary link* (x, y). *The node* x *modifies the value of the variable* S_x *infinitely many times, if and only if the node* y *modifies the value of the variable* S_y *infinitely many times.*

Proof. If node x modifies S_x finitely many times, then after a finite number of moves the value of S_x is not modified anymore. The next complete loop of node y after the last modification of S_x, makes $LS_y(x) = S_x$ and $LS_y(x)$ is not be modified by subsequent moves. Hence, after at most one modification, S_y remains unchanged. Conversely, if node y modifies S_y finitely many times then after a finite number of moves the value of S_y is not modified anymore. The next complete loop of node x after the last modification of S_y, makes $LS_x(y) = S_y$ and $LS_x(y)$ is not be modified by subsequent moves. Hence, after at most one modification, S_x remains unchanged.

Lemma 5. *For any fair execution, variable* S_r *at root node* r *is modified infinitely many times.*

Proof. By contradiction. Assume that the value of S_r is modified finitely many times. Then, after a finite number of moves, the value S_y for each child y of r will not be modified anymore, Lemma 4. Repeating the argument, after a finite number of moves no S or LS variables for any node in the tree may be modified. Consider now the leaf nodes. Since the execution is fair, the condition $S_z = LS_z(P_z)$ must be met for each leaf node z. If this condition is met for leaf nodes it must be met for the parents of leaf nodes too. Repeating the argument, the condition must be met by all nodes in the tree. But, if this condition is met, the root node r modifies its S_r variable in its next move, which is a contradiction.

Lemma 6. *For any fair execution and for any node* x, *the variable* S_x *is modified infinitely many times.*

Proof. If node x modifies its variable S_x finitely many times, its parent, say node z, must modify its variable S_z only finitely many times, Lemma 4. Repeating the argument, the root node also modifies its variable S_r finitely many times, which contradicts Lemma 5.

Lemma 7. *Consider an arbitrary node z ($\neq r$). If all links in the path from r to z are in a legitimate state, then these links remain in a legitimate state after an arbitrary move by any node in the system.*

Proof. Let (x, y) be an link in the path from r to z. Since (x, y) is in a legitimate state, we have $S_x \succeq LS_y(x) \wedge LS_y(x) \succeq S_y \wedge S_y \succeq LS_x(y)$ and at most one pair of successive variables in the sequence $(S_x, LS_y(x), S_y, LS_x(y))$ are unequal. We need to consider only those system moves (executed by nodes x and y) that can change the variables S_x, $LS_y(x)$, S_y, $LS_x(y)$. Considering each of these moves, we check that legitimacy of the link state is preserved in the resulting system state. The read moves (that update the variables $LS_y(x)$, $LS_x(y)$) obviously preserve legitimacy. To consider the move W_x, we look at two cases differently:

    Case 1 ($\mathbf{x} = r$): When W_x is executed, S_x can be modified (incremented by 1) only when $S_x = LS_x(y)$. Thus, since the state is legitimate, a W_x move can increment S_x only under the condition $S_x = LS_y(x) = S_y = LS_x(y)$ and after the move the link (x, y) remains in a legitimate state.

    Case 2 ($\mathbf{x} \neq r$): Since the link (t, x) (where node t is the parent of x, i.e., $= P_x$) is in a legitimate state, we have $LS_x(t) = S_x$ or $LS_x(t) = (S_x+1) \bmod 4$. When W_x is executed, S_x can be modified only by setting its value equal to $LS_x(t)$; hence, after the move the link (x, y) remains in a legitimate state.

Lemma 7 shows that if a path of legitimate links from the root to a node is created the legitimacy of the links in this path is preserved for all subsequent states. The next lemma shows that a new link is added to such a path in finite time.

Lemma 8. *Let (x, y) be a link in the tree. If all links in the path from root node to node x are in a legitimate state, then the link (x, y) becomes legitimate in a finite number of moves.*

Proof. First, we observe that the node y modifies the variable S_y infinitely many times, Lemma 6. Then, we use the same argument as in the proof of Theorem 3 to show that the link (x, y) becomes legitimate after a finite number of moves.

Theorem 5. *Starting from an arbitrary state, the system reaches a legitimate state in finite time (in finitely many moves).*

Proof. The first step is to prove that all links from the root node to its children become legitimate after a finite number of moves. This follows from the observation that each child x of the root node modifies its S variable infinitely many times and from an argument similar to the argument used in the proof of Theorem 3. Using Lemma 8, the theorem follows.

References

1. M. Mizuno and H. Kakugawa. A timestamp based transformation of self-stabilizing programs for distributed computing environments. In *Proceedings of the 10th International Workshop on Distributed Algorithms (WDAG'96)*, volume 304–321, 1996.
2. S. Dolev, A. Israeli, and S. Moran. Self-stabilization of dynamic systems assuming only read/write atomicity. *Distributed Computing*, 7:3–16, 1993.
3. M. Raynal. *Algorithms for Mutual Exclusion*. MIT Press, Cambridge MA, 1986.
4. P. K. Srimani and S. R. Das, editors. *Distributed Mutual Exclusion Algorithms*. IEEE Computer Society Press, Los Alamitos, CA, 1992.
5. M. Flatebo, A. K. Datta, and A. A. Schoone. Self-stabilizing multi-token rings. *Distributed Computing*, 8:133–142, 1994.
6. S. T. Huang and N. S. Chen. Self-stabilizing depth-first token circulation on networks. *Distributed Computing*, 1993.
7. E. W. Dijkstra. Solution of a problem in concurrent programming control. *Communication of the ACM*, 8(9):569, September 1965.
8. L. Lamport. A new solution of Dijkstra's concurrent programming problem. *Communications of the ACM*, 17(8):107–118, August 1974.
9. L. Lamport. The mutual exclusion problem: Part II – statement and solutions. *Journal of the ACM*, 33(2):327–348, 1986.
10. H. S. M. Kruijer. Self-stabilization (in spite of distributed control) in tree-structured systems. *Inf. Processing Letters*, 8(2):91–95, 1979.

Irreversible Dynamos in Tori *

P. Flocchini[1], E. Lodi[2], F. Luccio[3], L. Pagli[3], N. Santoro[4]

[1] Université de Montréal, Canada
[2] Università di Siena, Italy
[3] Università di Pisa, Italy
[4] Carleton University, Canada

Abstract. We study the dynamics of majority-based distributed systems in presence of permanent faults. In particular, we are interested in the patterns of initial faults which may lead the entire system to a faulty behaviour. Such patterns are called *dynamos* and their properties have been studied in many different contexts. In this paper we investigate dynamos for meshes with different types of toroidal closures. For each topology we establish tight bounds on the number of faulty elements needed for a system break-down, under different majority rules.

1 Introduction

Consider the following repetitive process on a *synchronous* network G: initially each vertex is in one of two states (colors), *black* or *white*; at each step, all vertices simultaneously (re)color themselves either black or white, each according to the color of the "majority" of its neighbors (majority rule). Different processes occur depending on how majority is defined (e.g., simple, strong, weighted) and on whether or not the neighborhood of a vertex includes that vertex. The problem is to study the initial configurations (assignment of colours) from which, after a finite number of steps, a monochromatic fixed point is reached, that is, all vertices become of the same colour (e.g., black). The initial set of black vertices is called *dynamo* (short for "dynamic monopoly") and their study has been introduced by Peleg [17] as an extension of the study of monopolies.

The dynamics of majority rules have been extensively studied in the context of cellular automata, and much effort has been concentrated on determining the asymptotic behaviors of different majority rules on different graph structures. In particular, it has been shown that, if the process is periodic, its period is at most two. Most of the existing research has focused, for example, on the study of the period-two behavior of symmetric weighted majorities on finite $\{0,1\}-$ and $\{0,\ldots,p\}$-colored graphs [9,19], on the number of fixed points on finite $\{0,1\}$-colored rings [1,2,10], on finite and infinite $\{0,1\}$-colored lines [12,13], on the behaviors of infinite, connected $\{0,1\}$-colored graphs [14]. Furthermore, dynamic majority has been applied to the immune system and to image processing [1,8].

Although the majority rule has been extensively investigated, not much is known regarding dynamos. Some results are known in terms of *catastrophic fault*

---

* This work has been supported in part by MURST, NSERC, and FCAR.

patterns which are dynamos based on "one-sided" majority for infinite chordal rings (e.g., [5, 15, 20]). Further results are known in the study of *monopolies*, that is dynamos for which the system converges to all black in a single step [3, 4, 16].

Other more subtle definitions have been posed. An *irreversible* dynamo is one where the initial black vertices do not change their colour regardless of their neighbourhood. This is opposed to *reversible* dynamos, for which a vertex may switch colour several times according to a changing neighborhood. Among the latter, *monotone* reversible dynamos are the ones for which black vertices remain always black because the neighborhood never forces them to turn white. Recently, some general lower and upper bounds on the size of monotone dynamos have been estibilished in [17], and a characterization of irreversible dynamos has been given for chordal rings in [7].

In this paper we consider irreversible dynamos in tori and we focus on their dimension, that is, the minimum number of initial black elements needed to reach the fixed point. The motivation for irreversible dynamos comes from fault-tolerance. Initial black vertices correspond to permanent faulty elements, and the white correspond to non-faulty. The faulty elements can induce a faulty behavior in their neighbors: if the majority of its neighbors is faulty (or has a faulty behavior), a non-faulty element will exhibit a faulty behavior and will therefore be indistinguishable from a faulty one. Irreversible dynamos are precisely those patterns of permanent initial faults whose occurrence leads the entire system to a faulty behavior (or catastrophe). In addition to its practical importance and theoretical interest, the study of irreversible dynamos gives insights on the class of monotone dynamos. In particular, all lower bounds established on the size of irreversible dynamos are immediately lower bounds for the monotone case.

The torus is one of the simplest and most natural way of connecting processors in a network. We consider different types of tori: the *toroidal mesh* (the classical architecture used in VLSI), the *torus cordalis* (also known as double-loop interconnection networks), and the *torus serpentinus* (e.g., used by ILIAC IV). For each of these topologies we derive lower and upper bounds on the dimensions of dynamos. For a summary of results see table 1. The upper bounds are constructive, that is we derive the initial black vertices constituting the dynamo, and we also analyze the completion time (i.e., the time necessary to reach the fixed point).

Limiting the discussion to meshes with toroidal connections avoids to examine border effects for some vertices, as it would occur in simple meshes. Our results and techniques can be easily adapted to simple meshes.

In the following, due to space limitations, all the proofs have been omitted; the interested reader is referred to [6].

2 Basic Definitions

Let us consider an $m \times n$ mesh M and denote with $\mu_{i,j}$, $0 \leq i \leq m - 1$, $0 \leq j \leq n - 1$, a vertex of M. The differences among the considered topologies consist only in the way that border vertices (i.e., $\mu_{i,0}, \mu_{i,n-1}$ with $0 \leq i \leq m - 1$

Table 1. Bounds on the size of irreversible dynamos for tori of $m \times n$ vertices, with $M = max\{m,n\}$, $N = min\{m,n\}$, and $H, K = m, n$ or $H, K = n, m$ (choose the alternative that yields stricter bounds).

	Simple majority		Strong majority	
	Lower Bound	*Upper Bound*	*Lower Bound*	*Upper Bound*
Toroidal mesh	$\lceil \frac{m+n}{2} \rceil - 1$	$\lceil \frac{m+n}{2} \rceil - 1$	$\lceil \frac{mn+1}{3} \rceil$	$\lceil \frac{H}{3} \rceil (K+1)$
Torus cordalis	$\lceil \frac{n}{2} \rceil$	$\lfloor \frac{n}{2} \rfloor + 1$	$\lceil \frac{mn+1}{3} \rceil$	$\lceil \frac{m}{3} \rceil (n+1)$
Torus serpentinus	$\lfloor \frac{N}{2} \rfloor$, for $\lceil \frac{M}{3} \rceil \geq \lfloor \frac{N}{2} \rfloor$ $\lfloor \frac{M}{3} \rfloor$, for $\lceil \frac{M}{3} \rceil < \lfloor \frac{N}{2} \rfloor$	$\lfloor \frac{N}{2} \rfloor + 1$	$\lceil \frac{mn+1}{3} \rceil$	$\lceil \frac{H}{3} \rceil (K+1)$

and $\mu_{0,j}, \mu_{m-1,j}$ with $0 \leq j \leq n-1$) are linked to other processors. The vertices $\mu_{i,n-1}$ on the last column are usually connected either to the opposite ones on the same rows (i.e. to $\mu_{i,0}$), thus forming ring connections in each row, or to the opposite ones on the successive rows (i.e. to $\mu_{i+1,0}$), in a snake-like way. The same linking strategy is applied for the last row.

In the toroidal mesh rings are formed in rows and columns; in the torus cordalis there are rings in the columns and snake-like connections in the rows; finally, in torus serpentinus there are snake-like connections in rows and columns. Formally, we have:

Definition 1 Toroidal Mesh
A toroidal mesh of $m \times n$ vertices is a mesh where each vertex $\tau_{i,j}$, with $0 \leq i \leq m-1, 0 \leq j \leq n-1$ is connected to the four vertices $\tau_{(i-1) \bmod m, j}$, $\tau_{(i+1) \bmod m, j}$, $\tau_{i,(j-1) \bmod n}$, $\tau_{i,(j+1) \bmod n}$ (mesh connections).

Definition 2 Torus Cordalis
A torus cordalis of $m \times n$ vertices is a mesh where each vertex $\tau_{i,j}$, with $0 \leq i \leq m-1, 0 \leq j \leq n-1$ has mesh connections except for the last vertex $\tau_{i,n-1}$ of each row i, which is connected to the first vertex $\tau_{(i+1) \bmod m, 0}$ of row $i+1$.

Notice that this torus can be seen as a chordal ring with one chord.

Definition 3 Torus Serpentinus
A torus serpentinus of $m \times n$ vertices is a mesh where each vertex $\tau_{i,j}$, with $1 \leq i \leq m-1, 0 \leq j \leq n-1$ has mesh connections, except for the last vertex $\tau_{i,n-1}$ of each row i which is connected to the first vertex $\tau_{(i+1) \bmod m, 0}$ of row $i+1$, and for the last vertex $\tau_{m-1,j}$ of each column j which is connected to the first vertex $\tau_{0,(j-1) \bmod n}$ of column $j-1$.

Majority will be defined as follows:

Definition 4 Irreversible-majority rule. *A vertex* v *becomes black if the majority of its neighbours are black. In case of tie* v *becomes black* (simple majority), *or keeps its color* (strong majority).

In the tori, simple (or strong) irreversible-majority asks for at least two (or three) black neighbours. We can now formally define dynamos:

Definition 5 *A* simple *(respectively:* strong*) irreversible dynamo is an initial set of black vertices from which an all black configuration is reached in a finite number of steps under the simple (respectively: strong) irreversible-majority rule.*

Simple and strong majorities will be treated separately because they exhibit different properties and are treated by different techniques.

3   Irreversible Dynamos with Simple Majority

Network behaviour changes drastically if we pass from simple to strong majority. We start our study from the former case.

3.1   Toroidal Mesh

Consider a toroidal mesh with a set T of $m \times n$ vertices, $m, n > 2$. Each vertex has four neighbors, then two black neighbors are enough to color black a white vertex. Let $S \subseteq T$ be a generic subset of vertices, and R_S be the *smallest rectangle* containing S. The size of R_S is $m_S \times n_S$. If S is all black, a *spanning set* for S (if any) is a *connected* black set $\sigma(S) \supseteq S$ derivable from S with consecutive applications of the simple majority rule. We have:

Proposition 1 *Let S be a black set, $m_S < m - 1$, $n_S < n - 1$. Then, any (non necessarily connected) black set B derivable from S is such that $B \subseteq R_S$.*

Proposition 2 *Let S be a black set. The existence of a spanning set $\sigma(S)$ implies $|S| \geq \lceil \frac{(m_S + n_S)}{2} \rceil$.*

From Propositions 1 and 2 we immediately derive our first lower bound result:

Theorem 1 *Let S be a simple irreversible dynamo for a toroidal mesh $m \times n$. We have:*
(i) $m_S \geq m - 1, n_S \geq n - 1$;
(ii) $|S| \geq \lceil \frac{(m+n)}{2} \rceil - 1$.

To build a matching upper bound we need some further results.

Proposition 3 *Let S be a black set, such that a spanning set $\sigma(S)$ exists. Then a black set R_S can be built from S. (note: R_S is then a spanning set of S).*

Definition 6 *An* alternating chain C *is a sequence of adjacent vertices starting and ending with black. The vertices of* C *are alternating black and white, however, if* C *has even length there is exactly one pair of consecutive black vertices somewhere.*

Theorem 2 *Let* S *be a black set consisting of the black vertices of an alternating chain* C, *with* $m_S = m - 1$ *and* $n_S = n - 1$. *Then, the whole torus can be colored black starting from* S, *that is* S *is a simple irreversible dynamo.*

An example of alternating chain of proper length, and the phases of the algorithm, are illustrated in Figure 1. From Theorem 2 we have:

```
o o o o o o o      o o o o o o o      o o o o o o o
● o ● o o o o      ● ● ● o o o o      ● ● ● ● ● ● o
o o o o o o o      ● o o o o o o      ● ● ● ● ● ● o
● o o o o ● o      ● o o o o ● o      ● ● ● ● ● ● o
o o o o o o o      ● o o o o ● o      ● ● ● ● ● ● o
● ● o ● o ● o      ● ● ● ● o ● o      ● ● ● ● ● ● o
o o o o o o o      o o o ● o ● o      ● ● ● ● ● ● o
o o o ● o ● o      o o o ● ● ● o      ● ● ● ● ● ● o

      S                 σ(S)                R_S
```

Fig. 1. The initial set S, the spanning set $\sigma(S)$ and the smallest rectangle R_S containing S.

Corollary 1 *Any* $m \times n$ *toroidal mesh admits a simple irreversible dynamo* S *with* $|S| = \lceil \frac{(m+n)}{2} \rceil - 1$.

Note that the lower bound of Theorem 1 matches the upper bound of Corollary 1. From the proof of this corollary we see that a dynamo of minimal cardinality can be built on an alternating chain. Furthermore we have:

Corollary 2 *There exists a simple irreversible dynamo* S *of minimal cardinality starting from which the whole toroidal mesh can be colored black in* $\lceil \frac{(m+n)}{2} \rceil$ *steps.*

An example of such a dynamo is shown in Figure 3.1.

An interesting observation is that the coloring mechanism shown in corollary 2 works also for asynchronous systems. In this case, however, the number of steps looses its significance.

3.2 Torus Cordalis

If studied on a torus cordalis, simple irreversible-majority is quite easy. We now show that any simple irreversible dynamo must have size $\geq \lceil \frac{n}{2} \rceil$, and that there exist dynamos with almost optimal size.

```
                              o o o o o
                              o ● o o o
                              o o o o o
                              o ● o o o
              o o o o o       o o ● o o
              o ● o o o       o o o ● o
              o o ● o o       o o o o ●
              o o o ● o       o o o o o
              o o o o ●       o o o o ●
```

Fig. 2. Examples of dynamos requiring $\lceil \frac{(m+n)}{2} \rceil$ steps.

Theorem 3 *Let S be a simple irreversible dynamo for a torus cordalis $m \times n$. We have: $|S| \geq \lceil \frac{n}{2} \rceil$.*

Theorem 4 *Any $m \times n$ torus cordalis admits a simple irreversible dynamo S with $|S| = \lfloor \frac{n}{2} \rfloor + 1$. Starting from S the whole torus can be colored black in $\lfloor \frac{m-3}{2} \rfloor n + 3$ steps.*

An example of such dynamos is given in Figure 3.

```
      o o o o o o o o       o o o o o o o o o
      o o o o o o o o       o o o o o o o o o
      ● o ● o ● o ● o       ● o o ● o ● o ● o
      o ● o o o o o o       o ● o o o o o o o
      o o o o o o o o       o o o o o o o o o
      o o o o o o o o       o o o o o o o o o
      o o o o o o o o       o o o o o o o o o
                            o o o o o o o o o
```

Fig. 3. Simple irreversible dynamos of $\lfloor \frac{n}{2} \rfloor + 1$ vertices for tori cordalis, with n even and n odd.

3.3 Torus Serpentinus

Since the torus serpentinus is symmetric with respect to rows and columns, we assume without loss of generality that $m \geq n$, and derive lower and upper bounds to the size of any simple irreversible dynamo. For $m \leq n$ simply exchange m with n in the expressions of the bounds.

Consider a *white cross*, that is a set of white vertices arranged as in Figure 5, with height m and width n. The parallel white lines from the square of nine vertices at the center of the cross to the borders of the mesh are called *rays*. Note that each vertex of the cross is adjacent to three other vertices of the same cross, thus implying:

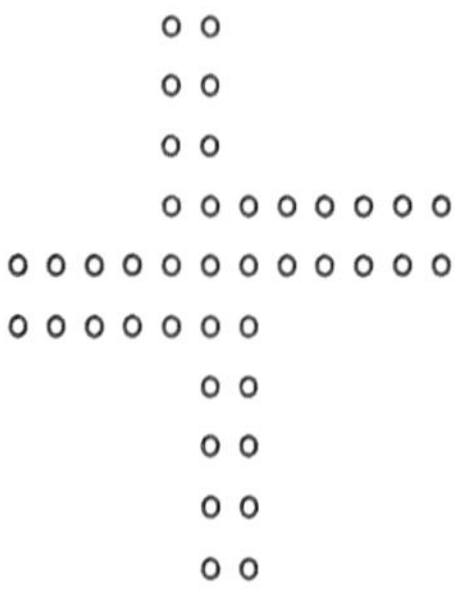

```
                  o o
                  o o
                  o o
                  o o o o o o o
      o o o o o o o o o o o o
      o o o o o o o
                  o o
                  o o
                  o o
                  o o
```

Fig. 4. The white cross configuration.

Proposition 4 *In a torus serpentinus the presence of a white cross prevents a simple irreversible dynamo to exist.*

We then have:

Proposition 5 *If a torus serpentinus contains three consecutive white rows, any simple irreversible dynamo must contain $\geq \lfloor \frac{n}{2} \rfloor$ vertices.*

Based on Proposition 5 we obtain the following lower bounds:

Theorem 5 *Let S be a simple irreversible dynamo for a torus serpentinus $m \times n$. We have:*

1. $|S| \geq \lfloor \frac{n}{2} \rfloor, \quad for \lceil \frac{m}{3} \rceil \geq \lfloor \frac{n}{2} \rfloor;$
2. $|S| \geq \lceil \frac{m}{3} \rceil, \quad for \lceil \frac{m}{3} \rceil < \lfloor \frac{n}{2} \rfloor.$

The upper bound for the torus serpentinus is identical to the one already found for the torus cordalis. In fact we have:

Theorem 6 *Any $m \times n$ torus serpentinus admits a simple irreversible dynamo S with $|S| = \lfloor \frac{n}{2} \rfloor + 1$. Starting from S the whole torus can be colored black in $\lfloor \frac{m-3}{2} \rfloor n + 3$ steps.*

4   Irreversible Dynamos with Strong Majority

A strong majority argument allows to derive a significant lower bound valid for the three considered families of tori (simply denoted by tori). Since these tori have a neighbourhood of four, three adjacent vertices are needed to color black a white vertex under strong majority. We have:

Theorem 7 *Let S be a strong irreversible dynamo for a torus $m \times n$. Then $|S| \geq \lceil \frac{mn+1}{3} \rceil.$*

We now derive an upper bound also valid for all tori.

Theorem 8 *Any $m \times n$ torus admits a strong irreversible dynamo S with $|S| = \lceil \frac{m}{3} \rceil (n+1)$. Starting from S the whole torus can be colored black in $\lfloor \frac{n}{2} \rfloor + 1$ steps.*

An example of such a dynamo is shown in Figure 5.

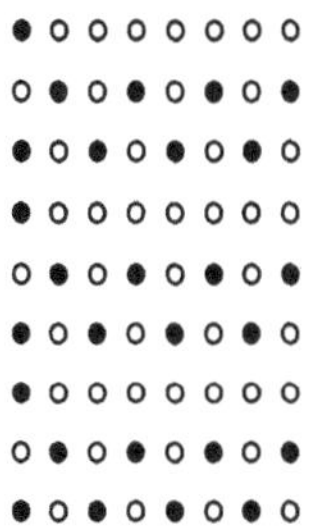

$$\bullet \circ \circ \circ \circ \circ \circ \circ$$
$$\circ \bullet \circ \bullet \circ \bullet \circ \bullet$$
$$\bullet \circ \bullet \circ \bullet \circ \bullet \circ$$
$$\bullet \circ \circ \circ \circ \circ \circ \circ$$
$$\circ \bullet \circ \bullet \circ \bullet \circ \bullet$$
$$\bullet \circ \bullet \circ \bullet \circ \bullet \circ$$
$$\bullet \circ \circ \circ \circ \circ \circ \circ$$
$$\circ \bullet \circ \bullet \circ \bullet \circ \bullet$$
$$\bullet \circ \bullet \circ \bullet \circ \bullet \circ$$

$$m = 9, \ n = 8, \ |S| = \lceil \tfrac{m}{3} \rceil (n+1) = 27$$

Fig. 5. A strong irreversible dynamo for toroidal mesh, torus cordalis or torus serpentinus.

For particular values of m and n the bound of Theorem 8 can be made stricter for the toroidal mesh and the torus serpentinus. In fact these networks are symmetrical with respect to rows and columns, hence the pattern of black vertices reported in figure 5 can be turned of 90 degrees, still constituting a dynamo. We immediately have:

Corollary 3 *Any $m \times n$ toroidal mesh or torus serpentinus admits a strong irreversible dynamo S with $|S| = max\{\lceil \frac{m}{3} \rceil (n+1), \lceil \frac{n}{3} \rceil (m+1)\}$.*

References

1. Z. Agur. Fixed points of majority rule cellular automata with application to plasticity and precision of the immune system. *Complex Systems*, 2:351–357, 1988.
2. Z. Agur, A. S. Fraenkel, and S. T. Klein. The number of fixed points of the majority rule. *Discrete Mathematics*, 70:295–302, 1988.
3. J-C Bermond, D. Peleg. The power of small coalitions in graphs. In *Proc. of 2nd Coll. on Structural Information and Communication Complexity*, 173-184, 1995.
4. J-C Bermond, J. Bond, D. Peleg, S. Perennes. Tight bounds on the size of 2-monopolies In *Proc. of 3rd Coll. on Structural Information and Communication Complexity*, 170-179, 1996.
5. R. De Prisco, A. Monti, L. Pagli. Efficient testing and reconfiguration of VLSI linear arrays. To appear in *Theoretical Computer Science*.
6. P.Flocchini, E. Lodi, F.Luccio, L.Pagli, N.Santoro. Irreversible dynamos in tori. Technical Report n. TR-98-05, Carleton University, School of Compiter Science, 1998.

7. P. Flocchini, F. Geurts, N. Santoro. Dynamic majority in general graphs and chordal rings. *Manuscript*, 1997.

8. E. Goles, S. Martinez. Neural and automata networks, dynamical behavior and applications. *Maths and Applications*, Kluwer Academic Publishers, 1990.

9. E. Goles and J. Olivos. Periodic behavior of generalized threshold functions. *Discrete Mathematics*, 30:187–189, 1980.

10. A. Granville. On a paper by Agur, Fraenkel and Klein. *Discrete Mathematics*, 94:147–151, 1991.

11. N. Linial, D. Peleg, Y. Rabinovich, M. Sachs. Sphere packing and local majority in graphs In *Proc. of 2nd ISTCS*, IEEE Comp. Soc. Press. pp. 141-149, 1993.

12. G. Moran. Parametrization for stationary patterns of the r-majority operators on 0–1 sequences. *Discrete Mathematics*, 132:175–195, 1994.

13. G. Moran. The r-majority vote action on 0–1 sequences. *Discrete Mathematics*, 132:145–174, 1994.

14. G. Moran. On the period-two-property of the majority operator in infinite graphs. *Transactions of the American Mathematical Society*, 347(5):1649–1667, 1995.

15. A. Nayak, N. Santoro, R. Tan. Fault intolerance of reconfigurable systolic arrays. *Proc. 20th Int. Symposium on Fault-Tolerant Computing*, 202-209, 1990.

16. D. Peleg. Local majority voting, small coalitions and controlling monopolies in graphs: A review. In *Proc. of 3rd Coll. on Structural Information and Communication Complexity*, 152-169, 1996.

17. D. Peleg. Size bounds for dynamic monopolies In *Proc. of 3rd Coll. on Structural Information and Communication Complexity*, 151-161, 1997.

18. S. Poljak. Transformations on graphs and convexity. *Complex Systems*, 1:1021–1033, 1987.

19. S. Poljak and M. Sûra. On periodical behavior in societies with symmetric influences. *Combinatorica*, 1:119–121, 1983.

20. N. Santoro, J. Ren, A. Nayak. On the complexity of testing for catastrophic faults. In *Proc. of 6th Int. Symposium on Algorithms and Computation*, 188-197, 1995.

MPI-GLUE: Interoperable High-Performance MPI Combining Different Vendor's MPI Worlds

Rolf Rabenseifner

Rechenzentrum der Universität Stuttgart, Allmandring 30, D-70550 Stuttgart,
rabenseifner@rus.uni-stuttgart.de,
http://www.hlrs.de/people/rabenseifner/

Abstract. Several metacomputing projects try to implement MPI for homogeneous and heterogeneous clusters of parallel systems. MPI-GLUE is the first approach which exports nearly full MPI 1.1 to the user's application without losing the efficiency of the vendors' MPI implementations. Inside of each MPP or PVP system the vendor's MPI implementation is used. Between the parallel systems a slightly modified TCP-based MPICH is used, i.e. MPI-GLUE is a layer that combines different vendors' MPIs by using MPICH as a global communication layer. Major design decisions within MPI-GLUE and other metacomputing MPI libraries (PACX-MPI, PVMPI, Globus and PLUS) and their implications for the programming model are compared. The design principles are explained in detail.

1 Introduction

The development of large scale parallel applications for clusters of high-end MPP or PVP systems requires efficient and standardized programming models. The message passing interface MPI [9] is one solution. Several metacomputing projects try to solve the problem that mostly different vendors' MPI implementations are not interoperable. MPI-GLUE is the first approach that exports nearly full MPI 1.1 without losing the efficiency of the vendors' MPI implementations inside of each parallel system.

2 Architecture of MPI-GLUE

MPI-GLUE is a library exporting the standard MPI 1.1 to parallel applications on clusters of parallel systems. MPI-GLUE imports the native MPI library of the system's vendor for the communication inside of a parallel system. Parallel systems in this sense are MPP or PVP systems, but also homogeneous workstation clusters. For the communication between such parallel systems MPI-GLUE uses a portable TCP-based MPI library (e.g. MPICH). In MPI-GLUE the native MPI library is also used by the portable MPI to transfer its byte messages inside of each parallel system. This design allows any homogeneous and heterogeneous combination of any number of parallel or nonparallel systems. The details are shown in Fig. 1.

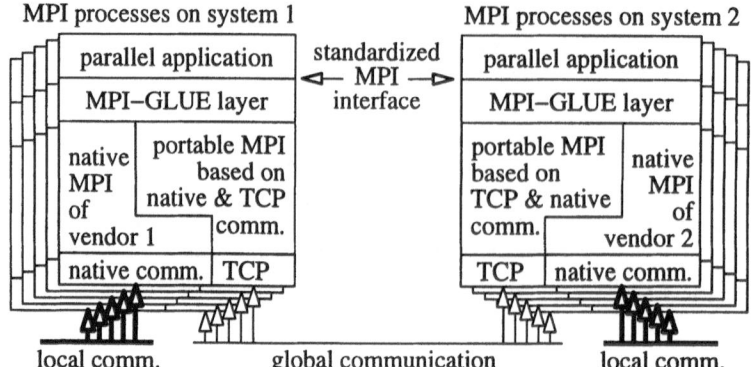

Fig. 1. Software architecture

3 Related Works

The **PACX-MPI** [1] project at the computing center of the University of Stuttgart has a similar approach as MPI-GLUE. At Supercomputing '97 PACX-MPI was used to demonstrate that MPI-based metacomputing can solve large scale problems. Two CRAY T3E with 512 processors each were combined to run a flow solver (URANUS) which predicts aerodynamic forces and high temperatures effecting on space vehicles when they reenter Earth's atmosphere [2]. Furthermore a new world record in simulating an FCC crystal with 1.399.440.00 atoms has been reached on that cluster [11]. PACX-MPI uses a special module for the TCP communication in contrast to the portable MPI in MPI-GLUE. The functionality of PACX-MPI is limited by this module to a small subset of MPI 1.1. The TCP communication does not directly exchange messages between the MPI processes. On each parallel system two special nodes are used as concentrators for incoming and outgoing TCP messages. PACX-MPI can compress all TCP messages to enlarge bandwidth.

PVMPI [4] has been developed at UTK, ORNL and ICL, and couples several MPI worlds by using PVM based bridges. The user interface does not meet the MPI 1.1 standard, but is similar to a restricted subset of the functionality of MPI_Comm_join in MPI 2.0 [10]. It is impossible to merge the bridge's inter-communicator into a global intra-communicator over all processes. Therefore, one can not execute existing MPI applications with PVMPI on a cluster of parallel systems.

In the **PLUS** [3] project at PC^2 at the University of Paderborn a bridge has been developed too. With this bridge MPI and PVM worlds can be combined. Inefficient TCP implementations of some vendors have been substituted by an own protocol based on UDP. It is optimized for wide area networks. As in PACX-MPI the communication is concentrated by a daemon process on each parallel system. The interface has the same restrictions as described for PVMPI, moreover, only a restricted subset of datatypes is valid. PLUS is part of the MOL (Metacomputer Online) project.

In the **Globus** [5] project and in the **I-WAY** [7] project MPICH [8] is used on top of the multi-protocol communication library NEXUS [6]. NEXUS is designed as a basic module for communication libraries like MPI. On the parallel system itself MPICH is used instead of the vendor's MPI library.

4 Design Decisions and Related Problems

Several decisions must be taken in the design of an MPI library usable for heterogeneous metacomputing. Major choices and the decisions in the mentioned projects are viewed in this section.

The glue approach versus the portable multi-protocol MPI implementation: The goals of high efficiency and full MPI functionality can be achieved by the following alternatives (Fig. 2): (a) The glue layer exports an MPI interface to the application and imports the native MPIs in each parallel system and a global communication module which allows communication between parallel systems. This gerneral glue approach is used in MPI-GLUE, PACX-MPI, PVMPI and PLUS. (b) In the portable multi-protocol MPI approach (used in Globus) each parallel system has its own local communication module. A common global communication module enables the communication between the parallel systems. The MPI library uses both modules. In this approach the global communication module is usually very small in comparison with the glue approach: it does only allow byte message transfers. The main disadvantage of this approach is that the performance of the native MPI library can not be used.

Global world versus dynamically created bridges: The standards MPI 1.1 and MPI 2.0 imply two metacomputing alternatives: (a) MPI_COMM_WORLD collects *all* processes (in MPI-GLUE, PACX-MPI and Globus) or (b) each partition of the cluster has a separate MPI_COMM_WORLD and inter-communicators connect the partitions. (doesn't conform to MPI 1.1, used in PVMPI and PLUS).

Global communication using an own module versus importing a portable MPI: The global communication module in the glue approach can be (a) a portable TCP-based MPI implementation (in MPI-GLUE), or (b) a special TCP-based module (in PACX-MPI, PVMPI and PLUS). With (a) its possible to implement the full MPI 1.1 standard without reimplementing most of the MPI's functionality inside of this special module.

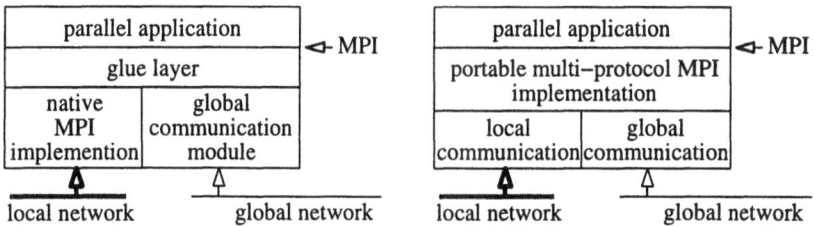

Fig. 2. The glue approach (left) and the multi-protocol MPI approach (right)

Using daemons or not using daemons: The global communication can be (a) concentrated by daemons or can be (b) done directly. In the *daemon-based approach* (a) the application processes send their messages on the local network (e.g. by using the native MPI) to a daemon, that transfers them on the global network (e.g. with TCP) to the daemon of the target system. From there the messages are received by the application processes within the local network of the target system. In the *direct-communication approach* (b) the work of the daemons is done by the operating system and each node communicates directly via the global network. With the daemon-based approach the application can fully parallelize computation and communication. The application is blocked only by native MPI latencies although in the background slow global communication is executed. Currently only the direct-communication approach is implemented in MPI-GLUE, but it would be no problem to add the daemon-based approach. A detailed discussion can be found in [12].

The first progress problem: All testing and blocking MPI calls that have to look at processes connected by the *local* network *and* at other processes connected by the *global* network should take into account that the latencies for probing the local and global network are different. On clusters of MPPs the ratio can be 1:100. Examples are a call to MPI_Recv with source=MPI_ANY_SOURCE, or a call to MPI_Waitany with requests handling local *and* global communication. More a workaround than a solution are asymmetric polling strategies: if polling is necessary then only in the nth round the central polling routine will inquire the state of the global interface, while in each round the local interface will be examined. n is given by the ratio mentioned above. Without this trick the latency for local communication may be expanded to the latency of the global communication. But with this trick the latency for global communication may be in the worst case twice the latency of the underlaying communication routine. This first progress problem arises in all systems that use no daemons for the global communication (MPI-GLUE, PVMPI configured without daemons, and Globus). In MPI-GLUE the asymmetric polling strategy has been implemented. In Globus it is possible to specify different polling intervals for each protocol.

The second progress problem: The restriction in MPI 1.1, page 12, lines 24 ff ("This document specifies the behavior of a parallel program assuming that only MPI calls are used for communication") allows that MPI implementations make progress on non-blocking and buffered operations *only inside* of MPI routines, e.g. in MPI_Wait... / _Test... / _Iprobe or _Finalize. This is a weak interpretation of the progress rule in MPI 1.1, page 44, lines 41ff. The rationale in MPI 2.0, page 142, lines 14-29 explicitly mentions this interpretation. This allows that a native MPI makes no progress, while the portable MPI is blocked by waiting for a message on the TCP link, and vice versa.

Therefore, in the glue approach the problem arises that the glue layer must not implement anything by blocking or testing only within the native MPI without giving the global communication module a chance to make progress, and vice versa. The only *hard problem* is the combination of using collective routines of the native MPI (because there isn't any non-blocking alternative) *and* using sends in the global MPI *and* these sends are buffered at the sender *and* the sender is blocked in the native collective operation. This problem can be solved in daemon-based systems by sending all messages (with a destination in another parallel system) immediately to the daemons and buffering them there (implemented in PACX-MPI and PLUS). It can also be solved by modifying the central probe and dispatching routine inside the native MPI.

The analogous problem (how to give the native MPI a chance to make progress, while the glue layer blocks in a global collective routine) usually only exists, if one uses an existing MPI as a global communication module, but in MPI-GLUE this problem has been solved, because the global MPI itself is a multi-protocol implementation which uses the local MPI as one protocol stack.

5 Design Details

The notation **"MPI"** is used for the MPI interface exported by this glue layer and for the glue layer software.

The notation **"MPX"** is used for the imported TCP-based MPI-1.1 library in which all names are modified and begin with "MPX_" instead of "MPI_". MPX allows the exchange of messages between all processes.

The notation **"MPL"** is used for the native (local) MPI 1.1 library. All names are modified ("MPL_" instead of "MPI_"). All MPL_COMM_WORLDs are disjoint and in the current version each process is member of exactly one MPL_-COMM_WORLD (that may have any size$\geq$1).

Details are given in the table below and in [12]. In the table **"X"** and **"L"** are abbreviations for MPX and MPL.

MPI constant, handle, routine	impl.ed by using	remarks
ranks	X	i.e. the MPI ranks are equal to the MPX ranks.
group handling	X	i.e. group handles and handling are impl. with MPX.
communicators	L and X	and *roots-communicator*, see Sec. *Opt. of collective op.*
datatypes	L and X	and a mapping of the L ranks to the X ranks and vice versa.
requests	L or X or delayed	MPI_IRECVs with MPI_ANY_SOURCE and communicators spawning more than one MPL region store all arguments of the IRECV call into the handle's structure. The real RECV is delayed until the test or wait calls.
sends, receives	L or X	(receives only with known source partition).
receives with ANY_SOURCE:		
- blocking	L and X	with asymmetric polling strategy.
- nonblocking	GLUE	postponed until the wait and test routines, see MPI requests above.
wait and test	L or X or L&X.	
MPI_Testall		Testing inside partial areas of the communicator may complete request handles, but the interface definition requires that only none or all requests are completed. Therefore (only for Testall), requests may be MPL/MPX-finished, but not MPI-finished.

The MPI process start and MPI_Init is mapped to the MPL process start, MPL_Init, MPX process start and MPX_Init. The user starts the application processes on the first parallel system with its MPL process start facility. MPI_Init first calls MPL_Init and then MPX_Init. The latter one is modified to reflect that the parallel partition is already started by MPL. Based on special arguments at the end of the argument list and based on the MPL ranks the first node recognizes inside of MPX_Init its role as MPICH's *big master*. It interactively starts the other partitions on the other parallel systems or uses a *remote shell* method (e.g. rsh) to invoke the MPL process start facility there. There as well the MPI_Init routine first invokes MPL_Init and then MPX_Init. Finally MPI_Init initializes the predefined handles. MPI_Finalize is implemented in the reverse order.

Optimization of collective operations: It is desirable that metacomputing applications do not synchronize processes over the TCP links. This implies that they should not call collective operations on a global communicators, especially those operations which make a barrier synchronization (barrier, allreduce, reduce_scatter, allgather(v), alltoall(v)). The following schemes help to minimize the latencies if the application still uses collective operations over global communicators. In general a collective routine is mapped to the corresponding routine of the native MPI *if the communicator is completely inside of one local MPL world.* To optimize MPI_Barrier, MPI_Bcast, MPI_Reduce and MPI_Allreduce *if the communicator belongs to more than one MPL world*, for each MPI-communicator, there is an additional internal *roots-communicator* that combines the first process of any underlying MPL-communicator by using MPX, e.g. MPI_Bcast is implemented by a call to native Bcast in each underlying MPL-communicator and one call to the global Bcast in the roots-communicator.

6 Status, Future Plans, and Acknowledgments

MPI-GLUE implements the full MPI 1.1 standard except some functionalities. Some restrictions result from the glue approach: (a) It is impossible to allow an unlimited number of user-defined operations. (b) Messages must be and need to be received with datatype MPI_PACKED if and only if they are sent with MPI_PACKED. Some other (hopefully minor) functionalities are still waiting to be implemented. A detailed list can be found in [12]. MPI-GLUE is portable. Current test platforms are SGI, CRAY T3E and Intel Paragon. Possible future extensions can be the implementation of MPI 2.0 functionalities, optimization of further collective routines, and integrating a daemon-based communication module inside the used portable MPI.

I would like to thank the members of the MPI-2 Forum and of the PACX-MPI project as well as the developers of MPICH who have helped me in the discussion which led to the present design and implementation of MPI-GLUE. And I would like to thank Oliver Hofmann for implementing a first prototype in his diploma thesis, Christoph Grunwald for implementing some routines during his practical course, especially some wait, test and topology functions, and Matthias Müller for testing MPI-GLUE with his P3T-DSMC application.

7 Summary

MPI-GLUE achieves interoperability for the message passing interface MPI. For this it combines existing vendors' MPI implementations losing neither full MPI 1.1 functionality nor the vendors' MPIs' efficiency. MPI-GLUE targets existing MPI applications that need clusters of MPPs or PVPs to solve large-scale problems. MPI-GLUE enables metacomputing by seamlessly expanding MPI 1.1 to any cluster of parallel systems. It exports a single virtual parallel MPI machine to the application. It combines all processes in one MPI_COMM_WORLD. All local communication is done by the local vendor's MPI and all global communication is implemented with MPICH directly, without using deamons. The design has only two unavoidable restrictions which concern user defined reduce operations and packed messages.

References

1. Beisel, T., Gabriel, E., Resch, M.: An Extension to MPI for Distributed Computing on MPPs. In Marian Bubak, Jack Dongarra, Jerzy Wasniewski (Eds.) Recent Advances in Parallel Virtual Machine and Message Passing Interface, LNCS, Springer, 1997, 75-83.
2. T. Bönisch, R. Rühle: Portable Parallelization of a 3-D Flow-Solver, Parallel Comp. Fluid Dynamics '97, Elsevier, Amsterdam, to appear.
3. Brune, M., Gehring, J., Reinefeld, A.: A Lightweight Communication Interface for Parallel Programming Environments. In Proc. High-Performance Computing and Networking HPCN'97, LNCS, Springer, 1997.
4. Fagg, G.E., Dongarra, J.J.: PVMPI: An Integration of the PVM and MPI Systems. Department of Computer Science Technical Report CS-96-328, University of Tennessee, Knoxville, May, 1996.
5. Foster, I., Kesselman, C.: Globus: A Metacomputing Infrastructure Toolkit. International Journal of Supercomputer Applications (to appear).
6. Foster, I., Geisler, J., Kesselman, C., Tuecke, S.: Managing Multiple Communication Methods in High-Performance Networked Computing Systems. J. Parallel and Distributed Computing, 40:35-48, 1997.
7. Foster, I., Geisler, J., Tuecke, S.: MPI on the I-WAY: A Wide-Area, Multimethod Implementation of the Message Passing Interface. Proc. 1996 MPI Developers Conference, 10-17, 1996.
8. Gropp, W., Lusk, E., Doss, N., Skjellum, A.: A High-Performance, Portable Implementation of the MPI Message Passing Interface Standard. Paralle Computing, 22, 1996.
9. MPI: A Message-Passing Interface Standard. Message Passing Interface Forum, June 12, 1995.
10. MPI-2: Extensions to the Message-Passing Interface. Message Passing Interface Forum, July 18, 1997.
11. Müller, M.: Weltrekorde durch Metacomputing. BI 11+12, Regionales Rechenzentrum Universität Stuttgart, 1997.
12. Rabenseifner, R.: MPI-GLUE: Interoperable high-performance MPI combining different vendor's MPI worlds. Tech. rep., Computing Center University Stuttgart. http://www.hlrs.de/people/rabenseifner/publ/mpi_glue_tr_apr98.ps.Z

High Performance Protocols for Clusters of Commodity Workstations

P. Melas and E. J. Zaluska

Electronics and Computer Science
University of Southampton, U.K.
pm95r@ecs.soton.ac.uk
ejz@ecs.soton.ac.uk

Abstract. Over the last few years technological advances in microprocessor and network technology have improved dramatically the performance achieved in clusters of commodity workstations. Despite those impressive improvements the cost of communication processing is still high. Traditional layered structured network protocols fail to achieve high throughputs because they access data several times. Network protocols which avoid routing through the kernel can remove this limit on communication performance and support very high transmission speeds which are comparable to the proprietary interconnection found in Massively Parallel Processors.

1   Introduction

Microprocessors have improved their performance dramatically over the past few years. Network technology has over the same period achieved impressive performance improvements in both Local Area Networks (LAN) and Wide Area Networks (WAN) (e.g. Myrinet and ATM) and this trend is expected to continue for the next several years. The integration of high-performance microprocessors into low-cost computing systems, the availability of high-performance LANs and the availability of common programming environments such as MPI have enabled Networks Of Workstations (NOWs) to emerge as a cost-effective alternative to Massively Parallel Processors (MPPs) providing a parallel processing environment that can deliver high performance for many practical applications.

Despite those improvements in hardware performance, applications on NOWs often fail to observe the expected performance speed-up. This is largely because the communication software overhead is several orders of magnitude larger than the hardware overhead [2]. Network protocols and the Operating System (OS) frequently introduce a serious communication bottleneck, (e.g. traditional UNIX layering architectures, TCP/IP, interaction with the kernel, etc). New software protocols with less interaction between the OS and the applications are required to deliver low-latency communication and exploit the bandwidth of the underlying intercommunication network [9].

Applications or libraries that support new protocols such as BIP [4], Active Messages [6], Fast Messages [2], U-Net [9] and network devices for ATM and

Myrinet networks demonstrate a remarkable improvement in both latency and bandwidth, which are frequently directly comparable to MPP communication performance.

This report investigates commodity "parallel systems", taking into account the internode communication network in conjunction with the communication protocol software. Latency and bandwidth tests have been run on various MPPs (SP2, CS2) and a number of clusters using different workstation architectures (SPARC, i86, Alpha), LANs (10 Mbit/s Ethernet, 100 Mbit/s Ethernet, Myrinet) and OS (Solaris, NT, Linux). The work is organised as follows: a review of the efficiency of the existing communication protocols and in particular TCP/IP from the NOW perspective is presented in the first section. A latency and bandwidth test of various clusters and an analysis of the results is presented next. Finally an example specialised protocol (BIP) is discussed followed by conclusions.

2 Clustering with TCP/IP over LANs

The task of a communication sub-system is to transfer data transparently from one application to another application, which could reside on another node. TCP/IP is currently the most widely-used network protocol in LANs and NOWs, providing a connection-oriented reliable byte stream service [8]. Originally TCP/IP was designed for WANs with relatively high error rate. The philosophy behind the TCP/IP model was to provide reliable communication between autonomous machines rather than a common resource [6].

In order to ensure reliability over an unreliable subnetwork TCP/IP uses features such as an in-packet end-to-end checksum, "SO_NDELAY" flag, "Time-to-live" IP field, packet fragmentation/reassembly, etc. All these features are useful in WANs but in LANs they represent redundancy and consume vital computational power [6].

Several years ago the bandwidth of the main memory and the disk I/O of a typical workstation was an order of magnitude faster than the physical network bandwidth. The difference in magnitude was invariably sufficient for the existing OS and communication protocol stacks to saturate network channels such as 10 Mbit/s Ethernet [3]. Despite hardware improvements workstation memory access time and internal I/O bus bandwidth in a workstation have not increased significantly during this period (the main improvements in performance have come from caches and a better understanding of how compilers can exploit the potential of caches). Faster networks have thus resulted in the gap between the network bandwidth and the internal computer resources being considerably reduced [3].

The fundamental design objective of traditional OS and protocol stacks at that time was reliability, programmability and process protection over an unreliable network. In a conventional implementation of TCP/IP a send operation involves the following stages of moving data: data from the application buffer are copied to the kernel buffer, then packet forming and calculation of the headers and the checksum takes place and finally packets are copied into the network interface for transmission. This requires extra context switching between applications and the kernel for each system call, additional copies between buffers and address spaces, and increased computational overhead [7].

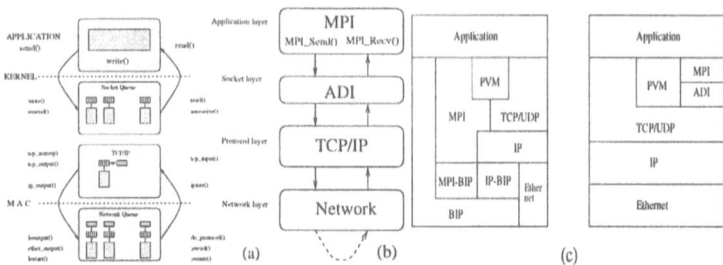

Fig. 1. a) Conventional TCP/IP implementation b)MPI on top of the TCP/IP protocol stack c)The BIP protocol stack approach

As a result, clusters of powerful workstations still suffer a degradation in performance even when a fast interconnection network is provided. In addition, the network protocols in common use are unable to exploit fully all of the hardware capability resulting in low bandwidth and high end-to-end latency. Parallel applications running on top of communication libraries (e.g. MPI, PVM, etc.) add an extra layer on top of the network communication stack, (see Fig. 1).

Improved protocols attempt to minimise the critical communication path of application-kernel-network device [4, 6, 2, 9]. Additionally these protocols exploit advanced hardware capabilities (e.g. network devices with enhanced DMA engines and co-processors) by moving as much as possible of their functionality into hardware. Applications can also interact directly with the network interface avoiding any system calls or kernel interaction. In this way processing overhead is considerably reduced improving both latency and throughput.

3 Latency and Bandwidth Measurements

In order to estimate and evaluate the effect of different communication protocols on clusters a ping-pong test measuring both peer-to-peer latency and bandwidth between two processors over MPI was written [2]. The latency test measures the time a node needs to send a sequence of messages and receive back the echo. The bandwidth test measures the time required for a sequence of back-to-back messages to be sent from one node to another. In both tests receive operations were posted before the send ones. The message size in tests always refers to the payload. Each test is repeated many times in order to reduce any clock jitter, first-time and warm-up effects and the median time from each measurement test presented in the results.

Various communication models [1, 5] have been developed in order to evaluate communication among processors in parallel systems. In this paper we follow a linear approach with a start-up time α (constant per segment cost) and a variable per-byte cost β. The message length at which half the maximum bandwidth is achieved $(n_{1/2})$ is an important indication as well.

$$t_n = \alpha + \beta n \qquad (1)$$

Parameters that can influence tests and measurements are also taken into account for each platform in order to analyse the results better, i.e. measurements on non-dedicated clusters have also been made to assess the effect of interference with other workload.

Tests on MPPs (SP2 and CS2) Latency and bandwidth tests have been run on the SP2 and CS2 in the University of Southampton and on the SP2 at Argonne National Laboratory. The results of the MPP test are used as a benchmark to analyse and compare the performance of NOW clusters. In both SP2 machines the native IBM's MPI implementation was used while on the CS2 the mpich 1.0.12 version of MPI was used. The difference in performance between the two SP2 machines is due to the different type of the high-performance switch (TB2/TB3). The breakpoint at 4 KB on the SP2 graph is due to the change in protocol used for sending small and large messages in the MPI implementation.

The NT cluster. The NT cluster is composed of 8 DEC Alpha based workstations and uses a dedicated interconnection network based around a 100Mbit/s Ethernet switch. The Abstract Device Interface makes use of the TCP/IP protocol stack that NT provides. Results on latency and bandwidth are not very impressive (Fig. 2), because of the experimental version of the MPI implementation used. The system was unable to make any effective use of the communication channel, even on large messages. An unnecessarily large number of context switches resulted in very low performance.

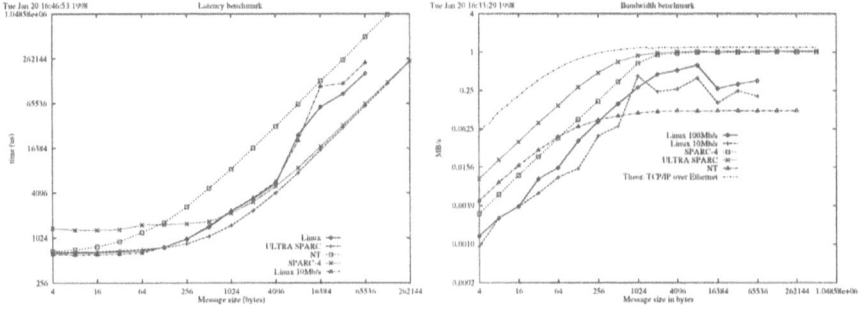

Fig. 2. Latency and bandwidth on NOW clusters

The Linux cluster. This cluster is build around Pentium-Pro machines running Linux 2.0.x. Tests have run using both the 10Mbit/s channel and FastEthernet with MPI version mpich 1.1. A poor performance was observed similar to the NT cluster, although in this case the non-dedicated interconnection network affected some of the results. Although its performance is slightly better than the NT cluster, the system fails to utilise the underlying hardware efficiently. In Fig.2 we can see a break point at the bandwidth plots close to 8 KB due to the interaction between TCP/IP and the OS.

The Solaris cluster. The Solaris cluster uses a non-dedicated 10Mbit/s Ethernet segment running SunOS 5.5.1 and the communication library is mpich 1.0.12. This cluster has the best performance among the clusters for 10Mbit/s Ethernet. Both SPARC-4 and ULTRA SPARC nodes easily saturated the network channel and push the communication bottleneck down to the Ethernet board. The non-dedicated intercommunication network had almost no affect on the results.

The Myrinet cluster. This cluster is build around Pentium-Pro machines running Linux 2.0.1, with an interconnection based on a Myrinet network. The communication protocol is the *Basic Interface for Parallelism* (BIP), an interface for network communication which is targeted towards message-passing parallel computing. This interface has been designed to deliver to the application layer the maximum performance achievable by the hardware [3]. Basic features of this protocol are a zero-copy protocol, user application level direct interaction with the network board, system call elimination and exploitative use of memory bandwidth. BIP can easily interface with other protocol stacks and interfaces such as IP or APIs, (see Fig. 1).

The cluster is flexible enough to configure the API either as a typical TCP/IP stack running over 10Mbit/s Ethernet, TCP/IP over Myrinet or directly on top of BIP. In the first case the IP protocol stack on top of the Ethernet network provides similar performance to the Linux cluster discussed above, although its performance is comparable to the Sun cluster. Latency for zero-size message length has dropped to 290 μs and the bandwidth is close to 1 MB/s for messages larger than 1 KB.

Changing the physical network from Ethernet to Myrinet via the TCP/BIP protocol provides a significant performance improvement. Zero size message latency is 171 μs and the bandwidth reaches 18 MB/s, with a $n_{1/2}$ figure below 1.5KB. Further change to the configuration enables the application (MPI) to interact directly with the network interface through BIP. The performance improvement in this case is impressive pushing the network board to the design limits. Zero length message latency is 11 μs and the bandwidth exceeds 114 MB/s with a $n_{1/2}$ message size of 8 KB.

Figure 3 shows the latency and bandwith graphs for those protocol stack configurations. A noticeable discontinuity at message sizes of 256 bytes reveals the different semantics between short and long messages transmission modes.

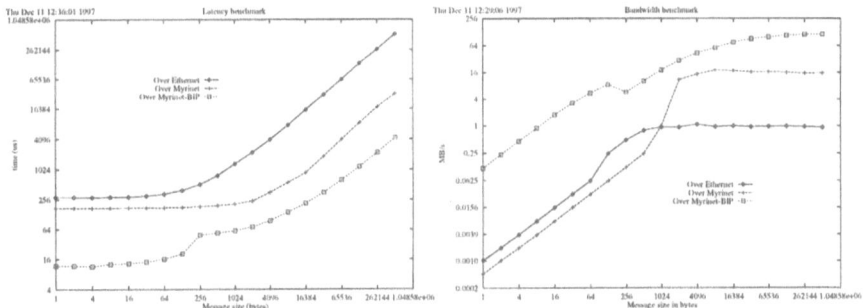

Fig. 3. Latency and bandwidth on a Myrinet cluster with page alignment

4   Analysis of the Results

The performance difference between normal NOW clusters and parallel systems is typically two or more orders of magnitude. A comparable performance difference can sometimes be observed between the raw performace of a NOW cluster and the performance delievered at the level of the user application (the relative latency and bandwidth performance of the above clusters is presented in Fig. 2). With reference to the theoretical bandwidth of a 10 Mbit/s Ethernet channel with TCP/IP headers we can see that only the Sun cluster approaches close to that maximum. The difference in computation power between the Ultra SPARC and SPARC-4 improves the latency and bandwidth of short messages.

On the other hand the NT and Linux cluster completely fail to saturate even a 10Mbit/s Ethernet channel and an order of magnitude improvement in the interconnection channel to FastEthernet does not improve throughput significantly. In both clusters the communication protocol implementation drastically limits performance. It is worth mentioning that both software and hardware for the Solaris cluster nodes comes from the same supplier, therefore the software is very well tuned and exploits all the features available in the hardware. By contrast the open market in PC-based systems requires software to be a more general and thus less efficient.

Similar relatively-low performance was measured on the Myrinet cluster using the TCP/IP protocol stack. The system exploits a small fraction of the network bandwidth and latency improvement is small. Replacing the communication protocol with BIP the Myrinet cluster is then able to exploit the network bandwidth and the application level performance becomes directly comparable with MPPs such as the SP2, T3D, CS2, etc.

As can be seen from Fig. 3, the Myrinet cluster when compared to those MPPs has significantly better latency features over the whole range of the measurements made. In bandwidth terms for short messages up to 256 bytes Myrinet outperforms all the other MPPs. Then for messages up to 4KB (which is the breakpoint of the SP2 at Argonne), the SP2 has a higher performance, but after this point performance of Myrinet is again better.

Table 1. Latency and Bandwidth results

Configuration	Cluster	H/.W	min Lat.	max BW	$n_{1/2}$
MPI over TCP/IP	NT/Alpha	FastEth.	673 μs	120 KB/s	100
MPI over TCP/IP	Linux/P.Pro	Ethernet	587 μs	265 KB/s	1.5 K
MPI over TCP/IP	Linux/P.Pro	FastEth.	637 μs	652 KB/s	1.5 K
MPI over TCP/IP	SPARC-4	Ethernet	660 μs	1.03 MB/s	280
MPI over TCP/IP	ULTRA	Ethernet	1.37 ms	1.01 MB/s	750
MPI over TCP/IP	Linux/P.Pro	Ethernet	280 μs	1 MB/s	300
MPI over TCP/BIP	Linux/P.Pro	Myrinet	171 μs	17.9 MB/s	1.5 K
MPI over BIP	Linux/P.Pro	Myrinet	11 μs	114 MB/s	8 K

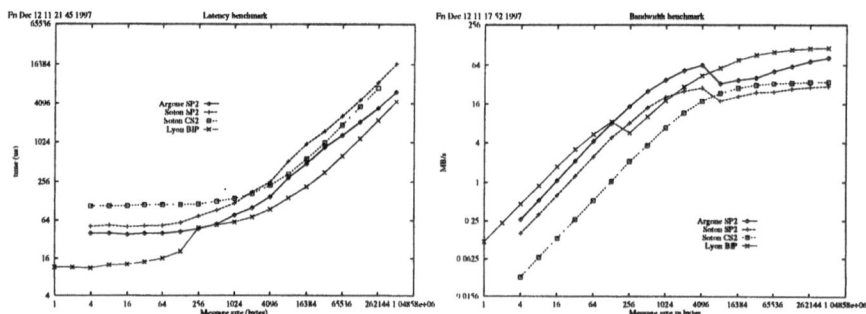

Fig. 4. Comparing latency and bandwidth between a Myrinet cluster and MPPs

5 Conclusion

In this paper the importance and the impact of some of the most common interconnecting network protocols for clusters has been examined. The TCP/IP protocol stack was tested with different cluster technologies and comparisons made with other specialised network protocols. Traditional communication protocols utilising the critical application-kernel-network path impose excessive communication processing cost and cannot exploit any advanced features in the hardware. Conversely communication protocols that enable applications to interact directly with network interfaces such as BIP, Fast Messages, Active Messages, etc can improve the performance of clusters considerable.

The majority of parallel applications use small-size messages to coordinate program execution and for this size of message latency overhead dominates the transmission time. In this case the communication cost cannot be hidden by any programming model or technique such as overlapping or pipelining. Fast communication protocols that deliver a drastically reduced communication latency are necessary for clusters to be used effectively in a wide range of scientific and commercial parallel applications as an alternative to MPPs.

ACKNOWLEDGEMENTS We acknowledge the support of the Greek State Scholarships Foundation, Southampton University Computer Services, Argonne National Laboratory and LHPC in Lyon.

References

1. J. J. Dongarra and T. Dunigan. Message-passing performance of various computers. Technical Report UT-CS-95-299, Department of Computer Science, University of Tennessee, July 1995.
2. Mario Lauria and Andrew Chien. MPI-FM: High performance MPI on workstation clusters. *Journal of Parallel and Distributed Computing*, 40(1):4–18, January 1997.
3. Bernard Tourancheau Loïc Prylli. Bpi: A new protocol design for high performance networking on myrinet. Technical report, LIP-ENS Lyon, September 1997.
4. Loic Prylli. Draft: Bip messages user manual for bip 0.92. Technical report, Laboratoire de l'Informatique du Parallelisme Lyon, 1997.
5. PARKBENCH Committee: Report-1, assembled by Roger Hockney (chairman), and Michael Berry secretary). Public international benchmarks for parallel computers. Technical Report UT-CS-93-213, Department of Computer Sci/ence, University of Tennessee, November 1993.
6. Steven H. Rodrigues, Thomas E. Anderson, and David E. Culler. High-performance local-area communication with fast sockets. In USENIX, editor, *1997 Annual Technical Conference, January 6–10, 1997. Anaheim, CA*, pages 257–274, Berkeley, CA, USA, January 1997. USENIX.
7. A. Chien S. Pakin, M. Lauria. High performance messaging on workstations: Illinois fast messages (fm) for myrinet. In *In Supercomputing '95*, San Diego, California, 1995.
8. W. R. Stevens. *TCP/IP Illustrated, Volume 1; The Protocols*. Addison Wesley, Reading, 1995.
9. Thorsten von Eicken, Anindya Basu, Vineet Buch, and Werner Vogels. U-net: a user-level network interface for parallel and distributed computing. In *Proceedings of the 15th ACM Symposium on Operating Systems Principles (SOSP)*, volume 29, pages 303–316, 1995.

Significance and Uses of Fine-Grained Synchronization Relations

Ajay D. Kshemkalyani

Dept. of Electrical & Computer Engg. and Computer Science
University of Cincinnati, P. O. Box 210030, Cincinnati, OH 45221-0030, USA
ajayk@ececs.uc.edu

Abstract. In a distributed system, high-level actions can be modeled by nonatomic events. Synchronization relations between distributed nonatomic events have been proposed to allow applications a fine choice in specifying synchronization conditions. This paper shows how these fine-grained relations can be used for various types of synchronization and coordination in distributed computations.

1 Introduction

High-level actions in several distributed application executions are realistically modeled by nonatomic poset events (where at least some of the component atomic events of a nonatomic event occur concurrently), for example, in distributed multimedia support, coordination in mobile computing, distributed debugging, navigation, planning, industrial process control, and virtual reality. It is important to provide these and emerging sophisticated applications a fine level of granularity in the specification of various synchronization/causality relations between nonatomic poset events. In addition, [20] stresses the theoretical and practical importance of the need for such relations. Most of the existing literature, e.g., [1, 2, 4, 5, 7, 10, 13, 15, 18–20] does not address this issue. A set of causality relations between nonatomic poset events was proposed in [11, 12] to specify and reason with a fine-grained specification of causality. This set of relations extended the hierarchy of the relations in [9, 14]. An axiom system on the proposed relations was given in [8]. The objective of this paper is to demonstrate the use of the relations in [11, 12].

The following poset event structure model is used as in [4, 9, 10, 12, 14–16, 20]. $(E, \prec)$ is a poset such that E represents points in space-time which are the primitive atomic events related by the causality relation $\prec$ which is an irreflexive partial ordering. Elements of E are partitioned into local executions at a coordinate in the space dimensions. In a distributed system, E represents a set of events and is discrete. Each local execution E_i is a linearly ordered set of events in partition i. An event e in partition i is denoted e_i. For a distributed application, points in the space dimensions correspond to the set of processes (also termed *nodes*), and E_i is the set of events executed by process i. Causality between events at different nodes is imposed by message communication.

High-level actions in the computation are nonempty subsets of E. Formally, if $\mathcal{E}$ denote the power set of E and $\mathcal{A}$ $(\neq \emptyset) \subseteq (\mathcal{E} - \emptyset)$, then each element A of $\mathcal{A}$ is termed an *interval* or a *nonatomic event*. It follows that if $A \cap E_i \neq \emptyset$, then $(A \cap E_i)$ has a least and a greatest event. $\mathcal{A}$ is the set of all the sets that represent a higher level grouping of the events of E that is of interest to an application.

Table 1. Some causality relations [9].

Relation r	Expression for $r(X,Y)$
$R1$	$\forall x \in X \forall y \in Y, x \prec y$
$R1'$	$\forall y \in Y \forall x \in X, x \prec y$
$R2$	$\forall x \in X \exists y \in Y, x \prec y$
$R2'$	$\exists y \in Y \forall x \in X, x \prec y$
$R3$	$\exists x \in X \forall y \in Y, x \prec y$
$R3'$	$\forall y \in Y \exists x \in X, x \prec y$
$R4$	$\exists x \in X \exists y \in Y, x \prec y$
$R4'$	$\exists y \in Y \exists x \in X, x \prec y$

The relations in [9] formed an exhaustive set of causality relations to express the possible interactions between a pair of linear intervals. These relations $R1$ - $R4$ and $R1'$ - $R4'$ from [9] are expressed in terms of the quantifiers over X and Y in Table 1. Observe that $R2'$ and $R3'$ are different from $R2$ and $R3$, respectively, when applied to posets. Table 1 gives the hierarchy and inclusion relationship of the causality relations $R1$ - $R4$. Each cell in the grid indicates the relationship of the row header to the column header. The notation for the inclusion relationship between causality relations is as follows. The inclusion relation "is a subrelation of" is denoted '$\sqsubseteq$' and its inverse is '$\sqsupseteq$'. For relations r_1 and r_2, we define $r_1 \parallel r_2$ to be $(r_1 \not\sqsubseteq r_2 \wedge r_2 \not\sqsubseteq r_1)$. The relations $\{ R1, R2, R3, R4 \}$ form a lattice hierarchy ordered by $\sqsubseteq$.

Table 2. Inclusion relationships between relations of Table 1 [9].

relation of row header to column header	$R1$	$R2$	$R3$	$R4$
$R1$	$=$	$\sqsubseteq$	$\sqsubseteq$	$\sqsubseteq$
$R2$	$\sqsupseteq$	$=$	$\parallel$	$\sqsubseteq$
$R3$	$\sqsupseteq$	$\parallel$	$=$	$\sqsubseteq$
$R4$	$\sqsupseteq$	$\sqsupseteq$	$\sqsupseteq$	$=$

The relations in [9] formed a comprehensive set of causality relations to express all possible interactions between a pair of linear intervals using only the $\prec$ relation between atomic events, and extended the partial hierarchy of relations of [14]. However, when the relations of [9] are applied to a pair of poset intervals, the hierarchy they form is incomplete. Causality relations between a pair of nonatomic poset intervals were formulated by extending the results [8, 9] to nonatomic poset events [11, 12]. The relations form a "comprehensive" set of causality relations between nonatomic poset events using first-order predicate logic and only the $\prec$ relation between atomic events, and fill in the existing partial hierarchy formed by relations in [9, 14]. A relational algebra for the relations in [11, 12] is given in [8]. Given any relation(s) between X and Y, the relational algebra allows the derivation of conjunctions, disjunctions, and negations of all other relations that are also valid between X and Y, as well as between Y and X.

Section 2 reviews the hierarchy of causality relations [11, 12]. Section 3 gives the uses of each of the relations. Some uses of the relations include modeling various forms of synchronization for group mutual exclusion, initiation of nested computing, termination of nested processing, and monitoring the start of a computation. Section 4 gives concluding remarks. The results of this paper are included in [8].

2   Relations between Nonatomic Poset Events

Previous work on linear intervals and time durations, e.g., [1, 2, 4, 5], identifies an interval by the instants of its beginning and end. Given a nonatomic poset interval, one needs to define counterparts for the beginning and end instants. These counterparts serve as "proxy" events for the poset interval just as the events at the beginning and end of linear intervals such as time durations serve as proxies for the linear interval. The proxies identify the durations on each node, in which the nonatomic event occurs. Two possible definitions of proxies are (i) $L_X = \{e_i \in X \mid \forall e'_i \in X, e_i \preceq e'_i\}$ and $U_X = \{e_i \in X \mid \forall e'_i \in X, e_i \succeq e'_i\}$, and (ii) $L_X = \{e \in X \mid \forall e' \in X, e \nsucc e'\}$ and $U_X = \{e \in X \mid \forall e' \in X, e \nprec e'\}$. Assume that one definition of proxies is consistently used, depending on context and application. Fig. 1 depicts the proxies of X and Y.

There are two aspects of a relation that can be specified between poset intervals. One aspect deals with the determination of an appropriate *proxy* for each interval. A proxy for X and Y can be chosen in four ways corresponding to the relations in $\{R1, R2, R3, R4\}$. From Table 1, it follows that these four relations form a lattice ordered by $\sqsubseteq$. The second aspect deals with how the atomic elements of the chosen proxies are related by causality. The chosen proxies can be related by the eight relations $R1$, $R1'$, $R2$, $R2'$, $R3$, $R3'$, $R4$, $R4'$ of Table 1, which are renamed a, a', b, b', c, c', d, d', respectively, to avoid confusion with their original names used for the first aspect of specifying the relations between poset intervals. The inclusion hierarchy among the six distinct relations forms a lattice ordered by $\sqsubseteq$; see Table 3.

The two aspects of deriving causality relations, described above, are combined to define the relations. The lattice of relations $\{ R1^*, R2^*, R3^*, R4^* \}$ between proxies of X and Y, and the lattice of relations $\{ a, a', b, b', c, c', d, d' \}$ between

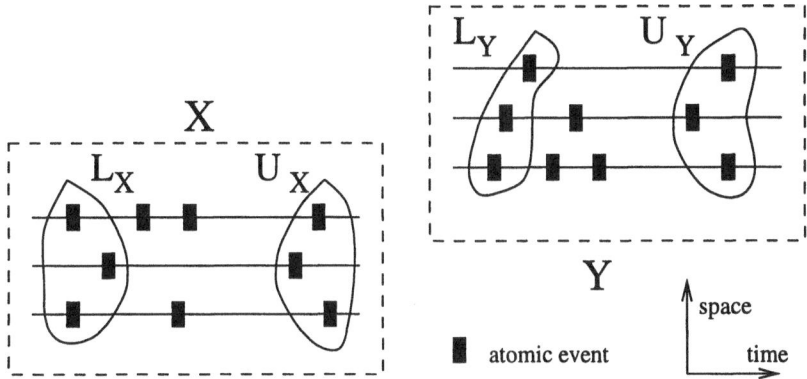

Fig. 1. Poset events X and Y and their proxies.

Table 3. Full hierarchy of relations of Table 1 [9]. Relations $R1$, $R1'$, $R2$, $R2'$, $R3$, $R3'$, $R4$, $R4'$ of Table 1 are renamed a, a', b, b', c, c', d, d', respectively. Relations in the row and column headers are defined between X and Y.

Relation names: its quantifiers for $x \prec y$	$R1,a$ (=$R1'$,a'): $\forall x \forall y$ (=$\forall y \forall x$)	$R2,b$: $\forall x \exists y$	$R2',b'$: $\exists y \forall x$	$R3,c$: $\exists x \forall y$	$R3',c'$: $\forall y \exists x$	$R4,d$ (=$R4'$,d'): $\exists x \exists y$ (=$\exists y \exists x$)
$R1,a$ (=$R1'$,a') : $\forall x \forall y (= \forall y \forall x)$	=	⊑	⊑	⊑	⊑	⊑
$R2,b$: $\forall x \exists y$	⊒	=	⊒	∥	∥	⊑
$R2',b'$: $\exists y \forall x$	⊒	⊑	=	∥	∥	⊑
$R3,c$: $\exists x \forall y$	⊒	∥	∥	=	⊑	⊑
$R3',c'$: $\forall y \exists x$	⊒	∥	∥	⊒	=	⊑
$R4,d$ (=$R4'$,d'): $\exists x \exists y (= \exists y \exists x)$	⊒	⊒	⊒	⊒	⊒	=

582

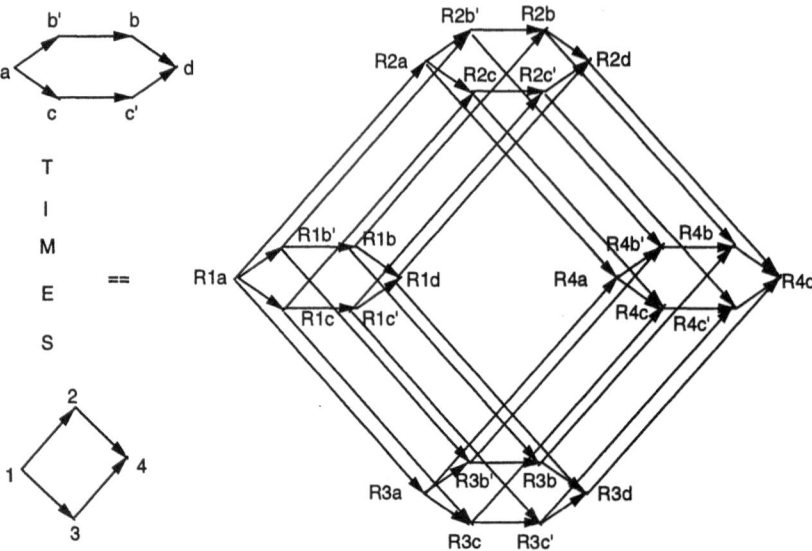

Fig. 2. Hierarchy of relations in [11, 12].

the elements of the proxies, give a product lattice of 32 relations over $A \times A$ to express $r(X, Y)$. The resulting set of poset relations, denoted $\mathcal{R}$, is given in Table 4. The relations in $\mathcal{R}$ form a lattice of 24 unique elements as shown in Fig. 2. $\mathcal{R}$ is comprehensive using first-order predicate logic and only the $\prec$ relation between atomic events. Relation $R?\#(X, Y)$ means that proxies of X and Y are chosen as per ?, and events in the proxies are related by $\#$. The two relations in [14], viz., $\longrightarrow$ and $-- \rightarrow$, correspond to $R1a$ and $R4d$, respectively, whereas the relations in [9] and listed in Table 1 correspond to the these relations as follows: $R1=R1'$, $R2$, $R2'$, $R3$, $R3'$, $R4=R4'$ correspond to $R1a$, $R2b$, $R2b'$, $R3c$, $R3c'$, $R4d$, respectively.

3 Significance of the Relations

The hierarchy of causality relations is useful for applications which use nonatomicity in reasoning and modeling and need a fine level of granularity of causality relations to specify synchronization relations and their composite global predicates between nonatomic events. Such applications include industrial process control applications, distributed debugging, navigation, planning, robotics, diagnostics, virtual reality, and coordination in mobile systems. The hierarchy provides a range of relations, and an application can use those that are relevant and useful to it. The relations and their composite (global) predicates provide a precise handle to *express* a naturally occurring or *enforce* a desired fine-grained level of causality or synchronization in the computation. A specific meaning of each relation and a brief discussion of how it can be enforced is now given. In

Table 4. Relations $r(X, Y)$ in $\mathcal{R}$ [11, 12].

Relation $r(X,Y)$	Relation definition specified by quantifiers for $x \prec y$, where $x \in X$, $y \in Y$	Relation $r(X,Y)$	Relation definition specified by quantifiers for $x \prec y$, where $x \in X$, $y \in Y$
R1a	$\forall x \in U_X \forall y \in L_Y$	R3a	$\forall x \in L_X \forall y \in L_Y$
R1a' (=R1a)	$\forall y \in L_Y \forall x \in U_X$	R3a' (=R3a)	$\forall y \in L_Y \forall x \in L_X$
R1b	$\forall x \in U_X \exists y \in L_Y$	R3b	$\forall x \in L_X \exists y \in L_Y$
R1b'	$\exists y \in L_Y \forall x \in U_X$	R3b'	$\exists y \in L_Y \forall x \in L_X$
R1c	$\exists x \in U_X \forall y \in L_Y$	R3c	$\exists x \in L_X \forall y \in L_Y$
R1c'	$\forall y \in L_Y \exists x \in U_X$	R3c'	$\forall y \in L_Y \exists x \in L_X$
R1d	$\exists x \in U_X \exists y \in L_Y$	R3d	$\exists x \in L_X \exists y \in L_Y$
R1d' (=R1d)	$\exists y \in L_Y \exists x \in U_X$	R3d' (=R3d)	$\exists y \in L_Y \exists x \in L_X$
R2a	$\forall x \in U_X \forall y \in U_Y$	R4a	$\forall x \in L_X \forall y \in U_Y$
R2a' (=R2a)	$\forall y \in U_Y \forall x \in U_X$	R4a' (=R4a)	$\forall y \in U_Y \forall x \in L_X$
R2b	$\forall x \in U_X \exists y \in U_Y$	R4b	$\forall x \in L_X \exists y \in U_Y$
R2b'	$\exists y \in U_Y \forall x \in U_X$	R4b'	$\exists y \in U_Y \forall x \in L_X$
R2c	$\exists x \in U_X \forall y \in U_Y$	R4c	$\exists x \in L_X \forall y \in U_Y$
R2c'	$\forall y \in U_Y \exists x \in U_X$	R4c'	$\forall y \in U_Y \exists x \in L_X$
R2d	$\exists x \in U_X \exists y \in U_Y$	R4d	$\exists x \in L_X \exists y \in U_Y$
R2d' (=R2d)	$\exists y \in U_Y \exists x \in U_X$	R4d' (=R4d)	$\exists y \in U_Y \exists x \in L_X$

the following discussion, the "X computation" and "Y computation" refer to the computation performed by the nonatomic events X and Y, respectively. A proxy of X is denoted $\hat{X}$.

We first consider the significance of the groups of relations $R^*a(X,Y)$, $R^*b(X,Y)$, $R^*b'(X,Y)$, $R^*c(X,Y)$, $R^*c'(X,Y)$, and $R^*d(X,Y)$. Each group deals with a particular proxy $\hat{X}$ and $\hat{Y}$.

- $R^*a(X,Y)$: All events in $\hat{Y}$ know the results of the X computation (if any,) upto all the events in $\hat{X}$. This is a strong form of synchronization between $\hat{X}$ and $\hat{Y}$.
- $R^*b(X,Y)$: For each event in $\hat{X}$, some event in $\hat{Y}$ knows the results of the X computation (if any,) upto that event in $\hat{X}$. The Y computation may then exchange information about the X computation upto $\hat{X}$, among the nodes participating in the Y computation.
- $R^*b'(X,Y)$: Some event in $\hat{Y}$ knows the results of the X computation (if any,) upto all events in $\hat{X}$. These relations are useful when it is sufficient for one node in $N_{\hat{Y}}$ to detect a global predicate across all nodes in $N_{\hat{X}}$. If the event in $\hat{Y}$ is at a node that behaves as the group leader of $N_{\hat{Y}}$, then it can either inform the other nodes in $N_{\hat{Y}}$ or make decisions on their behalf.
- $R^*c(X,Y)$: All events in $\hat{Y}$ know the results of the X computation (if any,) upto some common event in $\hat{X}$. This group of relations is useful when it is sufficient for one node in $N_{\hat{X}}$ to inform all the nodes in $N_{\hat{Y}}$ of its state,

such as when all the nodes in $N_{\hat{X}}$ have a similar state. If the node at which the event in $\hat{X}$ occurs has already collected information about the results/ states of the X computation upto $\hat{X}$ from other nodes in $N_{\hat{X}}$ (thus, that node behaves as the group leader of $\hat{X}$), then the events in $\hat{Y}$ will know the states of the X computation upto $\hat{X}$.

- $R^*c'(X,Y)$: Each event in $\hat{Y}$ knows the results of the X computation (if any,) upto some event in $\hat{X}$. If it is important to the application, then the state at each event in $\hat{X}$ should be communicated to some event in $\hat{Y}$.
- $R^*d(X,Y)$: Some event in $\hat{Y}$ knows the results of the X computation (if any,) upto some event in $\hat{X}$. The nodes under consideration at which the events in $\hat{Y}$ and $\hat{X}$, respectively, occur may be the group leaders of $N_{\hat{Y}}$ and $N_{\hat{X}}$, respectively. This group leader of $N_{\hat{X}}$ may have collected relevant state information from other nodes in $N_{\hat{X}}$, and conveys this information to the group leader of $N_{\hat{Y}}$, which in turn distributes the information to all nodes in $N_{\hat{Y}}$.

The above significance of each group of relations applies to each individual relation of that group. The specific use and meaning of each of the 24 relations in $\mathcal{R}$ is given next. We do not restrict the explanation that follows to any specific application.

$1^*(X,Y)$: This group of relations deals with U_X and L_Y. Each relation signifies different degree of transfer of control for synchronization, as in group mutual exclusion (*gmutex*), from the X computation to the Y computation.

- $R1a(X,Y)$: The Y computation at any node in N_Y begins only after that node knows that the X computation at each node in N_X has ended, e.g., a conventional distributed gmutex in which each node in N_Y waits for an indication from each node in N_X that it has relinquished control.
- $R1b(X,Y)$: For every node in N_X, the final value of its X computation is known by (or its mutex token is transferred to) some node in N_Y before that node in N_Y begins its Y computation. Thus, nodes in N_Y collectively (but not individually) know the final value of the X computation before the last among them begins its Y computation. This is a weak version of synchronization/gmutex.
- $R1b'(X,Y)$: Before beginning its Y computation, some node in N_Y knows the final value of the X computation at each node in N_X. This is a weak version of synchronization/gmutex (but stronger than $R1b$) with the property that at least one node in N_Y cannot begin its Y computation until the final value of the X computation at each node in N_X is known to it.
- $R1c(X,Y)$: The final value of the X computation at some node in N_X is known to all the nodes in N_Y before they begin their Y computation. This is a weak form of synchronization/gmutex which is useful when it suffices for a particular node in N_X to grant all the nodes in N_Y gmutex permission to proceed with the Y computation; this node in N_X may be the group leader of N_X, or simply all the nodes in N_X have the same final local state of the X computation within this application.

– $R1c'(X, Y)$: Each node in N_Y begins its Y computation only after it knows the final value of the X computation of some node in N_X. This is a weak form of synchronization/gmutex (weaker than $R1c$) which requires each node in N_Y to receive a final value (or gmutex token) from at least one node in N_X before starting its Y computation. This relation is sufficient for some applications such as those requiring that at most one (additional) process be admitted to join those in the critical section when one process leaves it.

– $R1d(X, Y)$: Some node in N_Y begins its Y computation only after it knows the final value of (or receives a gmutex token from) the X computation at some node in N_X. This is the weakest form of synchronization/gmutex.

$R2^*(X, Y)$: This group of relations deals with U_X and U_Y. The relations can signify various degrees of synchronization between the termination of computations X and Y, where X is nested within Y or X is a subcomputation of Y. Alternately, Y could denote activity at processes that have already spawned X activity in threads, and Y can complete only after X completes.

– $R2a(X, Y)$: The Y computation at any node in N_Y can terminate only after that node knows the final value of (or termination of) the X computation at each node in N_X. This is a strong synchronization before termination, between X and Y.

– $R2b(X, Y)$: For every node in N_X, the final value of its X computation is known by at least one node in N_Y before that node in N_Y terminates its Y computation. Thus, all the nodes in N_Y collectively (but not individually) know the final values of the X computation before they terminate their Y computation. This is a weak synchronization before termination.

– $R2b'(X, Y)$: Before terminating its Y computation, some node in N_Y knows the final value of the X computation at all nodes in N_X. This is a stronger synchronization before termination than $R2b$, wherein at least one node in N_Y cannot terminate its Y computation without knowing the final state of the X computation at all nodes in N_X. This suffices for all applications in which it is adequate for one node in N_Y to detect the termination of the X computation at each node in N_X before that node terminates its Y computation.

– $R2c(X, Y)$: The final value of the X computation at some node in N_X is known to all the N_Y nodes before they terminate the Y computation. This is a weak form of synchronization. The pertinent node in N_X could represent a critical thread in the X computation, or could be the group leader of N_X that represents the X computation at all nodes in N_X.

– $R2c'(X, Y)$: Each node in N_Y terminates its Y computation only after it knows the final value of the X computation at some node in N_X. This is a weak form of synchronization before termination (weaker than $R2c$), but is adequate when all the nodes in N_X are performing a similar X computation.

– $R2d(X, Y)$: Some node in N_Y terminates its Y computation after it knows the final value of the X computation at some node in N_X. This is a weak form of synchronization; however, if the concerned nodes in N_X and N_Y are the respective group leaders of the X and Y computations and, respectively, collect/distributed information from/to their groups, then a strong form of synchronization can be implicitly enforced because when Y terminates, it is known to each node in N_Y that the X computation has terminated.

$R3^*(X,Y)$: This group of relations deals with L_X and L_Y. The relations can signify various degrees of synchronization between the initiation of computations X and Y, where Y is nested within X or Y is a subcomputation of X. Alternately, X could denote activity at processes that have already spawned Y activity in threads.

- $R3a(X,Y)$: The Y computation at any node in N_Y begins after that node knows the initial values of the X computation at each node in N_X. This is a strong form of synchronization between the beginnings of the X and Y computations.
- $R3b(X,Y)$: For each node in N_X, the initial state of its X computation is known to some node in N_Y before that node in N_Y begins its Y computation. Thus, all the nodes in N_Y collectively (but not individually) know the initial state of the X computation. This synchronization is sufficient when the forked Y computations at each node in N_Y are only loosely coupled and should not know each others' initial states communicated by the X computation; while at the same time ensuring that the initial state of the X computation at each node in N_X is available to at least one node in N_Y before it commences its Y computation.
- $R3b'(X,Y)$: Before beginning its Y computation, some node in N_Y knows the initial state of the X computation at all the nodes in N_X. Thus the Y computation at this node can run a parallel X computation for fault-tolerance, or can be made an entirely deterministic function of the inputs to the X computation. The subject node in N_Y can coordinate the Y computation of the other nodes in N_Y. This synchronization is weaker than $R3a$ but stronger than $R3b$.
- $R3c(X,Y)$: The initial state of the X computation at some node in N_X is known to all the nodes in N_Y before they begin their Y computation. This is a weak synchronization; however, it is adequate when the subject node in N_X has forked all the threads that will perform Y, and behaves as the group leader of X that initiates the nested computation Y.
- $R3c'(X,Y)$: Each node in N_Y begins its Y computation only after it knows the initial state of the X computation at some node in N_X. Thus each node executing the computation Y has its Y computation forked or spawned by some node in N_X and its Y computation corresponds to a nested branch of X. The nodes in N_Y may not know each others' initial values for the Y computation; the X computations at (some of) the N_X nodes have semi-independently forked the Y computations at the nodes in N_Y.
- $R3d(X,Y)$: Some node in N_Y begins its Y computation only after it knows the initial state of the X computation at some node in N_X. This is a weak form of synchronization in which only one node in N_X and one node in N_Y coordinate their respective initial states of their local X and Y computations. However, if the node in N_X that initiated the X computation forks off

the main thread for the Y computation, then this form of synchronization between the initiations of X and Y is adequate to have Y as an entirely nested computation within X.

$R4^*(X,Y)$: This group of relations deals with L_X and U_Y. The relations signify different degrees of synchronization between a monitoring computation Y that knows the initial values with which the X computation begins, and then the monitoring computation Y terminates.

- $R4a(X,Y)$: The Y computation at any node in N_Y terminates only after that node knows the initial values of the X computation at each node in N_X. This is a strong form of synchronization between the start of X and the end of Y.
- $R4b(X,Y)$: For every node in N_X, the initial state of its X computation is known by at least one node in N_Y before that node in N_Y terminates its Y computation. Even if there is no exchange of information in the Y computation about the state of the X computation at individual nodes in N_X, this relation guarantees that when Y completes, the (initial) local states at each of the N_X nodes are collectively (but not individually) known by N_Y.
- $R4b'(X,Y)$: Before terminating its Y computation, some node in N_Y knows the initial state of the X computation at all the nodes in N_X. This node in N_Y can detect if an initial global predicate of the X computation across the nodes in N_X is satisfied, before it terminates its Y computation. If this node in N_Y is a group leader, it can then inform the other nodes in N_Y to terminate their Y computations.
- $R4c(X,Y)$: The initial state of the X computation at some node in N_X is known to all the nodes in N_Y before they terminate their Y computation. This weak synchronization is adequate for applications where all the N_X nodes start their X computation with similar values. Alternately, if the node in N_X behaves as a group leader, it can first detect the initial global state of the X computation and then inform all the nodes in N_Y.
- $R4c'(X,Y)$: Each node in N_Y terminates its Y computation only after it knows the initial state of the X computation at some node in N_X. This is a weaker form of synchronization than $R4c$ because the states of all nodes in N_X may not be observed before the nodes in N_Y terminate their Y computation. But this will be adequate for applications in which each node in N_X is reporting the same state/value of the X computation, and each node in N_Y simply needs a confirmation from some node in N_X before it terminates its Y computation. For example, a mobile host (an N_Y node) may simply need a confirmation from some base station (an N_X node) before it exits its Y computation.
- $R4d(X,Y)$: Some node in N_Y terminates its Y computation after it knows the initial state of the X computation at some node in N_X. This weak form of synchronization is sufficient when the group leader of X which is responsible for kicking off the rest of X informs some node (or the group leader) of the monitoring distributed program Y that computation X has successfully begun.

Consider enforcing group mutual exclusion (*gmutex*) between two groups of processes G_1 and G_2. Multiple processes of either group, but not both groups, are permitted to access some critical resources, such as distributed database records, at any time. Relations $R1*$ represent different degrees of gmutex that can be enforced, as explained for $R1^*(X, Y)$ earlier in this section. Also, the strongest form of gmutex, $R1a$, can also be enforced by $R1b'$, $R1c$, and $R1d$, if the communicating nodes in G_1 / G_2 are the respective group leaders. Thus, for $R1b'$, the nodes in G_1 communicate their states (gmutex tokens) to the group leader of G_2 which then collects all these states (gmutex tokens), and distributes them within G_2. For $R1c$, the group leader of G_1 collects all the gmutex tokens from G_1, then informs all the nodes in G_2. For $R1d$, the group leader of G_1 collects all the gmutex tokens from G_1, then informs the group leader of G_2 which then informs all the nodes in G_2. The above four ways of expressing the distributed mutual exclusion provide a choice in trade-offs of (i) knowledge of membership of G_1 and/or G_2, by members and/or group leaders, within each group and across groups (this is further complicated with mobile processes), (ii) different delay or the number of message exchange phases to achieve gmutex, (it is critical to have rapid exchange of access rights to the distributed database), (iii) different number of messages exchanged to achieve gmutex, (bandwidth is a constraint, particularly with the use of crypto techniques), and (iv) fault-tolerance implications (critical to sensitive applications).

4   Concluding Remarks

We showed the uses of fine-grained synchronization relations by applications that use nonatomicity in modeling actions and need a fine degree of granularity to specify synchronization relations and their composite global predicates. We showed the specific meaning and significance of each relation. Some uses of the relations in a distributed system include modeling various forms of synchronization for group mutual exclusion, initiation of nested computing, termination of nested computing, and monitoring the start of a computation. The synchronization between any X and Y computations can be performed at any "synchronization barrier" for the X and Y computations of any application. For example, $R2^*(X, Y)$ synchronization can be performed at a barrier to ensure that subcomputation X which is nested in subcomputation Y completes before subcomputation Y completes, following which $R3^*(Y, X)$ synchronization can be performed to kick off another nested subcomputation X within Y. This barrier synchronization may involve blocking of processes which need to wait for expected messages, analogous to the barrier synchronization for multiprocessor systems [17].

The synchronization relations provide a precise handle to express various types of synchronization conditions in first-order logic. Each synchronization performed satisfies a global predicate in the execution. (A classification of some types of global predicates is given in [3,6].) Complex global predicates can be formed by logical expressions on such synchronization relations. It is an interesting problem to classify the global predicates that can be specified using the synchronization relations.

Observe that performing a synchronization (corresponding to one of the relations) involves message passing, and hence also provides a direct way to *evaluate* global state functions and global predicates involving the nodes participating in the synchronization. Thus, performing the synchronization enables the detection of global predicates. Also, global predicates can be *enforced* by initiating the synchronization only when some local predicates become true. Identifying the types of global predicates that can be detected or enforced using the synchronization relations is an interesting problem. The design of algorithms to enforce global predicates is also a topic for study.

References

1. *Linear time, branching time, and partial orders in logics and models of concurrency,* J. W. de Bakker, W. P. de Roever, G. Rozenberg (Eds.), LNCS 354, Springer-Verlag, 1989.
2. J. V. Benthem, *The Logic of Time*, Kluwer Academic Publishers, (1ed. 1983), 2ed. 1991.
3. R. Cooper, K. Marzullo, Consistent detection of global predicates, *ACM/ONR Workshop on Parallel and Distributed Debugging*, 163-173, May 1991.
4. C. A. Fidge, Timestamps in message-passing systems that preserve partial ordering, *Australian Computer Science Communications*, Vol. 10, No. 1, 56-66, Feb. 1988.
5. P. C. Fishburn, *Interval Orders and Interval Graphs: A Study of Partially Ordered Sets,* J. Wiley & Sons, 1985.
6. V. Garg, B. Waldecker, Detection of weak unstable predicates in distributed programs, *IEEE Transactions on Parallel and Distributed Systems*, 5(3), 299-307, March 1994.
7. W. Janssen, M. Poel, J. Zwiers, Action systems and action refinement in the development of parallel systems, In J.C. Baeten, J.F. Groote, (Eds.) *Concur91*, LNCS 527, Springer-Verlag, 298-316, 1991.
8. A. Kshemkalyani, Temporal interactions of intervals in distributed systems, *TR-29.1933, IBM*, Sept. 1994.
9. A. Kshemkalyani, Temporal interactions of intervals in distributed systems, *Journal of Computer and System Sciences*, 52(2), 287-298, April 1996. (Contains some parts of [8]).
10. A. Kshemkalyani, Framework for viewing atomic events in distributed computations, *Theoretical Computer Science*, 196(1-2): 45-70, April 1998.
11. A. Kshemkalyani, Relative timing constraints between complex events, *8th IASTED Conf. on Parallel and Distributed Computing and Systems*, 324-326, Oct. 1996.
12. A. Kshemkalyani, Synchronization for distributed real-time applications, *5th Workshop on Parallel and Distributed Real-time Systems*, IEEE CS Press, 81-90, April 1997.
13. L. Lamport, Time, clocks, and the ordering of events in a distributed system, *CACM*, 558-565, 21(7), July 1978.
14. L. Lamport, On interprocess communication, Part I: Basic formalism, Part II: Algorithms, *Distributed Computing*, 1:77-101, 1986.

15. F. Mattern, Virtual time and global states of distributed systems, *Parallel and Distributed Algorithms*, North-Holland, 215-226, 1989.
16. F. Mattern, On the relativistic structure of logical time in distributed systems, In: Datation et Controle des Executions Reparties, *Bigre*, 78 (ISSN 0221-525), 3-20, 1992.
17. J. Mellor-Crummey, M. Scott, Algorithms for scalable synchronization on shared-memory multiprocessors, *ACM Transactions on Computer Systems*, 9(1): 21-65, Feb. 1991.
18. E.R. Olderog, *Nets, Terms, and Formulas*, Cambridge Tracts in Theoretical Computer Science, 1991.
19. A. Rensink, *Models and Methods for Action Refinement*, Ph.D. thesis, University of Twente, The Netherlands, Aug. 1993.
20. R. Schwarz, F. Mattern, Detecting causal relationships in distributed computations: In search of the holy grail, *Distributed Computing*, 7:149-174, 1994.

A Simple Protocol to Communicate Channels over Channels

Henk L. Muller and David May

Department of Computer Science, University of Bristol, UK
henkm@cs.bris.ac.uk, dave@cs.bris.ac.uk

Abstract. In this paper we present the communication protocol that we use to implement first class channels. Ordinary channels allow data communication (like CSP/Occam); first class channels allow communicating channel ends over a channel. This enables processes to exchange communication capabilities, making the communication graph highly flexible. In this paper we present a simple protocol to communicate channels over channels, and we show that we can implement this protocol cheaply and safely. The implementation is going to be embedded in, amongst others, ultra mobile computer systems. We envisage that the protocol is so simple that it can be implemented at hardware level.

1 Introduction

The traditional approach to communication in concurrent and parallel programming languages is either very flexible, or very restricted. A language like Occam [1], based on CSP [2], has a static communication graph. Processes are connected by means of channels. Data is transported over channels synchronously, meaning that input and output operations wait for each other before data can be communicated. In contrast, communication in concurrent object oriented languages is carried out via object identifiers. Once a process obtains an identifier of another object, it can communicate with that object, and any object can communicate with that object. Similarly, languages based on the π-calculus [3], use many-to-one communications.

We have recently developed a form of communication that lies between the purely static and very dynamic communication constructs. We enforce communication over point-to-point channels, just like Occam, but instead of having a fixed communication graph we allow channels to be passed between processes. This gives us a high degree of flexibility. A channel can be seen as a communication capability to or from a process. This is close to the NIL approach [4].

These flexible channels can be extremely useful in many application areas. Allowing channel ends to be passed like ordinary data gives us an elegant paradigm to encode the (mobile) software required for wearable computers. Multi-user games often require communication capabilities to be exchanged between nodes in the system. A similar type of channels has been successfully used for audio processing [5], and mobile channels appear to be a natural communication medium for continuous media.

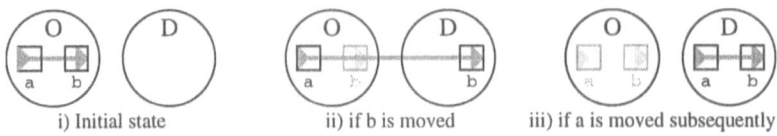

Fig. 1. A system, with two nodes 'D' and 'O', and a channel connecting ports a and b.
(i) the channel is local on 'O'; ii) port b is passed to node 'D', stretching the channel;
iii) port a is transferred to 'D', the channel snaps back.

In all of these applications, one wants to be able to pass around a communication capability from one part of a system to another. Having one-to-one communication channels as opposed to one-to-many communication models (which many concurrent OO languages offer as their default mechanism) makes it easier for the programmer to reason about the program behaviour, while it simplifies and speeds up the implementation. This is especially important when we are going to implement these protocols at hardware level.

In this paper we discuss the implementation of the mobile channel model. In Section 2 we first give a brief overview of the high level paradigm that we use to model these flexible channels. After that, we discuss the protocols that we have developed to implement the channels in Section 3. In Section 4 we present some performance figures of our implementation.

2 Moving Ports Paradigm

The programming paradigm that we have developed, Icarus, is designed to enable the development of mobile distributed software. In this section we give a brief description of the Icarus programming model, concentrating on the communication aspects; for an in-depth discussion of Icarus we refer to a companion paper [6]. Icarus processes communicate by means of channels. Like CSP and Occam channels, Icarus channels are point-to-point, and uni-directional. That means, at any moment in time, a channel has exactly one process that may input from the channel, and one process that may output to the channel. We call these channel ends *ports*, so each channel has an associated *input port* and *output port*. The two ports that are connected by a channel are called *companion* ports.

In contrast with CSP and Occam, Icarus ports are first-class. This means that one can declare a variable of the type "input port of integers", and assign it a channel end (the type system requires this to be the input-end of an integer channel). Alternatively one can declare a "channel of input port of integers", and communicate a port over this channel. Because an Icarus port can be passed around between processes, a flexible communication structure can be created.

There are two ways to look at mobile channels and ports. The easiest way to visualise mobile channels is to view a channel as a rubber pipe connecting two ports. Wherever the ports are moved, the channel connects them and transports data from the input port to the output port when required. The grey line in

```
{ chan !int d ; chan ?int c ;
  par { { chan z ;
          c ! z ; d ! z ;
        }
     || { ?int x ; int b ;
          c ? x ;
          x ? b ;
        }
     || { !int y ;
          d ? y ;
          y ! 1 ;
        }
      }
}
```

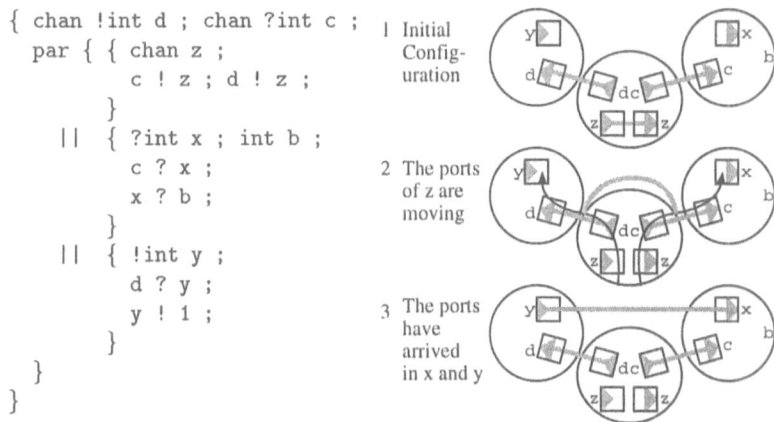

1 Initial
  Config-
  uration

2 The ports
  of z are
  moving

3 The ports
  have
  arrived
  in x and y

Fig. 2. Example Icarus program. Left: the code. Right: the execution, the processes before, during and after communication over c and d.

Figure 1(i) shows the channel between the two ports labelled **a** and **b**. Another way to view channels and ports is to enumerate them. If we assign a unique even numbers to each channel, say C, then we can define the ports of that channel to have numbers C and $C + 1$. Data output over port C will always arrive at its companion port $C + 1$; regardless the physical locations of the ports.

In order to preserve the point-to-point property of channels, ports are not *copied* when assigned or sent over a channel. Instead, they are *moved*. When reading a port variable, the variable is not just read, but read-and-destroyed in one atomic operation. These moving semantics guarantee that at any time each channel connects exactly two ports. A port-variable can thus either contain a port, that is, it is connected via a channel to a companion port, or it is unconnected, which we can represent by having no value in the port.

The syntax of Icarus allow us to declare an output port of something by using an exclamation mark: $!T$ is the type of a port over which values of type T can be output. The type $?T$ denotes the ports over which values of T can be input. So the type $!?$int is the type of port over which one can output a port over which one can input an integer.

Icarus has output (!) and input (?) operators, and a *select* statement to allow a choice between multiple possible inputs. Icarus communication is synchronous, which, as is shown later in this paper, simplifies the implementation dramatically. The example program in Figure 2 creates three processes. Process 1 sends an input port to process 2 and an output port to process 3. This establishes a channel between processes 2 and 3, over which they can exchange data.

3   The Protocol for Moving Ports

The protocol used to communicate ports over channels is simple. We present the protocol bottom up: we start with the basic communication of simple values, then

Port index	Companion port index	Remote Address	Port-state	More
12:	13	-	IDLE	...
13:	12	-	IDLE	...

Fig. 3. Part of the port data structure, with two companion ports.

we discuss sending ports itself, and we finally consider the difficult cases such as sending two companion ports simultaneously. We assume that the protocol is implemented on top of a reliable in-order packet delivery mechanism.

3.1 Communicating Single Values

The port data structure consists of an integer that denotes the index of the companion port, an optional field to denote the remote address of the companion port (not used for a port with a local companion), a field that gives the *state* of the port, and some additional fields for storing process identifications. Two companion ports are shown in Figure 3.

Communicating single values is implemented conventionally, using the same algorithms as used in, for example, the Transputer family [7]. The channel can be in one of three states, IDLE, INPUTTING and OUTPUTTING. This uniquely defines the state of both inputting and outputting processes.

The IDLE state denotes that no process wishes to communicate over the channel. The INPUTTING state denotes that a process has stopped for input and is waiting for communication, but that no process is ready for outputting yet. Conversely, the OUTPUTTING state denotes that a process is ready for output but that no process is ready for input. There is no state for two processes being ready for input and output, as that state is resolved immediately when the second process performs its input/output operation.

The state of a channel is represented as the state of its two ports. It is sufficient to store the state INPUTTING on the output port, and to store the state OUTPUTTING on the input port. This ensures that a check of the local port suffices to determine the state of the channel.

If we want to support a *select*-operation, we will need an extra state, SELECT-ING, which indicates that a process is willing to input. Upon communication the selecting process will perform some extra work to clean other ports which have been placed in the state SELECTING.

3.2 Communicating Simple Values Off-Node

Off-node communication is synchronous, but has been optimised to hide latency as much as possible. A process that can output data will immediately send the data on the underlying network, and then wait for a signal that the data has been consumed before continuing. Note that the process stops after outputting, until the data has been input; this is essential when implementing the full port-moving protocol.

Because the port-state of a remote companion port is not directly accessible, a port-state is stored at both nodes. Ports with a remote companion can be in the states REMOTE_IDLE, REMOTE_INPUTTING, REMOTE_OUTPUTTING and RE-MOTE_READY. The inputting port can be REMOTE_IDLE (no process willing to input at this moment), REMOTE_INPUTTING (a process is willing to input and waiting) or REMOTE_READY (no process is willing to input, but data has been sent and is waiting to be input). The outputting port can be in the state REMOTE_IDLE (no process is willing to output at this moment) or REMOTE_OUTPUTTING (a process has output data and is waiting for the data to be input).

When implementing a *select*-operation, a port may get into a state RE-MOTE_SELECTING to denote that a process is waiting for input from more than one source (note that our *select*-operation only allows selective input, we do not support a completely general *select*-operation [8])

3.3 Communicating Ports Off-Node

Locally, within a node, a port can be simply an index in an array. Within a node, ports can be freely moved around from one process to another simply by communicating this index value. We discuss transfers of ports to another node in four steps:

1. Transfer of a port from a source node to a destination node, where the companion port is on the source node (channel stretches).
2. Transfer of a port from a source node to a destination node, where the companion port is on a third node.
3. Transfer of a port from a source node to a destination node, where the companion port is on the destination node (channel snaps back)
4. Difficult cases (simultaneous transfers of two companion ports)

1. Transfer of a port from a source node to a destination node, where the companion port is on the source node. In this case the companion port is local, and stays behind. According to Section 3.1, the port that will be sent, must be in the state IDLE, INPUTTING, or OUTPUTTING. Because ports are associated with exactly one process, we can be sure that the port that is going to be sent is not active. That is, no process can be inputting or outputting on the port which is being sent: if a process was inputting or outputting on this port, then the process cannot send the port (it could only attempt to send the port over itself; the type system prevents this from happening).

Because the port being sent is not active, it is sufficient to send some identification of the companion port to the destination node. The identification that we send consists of a local index number, and the node address of the source (for example an IP-address and IP port-number). This tuple (local index, node address) is a globally unique identification of the port.

Upon receiving the identification, a new port is allocated, and the companion port index and node address are stored in this port. The port-state is initialised to REMOTE_IDLE, for the port was inactive. The second step is to inform the

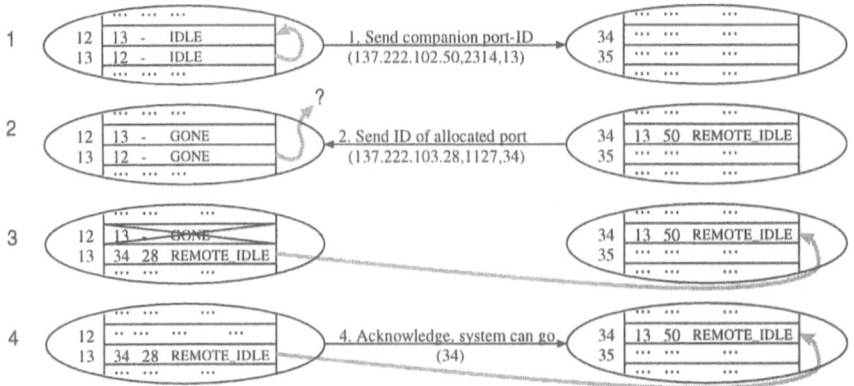

Fig. 4. Steps to send a port to a remote node. Grey arrows are channels.

companion node of the newly created port index, and the companion port's state is set to REMOTE_IDLE, REMOTE_INPUTTING, or REMOTE_OUTPUTTING. After the companion port is updated, step three is to delete the original port. Even though the port send was inactive, it was possible that at most one message was waiting on it, in that case the data is forwarded in order to preserve the invariant that the data has been sent to the remote node. Finally, in the fourth step an acknowledgement is sent to the destination node, signalling that the companion port is ready.

This four step process is summarised in Figure 4. Between steps 1 and 2 is a transitional period when the companion port is not connected to anything. For this reason, the companion port-state is set to GONE in step 1. Any I/O operation between steps 1 and 2 will wait until this transient state disappears.

2. Transfer of a port from a source node to a destination node, where the companion port is on a *different*, third, node. This case is slightly more difficult, and we need a generalised version of the previous protocol. This generalised protocol has four steps, outlined in Figure 5. In step 1 the global identification of the companion port is sent from 'S' to 'D'. In step 2 the newly allocated port-index is sent from the destination node 'D' to the companion node 'C'. This will allow the companion port to be updated. Step 3 will acknowledge to 'S' that the port has been transferred, allowing the source node to delete the original port. Finally step 4 is an acknowledgement to 'D' signalling that the port is ready for use.

Performing these four steps in this order will ensure that any message that might have been sent to the source node is collected and forwarded to the destination node. In the worst case, a message might be output on the companion port, just after the source node has initiated the transfer. This message will be delivered to the source node while the port is in state GONE. The data will be kept here until step 3 of the protocol, whereupon it is transferred to the destination node.

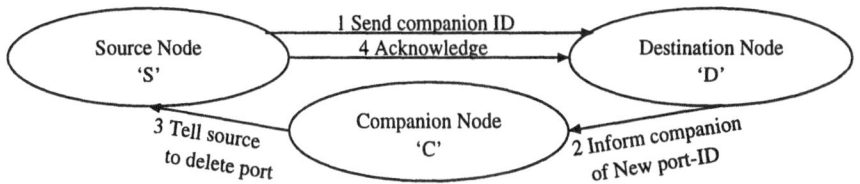

Fig. 5. Steps to send a port in a system involving three nodes.

Note that the case with two nodes discussed in the previous section is a special case of this four step protocol. If the companion port resides on the source node, then step 3 is a local operation.

3. Transfer of a port from a source node to a destination node, where the companion port is on the destination node. Because the port is being sent to the node where the companion resides, we must ensure that both companion ports are in a 'local' state, allowing data transfer to be optimised for local transport.

We use the same protocol as before. Because the destination node and the companion node are now one and the same, step 2, informing the companion node of its new status, is now a local activity. We still need to execute steps 3 and 4, telling the source node that the port can be deleted, and acknowledging the completion. These steps take care of the situation in which the port sent was originally in the state REMOTE_READY. The companion had sent data to the input-port, and the data was waiting to be picked up. Before the port is deleted, we forward the data to the destination node. Both companions are now on the destination node, so we change the input port-state to OUTPUTTING, and keep the data in the outputting process until a process is ready to input it. The data can be safely kept in the outputting process because it had not been allowed to proceed.

4. Difficult cases. There are two difficult cases worth discussing: two companion ports that are sent at the same time (to the same or different destination nodes); and the case where a port is transferring a port, which is in turn transferring a port (which in turn might be transferring a port).

When two ports are transferred simultaneously, the protocols moving the two ports might interfere with each other. The source of the interference is step 2, where the companion node is updated to reflect the new location of the port transferred. We overcome this problem by checking during step 2 whether the companion port is actually being moved, which is indicated by the port-state GONE. If this is the case, then the message is forwarded to the companion's destination node.

If a companion port is found in the state GONE, then the other port must find its companion GONE (because of the message ordering), and both nodes will forward the message of step 2. Eventually, both newly created ports will have a reference to each others' location, and both old ports are deleted. Note that if

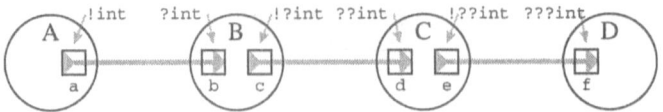

Fig. 6. Ports over ports over ports.

two companion ports are sent simultaneously, then both ports must have been IDLE, and no data will be sent until both have arrived at their destinations.

The most difficult situation is the case where a port is sent, and this port is actually a port of ports, carrying a port. In Figure 6 we have sketched a situation with 4 nodes (A, B, C and D) and 6 ports (a, b, c, d, e and f). The ports are linked up as denoted by the grey lines. Outputting port d over e is a normal output operation as discussed before. Similarly, outputting port b over c to d is a normal operation.

If the transfers of b and d happen simultaneously, then b is actually making a double hop, from node B via node C to node D. Indeed, there may be an arbitrarily long finite list of ports to be moved, only restricted by the number of '?' in the port type definition (the typing system guarantees it to be finite). This forwarding causes a serious complication in the protocol (around 20% of the code), and we are considering whether it is worthwhile to prohibit this.

3.4 Security

The protocol described above is not secure. Most notably, an attacker can forge a message for step 2 of the protocol, and deliver it to a node, in order to obtain a port in the process. The solution is to extend our protocol with an extra step. When executing step 1 of the process, we have to send a clearance message to the companion port, informing it that a move is imminent. The companion node will only execute step 2 of the protocol when the source node has sent this clearance. This provides security against unknown processes taking channel ends, but on the assumption that the underlying network enforces a unique naming scheme, and secure delivery of messages.

3.5 Discussion

The protocol has been optimised so that the frequent operations, such as sending ordinary data or ports, have high performance. There are communication patterns which will perform badly. The worst case is a program where a process outputs a large array over a port, while there is no process willing to input this data. If the input-port is now sent from one process to the next, then the data is shipped with it. If the input port were to be sent many times, we would waste bandwidth and increase latency. We could optimise this case by not sending data until required, but this would seriously impair performance in the more common case, when data is communicated between two remote nodes (incurring double latency on each transferral).

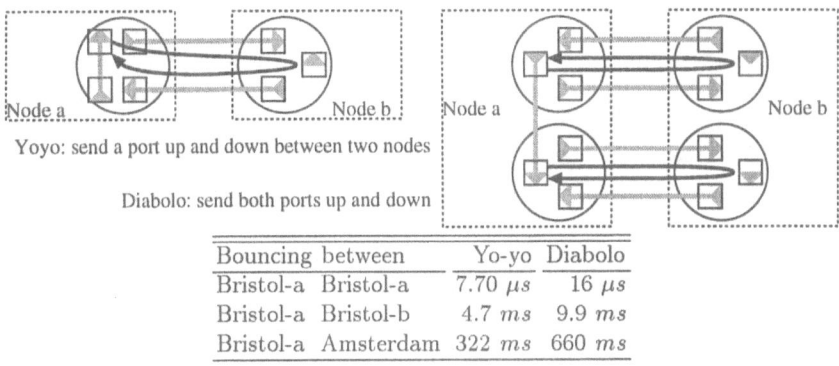

Yoyo: send a port up and down between two nodes

Diabolo: send both ports up and down

Bouncing between		Yo-yo	Diabolo
Bristol-a	Bristol-a	7.70 μs	16 μs
Bristol-a	Bristol-b	4.7 ms	9.9 ms
Bristol-a	Amsterdam	322 ms	660 ms

Fig. 7. The two test programs that we measured, with round trip times.

4 Results

The protocol has been implemented in C, to implement Icarus on a network of workstations. The implementation of the protocol takes less than 1000 lines of C, underlining its simplicity. We use UDP (User Data Protocol) for the underlying network, with an extra layer to guarantee reliable in-order delivery.

We have tested the protocol with many different applications. A class of students has written multi user games in Icarus. To convince the reader that our implementation works over the Internet, we show timings of the two test programs displayed in Figure 7. The first test program, Yo-yo, keeps one port at a fixed node, and sends the companion port forwards and backwards to a process on a second node. In this program the channel must stretch and snap back every time. The second test program, Diabolo, stretches the protocol with some more demanding communication. Two pairs of processes are running on two nodes, each pair of processes sends one of two companion forwards and backwards between two nodes.

We have measured three configurations: one where all processes are running within the same node (which measures communications inside the run-time support only), a configuration where we run the processes on two machines on the same LAN, and a configuration where we run the processes on two distant machines. The performance is linearly related to the average latency between the nodes involved, and to the number of messages that is needed for a round trip. This is clearly shown in the timings of Yo-yo versus Diabolo for the three configurations.

5 Conclusions

In this paper we presented a simple protocol that allows us to treat communication channels as first class objects. The programming paradigm allows processes to communicate over point-to-point channels. Values communicated can either be simple values (like integers or arrays of integers), or *ports* (ends of channels). Because the paradigm allows a synchronous implementation, we have been able to produce a simple implementation. Whenever a port is transferred, the port is known to be idle (because ports are point to point and synchronous), so we do not have to transfer any state around.

We have developed a four step protocol to move a communication port from one node to another. Step one carries the port identifier, creating a new port on the destination node. Step two informs the companion port of its new companion. Step three informs the source that the original port is to be deleted. Step four allows the newly created port to start communication. These steps may be local or remote, depending on the relative positions of the source, destination, and companion nodes. If the two connected ports end up in the same node, the channel snaps back to within the node, and the implementation switches to a highly efficient local communication protocol.

The applications of this protocol lie in many areas: multi user games (where players receive channels that allow them to communicate with other players), continuous media switchboards (where a channel carrying, for example, video signals is handed out to two terminals), and wearable computing (establishing communication channels between parts of a changing environment). We are currently integrating this protocol in a wearable computer system, and its applications in continuous media.

References

1. Inmos Ltd. *Occam-2 Reference Manual*. Prentice Hall, 1988.
2. C. A. R. Hoare. Communicating Sequential Processes. *Communications of the ACM*, 21(8):666–677, August 1978.
3. R. Milner, J. Parrow, and D. Walker. A Calculus of Mobile Processes, I. *Information and Computation*, 100(1):1–40, Sept. 1992.
4. R. Strom and S. Yemini. The NIL Distributed Systems Programming Language: A Status Report. *SIGPLAN notices*, 20(5):36–44, May 1985.
5. R. Kirk and A. Hunt. MIDAS–MILAN An Open Distributed Processing System for Audio Signal Processing. *Journal Audio Engineering Society*, 44(3):119–129, Mar. 1996.
6. D. May and H. L. Muller. Icarus language definition. Technical report, Department of Computer Science, University of Bristol, January 1997.
7. M. D. May, P. W. Thompson, and P. H. Welch, editors. *Networks, Routers & Transputers*. IOS Press, 1993.
8. G. N. Buckley and A. Silberschatz. An Effective Implementation for the Generalised Input-Output Construct of CSP. *ACM Transactions on Programming Languages and Systems*, pp 224, Apr. 1983.

SciOS: Flexible Operating System Support for SCI Clusters*

Povl T. Koch** and Xavier Rousset de Pina

SIRAC Laboratory, INRIA Rhône-Alpes,
ZIRST - 655, avenue de l'Europe
38330 Montbonnot Saint-Martin (France)
scios@inrialpes.fr

Abstract. The bottleneck for many parallel and distributed applications on networks of workstations is the high cost of communication on traditional network interfaces. Memory-mapped network interfaces provide latencies of a few microseconds and bandwidths close to the maximum of the local I/O bus. A major drawback with the current operating system support is that applications have to deal directly with data consistency and the allocation and placement of pinned physical memory. We propose a more elaborate interface, called SciOS, that provides a shared-memory abstraction where physical memory is treated like in a NUMA architecture. To lower the average memory access times, we use a relaxed memory model and dynamic page migration and replication combined with use of idle remote memory instead of disk swap. We describe the SciOS programming model and describe the issues for its implementation on an SCI cluster.

1   Introduction

Traditional network interfaces and operating systems often require that every communication is through expensive calls to the operating system, includes extensive buffer handling, and require memory copies between address spaces to enforce protection and provide flexibility and reliability. These interface designs typically result in high message latencies, bandwidth limitations, and high software overheads of communication. This hurts the performance of a wide range of parallel and distributed applications.

Many new networking technologies are based on virtual memory-mapped interfaces and have bandwidths in the range of Gigabits per second, message latencies of a few microseconds, and hardware-based reliable delivery and congestion control[1, 6, 8, 14]. After the setup of remote memory mappings, all communication is handled through the processor's normal load and store instructions

---

* This work was supported in part by the Region Rhône-Alpes in the framework of the Emergence Research Program (C.R. E901010001)

** Supported in part by the European Commission, the TMR Program. Also associated with University of Copenhagen (Denmark)

or system managed DMA with very little overhead on the sending and receiving processors. These types of interfaces have been shown to be well suited for message-passing interfaces [10] and parallel programming languages [9] because of their low latencies and the hardware support of zero-copy protocol implementations.

Our approach is to use the user-level load and store interface of memory-mapped networks as a means to build a Non-Coherent NUMA architecture and to provide a coherent shared memory system through operating system mechanism. A major performance issue for NUMA architectures is that applications have to deal with the high performance difference between local and remote memory references. On Cache-Coherent NUMA machines (CC-NUMA), remote accesses are typically 2-3 times more expensive than local access times. Therefore, the physical placement of data is crucial to performance of many applications. Each application has to deal explicitly with this locality issue which can become complex when dealing with dynamic sharing patterns. Mechanisms have been implemented to dynamically migrate and replicate shared memory pages [2, 3, 5, 16] to increase memory locality. For memory-mapped network interfaces, the difference between local and remote memory access is much higher than for CC-NUMA machines because the remote memory accesses have to pass by the I/O bus.

In Sect. 2, we describe the capabilities of memory-mapped networks together with the current operating system support provided for message-passing interfaces and its limitations. Section 3 describes the main features and implementation issues of our system, SciOS, based mainly on a coherent shared memory abstraction. Section 4 describes related work and Sect. 5 concludes the paper.

2 Memory-Mapped Network Implementations

The implementation of memory-mapped networks is based on a processor's user-level access to the network interface placed on the I/O bus. Although the I/O bus, e.g., PCI, is slower than the memory bus, it allows more general implementations that can be used on a wide range of architectures. We describe below the main mechanisms of the PCI-SCI adapter and the operating system support provided for it.

2.1 Dolphin's PCI-SCI Adapter

Dolphin's PCI-SCI adapter is based on an IEEE standard called Scalable Coherent Interface (SCI) [11] which specifies the implementation of a coherent shared memory abstraction and a low-level message passing interface. The cache coherence is optional and is typically not implemented on the I/O-based SCI implementations we are considering. SCI specifies a shared 64-bit physical address space. The high order 16 bits of an SCI address designates a node and the remaining 48 bits allow addressing of physical memory within each node. The

PCI-SCI adapter [6] provides a number of mechanisms to map remote memory, to send and receive low-level messages, generate remote interrupts, and to provide atomic operations on shared memory. SCI is based on a ring topology and Dolphin's implementation is based on 200 MB/s bidirectional point-to-point links and SCI switches which allow the interconnection of rings. The PCI-SCI adapter does not implement the cache consistency mechanisms in the SCI standard. Each application will therefore have to deal with issues such as when to flush or invalidate processor caches, buffers in the PCI-SCI adapter, etc. Below, we focus on the adapter's ability to map remote memory using a standard 32-bit PCI bus.

To share physical memory between nodes, a process can allocate local physical memory. The physical memory is accessed locally through a normal virtual-to-physical memory mapping. For a process on a remote node to access the shared memory, the process must set up two mappings using the operating system. (1) The remote memory is mapped into the address space of the local PCI bus. This is done by updating an address translation table (ATT) in the PCI-SCI adapter with the node ID and physical address of the remote memory. (2) The operating system then maps the specific address range of the PCI bus into its virtual address space through manipulation of its page tables. This mapping also enforces protection since the process can only access mapped memory and only in the right mode (read/write/execute). The mappings are shown in Fig. 1 where Process A has remotely mapped a physical memory page allocated and locally mapped by Process B on another node.

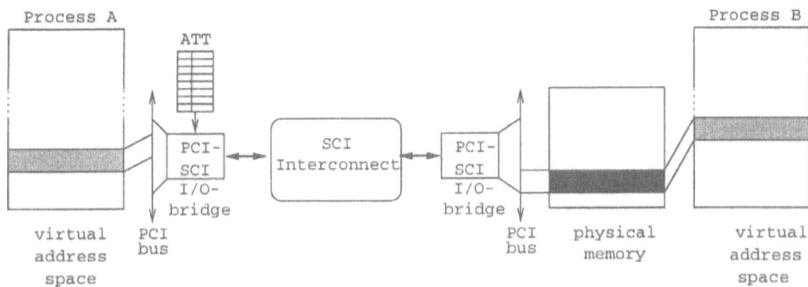

Fig. 1. Memory mappings for physical memory and PCI-SCI adapter.

2.2 Driver Support for the PCI-SCI Adapter

The driver support for the PCI-SCI adapter is structured as an operating system independent module which has different interfaces for each operating system and adapter versions. The main primitives directly represent the hardware functions presented in Sect. 2.1. The management of physical memory and memory mappings is implemented using device files. DMA functionality is provided through standard **read** and **write** calls. A number of reliability functions are provided to shield nodes from hardware failures.

All physical memory for a segment is marked as pinned so it is not placed on disk by the operating system. This is because the PCI bus only accesses the memory using physical addresses and the PCI-SCI adapter does not have a TLB functionality for incoming requests.

An application can create a shared segment between 8 KB and 512 KB which is allocated in contiguous physical memory. The allocation follows certain alignment requirements because the address translation table entries in the PCI-SCI adapter are aligned at 512 KB boundaries. The number of address translation table entries is 8,192 which can map up to 4 GB of remote memory, which is more than sufficient. But the main limitation is how much address space the adapter is allowed to use on the PCI bus. Currently, 128 MB is reserved on the PCI bus for the PCI-SCI adapter, which corresponds to only 256 active ATT entries. For the mapping of small memory blocks, e.g., for special synchronization mappings, a whole ATT entry must be used to map only a few physical pages. ATT entries must therefore be considered a scarce resource which limits the number of remote mappings. After some allocation and de-allocation of physical memory, it can be hard for the operating system to find contiguous and correctly aligned physical memory. Depending on the operating system, it can mean that only a limited number of segments can be created. For memory not perfectly allocated, more ATT entries are therefore needed. (The next generation PCI-SCI adapter has more ATT entries and allows mapping at a granularity smaller than 512 KB, thereby reducing the alignment requirements.)

Each segment has a unique key which is used to name the segment when it is mapped from a remote node. When the segment is mapped to the local PCI bus, the mmap call can map the segment into the virtual address space of a process. The processor cache is disabled for the region of the PCI bus that is used by the PCI-SCI adapter. This means that all load and store operations have to pass through the PCI-SCI adapter, i.e., loading a remote location recently stored to by the same processor results in PCI and SCI traffic. Since writes can arrive out of order special flush operations or remote reads can be used as a store barriers to guarantee some memory consistency.

Although the allocation, deallocation, mapping, and unmapping of segments is completely dynamic, the location of a segment is *fixed* during its lifetime. This is useful for simple, fixed sharing patterns, e.g., when implementing a message-passing interface. Since the bandwidth of remote writes much higher than those for remote reads, a receive buffer should be fixed on the receiving node to obtain the best performance. This performance may be hurt seriously for segments that have a dynamic sharing pattern.

3 The SciOS Prototype

The SciOS prototype provides a flexible coherent shared memory abstraction on a cluster of workstations connected by a load/store-based memory-mapped network interface. The goal is to minimize the cost of all memory references for different sharing patterns.

3.1 The SciOS Programming Model

SciOS provides a large, flat, shared address space. The abstraction for sharing is a *memory block* which is a contiguous range of virtual memory. To allow efficient implementations of the coherent shared memory, SciOS model is based on lazy release consistency [12]. This is a relaxed consistency model that allows many low-level optimizations such as the delaying of remote writes and delivery of remote writes out of order. The observation behind release consistency is that correctly synchronized applications only need to have shared memory coherent while inside critical sections. For an application to see a sequentially consistent memory, as in uniprocessor systems, all accesses to shared data are required to be bracketed with appropriate *acquire* and *release* operations when respectively entering and leaving a critical section. These operations are provided by the SciOS implementation. Traditional synchronization primitives such as locks can be implemented directly using the corresponding acquire/release operations while a barrier entry can use the release operation and a barrier departure, the acquire operation.

To allow the application to exercise some control over the underlying memory system, SciOS supports five different memory protocols. The protocols range from a simple, fixed memory placement as described in Sect. 2.2 to a more complex implementation with migration and replication of shared memory blocks:

FIXED_INIT Physical pages are pinned and fixed on the node that allocates it. Virtual addresses are coordinated so the same memory block is mapped at the same addresses in all sharing processes. Implementation of message passing interfaces and applications with multiple-producers/single-consumer patterns can benefit from using this memory protocol.

FIRST_TOUCH Physical pages are allocated and pinned on the first node that accesses them. After being allocated, a physical page is fixed on the node, as with the FIXED_INIT protocol. This protocol is targeted at parallel applications with a mostly-write sharing pattern or little active sharing.

MIGRATION Physical pages are allocated as in the FIRST_TOUCH protocol, but they may migrate individually between the nodes depending on the sharing pattern. Mechanisms ensure that thrashing (ping-pong) is minimized. Sequential sharing patterns can benefit from this protocol.

MIG_REP This is an extension of the MIGRATION protocol where physical pages are replicated when read by multiple nodes. Coherency mechanisms assure that the memory model is respected. Sharing patterns with mostly-read shared data can use this protocol.

GLOBAL This is similar to the MIG_REP protocol, but to globally minimize swap activity, decisions are coordinated with the virtual memory system and idle remote memory is used as swap space instead of a local disk. This protocol is primarily targeted at applications where each node is memory-bound ("out-of-core" applications).

3.2   Our SciOS Testbed

We are currently implementing the SciOS prototype on a cluster of a total of six uniprocessor Intel nodes running the Linux operating system. The nodes all have a standard 32-bit PCI bus. Each node has a single PCI-SCI adapter connected to our SCI switches. When the machines are connected back-to-back, we observe that the latency for remote 4-byte, non-cached accesses is 2.4 microseconds for writes and 4.6 microseconds for reads. Remote access bandwidth is measured to approximately 10 MB/s for reads and 35-70 MB/s for writes for packets sizes as small as 512 bytes and when using all hardware optimizations. Detailed performance analysis of this adapter type can be found in [15].

Dolphin's PCI-SCI driver runs as a Linux kernel module. SciOS is being implemented as another module that uses a few basic functions in the driver module. For SciOS's memory blocks, we use the normal UNIX file abstraction that can be implemented using Linux's Virtual File System (VFS) facility. After the SciOS module has been loaded and a directory mounted, an application can use the standard UNIX system calls **open**, **ioctl**, **mmap**, **munmap**, and **close** on memory blocks. The kernel directs these system calls — along with other kernel events such as page faults, swap in, and swap out — for all SciOS files to our SciOS module. This way, we can implement SciOS at the kernel level with full control of the memory blocks and the associated mappings and physical memory. It also allows us to manage the memory blocks with standard UNIX commands like **touch**, **rm**, and **ls** commands.

3.3   Issues for Implementing SciOS

In this section we will discuss the implementation issues for SciOS on our Linux cluster. The main SciOS data structure is a global table which holds information about each node and a directory for all the SciOS files. When a file is created, a page table holds information about the state of the physical memory allocated for it. With the SciOS implementation, we are dealing with the following four issues: (1) dynamic management of the scarce ATT entries in the PCI-SCI adapter, (2) allowing the use of normal, swappable physical memory for SciOS files, (3) the implementation of NUMA techniques for migration and replication of shared memory, and (4) the integration of the NUMA techniques with a remote swap mechanism. These issues are discussed below.

The PCI-SCI driver described in Sect. 2.2 simply refuses more remote mappings when there are no more ATT entries available. In SciOS, applications are allowed to map remote memory because ATT entries are shared dynamically, like the entries in the processor's TLB. When there are no ATT entries available for a new remote mapping, SciOS will invalidate all the virtual memory mappings for an already used ATT entry and reuse it. When the old mapping is used again, an ATT entry will be allocated in the same way. Since the PCI-SCI adapter does not provide statistics on the extent to which the ATT entry is used, we are forced to use a simple FIFO policy.

To alleviate the problem of allocation and alignment of pinned physical memory for the PCI-SCI adapter, it is possible during the startup of the Linux kernel to reserve a large part of physical memory. Since a *static* reservation is hard to make optimally and it does not allow the sharing of the physical memory with other applications, SciOS uses normal swappable physical memory. When the operating system is running out of physical memory, a "least recently used" page will be placed on disk so the page can be used by another process. SciOS is notified about such events (swapout) for the pages it has allocated. All remote mappings for the page must then be invalidated to avoid wrongly accessing the data of another application. An invalidation request must therefore be sent to all nodes that map the page and the page cannot be placed on disk until all acknowledgments have been received. When a remote process accesses an invalidated remote mapping, the page must be brought back from disk into the physical memory and the remote mapping must then be reestablished.

The NUMA techniques are reflected in SciOS's MIGRATION and MIG_REP protocols. During a page fault, two possibilities exist for the MIGRATION protocol: establish a remote mapping or migrate the page to allow local accesses. If the page has migrated much in the past, it is *frozen* for a period of time [5] and a remote mapping is established to avoid a ping-pong effect. Otherwise, the page is migrated to the node that accesses it. For the MIG_REP protocol, a third possibility exists: a page that is not frozen can be replicated on a read access. Replication is preferred over migration if the page has not been modified recently. On write accesses, page replicas must be invalidated to maintain memory coherency. In addition to the page fault decisions, a daemon periodically collects sharing information and may redo migration/replication decisions. Normal NUMA replication protocols eagerly invalidate all replicas before a write access can be allowed to enforce sequential consistency. With lazy release consistency [12], the invalidation of remote replicas can be postponed until the remote nodes perform an acquire operation. We will use this lazy technique for implementing the MIG_REP protocol.

On current architectures, access to remote memory is much faster than to local disk [7] even with traditional network. By avoiding disk swap activity, performance can be gained for applications with high memory requirements or when multiprogramming each SciOS node. With the GLOBAL protocol, SciOS takes into consideration the amount of idle physical memory on each node when pages are swapped out and while deciding when to migrate and replicate pages. If there is no idle physical memory on a node, SciOS can decide to map a page remotely instead of migrating or replicating the page. When the operating system wants to swap out a page, SciOS can simply discard replica pages. Instead of placing the page on a disk, it can be migrated to a node with a remote mapping and idle physical memory. In cases where the mapping nodes have no idle memory or when the page is not remotely mapped, the page can be migrated to other nodes with idle memory. Only when all physical memory is used on all nodes, SciOS will place pages on disk. To support the GLOBAL protocol, each node places information about its amount of idle memory in the global table.

4 Related Work

Our work draws on the research in the area of operating system support for distributed shared memory (DSM) abstractions for workstation clusters and NUMA architectures. Numerous DSM systems implement a coherent shared memory abstraction on workstation clusters without any hardware support for remote memory access. Munin [4] implements multiple memory protocols to support different sharing patterns like we will do in SciOS. The remote load/store capabilities of the PCI-SCI adapter widens the design space for SciOS compared traditional DSM system. The use of page migration and replication can lower memory access time and reduce load on the interconnect [3, 13]. Ibel *et al.* have implemented a global address space at the user level where application manages a pool of *preallocated* pinned segments and mappings of remote segments. Because SciOS is implemented in the kernel, we can share the physical memory and remote mappings between applications/processes. Verghese *et al.* [16] also takes the amount of idle physical memory into consideration when replicating pages like SciOS's GLOBAL protocol. Feeley *et al.*'s Global Memory Service [7] uses idle physical memory to reduce expensive disk swap but they do not deal with consistency of write-shared pages as SciOS's GLOBAL protocol.

5 Summary

Instead of using the memory-mapped network interfaces as a means to provide efficient message passing, we use it as a non-coherent NUMA architecture. We have presented SciOS, a system which provides a programming model based on a coherent shared memory that facilitates the programming of an SCI cluster. Since the placement of physical memory is important for performance, SciOS provides five different protocols so the programmer is able to control the behavior of the underlying memory system. Mechanisms such as migration and replication of shared pages will lower the average memory access times. A contribution of SciOS is a global memory protocol that tightly integrates the shared memory abstraction with the swap functionality of the virtual memory system. This will provide much better performance for applications with high memory requirements because the physical memory on all nodes is used as cache before pages are placed on a disk. By only using normal swappable physical memory and dynamically sharing address translation table entries, the SciOS implementation will remove many of the restrictions found in the current driver support. We are currently implementing SciOS on an Intel cluster using Dolphin's PCI-SCI adapter.

Acknowledgments

Kaare Lochsen and Hugo Kohmann from Dolphin Interconnect Solutions (Norway) willingly granted us access to the source codes for the PCI-SCI adapter and answered all of our questions. Roger Butenuth from Universität Paderborn (Germany) kindly gave us access to his Linux port of the PCI-SCI driver.

References

1. Matthias A. Blumrich, Kai Li, Richard Alpert, Cezary Dubnicki, and Edward W. Felten. Virtual memory mapped network interface for the SHRIMP multicomputer. In *Proceeding of the 21st Annual International Symposium on Computer Architecture*, pages 142–153, April 1994.
2. William J. Bolosky, Robert P. Fitzgerald, and Michael L. Scott. Simple but effective techniques for NUMA memory management. In *Proceedings of the 12th ACM Symposium on Operating System Principles*, pages 19–31, December 1989.
3. Edouard Bugnion, Scott Devine, and Mendel Rosenblum. Disco: Running commodity operating systems on scalable multiprocessors. In *Proceedings of the 16th ACM Symposium on Operating System Principles*, pages 143–156, October 1997.
4. John B. Carter, John K. Bennett, and Willy Zwaenepoel. Implementation and performance of Munin. In *Proceedings of the 13th ACM Symposium on Operating System Principles*, October 1991.
5. A. Cox and R. Fowler. The implementation of a coherent memory abstraction on a NUMA multiprocessor: Experiences with PLATINUM. In *Proceedings of the 12th ACM Symposium on Computer Architecture*, December 1990.
6. Dolphin Interconnect Solutions. PCI-SCI cluster adapter specification, May 1996. Version 1.2. See also http://www.dolphinics.no.
7. Michael J. Feeley, William E. Morgan, Frederic H. Pighin, Anna R. Karlin, Henry M. Levy, and Chandramohan A. Thekkath. Implementing global memory management in a workstation cluster. In *Proceedings of the 15th ACM Symposium on Operating System Principles*, pages 201–212, December 1995.
8. Marco Fillo and Richard B. Gillett. Architecture and implementation of MEMORY CHANNEL. *Digital Technical Journal*, 9(1):27–41, 1997.
9. Richard B. Gillett. Memory channel network for PCI. In *IEEE Micro*, volume 16, pages 12–18. February 1996.
10. Maximilian Ibel, Klaus E. Schauser, Chris J. Scheiman, and Manfred Weis. High-performance cluster computing using SCI. In *Hot Interconnects Symposium V*, August 1997.
11. IEEE. *IEEE Standard for Scalable Coherent Interface (SCI)*. 1992. Standard 1596.
12. Peter Keleher, Alan L. Cox, Sandhya Dwarkadas, and Willy Zwaenepoel. Tread-Marks: Distributed shared memory on standard workstations and operating systems. In *Proceedings of the 1994 Winter USENIX Conference*, pages 115–132, January 1994.
13. Richard P. Larowe Jr., Carla Schlatter Ellis, and Laurence S. Kaplan. The robustness of NUMA memory management. In *Proceedings of the 13th ACM Symposium on Operating System Principles*, pages 137–151, October 1991.
14. Evangelos P. Markatos and Manolis G.H. Katevenis. Telegraphos: High-performance networking for parallel processing on workstation clusters. In *Proceedings of the Second International Symposium on High-Performance Computer Architecture (HPCA)*. February 1996.
15. Knut Omang. Performance of a cluster of PCI based UltraSparc workstations interconnected with SCI. In *Proceedings of Network-Based Parallel Computing, Communication, Architecture, and Applications (CANPC'98)*, number 1362 in Lecture Notes in Computer Science, pages 232–246, January 1998.
16. Ben Verghese, Scott Devine, Anoop Gupta, and Mendel Rosenblum. Operating system support for improving data locality on CC-NUMA compute servers. In *Proceedings of the 7th Symposium on Architectural Support for Programming Languages and Operating Systems*, pages 279–289, October 1996.

Indirect Reference Listing:
A Robust Distributed GC*

José M. Piquer, Ivana Visconti

Universidad de Chile, Casilla 2777, Santiago, Chile
jpiquer@dcc.uchile.cl

Abstract. Reference Listing is a distributed Garbage Collection (GC) algorithm which replaces Reference Counts by a list of sites holding references to a given object. It has been successfully implemented on many systems during the last years, and some extensions to it have been proposed to collect cycles. However, Reference Listing is difficult to extend to an environment supporting migration and must remain blocked during a reference duplication to avoid race conditions on the reference lists.

On the other hand, Indirect GCs are a family of algorithms which use an inverse tree to group all the references to an object, supporting migration and in-transit references.

This paper presents a new algorithm, called Indirect Reference Listing which extends the Reference Listing to the tree, supporting migration, in-transit references (without blocking) and some degree of fault-tolerance. The GC has been implemented on a distributed Lisp system and the execution time overhead (compared to a Reference Count) is under 3%.

1 Introduction

Distributed Reference Counting algorithms (Lermen and Maurer [8]), (Bevan [1]), (Watson and Watson [21]) and (Piquer [10]) are being replaced by Reference Listing algorithms (Shapiro, Dickman and Plainfossé [17]; Birrel et al. [2]) in order to tolerate failures, as the algorithm is more robust. On the other hand, some extensions to Reference Listing have been proposed to recover cycles based on back tracing(Fuchs [5]; Maheshwari and Liskov [9]) and in partial tracing (Rodrigues and Jones [15]).

As in Reference Count, migration, in-transit references and the mixture of increment and decrement messages (now called add and delete) can cause problems.

Indirect Garbage Collection (Piquer [12]) is a family of algorithms based on a diffusion tree of the references, eliminating the need of two messages (only decrements exist) and supporting migration. However, the tree structure was very fragile and did not support any kind of failure.

---

* This work has been partially funded by Dirección de Investigación, Fac. Cs. Físicas y Matemáticas, U. de Chile

The environment in which our distributed garbage collector is designed to run is a loosely-coupled multi-processor system with independent memories, and a reliable point-to-point message passing system.

In general, we will suppose that message passing is very expensive and remote pointers already have many fields of information concerning the remote processor, object identifier, etc. In this kind of system, we can accept to lose memory (e.g., adding fields to the remote pointers) if extra messages can be avoided when requiring remote access, or when doing garbage collection.

In this paper, we present an new GC, integrating Reference Listing with the Indirect GC family. This algorithm has been implemented on a distributed Lisp environment (Piquer [11]) showing a minimum overhead. Furthermore, we will discuss how Reference Listing can be used to enhance the failure tolerance of the system.

The paper is organized as follows: section 2 presents the abstract model of a distributed system with remote pointers used throughout the paper. Section 3 presents the family of indirect garbage collectors, and the detailed algorithm of Indirect Reference Listing. Section 4 discusses the behavior of the algorithm in front of failures and section 5 compares it with other related work. Finally, section 6 presents the conclusions.

2   The Model

The basic model we will use in this paper is composed of the set of processors P, communicating through a reliable message-passing system, with finite but not bounded delays.

The set of all objects is noted O. Given an object $o \in O$, it always resides at one and only one processor $p_i \in P$: p_i is called the *owner* of o. Every object has always one and only one valid *owner* at a given moment in time.

The set of all the remote pointers to an object o is denoted $\mathbf{RP}(o)$. It is assumed that an object can migrate, thus changing its *owner*. However, this operation is considered an atomic operation, and objects are never in transit between two processors. This assumes a strong property in the underlying system, but in general a distributed system blocks every access to a migrating object, and the effect is the same. During a migration operation, a local object is transformed into a remote one and vice-versa. All the other remote pointers in the system remain valid after a migration as they refer to the object, not to a processor.

A remote pointer to an object o is called an o-reference[1]. For every object $o \in O$, $\mathbf{RP}(o)$ includes all of the existing o-references, including those contained in messages already sent but not yet received. This means that asynchronous remote pointer sending is allowed. In-transit remote pointers are considered to be always accessible. An o-reference is a *symbolic* pointer, usually implemented via an object identifier. It points to the object o, not to its *owner*.

---

[1] The o-reference terminology and the basis of this model were proposed by Lermen and Maurer ([8]), Piquer ([10]) and Tel and Mattern ([20]).

Only remote pointers are considered in this model, so the local pointers to o are not included in $\mathbf{RP}(o)$ and they are not called o-references. In fact, this paper just ignores local references to objects, assuming that only distant pointers exist. The only acceptable values inside an object are remote pointers.

The model also requires that, given an object o, every processor can hold at most one o-reference. Remote pointers are usually handled this way: if there are multiple pointers at a processor to the same remote object o, they pass by local indirection. This indirection represents for us the only o-reference at the processor. This is just for simplicity of the model, but it is not a requisite for the algorithms presented, although it simplifies some implementations. In our model then, an object containing a remote pointer to another object o is seen as an object containing a local pointer to an o-reference.

The four primitive operations supported are:

1. Creation of an o-reference
   A processor p_i, *owner* of an object o, transmits an o-reference to another processor p_j. The operation happens each time that the *owner* of o sends a message with an o-reference to another processor.
2. Duplication of an o-reference
   A processor p_j, which already has an o-reference to an object o (not being its *owner*), transmits the o-reference to another processor p_k.
3. Deletion of an o-reference
   A processor p_j, discards an o-reference. This means that the o-reference is no longer locally accessible[2].
4. Migration of an object o
   A processor p_j, holding an o-reference to an object o located at a distant processor p_i, becomes the new *owner* of o. In general, this operation implies the following actions: the *old owner* transforms its local object o into an o-reference and the *new owner* transforms its o-reference into a local object o. This operation preserves the number of elements of $\mathbf{RP}(o)$ since it just swaps an object o and an o-reference. A migration also generates many Duplication operations: one for each remote pointer contained in the migrating object, from the old to the new *owner*.
   The migration protocol is independent of the GC algorithm. However, the GC protocol must support asynchronous owner changes, with in-transit messages, without losing self-consistency.

We suppose that every o-reference creation is performed via one of the above operations. Remote pointer redirection is not supported, so an o-reference always points to the same object o.

The minimal set of primitive operations supported is powerful enough to model most of the existing distributed systems. Our simplified model enables a very simple exposition of the main ideas behind the algorithms but not their detailed implementation, which is usually very complex and system dependent. In fact, the model includes an underlying local GC: objects contain every remote

---

[2] In real implementations, this operation is usually invoked by the local GC.

pointer accessible from them. The Delete operation provides the interface to be used by the local GC to signal a remote pointer that is no longer locally accessible.

3 Indirect Reference Listing

In general, a distributed GC uses the information kept in the pointers by the underlying system to send its messages to the referenced object. This can be performed using the object finder, or just using the *owner* field of the pointer, and using forward pointers (Fowler [4]).

An Indirect GC is any GC modified to use a new data structure, exclusively managed and used by the GC, to send its messages: every object o has a distributed inverted tree associated to it, containing every host containing an o-reference. The o-references are extended with a new field **Parent** for the tree pointers. The object o is always at the root of the tree (see Fig. 1).

3.1 The GC Family

Any GC can be transformed into an Indirect GC, by simply replacing the messages sent directly to the object with messages sent through the inverted tree. This has two advantages: the GC is independent from the object manager, and it supports migrations, as will be shown later.

The inverted tree structure represents the diffusion tree of the pointer throughout the system. This structure contains every o-reference and the object o itself. When an o-reference is created or duplicated, the destination node is added as a child of the sender. If the destination node already had a valid **Parent** for this o-reference, it refuses the second **Parent** to avoid cycles in the structure.

When an o-reference is deleted, the corresponding GC algorithm will delete the o-reference from the tree only if it was a leaf. If not, a special o-reference (called a *zombie o*-reference) with all the GC fields, but no object access fields, is kept in the structure until it becomes a leaf. The *zombie o*-references are a side effect of the inverted tree: if one leaf of the sub-tree is still locally accessible, the GC needs *zombie o*-references to reach the object at the root. Obviously, when the root is the only node in the tree, there are no more o-references in the system, and the object itself is garbage.

The **Parent** field contains a processor identifier[3].

If the tree structure is always preserved (during every remote pointer operation), the GC can work on it without depending on the object finder. It is enough to use the **Parent** fields to send the GC messages to the ancestors in the tree, and following them recursively we can reach the object at the root, without using the underlying system.

---

[3] This optimization cannot be applied if we accept multiple o-references to the same object from the same processor. The **Parent** field is in fact a pointer to a remote pointer, but our simplified model allows this to be implemented as a pointer to a processor.

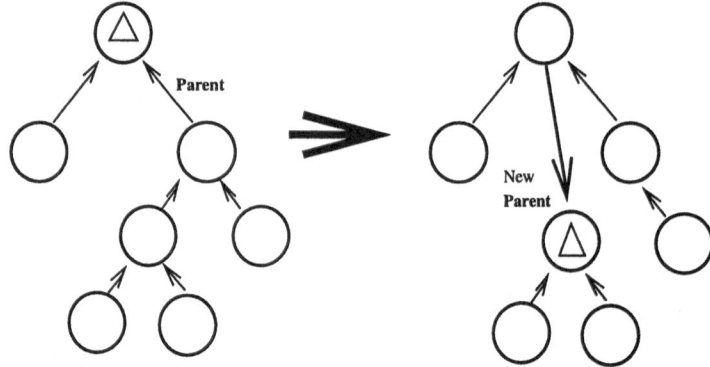

Fig. 1. A migration in the inverted tree

Migration is also allowed, by simply making a root change in the tree structure, which is trivial in an inverted tree (see Fig. 1).

On the other hand, the object finder itself does not need to use the **Parent** field. It can access the object by means of the *owner* fields, or whatever other fields are convenient.

3.2 Indirect Reference Listing

Indirect Reference Counting (IRC) was originally presented in (Piquer [10]). In a different context, this algorithm was also independently discovered at the same time by other authors (Ichisugi and Yonezawa [6]), (Rudalics [16]). In (Tel and Mattern [20]) it is shown that IRC is equivalent to the Dijkstra and Scholten ([3]) termination detection algorithm. Basically, IRC installs the reference counts at each node of the diffusion tree, counting the children of each one.

To implement Indirect Reference Listing (IRL), we will simply replace the counters by child lists, containing the site identifier of each node to which a reference has been sent. The changes to IRC in the implementation are minimal:

- Creation and Duplication
    - at p_i, when an o-reference is sent to p_j, p_j is added to **Ref_list**(x) (x being the object o for a Creation or the original o-reference for a Duplication)[4]
    - at p_j, upon reception of an o-reference from p_i, if it was already known, a *delete* message is sent to p_i. If it is newly installed, the o-reference is created with **Parent** $\leftarrow p_i$ and **Ref_list** $\leftarrow NIL$.
- Deletion
    - When an o-reference is deleted, it is marked as *Deleted*.

---

[4] If p_j was already in **Ref_list**(x), this operation will generate a duplicate. It is safe to do so, because a *delete* message could be in transit from p_j. It is a transient situation only, because p_j will refuse a second **Parent** for o, generating a *delete* message.

- Upon reception of a *delete* message for x from p_j, p_j is deleted from **Ref_list**(x).
- When an o-reference is marked as *Deleted* and its **Ref_list** $= NIL$, a *delete* message is sent to **Parent**(o-reference), and the o-reference is reclaimed. This test must be made upon reception of a *delete* message for an o-reference and upon o-reference deletion.

Since we assume in the model that every processor holds at most one o-reference per object o, a deleted o-reference with a non-null **Ref_list** is a *zombie* o-reference. If the same o-reference is received again in a message, a *delete* message must be sent to the o-reference sender, because the *zombie* already has a valid **Parent**.

- Migration

  The migration of an object o from p_i to p_j means a change of the root in the diffusion tree. This operation is trivial on an inverted tree if the old root is known. For IRL, the lists must also be kept consistent, so the new *owner* (p_j) has a new child (adding p_i to its **Ref_list**) and has no **Parent** (sending a *delete* to its old **Parent**). The migration costs one *delete* message, plus some work to be done locally at the respective processors, p_i and p_j.

When the **Ref_list** of an object is NIL, the object can be collected. An inverted tree with reference lists is shown in Fig. 2.

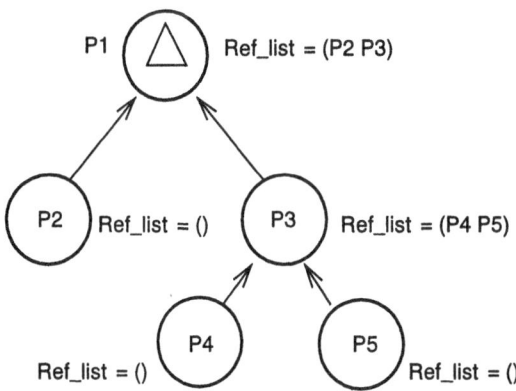

Fig. 2. Indirect Reference Listing

IRL is very straightforward to implement, but like Reference Count, it lacks the ability to collect cycles. However, the extensions proposed for Reference Listing to collect cycles, could also be applied here. Back tracing could be implemented following the reference lists through the tree.

IRL uses the inverted tree to send messages, so even as the **Parent** pointers are modified during migration (the other operations are only allowed to initialize them, not to change them), the reference lists never move. Objects migrate alone and they get the local o-reference **Ref_list**. Therefore, any in-transit *delete* message will always arrive at the correct destination.

IRL was implemented in a distributed Lisp System, an compared to IRC it only adds a 3% overhead on execution time. Of course, the main overhead is in space, depending on the total number of o-references. However, an interesting point, worth noting, is that IRL is backward-compatible with IRC (if *decrement* and *delete* messages are considered equivalent), and thus some sites could be using IRL while others use IRC in a correct Distributed GC.

4 Failures

Using Reference Lists, it is possible to resist some failures. IRL is a little different, so we will examine in turn: a node wanting to leave the system, a node crash and message loss or duplication.

4.1 Shutdown

One important problem for Indirect GCs is that it is not obvious for a node to leave the system, once it has diffusion trees traversing it. In large distributed systems, it is common for a node to need a shutdown, for maintainance or load reasons. It is important to be able to execute a shutdown at the node, deleting all its o-references and migrating all its remote-pointed objects. For Indirect GCs, *zombie o*-references must also be eliminated.

In IRC, a *zombie o*-reference could not be deleted, because its children need it to exist. On the other hand, we knew how many children it had, but not who they were. With IRL, we know the children identifiers, so we can start an intermediate node deletion algorithm (suppose p_i is shutting down):

- foreach x *zombie o*-reference
  - p_i sends (*adopt*, **Ref_list**(x)) to **Parent**(x)
- at p_j, upon reception of (*adopt*, list) for x,
  foreach p in list
  - Add p to **Ref_list**(x), send (*new_parent*, o-reference) to p
- At p_k, upon reception of *new_parent* for x from p_j:
  - send *deletion* to **Parent**(x), set **Parent**$(x) \leftarrow p_j$.

Processor p_i waits until there are no more *zombie o*-references before finally shutting down.

Any processor implementing IRL can shutdown cleanly using this protocol. If the other nodes are only using IRC, this protocol still works. If the node willing to leave does not implement IRL, this protocol cannot be engaged.

Another possible use of this protocol is to implement intermediate node deletion in the tree, avoiding *zombie o*-references completely: instead of marking an o-reference as deleted, the Delete operation could send an *adopt* message to the **Parent**.

4.2   Crashes

Supposing that every node is informed of a node crash, we would like to have a protocol to rebuild a consistent tree for every object accessed through the failed site. Every site can search into its **Ref_lists** the failed site, and they can delete the references coming from it. However, indirect references could also exist, and the failed site could also be **Parent** of some subtrees. These subtrees need to be rebuilt. One alternative is to generate an error on access, and to need to ask for the object again, or try to move the subtrees to the root. Anyway, the involved nodes now can detect this situation and handle it.

4.3   Message Loss/Duplication

Reference Lists are usually used to tolerate message loss and duplication. IRL does not support these behaviors, because we accept multiple copies of a site in a **Ref_list** to protect in-transit references.

   If it is necessary to deal with unreliable transports, exchanging the **Ref_lists** as proposed by Shapiro, Dickman and Plainfossé [17] is probably the best solution.

5   Related Work

- A reference listing algorithm using indirect references to provide fault-tolerance is SSP (Stub-Scion Pairs)(Shapiro, Dickman and Plainfossé [17]; Plainfossé [14]). This algorithm is based on replacing the reference counts by a site list, containing every site where a duplication has been sent. The reference lists are exchanged between sites to keep the information up to date. This extra information provides tolerance to duplication and disorder if a timestamp is added to the messages. A prototype implementation has shown a 10% overhead compared to plain IRC (Plainfossé and Shapiro [13]), but supporting message failures.
  Indirect Reference Listing is simpler and more efficient, being also more fragile in front of failures.
- Using Reference Lists to tolerate failures was originally proposed by Birrel et al [2] and it has been implemented into Sun's Remote Method Invocation for Java (Sun [19]). It does not support migration and must block the sender of a duplicated reference until an acknowledgment is received from the owner. IRL is not so strong in front of failures, but provides a more efficient solution to the problem.
- Fuchs [5] and Maheshwari and Liskov [9] propose back tracing techniques based on Reference Listing. Basically, they perform a tracing GC backwards, from the objects to their roots. Reference Lists are useful for that, because they give us the list of sites pointing to our site. However, inside local spaces the problem is harder, because an o-reference does not know which object contains it. Both papers propose techniques to deal with that issue. These

techniques can be equally applied to IRL, because following the **Ref_list** pointers the tree can be traversed from the root to the leaves (which is a back trace in this case).

6 Conclusions

We propose a new distributed GC algorithm, mixing Reference Listing with Indirect Garbage Collection: IRL. The algorithm is very simple and straightforward to implement, and backward compatible with IRC. The advantage is that a node can leave the system using a shutdown protocol, *zombie o*-references can be deleted and site crashes can be handled. Furthermore, back tracing can be applied to IRL, extending it to collect cycles (which is not possible with plain IRC).

An implementation in a distributed Lisp system shows an execution overhead under 3% and demonstrates the ability to inter-operate between IRC and IRL.

Compared with related work, IRL is simpler, supports migration and in-transit references and does not block a site during duplication. In the other hand it is more robust than plain IRC.

References

1. Bevan, D. I.: "Distributed Garbage Collection Using Reference Counting," LNCS 259, *PARLE'87 Proceedings Vol. II*, Eindhoven, Springer Verlag, June 1987.
2. Birrel, A., Evers D., Nelson G., Owicki S. and Wobber E.: "Distributed Garbage Collection for Network Objects," Technical Report 116, DEC-SRC, 1993.
3. Dijkstra, E. W. and Scholten, C. S.: "Termination Detection for Diffusing Computations," *Information Processing Letters*, Vol. 11, N. 1, August 1980.
4. Fowler, R. J.: "The Complexity of Using Forwarding Addresses for Decentralized Object Finding," in *Proc. 5th Annual ACM Symp. on Principles of Distributed Computing*, pp. 108–120, Alberta, Canada, August 1986.
5. M. Fuchs, "Garbage Collection on an Open Network," LNCS 986, *Proc. 1995 International Workshop on Memory Management (IWMM'95)*, Springer Verlag, pp. 251–265, Kinross, UK, September 1995.
6. Ichisugi, Y. and Yonezawa, A.: "Distributed Garbage Collection using group reference counting," Tech. Report 90-014, Dept Information Science, Univ., Of Tokyo, 1990.
7. Lang, B., Queinnec, C. and Piquer, J.: "Garbage Collecting the World," *19th ACM Conference on Principles of Programming Languages 1992*, Albuquerque, New Mexico, January 1992, pp. 39–50.
8. Lermen, C. W. and Maurer, D.: "A Protocol for Distributed Reference Counting," *Proc. 1986 ACM Conference on Lisp and Functional Programming*, Cambridge, Massachussets, August 1986.
9. Maheshwari, U., Liskov, B.: "Collecting Distributed Garbage Cycles by Back Tracing," *Proc. 1997 ACM Symp. on Principles of Distributed Computing (PODC'97)*, Santa Barbara, California, August 1997, pp. 239–248.
10. Piquer, J.: "Indirect Reference Counting: A Distributed GC," LNCS 505, *PARLE '91 Proceedings Vol I*, pp. 150–165, Springer Verlag, Eindhoven, The Netherlands, June 1991.

11. Piquer, J.: "A Re-Implementation of TransPive: Lessons from the Experience," *Proc. Parallel Symbolic Languages and Systems (PSLS'95)*, LNCS 1068, Vol I, pp. 310–329, Springer Verlag, Beaune, France, October 1995.

12. Piquer, J.: "Indirect Distributed Garbage Collection: Handling Object Migration," *ACM Trans. on Programming Languages and Systems*, V. 18, N. 5, September 1996, pp. 615–647.

13. Plainfossé, D. and Shapiro, M.: "Experience with a Fault-Tolerant Garbage Collector in a Distributed Lisp System," LNCS 637, *Proc. 1992 International Workshop on Memory Management (IWMM'92)*, Springer Verlag, pp. 116–133, St-Malo, France, September 1992.

14. Plainfossé, D.: "Distributed Garbage Collection and Referencing Management in the Soul Object Support System," PhD Thesis, U de Paris VI, France, June 1994.

15. Rodrigues, H. and Jones R,: "A Cyclic Distributed Garbage Collector for Network Objects," LNCS 1151, *Proc. 10th Workshop on Distributed Algorithms (WDAG'96)*, Springer-Verlag, pp. 123–140, Bologna, Italy, October 1996.

16. Rudalics, M.: "Implementation of Distributed Reference Counts," Tech. Report, RISC, J. Kepler Univ, Linz, Austria, 1990.

17. Shapiro, M., Dickman, P. and Plainfossé, D.: "Robust, Distributed References and Acyclic Garbage Collection," *ACM Symposium on Principles of Distributed Computing*, Vancouver, Canada, August 1992.

18. Shapiro, M., Dickman, P. and Plainfossé, D.: "SSP Chains: Robust, Distributed References Supporting Acyclic Garbage Collection," INRIA Res. Report 1799, November 1992.

19. Sun Microsystems Computer Corporation: "Java Remote Method Invocation Specification", http://sunsite.unc.edu/java, November, 1996.

20. Tel, G. and Mattern, F.: "The Derivation of Distributed Termination Detection Algorithms from Garbage Collection Schemes," *ACM Trans. on Programming Languages and Systems*, Vol. 15, N. 1, January 1993, pp. 1–35.

21. Watson, P. and Watson, I.: "An Efficient Garbage Collection Scheme for Parallel Computer Architectures," LNCS 259, *PARLE Proceedings Vol. II*, Eindhoven, Springer Verlag, June 1987.

Active Ports: A Performance-Oriented Operating System Support to Fast LAN Communications

G. Chiola, G. Ciaccio

DISI, Università di Genova, via Dodecaneso 35, 16146 Genova, Italy
{chiola,ciaccio}@disi.unige.it
http://www.disi.unige.it/project/gamma/

Abstract. The Genoa Active Message MAchine (GAMMA) is an efficient communication layer for 100base-T clusters of Personal Computers running Linux. It is based on Active Ports, a communication mechanism derived from Active Messages. GAMMA Active Ports deliver excellent communication performance at user level (latency 12.7 μs, maximum throughput 12.2 MByte/s, half-power point reached with 192 byte long messages), thus enabling cost-effective cluster computing on 100base-T. Despite being implemented at kernel level in the Linux OS, the performance numbers of GAMMA Active Ports are much better than many other LAN-oriented communication layers, including so called "user-level" ones (e.g. U-Net).

1 Introduction

Networks of workstations (NOWs) or, even better, clusters of Personal Computers (PCs) networked by inexpensive commodity interconnects, potentially offer a cost-effective support to parallel processing. The only obstacle to overcome is the inefficiency caused by the OS layers that sedimented in the communication architecture during the past evolutive eras of computer and communication technology. Thus the issue of enhancing or by-passing OS kernels with performance-oriented support to inter-process communication has become crucial. So far this issue has been addressed by following two ways, that we call the "efficient OS support" approach and the "user-level" approach.

In the first approach the messaging system is supported by the OS kernel with a small set of flexible and efficient low-level communication mechanisms and by a simplified communication protocols carefully designed according to a performance-oriented approach. Issues like choosing the right communication abstraction and implementing such abstraction efficiently are of primary concern here. The "efficient OS support" architecture fits nicely into the structure of modern OSs providing protected access to the communication devices. This way multi-users, multitasking features need not be limited or modified.

The "user-level" approach aims at improving performance by minimizing the OS involvement in the communication path in order to obtain a closer integration between the application and the communication device: The use of system calls is minimized allowing direct access to the network interface whenever possible. Here the challenge is to provide direct access to the communication devices without violating the OS protection model. In order for such approach not to compromise protection in the access to the communication devices, either a single-user environment or some form of gang scheduling are usually required. These two alternatives have their own obvious drawbacks. A third solution is to leverage programmable NICs which can run the necessary support for device multiplexing in place of the OS kernel. Currently this implies much higher hardware costs.

The best known representative of the "user-level" approach is U-Net. With U-Net, user processes are given direct protected access to the network device with no virtualization. Any communication layer as well as standard interface must move to user-level programming libraries. U-Net runs on an ATM network of SPARCstations [7], where a communication co-processor located on the NIC executes proper firmware in order to multiplex a pre-defined number of virtual "endpoints" over the NIC itself, with no OS involvement. In the Fast Ethernet emulation of U-Net [8], the low-cost NIC does not provide a programmable co-processor. The host CPU itself multiplexes the endpoints over the NIC by means of proper OS support, thus actually following an "efficient OS support" rather than a "user-level" architecture.

The "user-level" architecture is believed to deliver better communication performance than the "efficient OS support" architecture by avoiding the OS involvement. Such belief is now contradicted by our experience, at least in the case of commodity off-the-shelf devices.

2   Active Ports in GAMMA

The Genoa Active Message MAchine (GAMMA) [3] is a fast messaging system running on a 100base-T cluster of Pentium PCs running Linux and equipped with either 3COM 3C595-TX Fast Etherlink or 3COM 3C905-TX Fast Etherlink XL PCI 100base-T adapters. Porting GAMMA to DEC "tulip"-based and Intel EtherExpressPro NICs is in progress.

Following the "efficient OS support" principle, with GAMMA the Linux kernel has been enhanced with a communication layer implemented as a small set of additional *light-weight* system calls (that is, system calls with no intervention of the scheduler upon return) and a custom NIC driver. Most of the communication layer is embedded in the Linux kernel at the NIC driver level, the remaining part being placed in a user-level programming library. All the multi-user, multi-tasking functionalities of the Linux kernel have been preserved. All the communication software of GAMMA has been carefully developed according to a performance-oriented approach, extensively described in [3]. The adoption of a communication mechanism that we call *Active Ports* allowed a "zero copy" protocol, with no intermediate copies of messages along the whole user-to-user communication path.

With GAMMA, each process owns a number of (currently 255) *active ports*. GAMMA active ports are inspired to the Thinking Machines' CMAML Active Message library for the CM-5. Each active port can be used to exchange messages to another process in the same as well as a different process group. GAMMA directly supports SPMD parallel programming with automatic replication of the same process on different computation nodes. However plain MIMD programming is allowed as well since sender and receiver need not share code address space. Indeed the message handlers as used in active ports are designated by the receiver, not by the sender as in Generic Active Messages [6].

Prior to using an active port for outbound communications, the owner process must bind it to a destination, in the form of a triple (process group, process instance, destination active port). To send a message throughout an active port, the process has to invoke a "send" light-weight system call which traps to the GAMMA device driver. Each message is then transmitted as a sequence of one or more Ethernet frames. Following a "zero copy" policy, frames are arranged directly into the NIC's transmit FIFO by DMA of each frame body directly from the sender process' data structure with no intermediate buffering in kernel memory space. Frame headers are pre-computed at port binding time. As soon as the last frame of a message has been written to the NIC's transmit FIFO the "send" system call returns and the sender process resumes user-level activity.

Active ports can be used to broadcast messages as well. The Ethernet broadcast service is exploited directly, so that one single physical transmission occurs. Efficient barrier synchroniziation is also implemented using this facility [2].

For inbound communications, prior to use a given active port L the owner process must bind L to an application-defined function called *receiver handler* as well as to an application-defined final destination data structure (spanning a contiguous virtual memory region) where each message incoming through port L is to be stored. The final destination must be pinned down in physical RAM. At this point the local instance of the GAMMA driver knowns in advance the final destination in user space for messages incoming through port L. Therefore, when the GAMMA device driver detects frame arrivals for port L, the frame payloads can be copied directly in their final destination, with no temporary storage in kernel memory space and regardless of the schedule state of the receiver process at the arrival time. Moreover, as soon as the last piece of message has been delivered to port L, the GAMMA driver can immediately run the receiver handler bound to L so as to allow a real-time management of the message itself. The receiver handler is run on the interrupt stack. If the receiver process is not currently running, a temporary context switch is performed by the GAMMA driver so as to ensure that the handler is run in the receiver context. The possibility of running the receiver handler in real time is crucial in order to avoid that subsequent messages incoming to port L overlap the current one if this is to be avoided. Indeed a typical receiver handler: 1) integrates the current message into the main thread of the destination process; 2) notifies the message reception to the main thread itself; 3) re-binds port L to a fresh final destination in user space for the next incoming message.

The GAMMA implementation of Active Ports relies on a very simplified communication protocol with neither error recovery nor flow control. However the only source of frame corruption in modern LANs is due to frame collisions in shared networks. This may be avoided either using switched networks or explicitly scheduling communications at the application level. Flow control is unnecessary as long as the network is the throughput bottleneck. The flexibility of GAMMA allows either to use its simplified and efficient protocol "as is", or to build more complex and reliable higher level protocols using the GAMMA error detection features [3].

2.1 Error Handlers for User-Defined Error Recovery

As a by-product of the simplification of the communication protocol and of the elimination of intermediate copies of messages GAMMA provides neither message acknowledgement nor explicit flow control. However:

- In a LAN environment with good quality wiring, the only source of frame corruption is collision with other frames in case of shared Ethernet. Such possibility could be eliminated by using a switch instead of a repeater hub. In any case frames corrupted by collision have bad CRC and are discarded by the receiver NIC driver.
- If the receiver handler on the receiver side of a communication runs quickly enough (as usually is the case) both the RAM-to-NIC transfer rate on the sender side and the NIC-to-RAM transfer rate on the receiver side are much greater than the Fast Ethernet transfer rate. Hence, flow is implicitly controlled by simply testing whether there is enough room in the transmit FIFO of the sender's NIC before writing a frame.

The choice of GAMMA is to implement some error detection mechanisms and to allow applications to have their own error management policies if required. This is supported by allowing applications to bind each active port with an *error handlers*. An error handler is similar to a receiver handler but runs in case of communication errors rather than on message arrivals. Error handlers can implement higher level reliable communication protocols, if really needed for a given application.

3 Communication Performance and Conclusions

According to our average estimates based on a "ping-pong" microbenchmark, the GAMMA messaging system exhibit an unprecedented one-way user-to-user latency (delay for a zero byte message) as low as 12.7 μs, with a maximum throughput of 12.2 MByte/s (98% of the Fast Ethernet raw asymptotic bandwidth) with long (500 kByte) messages. The half-power point is achieved with messages as short as 192 bytes. Indeed such numbers enable cost-effective parallel computing on 100base-T clusters of PCs [1, 4, 5].

The improvement achieved by GAMMA with respect to Linux TCP/IP sockets on similar hardware (latency 150 μs, maximum throughput 7 MByte/s) is more than one order of magnitude for latency and 71% for bandwidth.

A latency improvement of more than 2 times has been obtained with respect to the U-Net "user-level" on similar or superior hardware (U-Net latency is 30 μs on 100base-T and 35.5 μs on 140 Mbps ATM). Therefore GAMMA delivers better communication performance than U-Net while providing a higher level and more flexible communication abstraction with same quality of service.

As a conclusion, we contradict the common belief that only a "user-level" communication architecture may yield a satisfactory answer to the demand for efficient inter-process communication in a low-cost LAN environment: the improved applications-to-device integration yielded by the "user-level" approach, and the corresponding performance gain, is negligible given that communication hardware cannot be closely integrated with the host CPU in commodity-based NOWs. Much better results can be obtained by optimizing the communication protocol and choosing a suitable communication abstraction. Indeed the GAMMA Active Ports are a flexible communication abstraction which preserves the Linux multi-user multi-tasking environment while still allowing best communication performance, based on traditional kernel-level protection mechanisms and low-cost commodity interconnects.

References

1. G. Chiola and G. Ciaccio. Architectural Issues and Preliminary Benchmarking of a Low-cost Network of Workstations based on Active Messages. In *Proc. 14th conf. on Architecture of Computer Systems (ARCS'97)*, Rostock, Germany, Sept. 1997.
2. G. Chiola and G. Ciaccio. Fast Barrier Synchronization on Shared Fast Ethernet. *2nd Int. Workshop on Communication and Architectural Support for Network-Based Parallel Computing (CANPC'98), LNCS 1362*, pages 132–143, Feb. 1998.
3. G. Ciaccio. Optimal Communication Performance on Fast Ethernet with GAMMA. In *Proc. Workshop PC-NOW, IPPS/SPDP'98*, pages 534–548, Orlando, Fl., April 1998. LNCS 1388.
4. G. Ciaccio and V. Di Martino. Porting a Molecular Dynamics Application on a Low-cost Cluster of Personal Computers running GAMMA. In *Proc. Workshop PC-NOW, IPPS/SPDP'98*, pages 524–533, Orlando, Fl., April 1998. LNCS 1388.
5. G. Ciaccio, V. Di Martino, and P. Lanucara. Porting the Flame Front Propagation Problem on GAMMA. In *Proc. HPCN Europe '98*, Amsterdam, The Netherlands, April 1998.
6. D. Culler, K. Keeton, L.T. Liu, A. Mainwaring, R. Martin, S. Rodriguez, K. Wright, and C. Yoshikawa. Generic Active Message Interface Specification. Tech. Report White Paper of the NOW Team, CS Dept., U. California at Berkeley, 1994.
7. T. von Eicken, A. Basu, V. Buch, and W. Vogels. U-Net: A User-Level Network Interface for Parallel and Distributed Computing. *15th ACM Symp. on Operating Systems Principles (SOSP'95)*, Copper Mountain, Co., Dec. 1995. ACM Press.
8. M. Welsh, A. Basu, and T. von Eicken. Low-latency Communication over Fast Ethernet. In *Proc. Euro-Par'96*, Lyon, France, August 1996.